HUMAN BIOLOGY

FOURTH EDITION

HUMAN BIOLOGY

FOURTH EDITION

CECIE STARR
Belmont, California

BEVERLY McMILLAN
Gloucester, Virginia

BROOKS/COLE

THOMSON LEARNING

Australia • Canada • Mexico • Singapore • Spain • United Kingdom • United States

BROOKS/COLE

THOMSON LEARNING

BIOLOGY PUBLISHER: Jack C. Carey

PROJECT DEVELOPMENT EDITOR: Kristin Milotich

MEDIA PROJECT MANAGER: Pat Waldo

EDITORIAL ASSISTANT: Karen Hansten

DEVELOPMENTAL EDITOR: Mary Arbogast

MARKETING TEAM: Rachel Alvelais, Mandie Houghan, Laura Hubrich, Carla Martin-Falcone

ASSISTANT EDITOR: Daniel Lombardino

PRODUCTION COORDINATOR: Mary Anne Shahidi

PRODUCTION SERVICE: Lachina Publishing Services, Inc.

PERMISSIONS EDITOR: Sue Ewing

TEXT DESIGN: Gary Head, Gary Head Design

COVER PHOTO: Photodisc

COVER DESIGN: Denise Davidson

INTERIOR ILLUSTRATION: Lisa Starr, Raychel Ciemma, Precision Graphics, Preface Inc., Robert Demarest, Jan Flessner, Darwen Hennings, Vally Hennings, Betsy Palay, Nadine Sokol, Kevin Somerville, Lloyd Townsend

PHOTO RESEARCHER: Lachina Publishing Services, Inc.

PRINT BUYER: Vena M. Dyer

COMPOSITION: Lachina Publishing Services, Inc.

COLOR PROCESSING: H&S Graphics (Tom Andersen, Michelle Kessel, Laurie Riggle, and John Deady)

PRINTING AND BINDING: Von Hoffmann Press

Printed in the United States of America

10 9 8 7 6 5 4 3 2

Library of Congress Cataloging-in-Publication Data
Starr, Cecie.
    Human biology / Cecie Starr, Beverly McMillan.—4th ed.
      p. cm.
    Includes bibliographical references and index.
    ISBN 0-534-37713-0
    1. Human biology. I. McMillan, Beverly. II. Title.
QP34.5 .S727 2000
612—dc21   00-045458

For more information about this or any other Brooks/Cole products, contact:
BROOKS/COLE
511 Forest Lodge Road
Pacific Grove, CA 93950 USA
www.brookscole.com
1-800-423-0563 (Thomson Learning Academic Resource Center)

For permission to use material from this work, contact us by
    www.thomsonrights.com
    Fax: 1-800-730-2215
    Phone: 1-800-730-2214

CONTENTS IN BRIEF

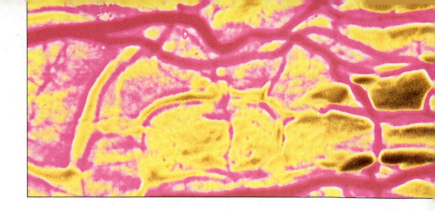

DETAILED CONTENTS

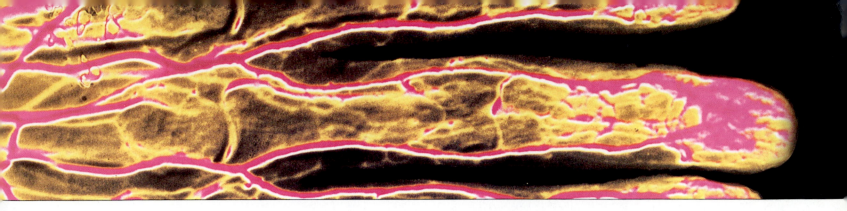

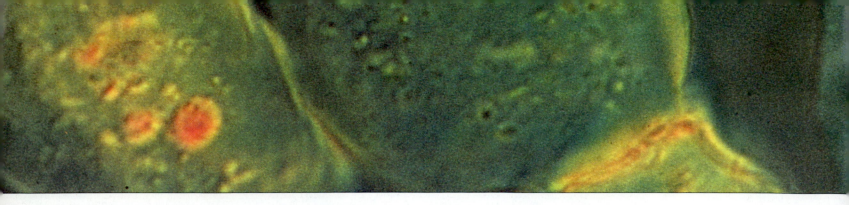

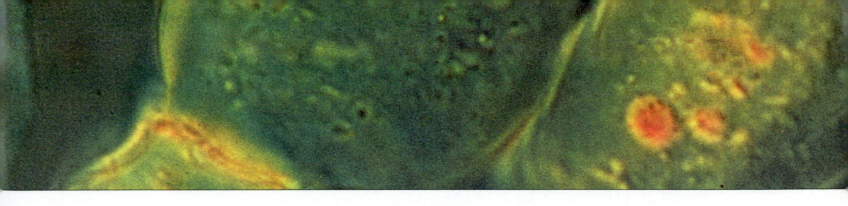

PREFACE

Teachers of introductory human biology know well the phenomenon of running as fast as one can merely to stay in the same place. New and updated information from scores of biology subdisciplines piles up daily, and somehow teachers are expected to distill it into a one-term course that speeds through the high points of human biology.

This book is a coherent account of the diverse topics in biology as they apply to humans. It is deliberately built to highlight concepts while presenting current knowledge and exposing students to major research trends. We believe that the biological perspective on human life is one of education's most valuable gifts. With it, students can navigate through the myriad medical, environmental, and social issues that confront us as our world becomes an ever more complex place in which to live.

This fourth edition of *Human Biology* is the culmination of a two-year process of reevaluating every element in the book, from its overall coverage and organization to the subtleties of how particular word choices affect the clarity and effectiveness of the writing. The outcome is a book that retains the strengths of previous editions while making some needed improvements and inaugurating some exciting changes.

Like previous editions, this book opens with an overview of basic concepts and methods, followed by principles of biochemistry and cell structure and function. Next come units on body systems, reproduction, and the principles of inheritance. A final unit discusses principles of evolution and ecology that are essential background for an informed citizenry. This conceptual organization parallels the levels of biological organization. Within that familiar framework, though, we did some important remodeling.

The changes begin with the simple but logical decision to call the introductory chapter on principles and methods Chapter 1. We also placed the chapters on reproduction and development in their own unit and included in it a new and timely chapter—one that addresses sexually transmitted diseases within the larger context of infectious disease (Chapter 16). To help students connect genetics concepts with some of their most intriguing applications, we moved the chapter on recombinant DNA technology and genetic engineering to directly follow the DNA chapter. An updated chapter on cancer rounds out the unit. And because most human biology courses lack the time to cover two evolution chapters, we integrated material on human evolution into the general presentation of evolutionary concepts (Chapter 23). This revision also helps students better relate the human evolutionary journey to the larger picture of evolution.

As before, we explain the structure and functioning of tissues, organs, and organ systems—the anatomy and physiology of the human body—in enough detail that students can build a working vocabulary about life's parts and processes. The theme that structure follows function is reinforced throughout. Working with dedicated users and reviewers, we refined and updated every part of the book, including the review material at the end of each chapter, the index, and the glossary. A major addition is a new Appendix I, which provides additional information on topics in cell metabolism and aerobic respiration. Students who want a deeper understanding of energy dynamics, enzyme function, and the steps of glycolysis, the Krebs cycle, and ATP production in mitochondria will find that information in this new appendix.

CONCEPT SPREADS

Throughout this book, we keep the story line in focus for students by subscribing to the question *"How do you eat an elephant?"* and its answer, *"One bite at a time."* Hence, here students will find descriptions, art, and supporting evidence for each concept organized on two facing pages, at most. Each "concept spread" starts with a numbered tab and ends with boldface, summary statements of key points. Students can use the statements to check whether they understand the concepts of one spread before starting on another. The clear demarcation also gives instructors greater flexibility in assigning or skipping topics within a chapter.

Restricting the space available for each concept forced us to avoid superfluous detail. Also, ongoing feedback from instructors has guided our decisions about when to leave core material alone and when to loosen it up with applications. Within each spread, headings and subheadings help students track the hierarchy of information. Careful transitions beween spreads help them keep the larger story in focus and discourage memorization for its own sake.

Based on extensive user feedback, we know that our clearly defined organization helps students find assigned topics easily and translates into improved test scores.

BALANCING CONCEPTS WITH APPLICATIONS

Each chapter starts with a lively or sobering application and an adjoining list of key concepts (the chapter's advance organizer). We also provide a brief outline of the main chapter sections. Strategically placed examples of applications parallel the core material—not so many as to be distracting, but enough to keep minds perking along with the conceptual development. Many brief applications are integrated into the text. Others are in boxed essays that provide more depth on health, medical, environmental, and social issues for interested students but do not interrupt the text. The book's last few pages separately index all applications for quick reference.

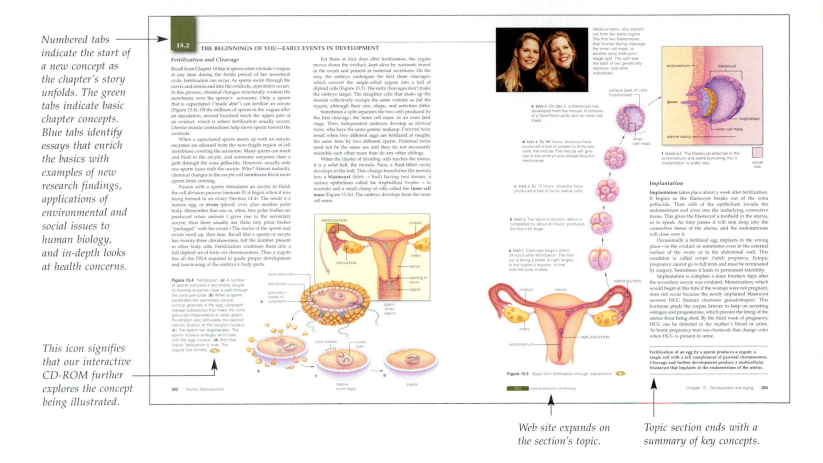

Numbered tabs indicate the start of a new concept as the chapter's story unfolds. The green tabs indicate basic chapter concepts. Blue tabs identify essays that enrich the basics with examples of new research findings, applications of environmental and social issues to human biology, and in-depth looks at health concerns.

This icon signifies that our interactive CD-ROM further explores the concept being illustrated.

Web site expands on the section's topic.

Topic section ends with a summary of key concepts.

FOUNDATIONS FOR CRITICAL THINKING

To help students increase their capacity for critical thinking, we selected certain text discussions and chapter introductions to show students how biologists use critical thinking to solve problems. The introductions to immunology (Chapter 9), the nervous system (Chapter 11), and DNA structure and function (Chapter 17) are examples. *Science Comes to Life* essays provide optional, more detailed examples. For example, one of these describes efforts to develop artificial skin, bone, and other tissues (Section 4.5). Another (Section 15.10) shows how research—in this case, increasingly sophisticated techniques of prenatal diagnosis—can bring an emerging story into sharp focus while it also poses new questions.

Each chapter concludes with *Critical Thinking* questions. Many reviewers suggested improvements or new topics for these questions and we often took their advice. Daniel Fairbanks developed many of the *Genetics Problems* in Chapters 18 and 19 that help students grasp the principles of inheritance.

VISUAL OVERVIEWS OF CONCEPTS

We simultaneously develop the text and art as inseparable parts of the same story. We also give visual learners a means to work their way through a visual overview of a major process before reading the corresponding text, which thereby becomes less intimidating. Students repeatedly let us know how much they appreciate this approach.

Our overview illustrations have step-by-step, written descriptions of biological parts and processes. Instead of wordless diagrams we break down information into a series of illustrated steps. For example, the descriptions of Figure 6.11, integrated with the art, walk students through the stages by which nutrient molecules are absorbed into the body, one step at a time.

Many of the anatomical drawings form integrated overviews of structure and function. Students need not jump back and forth from text, to tables, to illustrations and back again in order to comprehend how an organ system is put together and what its parts do. Individual descriptions of parts are hierarchically arranged to reflect the structural and functional organization of the system.

ZOOM SEQUENCES

Many illustrations progress from macroscopic to microscopic views of the same subject. For instance, Figure 5.16 is an overview illustration of skeletal muscle structure and function that begins with the biceps muscle in a human arm and culminates with the organization of a sarcomere, the basic unit of muscle contraction. Figure 17.10 diagrams the organization of DNA in chromosomes, showing step by step how the DNA molecule becomes increasingly condensed and coiled.

COLOR CODING

Visual consistency throughout the book helps students track complex parts and processes. Our line illustrations consistently use the same colors for molecules and cell structures:

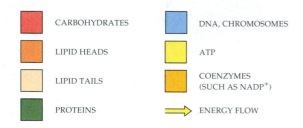

■ CARBOHYDRATES	■ DNA, CHROMOSOMES
■ LIPID HEADS	■ ATP
■ LIPID TAILS	■ COENZYMES (SUCH AS NADP⁺)
■ PROTEINS	→ ENERGY FLOW

ICONS

Small diagrams next to an illustration help students relate a topic to the big picture. For example, a simple representation of a mitochondrion subtly reminds students of where ATP production takes place.

A multimedia icon directs students to art in a CD-ROM enclosed in their book.

And information running along the bottom of each spread directs them to supplemental material on the Web.

Further reading: Student Guide to InfoTrac on web site ➝

1.3   www.brookscole.com/biology

END-OF-CHAPTER STUDY AIDS
AND APPENDICES

Each chapter ends with a summary in list form, review questions, a self-quiz, critical thinking questions, selected key terms, and a list of readings. Italicized page numbers tie the review questions and key terms to relevant text pages.

At the book's end, a glossary includes the boldfaced terms in the text, with pronunciation guides and word origins when such information can make formidable words less so. Our index is detailed enough to help students find a door to the text more quickly.

Appendix I offers new enrichment material on energy and cell metabolism, including visual overview illustrations. Students can use the periodic table of the elements in Appendix II for reference purposes. Appendix III includes metric-English conversion charts. The fourth appendix has detailed answers to genetics problems, the fifth has answers to self-quizzes, and the sixth provides hints that assist students in developing plausible solutions to the critical thinking questions.

CONTENT REVISIONS

For this edition we refined text and art in each concept spread and created a more balanced text-to-art ratio. Overall, we tightened the writing, but we also added applications and expanded topics that can be confusing if they are presented with too little explanation. In the chapters on anatomy and physiology, we strengthened the emphasis on ways that body systems contribute to homeostasis. Some adjustments reflect research trends. For example, in keeping with an emerging consensus, we adopted a six-kingdom classification scheme in Chapter 1; in Chapter 6 we noted changing views on aspects of nutrition.

PART I: FOUNDATIONS   Chapter 1, the student's introduction to concepts and methods in human biology, features a rewritten, expanded discussion of scientific methods. New text and art enliven the discussion of experimental design, and this edition describes and illustrates the important concept of sampling error. An updated example of the scientific method in action discusses and illustrates current research on the role of free radicals in photoaging. In this chapter and throughout the book, we fine-tuned *Critical Thinking* questions and added more of them. In Chapter 2, paragraph-by-paragraph rewriting simplified the discussion of basic chemistry, and in several sections we added photographs to enliven structural diagrams of molecules. We revised the section on acids, bases, and buffer systems, and developed a new diagram of the pH scale that links pH values with substances that are familiar to students. A new *Focus on Health* essay helps students understand free radicals and how antioxidants counteract them.

Typically, introductory textbook chapters on the cell start with the nucleus and present other cell parts in a rather "listy" way. We decided to reorganize this book's Chapter 3 on cell structure and function, juxtaposing topics and concepts that logically relate. So, after an overview, sections on the plasma membrane and cytomembrane system lead to the discussion of transport mechanisms (including diffusion and osmosis) across membranes. Then the other major organizing element of cells, the cytoskeleton, is described, followed by the nucleus and other organelles and their functions.

Chapter 4, on tissues, organs, and homeostasis, is a proven, workable introduction to those subjects and, beyond some streamlining in the discussion of connective tissues, for the most part we left it alone.

PART II: BODY SYSTEMS AND FUNCTIONS   Chapter 5 has revised sections on bone and on the formation of ATP in muscle tissue. We also revised the *Science Comes to Life*, shifting its focus to joint replacement in general. We updated the discussion of anabolic steroids in the *Focus on Health* focus essay. In the end-of-chapter review section, we added a table summarizing the anatomy and function-ing of bone and muscle tissue. Chapter 6 has a reworked opening vignette on diet, obesity, and weight control, timely subjects that are revisited in later chapter sections that bring students up to date on research findings about genetic influences on body weight. We also added vivid color diagrams and micrographs that help clarify the small intestine's role in nutrient absorption.

Chapter 7 incorporates revised illustrations of blood components and the cardiac cycle. We reworked the art and text devoted to cardiovascular disorders, updating and judiciously expanding some parts while consolidating others. Chapter 8 on immunology has new art throughout. We also included a bit more on inflammation's chemical mediators and revised the discussion of the immune response, adding short explanations that result in better integration of the text and art. In Chapter 9 the section on gas exchange and transport is rewritten. There also is more on vocal cords and speech, and a revised section on breathing mechanisms.

Chapter 10 provides an improved discussion of the kidneys' role in acid-base homeostasis. The discussion of the nervous system in Chapter 11 now begins with the compelling story of modern research on Parkinson's disease. With this edition's rewriting, students should find the going easier in reworked discussions of neuron function and action potentials. There is new art on action potential propagation along sheathed neurons. The section on memory is revised, as are the discussions of psycho-active drugs and drug abuse. We also consolidated the latter two sections, making for a clearer, more coherent treatment of those topics.

Chapter 12 on sensory systems features new art, with revised sections on somatic sensations, hearing, balance, and vision. Chapter 13 has revised text and art on the feedback control of adrenal hormones.

PART III: REPRODUCTION AND DEVELOPMENT   This new unit presents reproductive systems, development, aging, and threats to life and health from infectious dis-eases. Chapter 14 has updated text and new art in the

sections on birth control and infertility. In Chapter 15, new art helps illustrate early embryonic development and mor-phogenesis. The section on maternal lifestyle and nutrition is revised; the *Science Comes to Life* essay on technologies of prenatal diagnosis is updated, with a new illustration of fetoscopy. Other updating in this chapter includes the discussion of Alzheimer's disease, and new *Critical Thinking* questions.

Chapter 16 surveys the sources, transmission, and health effects of infectious diseases, including sexually transmitted ones. The previous edition's discussions of HIV and AIDS have been streamlined and updated. Antibiotic resistance is presented in the context of the biology of viruses, bacteria, viroids, prions, and patho-genic protozoa. The chapter also expands students' perspective with material on globally important threats such as malaria, tuberculosis, and diarrheal diseases.

PART IV: PRINCIPLES OF INHERITANCE   Throughout this unit we rewrote for clarity and accessibility, adding crisper definitions of terms and, where needed, additional bits of explanation. New Section 17.7 introduces chromo-some structure earlier in the unit, expanding the discus-sion of the cell cycle to set the stage for later sections. The illustrations of mitosis and meiosis are revised a bit; new art illustrates gametogenesis and the separation of sister chromatids during mitosis. In Chapter 18, new art on the health effects of sickle cell anemia illustrates pleiotropy. The chapter's *Science Comes to Life* on genomic imprinting is updated. Chapter 19, on chromosome variations and medical genetics, features a revised introduction that includes text and art on the Philadelphia chromosome and cancer. Sections on karyotyping, linkage, and cross-ing over are rewritten and illustrated with new color diagrams.

Chapter 20 has a revised discussion of DNA replica-tion, a new tRNA model and icon for the art on protein synthesis, and new, clearer illustrations of transcription and translation. Revised art and text examples make the mutation discussion easier to understand. Chapter 21 is reorganized, rewritten, and illustrated with striking new art. The changes begin with an opening vignette describing how researchers have traced the descendants of Thomas Jefferson and Sally Hemings by analyzing the Y chromosome of the Jefferson line. Other changes include new material on DNA fingerprinting, automated DNA sequencing, gene isolation, mapping the human genome, and safety issues.

Rounding out this unit, Chapter 22 offers an updated discussion of the causes, characteristics, and major types of cancer.

PART V: EVOLUTION AND ECOLOGY   Chapter 23 has reorganized, revised, and updated discussions on the fossil record, biogeography, and comparative anatomy and biochemistry. New to this edition is the integration of material on human evolution in this principles chapter, providing a single, more cohesive presentation of evolution for one-semester courses in human biology.

Chapter 24 has new art on the hydrologic, phosphorus, carbon, and nitrogen cycles; and updates on greenhouse gases and global warming. In Chapter 25 we updated the discussion and art on human population growth, and revised the age structure diagrams for baby boomers. The chapter's final sections include an update on Chernobyl's aftermath, a more thoughtful treatment of alternative energy sources, and a look at Wangari Maathai's work on reforestation.

SUPPLEMENTS FOR STUDENTS

Interactive Concepts in Biology. Packaged free with all student copies, this is the first CD-ROM to address the full sweep of biology. The cross-platform CD-ROM covers all concept spreads in the book. Because students can learn by doing, it encourages them to manipulate the book's art. Each illustration in the book that is on the CD has an icon in its caption. A combination of text, graphics, photographs, animations, video, and audio enhance each book chapter. Chapter-by-chapter interactive exercises and quizzes are also provided.

InfoTrac College Edition. A subscription to this online library is available FREE with each new copy of *Human Biology.* It gives students access to full articles—not simply abstracts—from more than 700 scholarly and popular periodicals dating back as far as four years. Some 700,000 articles are available through InfoTrac's impressive database, which has periodicals such as *Discover, Science News,* and *Health,* and many pamphlets on health concerns such as AIDS, STDs, and diabetes.

Student Guide to InfoTrac College Edition. This guide is on the Brooks/Cole Biology Resource Center site on the World Wide Web. It has an introduction to InfoTrac plus a critical thinking question for each text chapter and a set of electronic readings from InfoTrac that invite deeper examination of the issues.

Biology Resource Center. This is Brooks/Cole's web site for biology, which contains a home page for *Human Biology.* At present, all information is arranged by the fourth edition's chapters. Every month it presents new BioUpdates on relevant applications as well as hyperlinks and practice quiz questions for each chapter. It includes descriptions of college degrees and careers in biology, a student feedback site, and cool clip art. It also includes flashcards for all glossary terms, critical thinking exercises, news groups, surfing lessons (including biology surfing), a variety of search engines, biological games, and a BioTutor. Perhaps

most importantly, it has Internet exercises for each chapter to help students learn to extract information from the Web. The address for the Biology Resource Center is:

http://www.brookscole.com/biology

Study Guide and Workbook: An Interactive Approach. With sections numbered to match the concept spreads in the text, this guide allows students to focus on smaller amounts of material. Understanding and retention are increased as students respond to all questions by writing in the spaces provided. (ISBN: 0-534-38147-2)

Building Your Life Science Vocabulary. This booklet helps students learn biological terms by explaining root words and their applications. (ISBN: 0-534-26214-7)

Essential Study Skills for Science Students. Daniel Chiras's guide explains how to develop good study habits, sharpen memory, learn more quickly, get the most out of lectures, prepare for tests, produce excellent term papers, and improve critical thinking skills. (ISBN: 0-534-37595-2)

Also note that just inside the cover of instructors' examination copies is a section describing the comprehensive package of supplements available to instructors.

A COMMUNITY EFFORT

Authors can write accurately and often well about their field of interest, but it takes more than this to deal with the full breadth of biology as it relates to the human body. For us, it takes an educational network of more than 2000 teachers, researchers, and photographers in the United States, Canada, England, Germany, France, Australia, Sweden, and elsewhere. We list those reviewers whose recent contributions continue to shape our thinking. There simply is no way to describe the thoughtful effort that these individuals and other reviewers before them gave to our books. We can only salute their commitment to quality in education. Their insight and attention to detail are reflected on every page. For this edition, particular thanks go to Daniel Fairbanks, who rewrote Chapter 21 on genetic engineering and recombinant DNA technology.

Gary Head designed the book. Jeff Lachina and his staff at Lachina Publishing Services once again shepherded the book from our quirky Quark files to the printed page. We thank them for unstinting professionalism and grace under pressure. Mary Arbogast once again demonstrated that she is an editorial wizard with endurance and a sense of humor. Mary Anne Shahidi more than ably guided the production process at Brooks/Cole, and Daniel Lombardino and Karen Hansten handled myriad editorial tasks and details with amazing patience. Pat Waldo and her band of electronic *wunderkinder* created the remarkable package of CD-ROM and Web resources that accompany this book.

This fourth edition of *Human Biology* is much improved for the careful copyediting of Nancy Tenney. Thanks also to Lachina Publishing Services, who organized, analyzed, and finally assembled the remarkable illustration program in the following pages.

Two other people deserve immense credit for helping to bring this edition to fruition. One, biologist and philosopher Jack Musick, provided vital support and scientific insight, sometimes very late at night. The other is our publisher Jack Carey, who unfailingly supports our vision for what this book can achieve as an educational tool. Without Jack's guidance and unwavering friendship, this book would not be possible.

Cecie Starr
Beverly McMillan

REVIEWERS

ALCOCK, JOHN, *Arizona State University*
ALFORD, DONALD K., *Metropolitan State College of Denver*
ALLISON, VENITA F., *Southern Methodist University*
ANDERSON, D. ANDY, *Utah State University*
ANDERSON, DAVID R., *Pennsylvania State University–Fayette*
ARMSTRONG, PETER, *University of California, Davis*
AULEB, LEIGH, *San Francisco State University*
BAKKEN, AIMÉE H., *University of Washington, Seattle*
BARNUM, SUSAN R., *Miami University*
BEDECARRAX, EDMUND E., *City College of San Francisco*
BENNETT, JACK, *Northern Illinois University*
BENNETT, SHEILA K., *University of Maine at Augusta*
BOHR, DAVID F., *University of Michigan, Ann Arbor*
BOOTH, CHARLES E., *Eastern Connecticut State University*
BRAMMER, J.D., *North Dakota State University*
BROADWATER, SHARON T., *College of William and Mary*
BROWN, MELVIN K., *Erie Community College–City Campus*
BRUSLIND, LINDA D., *Oregon State University*
BUCHANAN, ALFRED G., *Santa Monica College*
BURKS, DOUGLAS J., *Wilmington College*
CHRISTENSEN, A. KENT, *University of Michigan*
CHRISTENSEN, ANN, *Pima Community College*
CHOW, VICTOR, *City College of San Francisco*
CONNELL, JOE, *Leeward Community College*
COX, GEORGE W., *San Diego State University*
COYNE, JERRY, *University of Chicago*
CROSBY, RICHARD M., *Treasure Valley Community College*
DANKO, LISA, *Mercyhurst College*
DELCOMYN, FRED, *University of Illinois at Urbana-Champaign*
DENNER, MELVIN, *University of Southern Indiana*
DENNISTON, KATHERINE J., *Towson State University*
DENTON, TOM E., *Auburn University Montgomery*
DLUZEN, DEAN, *Northeastern Ohio University's College of Medicine*
EDLIN, GORDON J., *University of Hawaii*
EMSLEY, MICHAEL, *George Mason University*
ERICKSON, GINA, *Highline Community College*
FAIRBANKS, DANIEL J., *Brigham Young University*
FALK, RICHARD H., *University of California, Davis*
FREY, JOHN E., *Mankato State University*
FROEHLICH, JEFFERY, *University of New Mexico*
FULFORD, DAVID E., *Edinboro University of Pennsylvania*
GARNER, JAMES G., *Long Island University, C.W. Post Campus*
GENUTH, SAUL M., *Case Western Reserve University*
GIANFERRARI, EDMUND A., *Keene State College of the University of New Hampshire*
GOODMAN, H. MAURICE, *University of Massachusetts Medical School*
GORDON, SHELDON R., *Oakland University*
GREGG, KENNETH W., *Winthrop University*
HAHN, MARTIN, *William Paterson College*
HALL, N. GAIL, *Trinity College*
HARLEY, JOHN P., *Eastern Kentucky University*
HASSAN, ASLAM S., *University of Illinois*
HENRY, MICHAEL, *Santa Rosa Junior College*
HERTZ, PAUL E., *Barnard College*
HILLE, MERRILL B., *University of Washington*
HOEGERMAN, STANTON F., *College of William and Mary*
HOHAM, RONALD W., *Colgate University*
HOSICK, HOWARD L., *Washington State University*
HUCCABY, PERRY, *Elizabethtown Community College*
HUNT, MADELYN D., *Lamar University*
HUPP, EUGENE W., *Texas Woman's University*
JOHNS, MITRICK A., *Northern Illinois University*
JOHNSON, LEONARD R., *University of Tennessee College of Medicine*
JOHNSON, TED, *St. Olaf College*
JOHNSON, VINCENT A., *St. Cloud State University*
JONES, CAROLYN K., *Vincennes University*

JOSEPH, JANN, *Grand Valley State University*
KAREIVA, PETER, *University of Washington*
KAYE, GORDON I., *Albany Medical College*
KENYON, DEAN H., *San Francisco State University*
KEYES, JACK, *Linfield College, Portland Campus*
KLEIN, KEITH K., *Mankato State University*
KENNEDY, KENNETH A.R., *Cornell University*
KIMBALL, JOHN W.
KROHNE, DAVID T., *Wabash College*
KRUMHARDT, BARBARA, *Des Moines Area Community College, Urban Campus*
KUPCHELLA, CHARLES E., *Southeast Missouri State University*
KUTCHAI, HOWARD, *University of Virginia*
LAMBERT, DALE, *Tarrant County College*
LAMMERT, JOHN M., *Gustavus Adolphus College*
LAPEN, ROBERT, *Central Washington University*
LASSITER, WILLIAM E., *University of North Carolina–Chapel Hill School of Medicine*
LEVY, MATTHEW N., *Mt. Sinai Hospital*
LITLLE, ROBERT C., *Medical College of Georgia*
LUCAS, CRAN, *Louisiana State University–Shreveport*
MANIA-FARRELL, BARBARA, *Purdue University Calumet*
MANN, ALAN, *University of Pennsylvania*
MANN, NANCY J., *Cuesta College*
MARCUS, PHILIP I., *University of Connecticut*
MARTIN, JOSEPH V., *Rutgers University Camden*
MATHIS, JAMES N., *West Georgia College*
MATTHEWS, PATRICIA, *Grand Valley State University*
MAYS, CHARLES, *DePauw University*
McCUE, JOHN F., *St. Cloud State University*
McMAHON, KAREN A., *University of Tulsa*
McNABB, F.M. ANNE, *Virginia Polytechnic Institute and State University*
MILLER, G. TYLER
MITCHELL, JOHN L.A., *Northern Illinois University*
MITCHELL, ROBERT B., *Pennsylvania State University*
MOHRMAN, DAVID E., *University of Minnesota, Duluth*
MOISES, HYLAN, *University of Michigan*
MORK, DAVID, *St. Cloud University*
MORTON, DAVID, *Frostburg State University*
MOTE, MICHAEL I., *Temple University*
MOWBRAY, ROD, *University of Wisconsin–LaCrosse*
MURPHY, RICHARD A., *University of Virginia Health Sciences Center*
MYKLES, DONALD L., *Colorado State University*
NORRIS, DAVID O., *University of Colorado at Boulder*
PARSON, WILLIAM, *University of Washington*
PETERS, LEWIS, *Northern Michigan University*
PIPERBERG, JOEL B., *Millersville University*
PORTEOUS-GAFFARD, SHIRLEY, *Fresno City College*
POZZI-GALLUZI, G., *Dutchess Community College*
QUADAGNO, DAVID, *Florida State University*
REZNICK, DAVID, *University of California, Riverside*
ROBERTS, JANE C., *Creighton University*
RUBENSTEIN, ELAINE, *Skidmore College*
SCALA, ANDREW M., *Dutchess Community College*
SAPP OLSON, SALLY
SHEPHERD, GORDON M., *Yale University School of Medicine*
SHERMAN, JOHN W., *Erie Community College–North*
SHERWOOD, LAURALEE, *West Virginia University*
SHIPPEE, RICHARD H., *Vincennes University*
SLOBODA, ROGER D., *Dartmouth College*
SMIGEL, BARBARA W., *Community College of Southern Nevada*
SMITH, ROBERT L., *West Virginia University*
SOROCHIN, RON, *State University of New York; College of Technology at Alfred*
STEELE, CRAIG W., *Edinboro University of Pennsylvania*

STEUBING, PATRICIA M., *University of Nevada, Las Vegas*
STEWART, GREGORY J., *State University of West Georgia*
STONE, ANALEE G., *Tunxis Community College*
SULLIVAN, ROBERT J., *Marist College*
TAUCK, DAVID, *Santa Clara University*
THIEMAN, WILLIAM, *Ventura College*
THOMPSON, ED W., *Winona State University*
TIZARD, IAN, *Texas A&M University*
TROTTER, WILLIAM, *Des Moines Area Community College Sciences Center*
TUTTLE, JEREMY B., *University of Virginia Health Sciences Center*
VALENTINE, JAMES W., *University of California, Berkeley*
VAN DE GRAAFF, KENT M., *Weber State University*

VAN DYKE, PETE, *Walla Walla Community College*
VARKEY, ALEXANDER, *Liberty University*
WALSH, BRUCE, *University of Arizona*
WARNER, MARGARET R., *Purdue University*
WEISBROT, DAN R., *William Paterson University*
WEISS, MARK L., *Wayne State University*
WHIPP, BRIAN J., *St. George's Hospital Medical School*
WHITTEMORE, SUSAN, *Keene State College*
WILLIAMS, ROBERTA B., *University of Nevada, Las Vegas*
WISE, MARY, *Northern Virginia Community College*
WOLFE, STEPHEN L., *University of California, Davis (Emeritus)*
YONENAKA, SHANNA, *San Francisco State University*

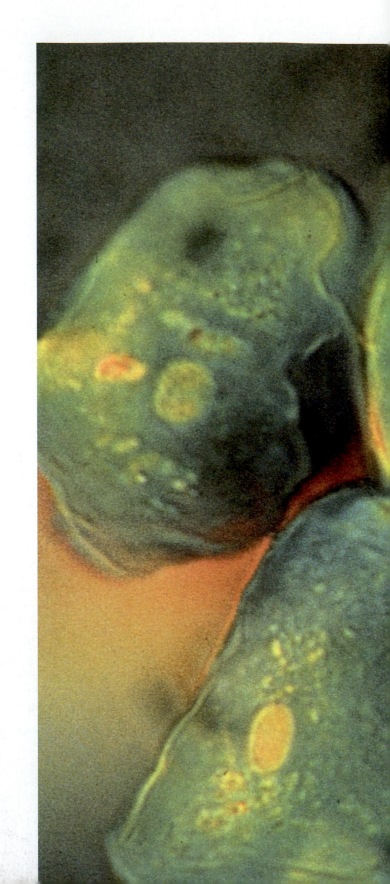

Human cheek cells magnified 200 times. Each cell is a tiny but highly ordered unit, the fundamental unit of life.

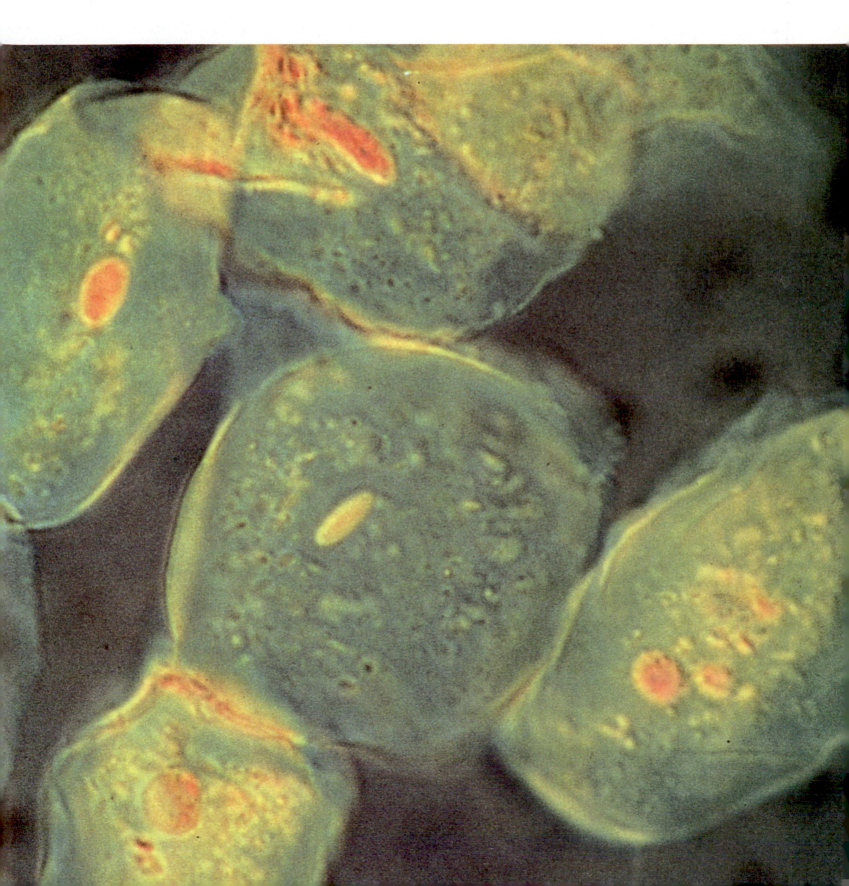

1 CONCEPTS AND METHODS IN HUMAN BIOLOGY

Meet Wonderful You

Buried somewhere in your brain are memories of discovering your own hands and feet, your family, the change of seasons, the smell of grass. The memories include early introductions to a great, disorganized parade of insects, flowers, frogs, and furred things—mostly living, sometimes dead. There are memories of questions: "What is life?" and "Where do I fit in the world around me?" There are memories of answers, some satisfying, others less so.

By observing, asking questions, and accumulating answers, you have gradually built up knowledge about life (Figure 1.1). Experience and education have been refining your questions, and most likely some answers are difficult to come by.

Think of a college student, twenty years old, whose motorcycle skidded into a truck. Now he is comatose, and doctors say his brain is functionally dead. His breathing and other basic functions will continue only as long as he stays hooked up to a respirator and other mechanical support systems. Would you say that this unfortunate young person is still "alive"?

Think of the world's forests—once vast tracts on every continent, now mere remnants due to logging, conversion for agriculture, the expansion of cities, and other activities that help keep many people sheltered, warm, and fed. In 1999 the human population surpassed 6 billion, and it is still growing rapidly as we begin the 21st century. With so few forests remaining, where will humanity find more usable land and forest products?

Or think of an embryo, a cluster of cells growing inside a pregnant woman. At what point does it become a human life? If you learn more about how an embryo actually develops, will the knowledge influence your thoughts about such incendiary issues as birth control and abortion?

If questions like these have ever crossed your mind, without knowing it you have been informally studying human biology. In formal terms, **biology** is the scientific study of life, here focused on the extraordinary human body. On a personal level, you probably already know a good deal about the workings of your own body—not to mention some of its quirks, too. This course is a chance to expand that base of knowledge. We hope you will come away from this course with a deeper understanding of just how fascinating and remarkable your body is.

Every biology textbook starts out by considering the often-perplexing question of what "life" is. This question opens a story that has probably been unfolding for more than 3.8 billion years. From that perspective, "life" is an outcome of ancient events in which nonliving matter—atoms and molecules—became assembled into the first living cells. Chapter 23 presents current ideas on the chemical origin of life on Earth. Once living cells existed, there began an ongoing process of biological **evolution**—change in many details of the body plan and functions of organisms through the generations. In the course of evolution major groups of life forms emerged, including animals like ourselves.

Humans, apes, and some other closely related animal species are primates. As primates we are also mammals, and all mammals, including humans, are vertebrates—animals with "backbones." Of course, we aren't alone on our planet—we share it with millions of other animal species, as well as with plants, fungi, countless bacteria, and other kinds of living things. Figure 1.2 sketches out a picture of the place our human species, *Homo sapiens*, occupies in the living world.

Figure 1.1 Think back on all you have known and seen; this is the foundation for your deeper study of human biology.

This textbook is about the biology of humans, but no matter where we look in the living world we find that all living things share some basic features. They capture and use energy and materials. They sense and respond to specific changes in the environment. Living things have the capacity to grow, develop, and reproduce.

A good deal of this textbook explores the body structures and functions through which your body maintains its living state. A partial list of those life-support systems includes the brain and other parts of our nervous system, a digestive system that extracts nutrients from the food we eat, systems for respiration and blood circulation, which keep the body supplied with oxygen, a urinary system that helps rid the body of wastes, and an immune system that defends it against countless threats. The body also has a hormone-based communication system, a skeleton and muscles for support and movement, and organs for reproduction.

This chapter introduces themes that are cornerstones of the study of human biology. One is our evolutionary heritage. Another theme is the state of internal flux called *homeostasis*. Living cells can function properly only within narrow environmental limits, and each of the body systems you will learn about helps maintain those conditions. As you read, you will also find essays that relate basic biological concepts to health topics, environmental questions, and social issues that concern each of us throughout our lives.

KEY CONCEPTS

1. Humans exhibit the basic characteristics of life. The human body consists of substances put together according to the laws that govern matter and energy. Humans obtain and use energy and materials from their environment, and they sense and respond to changes in their environment. They also grow and reproduce, based on instructions contained in DNA.

2. To sustain life, body systems must maintain a dynamic state of internal equilibrium called homeostasis.

3. As with other forms of life, the structures of the human body, and their functions, have been shaped by evolution.

4. Biology, like other branches of science, is based on systematic observations, hypotheses, predictions, and relentless testing. The external world, not internal conviction, is the testing ground for scientific theories.

CHAPTER AT A GLANCE

1.1 DNA, Energy, and Life

1.2 Two More Life Characteristics: Reproduction and Inheritance

1.3 Energy and Life's Organization

1.4 What Is Science?

1.5 A Scientific Method in Action: Probing a Burning Question

1.6 *How Do You Define Death?*

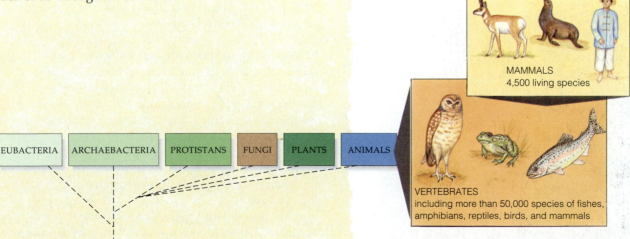

EUBACTERIA ARCHAEBACTERIA PROTISTANS FUNGI PLANTS ANIMALS

MAMMALS
4,500 living species

VERTEBRATES
including more than 50,000 species of fishes, amphibians, reptiles, birds, and mammals

Figure 1.2 Six kingdoms of life.

Human beings belong to one of well over a million species in the Animal kingdom. The other five kingdoms of life include more than 400,000 species of plants; various fungi; and vast numbers of single-celled organisms, including Protistans such as amoebas and two kingdoms of bacteria (Eubacteria and Archaebacteria).

DNA, ENERGY, AND LIFE

Nothing Lives without DNA

DNA AND THE MOLECULES OF LIFE   Picture a climber on a rock face, inching toward the top (Figure 1.3). Without even thinking about it, you know the climber is alive and the rock is not. Would you be able to explain why? At a fundamental level, both consist of atoms and their parts—components that Chapter 2 will examine in some detail. Atoms are the building blocks of larger structures called molecules. At the level of molecules, differences between living and nonliving things begin to emerge.

One molecule you will never find in a chunk of rock is **DNA**—short for *deoxyribonucleic acid*. However, every living organism—including you—*does* have DNA, which contains the instructions for using nonliving substances like carbon to build and operate the organism. DNA is rather like the instructions you might use to turn a heap of ceramic tiles into ordered patterns such as those shown in Figure 1.4.

The **cell** is the smallest unit of matter having the capacity for life. We define a cell as an organized unit that can survive and reproduce by itself, given instructions in DNA and the necessary energy and materials. No nonliving object can ever meet these specifications.

Figure 1.3 A climber cautiously chooses her route up a rock face. Although she is alive and the rock is not, both are made up of the same fundamental components.

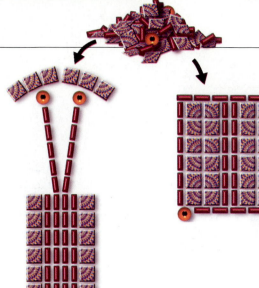

Figure 1.4 Examples of objects built from the same materials but with different assembly instructions.

Nothing Lives without Energy

ENERGY AND METABOLISM   Along with a set of DNA "directions" and certain raw materials, life requires an input of energy. **Energy** is most simply defined as the capacity to do work. In the realm of life, every living cell can (1) extract energy from its surroundings and (2) use that energy to maintain itself, grow, and make more cells. The general term for this capacity is **metabolism**. For now, all you need to know about metabolism is that it involves energy transfers. Those energy transfers are accomplished by an energy carrier called ATP (a topic of Chapter 3). They power the building of a cell's parts and specialized cellular work—for instance, the contraction of a muscle cell. Energy transfers also make possible other essential biochemical reactions. Cells even store some energy.

Like other living things, humans take in energy for metabolism from the external world. This is one reason why a person such as the child in Figure 1.5 must eat food. You will learn more about how the body extracts energy from food in Chapter 6.

SENSING AND RESPONDING TO ENERGY   Like all living cells, cells of your body can sense changes in their environment and make controlled responses. They manage this vital monitoring task with the help of **receptors**—molecules and structures that can detect stimuli. A **stimulus** is a specific form of energy detected by receptors. Examples are the heat energy from a burning match detected by receptors in your skin, or the mechanical energy in sound waves, detected by receptors in your ears.

Cells adjust their activities in response to signals from receptors. For example, the speed of your heartbeat is controlled by different nerves that either stimulate a more rapid beat or slow the heart rate down—all depending on some external factor, such as whether you are exercising or resting. Likewise, each cell in your body can withstand only so much heat or cold. It must rid itself of harmful substances. It requires certain nutrients in certain

Further reading: Student Guide to InfoTrac on web site →

you skip breakfast, then lunch, and the level of the sugar glucose in your blood—"blood sugar"—falls. Glucose is the body's main fuel molecule. A hormone then signals cells in your liver to dig into their stores of energy-rich molecules. Those complex molecules (glycogen) are broken down to the sugar glucose, which is released into the bloodstream, and your blood-sugar level returns to normal. This information feeds back to the hormone-secreting cells. Their activity then slows, and your liver stops breaking down glycogen. After you eat, glucose enters your bloodstream, your blood sugar rises, and cells of the pancreas increase their secretion of the hormone insulin. Most of your body cells have receptors for insulin, which prompts them to take up glucose—and your blood-sugar level again returns to normal.

In chapters to follow you will study body systems, such as the digestive and nervous systems, that each help maintain homeostasis. Those body systems, of course, are built of cells. The following diagram summarizes this fundamental, interdependent relationship:

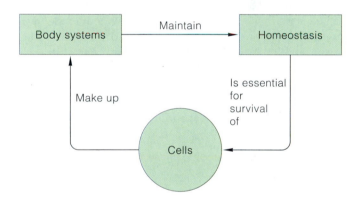

In Chapter 4 we'll take up the topic of homeostasis once more, with particular attention to the feedback mechanisms by which it operates.

Figure 1.5 This child is taking in energy from food. The food energy is transferred to ATP, which in turn makes it available for metabolic functions required to keep the child's cells, and the whole child, alive.

amounts. Yet temperature shifts and harmful substances are unavoidable, and at times you might gorge on junk food or fruit or skip a meal. All of these are examples of events that change the internal environment of your body, and they trigger responses that can restore the body to a state in which its internal environment is maintained within tolerable ranges. This state, called **homeostasis**, is one of the most important concepts in biology that you will encounter in this book; consider the following example.

A CLOSER LOOK AT HOMEOSTASIS The term homeostasis means "staying the same." But what exactly stays the same in homeostasis? The answer is, the fluids that make up the internal environment surrounding cells, including your blood. With homeostasis, healthy cells can survive; without it, they die. For instance, suppose

The structure and organization of living things begin with instructions contained in DNA molecules. Only living things have DNA.

Life emerges at the level of the cell.

Only cells have the capacity for metabolism, in which they take in and use energy to stockpile, tear down, and dispose of materials in ways that promote the cells' survival.

Cells respond to changes in the environment in ways that help maintain the stable internal state called homeostasis. When homeostasis is maintained, operating conditions inside the cell or the whole body favor survival.

TWO MORE LIFE CHARACTERISTICS: REPRODUCTION AND INHERITANCE

Like other organisms, humans are part of an ongoing journey that began billions of years ago. When a human egg and sperm join, the result is the first cell of a new human being. But the cell only exists because the sperm and egg formed earlier, in processes directed by DNA instructions that were passed down through countless generations. And with the time-tested instructions in DNA, a new human body develops in ways that will prepare it to help produce a new generation. By way of **reproduction**— the production of offspring by parents— the journey of life continues.

What "Inheritance" Means

Parents transmit to their offspring instructions for duplicating their body form and other traits. In biology, this passing of traits from generation to generation is what we mean by **inheritance**. Molecules of DNA contain the instructions that assure that offspring will resemble their parents. The instructions also allow the details of traits to vary (Figure 1.6). For instance, having five fingers on each hand is a human trait, but a few people are born with six fingers on each hand. How do such variations come about? Basically, variations in traits arise through **mutations**—changes in the structure or number of DNA molecules. Mutations that are present in reproductive cells, the sperm and eggs, can be inherited by offspring.

Most mutations are harmful because the separate bits of information in DNA are part of a coordinated whole.

For example, a single mutation in human DNA may lead to hemophilia, a genetic disorder in which blood cannot clot properly after the body is cut or bruised (Chapter 19).

Adaptive Traits and Human Evolution

Even though mutations usually cause problems, some mutations may prove harmless, or even beneficial, under prevailing environmental conditions. For example, the disease sickle-cell anemia is caused by a DNA change that results in a defective form of hemoglobin, a blood protein. The sickle-cell trait is most prevalent in parts of the world where malaria is common. People who inherit DNA with the mutation from both parents suffer the full-blown disease. But, for complex reasons, people who inherit the mutation from only one parent are resistant to malaria. For them, the mutation is adaptive; it *increases* their chances of survival.

An **adaptive trait** helps an organism survive and reproduce in a given environment. In the course of human evolution, countless DNA mutations were tested in the environments in which our ancestors lived. Through the ages, adaptive, heritable DNA changes have given rise to our large, complex brains and other components of the living world's most elaborate nervous system. Other adaptive changes paved the way for efficient mechanisms for taking in and distributing oxygen and food molecules—our systems for respiration, blood circulation, and food digestion, respectively. As subsequent chapters will describe, these are only some of the characteristics that enable us to live the biologically complex life of a human being. Chapter 23 delves into the various mechanisms by which the saga of human evolution occurs.

Figure 1.6 Reproduction is a life characteristic. Instructions in DNA assure that offspring will resemble parents—and they also permit variations in the details of traits, as the photograph demonstrates.

DNA is the molecule of inheritance. Its instructions for reproducing traits are passed on from parents to offspring.

Mutations introduce variations in heritable traits.

An adaptive trait helps an organism survive and reproduce in a particular environment.

1.3 ENERGY AND LIFE'S ORGANIZATION

As a whole, the metabolic activities of single cells and multicelled organisms such as humans maintain the great pattern of organization in nature. Notice that the levels of organization in Figure 1.7 begin with nonliving components and continue on through populations (such as the Earth's whole human population), ecosystems, and the biosphere. The term **biosphere** refers to all parts of the Earth's waters, crust, and atmosphere in which organisms live. Chapters 24 and 25 return to these concepts when we discuss the principles of ecology and human impacts on ecosystems.

Organisms Are Interdependent

In general, a flow of energy from the sun maintains the pattern of organization in nature. Plants and some other organisms that capture solar energy are the entry point for this energy. They are food producers for the living

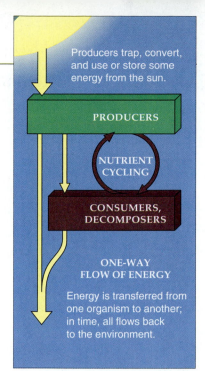

Figure 1.8 The flow of energy and the cycling of materials in the biosphere.

Producers trap, convert, and use or store some energy from the sun.

PRODUCERS

NUTRIENT CYCLING

CONSUMERS, DECOMPOSERS

ONE-WAY FLOW OF ENERGY

Energy is transferred from one organism to another; in time, all flows back to the environment.

world. Animals, including humans, are consumers; directly or indirectly, they feed on energy stored in plant parts. Thus you tap directly into this stored energy when you eat a banana, and you tap into it indirectly when you eat hamburger made from cattle that fed on grass or grain. Bacteria and fungi are decomposers. When they feed on tissues or remains of other organisms, they break down biological molecules to simple raw materials, which can be recycled back to producers. In time, all the energy that the producers initially captured from the sun's rays returns to the environment (as heat). For the moment, the simple, powerful message to remember is: Every part of the living world, including each of us, is linked to every other part.

For now, you only need to keep in mind that living things connect with one another by a one-way flow of energy through them and a cycling of materials among them, as Figure 1.8 diagrams. The interconnectedness of organisms on our planet affects the structure, size, and composition of the Earth's populations and communities. It affects ecosystems, even the whole biosphere. Once you understand the extent of these interactions and how they are affected by the activities of more than 6 billion human beings, you will gain insight into the thinning of the ozone layer, global warming, acid rain, and many other modern-day problems.

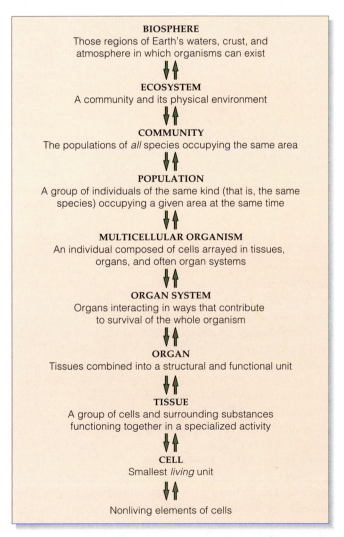

BIOSPHERE
Those regions of Earth's waters, crust, and atmosphere in which organisms can exist

ECOSYSTEM
A community and its physical environment

COMMUNITY
The populations of *all* species occupying the same area

POPULATION
A group of individuals of the same kind (that is, the same species) occupying a given area at the same time

MULTICELLULAR ORGANISM
An individual composed of cells arrayed in tissues, organs, and often organ systems

ORGAN SYSTEM
Organs interacting in ways that contribute to survival of the whole organism

ORGAN
Tissues combined into a structural and functional unit

TISSUE
A group of cells and surrounding substances functioning together in a specialized activity

CELL
Smallest *living* unit

Nonliving elements of cells

Figure 1.7 An overview of levels of organization in nature.

Levels of organization in the living world begin with subatomic particles, atoms, and molecules. Cells, multicellular organisms, and whole ecosystems are part of this continuum of increasing complexity.

All organisms, including humans, are part of webs of organization in nature, in that they depend directly or indirectly on one another for energy and raw materials.

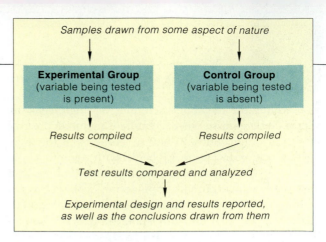

Samples drawn from some aspect of nature

| Experimental Group (variable being tested is present) | Control Group (variable being tested is absent) |

Results compiled — Results compiled

Test results compared and analyzed

Experimental design and results reported, as well as the conclusions drawn from them

Figure 1.9 Generalized sequence of steps involved in an experimental test of a prediction based on a hypothesis.

1.4 WHAT IS SCIENCE?

The preceding sections sketched out major concepts in human biology. Those concepts are firmly grounded in scientific research. But why should you accept that such facts have merit? You might even find yourself asking "What exactly *is* 'science'"? Good questions! Fortunately, there's nothing mysterious about science. Basically, it is an approach—a very powerful and productive one—to solving problems and answering questions about aspects of our world. This scientific approach emphasizes logic and the testability of conclusions. Its nuts and bolts are the practices considered next.

Observations, Hypotheses, and Tests

No matter what the topic, someone who wants to "do science" is guided by the following practices:

1. Observe some aspect of nature (including human biology), carefully check what others have learned about it, and then pose a question or identify a problem that your observation raises.

2. Develop a **hypothesis**, an educated guess, about possible answers to questions or solutions to problems.

3. With a hypothesis as a guide, make a **prediction**—that is, a statement of what you should observe about a body structure or function if you were to go looking for it. This is often called the "if-then" process. (*If* HIV leads to AIDS, *then* we should be able to observe a higher rate of AIDS among HIV-positive individuals.)

4. Devise ways to **test** the accuracy of your predictions, as by making systematic observations, building models, and doing experiments. (A **model** is a theoretical, detailed description or analogy that helps us visualize something that hasn't yet been observed directly.)

5. If the tests do not confirm the prediction, check for what might have gone wrong. Maybe you overlooked a factor that influenced the test results, or maybe the hypothesis is not a good one.

6. Repeat the tests or devise new ones—the more the better. Hypotheses supported by many different tests are more likely to be correct.

7. Objectively analyze and report the test results and conclusions you have drawn from them.

You might hear someone call these procedures "the scientific method," as if all scientists followed the same script in their work. They don't. Many observe, describe, and report on a topic and stop there, leaving subsequent steps to others. Scientists also can stumble onto findings they are not even looking for. The antibiotic penicillin was discovered by physician-researcher Alexander Fleming when the penicillin-producing mold strayed into a laboratory dish in which bacteria were growing.

When you use **logic,** you draw a conclusion that does not contradict the evidence used to support it. Lick a lemon slice and you notice that it is very sour. After more tries with the same result, you conclude that all lemons are very sour. You have correlated one specific (lemon) with another (sour). In this pattern of thinking, called *inductive logic,* a person derives a general statement from specific observations.

Express the general statement in "if-then" terms, and you have a hypothesis: "If you lick a lemon slice, then you will get a very sour taste in your mouth." This pattern of thinking is called *deductive logic*: you make inferences about specific consequences or specific predictions that *must follow* from a given hypothesis.

Suppose you decide to test your hypothesis by sampling as many lemon varieties as possible. As you do, you find that one, the Meyer lemon, is only slightly sour. You also discover that some people cannot taste lemons at all. Now, you must modify your original hypothesis: "If most people lick a lemon—except a Meyer lemon—they will get an extremely sour taste in their mouth." After sampling every known variety of lemon in the world, you conclude that the modified hypothesis is a good one. You can never prove it beyond all shadow of a doubt because there might be still other lemon trees growing in places people don't know about. But there is a high probability your hypothesis is right.

Comprehensive observations are a logical way to test the predictions that flow from a hypothesis. So are **experiments**. These tests simplify observation in nature or the laboratory by manipulating and controlling the conditions under which observations are made. Suitably designed observational and experimental tests allow you to predict that something will happen if a hypothesis isn't wrong (or won't happen if it *is* wrong). Figure 1.9 gives a general idea of the steps involved.

Every experiment is based on the premise that any aspect of nature has one or more underlying causes. This premise sets science apart from faith in the supernatural (meaning "beyond nature"). Every scientific hypothesis is potentially falsifiable—that is, it can be tested in the natural world in ways that might disprove it.

Further reading: Student Guide to InfoTrac on web site ➤

a Natalie, blindfolded, randomly plucks a jellybean from a jar of 120 green and 280 black jellybeans. That is a ratio of 30 to 70 percent.

b The jar is hidden before she removes her blindfold. She observes only a single green jellybean in her hand and assumes the jar holds only green jellybeans.

c Still blindfolded, Natalie randomly plucks 50 jellybeans from the jar and ends up with 10 green and 40 black ones.

d The larger sample leads her to assume one-fifth of the jar's jellybeans are green and four-fifths are black (a ratio of 20 to 80). Her larger sample more closely approximates the jar's green-to-black ratio. The more times Natalie repeats the sampling, the greater the chance she will come close to knowing the actual ratio.

Figure 1.10 A simple demonstration of sampling error.

EXPERIMENTAL DESIGN To get conclusive test results, experimenters follow certain practices. They refine their test design by reviewing information (published by other scientists) that may bear on their inquiry. They also design experiments that will test only one prediction of a hypothesis at a time. For each test, they set up a **control group**: a standard to which they can compare one or more experimental groups. Ideally, the control group is identical to the experimental group in every way *except* the variable being studied. A **variable** is an aspect of an object, individual, or event that may differ or change over time. For example, in experiments to identify the risk factors for lung cancer in women over age 50, one key variable is whether a woman smokes cigarettes.

SAMPLING ERROR It's rare when experimenters can observe all the individuals of a group under study. For instance, it would probably not be feasible to include all female smokers in North America in research on lung cancer. Instead, researchers use a sample group that is large enough to be representative of the whole. However, the larger the sample, the less likely it will be that any differences among the research subjects will distort the findings (Figure 1.10).

What Is a Scientific "Theory"?

Often, the first tests of a hypothesis are only a beginning, and its predictive power may undergo years of rigorous tests. If, after all this scrutiny, the hypothesis has not been disproved, scientists may use it to explain more data or observations, which could involve still more hypotheses. Only when a hypothesis meets such high standards can it become accepted as a **scientific theory**—an explanation of a broad range of related phenomena in the natural world. A scientific theory always remains open to further tests; it can be rejected if test results call it into question.

The Limits of Science

The call for objective testing strengthens the theories that emerge from scientists' work. Yet it also limits the kinds of studies scientists can do. For instance, science can't explain the "meaning of life." It can't tell us why each of us dies at one moment and not another. Answers to such questions are subjective: they come from within us, shaped by all our experiences and beliefs. And because people differ so much in this regard, questions that have subjective answers don't lend themselves to scientific analysis. The *Choices* essay on page 11 considers a few situations in which scientific information is only one of several factors in a complex decision-making process.

Outside the scientific arena, subjective answers can be both useful and enriching. For example, every culture and society has its own standards of morality and esthetics, and there are hundreds or thousands of different sets of religious beliefs. All guide their members in deciding what is important and good and what is not. All attempt to give meaning to what we do.

Every so often, scientists stir up controversy when they question or explain part of the world that was considered beyond natural explanation—that is, belonging to the supernatural. On occasion, a new, natural explanation runs counter to supernatural belief or to some other widely held, nonscientific view. This doesn't mean the scientists who raise the questions are any less moral, less law-abiding, less sensitive, or less caring than anyone else. It simply means one more standard guides their work: The external world, not internal conviction, must be the testing ground for scientific beliefs.

A scientific approach to studying human biology is based on asking questions, formulating hypotheses, making predictions, devising tests, and then objectively reporting the results.

A scientific theory is a testable explanation about the cause or causes of a broad range of related phenomena. It remains open to tests, revision, and tentative acceptance or rejection.

Heavy sun exposure is a risk factor for DNA damage that can lead to skin cancer (Figure 1.11), and these days many people slather on the sunscreen before they head outdoors. But cancer is only part of the equation when it comes to sun-related skin damage. Even after the link between sunlight and skin cancer was rock-solid, some biologists were puzzling over another question: What causes the leathery wrinkling, or *photoaging*, of the skin? Photoaging carves wrinkles into any face that has spent years soaking up sunshine, even in people who religiously use sunscreen.

Several years ago researchers Kerry Hanson and John Simon decided to probe this everyday observation more deeply. From other research, they knew that the rogue oxygen molecules called *free radicals* cause the physical damage to skin that shows up as photoaging (Figure 1.11*a*; see also Section 2.5). They also knew that free radicals can be a by-product of chemical events in skin cells—events that are triggered when light-sensitive chemicals are exposed to ultraviolet light. Several such chemicals have been identified, but Hanson and Simon thought that one in particular, a form of urocanic acid—UA, for short—might be the culprit. They focused on UA partly because it absorbs UV-A light—the ultraviolet wavelengths in sunlight that common sunscreens don't block. Another clue was the fact that UA had already been implicated in some other kinds of skin problems. Notice, however, that other hypotheses would have been reasonable, too. For instance, Hanson and Simon might have hypothesized that photoaging is triggered by poor protection against UV-B light—the wavelengths that most sunscreens *do* block to some extent. *In science, alternative hypotheses are the rule, not the exception.*

Moving from Hypothesis to Prediction

Armed with the hypothesis that UA responds to UV-A light in a way that promotes photoaging of skin cells, Hanson and Simon had to make a prediction that they and other scientists could readily test. They knew that in the 1980s researchers had shown that UV-A light could cause mouse skin to sag—a sign of photoaging—and they also knew that when UA absorbs light in one part of the UV-A spectrum, its chemical structure briefly changes. This fleeting change, which can be monitored using sophisticated measuring devices, might be a tip-off that something else was happening, too; for instance, it might signal events leading to the formation of free radicals. Following this reasoning, Hanson and Simon predicted that if they exposed UA to ultraviolet light, they would find that UV-A wavelengths associated with photoaged mouse skin would also be implicated in the chemical change in the structure of urocanic acid—a change that should be related to the production of free radicals.

 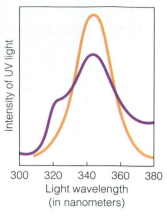

Figure 1.11 (**a**) This person's facial skin has the kind of wrinkles caused by photoaging. (**b**) Graph comparing the intensity of UV-A light associated with photoaging of the skin of mice (*purple* line), and with the formation of free radicals in a solution containing urocanic acid (*orange* line).

Testing a Prediction by Experiment

Some hypotheses can be tested by direct observation. For example, you can watch birds visiting a feeder to see whether one type of bird feeds more aggressively than others. Hanson and Simon's prediction required experimentation, using the experimental method described in Section 1.4. They began with experiments to uncover the cause of UA's brief structural change in response to UV-A light, and they found that the change occurred when UA absorbed and briefly stored light energy from a specific part of the UV-A spectrum. More study explained what happens when urocanic acid gives up its briefly stored light energy: it resumes its "normal" structure *and free radicals form.* New scientific understanding is always exciting, but the team still had to see whether the new information would support or contradict their prediction. The answer was clear when they graphed their data and compared it to the results on UV-A and photoaging in mice: as you can see in Figure 1.11*b*, exactly the same UV-A wavelengths are associated with both kinds of changes.

The research by Kerry Hanson and John Simon made headlines because it provided evidence of a direct link between the effects of UV-A light on urocanic acid in skin cells, the creation of free radicals, and photoaging. With aging a hot topic these days, it may not be long before we read about more such discoveries, each of them achieved through a carefully applied scientific method.

Every day, scientists develop hypotheses and make predictions that are testable, often through experimentation.

1.6

HOW DO YOU DEFINE DEATH?

In an austere hospital room, a young mother and father face the most anguished moment of their lives. Their baby has been born with no brain, except for a small portion of brain stem. For this child, none of the qualities that we associate with human life—the potential to think, feel joy or pain, learn, or speak—will ever be possible. For the moment, however, that bit of brain tissue controls lung and heart functions, so the baby's heart beats sporadically and its lungs occasionally take in air. Within hours or days, even those halting functions will cease.

In the United States, about one baby in a thousand is born with this condition, called *anencephaly*. For parents and physicians alike, the situation is agonizing. During the short period while the heart and lungs minimally function, other organs (such as the liver and kidneys) receive enough oxygen-carrying blood to keep them reasonably healthy. If the child is declared legally dead during that time, those organs can be transplanted and can bring the gift of life to others. If doctors wait until the nubbin of brain stem gives out, potentially transplantable organs will be irreversibly damaged by lack of oxygen and will be useless.

If this were your child, what course would you follow? You might not have a choice, if the matter went to court.

Many states have adopted a strict legal standard for such cases. A person may be declared legally dead only if *all* of the brain, including the brain stem, no longer functions. Some ethicists prefer this approach because it does not put society in the position of determining that some parts of the brain—but not others—make a person truly alive. Other people disagree strongly in cases in which the patient is in a "persistent vegetative state" (has no higher brain function) and there is no hope of recovery. In particular, advocates for a less strict definition of brain death point to the shortage of organs for transplants and the potential for saving other lives. They believe that, if all appropriate consents have been given, physicians should be allowed to terminate life support for "hopeless" patients and remove usable tissues and organs.

As individuals, we deal with this kind of issue in many ways. Some people hold religious beliefs requiring medical care as long as the heart can beat (Figure 1.12). Others draw up "living wills" designed to convey their wishes that life-support equipment, such as respirators and other high-tech machinery, not be used to prolong life artificially. Many people ignore the issue altogether, hoping that they or their loved ones will never have to grapple with it.

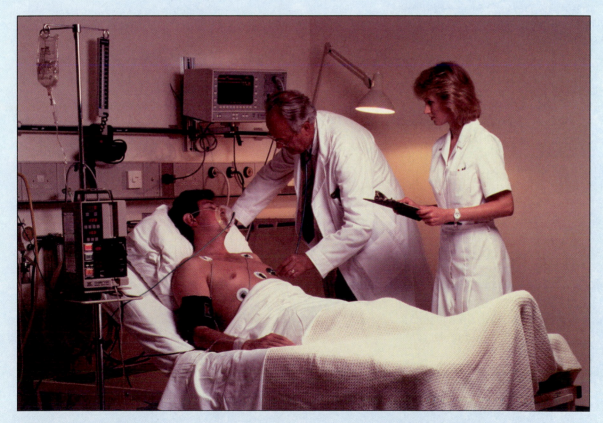

Figure 1.12 A critically ill hospital patient. Families of such individuals often find themselves confronting wrenching issues about the type of care their loved one should receive.

SUMMARY

1. Humans are living organisms. As such, they have the following characteristics:

   a. Their structure, organization, and interactions arise from the basic properties of matter and energy.

   b. Processes of metabolism and homeostasis maintain them in the living state.

   c. They have the capacity for growth, development, and reproduction, based on instructions contained in their DNA.

2. Diversity in features and characteristics arises through mutation. Mutations introduce changes in the DNA. The changes may lead to variations in the form, functioning, or behavior of individual offspring.

3. Individuals vary in their heritable traits (the traits that parents transmit to offspring). Such variations influence an organism's ability to survive and reproduce. Under prevailing conditions, some varieties of a given trait may be more adaptive than others; such traits will be "selected," whereas others will be eliminated through successive generations. Thus the population changes over time; it evolves. These points are central to the theory of evolution by natural selection.

4. There are various scientific methods, corresponding to many different fields of inquiry. The following key terms are important in all of those fields:

   a. Theory: an explanation of a broad range of related phenomena. An example is the theory of evolution by natural selection.

   b. Hypothesis: a possible explanation of a specific phenomenon.

   c. Prediction: a claim about what an observer can expect to see in nature if a theory or hypothesis is correct.

   d. Test: an effort to gather actual observations that may (or may not) match predicted or expected observations.

   e. Conclusion: a statement about whether a hypothesis (or theory) should be accepted, rejected, or modified, based on tests of the predictions derived from it.

5. Logic is a pattern of thought by which a person draws a conclusion that does not contradict evidence used to support the conclusion. A person using inductive logic derives a general statement from specific observations. With deductive logic, a person makes inferences about particular consequences or predictions that must follow from a hypothesis. This pattern of thinking is often expressed in "if-then" terms.

6. Predictions that flow from hypotheses can be tested by comprehensive observations or by experiments in nature or in the laboratory.

7. Experimental tests simplify observations in nature or the laboratory because the conditions under which the observations are made are manipulated or controlled. They are based on the premise of cause and effect: the phenomenon of interest has one or more underlying causes, which may or may not be obvious. A hypothesis is only scientific if it can be tested in ways that might disprove it.

8. A control group provides a standard against which one or more experimental groups (test groups) are compared. Ideally, it is the same as each experimental group except in the one variable being investigated.

9. A variable is a specific aspect of an object or event that might differ over time and between subjects. Experimenters directly manipulate the variable they are studying to support or disprove their prediction.

10. Test results are less likely to be distorted by sampling error when the sample is large and when sampling is repeated.

11. Systematic observations, hypotheses, predictions, and experimental tests are the foundation of scientific theories. The external world, not internal conviction, is the testing ground for those theories.

Review Questions

1. For this and subsequent chapters, make a list of the boldface terms that occur in the text. Write a definition next to each, and then check it against the one in the text.

2. Why is it difficult to give a simple definition of life? (For this and subsequent chapters, *italic numbers* following review questions indicate the sections in which the answers may be found.) *CI*

3. As a human, you are a living organism. List the characteristics of life that you exhibit. *1.1, 1.2, 1.3*

4. What is energy? Why is the concept of energy relevant to the study of human biology? *1.3*

5. Define metabolic activity and give an example of a metabolic event. *1.3*

6. What is the difference between reproduction and inheritance? *1.2*

7. What is an adaptive trait? *1.2*

8. Study Figure 1.7. Then, on your own, summarize what is meant by biological organization.

9. Describe the one-way flow of energy and the cycling of materials through the biosphere. *1.3*

10. Define and distinguish between: *1.4*

   a. hypothesis or scientific theory
   b. observational test and experimental test
   c. inductive and deductive logic

Self-Quiz (Answers in Appendix V)

1. The complex patterns of structural organization characteristic of life are based on instructions contained in _____.

2. _____ is the ability of cells to extract and transform energy from the environment and use it to maintain themselves, grow, and reproduce.

3. _____ is a state in which the body's internal environment is being maintained within a tolerable range. This state depends on _____, which are cells or structures that detect specific aspects of the environment.

4. Diverse structural, functional, and behavioral traits are considered to be _____ to changing conditions in the environment.

5. The capacity to evolve is based on variations in traits, which originally arise through _____.

6. A scientific approach to explaining some aspect of the natural world includes all of the following except _____.
   a. hypothesis   c. faith and public consensus
   b. testing   d. systematic observations

7. A related set of hypotheses that collectively explain some aspect of the natural world is a scientific _____.
   a. prediction   d. authority
   b. test   e. observation
   c. theory

8. The diagram below depicts the concept of:
   a. metabolism
   b. inheritance
   c. levels of organization
   d. energy transfers in the living world

parents

instructions in DNA

offspring

Critical Thinking: You Decide (Key in Appendix VI)

1. Court witnesses are asked "to tell the truth, the whole truth, and nothing but the truth." What are some problems inherent in the question? Is there a better alternative?

2. Design a test (or series of tests) to support or refute this hypothesis: A diet high in salt is associated with hypertension (high blood pressure), but hypertension is more common in people with a family history of the condition.

3. In popular magazine articles on diet, exercise, and other health-related topics, the authors often recommend a specific diet or dietary supplement. What kinds of evidence should the articles describe so you can decide whether you can accept their recommendations?

4. A scientific theory about some aspect of nature rests upon inductive logic. The assumption is that, because an outcome of some event has been observed to happen quite regularly, it will happen again. We can't know this for certain, however, because there is no way to account for all possible variables that may affect the outcome. To illustrate this point, Garvin McCain and Erwin Segal offer a parable:

> Once there was a highly intelligent turkey. It lived in a pen attended by a kind, thoughtful master, and had nothing to do but reflect on the world's wonders and regularities. Morning always began with the sky getting light, followed by the clop, clop, clop of its master's friendly footsteps, then by the appearance of delicious food. Other things varied—sometimes the morning was warm and sometimes cold—but food always followed footsteps. The sequence of events was so predictable, it became the basis of the turkey's theory about the goodness of the world. One morning, after more than one hundred confirmations of the theory, the turkey listened for the clop, clop, clop, heard it, and had its head chopped off.

The turkey learned the hard way that explanations about the world only have a high or low probability of not being wrong. Today, some people take this uncertainty to mean that "facts are irrelevant—facts change." If that is so, should we just stop doing scientific research? Why or why not?

Selected Key Terms

adaptive trait *1.2*	hypothesis *1.4*
ATP *1.1*	inheritance *1.2*
cell *1.1*	metabolism *1.1*
control group *1.4*	mutation *1.2*
DNA *1.1*	prediction *1.4*
energy *1.1*	reproduction *1.2*
experiment *1.4*	theory *1.4*
homeostasis *1.1*	variable *1.4*

Readings

Carey, S. 1994. *A Beginner's Guide to the Scientific Method*. Belmont, California: Wadsworth. Paperback.

Moore, J. 1993. *Science as a Way of Knowing—The Foundations of Modern Biology*. Cambridge, Massachusetts: Harvard University Press.

2 THE CHEMISTRY OF LIFE

Some Personal Chemistry

Right now you are breathing in oxygen. You would die without it. Two centuries ago, no one had a clue to what oxygen is, where it comes from, and how it helps keep people alive. Then researchers started unlocking the secrets of this chemical substance and others. With time, they began to innovate, developing things like fertilizers, nylons, cosmetics, aspirin, antibiotics, and plastic—a material that is everywhere we turn today.

Such chemical "magic" brings both benefits and problems. For example, the scale of agriculture required to sustain the human population, which now surpasses 6 billion, is astounding. Synthetic fertilizers boost crop yields by providing plants with nitrogen. The upside is that fertilizer-fed crops can help keep people from starving to death—a very real possibility in some parts of our world. The downside is that crop plants don't use all of the fertilizer that is applied to farm fields. Some nutrients are carried off in water from irrigation or rain and enter lakes, rivers, and seas where they nourish massive, detrimental "blooms" of algae and organisms that can sicken fish and shellfish.

Figure 2.1 At experimental wetlands in Corcoran, California, wild grasses help clean up the water from agricultural runoff when they take up selenium.

Sometimes "natural" chemicals become a problem when human activities alter the scheme of things. For instance, the chemical selenium is present in soil, and the tiny amounts of it in human food contribute to our health in important ways. In recent years, however, selenium has posed a dilemma. It can build up in the holding ponds that farmers often construct to receive agricultural runoff so that it will not enter streams and rivers. If enough of it accumulates, the pond water becomes toxic to livestock and birds. Inevitably, this kind of situation raises questions about possible harm to humans, and for good reason: people who consume excess selenium can develop various health problems, including vomiting, fatigue, and hair loss. In an effort to solve the dilemma, in 1996 biologist Norman Terry and his coworkers experimented with growing certain types of grasses in wetlands contaminated by selenium (Figure 2.1). As they predicted, the plants removed a significant amount of selenium from the water before it trickled away into the surrounding environment. Their research has pointed the way to an innovative chemical solution for a serious chemical problem.

In the final analysis, *everything in and around you is chemistry*. Every solid, liquid, or gaseous substance in and around you consists of one or more elements. Chemists define an **element** as a fundamental form of matter that has mass and takes up space.

Ninety-two elements occur naturally on earth; additional artificial ones have been created in laboratories. As is true of all organisms, most of the human body consists of four elements: oxygen, carbon, hydrogen, and nitrogen (Figure 2.2). Your body also contains some calcium, phosphorus, potassium, sulfur, sodium, and chlorine, as well as tiny but important amounts of so-called trace elements. A **trace element** is one that, like selenium, represents less than 0.01 percent of body weight.

The smallest unit that retains the properties of a given element is an **atom**, and atoms are the starting point for biological organization. Atoms of elements are building blocks of larger structures, called molecules. A molecule that consists of two or more elements is a compound. **Organic compounds** are built on a framework of linked atoms of the element carbon. In this chapter you'll learn more about how atoms are joined together, and about the four central types of organic compounds in living things—proteins, lipids (fats and related substances), carbohydrates, and nucleic acids (including DNA, the genetic material). In the following pages you may also gain a greater appreciation for water's role in your body's chemistry, and for the chemical processes that provide energy to power body functions.

HUMAN	
Oxygen	65
Carbon	18
Hydrogen	10
Nitrogen	3
Calcium	2
Phosphorus	1.1
Potassium	0.35
Sulfur	0.25
Sodium	0.15
Chlorine	0.15
Magnesium	0.05
Iron	0.004
Iodine	0.0004

EARTH'S CRUST	
Oxygen	46.6
Silicon	27.7
Aluminum	8.1
Iron	5.0
Calcium	3.6
Sodium	2.8
Potassium	2.6
Magnesium	2.1
Other elements:	1.5

Figure 2.2 Proportions of different elements in the Earth's crust and in the human body, as percentages of the total weight of each.

KEY CONCEPTS

1. All substances consist of one or more elements, such as hydrogen and carbon. Each element is composed of atoms, which are the smallest units that display the element's properties. An atom is composed of protons, electrons, and (except for hydrogen) neutrons.

2. An element's atoms all have the same number of protons and electrons, but they may vary slightly from one another in their number of neutrons. The variant forms of an element's atoms are called isotopes.

3. Atoms do not have an overall electric charge unless they become ionized—that is, unless they gain or lose electrons. An ion is an atom or molecule that has gained or lost one or more electrons and thereby has acquired an overall positive or negative charge.

4. When the arrays of electrons of two or more atoms interact in ways that unite them, the union is a chemical bond. The three types of chemical bonds are ionic bonds, covalent bonds, and hydrogen bonds.

5. Life on Earth is adapted to the properties of water. Especially important in human biology are water's ability to stabilize temperature and its capacity to mix with some substances and not others.

6. An organic compound has a framework of one or more carbon atoms. Other atoms are attached to the framework as cells build the organic compounds known as carbohydrates, lipids, proteins, and nucleic acids.

7. Cells use carbohydrates and lipids as building blocks and as their major energy sources. Many proteins are structural materials or enzymes (which speed up specific chemical reactions). Others transport substances or function in movements, changes in cell activities, and body defenses. Two nucleic acids, DNA and RNA, are the basis of inheritance and reproduction.

CHAPTER AT A GLANCE

The Structure of Atoms

An atom is the smallest unit that retains the properties of a given element. Small as atoms are, however, each one is composed of more than a hundred kinds of subatomic particles. The only ones you will need to consider in this book are **protons**, **electrons**, and **neutrons** (Figure 2.3).

All atoms have one or more protons, which carry a positive charge (p^+). Except for hydrogen, atoms also have one or more neutrons, which have no charge. Protons make up the atom's core region, the atomic nucleus. Electrons carry a negative charge (e^-), but they have no mass. They travel rapidly around the nucleus, and they occupy most of the atom's volume. Each atom has just as many electrons as protons.

Each element has a unique **atomic number**, which refers to the number of protons in its atoms. As Table 2.1 indicates, the atomic number is 1 for the hydrogen atom, which has one proton. It is 6 for the carbon atom, which has six protons.

Also, each element has a **mass number**, which is the combined number of protons and neutrons in the nucleus. A carbon atom, with six protons and six neutrons, has a mass number of 12.

You may also have heard the term "atomic weight." It is used in chemistry to refer to the relative masses of atoms, even though, technically speaking, mass is not quite the same thing as weight.

Why be concerned with atomic numbers and mass numbers? Such information gives us an idea of whether and how substances will interact. Equally important, it helps us predict how the substances of life will behave in cells, in the human body, and in the environment.

Isotopes—Varying Forms of Atoms

All atoms of an element have the same number of protons and electrons, but they may *not* have the same number of neutrons. When an atom of an element has more or fewer neutrons than the most common number, it is an **isotope**. For example, "a carbon atom" might be carbon 12 (six protons, six neutrons) or an isotope of carbon, carbon 14 (six protons, eight neutrons). These also can be written out as ^{12}C and ^{14}C. The prefix *iso-* means same, and all

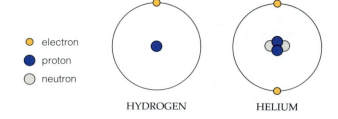

electron
proton
neutron

HYDROGEN   HELIUM

Figure 2.3 The hydrogen atom and helium atom according to one model of atomic structure. The model is simplified; the nuclei of these two representative atoms would be invisible specks at the scale used here.

isotopes of an element interact with other atoms in the same way. And because they do so, cells are able to use any isotope of an element for their metabolic activities.

Have you heard of radioactive isotopes? A French physicist, Henri Bequerel, discovered them in 1896, after he had placed a heavily wrapped rock atop an unexposed photographic plate in a desk drawer. The rock contained isotopes of uranium, which emit energy. A few days after the plate was exposed to the uranium emissions, a faint image of the rock appeared on it. Marie Curie, Bequerel's coworker, gave the name "radioactivity" to the substance's chemical behavior.

As we now know, a **radioisotope** is an isotope that has an unstable nucleus and that stabilizes itself by spontaneously emitting energy and particles. (Those particles are much smaller than protons, electrons, and neutrons.) This process, radioactive decay, transforms a radioisotope into an atom of a different element at a known rate. For instance, over a predictable time span, carbon 14 becomes nitrogen 14. This chapter's *Science Comes to Life* describes some of the ways radioisotopes are used in research and medicine.

Table 2.1	Atomic Number and Mass Number of Elements Common in Living Things		
Element	Symbol	Atomic Number	Most Common Mass Number
Hydrogen	H	1	1
Carbon	C	6	12
Nitrogen	N	7	14
Oxygen	O	8	16
Sodium	Na	11	23
Magnesium	Mg	12	24
Phosphorus	P	15	31
Sulfur	S	16	32
Chlorine	Cl	17	35
Potassium	K	19	39
Calcium	Ca	20	40
Iron	Fe	26	56
Iodine	I	53	127

An atom is the smallest unit of matter having the properties of a particular element. Each atom consists of subatomic particles, including one or more positively charged protons, negatively charged electrons, and (except for hydrogen) neutrons. Protons and neutrons make up the atom's nucleus.

Atoms of an element have the same number of electrons and protons, but the number of their neutrons can vary. The variant forms of atoms of the same element are called isotopes.

USING RADIOISOTOPES TO TRACK CHEMICALS AND SAVE LIVES

To a person who is ill, radioisotopes can be of more than passing interest. For example, they can allow a physician to diagnose disease with exquisite precision—and without subjecting the patient to exploratory surgery. Radioisotopes also are part of the actual treatment of certain cancers. Regardless of the use, radioisotopes always are handled with great care. For safety's sake, clinicians use only radioisotopes with extremely short half-lives. **Half-life** is the time it takes for half of a quantity of a radioisotope to decay into a different, more stable isotope.

TRACKING TRACERS   Various devices can detect radioisotope emissions. Thus, radioisotopes can be employed in tracers. A **tracer** is a substance with a radioisotope attached to it, rather like a shipping label, that a physician can administer to a patient. The tracking device then follows the tracer's movement through a pathway or pinpoints its destination.

SAVING LIVES   In nuclear medicine, radioisotopes are used under carefully controlled conditions to diagnose and treat diseases. Serious illnesses involving the thyroid are a case in point. The thyroid is the only gland in the human body that takes up iodine. After a tiny amount of a radioactive isotope of iodine, iodine 123, is injected into a patient's bloodstream, the thyroid can be scanned, producing the kinds of images you see in Figure 2.4.

Treatments for some cancers rely on radiation from radioisotopes to destroy or impair the activity of cells that are not functioning properly. For instance, such **radiation therapy** can be used to target a small, localized cancer—one that has not spread—and bombard it with radium 226 or cobalt 60.

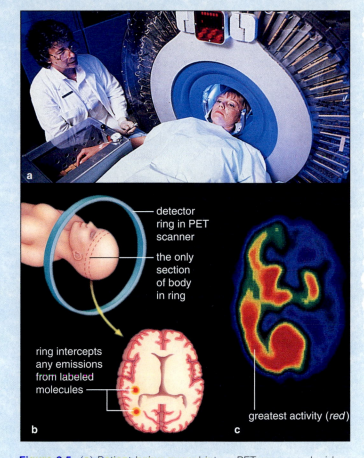

Figure 2.5 (**a**) Patient being moved into a PET scanner. Inside (**b**), a ring of detectors intercepts the radioactive emissions from labeled molecules that were injected into the patient. Computers analyze and color-code the number of emissions from each location in the scanned body region. (**c**) Brain scan of a child who has a neurological disorder. Different colors in a brain scan signify differences in metabolic activity. The right half of this brain shows very little activity. By comparison, cells of the left half absorbed and used the labeled molecule at expected rates.

In the figure, the labels read: "detector ring in PET scanner", "the only section of body in ring", "ring intercepts any emissions from labeled molecules", "greatest activity (*red*)".

normal   *enlarged*

cancerous

Figure 2.4 Scans of the thyroid gland from three patients who have ingested radioactive iodine, which is taken up by the thyroid.

Patients with irregular heartbeats use artificial pacemakers powered by energy emitted from plutonium 238. (This dangerous radioisotope is sealed in a case to prevent emissions from damaging body tissues.) Used with the imaging technology called positron-emission tomography (PET), radioisotopes provide information about abnormalities in the metabolic functions of specific tissues. The radioisotopes are attached to the sugar glucose or to some other biological molecule, then injected into a patient, who is moved into a PET scanner (Figure 2.5*a,b*). When cells in a target tissue absorb glucose, emissions from the radioisotope can be used to produce a vivid image of changes in metabolic activity. PET has been extremely useful for studying brain activity (Figure 2.5*c*).

WHEN ATOM BONDS WITH ATOM

Electrons and Energy Levels

Life requires chemical reactions, and in every chemical reaction, atoms interact. Three kinds of interactions are possible: A given atom may share one or more electrons, accept extra ones, or donate electrons to another atom. *Which* one of these events occurs depends on how many electrons an atom has and how they are arranged.

If you have ever played with magnets you know that like charges (++ or - -) repel each other and unlike charges (+ -) attract each other. Electrons carry a negative charge. In an atom, they repel each other, as you would predict. As you would also predict, electrons are attracted to the positive charge of protons. An atom's electrons spend as much time as possible near its protons—and as far away as possible from other electrons—by moving around the atomic nucleus in volumes of space. You can visualize these regions fairly accurately if you imagine the electrons contained inside shells. Think of a hollow Ping-Pong ball suspended inside a hollow baseball, which is suspended inside a hollow soccer ball, and you begin to get the idea. Hydrogen, the simplest atom, has just one electron in a single shell, as sketched in Figure 2.6. In atoms of every other element, this shell holds two electrons. Additional electrons are in shells farther from the nucleus, as shown in Figure 2.7.

The shells around an atom's nucleus are equivalent to energy levels. The one closest to the nucleus represents the *lowest energy level* in the atom, and shells farther from the nucleus are at progressively *higher* energy levels. It stands to reason, then, that since the atoms of different elements have different numbers of electrons, they also have different numbers of shells that electrons can occupy. The main point to remember, however, is that, at most, only eight electrons can be in a shell. This means that larger atoms have many shells. At the back of this book, Appendix II lists the known elements, including some large, extremely complex ones whose atoms have lots of electrons occupying shell after shell.

Chemical Bonds

Why be interested in how electrons are organized in atoms? Because that information is the heart of one of

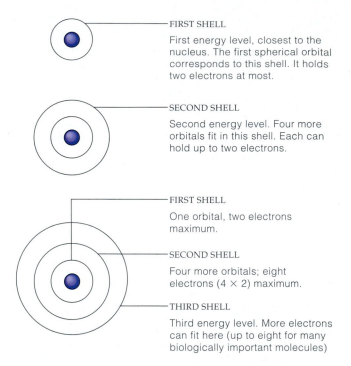

FIRST SHELL

First energy level, closest to the nucleus. The first spherical orbital corresponds to this shell. It holds two electrons at most.

SECOND SHELL

Second energy level. Four more orbitals fit in this shell. Each can hold up to two electrons.

FIRST SHELL

One orbital, two electrons maximum.

SECOND SHELL

Four more orbitals; eight electrons (4 × 2) maximum.

THIRD SHELL

Third energy level. More electrons can fit here (up to eight for many biologically important molecules)

Figure 2.7 Shell model of electron distribution in atoms. Three of the many possible energy levels (shells) are shown.

the most important events in living systems, chemical bonding. We define a **chemical bond** as a union between the electron structures of atoms. Having a filled outer shell is the most stable state for atoms. However, many atoms do not have enough electrons to fill up their outermost shell. And atoms with an unfilled outer shell tend to form chemical bonds with other atoms in order to fill their outer shell. Atoms of oxygen, carbon, hydrogen, and nitrogen—the four most abundant elements in the human body—are this way (see Figure 2.2). The single shell of hydrogen and helium atoms is full when it contains two electrons. Other kinds of atoms that have unfilled outer shells follow the *octet rule*—they take part in chemical bonds that result in a filled outer shell containing eight electrons. To predict whether two atoms will interact, you can check for electron vacancies in their outermost shell. Vacancies mean that, under the right conditions, an atom might give up, gain, or share electrons. In that event, the number or distribution of its electrons will change.

In Figure 2.8, which gives the electron distribution for several atoms, electrons are assigned to an energy level (a circle, representing a shell). You can easily count the electron vacancies in the outermost shell of each atom. As the table shows, helium and neon have no vacancies. They are among the *inert* atoms, which generally do not take part in chemical reactions.

———— electron orbital

———— nucleus

HYDROGEN ATOM

Figure 2.6 The electron structure of a hydrogen atom. A single electron occupies the lone spherical orbital.

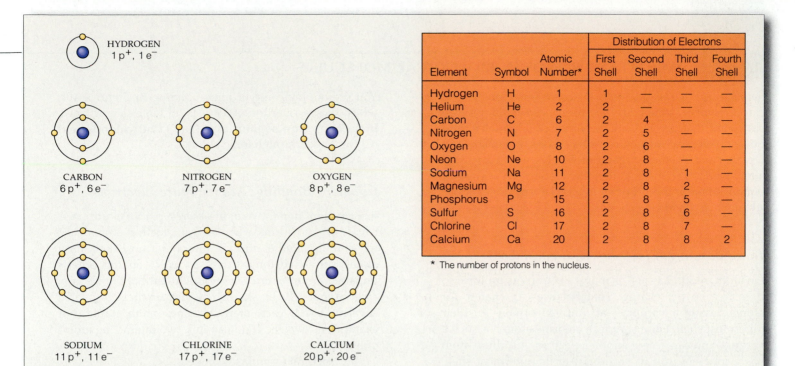

Element	Symbol	Atomic Number*	Distribution of Electrons			
			First Shell	Second Shell	Third Shell	Fourth Shell
Hydrogen	H	1	1	—	—	—
Helium	He	2	2	—	—	—
Carbon	C	6	2	4	—	—
Nitrogen	N	7	2	5	—	—
Oxygen	O	8	2	6	—	—
Neon	Ne	10	2	8	—	—
Sodium	Na	11	2	8	1	—
Magnesium	Mg	12	2	8	2	—
Phosphorus	P	15	2	8	5	—
Sulfur	S	16	2	8	6	—
Chlorine	Cl	17	2	8	7	—
Calcium	Ca	20	2	8	8	2

* The number of protons in the nucleus.

HYDROGEN
$1 p^+, 1 e^-$

CARBON
$6 p^+, 6 e^-$

NITROGEN
$7 p^+, 7 e^-$

OXYGEN
$8 p^+, 8 e^-$

SODIUM
$11 p^+, 11 e^-$

CHLORINE
$17 p^+, 17 e^-$

CALCIUM
$20 p^+, 20 e^-$

Figure 2.8 Examples of the shell model of the distribution of electrons in atoms. Hydrogen, carbon, and other atoms having electron vacancies (unfilled orbitals) in their outermost shell tend to give up, accept, or share electrons. The electron structures of helium and other atoms with no electron vacancies in the outermost shell are much more stable. Such atoms tend not to react with other atoms.

We use symbols for elements when writing *formulas*, which identify the composition of compounds. For example, water has the formula H_2O. Symbols and formulas are used in *chemical equations*, which are representations of reactions among atoms and molecules.

In written chemical reactions, an arrow means "yields." Substances entering a reaction (reactants) are to the left of the arrow. Reaction products are to the right. For example, the reaction between hydrogen and oxygen that yields water is summarized this way:

$$2 H_2 \quad + \quad O_2 \quad \longrightarrow \quad 2 H_2O$$

4 hydrogens   2 oxygens   4 hydrogens, 2 oxygens

Note that there are as many atoms of each element to the right of the arrow as there are to the left. Although atoms are combined in different forms, none is consumed or destroyed in the process. The total mass of all products of any chemical reaction equals the total mass of all its reactants. All equations used to represent chemical reactions, including reactions in cells, must be balanced this way.

Figure 2.9 Chemical bookkeeping.

Combinations of Atoms

When chemical bonding joins atoms, the new chemical structure is a **molecule**. Many molecules contain atoms of only one element. Molecular nitrogen (N_2), with its two nitrogen atoms, is an example. Figure 2.9 explains how to read the notation used in representing chemical reactions that occur between atoms and molecules.

Many other kinds of molecules are **compounds**. That is, they consist of two or more elements in proportions that never vary. For example, water is a compound. Every water molecule has one oxygen atom bonded with two hydrogen atoms. No matter where water molecules are—in rain clouds or in a lake or in your bathtub—they *always* have twice as many hydrogen as oxygen atoms.

In a **mixture**, two or more kinds of molecules simply mingle, in proportions that *can* vary. To take an example, the sugar sucrose is a compound of carbon, hydrogen, and oxygen. Swirl together molecules of sucrose and water and you get a mixture—sugar-sweetened water. Increase the proportion of sucrose in the mixture and it will still be a mixture—just an intensely sweet one, such as syrup.

In an atom, electrons occupy orbitals—volumes of space around the nucleus. In a simple model, orbitals are arranged as a series of shells. Successive shells in an atom's electron structure correspond to levels of energy. The farther away an energy level is from the nucleus, the greater its energy.

Only one or two electrons can be in an orbital. Atoms with unfilled orbitals in their outermost shell tend to interact with other atoms; those with no vacancies do not.

In molecules of an element, all of the atoms are of the same kind. In molecules of a compound, atoms of two or more elements are bonded together, in proportions that stay the same. In a mixture, the proportions can vary.

2.4 IMPORTANT BONDS IN BIOLOGICAL MOLECULES

Eat more fruit! Drink your milk! Probably for as long as you can remember, somebody has been been telling you to eat foods that are rich in carbohydrates, proteins, and other "biological molecules." Only living organisms put together and use these organic molecules, which consist of a few kinds of atoms held together by only a few kinds of bonds. Mainly, these bonds are those known as ionic, covalent, and hydrogen bonds.

Ionic Bonding: Electrons Gained or Lost

An atom, recall, has just as many electrons as protons, so it carries no net charge. That balance can change for atoms having a vacancy—an unfilled orbital—in their outermost shell. For example, a chlorine atom has such a vacancy and can accept another electron. A sodium atom has a lone electron in an orbital in its outermost shell, and that electron can be knocked out or pulled away from the orbital. Any atom that has either lost or gained one or more electrons is an **ion**. The balance between its protons and its electrons has shifted, so the atom has become ionized; it has become positively or negatively charged (Figure 2.10a).

In living cells, neighboring atoms commonly accept or donate electrons among one another. When one atom loses an electron and one gains, both become ionized. Depending on cellular conditions, the ions may separate, or they may remain together as a result of the mutual attraction of their opposite charges. An association of two ions that have opposing charges is called an **ionic bond**.

You see one outcome of ionic bonding in Figure 2.10b, which shows a portion of a crystal of table salt, or NaCl. In such crystals, sodium ions (Na^+) and chloride ions (Cl^-) interact through ionic bonds.

Covalent Bonding: Atoms Share Electrons

In a **covalent bond**, two atoms *share* a pair of electrons. A covalent bond forms when two interacting atoms each have an unpaired electron in their outermost shell. Each atom's attractive force exerts a "pull" on the other's unpaired electron, but not enough to attract the electron away completely. By sharing the unpaired electrons in a covalent bond, each atom becomes more stable. For example, a covalent bond can link two interacting atoms of hydrogen.

In structural formulas, a single line between two atoms means they share a *single* covalent bond. Molecular hydrogen (a molecule made up of two hydrogen atoms) has such a bond, which can be written as H—H. In a *double* covalent bond, two atoms share two pairs of electrons. Molecular oxygen (O=O) is like this. In a *triple* covalent bond, two atoms share three pairs of electrons. An example is molecular nitrogen (N≡N). All three examples happen to be gaseous molecules. Every time you take a breath, you inhale H_2, O_2, and N_2 molecules.

MOLECULAR HYDROGEN (H_2)

Covalent bonds are polar or nonpolar. In a *nonpolar* covalent bond, the participating atoms exert the same pull on electrons and share them equally. The term "nonpolar" refers to the fact

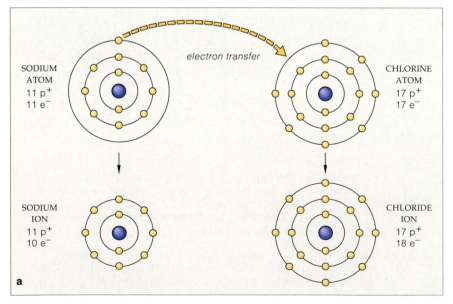

Figure 2.10 (a) Ionization by way of an electron transfer. In this case, a sodium atom donates the lone electron in its outermost shell to a chlorine atom, which has an unfilled orbital in *its* outer shell. This interaction results in a sodium ion (NA^+) and a chloride ion (Cl^-). (b) In each crystal of table salt, or NaCl, many sodium and chloride ions stay together because of the mutual attraction of their opposite charges. Their interaction is an example of ionic bonding.

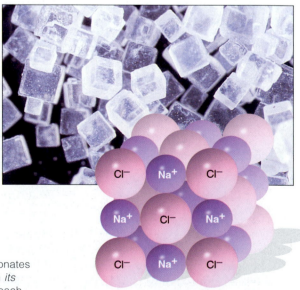

b Crystals of sodium chloride (NaCl).

that there is no difference in charge at the two ends of the bond (that is, at its two "poles"). Molecular hydrogen is a simple example. Its two hydrogen atoms, each with one proton, attract the shared electrons equally.

In a *polar* covalent bond, nuclei of different atoms (which have different numbers of protons) do not exert the same pull on shared electrons. One exerts more pull, and that atom ends up with a slight negative charge; the atom is "electronegative." Its effect is balanced out by the other atoms, whose nuclei exert less pull and end up with a slight positive charge. In other words, taken together, the atoms interacting in a polar covalent bond have no *net* charge—but the charge is distributed unevenly between the bond's two ends.

As an example, a water molecule has two polar covalent bonds: H—O—H. In this molecule, electrons are less attracted to the hydrogens than to the oxygen, which has more protons. A water molecule carries no net charge, but you will see shortly that its polarity makes it weakly attract neighboring polar molecules and ions.

Hydrogen Bonding

Covalent bonds are strong, holding atoms together in molecules in specific arrangements. In some cases, those arrangements position atoms in ways that give rise to weak attractions and repulsions between the charged regions of large molecules, or between molecules and ions. Such interactions break and form easily, like the connections between skydivers performing elaborate stunts (Figure 2.11). Yet such interactions are important in the structure and functioning of biological molecules.

Figure 2.11 Like skydivers who briefly clasp hands to form an orderly pattern, weak attractions within and between molecules (and between ions and molecules) can form and break easily.

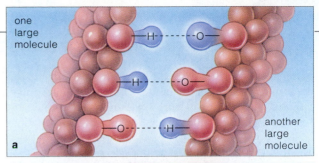
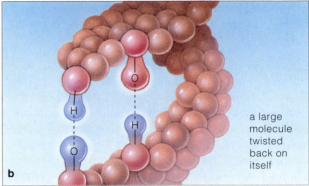

Figure 2.12 Two examples of hydrogen bonding. Compared with a covalent bond, a hydrogen bond is easier to break. Collectively, however, hydrogen bonds are important in water, DNA, proteins, and many other substances.

For example, a **hydrogen bond** is a weak attraction between an electronegative atom (such as an oxygen or nitrogen atom taking part in a polar covalent bond) and a hydrogen atom taking part in a second polar covalent bond. Hydrogen's slight positive charge weakly attracts the atom with the slight negative charge (Figure 2.12*a*).

Hydrogen bonds may form between two or more molecules. They also may form between different parts of the same molecule, where it twists and folds back on itself (Figure 2.12*b*). For example, DNA, the genetic material, is built of two parallel strands of chemical units, and the strands are held together by numerous hydrogen bonds. Individually, the hydrogen bonds break easily, but collectively they help stabilize DNA's structure. Similarly, hydrogen bonds form between the molecules that make up water. As you will read in Section 2.6, they contribute to water's life-sustaining properties.

In an ionic bond, two ions of opposite charge attract each other and remain together. Ions form when atoms gain or lose electrons, acquiring a net positive or negative charge.

In a covalent bond, atoms share electrons. If the atoms share electrons equally, the bond is nonpolar. If the sharing is not equal, the bond itself is polar (slightly positive at one end, slightly negative at the other).

In a hydrogen bond, a small, highly electronegative atom or molecule interacts weakly with a neighboring hydrogen atom that is already taking part in a polar covalent bond.

ANTIOXIDANTS—FIGHTING FREE RADICALS

Stop electron transfers in your body cells and your life will stop—*blam!* Plain and simple, your survival depends on all that chemical swapping. Yet there's another, not-so-great side to one sort of transfer, the kind called oxidation.

THE FREE RADICAL FACTORY *Oxidation* is the name chemists give to the process by which an atom or molecule *loses* one or more electrons to another atom or molecule. It's what causes a match to burn and an iron nail to rust, and it's a chemical workhorse in all kinds of important metabolic events in body cells. For example, oxidation is a crucial part of the chemistry that generates the body's energy (a topic of Chapter 3). Unfortunately, however, the countless oxidations that go on in a person's cells throughout life also release highly unstable molecules called **free radicals**. Each one is a molecule (such as O_2^-) that includes an oxygen atom that lacks a full complement of electrons in its outer shell. Because a free radical is shy an electron, it has powerful potential to oxidize: to fill its empty electron slot, it can readily "steal" an electron from another, stable molecule. Like stealing a tire from a car, the oxidative theft will disrupt both the structure of the affected molecule and its ability to operate normally. Depending on how a molecule is affected, the oxidation can even be a fatal blow.

Like a lot of other chemical scenarios in the body, the real issue with free radicals is balance. The body has uses for a small supply of them, but the numbers produced by normal cell functions far exceed this "healthy" amount. And in large numbers, free radicals pose a serious threat to essential molecules, including a cell's DNA. Other chemical assaults on the body, such as cigarette smoke and the ultraviolet radiation in sunlight, produce additional free radicals, making a problematical situation worse. Elsewhere in this book you will discover that current research on heart disease, aging, lung cancer, and some other health concerns is turning up evidence that free radical damage is at least partly to blame.

THE ANTIOXIDANT ARMY For all of the foregoing reasons, the shelves of supermarkets and "health food" stores are loaded with vitamins and other substances touted as "antioxidants." Behind the hype there's some interesting science. An **antioxidant** is a chemical that can give up an electron to a free radical before the rogue does damage to a DNA molecule or some other cell constituent. The body manufactures some antioxidants, including the hormone melatonin (Chapter 12), which give up electrons to free radicals and thereby neutralize them. This home-grown antioxidant army isn't enough to balance the ongoing production of free radicals, however, and we now know of other sources of additional antioxidants.

For instance, ascorbic acid—also known as vitamin C—is an antioxidant, in addition to serving other crucial roles in cell operations (Chapter 6). A molecule of ascorbic acid

Figure 2.13 Antioxidants occur in many plant-derived foods—just one more reason why a healthful diet includes lots of orange and green vegetables and fruits.

has two hydrogen atoms positioned so that they can be stripped off the molecule (making it dehydroascorbic acid). (Because each hydrogen atom has one electron, for all intents and purposes removing it is the same chemically as removing an electron.) Once a cell has enough vitamin C to meet its other needs, every "extra" molecule of vitamin C can give up two electrons, neutralizing two free radical molecules that might otherwise cause damage. Consider the following case in point. Nitrite, a reactive compound built of a nitrogen atom and two oxygens, is a common food additive in processed meats. It is worrisome because it reacts with food components called amines to produce nitrosamines—substances that have been convincingly linked to some kinds of cancer. In some promising research, however, vitamin C has been shown to operate as an antioxidant, giving up an electron to nitrite. The reaction forms nitrate, a much more stable—and benign—compound having a nitrogen and *three* oxygens.

Vitamin E is another free radical scavenger. So are some carotenoids, which are pigments in orange and leafy green vegetables, among other foods (Figure 2.13). Alpha carotene (found in foods such as carrots, pumpkins, and some squashes) and beta cryptoxanthin are two of these; another is beta carotene, although it may not be as effective as other antioxidants, overall. In general, nutritionists recommend adding antioxidants to the diet by eating lots of the foods that contain them, using supplements only in moderation. Antioxidant-rich foods typically are low in fat, high in fiber, and just plain good for us in plenty of other ways as well.

PROPERTIES OF WATER THAT SUPPORT LIFE

No sprint through basic chemistry is complete without an introduction to H₂O—water. Life on our planet may have begun in water, and for us humans (and other organisms) it is indispensable. By weight, the body is roughly two-thirds water, which is essential to maintaining the shape and internal structure of body cells. Water makes up most of the fluid in blood, and a slew of life-sustaining chemical reactions either require water as a reactant or occur only after other substances dissolve in water. Water is suited for these and other pivotal roles in the living world because of its unusual properties.

Polarity: Hydrogen Bonding

You may recall that a water molecule has no net charge, but the charges it does carry are distributed unevenly. The water molecule's oxygen "end" is a bit negative and its hydrogen end is a bit positive (Figure 2.14*a*). This uneven distribution of electrical charge makes water molecules polar; the polarity in turn makes every water molecule attract other water molecules and form hydrogen bonds with them. Collectively, the bonds are so strong that they hold the water molecules close together; this is why water is a liquid at room temperature. Without such strong hydrogen bonding between H₂O molecules, there wouldn't be liquid water on earth—and we wouldn't be here, either.

Water attracts and hydrogen-bonds with other polar substances, such as sugars. Because polar molecules are attracted to water, they are termed **hydrophilic**, which means "water-loving." Conversely, water repels nonpolar substances, including oils. Hence nonpolar molecules are **hydrophobic**, or "water-dreading." You can observe this effect by shaking a bottle of water and salad oil, then letting it settle. Soon, new hydrogen bonds replace those that broke apart when you shook the bottle. As water molecules reunite they push aside the oil molecules, which cluster as droplets or a film at the water's surface.

More Key Properties of Water

Water's hydrogen bonds give it a high *heat capacity*—they enable it to absorb a great deal of heat energy before it warms significantly or evaporates. In a volume of water, a large amount of heat must be applied to break the many hydrogen bonds present. Water's ability to absorb significant heat before becoming hot is the reason it can be used to cool a hot automobile engine. Likewise, water helps stabilize the temperature inside cells, which are mostly water. The chemical reactions in cells constantly produce heat, yet cells must stay fairly cool because their proteins can't function properly outside narrow temperature limits.

When enough heat energy is present, hydrogen bonds between water molecules do break apart and they do not

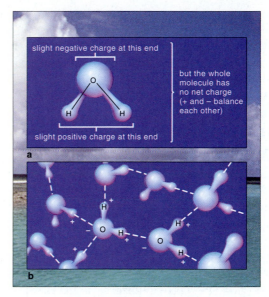

Figure 2.14 (**a**) Polarity of a water molecule. (**b**) Hydrogen bonds between molecules in liquid water (dashed lines).

re-form. Then liquid water evaporates—molecules at its surface escape into the air. When many water molecules evaporate, heat energy is lost. This is why sweating helps cool you off on a hot, dry day. Sweat is 99 percent water, and when it evaporates from the millions of sweat glands in your skin, heat leaves with it.

Finally, water is a superb *solvent*, which means that ions and polar molecules readily dissolve in it. Any dissolved substance is a **solute**. When a substance dissolves, clusters of water molecules surround its individual molecules or ions. This is what happens to solutes in cellular fluid, in blood, and in all other body fluids. Most chemical reactions in the body occur in water-based solutions.

Figure 2.15 Clusters of water molecules around ions.

Figure 2.15 shows what happens when you pour some table salt (NaCl) into a cup of water. After a while, the salt crystals separate into Na⁺ and Cl⁻. Each Na⁺ attracts the negative end of some of the water molecules while each Cl⁻ attracts the positive end of others.

A water molecule is polar. This polarity allows water molecules to hydrogen-bond with one another and with other polar (hydrophilic) substances. Water molecules tend to repel nonpolar (hydrophobic) substances.

Among other effects, the hydrogen bonds in water help it stabilize temperature in body fluids and allow it to dissolve many substances.

ACIDS, BASES, AND BUFFERS

A great variety of ions dissolved in the fluids inside and outside cells influence cell structure and functioning. Especially important are **hydrogen ions**, or H⁺. Hydrogen ions are the same thing as free (unbound) protons. They have far-reaching effects largely because they are chemically active and there are so many of them.

The pH Scale

When water is in its liquid form, at any given moment some water molecules are breaking apart into hydrogen ions and hydroxide ions (OH⁻). This ionization of water is the basis for the **pH scale**, as in Figure 2.16. Biologists use this scale to measure the concentration of H⁺ in water, blood, and other fluids. Pure water (not rainwater or tapwater) always contains just as many H⁺ as OH⁻ ions, and so can other fluids. Whenever this situation exists, the pH is said to be neutral. On the pH scale, which ranges from 0 (the highest H⁺ concentration) to 14 (the lowest), neutrality corresponds to a value of 7.

Starting at neutrality, each change by one unit of the pH scale corresponds to a tenfold increase or decrease in H⁺ concentration. An easy way to sense the differences is to dissolve a bit of baking soda (pH 9) on your tongue, then follow it with water (7), then lemon juice (2.3).

How Do Acids Differ from Bases?

When it dissolves in water, any substance categorized as an **acid** donates protons (H⁺) to other solutes or to water molecules. By contrast, any substance categorized as a **base** accepts H⁺ when it is dissolved in water. When either an acid or a base dissolves, OH⁻ then forms in the solution also. *Acidic* solutions, such as lemon juice, the gastric fluid in your stomach, and coffee, release more H⁺ than OH⁻; their pH is below 7. *Basic* solutions, such as seawater, baking soda, and egg white, release more OH⁻ than H⁺. Such solutions are also called *alkaline* fluids; they have a pH above 7.

The fluid inside most human cells is about 7 on the pH scale. Body cells also are surrounded by fluids, and the pH values of most of those fluids are slightly higher; they range between 7.3 and 7.5. The pH of the fluid portion of blood is in the same range.

Chemists characterize most acids as being either weak or strong. Weak ones, such as carbonic acid (H₂CO₃), don't readily donate H⁺. Depending on the pH, they just as easily accept H⁺ after giving it up, so they alternate between acting as an acid and acting as a base. On the other hand, strong acids totally give up H⁺ when they dissociate in water. Hydrochloric acid (HCl), nitric acid (HNO₃), and sulfuric acid (H₂SO₄) are examples.

High concentrations of strong acids or strong bases can play important roles in the body. For instance, when you consume food, cells in your stomach are stimulated to secrete HCl, which separates into H⁺ and Cl⁻ in water. The H⁺ ions make the fluid in your stomach more acidic, and the increased acidity switches on enzymes that can digest (chemically break down) food particles. It also helps kill harmful bacteria. However, eating a meal that contains too much of certain kinds of foods can lead to "acid stomach." Antacids, such as milk of magnesia, are strong bases. When milk of magnesia dissolves in your stomach, it releases magnesium ions and OH⁻ to neutralize the acid. OH⁻ combines with excess H⁺ in your stomach fluid, the reaction raises the pH, and your acid stomach goes away.

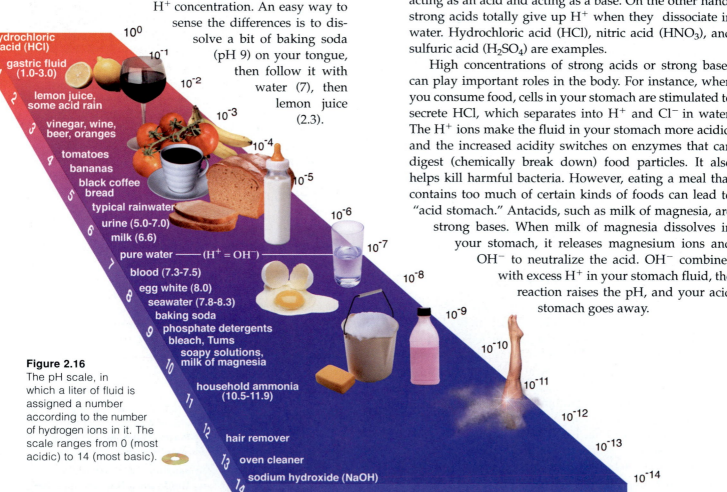

Figure 2.16
The pH scale, in which a liter of fluid is assigned a number according to the number of hydrogen ions in it. The scale ranges from 0 (most acidic) to 14 (most basic).

hydrochloric acid (HCl)
10^0
0
gastric fluid (1.0–3.0)
10^{-1}
1
lemon juice, some acid rain
10^{-2}
2
vinegar, wine, beer, oranges
10^{-3}
3
tomatoes
bananas
10^{-4}
4
black coffee
bread
10^{-5}
5
typical rainwater
urine (5.0–7.0)
10^{-6}
6
milk (6.6)
pure water ———— (H⁺ = OH⁻)
10^{-7}
7
blood (7.3–7.5)
egg white (8.0)
10^{-8}
8
seawater (7.8–8.3)
baking soda
phosphate detergents
bleach, Tums
10^{-9}
9
soapy solutions, milk of magnesia
10^{-10}
10
household ammonia (10.5–11.9)
10^{-11}
11
10^{-12}
12
hair remover
10^{-13}
13
oven cleaner
sodium hydroxide (NaOH)
10^{-14}
14

Further reading: Student Guide to InfoTrac on web site →

High concentrations of strong acids or bases also can disrupt the external environment and pose dangers to life. Read the labels on bottles of ammonia, drain cleaner, and other products that are often stored in households. Many can cause severe chemical burns. So can sulfuric acid in car batteries. Smoke from fossil fuels, exhaust from motor vehicles, and nitrogen fertilizers release strong acids, which alter the pH of rain (Figure 2.17). Acid rain is an ongoing environmental problem that is considered in more detail in Chapter 25.

Figure 2.17 Sulfur dioxide emissions from a coal-burning power plant. Camera lens filters revealed the otherwise invisible emissions. Sulfur dioxide and other airborne pollutants dissolve in water vapor to form acidic solutions. They are a major component of acid rain.

Buffers against Shifts in pH

Chemical reactions in cells are sensitive to even slight shifts in pH, for H^+ and OH^- can combine with many different molecules and alter their functions. Normally, control mechanisms minimize unsuitable shifts in pH, as they do when HCl enters the stomach in response to a meal. Many of the controls involve buffer systems.

A **buffer system** is a partnership between a weak acid and a base that forms when the acid dissolves in water. The two work as a pair to counter slight shifts in pH. Remember, when a strong base enters a fluid, the OH^- level rises. But a weak acid neutralizes part of the added OH^- by combining with it. By this interaction, the weak acid's partner forms. Later, if a strong acid floods in, the base will accept H^+ and thereby become its partner in the system.

Bear in mind, the action of a buffer system cannot make new hydrogen ions or eliminate ones that already are present. It can only bind or release them.

A variety of buffer systems operate in the blood and in tissue fluids—the internal environment of the body. For example, chemical reactions in the lungs and kidneys help control the acid–base balance of this environment, at levels suitable for life (Sections 9.5 and 10.8). For now, consider what happens when the blood level of H^+ decreases and the blood is not as acidic as it should be. At such times, carbonic acid that is dissolved in blood releases H^+ and becomes the partner base, bicarbonate:

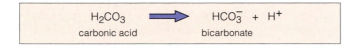

$$H_2CO_3 \longrightarrow HCO_3^- + H^+$$
carbonic acid   bicarbonate

When the blood becomes more acidic, more H^+ is bound to the base, thus forming the partner acid.

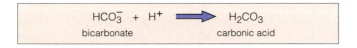

$$HCO_3^- + H^+ \longrightarrow H_2CO_3$$
bicarbonate   carbonic acid

Uncontrolled shifts in pH in body fluids are recipes for disaster. If blood's pH (7.3–7.5) declines to even 7, a person will enter a *coma*, a sometimes irreversible state of unconsciousness. An increase to 7.8 can lead to *tetany*, a potentially fatal condition in which the body's skeletal muscles contract uncontrollably. In *acidosis*, carbon dioxide builds up in the blood, too much carbonic acid forms, and blood pH severely decreases. The condition called *alkalosis* is an uncorrected increase in blood pH. Both acidosis and alkalosis weaken the body and can be lethal.

Salts

Salts are compounds that release ions *other than* H^+ and OH^- in solutions. Salts and water often form when a strong acid and a strong base interact. Depending on a solution's pH value, salts can form and dissolve easily. Consider how sodium chloride forms, then dissolves:

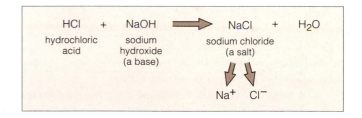

$$HCl + NaOH \longrightarrow NaCl + H_2O$$
hydrochloric   sodium   sodium chloride
acid   hydroxide   (a salt)
              (a base)

$$Na^+ \quad Cl^-$$

Many salts dissolve into ions that have key functions in cells. For example, the activity of nerve cells depends on ions of sodium, potassium, and calcium, and your muscles contract with the help of calcium ions.

Hydrogen ions (H^+) and other ions dissolved in the fluids inside and outside cells affect cell structure and function.

When dissolved in water, acidic substances release H^+, and basic (alkaline) substances accept them. Certain acid–base interactions, such as in buffer systems, help maintain the pH value of a fluid—that is, its H^+ concentration.

A buffer system counters slight shifts in pH by releasing hydrogen ions when their concentration is too low or by combining with them when the concentration is too high.

Salts are compounds that release ions other than H^+ and OH^-, and many of those ions have key roles in cell functions.

ORGANIC COMPOUNDS: BUILDING ON CARBON ATOMS

The Molecules of Life

Only living cells synthesize the molecules characteristic of life—the carbohydrates, lipids, proteins, and nucleic acids. Different classes of these biological molecules serve as packets of instantly available energy, energy stores, structural materials, metabolic workers, libraries of hereditary information, and signals between cells.

As this chapter's introduction noted, the molecules of life are organic compounds: they have hydrogen and often other elements bonded to carbon atoms (mainly by covalent bonds). At one time, chemists thought "organic" substances were the ones they obtained from animals and vegetables, as opposed to "inorganic" ones they got from minerals. The term is still used today, even though organic compounds now can be made in laboratories.

Carbon's Bonding Behavior

Like every living organism, the human body is mostly oxygen, hydrogen, and carbon. Much of the oxygen and hydrogen is in the form of water. Remove the water, and carbon makes up more than half of what's left.

Carbon's importance in life arises from its versatile bonding behavior. As shown in the sketch below, *each carbon atom can share pairs of electrons with as many as four other atoms.* Each covalent bond formed in this way is relatively stable because the carbon atoms share pairs of electrons equally. Such bonds link carbon atoms together in chains. These form a backbone to which hydrogen, oxygen, and other elements become attached.

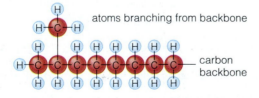
single covalent bond

carbon atom

The angles of the covalent bonds are the start of the three-dimensional shapes of organic compounds. A chain of carbon atoms, bonded covalently one after another, forms a backbone from which other atoms can project:

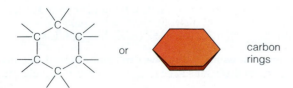
atoms branching from backbone

carbon backbone

The backbone can coil back on itself in a ring structure, which we can diagram in such ways as:

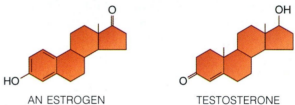
or

carbon rings

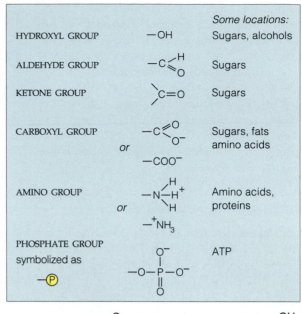

		Some locations:
HYDROXYL GROUP	—OH	Sugars, alcohols
ALDEHYDE GROUP	—C⟨H,O	Sugars
KETONE GROUP	⟩C=O	Sugars
CARBOXYL GROUP	—C⟨O,O⁻ or —COO⁻	Sugars, fats amino acids
AMINO GROUP	—N⟨H,H⁺,H or —⁺NH₃	Amino acids, proteins
PHOSPHATE GROUP symbolized as Ⓟ	—O—P—O⁻ with O⁻ and O	ATP

AN ESTROGEN   TESTOSTERONE

Figure 2.18 (**a**) A few examples of functional groups. (**b**) The sex hormones estrogen and testosterone have hydroxyl groups attached to different carbons.

Functional Groups

A carbon backbone with only hydrogen atoms attached to it is a hydrocarbon, which is a very stable structure. Besides hydrogen atoms, biological molecules also have **functional groups:** various kinds of atoms or clusters of them that are covalently bonded to the backbone and can influence the behavior of organic compounds.

Figure 2.18*a* shows a few functional groups. Sugars and other organic compounds classified as alcohols have one or more hydroxyl groups (—OH). Water hydrogen-bonds with hydroxyl groups, which is why sugars can dissolve in water. The backbone of a protein forms by reactions between amino groups and carboxyl groups. As you will see shortly, the backbone is the start of bonding patterns that produce the protein's three-dimensional structure. Amino groups also can combine with hydrogen ions and act as buffers against decreases in pH. And the functional groups shown in Figure 2.18*b*, those of sex hormones, are a molecular starting point for differences between males and females.

How do cells assemble the organic compounds they require for their structure and functioning? The details

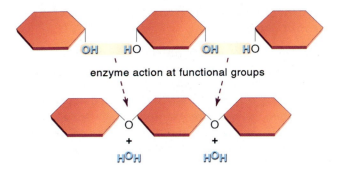

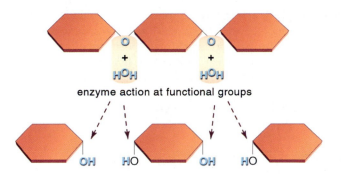

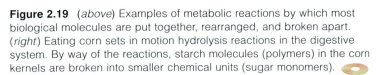
enzyme action at functional groups

a Two condensation reactions. Enzymes remove an —OH group and H atom from two molecules, which covalently bond as a larger molecule. Two water molecules form.

b Hydrolysis, a water-requiring cleavage reaction. Enzyme action splits a molecule into three parts, then attaches an H atom and an —OH group derived from a water molecule to each exposed site.

Figure 2.19 (*above*) Examples of metabolic reactions by which most biological molecules are put together, rearranged, and broken apart. (*right*) Eating corn sets in motion hydrolysis reactions in the digestive system. By way of the reactions, starch molecules (polymers) in the corn kernels are broken into smaller chemical units (sugar monomers).

can (and do) fill whole books, but here you only need to focus on a few basic concepts. First, whatever happens in a cell requires energy. This fundamental rule holds regardless of whether a cell is building, rearranging, or even breaking apart large organic compounds needed for its operations. Chemical reactions in cells also require a class of proteins called **enzymes**, which make specific metabolic reactions take place faster than they would on their own. Different enzymes facilitate different kinds of reactions. Two reactions that go on constantly in cells are those called condensation and hydrolysis.

Condensation and Hydrolysis Reactions

As a cell builds or alters organic compounds, a common chemical step is the **condensation** reaction. Often in this kind of reaction, enzymes remove a hydroxyl group from one molecule and an H atom from another, then speed the formation of a covalent bond between the two molecules at their exposed sites (Figure 2.19a). The discarded hydrogen and oxygen atoms may combine to form a molecule of water (H_2O). It's as if you and a friend have to remove your gloves, pin the matched pairs together, and set them aside before you can hold hands. Because this kind of reaction often forms water as a by-product, condensation is sometimes called *dehydration* ("de-watering") *synthesis*. Cells can use whole series of condensation reactions to assemble polymers. A **polymer** is a large molecule built of three to millions of subunits. The individual subunits may be called **monomers**, and they may be identical or different.

Hydrolysis is like condensation in reverse (Figure 2.19b). Enzymes that recognize specific functional groups split molecules into two or more parts, then *attach* an —OH group and a hydrogen atom derived from a molecule of water to the exposed sites. With hydrolysis, cells can break apart large polymers into smaller units when these are required for building blocks or energy (Figure 2.19c).

Organic compounds have carbon backbones. The different bonding arrangements that can arise from the backbone help give organic compounds their three-dimensional shapes.

Functional groups that are covalently bonded to the carbon backbone of organic compounds tremendously increase their structural and functional diversity.

Carbohydrates, lipids, proteins, and nucleic acids are the main biological molecules—the organic compounds that only living cells can assemble.

Cells assemble, rearrange, and break down organic compounds mainly through reactions facilitated by enzymes. These reactions include the combining or splitting of molecules, as in condensation and hydrolysis reactions.

CARBOHYDRATES

Consider first the carbohydrates—the most abundant of all biological molecules, which cells use as structural materials and as transportable or storage forms of energy. Most **carbohydrates** consist of carbon, hydrogen, and oxygen in a 1:2:1 ratio. Differences in structure separate carbohydrates into three classes: the **monosaccharides**, **oligosaccharides**, and **polysaccharides**.

Monosaccharides: A Single Sugar Unit

"Saccharide" comes from a Greek word meaning sugar. A *mono*saccharide, meaning "one monomer of sugar," is the simplest carbohydrate. It has at least two —OH groups joined to the carbon backbone plus an aldehyde or a ketone group. Most monosaccharides taste sweet and dissolve readily in water. The most common ones have a backbone of five or six carbons; for example, deoxyribose, the sugar component of DNA, has five carbon atoms. The monosaccharide glucose has six. It is the main energy source for body cells, and it also is the precursor (parent molecule) for many compounds and a building block for larger carbohydrates (Figure 2.20*a*). Vitamin C (a sugar acid) and glycerol (an alcohol with three hydroxyl groups) are other examples of compounds derived from sugar monomers.

Oligosaccharides: Short Chain Carbohydrates

Unlike the simple sugars, an *oligo*saccharide is a short chain of two or more sugar monomers that are united by covalent bonds. (*Oligo-* means a few.) The type known as *di*saccharides consist of just two sugar units. Lactose, sucrose, and maltose are examples. Lactose (a glucose and a galactose unit) is a milk sugar. Sucrose, the most plentiful sugar in nature, consists of one glucose and one fructose unit (Figure 2.20*c*). You consume sucrose when you eat fruit, among other plant foods. Table sugar is sucrose crystallized from sugar cane and sugar beets.

Proteins and other large molecules often have oligosaccharides attached as side chains to their carbon backbone. Some chains take part in activities of cell membranes, as you will read in Chapter 3. Others have roles in the body's defenses against disease.

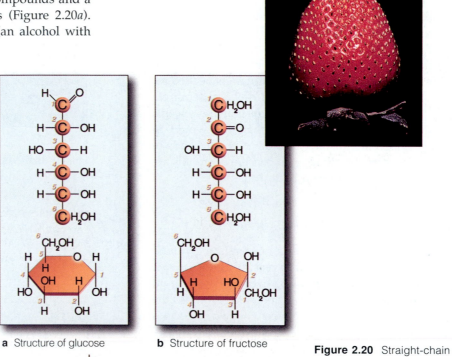

a Structure of glucose   **b** Structure of fructose

c Formation of a sucrose molecule from two simple sugars

glucose   fructose

+ H_2O

sucrose

Figure 2.20 Straight-chain and ring forms of (**a**) glucose and (**b**) fructose. For reference purposes, the carbon atoms of simple sugars are commonly numbered in sequence, starting at the end closest to the molecule's aldehyde or ketone group. (**c**) Condensation of two monosaccharides (glucose and fructose) into a disaccharide (sucrose).

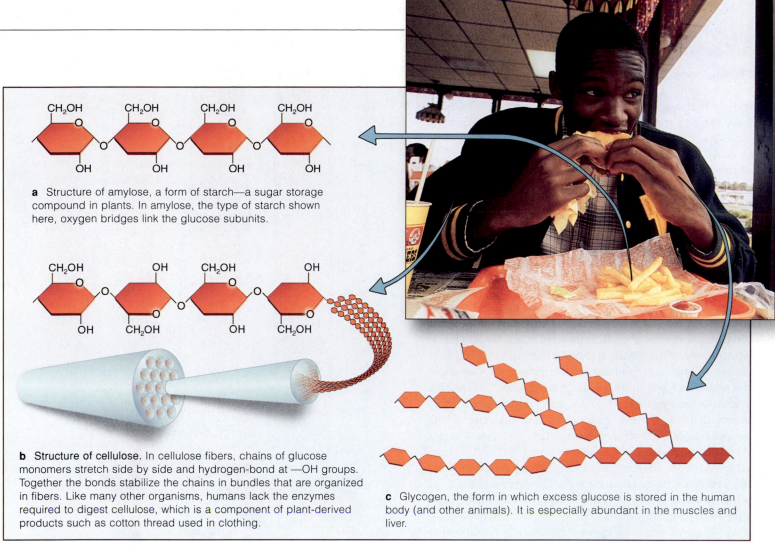

a Structure of amylose, a form of starch—a sugar storage compound in plants. In amylose, the type of starch shown here, oxygen bridges link the glucose subunits.

b Structure of cellulose. In cellulose fibers, chains of glucose monomers stretch side by side and hydrogen-bond at —OH groups. Together the bonds stabilize the chains in bundles that are organized in fibers. Like many other organisms, humans lack the enzymes required to digest cellulose, which is a component of plant-derived products such as cotton thread used in clothing.

c Glycogen, the form in which excess glucose is stored in the human body (and other animals). It is especially abundant in the muscles and liver.

Figure 2.21 Starch, cellulose, and glycogen. All three carbohydrates are built only of glucose monomers.

Polysaccharides: Complex Carbohydrates

The "complex carbohydrates," or *poly*saccharides, are straight or branched chains of many sugar monomers—often hundreds or thousands of the same or different kinds. Most of the carbohydrates you eat are in the form of polysaccharides. The most common ones—glycogen, starch, and cellulose—consist only of glucose.

When you eat meat you are consuming glycogen. Glycogen is a sugar storage form in animals—notably in muscle tissues and the liver. When blood-sugar levels fall, liver cells break down glycogen and release glucose units to the blood. During exercise, muscle cells tap into their glycogen stores for quick access to energy. Figure 2.21 shows just a few of a typical glycogen molecule's numerous branchings.

Foods such as potatoes, rice, wheat, and corn are all rich in starch (Figure 2.21a), which is a storage form of glucose in plant cells. In starch the glucose subunits are arranged in a linear fashion. Cellulose is a tough, insoluble structural material in plant cell walls (Figure 2.21b).

In cellulose, the links between neighboring glucose monomers have a different orientation than they do in glycogen and starch. Also, hydrogen bonds form between —OH groups of adjacent cellulose molecules; the result is a fine strand. Humans do not have digestive enzymes that can break down (hydrolyze) the cellulose in vegetables, whole grains, and other plant tissues. We benefit from it, however, as undigested "fiber" that adds bulk and so helps move wastes through the lower digestive tract. Chapter 6 talks more about the health benefits of dietary fiber.

Carbohydrates are either simple sugars (such as glucose) or molecules composed of many sugar units.

In order of their structural complexity, the three types of carbohydrates are monosaccharides, oligosaccharides, and polysaccharides.

All cells require carbohydrates, which the body uses for structural materials, to store energy, and as transportable packets of energy.

Oil and water don't mix—and a little biochemistry explains why. Oils are a type of **lipid**, and a lipid is mostly hydrocarbon. As a result, a lipid has a large non-polar region that makes it hydrophobic—the reason it tends not to dissolve in water. Lipids *do* readily dissolve in nonpolar substances, however, and all are greasy or oily to the touch. Cells have many uses for lipids: as primary energy reservoirs, as structural materials (as in cell membranes), and as signaling molecules. Here we focus on the neutral fats, phospholipids, and waxes, all of which have fatty acid components. We also consider the sterols, each with a backbone of four carbon rings.

Fats and Fatty Acids

Fats are lipids having one, two, or three fatty acids attached to glycerol. Each **fatty acid** has a backbone of up to thirty-six carbons, a carboxyl group (—COOH) at one end, and hydrogen atoms occupying most or all of the remaining bonding sites. It typically stretches out like a flexible tail. An *unsaturated* tail has one or more double bonds in its backbone. A *saturated* tail has only single bonds. Figure 2.22a shows examples.

Most animal fats have lots of saturated fatty acids, which pack together by weak bonding interactions. Like the visible fat in bacon, they are solid at temperatures at which most plant fats remain liquid. "Vegetable oils" such as canola, corn oil, and olive oil flow freely because the packing interactions in plant fats are not as stable due to rigid kinks in their fatty acid tails.

Butter, lard, plant oils, and other dietary fats consist mostly of **triglycerides**: neutral fats that have three fatty acid tails attached to a glycerol backbone (Figure 2.22b). Triglycerides are the body's most abundant lipids and its richest source of energy. Gram for gram, they yield more than twice as much energy when they are broken down, compared to complex carbohydrates like starches. This is because triglycerides have more removable electrons than carbohydrates do—and energy is released when electrons are removed. In the body, cells of fat-storing (adipose) tissues stockpile triglycerides as fat droplets.

The *trans fatty acids* are partially saturated, or "hydrogenated." Such molecules often are the main ingredient in solid margarines. Many people limit the amount of trans fatty acids in their diet because trans fatty acids have been linked to the development of some types of heart disease (Chapter 7).

One type of fatty acid is a precursor to eicosanoids, a class of molecules that carry signals between cells in specific body regions. Different eicosanoids bind to receptors on cells and help regulate many physiological processes, such as muscle contraction and message transmission through the nervous system.

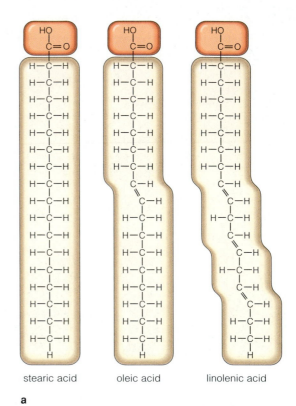

Figure 2.22 **(a)** Structural formulas for three fatty acids. Stearic acid's carbon backbone is fully saturated with hydrogens. Oleic acid, with its double bond in the carbon backbone, is unsaturated. Linolenic acid, with three double bonds, is a "polyunsaturated" fatty acid. **(b)** Condensation of fatty acids and glycerol into a triglyceride.

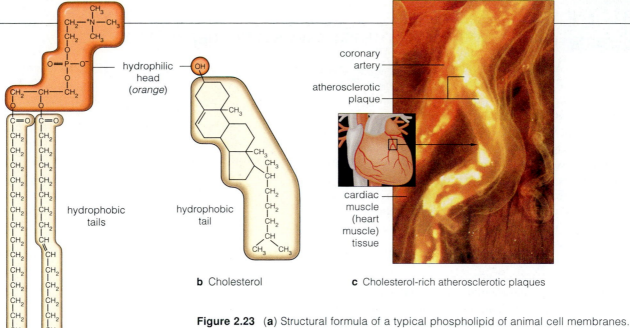

hydrophilic head (*orange*)

hydrophobic tails

hydrophobic tail

a One of the phospholipids

b Cholesterol

coronary artery

atherosclerotic plaque

cardiac muscle (heart muscle) tissue

c Cholesterol-rich atherosclerotic plaques

Figure 2.23 (**a**) Structural formula of a typical phospholipid of animal cell membranes. Are its hydrophobic tails saturated or unsaturated? (**b**) Structural formula of cholesterol, the major sterol in the human body. (**c**) The liver synthesizes enough cholesterol to meet body needs, so a fat-rich diet may result in excessive cholesterol in the blood—and the formation of abnormal masses of fatty material in certain blood vessels (arteries). Such atherosclerotic plaques may dangerously clog the arteries that deliver blood to the heart.

Phospholipids

A **phospholipid** has a glycerol backbone, two fatty acid tails, and a hydrophilic "head" with a phosphate group—a phosphorus atom bonded to four oxygen atoms—and another polar group (Figure 2.23*a*). Phospholipids are the main materials of cell membranes, which have two layers of lipids. The heads of one layer are dissolved in the cell's fluid interior, and the heads of the other layer are dissolved in the surroundings. Sandwiched between the two are all the fatty acid tails, which are hydrophobic.

Sterols and Substances Derived from Them

Sterols are among the lipids that have no fatty acid tails. Sterols differ in the number, position, and type of their functional groups, but they all have a rigid backbone of four fused-together carbon rings:

sterol backbone

Many people associate the sterol cholesterol with heart and artery disease. However, normal amounts of this sterol are crucial to the structure and proper functioning of cells. For instance, cholesterol is a vital component of membranes of every cell in your body. Some important

derivatives of cholesterol include vitamin D (essential for bone and tooth development), bile salts (which help with fat digestion in the small intestine), and steroid hormones such as estrogen and testosterone. Later chapters will tell more of the fascinating story of how steroid hormones influence reproduction, development, growth, and some other body functions.

Waxes

The lipids called **waxes** have long-chain fatty acids, tightly packed and linked to long-chain alcohols or to carbon rings. Waxes have a firm consistency and repel water. In humans, glands in the ear canal produce earwax, a yellow-brown substance that protects tissues from drying out. Sometimes the wax builds up and obstructs the canal, causing discomfort and impairing a person's ability to hear.

Lipids are hydrophobic, greasy or oily compounds. They include:

• Neutral fats (triglycerides), major reservoirs of energy

• Phospholipids, the main components of cell membranes

• Sterols (such as cholesterol), components of membranes and precursors of steroid hormones and other vital molecules

• Waxes, firm yet pliable components of water-repelling and lubricating substances such as earwax

---

AMINO ACIDS AND THE PRIMARY STRUCTURE OF PROTEINS

Have you ever wondered how a permanent wave makes a person's hair curly (Figure 2.24)? The characteristics of many body components, including chemically "waved" hair, start with the structure of proteins.

Of all the large biological molecules, **proteins** are the most diverse. Those called enzymes make metabolic reactions proceed much faster than they otherwise would. Structural proteins are major components of your bones, muscles, hair, and other body elements. Transport proteins help move substances into or out of cells or through body fluids. Regulatory proteins, including the protein hormones, help adjust body functions by serving as signals for change in cell activities. Without them, a host of bodily events and activities—waking, sleeping, going through puberty, and engaging in sexual activity, to cite just a few—would be impaired or impossible. Many other proteins function as weapons against harmful bacteria and other invaders. Most amazingly, perhaps, is the fact that body cells build all these different proteins from only twenty or so kinds of amino acids.

Structure of Amino Acids

An **amino acid** is a small organic compound that consists of an amino group, a carboxyl group (an acid), an atom of hydrogen, and one or more atoms called its R group. As you can see from the structural formula in Figure 2.25a, these parts generally are covalently bonded to the same carbon atom. Figure 2.25b shows two amino acids that we will consider later in the book.

Figure 2.24 The hair of actress Nicole Kidman (**a**) before and (**b**) after a permanent wave. The difference, as you will see from this section and the next, starts with strings of amino acids in polypeptide chains.

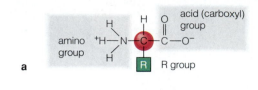
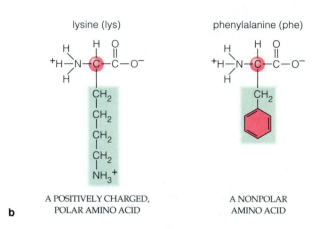

Figure 2.25 (**a**) Generalized structural formula for amino acids. (**b**) Structural formulas for two of the twenty common amino acids. Green boxes highlight the R groups, which are side chains with functional groups.

Primary Structure of Proteins

When a cell synthesizes a protein, amino acids become linked, one after the other, by *peptide* bonds. As Figure 2.26 shows, this is the type of covalent bond that forms between one amino acid's amino group (NH_3^+) and the carboxyl group ($—COO^-$) of the next amino acid.

When peptide bonds join two amino acids together, we have a dipeptide. When they join three or more amino acids, we have a **polypeptide chain**. The backbone of each chain incorporates nitrogen atoms in this regular pattern: —N—C—C—N—C—C—.

For each particular kind of protein, different amino acid units are added in a specific order, one at a time, from the twenty kinds available. As a later chapter describes more fully, genes determine the order in which amino acids are "chosen" to be added to the growing chain. Once the chain is completed, the order of its linked amino acids is the unique sequence for that particular kind of protein (Figure 2.27). Every other kind of protein in the body will have its own specific sequence of amino acids, linked one to the next like the links of a chain. This sequence is a protein's *primary* structure. It is determined by DNA, as described in Chapter 20. How many amino acids can be linked up this way? The primary structure of the largest known protein, which is a component of human muscle, is a string of some 27,000 amino acids!

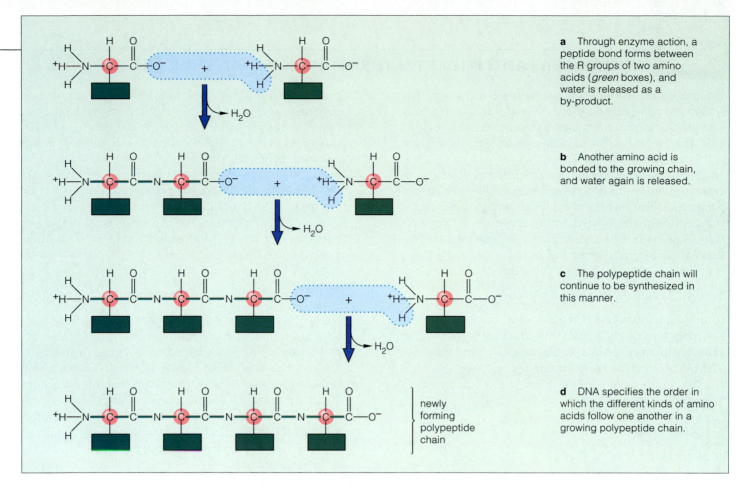

a Through enzyme action, a peptide bond forms between the R groups of two amino acids (*green* boxes), and water is released as a by-product.

b Another amino acid is bonded to the growing chain, and water again is released.

c The polypeptide chain will continue to be synthesized in this manner.

newly forming polypeptide chain

d DNA specifies the order in which the different kinds of amino acids follow one another in a growing polypeptide chain.

Figure 2.26 The formation of peptide bonds during protein synthesis. This is a condensation reaction.

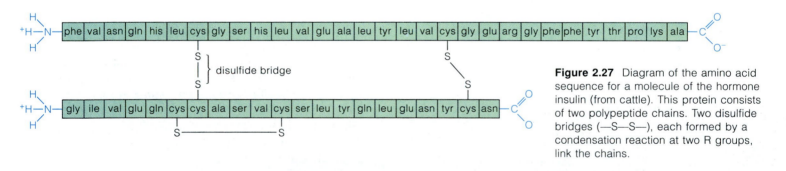

disulfide bridge

Figure 2.27 Diagram of the amino acid sequence for a molecule of the hormone insulin (from cattle). This protein consists of two polypeptide chains. Two disulfide bridges (—S—S—), each formed by a condensation reaction at two R groups, link the chains.

Now consider this: Different cells make thousands of different proteins. Many of the proteins are fibrous, with polypeptide chains organized as strands or sheets. Collectively, many such molecules contribute to the shape and internal organization of cells. Other kinds of proteins are globular, with one or more polypeptide chains folded into compact, rounded shapes. Most enzymes are globular proteins. So are the proteins that contribute to cell movement.

Regardless of the type of protein, its shape and its function arise from the primary structure—that is, from information built into its amino acid sequence. As you will see in the next section, that information dictates which parts of a polypeptide chain will coil, bend, or even interact with other chains nearby. And the type and

arrangement of atoms in the coiled, stretched out, or folded regions determine whether a protein will act as, say, an enzyme, a transporter, a receptor, or even be a possible target for a bacterium or virus. Said another way, a protein's primary structure dictates both its final structure and its chemical behavior.

A protein consists of one or more polypeptide chains, in which amino acids are joined together by peptide bonds.

A unique sequence of amino acids (which kind follows another) makes up each kind of protein; this sequence is the protein's primary structure. It dictates the protein's unique final structure, its chemical behavior, and its function.

HOW DOES A PROTEIN GET ITS THREE-DIMENSIONAL STRUCTURE?

The preceding section gave you a sense of how amino acids are strung together in a polypeptide chain, which is a protein's primary structure. Now we are going to look at a few examples of how primary structure gives rise to a protein's shape.

For the most part, the primary structure gives rise to a protein's shape in two ways. First, it allows hydrogen bonds to form between different amino acids in a polypeptide chain. Second, it puts R groups in positions that allow them to interact. Through their interactions, the chain is forced to bend and twist.

Second Level of Protein Structure

Hydrogen bonds form at regular, short intervals along a new polypeptide chain, and they give rise to a coiled or extended pattern known as the *secondary* structure of a

protein. Think of a polypeptide chain as a set of rigid playing cards joined by links that can swivel a bit. Each "card" is a peptide group (Figure 2.28*a*). Atoms on either side of it can rotate slightly around their covalent bonds and form bonds with neighboring atoms. For instance, in many chains a hydrogen bond readily forms after every third amino acid. The bonding pattern forces the peptide groups to coil helically, like a spiral staircase (Figure 2.28*b*). In other proteins, the hydrogen-bonding pattern holds two or more chains side by side, in a sheetlike array (Figure 2.28*c*).

Third Level of Protein Structure

Most polypeptide chains that have a coiled secondary structure undergo more folding, owing to the number and location of certain amino acids along their length. These amino acids bend a chain at certain angles and in certain directions to make the chain loop out. R groups far apart along its length interact and hold the loops in characteristic positions. A polypeptide chain folded as an outcome of its bend-producing amino acids and its R-group interactions has reached a third structural level, or *tertiary* structure.

Figure 2.29*a* shows how a polypeptide chain became folded into the compact tertiary structure of the protein globin. Hydrogen bonds between particular functional groups along the chain's length caused the folding.

Fourth Level of Protein Structure

Imagine that bonds form between *four* molecules of globin and that an iron-containing functional group, a heme group, is positioned near the center of each. The result is **hemoglobin**, an oxygen-transporting protein

a One peptide group

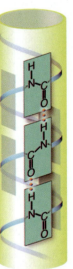

b

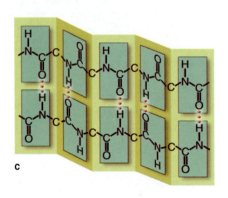

c

Figure 2.28 Major bonding patterns between (**a**) peptide groups of polypeptide chains. Extensive hydrogen bonding (the *dotted* lines) can bring about (**b**) coiling or (**c**) sheetlike chains.

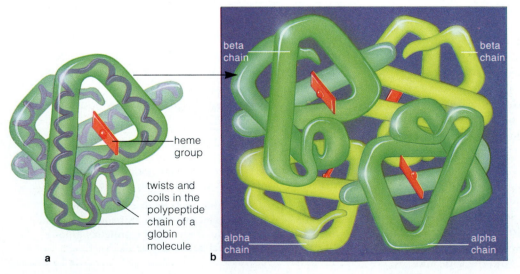

heme group

twists and coils in the polypeptide chain of a globin molecule

a

beta chain

beta chain

alpha chain

alpha chain

b

Figure 2.29 (**a**) Globin molecule. This coiled polypeptide chain associates with a heme group, an iron-containing group that strongly attracts oxygen. There is a whole family of globin molecules. The member molecules are identical in most parts of their amino acid sequences, but each is unique in other parts. In this diagram, the artist drew a transparent green "noodle" around each chain to help you visualize how it folds in three dimensions. (**b**) Hemoglobin, a protein that transports oxygen in blood, consists of four polypeptide chains (globin molecules) and four heme groups. Two chains, designated *alpha*, have a slightly different amino acid sequence than the other two (the *beta* chains). To keep this sketch from looking like a tangle of noodles, the chains in the background are tinted differently from those in the foreground.

Further reading: Student Guide to InfoTrac on web site →

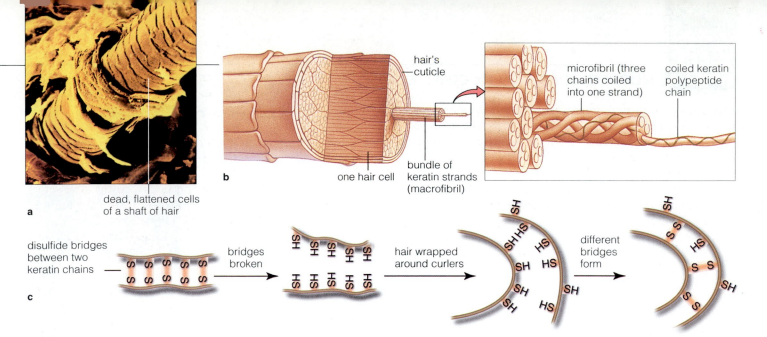

a dead, flattened cells of a shaft of hair

b one hair cell | hair's cuticle | bundle of keratin strands (macrofibril)

microfibril (three chains coiled into one strand) | coiled keratin polypeptide chain

c disulfide bridges between two keratin chains → bridges broken → hair wrapped around curlers → different bridges form

(Figure 2.29b). At this moment, each of the millions of red blood cells in your body is transporting a billion molecules of oxygen, bound to 250 million molecules of hemoglobin.

Hemoglobin is a fine example of the fourth level of protein structure, *quaternary* structure. In all proteins at this level of organization, two or more polypeptide chains have become joined together by numerous weak interactions (such as hydrogen bonds) and sometimes by disulfide bridges—covalent bonds between sulfur atoms of R groups. Hemoglobin has four polypeptide chains.

Hemoglobin is a globular protein; so are most enzymes. Many other proteins that have quaternary structure are fibrous. Keratin, a structural protein of hair (Figure 2.30), is like this. So is collagen, the most common protein in the body. Skin, bone, corneas, blood vessels, and other body parts depend on the strength of collagen.

Glycoproteins and Lipoproteins

Some proteins have other organic compounds attached to their polypeptide chains. For example, lipoproteins form when certain proteins circulating in blood combine with cholesterol, triglycerides, and phospholipids that were absorbed from the gastrointestinal tract after a meal. Similarly, most **glycoproteins** have linear or branched oligosaccharides bonded to them. Nearly all the proteins at the surface of your cells are glycoproteins. So are most proteins secreted from body cells (including protein hormones) and many proteins in blood.

Structural Changes by Denaturation

Breaking weak bonds of a protein or any other large molecule disrupts its three-dimensional shape, an event called **denaturation**. For example, weak hydrogen bonds are sensitive to increases or decreases in temperature and pH. If the temperature or pH exceeds a protein's

Figure 2.30 Structure of hair. (**a,b**) Hair cells develop from modified skin cells. They manufacture polypeptide chains of the protein keratin. Disulfide bridges link three chains together as fine fibers, which are bundled into larger, cablelike fibers. The cablelike fibers almost fill the cells, which eventually die. As you can see in in the photograph, dead, flattened cells form a tubelike cuticle around the developing hair shaft.

(**c**) For a permanent wave, hair is bathed with chemicals that break the disulfide bridges and wrapped around curlers that hold their polypeptide chains in new positions. Next, a chemical is applied that causes new disulfide bridges to form, but they form between different sulfur-bearing amino acids than before. The displaced bonding locks the hair in curled positions (compare to Figure 2.23).

range of tolerance, its polypeptide chains will unwind or change shape, and the protein will no longer function. Consider albumin, concentrated in the "egg white" of uncooked chicken eggs. When you cook eggs, the heat does not disrupt the strong covalent bonds of albumin's primary structure. But it destroys weaker bonds contributing to the three-dimensional shape. For some proteins, denaturation might be reversed when normal conditions are restored—but albumin isn't one of them. There is no way to uncook a cooked egg white.

The secondary structure of a protein is a coiled pattern or an extended, sheetlike pattern that arises by hydrogen bonding at short, regular intervals along a polypeptide chain.

At the third level of protein structure, bend-producing amino acids make a coiled chain loop at certain angles and in certain directions, and interactions among R groups in the chain hold the loops in characteristic positions.

At the fourth level of protein structure, numerous hydrogen bonds and other interactions join two or more polypeptide chains. Many proteins, including hemoglobin, are this structurally complex.

Nucleotides are small organic compounds with a sugar, at least one phosphate group, and a base. The sugar is either ribose or deoxyribose, both of which have a five-carbon ring structure. The only difference is that ribose has an oxygen atom attached to carbon 2 in the ring and deoxyribose does not. The bases have a single or double carbon ring structure that incorporates nitrogen:

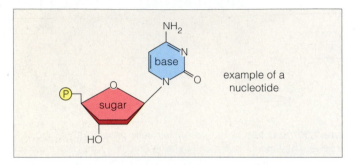

One nucleotide, **ATP** (for adenosine triphosphate), has three phosphate groups attached to its sugar:

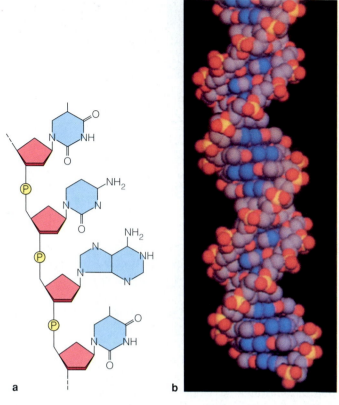

Figure 2.31 (**a**) Examples of bonds between nucleotides in a nucleic acid molecule. (**b**) Model of DNA. The molecule consists of two strands of nucleotides joined by hydrogen bonds and twisted into a double helix. The nucleotide bases are blue.

In every living cell, molecules of ATP couple chemical reactions that *release* energy with other reactions that *require* energy. This is because ATP molecules can readily transfer a phosphate group to many other molecules inside the cell. When that happens, the acceptor molecules have the energy they require to enter into a reaction. Thus ATP is central to metabolism.

Different nucleotides are building blocks for single- or double-stranded molecules classified as **nucleic acids**. In the backbones of such strands, each nucleotide's sugar component is covalently bonded to a phosphate group of the adjacent nucleotides (Figure 2.31*a*). In this book and in your daily life you will be hearing a great deal about **DNA** (deoxyribonucleic acid). This molecule is composed of two strands of nucleotides, twisted together helically (Figure 2.31*b*). Hydrogen bonds between the nucleotide bases join the strands. Genetic (heritable) information is encoded in the base sequences. In some parts of the molecule, the particular bases occur in a sequence that is unique to each species. Unlike DNA, the **RNAs** (short for ribonucleic acid) usually are single strands of nucleotides. Ever since living cells first appeared on Earth, RNAs have played key roles in processes by which genetic information is used to build proteins.

In chapters to come, you'll also be reading about some other nucleotides. For instance, certain nucleotides are subunits of coenzymes. A **coenzyme** is a molecule that accepts hydrogen atoms and electrons that are being removed from other molecules and transfers them to locations where they can be used for other reactions. Examples of coenzymes include certain vitamins and compounds abbreviated NAD^+ and FAD. Other nucleotides have roles as chemical messengers inside and between cells. One such messenger nucleotide, called cAMP (for cyclic adenosine monophosphate), is a major player in events by which some hormones exert their effects on the body.

Nucleotides are small, nitrogen- and phosphorus-containing organic compounds. They are building blocks for DNA and RNA. Some, including ATP, have central roles in metabolism.

DNA is a double-stranded nucleic acid. Genetic information is encoded in its sequence of nucleotide bases. RNAs are single-stranded nucleic acids with roles in the processes by which DNA's genetic information is used to build proteins.

Further reading: Student Guide to InfoTrac on web site →

2.14

FOOD PRODUCTION AND A CHEMICAL ARMS RACE

The next time you shop for groceries, consider what it takes to provide you with your daily supply of organic compounds. For example, the lettuce for your salad most likely grew in fertilized cropland, and the grower may well have been concerned about invading weeds and attacks by insects. Each year, these food pirates and others ruin or gobble up nearly half of the food that people all over the world try to grow.

People—and plants—marshal various chemical defenses against the attackers. For instance, the tissues of many plants contain toxins that repel or kill insects or harmful fungi or other organisms. A toxin is an organic compound, a normal metabolic product of some species, but its chemical effects can harm or kill members of certain other species. Humans encounter natural plant toxins in a wide range of foods—chili peppers, potatoes, figs, celery, and alfalfa sprouts, for instance. By and large, however, our bodies seem to be able to cope with those chemicals just fine—possibly because we have evolved our own biochemical ways of neutralizing them.

In 1945, the human race took a cue from the plant world as chemists began developing toxins that could improve our ability to protect crop yields, stored grains, public water supplies, and even our pets and ornamental plants. The arsenal of synthetic toxins came to include three broad groups of potent chemicals. The *herbicides* kill weeds by disrupting a target plant's metabolism and growth. Most *insecticides* cause a target insect to suffocate, prevent its nerves and muscles from functioning properly, or block its reproduction. *Fungicides* work against harmful fungi, including a mold that makes aflatoxin, one of the deadliest toxins. By 1955, each year people were spraying or spreading more than 1.25 billion pounds of these toxins though agricultural fields, gardens, homes, and industrial sites (Figure 2.32).

What about the drawbacks of pesticides? Along with pests, some of our toxic organic compounds kill birds and other predators which, in natural settings, help control pest populations. Certain ones, such as DDT, stay active for years. Also, as we have learned—and forget at our peril—pesticides cannot be released haphazardly into the environment, for people can inhale them, ingest them along with food, or absorb them through the skin. Some trigger rashes, hives, sickening headaches, asthma, and joint pain, affecting millions. In susceptible individuals, some can even trigger life-threatening allergic reactions.

At this time, long-lived pesticides such as DDT are banned in the United States (although they are still used in certain countries from which we import fruits and vegetables). Even those that break down soon after being applied, such as malathion

Figure 2.32 Lettuce, a common crop plant, and a low-flying crop duster with its rain of pesticides.

and other organophosphates, must be used according to strict application procedures and are subjected to continuing safety tests.

Yet despite decades of refinements, the pressure to develop new weapons against pests never stops. Crop-attacking insects in particular can and have developed resistance to some chemicals in current use, forcing chemists to look for new, more potent insecticides, always trying to balance effectiveness with safety. It's a difficult assignment. Trying a different tack, some seed companies have genetically engineered crop plants that themselves produce substances that kill insect predators—but as you will read later in this book, those efforts are controversial because some researchers fear potential harm to "good" insects such as butterflies. There will be few easy answers as the search for effective countermeasures goes on.

SUMMARY

1. Protons, neutrons, and electrons are the basic constituents of atoms. An element's atoms have the same number of protons and electrons but may vary in the number of neutrons. The various forms are isotopes. Whether an atom interacts with others depends on the number and arrangement of its electrons.

2. In one model, electrons occupy orbitals (volumes of space) inside a series of shells around the nucleus of an atom. Atoms with one or more unfilled orbitals in their outermost shell tend to take part in chemical bonds.

3. Hydrogen, oxygen, carbon, and nitrogen are the most abundant elements in organisms. They all tend to form bonds with other atoms.

4. Atoms have no net charge, but may gain or lose one or more electrons and become ions, with an overall positive or negative charge.

5. In ionic bonds, positive and negative ions stay together by the mutual attraction of their opposite charges. In covalent bonds, atoms share one or more electrons. In hydrogen bonds, a small, electronegative atom interacts weakly with a weakly positive hydrogen atom that is already part of a polar covalent bond.

6. A pH value refers to the concentration of hydrogen ions in a fluid. Acids release hydrogen ions (H^+); and bases release hydroxide ions (OH^-) that can combine with H^+. At pH 7, the H^+ and OH^- concentrations in a solution are equal. Buffers maintain pH values of blood, tissue fluids, and the fluid inside cells.

7. Water is crucial for the metabolic activity, shape, and internal organization of cells. Due to hydrogen bonds between its polar molecules, water has a range of special properties. These include its ability to resist temperature changes and to dissolve other polar substances.

8. A carbon atom forms up to four covalent bonds with other atoms. Carbon atoms bonded together in linear or ring structures are the backbone of organic compounds. The chemical and physical properties of many of those compounds depends largely on functional groups (atoms attached to the backbone).

9. Cells assemble, rearrange, and break apart most organic compounds by five kinds of enzyme-mediated reactions: functional-group transfers, electron transfer, internal rearrangements, condensation reactions, and cleavages (including hydrolysis).

10. Cells have pools of dissolved sugars, fatty acids, amino acids, and nucleotides. These are all small organic compounds that include no more than about 20 carbon atoms. They are building blocks for the larger biological molecules—the polysaccharides, lipids, proteins, and nucleic acids (Table 2.2).

Review Questions

1. Carbohydrates, lipids, proteins, and nucleic acids are assembled from families of small organic molecules. What are the families? 2.4

2. Which of the following is the carbohydrate, the fatty acid, the amino acid, and the polypeptide? 2.9, 2.10, 2.11
   a. $^+NH_3$ ⟍ CHR ⟍ COO^- c. $(glycine)_{20}$
   b. $C_6H_{12}O_6$ d. $CH_3(CH_2)_{16}COOH$

3. Describe the four levels of protein structure. How do a protein's side groups influence its interactions with other substances? What is denaturation? 2.11, 2.12

4. Distinguish among the following:
   a. monosaccharide, polysaccharide, disaccharide 2.9
   b. peptide bond, polypeptide 2.11
   c. glycerol, fatty acid 2.10
   d. nucleotide, nucleic acid 2.13

Self-Quiz (Answers in Appendix V)

1. The backbone of organic compounds forms when _____ atoms are covalently bonded into chains and rings.

2. Each carbon atom can form up to _____ bonds with other atoms.
   a. four   c. eight
   b. six   d. sixteen

3. All of the following except _____ are small organic molecules that serve as the main building blocks or energy sources in cells.
   a. fatty acids   c. lipids   e. amino acids
   b. simple sugars   d. nucleotides

4. Which of the following is not a carbohydrate?
   a. glucose molecule   c. fat
   b. simple sugar   d. polysaccharide

5. _____, a class of proteins, make metabolic reactions proceed much faster than they would on their own.
   a. Nucleic acids   c. Fatty acids
   b. Amino acids   d. Enzymes

6. Examples of nucleic acids, the basis of inheritance, are
   a. polysaccharides   c. proteins
   b. DNA and RNA   d. simple sugars

7. Which phrase best describes what a functional group does?
   a. assembles large organic compounds
   b. influences the behavior of organic compounds
   c. splits molecules into two or more parts
   d. speeds up metabolic reactions

8. In _____ reactions, small molecules become covalently linked, and water can also form.
   a. symbiotic   c. condensation
   b. hydrolysis   d. ionic

9. Match each type of molecule with the correct description.

   _____ chain of amino acids   a. carbohydrate
   _____ energy carrier   b. phospholipid
   _____ glycerol, fatty acids,   c. protein
        phosphate   d. DNA
   _____ chain of nucleotides   e. ATP
   _____ one or more sugar units

10. What kinds of bonds often control the shape (or tertiary form) of large molecules such as proteins?
    a. hydrogen   c. covalent   e. single
    b. ionic   d. inert

Table 2.2 Summary of the Main Carbon Compounds In Living Things

Category	Main Subcategories	Some Examples and Their Functions	
Carbohydrates *contain an aldehyde or a ketone group and one or more hydroxyl groups*	**Monosaccharides (simple sugars) Oligosaccharides**	Glucose Sucrose (a disaccharide)	Structural roles, energy source Form of sugar transported in plants
	Polysaccharides (complex carbohydrates)	Starch Cellulose	Energy storage Structural roles
Lipids *are largely hydrocarbon, generally do not dissolve in water but dissolve in nonpolar solvents*	**Lipids with fatty acids:** *Glycerides:* one, two, or three fatty acid tails attached to glycerol backbone *Phospholipids:* phosphate group, another polar group, and (often) two fatty acids attached to glycerol backbone *Waxes:* long-chain fatty acid tails attached to alcohol **Lipids with no fatty acids:** *Sterols:* four carbon rings; the number, position, and type of functional groups vary	Fats (e.g., butter) Oils (e.g., corn oil) Phosphatidylcholine Waxes in cutin Cholesterol	Energy storage Key component of cell membranes Water retention by plants Component of animal cell membranes, can be rearranged into other steroids (e.g., vitamin D, sex hormones)
Proteins *are polypeptides (up to several thousand amino acids, covalently linked)*	**Fibrous proteins:** Individual polypeptide chains, often linked into tough, water-insoluble molecules **Globular proteins:** One or more polypeptide chains folded and linked into globular shapes; many roles in cell activities	Keratin Collagen Enzymes Hemoglobin Insulin Antibodies	Structural element of hair, nails Structural element of bones and cartilage Increase in rates of reactions Oxygen transport Control of glucose metabolism Tissue defense
Nucleic Acids (and Nucleotides) *are chains of units (or individual units) that each consist of a five-carbon sugar, phosphate, and a nitrogen-containing base*	**Adenosine phosphates Nucleotide coenzymes** **Nucleic acids:** Chains of thousands to millions of nucleotides	ATP NAD^+, $NADP^+$ DNA, RNAs	Energy carrier Transport of protons (H^+) and electrons from one reaction site to another Storage, transmission, translation of genetic information

Critical Thinking: You Decide (Key in Appendix VI)

1. Black coffee has a pH of 5, whereas milk of magnesia has a pH of 10. Is coffee twice as acidic as milk of magnesia?

2. Your cotton shirt has stains from whipped cream and strawberry syrup, and your dry cleaner says that two separate cleaning agents will be needed to remove the stains. Explain why two agents are needed and what different chemical characteristic each would have.

3. A store clerk tells you that "natural" vitamin C extracted from rose hips is better for you than synthetic vitamin C. Based on what you know of the structure of organic compounds, does this claim seem credible? Why or why not?

4. Use the Internet to find three examples of acid rain damage and efforts to combat the problem. You might start with the United States Environmental Protection Agency's acid rain home page.

Selected Key Terms

acid *2.7*
amino acid *2.11*
atom *CI*
ATP *2.13*
base *2.7*
buffer system *2.7*
carbohydrate *2.9*
chemical bond *2.3*
ion *2.4*
ionic bond *2.4*
isotope *2.1*
lipid *2.10*
molecule *2.3*
monomer *2.8*
monosaccharide *2.9*
neutron *2.1*
coenzyme *2.13*
compound *2.3*
condensation *2.8*
covalent bond *2.4*
denaturation *2.12*
DNA *2.13*
electron *2.1*
element *CI*
enzyme *2.8*
fatty acid *2.10*
functional group *2.8*
glycoprotein *2.11*
hydrocarbon *2.4*
hydrogen bond *2.4*
hydrolysis *2.8*
hydrophilic *2.6*
hydrophobic *2.6*
nucleic acid *2.13*
nucleotide *2.13*
oligosaccharide *2.9*
organic compound *CI*
pH scale *2.7*
phospholipid *2.10*
polymer *2.8*
polypeptide chain *2.11*
polysaccharide *2.9*
protein *2.11*
proton *2.1*
RNA *2.13*
salt *2.7*
solute *2.6*
sterol *2.10*
triglyceride *2.10*
wax *2.10*

Readings

Goodsell, D. September–October 1992. "A Look Inside the Living Cell." *American Scientist.* Models of biological molecules.

Wolfe, S. 1995. *Introduction to Molecular and Cellular Biology.* Belmont, Calif.: Wadsworth.

Zimmer, C. October 1992. "Wet, Wild, and Weird." *Discover.* Water is nature's "hardest" liquid, and other intriguing discoveries about this fluid crucial for life.

3 CELLS

"Very Small Animalcules . . . "

Early in the 17th century, a scholar named Galileo Galilei arranged two glass lenses inside a cylinder. With this instrument he chanced to look at an insect, and he noticed the stunning geometric patterns of its tiny eyes. Thus Galileo, who was not a biologist, was among the first to record a biological observation made through a microscope. The study of the cellular basis of life was about to begin.

Before long, Robert Hooke, Curator of Instruments for the Royal Society of England, was using a microscope to view thinly sliced cork from a tree. Viewing tiny compartments (Figure 3.1*a*), he gave them the Latin name *cellulae*, meaning small rooms—which is the origin of the biological term "cell." Another pioneer in microscopy was Antony van Leeuwenhoek, a Dutch shopkeeper. He had exceptional skill in constructing lenses and possibly the keenest vision of all three men. Intensely curious, by the late 1600s van Leeuwenhoek was finding natural wonders everywhere, including "many very small animalcules" in tartar scraped from his own teeth.

Eventually microscopy yielded three insights that together make up the **cell theory**: First, every organism is composed of one or more cells. Second, the cell is the smallest unit having the properties of life. Third, the continuity of life arises directly from the growth and division of single cells.

Modern microscopy is one of the most technically advanced tools of biology (Figure 3.1*b*), and it has allowed us to learn a great deal about cells in the human body. A photograph formed using a microscope is called a micrograph; those in Figure 3.2 compare the sorts of detail different types of microscopes can reveal. For example, the red blood cells in Figure 3.2*a* were viewed using a *compound light microscope*, in which two or more glass lenses bend (refract) incoming light rays to form an enlarged image of a specimen. With this method, the cell must be small or thin enough for light to pass through, and its parts must differ in color or optical density from their surroundings. Unfortunately, most cell parts are nearly colorless and are uniform in density. For this reason, prior to viewing cells through a light microscope, researchers expose the cells to dyes that react with some cell parts but not others. Although useful, this "staining" can alter cell parts and kill the cell. Dead cells begin to break down at once, so they are sometimes preserved before staining. However, no matter how good a glass lens system is, when the diameter of the object being viewed is magnified by 2,000 times or more, cell parts appear larger but are not clearer. The properties of light waves passing through glass lenses limit the resolution of smaller details.

Electron microscopy was developed to solve this problem. Although electrons are particles, they behave like waves, and an electron microscope accelerates the

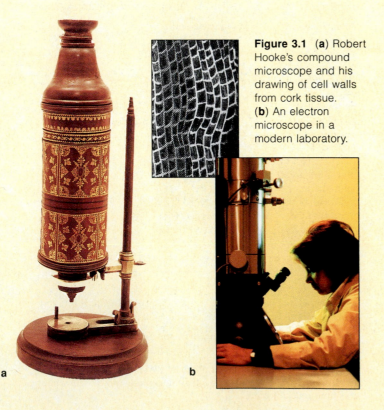

Figure 3.1 (**a**) Robert Hooke's compound microscope and his drawing of cell walls from cork tissue. (**b**) An electron microscope in a modern laboratory.

a   b

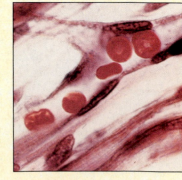

Figure 3.2 Comparison of how different types of microscopes reveal cellular details. The cells are human red blood cells.

a Red blood cells inside a small blood vessel, as revealed by a light microscope.

flow of streams of electrons that have wavelengths about 100,000 times shorter than those of visible light. This means that electron microscopes can achieve much greater resolution of smaller details than even the best light microscopes can, allowing the viewer to obtain remarkably sharp images of cells or other specimens.

A *transmission electron microscope* uses a magnetic field as the "lens" that bends the electron stream and focuses it into an image, which then is magnified. With a *scanning electron microscope*, a narrow beam of electrons is directed back and forth across a specimen thinly coated with metal. The metal responds by emitting some of its own electrons. A detector connected to electronic circuitry transforms the electron energy into an image of the specimen's surface on a television screen. Most of the images have fantastic depth, as you can see in Figure 3.2c below.

This chapter gives an overview of the structure and functioning of cells, and as you read, you will discover just how remarkable your body cells are. They generally are built so that they can bring in certain substances, release or keep out others, and conduct their internal activities with great precision. Every cell also is adapted to function under particular environmental conditions, which mechanisms of home-ostasis must maintain. The sum total of these activities is called metabolism, and you will also learn here how cells obtain the energy that keeps the body's metabolic fires burning.

b Electron micrographs. (*left*) This transmission electron micrograph (TEM) shows the inside of mature red blood cells, which is packed with hemoglobin. (*right*) Scanning electron micrograph (SEM) with color added shows the "concave disk" shape of red blood cells.

 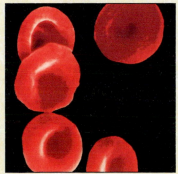

KEY CONCEPTS

1. A cell is the smallest living unit.

2. Cells have an outer, double-layered membrane. This membrane separates the interior of a cell from the environment outside it. Cells contain cytoplasm—an organized internal region where activities necessary for survival take place. The cells of humans and other complex (eukaryotic) organisms also contain organelles. These are compartments that physically separate different metabolic reactions and allow them to take place in an orderly way.

3. DNA is located in an organelle called the nucleus.

4. Cell membranes consist mostly of phospholipids and proteins. The phospholipids form two adjacent layers that give the membrane its basic structure and prevent water-soluble substances from freely crossing it. The proteins perform most other membrane functions.

5. Whether a given molecule or ion crosses a membrane depends partly on its concentrations on both sides of that membrane. Substances tend to diffuse from regions of higher to lower concentration. Membrane proteins help or hinder this movement.

6. Cells harness and use energy for building, storing, breaking apart, and eliminating substances in ways that help them survive and reproduce. These activities are called metabolism. Inside cells, enzymes greatly increase the rate of specific biochemical reactions.

7. ATP carries usable energy in chemical form from one reaction site to another. Most of the ATP used by human cells is generated by aerobic (with oxygen) respiration.

CHAPTER AT A GLANCE

AN INTRODUCTION TO CELLS

In your body live trillions of cells, each one a highly organized bit of life. In a complex organism such as a human, cells live in interdependency. Yet, in a laboratory setting each one still has the capacity to survive on its own, and the natural world teems with single-celled organisms such as bacteria. In other words, no matter what kind of organism we consider, the cell is still the basic unit of life. Its structure is highly organized for metabolism. It senses and responds to its environment. And, based upon inherited instructions in its DNA, it has the potential to reproduce.

Structural Organization of Cells

Cells differ enormously in size, shape, and activities, but all living cells are alike in three respects. All start out life with a plasma membrane, a region of DNA, and a region of cytoplasm:

1. **Plasma membrane**. This thin, outer membrane maintains the cell as a distinct entity. By doing so, it allows metabolic events to proceed apart from random events in the environment outside a cell. A plasma membrane does not *isolate* the cell interior. Substances move across it in highly controlled ways.

2. **DNA-containing region**. DNA, with its heritable instructions, occupies part of the interior, along with molecules that can copy or read the instructions.

3. **Cytoplasm**. Cytoplasm is everything between the plasma membrane and the region of DNA. It consists of a semifluid matrix and other components. For example, it contains **ribosomes**, which are structures upon which proteins are built.

There are two fundamentally different kinds of cells. In a **prokaryotic cell** (loosely, prokaryotic means "before the nucleus") nothing separates the region of DNA from other internal cell parts. Bacteria, like the one shown in Figure 3.3*a*, are the only prokaryotic cells.

By contrast, all other cells are **eukaryotic cells** ("true nucleus"). Their cytoplasm includes **organelles**, tiny sacs and compartments bounded by membranes (Table 3.1). One organelle, the nucleus, is where the DNA of a eukaryotic cell is housed. Nuclei are clearly visible in the cells pictured in Figure 3.4*a*.

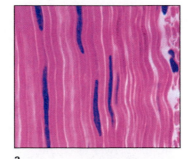

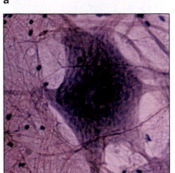

Figure 3.4 Two types of cells in the human body. (**a**) Smooth muscle cells, found in the walls of hollow organs such as the stomach, have an elongated shape. (**b**) A motor neuron, a type of nerve cell, has threadlike extensions.

Cell Size and Shape

You may be wondering how small cells really are. Can any be observed with the unaided eye? There are a few, including the yolks of bird eggs, cells in the red part of watermelons, and the fish eggs we call caviar. However, most cells can only be observed with a microscope. For instance, each of your red blood cells is about 8 millionths of a meter across—so tiny that you could line up 2,000 of them across your thumbnail.

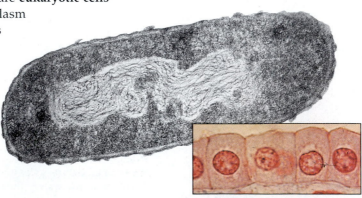

Figure 3.3 (**a**) Micrograph of a bacterium, *Escherichia coli*. This prokaryote is a normal inhabitant of the human gut. The light, spaghetti-like material in the center of the cell is its genetic material, DNA. (**b**) Cross section of cells inside a tiny tube in a kidney. A dye gives the membrane-bound nucleus a reddish hue.

Table 3.1 Eukaryotic and Prokaryotic Cells Compared

	Eukaryotic	Prokaryotic
Plasma membrane	yes	yes
DNA-containing region	yes	yes
Cytoplasm	yes	yes
Nucleus bound by a membrane	yes	no
Other organelles	yes	no

Why are most cells so small? The **surface-to-volume ratio** constrains increases in their size. Simply put, this is a physical relationship that dictates that as the diameter of a growing cell expands, the cell's volume increases faster than its surface area. As a practical matter, it means that the surface of a very large, round cell would not provide enough area for the necessary "imports" and "exports" of substances that would have to occur for the cell to survive. The inward flow of nutrients and the outward flow of wastes would not be fast enough, and the cell would die.

A large, round cell also would have trouble moving materials through its cytoplasm. In small cells, though, random, tiny motions of molecules easily distribute materials. If a cell isn't small, it probably is long and thin or has folds that increase its surface area relative to its volume. The smaller or narrower or more frilly the cell, the more efficiently materials can cross its surface and become distributed inside it. Figure 3.4b shows two of the many shapes of cells in your own body. One of the examples depicts long, slender cells in smooth muscle. The cells of skeletal muscle (muscle that attaches to the skeleton, as described in Chapter 5) also are slender and can be very long indeed. For instance, in the biceps of your upper arm, they are many inches long—as long as the muscle itself.

Cell Membranes

Shortly you will be reading more about how crucial a eukaryotic cell's membranes are to its survival. Those membranes consist mostly of phospholipids, and their functioning depends on this structure. Recall that a phospholipid has a hydrophilic (water-loving) head and two fatty acid tails, which are hydrophobic (water-dreading). This arrangement is pictured in Section 2.10. Research has shown that when a large number of

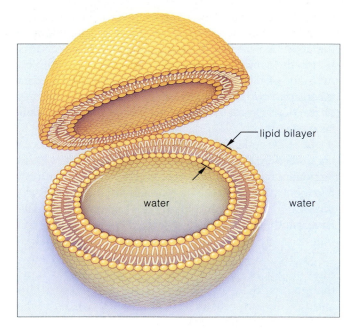

Figure 3.5 Phospholipids arranged in a lipid bilayer. Lipids placed in liquid water may spontaneously arrange themselves this way.

phospholipids are immersed in water, they interact with the water molecules and with one another until they spontaneously cluster at the water's surface. Their movements may even force them beneath the water's surface, where they become organized into two layers—with all the hydrophobic tails sandwiched between all the heads. This heads-out, tails-in arrangement is called a **lipid bilayer**. Sketched in Figure 3.5, it is the basic structure of cell membranes.

All cells have an outer plasma membrane, an internal region of DNA, and a region of cytoplasm. This is true of both prokaryotic and eukaryotic cells. The cells of humans and all other complex organisms are eukaryotic.

Besides the plasma membrane, eukaryotic cells have internal membrane-bound compartments called organelles. The cell's organelles include a nucleus that contains the genetic material DNA.

The size and shape of a growing cell is constrained by the surface-to-volume ratio. There must be enough surface area to allow proper inward flow of nutrients and outward flow of wastes.

Each cell membrane consists largely of phospholipids arranged in a bilayer structure. The hydrophobic parts are sandwiched between the hydrophilic parts, which are dissolved in the watery fluids inside and outside a cell.

EUKARYOTIC CELLS AND THEIR COMPONENTS

In the tiny space of a eukaryotic cell, at any given moment, a vast number of chemical reactions go on. Many of the reactions would be incompatible if they occurred within the same compartment of the cell. For example, a molecule of fat can be assembled by some reactions and taken apart by others, but a cell gains nothing if both sets of reactions proceed at the exact same time on the same fat molecule.

In eukaryotic cells such as those of the human body, organelles are the solution to this dilemma (Table 3.2). An organelle, remember, is a membrane-bound sac or compartment, and organelle membranes are the barriers that physically separate incompatible reactions. Organelles have another function, too: They allow compatible and interconnected reactions to proceed at different times.

Figure 3.6 shows where organelles and other structures might be located in a body cell. Keep in mind, though, that there are major differences in the structures and functions of cells in different tissues. Much of what we know about those features has been gained through the methods of microscopy outlined at the beginning of this chapter.

Table 3.2 Common Features of Eukaryotic Cells	
ORGANELLES AND THEIR MAIN FUNCTIONS:	
Nucleus	Localizing the cell's DNA
Endoplasmic reticulum	Routing and modifying the newly formed polypeptide chains; also, synthesizing lipids
Golgi body	Modifying polypeptide chains into mature proteins; sorting and shipping proteins and lipids for secretion or for use inside cell
Various vesicles	Transporting or storing a variety of substances; digesting substances and structures in the cell; other functions
Mitochondria	Producing many ATP molecules in highly efficient fashion
NON-MEMBRANOUS STRUCTURES AND THEIR FUNCTIONS:	
Ribosomes	Assembling polypeptide chains
Cytoskeleton	Imparting overall shape and internal organization to cell; moving the cell and its internal structures

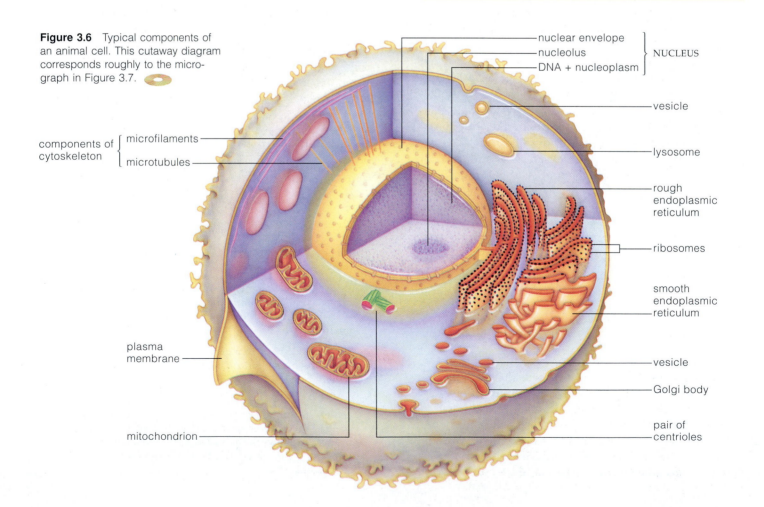

Figure 3.6 Typical components of an animal cell. This cutaway diagram corresponds roughly to the micrograph in Figure 3.7.

nuclear envelope
nucleolus
DNA + nucleoplasm
} NUCLEUS

vesicle

lysosome

rough endoplasmic reticulum

ribosomes

smooth endoplasmic reticulum

vesicle

Golgi body

pair of centrioles

components of cytoskeleton { microfilaments
microtubules

plasma membrane

mitochondrion

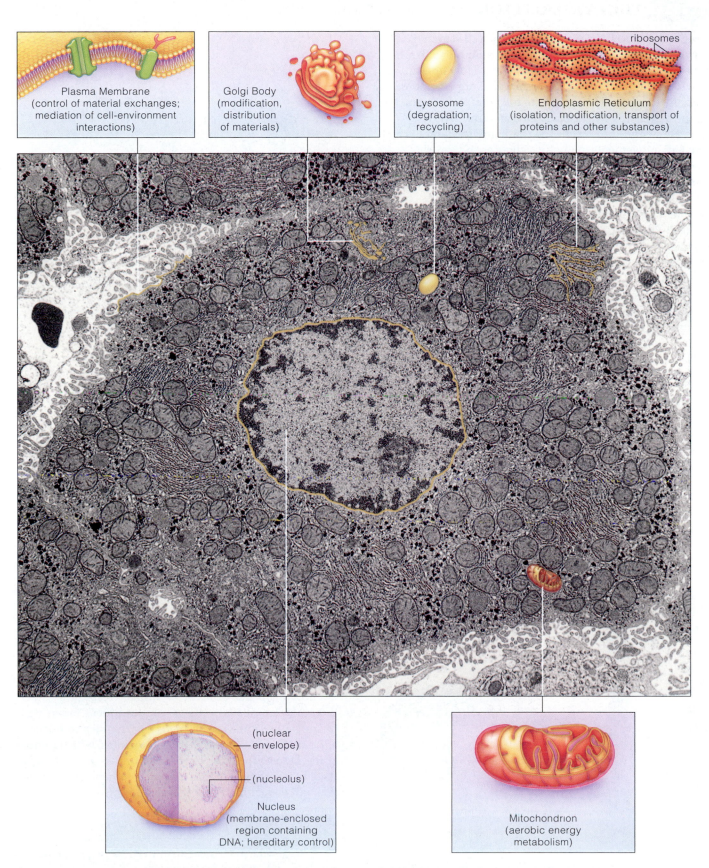

Plasma Membrane
(control of material exchanges;
mediation of cell-environment
interactions)

Golgi Body
(modification,
distribution
of materials)

Lysosome
(degradation;
recycling)

ribosomes

Endoplasmic Reticulum
(isolation, modification, transport of
proteins and other substances)

(nuclear
envelope)

(nucleolus)

Nucleus
(membrane-enclosed
region containing
DNA; hereditary control)

Mitochondrion
(aerobic energy
metabolism)

Figure 3.7 Transmission electron micrograph of a liver cell in cross section. Although this is a rat liver cell, it is representative of liver cells in humans.

THE CYTOSKELETON

We begin our close-up look at eukaryotic cell parts with the **cytoskeleton**, a system of interconnected fibers, threads, and lattices in the fluid part of the cytoplasm. The system gives cells their shape, internal organization, and ability to move. The composite micrograph in Figure 3.8 shows a cell's cytoskeleton, emphasizing two of its major elements: **microtubules** and **microfilaments**. Nearly all movements of body cells depend on these two classes of elements. Microtubules consist of protein subunits, called tubulins, that form a hollow cylinder. Microfilaments consist of two chains of subunits of the protein actin, twisted together. Some kinds of cells also have **intermediate filaments**, a class of ropelike cytoskeletal elements that mechanically strengthen cells and tissues. Figure 3.9 shows the basic structure of all three cytoskeletal elements.

Various accessory proteins can be joined to tubulin and actin, allowing them to serve different functions. Examples include subunits of "motor proteins" such as those called myosin and dynein, which typically project from microtubules or microfilaments that have roles in cell movement. Myosin also is part of the machinery by which

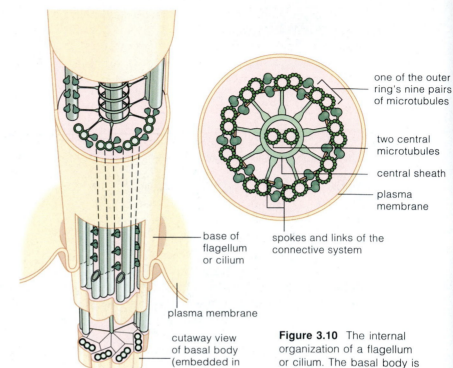

one of the outer ring's nine pairs of microtubules

two central microtubules

central sheath

plasma membrane

base of flagellum or cilium

spokes and links of the connective system

plasma membrane

cutaway view of basal body (embedded in cytoplasm)

Figure 3.10 The internal organization of a flagellum or cilium. The basal body is formed from a centriole.

muscle cells contract. "Crosslinking proteins" splice microtubules or microfilaments together. Certain kinds take part in forming the cell cortex, which is an extensive mesh of microfilaments and other proteins underneath the plasma membrane. The cortex provides structural reinforcement for the cell surface. It also has roles in cell movements and changes in shape.

A **flagellum** (plural: flagella) or **cilium** (plural: cilia) is an example of cytoskeletal organization. Nine pairs of microtubules ring a central pair (Figure 3.10). A system of spokes and links holds this "9 + 2 array" together. The flagellum or cilium bends when pairs of microtubules in the ring slide over each other. Like whiplike tails, flagella propel human sperm.

Cilia are shorter than flagella, and usually there are more of them per cell. In the human respiratory tract, thousands of ciliated cells whisk out mucus laden with dust or other undesirable material. The microtubules of a flagellum or cilium arise from **centrioles**, which remain at the base of the completed structure, as a **basal body**. A centriole is a type of microtubule-producing center, and you will be reading about it and some others later in this book.

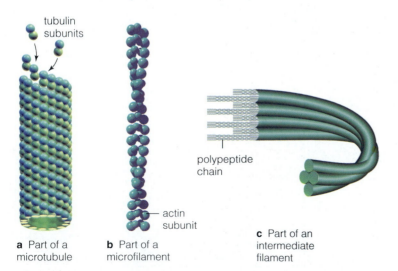

Figure 3.8 An animal cell's cytoskeleton. Its microtubules are tinted green and two kinds of microfilaments are blue and red.

tubulin subunits

polypeptide chain

actin subunit

a Part of a microtubule

b Part of a microfilament

c Part of an intermediate filament

Figure 3.9 Structural organization of (**a**) a microtubule, (**b**) a microfilament, and (**c**) an intermediate filament.

The cytoskeleton gives a eukaryotic cell its shape, internal structure, and capacity for movement.

THE PLASMA MEMBRANE: A "FLUID MOSAIC"

The plasma membrane is not a solid, rigid barrier between a cell's cytoplasm and the fluid outside (called extracellular fluid). If it were, needed substances couldn't enter the cell and wastes couldn't leave it. Instead, the membrane's lipid bilayer structure—the phospholipid "sandwich" described in Section 3.2—makes a cell membrane somewhat fluid, a structural quality that is vital to its functioning, as you'll see. As researchers learned more about the membrane's physical makeup, they developed what is called the **fluid mosaic model** of cell membrane structure.

Figure 3.11 shows a bit of membrane corresponding to the fluid mosaic model. In this model, cell membranes have a mixed composition—they are a "mosaic" of proteins and various lipids: phospholipids, glycolipids, and, in the cells of humans (and other animals), the lipid cholesterol. The proteins are embedded in the bilayer or positioned at its outer or inner surface. Different events take place at the inner and outer surfaces, which have different numbers and kinds of lipids and proteins, arranged in different ways.

Why is the membrane somewhat "fluid"? The model tells us that this quality is due to the movements and interactions of the membrane's components. Most phospholipids can spin on their long axis, move sideways, and flex their tails—all of which help keep neighboring molecules from packing into a solid layer. Short or kinked (unsaturated) hydrophobic tails also contribute to the membrane's fluid nature.

The proteins embedded in a lipid bilayer or attached to one of its surfaces carry out most of a cell membrane's functions. Many of the proteins are enzymes that serve in a cell's metabolic machinery. Others are *transport* proteins that span the bilayer and allow water-soluble substances to move through their interior. They bind molecules or ions on one side of the membrane, then release them on the other side. *Receptor* proteins bind extracellular ("outside the cell") substances, such as hormones, that trigger changes in cell activities. For example, when a young person reaches puberty, certain enzymes that "crank up" the machinery for cell growth and division are switched on when growth hormone binds with receptors for it. Different cells have different combinations of receptors. Diverse *recognition* protein at the cell surface are like molecular fingerprints; they identify a cell as being of a specific type. Finally, *adhesion* proteins help cells of the same type stick together and stay positioned in their proper tissues.

A cell membrane is a "fluid mosaic" of phospholipids and proteins. Its fluid quality is due to the movements and interactions of the membrane's components.

Proteins associated with the bilayer carry out most membrane functions. Many types transport substances across the bilayer or serve as receptors for extracellular substances. Other types of proteins function in cell-to-cell recognition or adhesion.

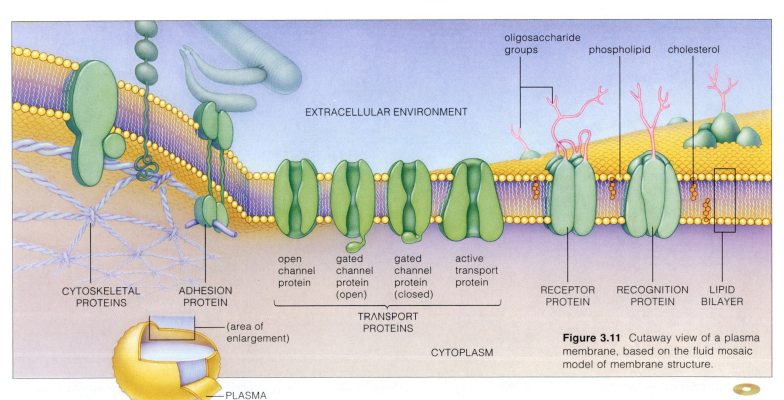

Figure 3.11 Cutaway view of a plasma membrane, based on the fluid mosaic model of membrane structure.

THE CYTOMEMBRANE SYSTEM

The **cytomembrane system** is a series of organelles in which lipids are assembled and new polypeptide chains are modified into final proteins. Its products are sorted and shipped to different destinations. Figure 3.12 shows how its organelles—the ER, Golgi bodies, and various vesicles—functionally connect with one another.

Endoplasmic Reticulum

The functions of the cytomembrane system begin with **endoplasmic reticulum**, or **ER**. The ER is continuous with the cell membrane enclosing the nucleus and curves through the cytoplasm. ER appears rough or smooth, depending mainly on whether ribosomes are attached to the membrane facing the cytoplasm.

We typically observe *rough* ER arranged into stacks of flattened sacs with many ribosomes attached (Figure 3.13a). Every new polypeptide chain is synthesized on ribosomes. But only the newly forming chains having a built-in signal can enter the space inside rough ER or be incorporated into ER membranes. (The signal is a string of fifteen to twenty specific amino acids.) Once the chains are in rough ER, enzymes may attach oligosaccharides and other side chains to them. Many specialized cells secrete the final proteins, and rough ER is abundant in such cells. For example, in your pancreas, ER-rich gland cells make and secrete enzymes that end up in the small intestine and help digest your meals.

Smooth ER has no ribosomes and curves through the cytoplasm like connecting pipes (Figure 3.13b). Many cells assemble most lipids inside the pipes. In liver cells, smooth ER inactivates certain drugs and harmful by-products of metabolism. In skeletal muscle cells a type of smooth ER called sarcoplasmic reticulum stores and releases calcium ions essential in muscle contraction.

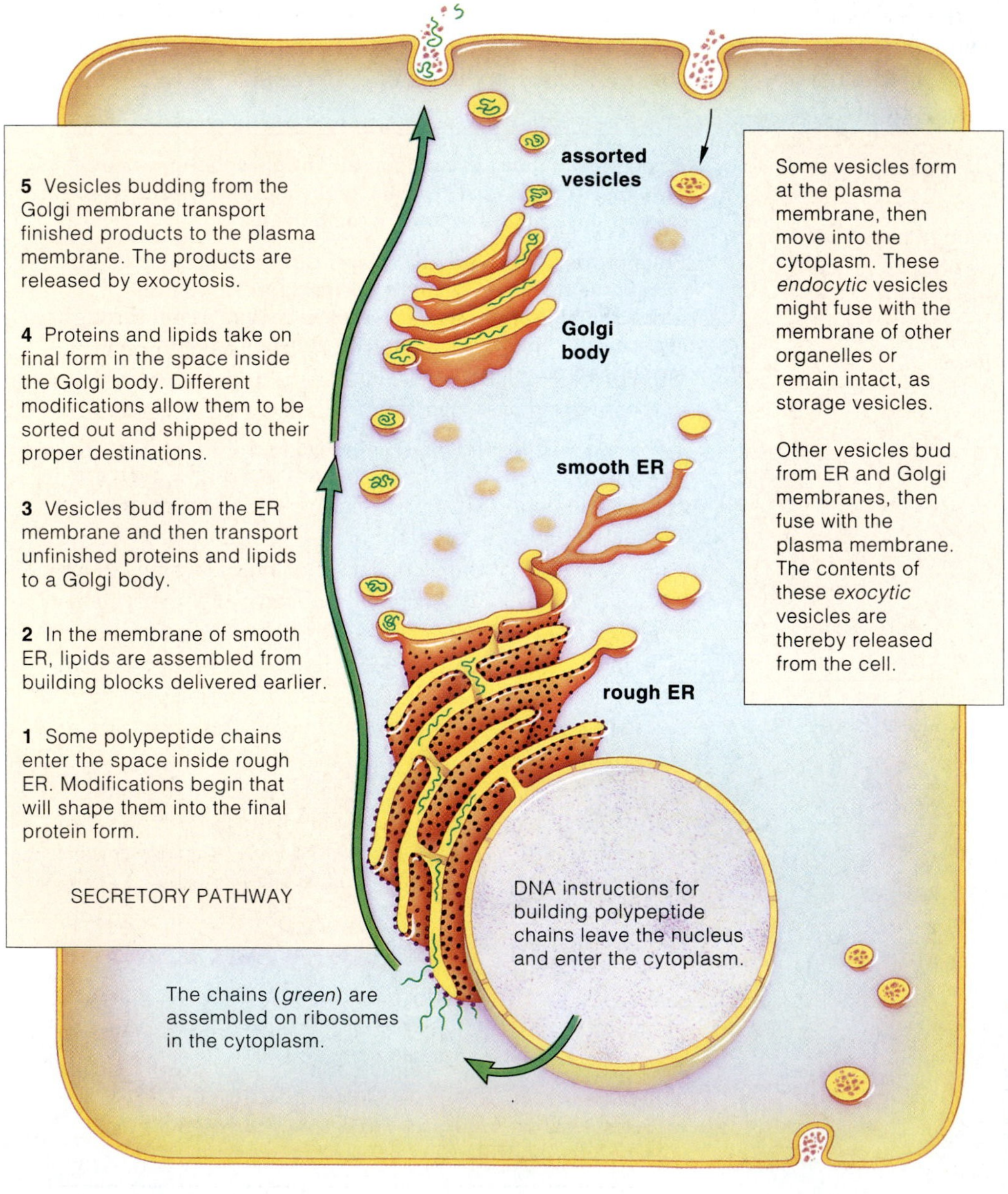

Golgi Bodies

In **Golgi bodies** enzymes put the finishing touches on proteins and lipids, sort them out, and package them inside vesicles for shipment to specific locations. For example, an enzyme in one Golgi region might attach a phosphate group to a new protein, thereby giving it a "mailing tag" to its proper destination.

Commonly, a Golgi body looks vaguely like a stack of pancakes; it is composed of a series of flattened membrane sacs (Figure 3.14). In functional terms, the final portion of a Golgi body corresponds to the top pancake. Here, vesicles form as patches of the membrane bulge out, then break away into the cytoplasm.

5 Vesicles budding from the Golgi membrane transport finished products to the plasma membrane. The products are released by exocytosis.

4 Proteins and lipids take on final form in the space inside the Golgi body. Different modifications allow them to be sorted out and shipped to their proper destinations.

3 Vesicles bud from the ER membrane and then transport unfinished proteins and lipids to a Golgi body.

2 In the membrane of smooth ER, lipids are assembled from building blocks delivered earlier.

1 Some polypeptide chains enter the space inside rough ER. Modifications begin that will shape them into the final protein form.

SECRETORY PATHWAY

assorted vesicles

Golgi body

smooth ER

rough ER

Some vesicles form at the plasma membrane, then move into the cytoplasm. These *endocytic* vesicles might fuse with the membrane of other organelles or remain intact, as storage vesicles.

Other vesicles bud from ER and Golgi membranes, then fuse with the plasma membrane. The contents of these *exocytic* vesicles are thereby released from the cell.

DNA instructions for building polypeptide chains leave the nucleus and enter the cytoplasm.

The chains (*green*) are assembled on ribosomes in the cytoplasm.

Figure 3.12 Cytomembrane system, a membrane system in the cytoplasm that assembles, modifies, packages, and ships proteins and lipids. Green arrows highlight a secretory pathway by which certain proteins and lipids are packaged and released from many types of cells, including gland cells that secrete mucus, sweat, and digestive enzymes.

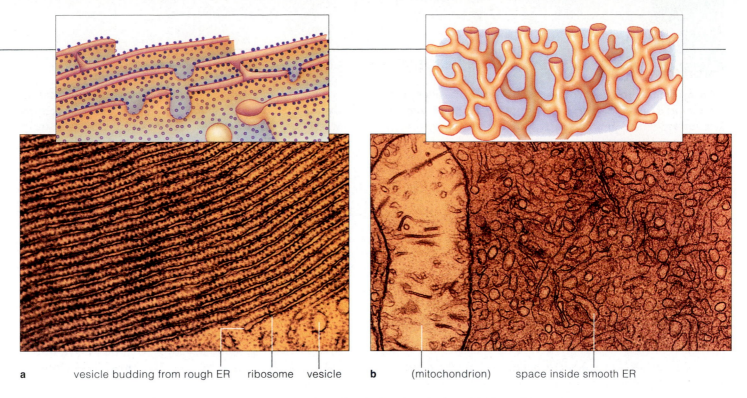

a   vesicle budding from rough ER   ribosome   vesicle   b   (mitochondrion)   space inside smooth ER

Figure 3.13 Transmission electron micrographs and sketches of endoplasmic reticulum. (**a**) Many ribosomes dot the flattened surfaces of rough ER that face the cytoplasm. (**b**) This section reveals the diameters of the many interconnected, pipelike regions of smooth ER.

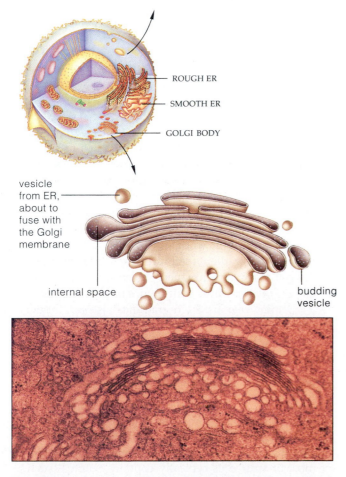

ROUGH ER

SMOOTH ER

GOLGI BODY

vesicle from ER, about to fuse with the Golgi membrane

internal space

budding vesicle

Figure 3.14 Sketch and micrograph of a Golgi body.

A Variety of Vesicles

Vesicles are tiny, membraneous sacs that move through the cytoplasm or take up positions in it. A common type, the lysosome, buds from Golgi membranes. A **lysosome** is an organelle in which intracellular digestion takes place. Enzymes in lysosomes are a potent brew that speeds the breakdown of proteins, complex sugars, nucleic acids, and some lipids. Lysosomes may even digest whole cells or cell parts. Often, lysosomes fuse with vesicles that formed at the plasma membrane. Such vesicles typically hold molecules, bacteria, or other items that attached to the plasma membrane. As Chapter 8 describes, certain white blood cells of the immune system take in foreign material and dispose of it this way.

Peroxisomes, another type of vesicle, are tiny sacs of enzymes that break down fatty acids and amino acids. The reactions produce hydrogen peroxide, a potentially harmful substance. But before hydrogen peroxide can injure the cell, another enzyme in peroxisomes converts it to water and oxygen or uses it to break down alcohol. After someone drinks alcohol, nearly half of it is broken down in peroxisomes of liver and kidney cells.

In the ER and Golgi bodies of the cytomembrane system, many proteins take on final form and lipids are synthesized.

Lipids, proteins, and other substances become packaged in vesicles destined for export, storage, membrane building, intracellular digestion, and other cell activities.

Picture a cell membrane, with water bathing both sides of its bilayer. Lots of substances are dissolved in the water, but the kinds and amounts are not the same on the two sides. The membrane itself helps establish and maintain those differences, which are essential for cell functioning. How does the cell membrane accomplish this feat? It shows **selective permeability**. *Because of its molecular structure, the membrane allows some substances but not others to cross it in certain ways, at certain times.*

Carbon dioxide, molecular oxygen, and other small, nonpolar solutes can readily cross a cell membrane's lipid bilayer, and so can water molecules:

Although water molecules are polar, they might move through gaps that open up in the bilayer, then close up again. Glucose and other large, polar molecules almost never cross the bilayer independently. Neither do ions:

Such water-soluble substances must cross the bilayer through the interior of transport proteins, as you will read in Section 3.7. Why does a solute move one way or another at any given time? The answer starts with concentration gradients.

Concentration Gradients and Diffusion

"Concentration" refers to the number of molecules of a substance in a specified volume of fluid. "Gradient" means that the number in one region is not the same as it is in another. Therefore, a **concentration gradient** is a difference in the number of molecules or ions of a given substance in two adjoining regions.

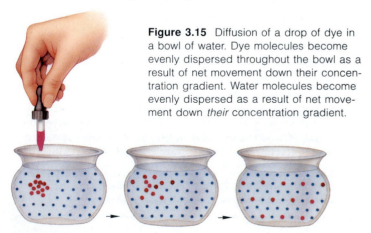

Figure 3.15 Diffusion of a drop of dye in a bowl of water. Dye molecules become evenly dispersed throughout the bowl as a result of net movement down their concentration gradient. Water molecules become evenly dispersed as a result of net movement down *their* concentration gradient.

Unless other forces come into play, a substance moves from a region where it is more concentrated to a region where it is less concentrated. The energy inherent in its molecules, which keeps them in constant motion, drives the directional movement. While the molecules collide randomly and career back and forth, millions of times a second, the *net* movement is away from the place of greater concentration (and the most collisions).

Diffusion is the name for the net movement of like molecules or ions down a concentration gradient. It is a key factor in the movement of substances across cell membranes and through cytoplasmic fluid. In the body, diffusion moves substances to and from cells, into and out of the fluids bathing them.

If a solution contains more than one solute, each solute still diffuses according to its *own* concentration gradient. For example, if you put a drop of dye in one side of a bowl of water, the dye molecules diffuse to the region where they are less concentrated. Likewise, the water molecules move in the opposite direction, to the region where *they* are less concentrated (Figure 3.15).

Diffusion is faster when the gradient is steep. In the region of greatest concentration, more molecules are moving outward, compared to the number that are moving in. As the gradient decreases, the difference in the number of molecules moving either way declines. When the gradient finally disappears, molecules are still in motion. However, now the total number moving one way or the other during a specified interval is about the same. When the net distribution of molecules becomes nearly uniform throughout two adjoining regions, we call this "dynamic equilibrium."

For charged molecules, transport is influenced by both the concentration gradient and the *electric gradient*— a difference in electric charge across the cell membrane. The diffusion of ions down concentration and electric gradients is vital for cell functions. Many processes, such as the flow of information through your nervous system, depend on the combined influences of electric and concentration gradients to attract specific substances across cell plasma membranes.

A *pressure gradient* is a difference in pressure between two adjoining regions. Such gradients also can influence the speed and direction in which molecules diffuse. Breathing is an essential body function that depends on pressure gradients, as you will read in Chapter 9.

Osmosis

Because the plasma membrane is selectively permeable, the solute concentrations can increase on one side of the membrane and not the other. For example, the cytoplasm of most of your cells typically contains solutes (such as

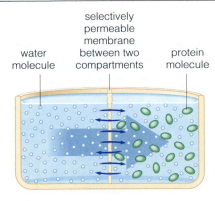

water molecule / selectively permeable membrane between two compartments / protein molecule

Figure 3.16 How a solute concentration gradient affects osmotic movement.

Start with a container divided by a membrane that water but not proteins can cross. Pour water into the left compartment. Pour the same volume of a protein-rich solution into the right compartment. There, proteins occupy some of the space. The net diffusion of water in this example is from left to right (large blue arrow).

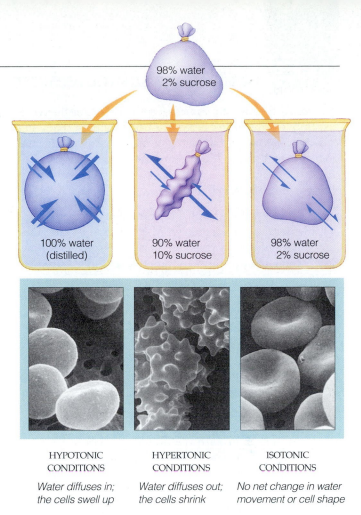

98% water
2% sucrose

100% water (distilled) | 90% water 10% sucrose | 98% water 2% sucrose

HYPOTONIC CONDITIONS | HYPERTONIC CONDITIONS | ISOTONIC CONDITIONS

Water diffuses in; the cells swell up | *Water diffuses out; the cells shrink* | *No net change in water movement or cell shape*

Figure 3.17 Tonicity and the diffusion of water. In the sketches, membranous bags through which water but not sucrose can move are placed in hypotonic, hypertonic, and isotonic solutions. In each container, arrow width represents the relative amount of water movement. The sketches show what happens when red blood cells—which cannot actively take in or expel water—are placed in comparable solutions.

proteins) that cannot cross the plasma membrane by simple diffusion. Such solutes may become concentrated within a cell, producing solute concentration gradients that affect the diffusion of water across the plasma membrane. **Osmosis** is the name for the diffusion of water across a selectively permeable membrane in response to solute concentration gradients (Figure 3.16).

A characteristic called *tonicity* influences osmotic water movements. Tonicity refers to the relative solute concentrations in two fluids. When two fluids on opposing sides of a cell membrane contain equal concentrations of solutes, we say they are *isotonic* (*iso-* means same) and there is no net movement of water in either direction across the membrane. When the solute concentrations are not equal, one fluid is *hypotonic* (has fewer solutes) and the other is *hypertonic* (has more solutes). Figure 3.17 shows how the tonicity of a fluid affects red blood cells. The important point to remember is that water always tends to move from a hypotonic solution to a hypertonic one because it moves down its concentration gradient.

Osmotic water movements across a cell's plasma membrane produce *osmotic pressure*. This term refers to water's tendency to pass from a hypotonic solution to a hypertonic one when the two solutions are separated by a selectively permeable membrane. The greater the difference in solute concentration between the solutions, the more osmotic pressure builds, because water "wants" to move into the region where solutes are more concentrated. In certain situations osmotic pressure can be counterbalanced by *hydrostatic pressure*. In a cell, this is a mechanical pressure that various factors (such as the plasma membrane's resistance to stretching) exert on the cytoplasm. If the opposing hydrostatic pressure becomes great enough, it can stop or even reverse the osmotic flow of water. Blood is largely water and it also exerts hydrostatic pressure in the body. Chapter 7 explores some important physiological implications of this fact.

Moment to moment, cell activities and other events alter the factors that influence the solute concentrations of body fluids and water movements between those fluids. If cells did not have mechanisms for adjusting to such differences, they would shrivel or burst, as Figure 3.17 illustrates. In subsequent chapters we will return to some key ways in which osmotic water movements help maintain the body's proper water balance.

Diffusion is the net movement of like molecules or ions from a region of higher concentration to a region of lower concentration.

Osmosis is the net diffusion of water between two solutions that differ in water concentration and that are separated by a selectively permeable membrane. The more molecules and ions dissolved in a solution, the lower its water concentration will be.

The fluid pressure that a solution exerts against a membrane also influences the osmotic movement of water.

Besides diffusion and osmosis, some other mechanisms also move substances across cell membranes. They are called passive transport, active transport, exocytosis, and endocytosis, and we'll now consider them in turn.

Transport through Proteins

Section 3.4 introduced transport proteins, which span the cell membrane's lipid bilayer. Many of them allow ions and other solutes to freely diffuse across a membrane, down their concentration gradients. **Passive transport** is the name for this kind of flow of solutes through the interior of transport proteins (Figure 3.18). This directional movement (sometimes called "facilitated diffusion") is "passive" because it does not require energy from ATP, the energy-delivering nucleotide introduced in Chapter 2.

Two features allow a transport protein to fulfill its role. First, its interior can open on both sides of a cell membrane. Second, when the protein interacts with a solute, its shape changes, then changes back again. The changes shunt the solute through the protein's interior, from one side of the lipid bilayer to the other. While transport proteins allow solutes to move both ways across a membrane, they are choosy about *which* solutes travel through them. For example, the protein that transports amino acids will not transport glucose.

As cells use and produce substances in their activities, solute concentrations on either side of their membranes are constantly changing. Cell operations also require a cell to actively move certain solutes to and from its interior. Action requires energy, and so cells have energy-using mechanisms called "membrane pumps" that make solutes cross membranes *against* concentration gradients. The name for these mechanisms is **active transport** (Figure 3.19). ATP provides most of the energy for active transport, and membrane pumps can continue until the solute is *more* concentrated on the side of the membrane where it is being pumped. For example, the calcium pump helps keep the calcium concentration in cells at least a thousand times lower than the concentration outside, laying the chemical groundwork for muscle contraction and some other essential physiological processes.

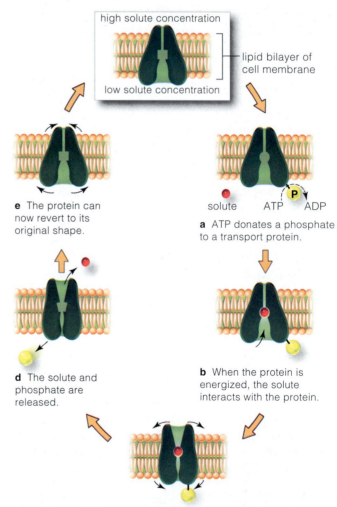

e The protein can now revert to its original shape.

d The solute and phosphate are released.

a ATP donates a phosphate to a transport protein.

solute   ATP   P   ADP

b When the protein is energized, the solute interacts with the protein.

c The protein's shape changes, and the solute ends up facing the opposite side of the membrane.

Figure 3.19 Active transport across a cell membrane. ATP provides the energy for this mechanism when it transfers a phosphate group to a transport protein. The transfer triggers reversible changes in the protein's shape.

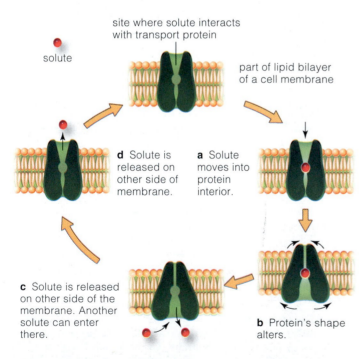

site where solute interacts with transport protein

solute

part of lipid bilayer of a cell membrane

d Solute is released on other side of membrane.

a Solute moves into protein interior.

c Solute is released on other side of the membrane. Another solute can enter there.

b Protein's shape alters.

Figure 3.18 Passive transport across a cell membrane. A solute can move in both directions through transport proteins. In passive transport, the *net* movement will be down a solute's concentration gradient (from higher to lower concentration) until its concentrations are the same on both sides of the membrane.

Further reading: Student Guide to InfoTrac on web site →

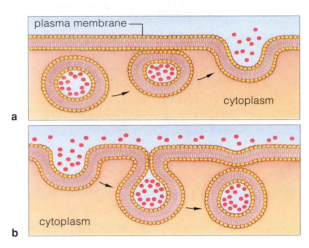

a

b

Figure 3.20 (**a**) Exocytosis. Cells release substances when an exocytic vesicle's membrane fuses with the plasma membrane. (**b**) Endocytosis. A bit of plasma membrane balloons inward beneath water and solutes outside, then it pinches off as an endocytic vesicle that moves into the cytoplasm.

Transport by Vesicles

Transport proteins can only move small molecules and ions into or out of cells. To take in or expel large molecules or particles, cells use vesicles that form through exocytosis and endocytosis.

In **exocytosis** ("moving outside a cell"), a vesicle moves to the cell surface and the protein-studded lipid bilayer of its membrane fuses with the plasma membrane (Figure 3.20*a*). While this exocytic vesicle is losing its identity, its contents are released to the surroundings.

In **endocytosis** ("coming inside a cell"), a cell takes in substances next to its surface. A small indentation forms at the plasma membrane, balloons inward, and pinches off. The resulting endocytic vesicle transports its contents or stores them in the cytoplasm (Figure 3.20*b*). When endocytosis brings organic matter into the cell, the process is called *phagocytosis*, which literally means "cell eating." Certain cells of the immune system engage in phagocytosis when they defend the body against harmful viruses, bacteria, and other threats.

Diseases that upset the normal balance of solutes and water in cells can be deadly, as this chapter's *Focus on Our Environment* explains.

In passive transport, a solute diffuses down its concentration gradient through the interior of a transport protein. In active transport, membrane pumps move solutes against their concentration gradient into or out of a cell. ATP provides much of the energy for active transport. Exocytosis and endocytosis move large molecules or particles across the membrane.

REVENGE OF EL TOR

In many parts of our world, cholera is more dreaded than AIDS. A person can wake up one morning with symptoms—mainly, overwhelming diarrhea—and by nightfall be dead, his or her body cells literally drained of water. The cause is a poison, cholera exotoxin, that is produced by marauding armies of the bacterium *Vibrio cholerae*. Today, a strain of the bacterium called El Tor is marching around the globe, carried by contaminated water supplies (Figure 3.21), shellfish, and other foods. In the Western Hemisphere alone, it caused more than 700,000 cases during a cholera epidemic in the early 1990s. It is also a threat in Africa and Asia.

El Tor's toxin wages war on a person's cells by disrupting homeostasis—specifically, by triggering the uncontrolled loss of water by osmosis. It does so by raising the level of a signaling molecule that causes cells to dump chloride into the intestine. Other solutes, including sodium, follow. As the solute concentration in the extracellular fluid increases, the concentration of water falls. Therefore, osmosis pulls water out of cells and ultimately into the intestine, where the life-sustaining fluid leaves as a horrific, nearly clear diarrhea.

In the developed world, cholera victims are treated with antibiotics. Even in regions where antibiotics are not readily available, patients often can recover through *oral rehydration therapy* (ORT), in which a fluid containing glucose and other solutes is administered. The particular mix of solutes and water in ORT causes the body to take up water faster than it is lost, giving the person's immune system time to mount a counterattack on the invader. Meanwhile, researchers are working to improve a not-very-effective cholera vaccine, and public health officials worry about the next outbreak of El Tor.

Figure 3.21 In many places where public sanitation is primitive or lacking altogether, people run the risk of contracting cholera because drinking water and some food sources (especially fish) are contaminated by human sewage that carries the *Vibrio cholerae* bacterium.

THE NUCLEUS

The **nucleus** encloses a eukaryotic cell's DNA. The DNA contains instructions for building all of a cell's proteins and, through those proteins, for determining a cell's structure and function. Figure 3.22 shows the distinctive structure of the nucleus, and Table 3.3 lists its components. The nucleus has two key functions. *First*, it physically separates DNA from the complex metabolic machinery of the cytoplasm. Each body cell contains forty-six DNA molecules, and the separation of DNA makes it easier to sort out hereditary instructions when it comes time for a cell to divide. When a cell divides, DNA molecules must be assorted into parcels—one for each new cell that forms. *Second*, outer membranes of the nucleus form a boundary where the cell can control the passage of substances and signals to and from the cytoplasm.

The Nuclear Envelope

Unlike the cell itself, a nucleus has two outer membranes, one wrapped around the other. This double-membrane system is called a **nuclear envelope**. It consists of two lipid bilayers in which numerous protein molecules are embedded. It surrounds the fluid portion of the nucleus (the nucleoplasm).

Threadlike bits of protein attach to the innermost surface of the nuclear envelope. They anchor the DNA molecules to the envelope and help keep them organized. The envelope's outer surface is studded with ribosomes. All proteins are built either on such membrane-bound ribosomes or on ribosomes located in the cytoplasm.

As with all cell membranes, a nuclear envelope's lipid bilayers keep water-soluble substances from moving freely into and out of the nucleus. But many pores, each

Table 3.3	Components of the Nucleus
Nuclear envelope	Double-membraned, pore-riddled boundary between cytoplasm and the nucleus interior
Nucleolus	Dense cluster of the RNA and proteins used to assemble ribosome subunits
Nucleoplasm	Fluid portion of the nucleus interior
Chromosomes	DNA molecules and proteins attached to them
Chromatin	All DNA molecules and their associated proteins in the nucleus

composed of clusters of proteins, span both bilayers. They allow ions and small, water-soluble molecules to move freely across the nuclear envelope, but they control the passage of large molecules, such as ribosome subunits.

The Nucleolus

As a cell grows, one or more dense masses appear within its nucleus. Each mass is a **nucleolus** (plural: nucleoli). Here a great number of protein and RNA molecules are constructed. They are the subunits from which ribosomes are built, and they will pass through nuclear pores and into the cytoplasm. In times of protein synthesis, ribosomes form (each from two subunits) in the cytoplasm.

Chromosomes

When a eukaryotic cell is not dividing, its DNA looks like thin threads inside the nucleus. Many protein molecules are attached to the threads, a bit like beads on a string. Except when they are highly magnified, the threaded beads look grainy, as you see in Figure 3.22. However, when the cell is preparing to divide, it duplicates its DNA molecules so that each new cell will get all the required hereditary instructions. In addition, each DNA molecule becomes folded and twisted into a condensed structure, proteins and all.

Early microscopists bestowed the name *chromatin* on the seemingly grainy substance and *chromosomes* on the condensed structures. We now define **chromatin** as a cell's collection of DNA, together with all of the proteins associated with it. Each **chromosome** is one DNA molecule and its associated proteins, regardless of whether it is threadlike or condensed. In other words, a chromosome doesn't always look the same during the life of a cell.

The nucleus, an organelle with two outer membranes, keeps a cell's DNA molecules separated from the metabolic machinery of the cytoplasm. The separation makes it easier to organize the DNA and to copy it before a parent cell divides into daughter cells.

Pores across the nuclear envelope help control the passage of many substances between the nucleus and cytoplasm.

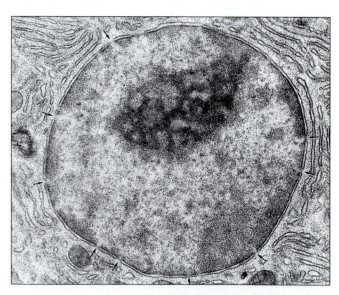

Figure 3.22 The nucleus of an animal cell. Arrows point to pores in the nuclear envelope.

Further reading: Student Guide to InfoTrac on web site →

THE MITOCHONDRION

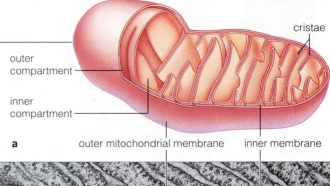

cristae

outer compartment

inner compartment

a

outer mitochondrial membrane

inner membrane

For as long as a cell lives, it hums with chemical reactions. The vast majority of those reactions require energy, which is provided by ATP—the premier energy carrier described in Section 2.13. Energy associated with ATP's phosphate groups can be delivered to nearly all reaction sites in a cell, and drives nearly all cell activities. Many ATP molecules form when organic compounds are completely broken down to carbon dioxide and water in a **mitochondrion** (plural: mitochondria).

Only eukaryotic cells contain these organelles. The example in Figure 3.23 gives you an idea of their structure. The kind of ATP-forming reactions that occur in mitochondria extract far more energy from organic compounds than can be done by any other means. They cannot run to completion without plenty of oxygen. Like other land-dwelling vertebrates, every time you breathe in, you take in oxygen mainly for mitochondria in cells—in your case, trillions of them.

Each mitochondrion has a double-membrane system. As shown in the sketch at the upper right, the outermost membrane faces the cytoplasm. In most cases, the inner membrane folds back on itself repeatedly. Each fold is a crista (plural: cristae).

What is the function of the intricate membrane system? It forms two distinct compartments within a mitochondrion. Enzymes and other proteins stockpile hydrogen ions in the outer compartment. Electron transfers drive the stockpiling, and oxygen helps keep the machinery running by binding and thus removing the spent electrons. Hydrogen ions flow out of the compartment in controlled ways. Energy inherent in the flow drives ATP formation, as Section 3.14 describes.

All eukaryotic cells have at least one mitochondrion. If you look back at Figure 3.7, you can see mitochondria all through the cytoplasm of that one thin slice from one liver cell. The message in that micrograph is that the liver is an exceptionally active, energy-demanding organ.

Mitochondria have intrigued biologists because they resemble bacteria in size and in many details of their biochemistry. They even have their own DNA and some ribosomes, and they divide on their own. Many biologists believe mitochondria evolved from ancient bacteria that were consumed by another ancient cell, yet did not die. Perhaps they were able to reproduce inside the predatory cell and its descendants. If they became permanent, protected residents, they might have lost structures and functions required for independent life while they were

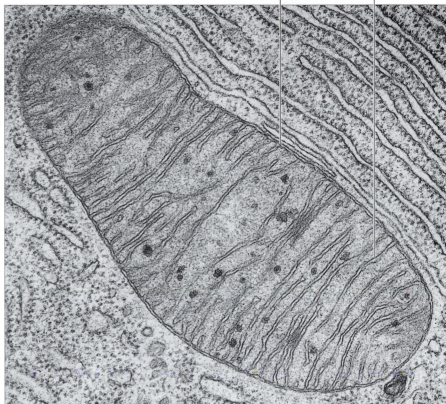

b

Figure 3.23 (**a**) Sketch and (**b**) transmission electron micrograph of a thin slice through a typical mitochondrion. Reactions inside this organelle produce ATP, which is the major energy carrier in cells.

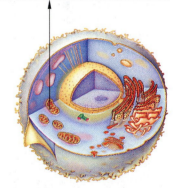

becoming mitochondria—the ATP-producing organelles without which we humans could not survive.

The organelles called mitochondria are the ATP-producing powerhouses of eukaryotic cells.

Energy-releasing reactions take place in the compartmented, internal membrane system of mitochondria. The reactions, which require oxygen, produce far more ATP than can be made by other cellular reactions.

The 65 trillion living cells in your body require energy for their operations: building, storing, dismantling, and releasing substances in controlled ways. This demanding cellular work is called **metabolism**. The "raw" energy for it comes from organic compounds in food, which a cell's mitochondria convert to ATP—a chemical form the cell can use. To set the stage for that story, we look first at key events and players in metabolism, beginning with ATP.

ATP—The Cell's Energy Currency

Some reactions in cells release energy, others require an energy input. ATP couples the two kinds of reactions, carrying energy from one reaction site to another.

Remember from Section 2.13 that ATP is short for adenosine triphosphate, one of the organic compounds called nucleotides. Each ATP molecule consists of the five-carbon sugar ribose to which adenine (a nucleotide base) and three phosphate groups are attached.

Hundreds of different enzymes can break the covalent bond between the two outermost phosphate groups of the ATP molecule's tail (shown at left in yellow). The enzymes then can attach this phosphate group to another substance. Any transfer of a phosphate group to a molecule is a **phosphorylation**. The energy transferred can activate hundreds of different molecules and drive hundreds of cellular activities. For example, in a contractile unit of a muscle cell, ATP transfers a phosphate group to a transport protein. Energy associated with the transfer makes the protein's shape change in a way that pumps calcium ions out of the cell.

Cells renew their ATP supply. At certain steps of many metabolic processes, a phosphate atom (symbolized by P_i) or a phosphate group that has been split off from some substance is attached to ADP, adenosine diphosphate (the prefix *di-* indicates *two* phosphate groups). The resulting molecule, with three phosphates, is ATP. And when ATP transfers a phosphate group elsewhere, it reverts to ADP. In this way it completes the **ATP/ADP cycle**:

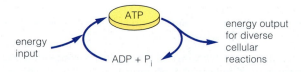

Like currency earned in an economy, ATP is earned in reactions that yield energy and spent in reactions that require it. That is why textbooks often use a cartoon coin to symbolize ATP.

Metabolic Pathways

At any instant, thousands of reactions are transforming thousands of substances inside a cell. Most of these reactions are organized as **metabolic pathways**, in which reactions proceed one after another, in orderly steps that enzymes mediate. There are two main types of metabolic pathways: biosynthetic ones and degradative ones.

In a biosynthetic pathway, small molecules are assembled into more complex ones in which the energy content of the chemical bonds is increased. The pathways that build complex carbohydrates, proteins, and other large molecules are all biosynthetic; the energy stored in their bonds is a major reason why we can use them as food. In a degradative pathway, large molecules are broken down to products of lower energy content. Sometimes biosynthetic activity is called **anabolism**, and degradative activity is called **catabolism**.

A **substrate** is any substance that enters a reaction. Substrates are also called reactants or precursors. Any substance that forms between the start and end of a pathway is an *intermediate*. Those remaining at the end of a reaction or a pathway are the *end products*. The *energy carriers* are ATP and a few other compounds that can activate (donate energy to) substances by transferring functional groups to them. Most *enzymes* are proteins that speed specific reactions. *Cofactors* are organic molecules or metal ions that assist enzymes or pick up electrons, atoms, or functional groups at one reaction site and taxi them to a different site. *Transport proteins* are proteins that span a cell membrane and let substances across it in controlled ways. They help adjust concentrations of substances on both sides of the membrane. In this way they influence metabolic reactions.

Many metabolic pathways advance step by step in linear fashion, from substrates to end products. Many others are cyclic; the steps proceed in a circle, with end products serving as reactants to start things over. Many times, the intermediates or end products of one pathway also can enter different metabolic pathways.

Quite a few metabolic pathways have steps that are linked by a transfer of electrons: One substance gives up one or more electrons and another substance accepts them. In other words, the reactions are *coupled*. As they transfer electrons, coupled reactions in mitochondria release the energy that is captured in ATP.

Four Features of Enzymes

Enzymes are catalytic molecules: They greatly speed the rate at which specific reactions occur. Again, nearly all

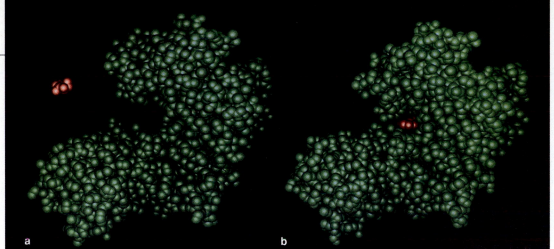

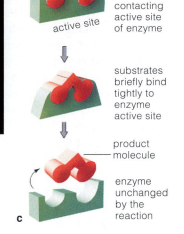

two substrate molecules

substrates contacting active site of enzyme

active site

substrates briefly bind tightly to enzyme active site

product molecule

enzyme unchanged by the reaction

c

Figure 3.24 Model of an enzyme, hexokinase, at work. Hexokinase catalyzes phosphorylation of glucose—shown above (**a**) as a small red molecule heading toward the active site in the enzyme (green). (**b**) When the glucose molecule contacts the enzyme's active site, parts of the enzyme briefly close around it and prompt the enzyme to enter the reaction. (**c**) The enzyme-substrate complex is short-lived, partly because it is held together by weak bonds. The enzyme resumes its prebinding shape as a product molecule is released.

enzymes are proteins (the exceptions are certain RNA molecules). All enzymes share four features. First, they usually make reactions occur hundreds to millions of times faster than would otherwise happen. Second, reactions do not permanently alter or use up enzyme molecules; the same one may act repeatedly. Third, the same type of enzyme usually works for the forward and reverse directions of a reaction. Fourth, each enzyme only interacts with specific substrates. The substrates are substances that it can chemically recognize, bind, and modify in certain ways. An example is thrombin, one of the enzymes required to clot blood. It only recognizes a side-by-side alignment of arginine and glycine, two amino acids, in a protein. When it does so, it cleaves the peptide bond between them. An enzyme has at least one **active site** (Figure 3.24), a surface crevice where the enzyme and a substrate interact.

Each type of enzyme functions best within a certain temperature range. For example, when body temperature rises too high, the increased heat energy disrupts bonds holding an enzyme in its three-dimensional shape. This alters the active site so substrates cannot bind to it, and metabolic activities are thrown into turmoil. This is what happens when sick people develop dangerously high fevers. They will usually die if their internal temperature reaches 44 °C (112 °F). Enzymes also function best within a certain pH range—in the body, from pH 7.35 to 7.4. Above or below this pH range most enzymes cannot function properly.

During many reactions, enzymes speed the transfer of electrons, atoms, or functional groups from one substrate to another. Cofactors that help with the reactions include organic molecules called **coenzymes**. Many coenzymes, including **NAD+** (nicotinamide adenine dinucleotide) and **FAD** (flavin adenine dinucleotide) are derived from vitamins. As you will read shortly, they play central roles in reactions by which cells make ATP.

How Is Enzyme Activity Controlled?

Each cell controls its enzyme activity. By coordinating control mechanisms, it maintains, lowers, or raises the concentrations of substances. Controls that adjust how fast enzyme molecules are synthesized affect how many are available for a metabolic pathway. Other controls boost or slow the action of enzymes that were synthesized earlier.

In your own body, the signaling agents called hormones have major enzyme-regulating effects. As Chapter 13 describes, specialized cells release hormones into the bloodstream and any cell with receptors for a given hormone will take it up. Then, the cell's program for a specific activity will be changed. For example, when you eat, food arriving in your stomach causes gland cells in the stomach to secrete the hormone gastrin into your bloodstream. Stomach cells with receptors for gastrin respond in several ways, such as making and secreting the ingredients of "gastric juice"—including enzymes that break down food proteins.

Most chemical reactions in cells are organized in the orderly steps of metabolic pathways.

An enzyme speeds up the rate at which a specific reaction occurs. Enzymes act only on specific substrates, and a given enzyme may catalyze the same reaction repeatedly, as long as substrates are available.

Enzymes function best within limited ranges of temperature and pH. Cofactors (coenzymes and metal ions) help catalyze reactions or transfer electrons, atoms, and functional groups from one substrate to another.

Control mechanisms influence the synthesis of new enzymes and stimulate or inhibit existing ones.

MAKING ATP: THE FIRST TWO STAGES

The chemical reactions that sustain cells—and, of course, the whole body—depend on the energy that cells capture when they synthesize ATP. Cells make ATP by breaking covalent bonds of carbohydrates (especially glucose), lipids, and proteins. During the breakdown reactions, electrons are removed from intermediate compounds. Energy associated with those liberated electrons drives the formation of ATP, as you'll soon read.

Cells of the human body typically form ATP by way of **aerobic respiration**. "Aerobic" means the pathway cannot be completed without oxygen. With every breath, you are providing your cells with oxygen. Other energy-releasing pathways are *anaerobic*, which means that they can be completed without using oxygen. The most common anaerobic pathways are called **fermentation pathways**. Cells of some human tissues, such as skeletal muscle, can briefly use a fermentation pathway, but only when they are not receiving enough oxygen.

Both types of energy-releasing pathways start with the same set of reactions, which take place in the cell cytoplasm. These initial reactions are called **glycolysis**. In glycolysis, enzymes cleave and rearrange each glucose molecule into two molecules of **pyruvate**, each with a backbone of three carbon atoms. Notice that oxygen has no role in glycolysis.

Once glycolysis is completed, the next events in the aerobic pathway for making ATP take place in a cell's mitochondria. There, oxygen serves as the final acceptor of electrons used in the reactions. This is why aerobic respiration requires oxygen. In anaerobic pathways such as fermentation, a substance other than oxygen is the final electron acceptor.

As you examine the energy-releasing pathways in this section and the next, keep in mind that enzymes catalyze each reaction step and intermediate molecules formed at one step serve as substrates for the next enzyme in the pathway.

The Breakdown of Glucose Begins

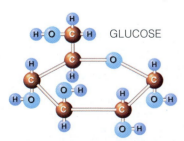

Glucose, recall, is a simple sugar. Each glucose molecule consists of six carbon, twelve hydrogen, and six oxygen atoms, all joined by covalent bonds. The carbons make up the backbone. In glycolysis, a glucose molecule (or another carbohydrate in the cytoplasm) is partially broken down, so that two molecules of the three-carbon compound pyruvate form:

$$glucose \rightarrow glucose\text{-}6\text{-}phosphate \rightarrow 2\ pyruvate$$

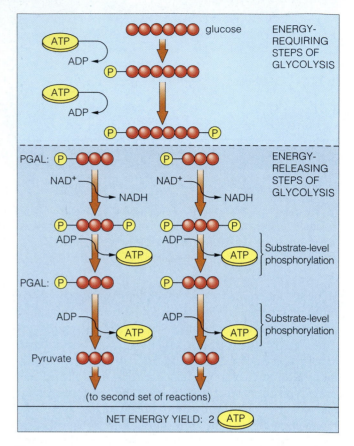

Figure 3.25 Reaction steps of glycolysis. The red circles represent the six carbon atoms of the glucose molecule.

The first steps of glycolysis require energy. As you can see in Figure 3.25, the steps advance only when two ATP molecules each transfer a phosphate group to glucose and so donate energy to it. Such transfers, remember, are "phosphorylations." In this case, they raise the energy content of glucose to a level high enough to enable the energy-releasing steps of glycolysis to begin.

The first energy-releasing step cleaves the activated glucose into two molecules, which we can call **PGAL** (phosphoglyceraldehyde). Each PGAL is converted to an unstable intermediate. Next, the two molecules of the intermediate each donate a phosphate group to ADP, forming ATP. The next intermediate in the sequence does the same thing.

Thus, a total of four ATP form by **substrate-level phosphorylation**. This is the direct transfer of a phosphate group from a substrate of a reaction to another molecule, such as ADP. Keep in mind, though, that two ATP were invested to start the reactions. So the *net* energy yield is only two ATP.

Further reading: Student Guide to InfoTrac on web site →

Meanwhile, the coenzyme NAD$^+$ picks up electrons and hydrogen ions released from each PGAL. In so doing, it becomes NADH. When NADH gives up its cargo at a different reaction site, it reverts to NAD$^+$. Said another way, NAD$^+$ is reusable, like all coenzymes.

The Krebs Cycle and Preparatory Steps

Now let's consider what happens when two molecules of pyruvate, formed by glycolysis, enter a mitochondrion. There, in the mitochondrion's inner compartment (the matrix), the aerobic pathway will be completed. That story unfolds during a stage of reactions in which a bit more ATP forms. Carbon and oxygen atoms depart, in the form of carbon dioxide and water. And coenzymes accept electrons and hydrogen removed from intermediates of the reactions (Figure 3.26).

In a few preparatory steps, an enzyme removes a carbon atom from each pyruvate molecule. A coenzyme called coenzyme A becomes acetyl-CoA by combining with the remaining two-carbon fragment. The fragment is transferred to oxaloacetate, the entry point for the **Krebs cycle.** The name of this cyclic pathway honors the biologist who began working out its details in the 1930s. Notice that *six* carbon atoms, three in each pyruvate backbone, enter this second stage of reactions. Six also leave (in six carbon dioxide molecules) during the preparatory reactions and the Krebs cycle. The CO_2 is transported in blood to the lungs and exhaled.

Second-stage reactions have three major functions. First, NAD$^+$ and FAD pick up H+ and electrons, to become NADH and FADH$_2$. Second, two molecules of ATP are produced, by substrate-level phosphorylations. Third, the reactions rearrange the Krebs cycle intermediates into oxaloacetate. Cells have only so much oxaloacetate, and it must be regenerated to keep the cyclic reactions going.

The two ATP that form do not add much to the small yield from glycolysis. However, for each original glucose molecule, a large number of coenzymes pick up hydrogen and electrons for transport to the sites of the third and final stage of the aerobic pathway:

Glycolysis:	2 NADH
Pyruvate conversion preceding Krebs cycle:	2 NADH
Krebs cycle:	2 FADH$_2$ + 6 NADH
Coenzyme sent to the third stage:	2 FADH$_2$ + 10 NADH

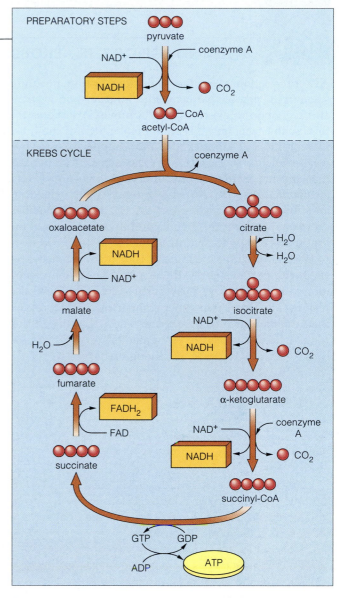

Figure 3.26 The Krebs cycle and a few reactions that precede it. For each three-carbon pyruvate molecule entering the cycle, three CO_2, one ATP, four NADH, and one FADH$_2$ molecules form. The steps shown proceed *twice* because the glucose molecule was broken down initially to *two* pyruvate molecules.

Glycolysis is an energy-releasing stage of reactions in which glucose or another carbohydrate is partially broken down to two molecules of pyruvate.

Two NADH and four ATP form. However, when we subtract the two ATP required to start the reactions, the *net* energy yield of glycolysis is two ATP.

The two pyruvate molecules from glycolysis enter a mitochondrion, where each gives up a carbon atom and the remnant enters the Krebs cycle. The carbon atoms end up in carbon dioxide.

The preparatory steps and the Krebs cycle yield two ATP. Oxaloacetate, the entry point for the cycle, is regenerated. Coenzymes pick up hydrogen and electrons removed from substrates, for delivery to the final stage of the pathway.

THIRD STAGE OF THE AEROBIC PATHWAY

ATP production goes into high gear during the third stage of the aerobic pathway. Electron transport systems and neighboring enzymes (ATP synthases) serve as the production machinery. They are embedded in the inner membrane that divides the mitochondrion into two compartments (Figure 3.27*a*). The actual ATP-making mechanism is **oxidative phosphorylation**. As it proceeds, electron transport systems interact with electrons and unbound hydrogen—that is, H^+ ions. Remember, coenzymes deliver this bounty from reaction sites of the first two stages of the pathway.

Briefly, electrons are transferred from one molecule of each transport system to the next in line. When certain molecules accept and then donate electrons, they also pick up H^+ ions in the inner compartment, then release them to the outer compartment. Their shuttling action sets up H^+ concentration and electric gradients across the inner mitochondrial membrane. Nearby in that membrane, the ions follow the gradients back into the inner compartment, through the interior of ATP synthases (Figure 3.27*b*).

The H^+ flow through these transport proteins drives formation of ATP from ADP and unbound phosphate. Free oxygen keeps ATP production going; without it, ATP production stops and cells die. The oxygen withdraws electrons at the end of the transport systems and then combines with H^+. Water is the result.

Net ATP Yield of Aerobic Respiration

In many types of cells, thirty-two ATP form in the third stage of aerobic respiration. Add these to the net yield

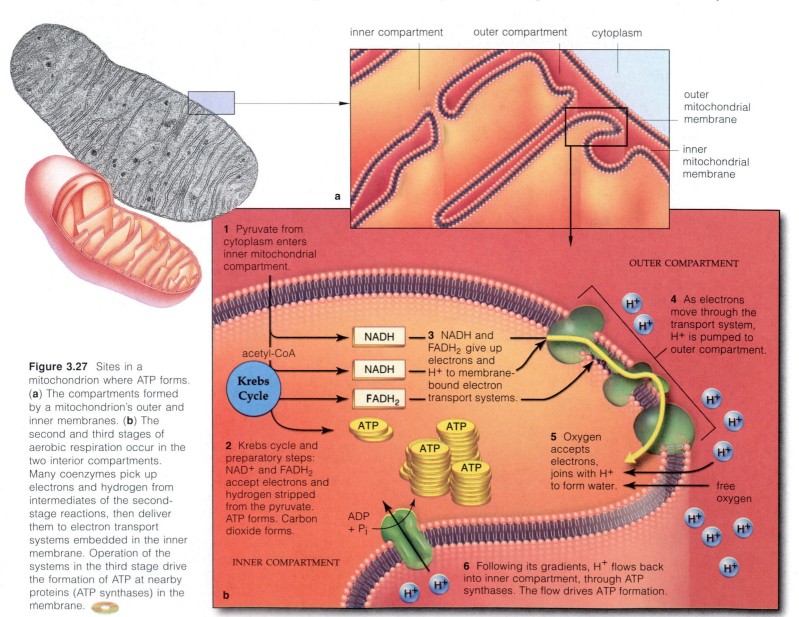

Figure 3.27 Sites in a mitochondrion where ATP forms. (**a**) The compartments formed by a mitochondrion's outer and inner membranes. (**b**) The second and third stages of aerobic respiration occur in the two interior compartments. Many coenzymes pick up electrons and hydrogen from intermediates of the second-stage reactions, then deliver them to electron transport systems embedded in the inner membrane. Operation of the systems in the third stage drive the formation of ATP at nearby proteins (ATP synthases) in the membrane.

inner compartment   outer compartment   cytoplasm

outer mitochondrial membrane

inner mitochondrial membrane

a

OUTER COMPARTMENT

1 Pyruvate from cytoplasm enters inner mitochondrial compartment.

acetyl-CoA

Krebs Cycle

NADH

NADH

FADH₂

3 NADH and FADH₂ give up electrons and H^+ to membrane-bound electron transport systems.

4 As electrons move through the transport system, H^+ is pumped to outer compartment.

ATP   ATP

ATP

ATP

2 Krebs cycle and preparatory steps: NAD^+ and $FADH_2$ accept electrons and hydrogen stripped from the pyruvate. ATP forms. Carbon dioxide forms.

5 Oxygen accepts electrons, joins with H^+ to form water.

free oxygen

ADP + P_i

INNER COMPARTMENT

6 Following its gradients, H^+ flows back into inner compartment, through ATP synthases. The flow drives ATP formation.

b

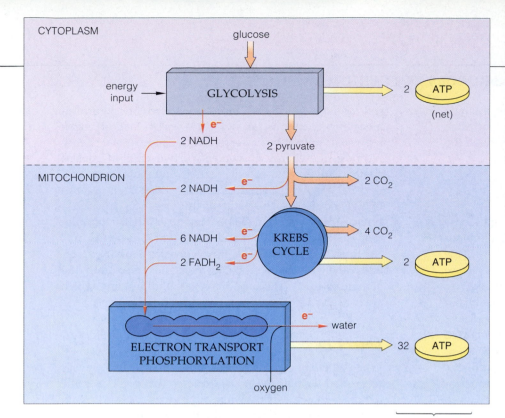

CYTOPLASM

glucose

energy input → **GLYCOLYSIS** → 2 **ATP** (net)

e⁻ → 2 NADH

2 pyruvate

MITOCHONDRION

2 NADH e⁻ → 2 CO_2

6 NADH e⁻
KREBS CYCLE → 4 CO_2
2 FADH₂ e⁻ → 2 **ATP**

ELECTRON TRANSPORT PHOSPHORYLATION e⁻ → water → 32 **ATP**

oxygen

Typical Energy Yield: 36 ATP

Figure 3.28 Summary of aerobic respiration. From start to finish, the typical net energy yield from each glucose molecule is thirty-six ATP. Only this pathway delivers enough energy to build and maintain a large, active, multicellular organism such as a human.

from the preceding stages, and the total harvest is thirty-six ATP from one glucose molecule. That's a lot! But while thirty-six ATP may be a typical yield, the actual amount depends on the conditions in a cell at a given moment—as when a cell requires a given intermediate elsewhere and pulls it out of the reaction sequence. The yield also depends on how particular cells use the NADH that formed in glycolysis. NADH produced in the cytoplasm can't enter a mitochondrion. It can only deliver electrons and hydrogen *to* proteins in that organelle's outer membrane, where proteins shuttle them across to NAD⁺ or to FAD molecules inside the mitochondrion. Then both coenzymes deliver the electrons to transport systems of the inner membrane. But FAD delivers them to a lower entry point in the system, so its deliveries produce *less* ATP. Figure 3.28 summarizes the ATP yield from all three stages of aerobic respiration. If you would like to learn more about these events, turn to Appendix I at the back of this book.

ATP from Anaerobic Pathways

As noted earlier, cells also can use glucose as a substrate for fermentation pathways. These anaerobic pathways do not use oxygen as the final acceptor of the electrons that ultimately drive the ATP-forming machinery. Various microorganisms can *only* use such pathways. Examples include bacteria that live in our intestine and those that cause botulism, tetanus, and some other diseases. On a more positive note, bacterial fermentation also is vital in the manufacture of some cheeses and therapeutic drugs.

Glycolysis is the first stage of fermentation pathways. Here again, enzymes split and rearrange the fragments of a glucose molecule into two pyruvate molecules. Two NADH form, and the net energy yield is two ATP. However, the reactions do not completely break down glucose to carbon dioxide and water; and they produce no more ATP beyond the glycolysis yield. The final steps serve to regenerate the coenzyme NAD⁺.

Lactate fermentation occurs in muscle cells. In this pathway, the pyruvate formed during glycolysis accepts hydrogen and electrons from NADH. With the transfer, each pyruvate is converted to lactate, a three-carbon compound. (Sometimes lactate is also called lactic acid.) Some types of animal cells can switch to lactate fermentation for a quick fix of ATP when they run low on oxygen. As already noted, when a muscle's demands for energy are intense but brief—say, during a short footrace—muscle cells use this anaerobic pathway. They can only do so briefly, though. When glucose stores are gone, muscles fatigue and can no longer contract.

In the final stage of the aerobic pathway, coenzymes deliver electrons to transport systems at the inner mitochondrial membrane. Electrons and H⁺ move through the systems. Oxygen is the final electron acceptor. ATP forms when H⁺ flows through enzymes that also span the inner membrane.

From start (glycolysis in the cytoplasm) to finish (in the mitochondrion) aerobic respiration typically has a net yield of thirty-six ATP for every glucose molecule.

So far, we have looked at what happens after a single glucose molecule enters an energy-releasing pathway. Now we can consider what cells do when they have too many or too few molecules of glucose.

Carbohydrate Breakdown in Perspective

When you eat, your body absorbs a great deal of glucose and other small organic molecules, which the bloodstream transports throughout your body. A rise in the glucose level in blood prompts the pancreas to release insulin, a hormone that stimulates cells to take up glucose faster. The cells convert the windfall of glucose to glucose-6-phosphate and so "trap" it in the cytoplasm. (When phosphorylated, glucose cannot be transported back out, across the plasma membrane.) Glucose-6-phosphate is the first activated intermediate of glycolysis.

If your glucose intake exceeds cellular demands for energy, ATP-producing machinery goes into high gear. Unless a cell is rapidly using ATP, its concentration of ATP can rise to a high level. Then, glucose-6-phosphate is diverted into a biosynthesis pathway that assembles glucose units into glycogen, a storage polysaccharide (Section 2.9). This is especially the case for muscle and liver cells, which maintain the largest glycogen stores.

When you are not eating, glucose is not entering your bloodstream and its level in the blood declines. If the decline were not counteracted, that would be a problem for the brain, the body's glucose "hog." The brain constantly takes up more than two-thirds of the glucose circulating in the bloodstream because its hundreds of millions of cells use glucose as their preferred energy source.

The pancreas responds to the decline by secreting glucagon, a hormone that prompts liver cells to convert glycogen back to glucose and send it back to the blood. The blood glucose level rises, and brain cells keep on functioning. Thus, hormones control whether the body's

cells use free glucose as an energy source or "squirrel it away"—typically, as fat, not glycogen. In fact, glycogen represents only about 1 percent of the body's total energy reserves, the energy equivalent of two cups of cooked pasta. Unless you eat at regular intervals, you will deplete the liver's small glycogen stores in less than twelve hours. Of the total energy reserves in, say, a typical adult American, 78 percent (about 10,000 kilocalories) is stored in body fat and 21 percent in proteins.

Table 3.4 summarizes how complex carbohydrates and other types of starting molecules feed into the aerobic pathway. The subunits of fats and proteins, which we consider next, enter at points beyond glycolysis.

Energy from Fats

When a person goes on a "diet," the aim nearly always is to lose fat. While Chapter 6 explores that topic in some detail, here we can ask: Just how does the body gain access to its huge reservoir of fats? Remember from Section 2.10 that a fat molecule has a glycerol head and one, two, or three fatty acid tails. Most of the fat stored in the body consists of triglycerides, which accumulate inside the fat cells in certain tissues (*adipose* tissues) of the buttocks and other strategic locations beneath the skin.

Between meals or during exercise, the body taps triglycerides as energy alternatives to glucose. Then, enzymes in fat cells break the bonds holding glycerol and fatty acids together. Next the breakdown products enter the bloodstream. When glycerol reaches the liver, enzymes convert it to PGAL—which, remember, is an intermediate of glycolysis. Most body cells take up the circulating fatty acids. Enzymes again break bonds in the carbon backbone of fatty acid tails, then convert the fragments to acetyl-CoA, which can enter the Krebs cycle (Figure 3.29). Each fatty acid tail has many more carbon-bound hydrogen atoms than glucose does, so breaking down a fatty acid yields much more ATP. In fact, when this pathway is operating, it can supply about half the ATP that your muscle, liver, and kidney cells require.

Energy from Proteins

A diet that lacks sufficient protein is unhealthy. Why? Because unlike fats, proteins are not stored in the body. Eat more proteins than your body requires, and enzymes simply split these "extra" proteins into amino acid units. Then they remove the amino group ($—NH_3^+$) from each unit and ammonia (NH_3) forms. Depending on the conditions in the cell, the leftover carbon backbones can be converted to fats or carbohydrates. Or they may enter the Krebs cycle, as shown in Figure 3.29, where coenzymes can pick up hydrogen and electrons removed from the carbon atoms. The ammonia that forms is converted

Table 3.4 Summary of Energy Sources in the Human Body		
Starting Molecule	Subunit	Entry Point into the Aerobic Pathway
Complex carbohydrate	Simple sugars (e.g., glucose)	Glycolysis
Fat	Fatty acids	Preparatory reactions for Krebs cycle
	Glycerol	Raw material for key intermediate in glycolysis (PGAL)
Protein	Amino acids	Carbon backbones enter Krebs cycle or preparatory reactions

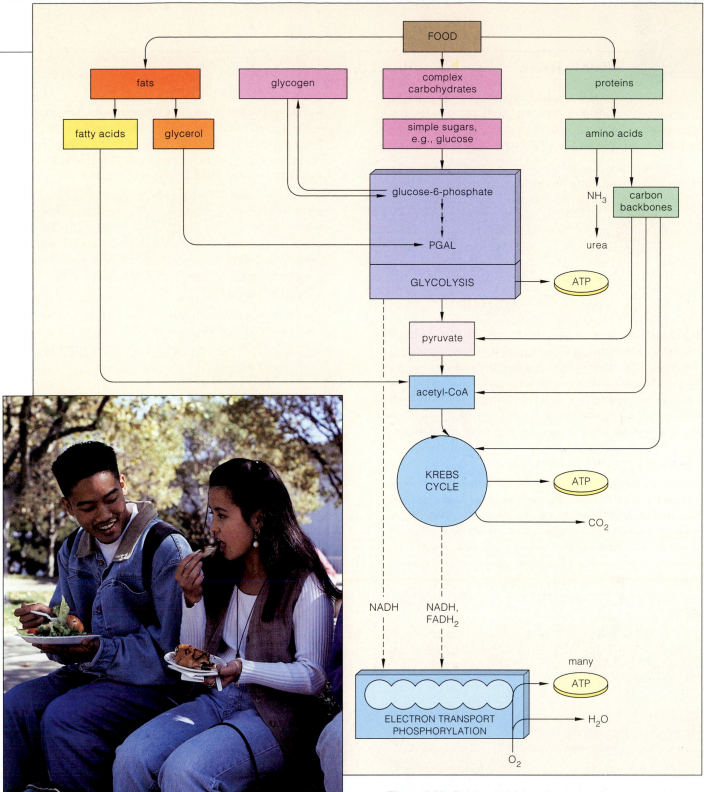

Figure 3.29 Points at which various organic compounds can enter the reaction stages of aerobic respiration.

to urea. This is a nitrogen-containing waste product that is toxic if too much of it accumulates. Normally your body excretes urea, in urine.

As you may have realized by now, maintaining and gaining access to the body's energy reserves is not a simple business. However, as will become clear in later chapters, providing all of your cells, organs, and organ systems with energy begins with the kinds, amounts, and proportions of various foods you put in your mouth.

Subunits of complex carbohydrates, fats, and proteins all can serve as energy sources in the human body.

Whether glucose or other organic compounds enter an energy-releasing pathway depends on the kinds and proportions of carbohydrates, fats, and proteins in the diet as well as on the type of cell.

SUMMARY

1. Each living cell has a plasma membrane surrounding an inner region called the cytoplasm. In eukaryotic cells, including cells of the human body, cell membranes divide the cell into functional compartments called organelles.

2. Organelle membranes separate metabolic reactions in the cytoplasm and allow different kinds to proceed in an orderly way. The largest organelle is the nucleus; the genetic material (DNA) is located there. The cytoskeleton functions in cell shape, internal organization, and movements. Mitochondria specialize in the oxygen-requiring reactions that produce ATP. The cytomembrane system includes the endoplasmic reticulum, Golgi bodies, and various vesicles. In this system new proteins are modified into final form and lipids are assembled.

3. Membranes are crucial to cell structure and function. Cell membranes consist mostly of lipids and proteins. The lipids are mainly phospholipids, arranged as two layers. This lipid bilayer gives structure to the membrane and bars water-soluble substances from freely crossing it. Various kinds of proteins in the bilayer or at one of its surfaces perform most membrane functions. Properties of a membrane's phospholipids and proteins make it only selectively permeable to substances.

4. Some membrane proteins allow water-soluble substances to cross the membrane. Others are receptors for hormones or other substances that trigger alterations in cell behavior. Still other proteins have short carbohydrate chains that function in cell-to-cell recognition. Adhesion proteins help cells stay together in their proper tissues.

5. Substances cross cell membranes by several transport mechanisms. Simple diffusion is the random, unassisted movement of solutes down a concentration gradient. Osmosis is the movement of water across a selectively permeable membrane in response to concentration gradients, a pressure gradient, or both. In passive transport, a solute simply moves down its concentration gradient, through the interior of a membrane transport protein. In active transport, a solute is pumped *against* its concentration gradient through the interior of a membrane protein. Active transport requires an energy boost, as from ATP.

6. A metabolic pathway is a stepwise sequence of reactions mediated by enzymes. Biosynthetic pathways build large, energy-rich organic compounds from smaller molecules of lower energy content. Degradative pathways break down molecules to smaller ones of lower energy content. Enzymes are catalytic molecules that greatly enhance the rate of metabolic reactions; most enzymes are proteins. Cofactors, including coenzymes, help catalyze reactions or carry electrons, hydrogen, or functional groups from a substrate to other sites.

7. Controls stimulate or inhibit enzyme activity at key steps in metabolic pathways. They help coordinate the kinds and amounts of substances available.

8. ATP is the main energy carrier in cells. The metabolic reactions of nearly all biosynthetic pathways run on energy it delivers. In body cells, aerobic respiration produces most ATP. This pathway releases chemical energy from glucose and other organic compounds.

9. In aerobically respiring cells, oxygen is the final acceptor of electrons removed from glucose. The pathway has three stages: glycolysis (in the cytoplasm), the Krebs cycle, and oxidative phosphorylation coupled with electron transport (in mitochondria). The net energy yield of aerobic respiration is commonly thirty-six ATPs.

10. Fermentation, an anaerobic pathway, has a small net yield (two ATP, from glycolysis) because glucose is not completely broken down. After the initial reactions (glycolysis), the pathway simply regenerates the NAD^+ required for glycolysis.

Review Questions

1. Describe the general functions of the following in a eukaryotic cell: the plasma membrane, cytoplasm, DNA, ribosomes, organelles, and cytoskeleton. *3.1, 3.2*

2. Which organelles are part of the cytomembrane system? *3.4*

3. Distinguish between the following pairs of terms:
   a. diffusion; osmosis *3.5*
   b. passive transport; active transport *3.6*
   c. endocytosis; exocytosis *3.6*

4. What is an enzyme? Describe its role in metabolic reactions. *3.10*

5. In aerobic respiration, which reactions occur only in the cytoplasm? Which occur only in a cell's mitochondria? *3.11*

6. For this diagram of the aerobic pathway, fill in the number of molecules of pyruvate, coenzymes, and end products, as well as the net ATP formed at each stage. *3.11, 3.12*

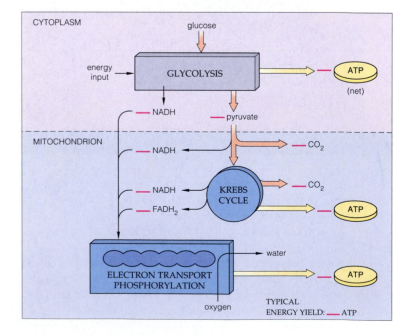

Self-Quiz (Answers in Appendix V)

1. The plasma membrane _____.
   a. surrounds an inner region of cytoplasm
   b. separates the nucleus from the cytoplasm
   c. separates cell interior from the environment
   d. acts as a nucleus in prokaryotic cells
   e. only a and c are correct

2. The _____ is responsible for a eukaryotic cell's shape, internal organization, and cell movement.

3. Cell membranes consist mainly of a _____.
   a. carbohydrate bilayer and proteins
   b. protein bilayer and phospholipids
   c. phospholipid bilayer and proteins

4. _____ carry out most membrane functions.
   a. Proteins   c. Nucleic acids
   b. Phospholipids   d. Hormones

5. The passive movement of a solute through a membrane protein down its concentration gradient is an example of _____.
   a. osmosis   c. diffusion
   b. active transport d. facilitated diffusion

6. Match each organelle with its correct function.
   ____protein synthesis   a. mitochondrion
   ____movement   b. ribosome
   ____intracellular digestion   c. smooth ER
   ____modification of new proteins d. rough ER
   ____lipid synthesis   e. nucleolus
   ____ATP formation   f. lysosome
   ____ribosome subunit assembly   g. flagellum

7. Which of the following statements is *not* true? Metabolic pathways _____.
   a. occur in stepwise series of chemical reactions
   b. are speeded up by enzyme activity
   c. may break down or assemble molecules
   d. always produce energy (i.e., ATP)

8. Enzymes _____.
   a. enhance reaction rates   c. act on specific substrates
   b. are affected by pH   d. all of the above are correct

9. Match each substance with its correct description.
   ____a coenzyme or metal ion   a. reactant
   ____substance formed at end of a b. enzyme
       metabolic pathway   c. cofactor
   ____mainly ATP   d. energy carrier
   ____substance entering a reaction e. end product
   ____catalytic protein

10. The pathway of aerobic respiration is completed in the ____
    a. nucleus   c. plasma membrane
    b. mitochondrion   d. cytoplasm

11. Match each type of metabolic reaction with its function:
    ____glycolysis   a. ATP, NADH, and CO_2 form
    ____Krebs cycle   b. glucose to two pyruvate molecules
    ____oxidative   c. H^+ flows through channel
        phosphorylation   proteins, ATP forms

12. In a mitochondrion, where are the electron transport systems and carrier proteins required for ATP formation located?

Critical Thinking: You Decide (Key in Appendix VI)

1. Using this chapter's opening vignette as a reference, suppose you want to observe the surface of a microscopic section of bone. Would you benefit most from using a compound light microscope, a transmission electron microscope, or a scanning electron microscope?

2. Jogging is considered aerobic exercise because the cardiovascular system (heart and lungs) can adjust its activity to supply the oxygen needs of working cells. In contrast, sprinting the 100-meter dash might be called "anaerobic" exercise, and golf "nonaerobic" exercise. Explain these last two observations.

3. The cells of your body never use nucleic acids as an energy source. Suggest why.

4. Your professor shows you an electron micrograph of a cell with large numbers of mitochondria and Golgi bodies. You notice that this particular cell also contains a great deal of rough endoplasmic reticulum. What kinds of activities would require the cell to have so many of the three kinds of organelles?

Selected Key Terms

active site 3.11
active transport 3.7
ADP/ATP cycle 3.11
aerobic respiration 3.12
centriole 3.3
chromosome 3.9
cilium 3.3
coenzyme 3.11
concentration gradient 3.6
cytomembrane system 3.5
cytoplasm 3.1
cytoskeleton 3.3
diffusion 3.6
endocytosis 3.7
endoplasmic reticulum (ER) 3.5
enzyme 3.11
eukaryotic cell 3.1
exocytosis 3.7
FAD 3.11
fermentation pathway 3.12
flagellum 3.3
fluid mosaic model 3.4
glycolysis 3.12
Golgi body 3.5

intermediate filament 3.3
Krebs cycle 3.12
lipid bilayer 3.1
lysosome 3.5
metabolic pathway 3.11
metabolism 3.11
microfilament 3.3
microtubule 3.3
mitochondrion 3.10
NAD^+ 3.11
nuclear envelope 3.9
nucleolus 3.9
nucleus 3.9
organelle 3.1
osmosis 3.6
oxidative phosphorylation 3.13
passive transport 3.7
PGAL 3.12
phosphorylation 3.11
plasma membrane 3.4
ribosome 3.1
substrate 3.11
vesicle 3.5

Readings

Inger, D. E. January 1998. "The Architecture of Life." *Scientific American*. A fascinating discussion of the "building rules" that seem to guide the design of cells and other organic structures.

Trefil, James. June 1992. "Seeing Atoms." *Discover*. How researchers use microscopes to study cell organelles, molecules, and even atoms.

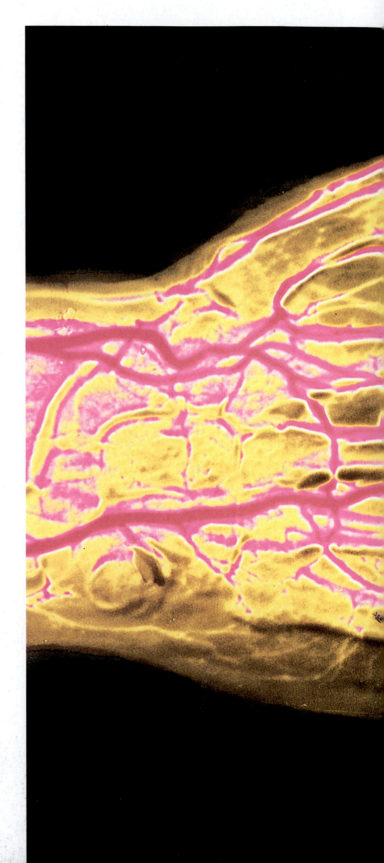

Intricate network of blood vessels in a human hand.

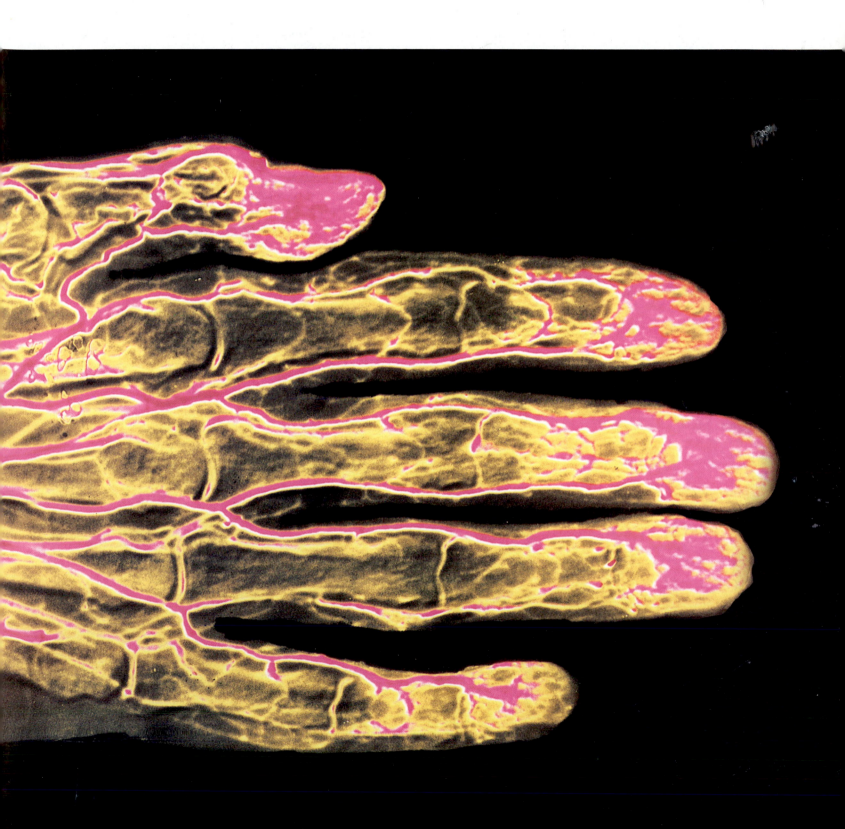

INTRODUCTION TO TISSUES, ORGAN SYSTEMS, AND HOMEOSTASIS

If You've Got It, Flaunt It!

Remember the levels of biological organization outlined in Chapter 1? Did you ever stop to think that you are an example of more than half of them? From nonliving atoms and molecules to the tiny, living dynamos that are your cells, one kind of structural "hardware" becomes the building block for the next (Figure 4.1*a*). At each step along the way, more elaborate hardware comes, in the lingo of computerese, with what some might think of as fancier "applications"—the functions that living structures can carry out, as long as they have the required energy and materials. Cells, of course, are just the starting points for the next levels of biological organization, which we consider here: tissues, organs, and organ systems.

We can probably safely assume that no computer or any other machine will ever be able to do all the complex things a person does. From reading a book to playing the flute or exercising in an aerobics class, we humans come equipped for some extraordinary

biological feats (Figure 4.1*b*). Part of what makes us so remarkable is that, like other complex animals, our amazingly intricate bodies are built of just four basic types of tissues. A **tissue** is a group of cells and inter-cellular substances, all interacting in one or more tasks.

For example, muscle is one of the four basic tissue types, and the specialized task of its contractile cells is moving body parts. Another basic tissue type is nerve tissue, which is made of cells specialized for detecting, processing, and coordinating information. Connective tissues, such as cartilage and bone, provide structural support and, in some cases, metabolic support as well. The fourth tissue category, epithelial tissue, includes tissues that cover the body surface—your skin—or line its cavities and tubes.

Combinations of tissues make up the body's organs. To be precise, an **organ**, such as the heart, consists of different tissues organized in specific proportions and patterns. Thus every human heart has predictable

MULTICELLULAR ORGANISM
An individual composed of cells arrayed in tissues, organs, and often organ systems

ORGAN SYSTEM
Organs interacting in ways that contribute to survival of the whole organism

ORGAN
Tissues combined into a structural and functional unit

TISSUE
A group of cells and surrounding substances functioning together in a specialized activity

CELL
Smallest *living* unit

Nonliving subcellular components

arrangements of epithelial, connective, muscle, and nerve tissues. An **organ system** consists of two or more organs that interact physically, chemically, or both, in performing a common task, as when interconnected arteries and other vessels transport blood through the body under the driving force of a beating heart.

Like cells, tissues, and organs, organ systems are specialized for particular functions. For instance, the digestive system brings needed nutrients into the body, and the circulatory system—your heart and blood vessels—efficiently transports them to cells. Working together, the circulatory and respiratory systems supply cells with the oxygen they need for metabolism and making ATP. Those systems also swiftly remove the wastes, such as carbon dioxide, produced by all that cellular activity.

Any number of cell activities change the body's **internal environment.** This term refers to the bloodstream and the fluids bathing body cells. As you know from Chapter 3, cells continually take up substances and produce others, and this biochemical give-and-take alters the composition of blood and tissue fluid. Unregulated, the changes would eventually kill cells. Fortunately, your kidneys and other elements of the urinary system work to keep this disaster at bay.

Governing all this activity are the body's command posts—the nervous system, and an endocrine system that operates by way of hormones.

With this whirlwind introduction, we turn now to a closer examination of the basic characteristics of the human body's tissues, organs, and organ systems. We will also take another look at the types of controls that affect the functions of organ systems in ways that help maintain homeostasis—stable operating conditions in the internal environment.

KEY CONCEPTS

1. Cells interact to form *tissues*, many of which are combined in *organs*, which are components of *organ systems*.

2. Organs are constructed of four basic classes of tissue: epithelial, connective, muscle, and nerve tissues.

3. Each cell engages in basic metabolic activities that assure its own survival. At the same time, cells, tissues, and organs perform activities that contribute to the survival of the body as a whole.

4. The combined contributions of cells, tissues, organs, and organ systems help maintain a stable internal environment, which is required for our body's survival. This concept helps us understand the functions of any organ system, regardless of its complexity.

5. Homeostasis is the formal name for stable operating conditions in the internal environment.

CHAPTER AT A GLANCE

Figure 4.1 Doing aerobics. As we'll see in this chapter, physical fitness is in part a reflection of the body's ability to maintain homeostasis, keeping internal operating conditions within life-sustaining limits.

Your skin. The rosy tissue lining your mouth. Each of these is an example of epithelial tissue, or **epithelium** (plural: epithelia). This tissue has a free surface, which faces either a body cavity or the outside environment. Epithelial cells nestle together, with little material in between them. They are arranged in one or more layers, and they are linked by specialized junctions that connect cells both structurally and functionally. Sometimes cells at an epithelium's free surface have slim protrusions of cytoplasm—the cilia described in Section 3.3, or **microvilli** (singular: **microvillus**), which are specialized for absorbing substances into the cell. The other surface of an epithelium adheres to a **basement membrane**, a layer between the epithelium and the underlying connective tissue (Figure 4.2*a*). A basement membrane does not contain cells but is densely packed with proteins and polysaccharides.

Two Types of Epithelia

Basically, there are only two types of epithelia. Their names are simple epithelium and stratified epithelium. Simple epithelium lines the body's cavities, ducts, and tubes—for example, the chest cavity, tear ducts, and the tubes in the kidneys, where urine is formed (Figure 4.2*b–d*). Simple epithelium is a single layer of cells. In general, these cells function in the diffusion, secretion, absorption, or filtering of substances across the layer.

Stratified epithelium has two or more cell layers (see the micrograph in Figure 4.2*a*). Protection is its typical function—for example, it is found at the surface of your skin, which is exposed to various kinds of bumps, abrasion, and so forth. *Pseudostratified* epithelium actually is simple epithelium in disguise. It is a single cell layer that looks like a multiple layer when it is viewed from the side because the nuclei of adjacent cells are staggered. Most of the cells have cilia. Pseudostratified epithelium occurs in your throat, nasal passages, reproductive tract, and other body locations where cilia sweep mucus or some other fluid across the epithelial surface.

The basic types of epithelium are subdivided into several categories, depending on the shape of cells at the tissue's free surface. A *squamous epithelium* has flattened cells, a *cuboidal epithelium* has cube-shaped cells, and a *columnar epithelium* has elongated cells. The different shapes correlate with different functions. For instance, oxygen and carbon dioxide easily diffuse across the thin layer of simple squamous epithelium making up the walls of fine blood vessels, as in Figure 4.2*b*. The plumper cells of cuboidal and columnar epithelia have apparatus for secreting substances. Table 4.1 summarizes the various types of epithelium and their roles. *Science Comes to Life* on page 74 describes experimental efforts to grow epithelium and other tissues outside the body.

Glands: Derived from Epithelium

A **gland** is a cell or a multicellular structure that is derived from epithelium. Often, a gland stays connected to the epithelium that gave rise to it. The biological function of glands is to make and secrete specific products. Mucus-secreting goblet cells, for instance, are embedded in pseudostratified epithelium that lines the trachea (your windpipe) and other tubes leading to the lungs. The epithelium of your stomach is blanketed with gland cells that secrete protective mucus and other substances.

Biologists often classify glands according to how their secretions reach the place where they are used. **Exocrine glands** secrete substances onto an epithelial surface through ducts or tubes. Mucus, saliva, earwax, oil, milk, and digestive enzymes all are exocrine secretions. Many exocrine glands simply secrete the substance they are specialized to make; salivary glands and most sweat glands are like this. In other cases, though, a gland's secretions include bits of the gland cells themselves. For instance, milk secreted from a nursing woman's mammary glands contains bits of the glandular epithelial tissue. In still other cases,

Table 4.1 Major Types of Epithelium		
Type	Shape	Typical Locations
Simple (one layer)	Squamous	Linings of blood vessels, lung alveoli (sites of gas exchange)
	Cuboidal	Glands and their ducts, surface of ovaries, pigmented epithelium of eye
	Columnar	Stomach, intestines, uterus
Stratified (two or more layers)	Squamous	Skin (keratinized), mouth, throat, esophagus, vagina (nonkeratinized)
	Cuboidal	Ducts of sweat glands
	Columnar	Male urethra, ducts of salivary glands
Pseudostratified	Columnar	Throat, nasal passages, sinuses, trachea, male genital ducts

Further reading: Student Guide to InfoTrac on web site ➝

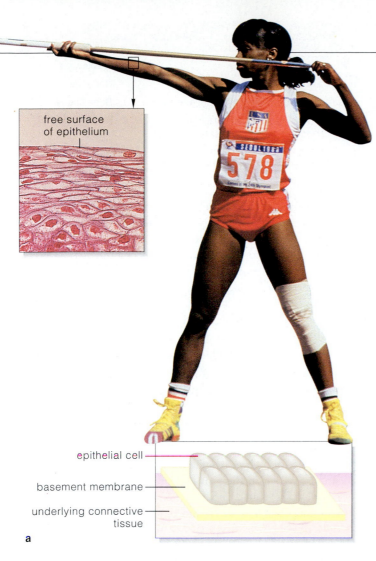

free surface
of epithelium

such as sebaceous (oil) glands in your skin, whole cells full of the material to be secreted are actually shed into the duct, where they burst and their contents spill out.

In contrast to exocrine glands, **endocrine glands** do not secrete substances through tubes or ducts. Their secretions, hormones, pour directly into the extracellular fluid bathing the glands. Typically, the bloodstream picks up hormones and distributes them to target cells elsewhere in the body. Examples of endocrine glands include the thyroid, the adrenals, the pituitary gland, and other glands that are discussed in Chapter 13.

Epithelia are sheetlike tissues with one free surface. Simple epithelium lines body cavities, ducts, and tubes. Stratified epithelium, like that at the skin surface, typically protects the underlying tissues.

Glands are secretory structures derived from epithelium, and they often remain connected to it. Some glands are single cells, while others are multicellular.

epithelial cell

basement membrane

underlying connective tissue

a

Figure 4.2 Some basic characteristics of epithelium. (**a**) All epithelia have a free surface. There is a basement membrane between the opposite surface and underlying connective tissue. The diagram shows simple epithelium, which consists of a single layer of cells. The micrograph shows the upper portion of stratified squamous epithelium, which has more than one layer of cells. The cells become more flattened toward the surface. (**b–d**) Examples of simple epithelium, showing the three basic cell shapes in this type of tissue.

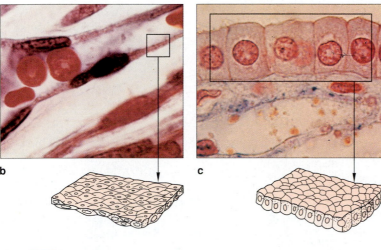

b

TYPE: Simple squamous

DESCRIPTION: Single layer of flattened cells

COMMON LOCATIONS: Blood vessel walls; air sacs of lungs

FUNCTION: Diffusion

c

TYPE: Simple cuboidal

DESCRIPTION: Single layer of cubelike cells; free surface may have microvilli (absorptive structures)

COMMON LOCATIONS: Salivary glands, sweat glands, and ducts of glands and kidney tubules

FUNCTION: Secretion, absorption

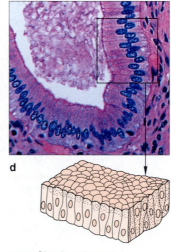

d

TYPE: Simple columnar

DESCRIPTION: Single layer of tall, slender cells; free surface may have microvilli

COMMON LOCATIONS: Part of lining of gut and respiratory tract

FUNCTION: Secretion, absorption

CONNECTIVE TISSUES

Of all body tissues, **connective tissues** are the most abundant and widely distributed. They range from soft connective tissues to specialized types, which include cartilage, bone, adipose tissue, and blood (Table 4.2). In all types of connective tissue except blood, cells secrete fibers of structural proteins—collagen and elastin. Plastic surgeons inject collagen fibers to "plump up" wrinkled skin, lips, and sunken acne scars. Elastin fibers are rubbery and make connective tissue stretchable; for example, elastin is abundant in the lungs, which must stretch and recoil as air moves in and out. Connective tissue cells also secrete modified polysaccharides, which accumulate between cells and fibers as the connective tissue's "ground substance."

Soft Connective Tissues

These tissues have mostly the same components, but in different proportions. **Loose connective tissue** has fibers and cells loosely arranged in a semifluid ground substance (Figure 4.3a). Often it serves as a support framework for epithelium. It contains fibroblasts (cells that produce and secrete fibers) and white blood cells that fight infections. For example, when bacteria enter skin through a cut, these white blood cells mount an early counterattack.

Dense, **irregular connective tissue** has fibroblasts and many fibers (mostly collagen-containing ones) that

Table 4.2 Categories of Connective Tissue	
SOFT	**SPECIALIZED**
Loose connective tissue	Cartilage
Dense, irregular connective tissue	Bone tissue
Dense, regular connective tissue (ligaments, tendons)	Adipose tissue
	Blood

are oriented every which way. This tissue is present in skin and forms protective capsules around organs that need not stretch much (Figure 4.3b). In **dense**, **regular connective tissue**, the fibroblasts occur in rows between many parallel bundles of fibers. Tendons, which attach skeletal muscle to bone, have this tissue. The bundles of collagen fibers help tendons resist being torn (Figure 4.3c). Dense, regular connective tissue also is present in elastic ligaments, which attach one bone to another. The elastic fibers allow movement around joints, such as the knee.

Specialized Connective Tissues

Like rubber, the intercellular material called **cartilage** is solid but pliable, and resists compression. Cells called chondroblasts made and released the material. As they matured (into chondrocytes) they became imprisoned in small cavities in their own secretions (Figure 4.3d). The cavities are called *lacunae*. Cartilage does not contain blood vessels and so has no blood supply; this is why it does not easily heal when it is injured.

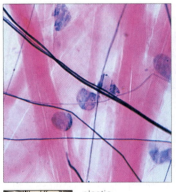
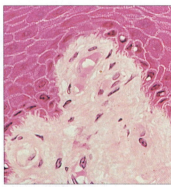
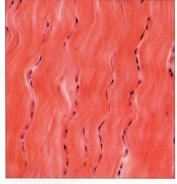
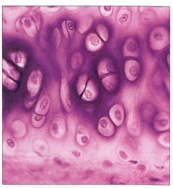

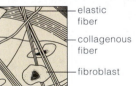

- elastic fiber
- collagenous fiber
- fibroblast

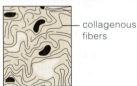

- collagenous fibers

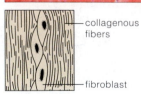

- collagenous fibers
- fibroblast

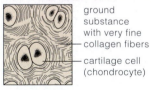

- ground substance with very fine collagen fibers
- cartilage cell (chondrocyte)

a

TYPE: Loose connective tissue
DESCRIPTION: Fibroblasts, other cells, plus fibers loosely arranged in semifluid ground substance
COMMON LOCATIONS: Under the skin and most epithelia
FUNCTION: Elasticity, diffusion

b

TYPE: Dense, irregular connective tissue
DESCRIPTION: Collagenous fibers, fibroblasts, less ground substance
COMMON LOCATIONS: In skin and capsules around some organs
FUNCTION: Support

c

TYPE: Dense, regular connective tissue
DESCRIPTION: Collagen fibers in parallel bundles, long rows of fibroblasts, little ground substance
COMMON LOCATIONS: Tendons, ligaments
FUNCTION: Strength, elasticity

d

TYPE: Cartilage
DESCRIPTION: Cells embedded in pliable, solid ground substance
COMMON LOCATIONS: Ends of long bones, nose, parts of airways, skeleton of vertebrate embryos
FUNCTION: Support, flexibility, low-friction surface for joint movement

Figure 4.3 Examples of soft connective tissues and specialized connective tissues.

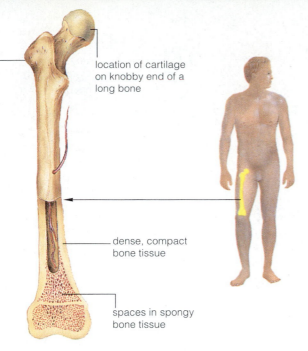

location of cartilage on knobby end of a long bone

dense, compact bone tissue

spaces in spongy bone tissue

Figure 4.4 Cartilage and bone tissue in a human leg bone. Spongy bone tissue has tiny, needlelike hard parts with spaces between. Compact bone tissue is much denser.

The most common type of cartilage in the human body is *hyaline cartilage*, in which the matrix is laced with many small collagen fibers. In mature bones that can move (such as leg bones), hyaline cartilage provides a friction-reducing cover where the bone ends articulate in joints. It makes up parts of the nose, ribs, and windpipe (trachea). In the developing skeleton of an embryo, hyaline cartilage is the forerunner of bone.

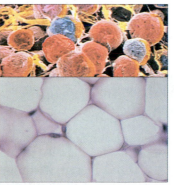

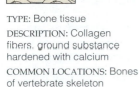

compact bone tissue

spaces that house living bone cells (osteocytes)

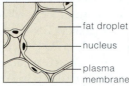

fat droplet

nucleus

plasma membrane

TYPE: Bone tissue

DESCRIPTION: Collagen fibers, ground substance hardened with calcium

COMMON LOCATIONS: Bones of vertebrate skeleton

FUNCTION: Movement, support, protection

TYPE: Adipose tissue

DESCRIPTION: Large, tightly packed fat cells occupying most of ground tissue

COMMON LOCATIONS: Under skin, around heart, kidneys

FUNCTION: Energy reserves, insulation, padding

e

f

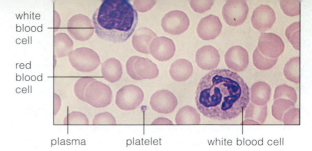

white blood cell

red blood cell

plasma   platelet   white blood cell

Figure 4.5 Some components of human blood. This tissue's straw-colored, liquid matrix (plasma) is mostly water in which a variety of substances are dissolved.

Elastic cartilage occurs in places where a flexible yet rigid structure is required. It also has both collagen and elastin fibers. You can bend your outer ear because it is built of elastic cartilage. So is the epiglottis, a flexible flap of tissue that closes off the opening of the larynx when you swallow. As a result, food and liquids enter the tube leading to your stomach, not the one to your lungs.

Sturdy and resilient *fibrocartilage* is packed with thick bundles of collagen fibers. It can withstand tremendous pressure, and it forms the cartilage "cushions" in joints such as the knee (Section 5.5) and in the disks that separate the vertebrae in the spinal column.

Bone tissue is mineral-hardened; its collagen fibers and ground substance are rich in calcium salts (Figures 4.3*e* and 4.4). It is the main tissue in bones. In the skeleton, different bones support and protect softer tissues and organs. Limb bones, such as the long bones of the legs, bear weight. They also interact with skeletal muscles to bring about movement. Bones also store the mineral calcium, and some produce blood cells.

Excess carbohydrates and proteins that cells do not use at once are converted to storage forms, especially fats. A few fat droplets collect in the cytoplasm of various cells. But cells in **adipose tissue**, which is located mainly beneath the skin, specialize in fat storage. Fat droplets practically fill their cytoplasm (Figure 4.3*f*). A rich blood supply in the tissue is an accessible medium through which fats move.

Blood is derived primarily from connective tissue, and that is why many biologists consider it a connective tissue. Blood's biological role is transport. Circulating in plasma, its fluid medium, are a great many red blood cells, white blood cells, and platelets, along with many dissolved ions, nutrients, hormones, and other substances (Figure 4.5). We take a closer look at this complex tissue in Section 7.1.

Diverse types of connective tissues bind together, support, strengthen, protect, and insulate other body tissues.

Soft connective tissues consist of protein fibers as well as a variety of cells arranged in a ground substance.

Cartilage, bone, blood, and adipose tissue are specialized connective tissues. Cartilage and bone are both structural materials. Blood is a fluid tissue with transport functions. Adipose tissue is a reservoir of stored energy.

MUSCLE TISSUE

In all **muscle tissues**, cells *contract* (that is, shorten) in response to stimulation from the outside, then lengthen and so return to their noncontracted state. These tissues have many long, cylindrical cells arranged in parallel arrays. Muscle layers and muscular organs contract and relax in coordinated fashion. Their action moves

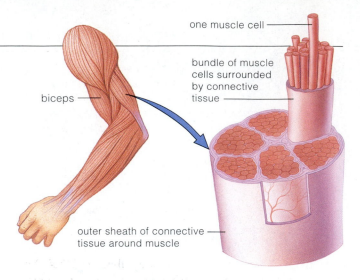

Figure 4.7 Location and general arrangement of muscle cells in a typical skeletal muscle, the biceps.

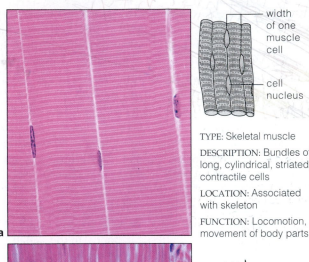
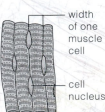

TYPE: Skeletal muscle

DESCRIPTION: Bundles of long, cylindrical, striated, contractile cells

LOCATION: Associated with skeleton

FUNCTION: Locomotion, movement of body parts

a

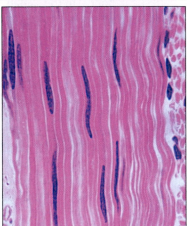

TYPE: Smooth muscle

DESCRIPTION: Contractile cells with tapered ends

LOCATION: Wall of internal organs, such as stomach

FUNCTION: Movement of internal organs

b

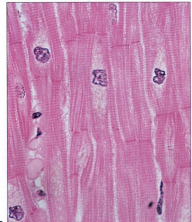

TYPE: Cardiac muscle

DESCRIPTION: Cylindrical, striated cells that have specialized end junctions

LOCATION: Wall of heart

FUNCTION: Pump blood within circulatory system

c

Figure 4.6 Characteristics and examples of skeletal muscle, smooth muscle, and cardiac muscle tissues.

the body through the environment, and it maintains and changes the positions of body parts. The three types of muscle tissue are called skeletal, smooth, and cardiac muscle tissues.

Skeletal muscle tissue is located in muscles that are securely fastened to your bones (Figure 4.6*a*). In a typical muscle, such as the biceps, striated skeletal muscle cells are bundled closely together, in parallel. (*Striated* means striped.) The bundles are called fascicles. A sheath of tough connective tissue encloses several fascicles, as you can see from Figure 4.7. This bundled arrangement is "a muscle." The structure and function of skeletal muscle tissue are topics of Chapter 5.

The contractile cells of **smooth muscle tissue** taper at both ends (Figure 4.6*b*). Specialized junctions hold the cells together (Section 4.5), and they are bundled together in a connective tissue sheath. The walls of internal organs—including blood vessels, the stomach, and the intestines—contain this type of muscle tissue. The contraction of smooth muscle is sometimes said to be "involuntary" because we usually cannot make it contract just by thinking about it (as we can with skeletal muscle).

Cardiac muscle tissue is a contractile tissue that is present only in the heart. Specialized junctions fuse the plasma membranes of cardiac muscle cells and make them stick together. Communication junctions at some fusion points allow the cells to contract as a unit. That is, when one muscle cell receives a signal to contract, its neighbors are stimulated to contract, too.

Muscle tissue, which can contract (shorten) in response to stimulation, helps move the body and specific body parts.

Skeletal muscle is the only muscle tissue attached to bones. Smooth muscle is a component of internal organs. Cardiac muscle makes up the contractile walls of the heart. Connective tissue sheathes all three types of muscle tissues.

NERVOUS TISSUE

Of all tissues, **nervous tissue** has the most control over the body's ability to respond to changing conditions. In your own body, tens of thousands of cells called **neurons** are centrally located in the brain. Millions of others extend throughout the body, some of them three or four feet. As the cells that transmit nerve impulses, neurons are the body's communication lines.

Neuron Structure and Functions

Each neuron consists of a cell body that contains the nucleus and cytoplasm. It also has two types of extensions, or cell processes (Figure 4.8a). Branched processes called dendrites pick up incoming chemical messages. Outgoing messages are conducted by an axon. Depending on the type of neuron, its axon may extend outward a few millimeters or more than a meter. If you would like to see a diagram of this structure, look ahead to Section 11.1.

A cluster of processes from several neurons forms a **nerve**. Nerves conduct messages from the central nervous system (the brain and spinal cord) to muscles and glands. They also carry messages from specialized sensory receptors back to the central nervous system.

As you might guess, as the body's communication lines, neurons are especially important in maintaining homeostasis. Some types of neurons detect specific changes in both internal and external environmental conditions. For example, as you will read in Chapter 12, millions of specialized sensory neurons in epithelium lining the upper reaches of the nose respond to odor molecules. They are vital to our senses of smell and taste. Sensory neurons in the retina of the eye are linked with receptors that respond to light, while different types in skin and muscle tissue are linked to receptors that detect pressure, stretching, and other stimuli.

A constant flow of information coordinates and regulates the activities of the body's billions of cells and the tissues and organs they make up. Certain neurons— mainly in the central nervous system—receive sensory input, integrate it with additional information, and then trigger responses by influencing the activity of still other neurons. In this way, they coordinate the body's immediate and long-term responses to change. For example, if you stand on one foot and your body starts to tilt precariously, the combined actions of sensory cells and neurons will "tell" you to put your other foot down or take some other action to keep from falling over. Yet other neurons are part of a peripheral nervous system that relays signals to muscles and glands that can carry out responses. When you put your foot down to regain your balance, the commands to your leg muscles arrived via the peripheral nervous system.

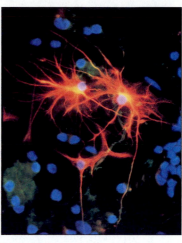

a   cell body of neuron   b

Figure 4.8 A sampling of the millions of cells in nerve tissue. (**a**) Motor neurons, which relay signals from the brain or spinal cord to muscles and glands. The large, dark regions are neuron cell bodies. (**b**) Astrocytes, a type of neuroglia. These and other kinds of neuroglia make up more than half the volume of nerve tissue and provide vital support and other services for neurons.

Neuroglia—Support Cells in Nervous Tissue

Neuroglia is the collective name for various types of accessory cells that make up more than half the volume of nervous tissue. The word *glia* means glue, and neuroglia were once simply thought to be the "mortar" that physically supported neurons. Today we know that many neuroglia do provide physical support, but they also have other important functions. The type called astrocytes ("astro" because they are star-shaped; Figure 4.8b) help shuttle nutrients to neurons, provide physical support, and perform other services. Another type removes debris, microorganisms, or other foreign matter. Still another type provides insulation—a vital function that helps speed nerve impulses through the body, as described in Chapter 11.

Neurons are the basic units of communication in nervous tissue. Different kinds detect specific stimuli, integrate information, and issue or relay commands for response.

Various types of neuroglia lend structural support to neurons, help nourish and protect them, or provide insulation.

WELCOME TO THE HUMAN BODY SHOP!

Need a new swatch of skin or a piece of cartilage? How about a new heart valve grown especially for you? Growing replacement tissues such as epithelium and cartilage is already a reality. On the horizon is the possibility of growing replacement blood vessels, and whole new organs such as livers, bladders, and even hearts. Such spare parts for the body can give new hope to the millions of people who each year lose chunks of tissue or whole organs to accidents or disease. They also can help the thousands of children born with certain birth defects.

Replacement Tissues

A new breed of scientists called tissue engineers have had tremendous success growing epithelial tissue that can be used as artificial skin. At least one type is already on the market. One method for growing such "designer" skin starts with collagen extracted from cattle. As described in Section 4.3, collagen is a major structural protein in the body. After the extracted collagen is purified, cells from the skin's inner layer (called the dermis) are "seeded" on it and allowed to multiply. About one week later, cells from skin's upper layer (epidermis) are added. In some cases, the cells to be cultured are taken from a patient's own body; more often, however, they are simply "generic" skin cells taken from cultures that are maintained in the laboratory.

Provided with the proper nutrients and growth environment, the mix of skin cells and collagen slowly becomes transformed into delicate pink sheets of "custom-made" skin. This part of the process takes about two weeks. When layers of the new skin are placed over,

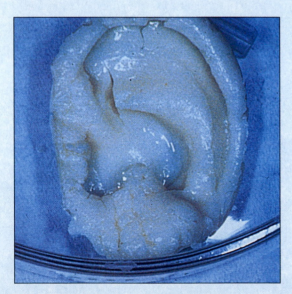

Figure 4.9 This bioengineered ear was grown using a plastic polymer mold spiked with cartilage cells.

say, an ulcer on the leg of a diabetes patient, blood vessels grow into the new tissue, providing it with blood's life-sustaining cargoes of oxygen and nutrients. With time, the once-gaping wound heals. Lab-grown epithelium can also be used to replace skin lost due to severe burns or other damage. Importantly, the artificial skin is not rejected by the recipient's immune system.

Using a similar method, cartilage and bone tissues are also being grown in the laboratory. In the body, these tissues help form organs and other body parts with a variety of shapes. To recreate a replacement tissue in a desired shape, tissue engineers start with a biodegradable model made of plastic. The model has tiny spaces in it, like the spaces between the crossbars of a garden trellis. As with epithelium, the cartilage or bone cells are seeded onto the model. Supplied with nutrients and any other needed chemical supplies, the cells multiply. As they do so, they take on the model's shape. After a while the plastic model biodegrades, leaving behind a usable piece of tissue such as the cartilage ear shown in Figure 4.9.

Laboratory-made materials called organoapatites now are being used to repair bone that has been damaged by disease or accident. Derived from mineral crystals that occur naturally in living bone (and teeth), organoapatites rapidly become integrated into the patient's healthy bone tissue. Blood vessels will even grow into the artificial material, helping to speed up natural processes of bone tissue replacement. Another option is to use a piece of coral as the model for growing bone. Corals are aquatic animals whose soft parts grow on a hard, mineralized framework that is laced with tiny holes—and the framework is ideal for "seeding" with bone cells. In one case, researchers used this technique to grow a new thumb-tip for a man who had accidentally cut his off. They fashioned a bit of coral into the right shape, seeded it with the man's own bone cells (which had been grown in culture), and implanted the new part on the stump of the patient's thumb. Like a plastic model, the coral framework will break down as the bone cells grow into and through it. If all goes well, the new thumb tip will function almost as well as the lucky patient's original one did.

New Organs

A "bioreactor" developed for space shuttle missions (Figure 4.10) has been used to grow heart, liver, and lung cells, among other cell types. All behave much as they would in intact organs. In this and similar devices, cells grow in extremely close proximity to one another, which reduces physical shear stresses that might separate the cells if they were grown in a culture dish. Nestled together in a bioreactor, cells presumably also can more easily exchange intercellular signals that are essential for tissues to form and function normally.

Further reading: Student Guide to InfoTrac on web site →

When it comes to growing an entire organ, the premise is the same as for growing tissues: Give cells the structural model and nutrients they require, and they will grow into the tissues of the desired organ—one that will function as it is supposed to. Thus far, researchers have succeeded in growing small artificial bladders, using cultured muscle and epithelial cells (for the bladder lining). For the time being, such artificial organs are only intriguing experiments, but they are the first steps toward a time when we may be able to grow replacements for aging or diseased hearts and other body parts.

The Challenge of Replacing Blood

Every three seconds, someone in the United States receives a blood transfusion. And the demand for blood for surgery patients and the treatment of severe injuries is growing dramatically. Why? A key reason is that the human population is aging, and older people are more likely to require surgical procedures—such as joint replacement or heart surgery—in which significant blood loss is normal. Blood is transfused in "units," each about one-half liter (roughly a pint). Worldwide, the demand for transfusable blood is exploding, with an annual *increase* of over 7 million units.

Blood donations aren't even close to keeping up with this incessant demand. And in the 1980s the world realized that donated blood, and products derived from it, could transmit HIV, the virus responsible for AIDS, as well as some other viral diseases. Although sensitive methods for testing donated blood have since become routine, many people deeply fear receiving transfused blood.

Enter researchers who are racing to develop synthetic blood that can safely perform blood's functions. Since the body can renew its blood supply (described in Chapter 7), a substitute would need only to fill the gap, so to speak, for a limited time. But the basic functions such a substance would have to carry out are absolutely crucial to the survival of cells, tissues, and organs. Most vital is transporting the oxygen that cells need to make ATP. At the very least, this means that a blood substitute must perform the tasks of hemoglobin—the iron-containing transport protein described in Section 2.12. In addition, the body must not identify it as "foreign," and thus trigger an immune response. Finally, it must be free of dangerous side effects.

One major type of whole blood substitute is a solution containing large carbohydrate molecules that help maintain the proper water content of the fluid, along with synthetic molecules that contain atoms of carbon and fluorine. This mixture, called a perfluorocarbon, or PFC, can bind roughly two-thirds the amount of oxygen that would be found in normal whole blood. Unfortunately, PFCs can have serious side effects, generally are eliminated from the body quickly, and are difficult to store for

Figure 4.10 A rotating cell culture device developed by NASA researchers creates laboratory conditions that encourage cells to differentiate into specialized cell types.

long periods. Even so, they may be an alternative when an individual's religious beliefs forbid transfusion of real blood. Another approach is to create a blood substitute using actual human hemoglobin. The hemoglobin is extracted from donated blood that has outlived its shelf life. It then is chemically modified in various ways.

For example, hemoglobin extracted from human blood is linked to other substances, mimicking conditions inside a red blood cell that allow hemoglobin to bind oxygen (in the lungs), then release it to cells. Other alterations cause hemoglobin molecules to stay in the recipient's body longer than they otherwise would, allowing more time for the person's own red blood cells to be regenerated. The hemoglobin molecules also may be reshaped so that they can carry more oxygen than usual. This characteristic can be useful in treating some kinds of cancer tumors, which are more susceptible to radiation if there is more oxygen in the cancerous tissue.

Genetic engineers have inserted the gene for human hemoglobin into pigs, which then produce the protein. The goal is to obtain quantities of purified hemoglobin, which can be used to improve the oxygen-carrying capacity of the blood in people suffering from a severe blood loss.

Blood substitutes all have pitfalls, ranging from high manufacturing costs to short retention in the body to the danger of transmitting disease. For example, products made from cow's blood could carry *bovine spongiform encephalitis*—otherwise known as "mad cow disease." Given their potential for improving health care and saving lives, however, it is only a matter of time before one or more safe, effective blood substitutes become available.

Look at the skin of your hand. Now try to imagine what would happen to that skin—and to all the other tissues in your body—if the cells of which they are composed did not stick together. Quite literally, you'd fall apart. In epithelium and other tissues, we observe three classes of cell junctions. By way of these contacts, cells adhere strongly to one another. Cell-to-cell contacts are especially plentiful in places where substances must not leak from one body compartment to another.

You can see examples of cell-to-cell contacts in Figure 4.11. **Tight junctions** (Figure 4.11a) are strands of protein that help stop substances from leaking across a tissue. The strands form gasketlike seals that prevent molecules from moving freely across the junction. In epithelium, for example, tight junctions allow the epithelial cells to control what enters the body. For instance, during food digestion various types of nutrient molecules can diffuse into epithelial cells or enter them selectively by active transport, but tight junctions keep those needed molecules from slipping *between* cells. Tight junctions also prevent the highly acidic gastric fluid in your stomach from leaking out and digesting proteins of your own body instead of those you consume in food. Actually, that is what happens in people who have peptic ulcers (Section 6.3).

Adhering junctions (Figure 4.11b) cement cells together. One type, sometimes called desmosomes, are like spot welds at the plasma membranes of two adjacent cells. They are anchored to the cytoskeleton in each cell and help hold cells together in tissues that are subject to stretching, such as epithelium of the skin, the lungs, and the stomach. Another type of adhering junction (also called zonula adherens junctions) form a tight collar around epithelial cells.

Gap junctions (Figure 4.11c) are channels connecting the cytoplasm of adjacent cells. They help cells communicate by promoting the rapid transfer of ions and small molecules between them.

Figure 4.11 Cell junctions. (**a**) A tight junction. In this kind of junction, protein strands form tight seals that ring each cell and seal it to its neighbors. The seals prevent substances from leaking across the free surface of some kinds of epithelia. The only way a substance can reach tissues below is to pass *through* the epithelial cells, which have built-in mechanisms that can control their passage. (**b**) Adhering junctions (desmosomes) are like spot welds that "cement" cells of epithelium (and all other tissues) together so that they function as a unit. They are abundant in the skin's surface layer and other tissues subjected to abrasion. (**c**) Gap junctions promote diffusion of ions and small molecules from cell to cell. They are abundant in the heart and other organs in which cell activities must be rapidly coordinated.

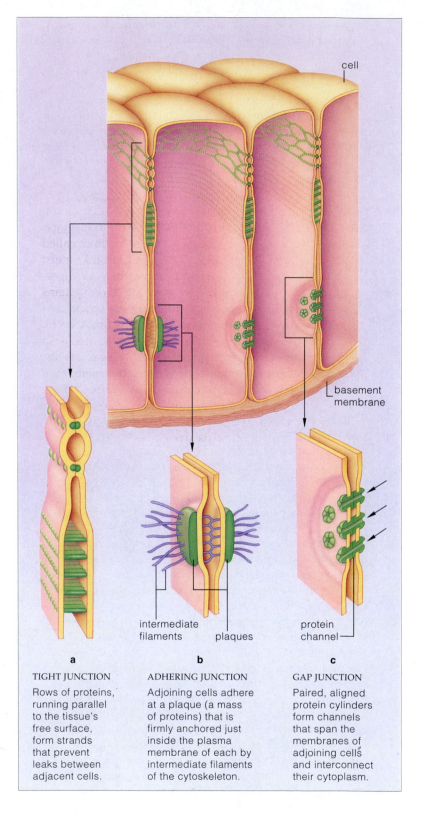

a	b	c
TIGHT JUNCTION	**ADHERING JUNCTION**	**GAP JUNCTION**
Rows of proteins, running parallel to the tissue's free surface, form strands that prevent leaks between adjacent cells.	Adjoining cells adhere at a plaque (a mass of proteins) that is firmly anchored just inside the plasma membrane of each by intermediate filaments of the cytoskeleton.	Paired, aligned protein cylinders form channels that span the membranes of adjoining cells and interconnect their cytoplasm.

MEMBRANES

The body's surfaces and cavities are covered by different kinds of thin, sheetlike membranes. Some mainly provide protection, while others both protect and lubricate organs. However, it is a membrane's structure, not its functions, that determines which of two categories it is assigned to. These categories are *epithelial membranes* and *connective tissue membranes*, and we'll now consider some examples of each.

Epithelial Membranes

Epithelial membranes consist of a sheet of epithelium together with underlying connective tissue. For instance, consider the body's *mucous membranes*, sometimes called mucosae (singular: mucosa). These are the pink, moist membranes lining the tubes and cavities of your digestive, respiratory, urinary, and reproductive systems (Figure 4.12a). Most mucous membranes are specialized to absorb or secrete substances (and sometimes both). And most, like the mucosa lining the stomach, contain glands including mucous glands that secrete mucus. Not all do, though. For instance, the mucous membrane lining the urinary tract (including the tubes through which urine flows out) does not. In chapters to come, you will read about many examples of how mucous membranes protect other tissues and secrete or absorb substances.

Serous membranes are another type of epithelial membrane. These membranes occur in paired sheets; imagine one paper sack inside another, with a narrow space between them, and you'll get the idea. Serous membranes don't have glands, but the layers do secrete a fluid that fills the space between them. Examples include the membranes that line the chest (thoracic) cavity and enclose the heart and lungs. Among other functions, serous membranes help anchor internal organs in place and provide lubricated smooth surfaces that prevent chafing or abrasion between adjacent organs or between organs and the body wall.

A third type of epithelial membrane is the *cutaneous membrane* (Figure 4.12c). You know this hardy, dry membrane as your skin. Its tissues also are part of one of the body's major organ systems, the integumentary system, which we will examine in some detail in Section 4.8.

Connective Tissue Membranes

A few membranes in the body have no epithelial cells, only connective tissue. These *synovial membranes* line the sheaths of tendons and the capsulelike cavities around certain joints. Cells in the membranes secrete fluid that lubricates the ends of moving bones or prevents friction between a moving tendon and the bone it is attached to.

Epithelial membranes consist of epithelium overlying connective tissue. They line body surfaces, cavities, ducts, and tubes. Different types include mucous, serous, and the cutaneous membrane of skin. Most epithelial membranes contain glands and epithelial cells specialized for secretion, absorption, or both.

Connective tissue membranes consist only of connective tissue. They line joint cavities.

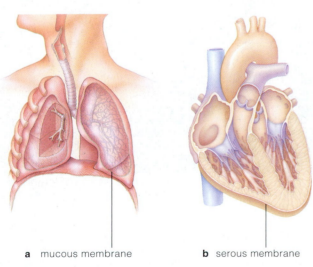   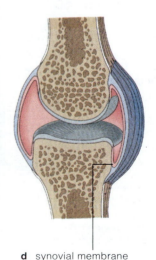

a mucous membrane   **b** serous membrane   **c** cutaneous membrane (skin)   **d** synovial membrane

Figure 4.12 Examples of body membranes.

ORGAN SYSTEMS

Figure 4.13 provides an overview of the human body's organ systems. An important point to keep in mind is that each organ system contributes to the survival of all living cells in the body. You may think this is stretching things a bit. After all, how could, say, bones and muscles be helping each microscopically small cell to stay alive? Yet, interactions between your skeletal and muscular systems allow you to move about—toward sources of nutrients and water, for example. Parts of those systems help keep blood circulating to cells, as when leg muscle contractions help move blood in veins back to the heart. Blood inside the circulatory system rapidly transports nutrients and other substances to cells and transports products and wastes away from them. The respiratory system swiftly delivers oxygen from the outside air to the circulatory system and takes up carbon dioxide wastes from it, skeletal muscles assist the respiratory system—and so it goes, throughout the entire body.

Where do the tissues of organ systems come from? When a man's sperm fertilizes a woman's egg, a tiny embryo soon forms. In short order, the embryo's cells become arranged as three primary tissues, called ectoderm, mesoderm, and endoderm. These three tissues

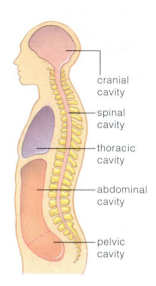

Figure 4.14 (**a**) The major cavities in the human body. (**b**) Directional terms and planes of symmetry for the human body. Notice how the midsagittal plane divides the body into right and left halves. The transverse plane divides it into superior (top) and inferior (bottom) parts. The frontal plane divides it into anterior (front) and posterior (back) parts.

cranial cavity

spinal cavity

thoracic cavity

abdominal cavity

pelvic cavity

a

INTEGUMENTARY SYSTEM	MUSCULAR SYSTEM	SKELETAL SYSTEM	NERVOUS SYSTEM	ENDOCRINE SYSTEM	CIRCULATORY SYSTEM
Protect body from injury, dehydration, and some pathogens; control its temperature, excrete some wastes; receive some external stimuli.	Move body and its internal parts; maintain posture; generate heat (by increases in metabolic activity).	Support and protect body parts; provide muscle attachment sites; produce red blood cells; store calcium, phosphorus.	Detect both external and internal stimuli; control and coordinate responses to stimuli; integrate all organ system activities.	Hormonally control body function; work with nervous system to integrate short-term and long-term activities.	Rapidly transport many materials to and from cells; help stabilize internal pH and temperature.

Figure 4.13 Overview of human organ systems and their functions.

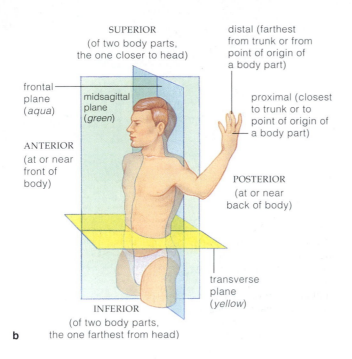

SUPERIOR
(of two body parts,
the one closer to head)

distal (farthest
from trunk or from
point of origin of
a body part)

frontal
plane
(*aqua*)

midsagittal
plane
(*green*)

proximal (closest
to trunk or to
point of origin of
a body part)

ANTERIOR
(at or near
front of
body)

POSTERIOR
(at or near
back of body)

transverse
plane
(*yellow*)

b

INFERIOR
(of two body parts,
the one farthest from head)

are embryonic forerunners of all tissues in an adult. Ectoderm gives rise to the skin's outer layer and to tissues of the nervous system. From mesoderm, tissues of the muscles, bones, and most of the circulatory, reproductive, and urinary systems will develop. Endoderm gives rise to the lining of the digestive tract and to organs derived from it. Chapter 15 describes these events.

Figure 4.14*a* shows the major body cavities in which many important organs are located. The **cranial cavity** and **spinal cavity** house the central nervous system— your brain and spinal cord. Your heart and lungs reside in the **thoracic cavity**—essentially, inside your chest. The diaphragm muscle separates the thoracic cavity from the **abdominal cavity**, which holds your stomach, liver, most of the intestine, and other organs. Your reproductive organs, bladder, and rectum are located within the **pelvic cavity.** Figure 4.14*b* lists some terms that are used to describe the positions of various organs.

The body's organ systems each serve a specialized function that contributes to the survival of all living body cells.

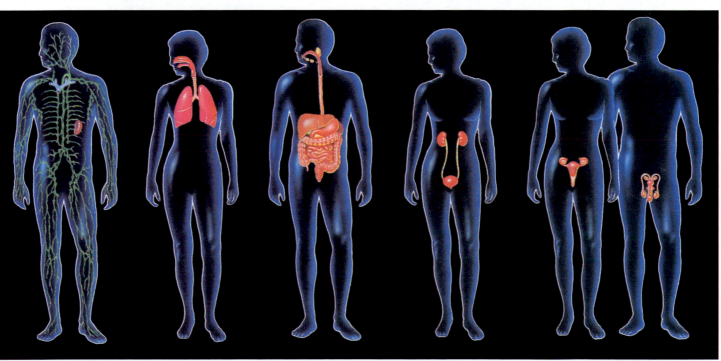

LYMPHATIC SYSTEM	RESPIRATORY SYSTEM	DIGESTIVE SYSTEM	URINARY SYSTEM	REPRODUCTIVE SYSTEM
Collect and return some tissue fluid to the bloodstream; defend the body against infection and tissue damage.	Rapidly deliver oxygen to the tissue fluid that bathes all living cells; remove carbon dioxide wastes of cells; help regulate pH.	Ingest food and water; mechanically, chemically break down food, and absorb small molecules into internal environment; eliminate food residues.	Maintain the volume and composition of internal environment; excrete excess fluid and blood-borne wastes.	*Female:* Produce eggs; after fertilization, afford a protected, nutritive environment for the development of new individual. *Male:* Produce and transfer sperm to the female. Hormones of both systems also influence other organ systems.

THE SKIN: EXAMPLE OF AN ORGAN SYSTEM

No garment even comes close to having the qualities of your skin. What body covering besides skin holds its shape after repeated stretchings and washings, blocks harmful rays from the sun, kills many bacteria on contact, holds in moisture, fixes small cuts and burns, *and* lasts as long as its owner? Indirectly, skin helps control internal body temperature, for the nervous system can rapidly adjust the flow of blood (which transports metabolic heat) to and from the skin's many tiny blood vessels. Also, signals from skin's sensory receptors help the brain assess what's going on in the outside world.

As if all this were not enough, your skin also makes vitamin D, or cholecalciferol. This steroid-like compound helps the body absorb calcium from food. It is produced when certain cells in skin absorb sunlight. The cells then release vitamin D to the bloodstream, which transports it to absorptive cells in the intestinal lining. This is a hormone-like action—which means that skin acts like an endocrine gland when exposed to sunlight.

Technically, skin is part of the body's **integument** (from Latin *integere*, meaning "to cover"). This organ system includes the skin and the structures (such as hair and nails) derived from epidermal cells of its outer tissue layers. If you are an average-sized adult, your skin weighs about 9 pounds (making it your body's largest organ). Except for places subjected to regular pounding or abrasion (such as the palms of the hands and soles of the feet), your skin is generally not much thicker than a piece of construction paper. On your eyelids and some other places, it is even thinner.

Skin has two regions—an outer **epidermis** and an underlying **dermis** (Figure 4.15). The dermis is mainly dense connective tissue containing many elastin fibers (which make the skin resilient) and collagen fibers (which make it strong). Together, the epidermis and dermis make up the cutaneous membrane mentioned in Section 4.6. Below the dermis is a subcutaneous ("under the skin") layer, the hypodermis. This is a loose connective tissue that anchors the skin while allowing it to move a bit. Fat stored in the hypodermis helps insulate the body and cushions some of its parts.

Like puff pastry, epidermis has layers; it is a stratified squamous epithelium. Cell junctions knit the epithelial cells together. The cells arise in deeper layers and are pushed toward the skin's surface as cell divisions produce new cells beneath them. (This efficient replacement is one reason why skin can mend minor damage so quickly.) Due to pressure from the continually growing cell mass and from normal wear and tear at the surface, older cells are dead and flattened by the time they reach the outer layers. There, they are abraded off or flake away.

Most cells of the epidermis are **keratinocytes**. These cells make keratin, a tough, water-insoluble protein. By the time they reach the skin surface and have died, all

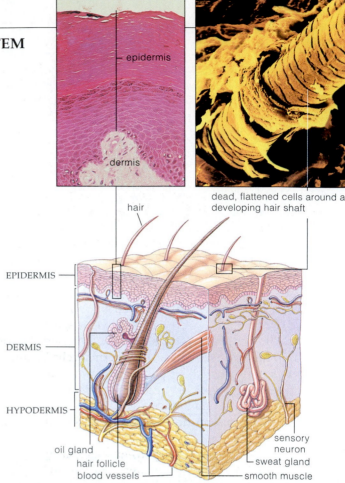

Figure 4.15 Structure of human skin. The uppermost layer is the epidermis (stratum corneum); deeper layers include the stratum basales (basal layer) and the dermis. The dark spots in the basal layer are epidermal cells to which melanocytes have passed pigment.

that remain are the keratin fibers, packed inside plasma membranes. This helps make skin's outermost layer—the stratum corneum—tough and waterproof.

In the deepest layer of epidermis, specialized cells called **melanocytes** produce a brown-black pigment called melanin and donate it to the keratinocytes. Melanin blocks harmful ultraviolet radiation. In general, all humans have the same number of melanocytes, but skin color varies due to differences in the distribution and activity of those cells. For example, albinos cannot produce all the enzymes necessary to produce melanin. The pale skin of Caucasians contains little melanin, so the pigment hemoglobin inside red blood cells shows through thin-walled blood vessels and the epidermis itself, both of which are transparent. There is more melanin in naturally brown or black skin. Skin color also is influenced by the yellow-orange pigment carotene in cells of the dermis.

Besides having cells that produce melanin and keratin, the epidermis also contains two other cell types that help protect the body against harm. *Langerhans cells* are phagocytes ("cell eaters") that develop in bone marrow and migrate to the skin, where they engulf bacteria or virus particles. Then they display surface proteins that can mobilize the immune system against the attack. *Granstein cells* are rather mysterious, but apparently they interact with white blood cells of the immune system, suppressing immune responses in the skin. By doing so, they may help keep the responses under control.

Dense connective tissue makes up most of the dermis, and it fends off damage from everyday stretching and other mechanical insults. This protection has limits, though. For example, the dermis tears when skin over a woman's abdomen is stretched too much during pregnancy, leaving white scars ("stretch marks"). If abrasion is persistent—as might happen if you wear a too-tight shoe—the epidermis separates from the dermis, the gap fills with a watery fluid, and you get a blister.

The dermis is laced with blood vessels, lymph vessels, and the receptor endings of sensory nerves. Nutrients from the bloodstream reach epidermal cells by diffusing through the dermal tissue. Sweat glands, oil glands, the husklike hair follicles, and fingernails and toenails are derived from epidermal tissue, although they are embedded to some degree in the dermis (Figure 4.15).

The body has about 2.5 million sweat glands. The fluid they secrete is 99 percent water; it also contains dissolved salts, traces of ammonia and other wastes, vitamin C, and other substances. A type of sweat gland that is plentiful in the palms of the hands, soles of the feet, forehead, and armpits functions mainly in temperature regulation. Another type is abundant in skin around the genitals. Stress, pain, and sexual foreplay all can increase their secretory activity.

Oil glands (or *sebaceous glands*) are everywhere except on the palms and soles of the feet. Their oily secretions soften and lubricate the hair and skin, and other secretions kill many harmful bacteria. *Acne* is a skin inflammation that develops after bacteria infect oil gland ducts.

A **hair** is a flexible structure of mostly keratinized cells, rooted in skin with a shaft above its surface. As cells divide near the base of the root, older cells are pushed upward, then flatten and die. The outermost layer of the shaft consists of flattened cells that overlap like roof shingles (Figure 4.15, *right*). These dead cells are what frizz out as "split ends." An average human scalp has about 100,000 hairs. However, both the growth and the density of a person's hair are influenced by genes, nutrition, hormones, and stress.

With its layers of keratinized and melanin-shielded epidermal cells, skin helps the body conserve water, avoid damage from ultraviolet radiation, and resist mechanical stress. Hairs, oil glands, sweat glands, and other structures associated with skin are derived from epidermal cells.

4.10 SUN, SKIN, AND THE OZONE LAYER

Exposure to ultraviolet (UV) radiation stimulates the melanin-producing cells of the epidermis. With prolonged sun exposure, melanin levels increase and light-skinned people become tanned. Tanning gives some protection against UV radiation, but over the years, it causes elastin fibers in the dermis to clump together. The skin loses its resiliency and begins to look leathery.

Excessive exposure to UV radiation also suppresses a person's immune system. For instance, getting a sunburn interferes with the infection-fighting phagocytes in the epidermis. This may be why sunburns can trigger the painful blisters called "cold sores" that announce the appearance of *herpes simplex*. Nearly everyone harbors this virus. Usually it becomes localized in a nerve ending near the skin surface, where it remains dormant. Stress factors, including sunburn, can activate the virus and trigger the skin eruptions.

Ultraviolet radiation from sunlight or from the lamps of tanning salons also can activate proto-oncogenes in skin cells (Section 22.2). These bits of DNA can trigger cancer, especially when some factor alters their structure (and hence their function).

Most of the UV radiation to which we are exposed comes from the sun. Until fairly recently, a layer of ozone in the stratosphere intercepted much of the potentially damaging UV radiation (specifically UV-B). Today, however—due in large part to human activities described in Chapter 25—the ozone layer over ever larger regions of the globe is being destroyed faster than natural processes can replace it. The rate of skin cancers now is rapidly increasing. Skin cancers are among the easiest to cure, but only if they are promptly removed surgically. As for the protective ozone layer, experts estimate that even if all ozone-depleting substances were banned tomorrow, it would take about 100 years for the planet to recover.

The Internal Environment

To stay alive, your cells must be continuously bathed in fluid that supplies them with nutrients and carries away metabolic wastes. In this they are no different from an amoeba or any other free-living, single-celled organism. The difference is, trillions of cells coexist in your body, and all of them must draw nutrients from and dump wastes into the same fifteen liters of fluid. That is less than four gallons.

The fluid not inside cells is **extracellular fluid** (in this context, "extra" means outside). Much of it is *interstitial*, meaning it fills spaces between cells and tissues. The rest is blood plasma, the fluid portion of blood. Interstitial fluid exchanges substances with the cells it bathes and with blood.

Functionally, changes in extracellular fluid cause changes in cells. That is why drastic changes in its composition and volume have drastic effects on cell activities. The type and number of ions are especially crucial, for they must be maintained at levels that are compatible with metabolism. Otherwise, the body cannot survive.

Look back at the simple diagram in Section 1.1, where you first read about the concept of homeostasis. The point of that diagram is that *the component parts of the body work together to maintain the stable fluid environment required by its living cells.* This concept has three key points:

1. Each body cell engages in metabolic activities that ensure its own survival.

2. At the same time, the cells of a given tissue perform one or more activities that contribute to the survival of the whole body.

3. The combined contributions of individual cells, tissues, organs, and organ systems help maintain the stable internal environment—the extracellular fluid—required for individual cell survival.

Mechanisms of Homeostasis

Homeostasis literally means "staying the same." It means stable operating conditions in the body's internal environment. Three components interact to maintain this state. They are called sensory receptors, integrators, and effectors. **Sensory receptors** are cells or cell parts that can detect a **stimulus**, which is a specific change in the environment. For example, when someone kisses you, there is a change in pressure on your lips. Receptors in the skin of your lips translate the stimulus into a signal, which can be sent to the brain. Your brain is an **integrator**, a control point where different bits of information are pulled together in the selection of a response. It can send signals to your muscles or glands (or both). Muscles and glands

are **effectors**—they carry out the response. In this case, the response might include kissing the person back. Of course, you cannot engage in a kiss indefinitely, because eventually you must eat and perform other tasks that maintain your body's operating conditions.

So how does the brain reverse the physiological changes induced by the kiss? Receptors provide it with information only about how things *are* operating. The brain also maintains information about how things *should be* operating—that is, information from "set points." When physical or chemical conditions deviate sharply from a set point, the brain functions to bring them back within an effective operating range. It does this by way of signals that cause specific muscles and glands to increase or decrease their activity. Set points are key elements of many physiological mechanisms, including those that influence eating behavior, breathing, thirst, and urination, to name a few.

NEGATIVE FEEDBACK Feedback mechanisms are controls that help keep physical and chemical aspects of the body within tolerable ranges. In **negative feedback**, some activity alters a condition in the internal environment, and this triggers a response that reverses the altered condition (Figure 4.16). To grasp how the mechanism operates, think of a furnace with a thermostat. The thermostat senses the air temperature and mechanically "compares" it to a preset point on a thermometer built into the furnace control system. When the temperature falls below the preset point, the thermostat signals a switching mechanism that turns on the heating unit. When the air becomes heated enough to match the prescribed level, the thermostat signals the switching mechanism, which shuts off the heating unit.

Similarly, feedback mechanisms help keep the body temperature of humans (and many other animals) within a normal range, even during extremely hot or cold weather or under other conditions that alter temperature

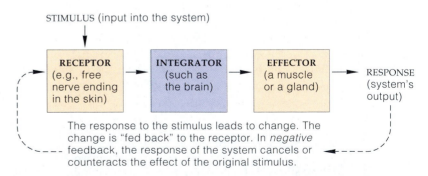

Figure 4.16 Components necessary for negative feedback at the organ level.

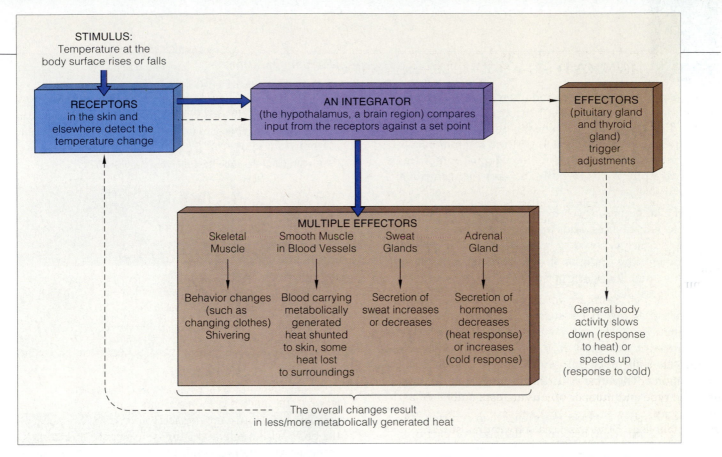

Figure 4.17 Homeostatic controls over the internal temperature of the human body. The dashed line shows how the feedback loop is completed. The *blue* arrows indicate the main control pathways.

(Figure 4.17). For example, when the body senses that its skin is getting too hot while you work outside in the summer sun, mechanisms are set in motion that slow down metabolic activity and overall body activity. You move more slowly and may seek the shelter of a shade tree. At the same time, blood flow to the skin increases and your sweat glands are prodded to secrete larger quantities of sweat. As water in sweat evaporates, more heat is lost from the body. These and other mechanisms counter overheating by curbing the body's heat-generating activities and giving up excess heat to the surroundings.

POSITIVE FEEDBACK   Sometimes a **positive feedback** mechanism operates. This type of mechanism sets in motion a chain of events that *intensify* a change from an original condition—and after a limited time, the intensification reverses the change. Positive feedback is associated with instability in a system. For example, during sexual intercourse, chemical signals from a female's nervous system can cause intense physiological responses to her sexual partner. Her responses stimulate changes in her partner that stimulate the female even more—and so on until she reaches an explosive, climax level of excitation. Normal conditions then return, and homeostasis prevails.

As another example, during childbirth, a fetus exerts pressure on the walls of its mother's uterus. Pressure stimulates production and secretion of oxytocin, a

hormone. Oxytocin causes wall muscles to contract and exert pressure on the fetus, which exerts more pressure on the uterine wall, and so on until the fetus is expelled.

What we have been describing here is a general pattern of monitoring and responding to a constant flow of information about the external and internal environments. During this activity, organ systems operate together in coordinated fashion.

Throughout this unit we will be asking the following questions about organ systems:

1. What physical or chemical aspect of the internal environment are organ systems working to maintain as conditions change?

2. How are organ systems kept informed of the various changes?

3. How do they process the incoming information?

4. What mechanisms are set in motion in response?

As you will see in chapters to follow, the operation of all organ systems is under precise neural and endocrine control.

Homeostatic control mechanisms maintain physical and chemical aspects of the internal environment within ranges that are most favorable for cell activities.

SUMMARY

1. A tissue is a group of cells and intercellular substances that perform a common task. An organ is a structural unit of different tissues combined in definite proportions and patterns that allow them to perform a common task. An organ system has two or more organs interacting chemically, physically, or both, in ways that contribute to the survival of the body as a whole.

2. Epithelial tissues cover external body surfaces and line internal cavities and tubes. These tissues have one free surface exposed to body fluids or the external environment; the opposite surface rests on a basement membrane that intervenes between it and an underlying connective tissue.

3. Different connective tissues bind together, support, strengthen, protect, and insulate other tissues. Most have fibers of structural proteins (especially collagen), fibroblasts, and other cells within a ground substance. They include connective tissue proper and specialized connective tissues such as cartilage, bone, and blood.

4. Muscle tissues contract (shorten), then return to the resting position. They help move the body or part of it. The three types of muscles are skeletal muscle, smooth muscle, and cardiac muscle.

5. Nerve tissue intercepts and integrates information about internal or external conditions, and governs the body's responses to change. Neurons of this tissue are the basic units of the nervous system.

6. Tissues, organs, and organ systems work together to maintain a stable internal environment (the extracellular fluid) required for survival. At homeostasis, conditions in the internal environment are balanced at levels most favorable for cell activities.

7. An example of an organ system is the integumentary system, or skin, which protects the rest of the body from abrasion, bacterial attack, ultraviolet radiation, and dehydration. It helps control internal temperature, contains cells that synthesize vitamin D, and serves as a blood reservoir for the rest of the body. Its receptors are essential in detecting environmental stimuli.

8. Feedback controls help maintain internal conditions. In negative feedback (the most common mechanism), a change in some condition triggers a response that reverses the change. In positive feedback, a response reverses a change in some condition by intensifying the change for a limited time.

9. Homeostasis depends on receptors, integrators, and effectors. Receptors detect stimuli, which are specific changes in the environment. Integrators such as the brain process information and direct muscles and glands (the body's effectors) to carry out responses.

Review Questions

1. Describe the characteristics of epithelial tissue in general. Then describe the various types of epithelial tissues in terms of specific characteristics and functions. *4.1*

2. List the major types of connective tissues; add the names and characteristics of their specific types. *4.2*

3. Identify and describe the following tissues: *4.1–4.3*

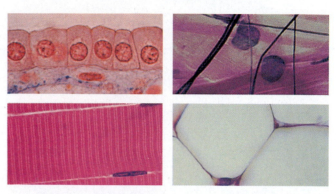

4. Identify this category of tissue and its characteristics: *4.4*

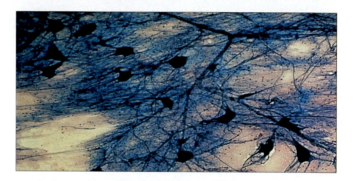

5. List the types of cell junctions and their functions. *4.6*

6. List the basic types of membranes in the body. *4.7*

7. Define tissue, organ, and organ system. List the eleven major organ systems of the human body. *CI, 4.8*

8. What are some of the functions of skin? *4.10*

9. Define homeostasis. *CI, 4.11*

10. How do negative feedback and positive feedback differ? *4.11*

Self-Quiz *(Answers in Appendix V)*

1. _____ tissues have closely linked cells and one free surface.
    a. Muscle   c. Connective
    b. Nerve   d. Epithelial

2. In most _____ cells secrete fibers of collagen and elastin.
    a. muscle tissues   c. connective tissues
    b. nerve tissue   d. epithelial tissues

3. _____ has a semifluid ground substance and occurs under most epithelia.
    a. Dense, irregular   c. Dense, regular
       connective tissue   connective tissue
    b. Loose connective tissue   d. Cartilage

4. _____ a specialized connective tissue, is mostly plasma with cellular components and various dissolved substances.
   a. Irregular connective tissue   c. Cartilage
   b. Blood   d. Bone

5. Components of _____ tissue coordinate information about environmental changes and control responses to those changes.
   a. muscle   c. connective
   b. nerve   d. epithelial

6. In your own body, _____ can shorten (contract).
   a. muscle tissue   c. connective tissue
   b. nerve tissue   d. epithelial tissue

7. After you eat too many carbohydrates and proteins, your body converts the excess to storage fats, which accumulate in _____ .

   a. connective tissue proper   c. adipose tissue
   b. dense connective tissue d. both b and c

8. Cells in the human body _____ .
   a. survive by their own metabolic activities
   b. contribute to the survival of the body as a whole
   c. help maintain extracellular fluid
   d. all of the above

9. In a state of _____, physical and chemical aspects of the body are being kept within tolerable ranges by controlling mechanisms.
   a. positive feedback   c. homeostasis
   b. negative feedback   d. metastasis

10. In negative feedback mechanisms, _____ .
   a. a detected change brings about a response that tends to return internal operating conditions to the original state
   b. a detected change suppresses internal operating conditions to levels below the set point
   c. a detected change raises internal operating conditions to levels above the set point
   d. a detected change causes fewer solutes to be fed back to the affected cells

11. Fill in the blanks: _____ detect specific environmental changes, an _____ pulls different bits of information together in the selection of a response, and _____ carry out the response.

12. Match the concepts:
   _____ muscles and glands   a. integrating center
   _____ positive feedback   b. response that reverses an
   _____ sites of body receptors   altered condition
   _____ negative feedback   c. eyes and ears
   _____ brain   d. effectors
                                   e. chain of events that intensifies the original condition

Critical Thinking: You Decide *(Key in Appendix VI)*

1. In people who have the genetic disorder *anhidrotic ectodermal dysplasia*, patches of tissue have no sweat glands. What kind of tissue are we talking about?

2. Adipose tissue and blood are often said to be "atypical" connective tissues. Compared to other connective tissues, which of their features are *not* typical?

3. The disease called scurvy results from a deficiency of vitamin C, which the body requires for collagen synthesis. Explain why (among other symptoms) scurvy sufferers tend to lose teeth, and why any wounds heal much more slowly than normal, if at all.

4. After graduating from high school, Jeff and Ryan set out on a trip through the desert roads of California and Arizona. One hot, dry morning in Joshua Tree National Monument, they saw an unusual rock formation that didn't appear to be too far from the road. They left the car and started hiking toward it. Their destination turned out to be farther away than they thought, and the sun was more relentless than they anticipated. When they finally reached the shade of the rocks, their canteen was nearly empty and they began to experience overwhelming thirst and other symptoms of dehydration. They knew they had to locate and drink water (or some other fluid), which is what the brain usually tells us to do when dehydration sets in.

   From what you read in this chapter, would you suspect that Jeff and Ryan's thirst behavior is part of a positive or negative feedback control mechanism?

5. Porphyria is a genetic disorder that shows up in about 1 in 25,000 people. Affected individuals lack certain enzymes that are part of a metabolic pathway leading to formation of heme, which is the iron-containing group in hemoglobin. Accumulating porphyrins, which are intermediates in the pathway, cause awful symptoms, especially if the person is exposed to sunlight. Lesions and scars form on the skin. Hair grows thickly on the face and hands. As the gums retreat from the sufferer's teeth, the canine teeth can begin to look like fangs. The symptoms get worse if the affected person consumes alcohol or garlic. The individual may be advised to avoid sunlight and aggravating substances and may also receive injections of heme from normal red blood cells.

   If you are familiar with vampire stories, which date from the Middle Ages or earlier, can you think of a reason why they may have evolved among superstitious folk who had no knowledge of the medical basis of porphyria?

6. Use the World Wide Web to find at least one free searchable database dealing with each of the body's eleven main organ systems.

Selected Key Terms

adipose tissue *4.2*	gland *4.1*
basement membrane *4.1*	homeostasis *4.11*
blood *4.2*	integrator *4.11*
bone *4.2*	integument *4.9*
cartilage *4.2*	muscle tissue *4.3*
connective tissue *4.2*	negative feedback mechanism *4.11*
dermis *4.9*	nervous tissue *4.4*
effector *4.11*	organ *CI*
endocrine gland *4.1*	organ system *CI*
epidermis *4.9*	positive feedback *4.11*
epithelium *4.1*	sensory receptor *4.11*
exocrine gland *4.1*	tissue *CI*
extracellular fluid *4.11*	

Readings

Mooney, D. J., and A. Mikos. April 1999. "Growing New Organs." *Scientific American*.

Nucci, M., and A. Abuchowski. February 1998. "The Search for Blood Substitutes." *Scientific American*.

THE MUSCULAR AND SKELETAL SYSTEMS

Row, Row, Row Your Boat

Betsy and Mary McCagg are twins—and two of the best competitive rowers in the world (Figure 5.1). Beginning in high school, they built a reputation as elite athletes. Among other honors, they were members of the United States national rowing team in the 1992 and 1996 Olympic Games. One year they were part of an eight-woman team that won the world championship for the distance of 2,000 meters in women's rowing.

The McCagg sisters have genetically determined characteristics that contribute to their prowess in their chosen sport. Both are tall (188 centimeters, or 6 feet 2 inches) with long thighbones in their legs. This feature

is important, because the longer a rower's thighbones (femurs), the more power the rower can generate with each stroke. In addition, however, Betsy and Mary put in grueling hours of work with exercise physiologists and other sport scientists to hone their bodies to peak condition. Like other world-class athletes, their main goal is to maximize the potential of the systems that support and move their bodies.

Long before each race, the McCagg twins follow a strict training regimen. The physical workouts increase the capacity of their musculoskeletal system to support and help move their bodies. How? To begin with,

rigorous exercise stresses bones, and the bone mass increases in response. Our muscle cells respond to training by becoming larger or loaded with mitochondria, the organelles that produce ATP energy during aerobic respiration. Also, more blood vessels develop, increasing blood flow for the transport of nutrients and wastes. These and other training-related changes can produce amazing increases in strength and stamina—qualities that are essential for competitive success in rowing, cycling, and other endurance sports.

In this chapter we examine the body's systems for structural support, protection, and movement. The skeleton is a strong, internal framework built of rigid, mineralized bones. Along with cartilages, joints, and ligaments, it makes up the **skeletal system** (Figure 5.2). Skeletal muscles, which attach to bones and pull on them to move the body and its parts, make up our **muscular system**. The two systems work so closely together that sometimes we refer simply to the "musculoskeletal system."

We take for granted that our bodies' softer parts have a supporting framework—one function, among several, that our hard, bony skeleton serves. But while bones are quite sturdy, much of the structural material inside them is surprisingly lightweight. This adaptation allows our muscles to move bones at a lower energy cost. Muscles also work with one another as well as with the skeleton. Guided and controlled by the nervous system, this coordinated organization enables a person to execute the movements and position changes required for the full range of our daily activities—from rowing a racing scull in the Olympic Games to turning the pages of a textbook.

Figure 5.1 Betsy and Mary McCagg giving a workout to their bodies' systems of support and movement—the skeleton and muscles.

KEY CONCEPTS

1. Bones are rigid, mineralized organs. They are the main component of the skeletal system.

2. Bones contribute to homeostasis in many ways. They protect and support soft organs, and they interact with muscles that move the body. They also store minerals. Some bones include tissues that produce blood cells.

3. Joints are sites of contact or near-contact between bones. Joints at the articulating ends of bones allow skeletal movements.

4. The muscular system consists of skeletal muscles, which attach to bones. Skeletal muscles pull on bones, and so move the body and its parts. Smooth muscle is found in the walls of organs and blood vessels; cardiac muscle is the main tissue of the heart.

5. A skeletal muscle contracts when it is stimulated by the nervous system. Such contractions are usually voluntary. The contractions of smooth muscle and cardiac muscle are involuntary.

CHAPTER AT A GLANCE

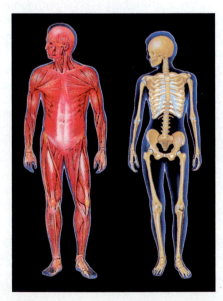

Figure 5.2 Overview of the human body's muscle system and skeletal system.

CHARACTERISTICS OF BONE

Bones are the main components of our human skeletal system. It can be easy to take them for granted simply as "girders" that support our soft, rather squishy flesh, but in fact our bones are complex organs that contribute to homeostasis in many ways (Table 5.1). For instance, bones that support and anchor skeletal muscles help maintain or change the positions of our body parts. Some form hard compartments that enclose and protect other organs; for example, the skull does this for the brain and the rib cage protects the lungs. Bones also serve as a "pantry" where the body can store calcium and phosphorus (as ions). Deposits and withdrawals of those minerals from bones help maintain blood levels of calcium and phosphorus, thus supporting metabolic activities that use or produce the minerals. Some bones have marrow cavities where blood cells are formed.

Human bones range in size from ear bones about the size of a watch battery to massive thighbones. Bone shapes vary, too. Long bones, like the thighbone in Figure 5.3, are longer than they are wide, while others, like the ankle bones, are short. Others, such as the breastbone (sternum), are flat, and still others, such as our spinal vertebrae, are "irregular." However, all bones are alike in some ways. They all contain connective tissues—including bone tissue and other connective tissue that lines the surfaces of bones and the cavities inside them. At joints, where one bone meets with another, bones have cartilage. Other tissue components of bones include nerve tissue and epithelium, which occurs in the walls of blood vessels that transport substances to and from bones.

In bone tissue itself, there are both living cells and nonliving elements. The living cells are **osteocytes** (*osteo* = bone). As a bone develops, precursors of osteocytes, called osteoblasts, secrete most of the materials that will make up the mature bone—collagen, some elastin, and a ground substance of proteins and carbohydrates. Eventually, the osteocytes are enclosed in spaces in the ground substance, called lacunae (*lacuna* = hole). With time, calcium salts are deposited and the ground substance hardens—that is, it becomes mineralized.

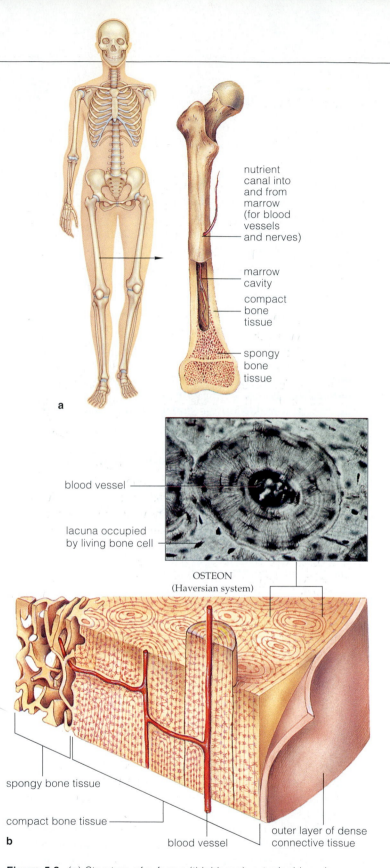

a

blood vessel

lacuna occupied by living bone cell

OSTEON
(Haversian system)

spongy bone tissue

compact bone tissue

blood vessel

outer layer of dense connective tissue

b

nutrient canal into and from marrow (for blood vessels and nerves)

marrow cavity

compact bone tissue

spongy bone tissue

Figure 5.3 (**a**) Structure of a femur (thighbone), a typical long bone. (**b**) The appearance of its spongy and compact bone tissue. Thin, dense layers of compact bone tissue form cylindrical, interconnected arrays around canals that contain blood vessels and nerves. Each array is an osteon (Haversian system). The blood vessel threading through its center transports substances to and from osteocytes, living bone cells in small spaces (lacunae) in the bone tissue. Small tunnels called canaliculi connect neighboring spaces.

Table 5.1 Functions of Bone

1. *Movement.* Bones interact with skeletal muscle to maintain or change the position of body parts.

2. *Support.* Bones support and anchor muscles.

3. *Protection.* Many bones form hard compartments that enclose and protect soft internal organs.

4. *Mineral storage.* Bones are a reservoir for calcium and phosphorus, the deposits and withdrawals of which help to maintain ion concentrations in body fluids.

5. *Blood cell formation.* Some bones contain regions where blood cells are produced.

The Two Kinds of Bone Tissue

There are two kinds of tissue in bone, compact bone and spongy bone. Figure 5.3 shows how these two kinds of tissue are arranged in one of the body's largest bones, the thighbone (femur). As you might guess from its name, **compact bone** is a dense tissue that looks solid and smooth. In a long bone, it forms the bone's shaft and the outer part of its two ends. A membrane called the periosteum covers the shaft. The collagen fibers in bone tissue give bones tensile strength—in a long bone, the strength to withstand mechanical stresses associated with standing, lifting, tugging, and so forth. The calcium salts in bone tissue make it hard.

When a bone is developing, compact bone tissue is laid down as thin, circular layers around small central canals. Each set of layers is called a Haversian system, or an **osteon**. The canals are interconnected. They serve as channels for blood vessels and nerves that transport substances to and from the osteocytes in compact bone. Osteocytes also send slender cell processes into narrow channels between lacunae. Through these channels, nutrients can move through the hard ground substance from osteocyte to osteocyte. Wastes can exit the same way.

The bone tissue *inside* a long bone's shaft and at its ends looks like a sponge. Tiny, flattened struts are fused together to make up this **spongy bone** tissue, which looks lacy and delicate but actually is quite firm and strong. In some bones, red **bone marrow** fills the spaces between the struts. Much of the body's supply of blood cells forms in red marrow in irregular bones such as the hipbone and in flat bones such as the sternum. This is because, as a person grows into adulthood, the red marrow in the shafts of most long bones is replaced by yellow marrow. Yellow marrow is mostly fat, although if need be it can convert to red marrow, which produces red blood cells.

How a Bone First Develops

The skeleton of a developing embryo consists only of cartilage and membranes. Yet, after only about two months of life in the womb, this pliable framework is completely transformed into a bony skeleton. Again, we see a good example of this remarkable process in the development of a long bone.

In a growing embryo, a cartilage "model" provides a template for each long bone (Figure 5.4). Once the outer membrane is in place on the bone model, it gives rise to the bone-forming osteoblasts. As these cells become active, a bony "collar" forms around the cartilage shaft. Then the cartilage inside the shaft calcifies, and an artery, a vein, and other elements (including some osteoblasts) infiltrate the forming bone. Soon, the marrow cavity forms.

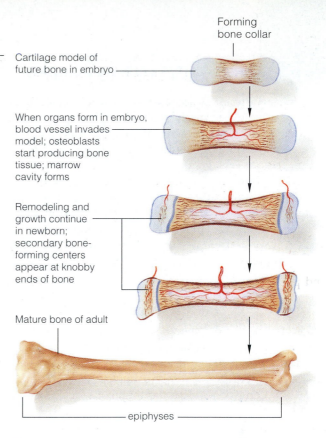

Cartilage model of future bone in embryo

Forming bone collar

When organs form in embryo, blood vessel invades model; osteoblasts start producing bone tissue; marrow cavity forms

Remodeling and growth continue in newborn; secondary bone-forming centers appear at knobby ends of bone

Mature bone of adult

epiphyses

Figure 5.4 How a long bone forms. The process begins when osteoblasts become active in a cartilage model (here, already formed in the embryo). The bone-forming cells are active first in the shaft, then at the knobby ends. In time, cartilage is left only in the epiphyses at the ends of the shaft.

Meanwhile, the osteoblasts secrete ground substance destined to become mineralized. When the osteoblasts are finally trapped by the matrix they secreted they become osteocytes—mature living bone cells.

A long bone has flaring ends, each one called an epiphysis (plural: epiphyses). In growing children and young adults epiphyses are separated from the bone shaft by an **epiphyseal plate** of cartilage. Human growth hormone (GH) triggers the development of such cartilage plates and also maintains them (Chapter 13). As long as the epiphyseal plate is present, the bone can lengthen as a person grows. When a person stops growing in late adolescence, bone replaces the cartilage plates.

Bones are mineralized organs composed of living cells, a nonliving mineralized matrix of collagen fibers and ground substance, and an outer covering of dense connective tissue.

Bones contribute to homeostasis by providing body support, enabling movement, storing minerals, and, in some cases, producing blood cells (in marrow).

Two kinds of marrow, red and yellow, fill the spaces in certain bones. Blood cells form in red marrow. Yellow marrow, which is mostly fat, replaces red marrow in the shafts of most long bones in adults. It can be converted to red marrow in time of need.

HOW THE SKELETON GROWS AND IS MAINTAINED

From the standpoint of homeostasis, it would be hard to overstate how important the mineral calcium is. Not only is it a vital structural element in bone, but physiological processes like muscle contraction and the operation of the nervous system depend on it. From childhood onward, calcium—and another key mineral, phosphate—are constantly deposited in and withdrawn from bone tissue. The name for this dynamic turnover of minerals is "remodeling." As you'll now read, remodeling is how bones grow, gain strength, take on the proper proportions, and are repaired when they break. Remodeling also is crucial in keeping the body's calcium supplies in balance.

Bone Remodeling

In remodeling, osteoblasts deposit bone and cells called *osteoclasts* break it down. Both processes depend on hormones. For example, when the blood level of calcium falls below a set point, the hormone PTH (parathyroid hormone) stimulates osteoclasts to secrete enzymes that break down bone tissue. In an adult, osteocytes also respond to PTH and withdraw calcium from bone. The released calcium enters interstitial fluid, then moves into the bloodstream. If there is more calcium in the blood than the body requires at a given time, another hormone, calcitonin, prompts osteoblasts to take up calcium from the interstitial fluid and use it to produce new bone. Chapter 13 has more on these hormonal controls.

Remodeling also helps keep bones themselves strong. This is because mechanical stress on mature bones triggers more bone deposits than withdrawals. Such stress comes partly from muscles pulling on the bone during body movements. Hence, brisk walking, jogging, lifting weights, and similar activities cause affected bones to become denser—and stronger—as more bone tissue is created. By the same token, people who are inactive or bedridden tend to lose bone mass because the mechanical stress on their bones is so slight.

Before adulthood, a person's bones grow by way of remodeling. During this time, the body requires ample calcium to meet the combined demands of bone growth and other metabolic needs for the minerals stored in our bones. Along with calcium in the diet, turnover helps meet the demand. For example, in a growing child the diameter of the thighbones increases as osteoblasts deposit calcium phosphate at the surface of each shaft. Simultaneously, however, osteoclasts break down a small amount of bone tissue *inside* the shaft. Thus the child's thighbones become thicker and stronger to support the increasing body weight, but they don't get too heavy.

Remodeling also helps heal broken bones. A patch of fibrocartilage forms at the break first, then osteoblasts and osteoclasts migrate to the site. Over time, the cartilage is

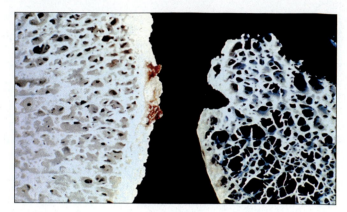

Figure 5.5 Osteoporosis. In normal tissue (**left**), mineral deposits continually replenish mineral withdrawals. After the onset of osteoporosis (**right**), replacements can't keep pace with withdrawals. The tissue gradually erodes, and bones become progressively hollow and brittle.

replaced by bone tissue that gains strength by remodeling as it is "stressed" by a person's gradually increasing physical activity during healing.

As we age, bone tissue may break down faster than it is renewed. This progressive bone deterioration is called *osteoporosis* (Figure 5.5), and when it occurs the backbone, pelvis (hip bones), and other bones lose mass. The spine can collapse and curve so much that the rib cage lowers (a condition called *lordosis*), and internal organs become crowded. Osteoporosis is especially common in older women, although men can be affected, too. Deficiencies of calcium and sex hormones, eating too much protein, smoking, and a sedentary lifestyle all may contribute to osteoporosis. On the other hand, getting plenty of exercise (to stimulate bone deposits) and taking in plenty of calcium can help minimize bone loss.

The Skeleton: A Preview

A fully formed human skeleton has 206 bones, which grow by way of remodeling until a person is about twenty. The bones are organized into an **axial skeleton** and an **appendicular skeleton** (Figure 5.6). Straplike **ligaments** connect them at joints. Ligaments are composed of dense, regular connective tissue with many elastic fibers, so it is stretchable and resilient. **Tendons** are cords or straps that attach muscles to bones or to other muscles. In contrast to ligaments, tendons are built of dense, regular connective tissue packed with collagen—which makes it strong.

Bones grow and become strong by way of remodeling, in which osteoblasts deposit bone and osteoclasts break it down. The fully formed human skeleton consists of 206 bones, in axial and appendicular divisions.

Further reading: Student Guide to InfoTrac on web site →

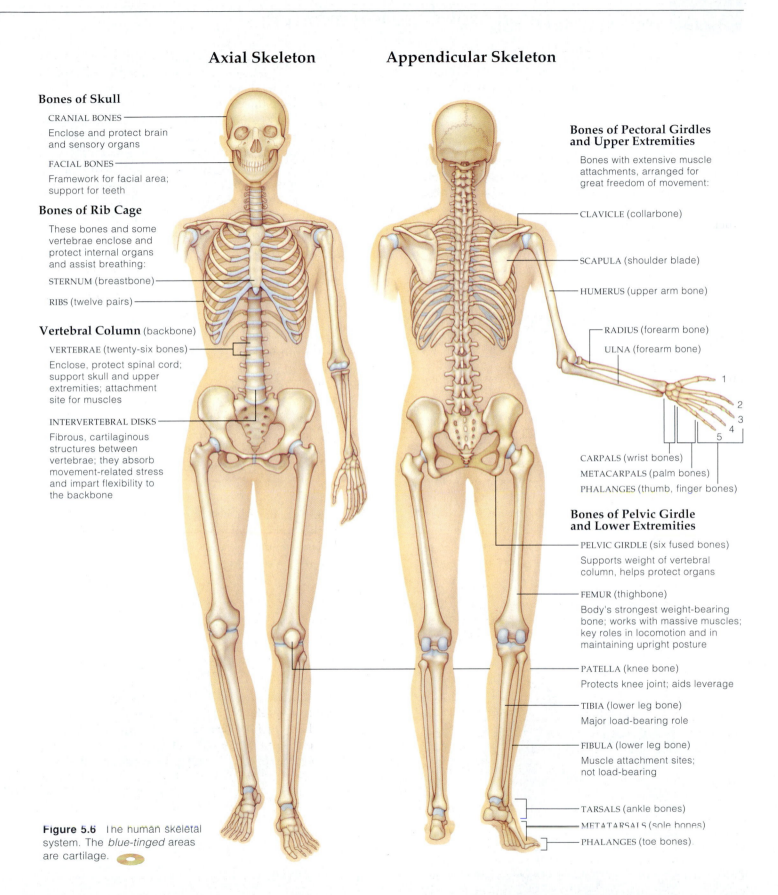

Axial Skeleton

Appendicular Skeleton

Bones of Skull

CRANIAL BONES

Enclose and protect brain
and sensory organs

FACIAL BONES

Framework for facial area;
support for teeth

Bones of Rib Cage

These bones and some
vertebrae enclose and
protect internal organs
and assist breathing:

STERNUM (breastbone)

RIBS (twelve pairs)

Vertebral Column (backbone)

VERTEBRAE (twenty-six bones)

Enclose, protect spinal cord;
support skull and upper
extremities; attachment
site for muscles

INTERVERTEBRAL DISKS

Fibrous, cartilaginous
structures between
vertebrae; they absorb
movement-related stress
and impart flexibility to
the backbone

Bones of Pectoral Girdles and Upper Extremities

Bones with extensive muscle
attachments, arranged for
great freedom of movement:

CLAVICLE (collarbone)

SCAPULA (shoulder blade)

HUMERUS (upper arm bone)

RADIUS (forearm bone)

ULNA (forearm bone)

1
2
3
4
5

CARPALS (wrist bones)

METACARPALS (palm bones)

PHALANGES (thumb, finger bones)

Bones of Pelvic Girdle and Lower Extremities

PELVIC GIRDLE (six fused bones)

Supports weight of vertebral
column, helps protect organs

FEMUR (thighbone)

Body's strongest weight-bearing
bone; works with massive muscles;
key roles in locomotion and in
maintaining upright posture

PATELLA (knee bone)

Protects knee joint; aids leverage

TIBIA (lower leg bone)

Major load-bearing role

FIBULA (lower leg bone)

Muscle attachment sites;
not load-bearing

TARSALS (ankle bones)

METATARSALS (sole bones)

PHALANGES (toe bones)

Figure 5.6 The human skeletal
system. The *blue-tinged* areas
are cartilage.

THE AXIAL SKELETON

The axial skeleton is made up of bones that roughly form the body's "head to toe" vertical axis. These bones include the skull, vertebral column (backbone), ribs, and sternum (the breastbone).

The Skull

You might be surprised to learn that your **skull** consists of more than two dozen bones. These bones are divided into several groupings, and while many of them are traditionally called by complex-sounding names derived from Latin, their roles are much simpler to grasp. For example, one grouping, the "cranial vault," or **brain case**, includes eight bones that together surround and protect your all-important brain. As Figure 5.7a shows, the *frontal bone* makes up the forehead and upper ridges of the eye sockets. It contains air spaces, called **sinuses**, that are lined with mucous membrane. Sinuses make the skull lighter, which means there is less weight for the backbone and neck muscles to support. But passages link them to the upper respiratory tract—and their ability to produce mucus can mean misery for anyone who has a head cold or pollen allergies. Bacterial infections in the nasal passages can spread to the sinuses, causing *sinusitis*. Figure 5.7c shows sinuses in the cranial and facial bones.

Temporal bones form the lower sides of the cranium and surround the ear canals. Each canal is a tunnel that leads from the outside to the middle and inner ear. Inside the middle ear are the tiny bones that function in hearing (Chapter 11). On either side of your head, just in front of each temporal bone, a *sphenoid bone* extends inward to form part of the inner eye socket. The *ethmoid bone* also

contributes to the inner socket and helps support the nose. A pair of *parietal bones* above and behind the temporal bones form a large part of the skull; they sweep upward and meet at the top of the head. An *occipital bone* forms the back and base of the skull. This bone also encloses a large opening, the *foramen magnum* ("large hole"). Here, the spinal cord emerges from the base of the brain and enters the spinal column (Figure 5.7b). Quite a few passageways run through and between various skull bones for nerves and blood vessels, especially at the base of the skull. For instance, the jugular veins, which carry blood leaving the brain, pass through openings between the occipital bone and each temporal bone.

Facial Bones

Figure 5.7 also shows facial bones, many of which you can easily feel with your fingers. The largest is your lower jaw, or **mandible**. The upper jaw consists of two *maxillary bones*. Two *zygomatic bones* form the middle of the protuberances we call "cheekbones" and the outer parts of the eye sockets. A small, flattened *lacrimal bone* fills out the inner eye socket. Tear ducts pass between this bone and the maxillary bones and drain into the nasal cavity—one reason why your nose runs when you cry. Tooth sockets in the upper and lower jaws also contain the teeth.

Palatine bones make up part of the floor and side wall of the nasal cavity. (Extensions of these bones, together with the maxillary bones, form the back of the hard palate, the "roof" of your mouth.) A *vomer bone* forms part of the nasal septum, a thin "wall" that divides the nasal cavity into two sections.

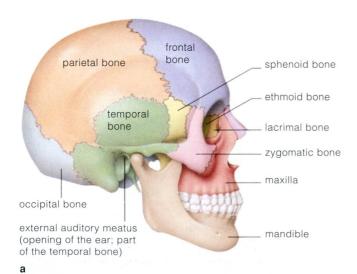

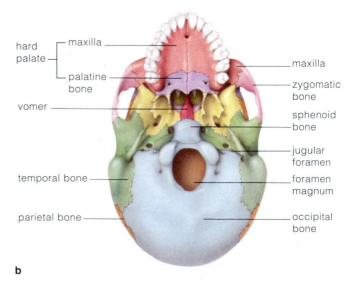

Figure 5.7 The skull. (**a**) The irregular junctions between different bones are called sutures. (**b**) An "inferior," or bottom-up, view of the skull. The large foramen magnum is situated atop the uppermost cervical vertebra. (**c**) Sinuses in bones associated with the nasal cavity.

Vertebral Column: The Backbone

The flexible, curved human vertebral column—your backbone or spine—extends from the base of the skull to the hipbones (pelvic girdle). This arrangement transmits the weight of a person's torso to the lower limbs. The vertebrae are stacked one atop the other. They have bony projections that form a protected channel for the delicate spinal cord. As sketched in Figure 5.8, humans have seven *cervical* vertebrae in the neck, twelve *thoracic* vertebrae in the chest area, and five *lumbar* vertebrae in the lower back. Counting these, there are thirty-three vertebrae, in all. In the course of human evolution, five other vertebrae have become fused to form the sacrum, and another four have become fused to form the coccyx, or "tailbone."

Roughly a quarter of your spine's length consists of **intervertebral disks**—compressible pads of fibrocartilage that are sandwiched between vertebrae. The disks are shock absorbers and flex points. They are thickest between cervical vertebrae and between lumbar vertebrae. Severe or rapid shocks, as well as changes due to aging, can cause a disk to "slip out," or *herniate*. If the slipped disk ruptures, its jellylike core material may squeeze out, making matters worse. And if the changes compress neighboring nerves or the spinal cord, the result can be excruciating pain and the loss of mobility that often comes with it. Depending on the situation, treatment can range from bed rest and use of painkilling drugs to surgery.

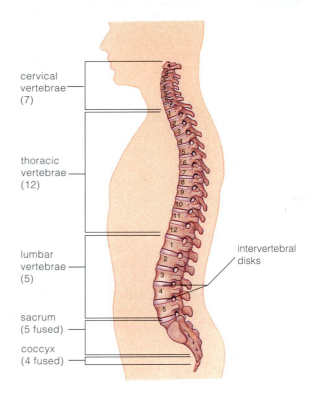

cervical vertebrae (7)

thoracic vertebrae (12)

lumbar vertebrae (5)

sacrum (5 fused)

coccyx (4 fused)

intervertebral disks

Figure 5.8 Side view of the vertebral column or backbone. The cranium balances on the column's uppermost vertebra.

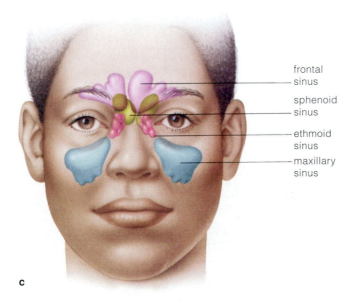

frontal sinus

sphenoid sinus

ethmoid sinus

maxillary sinus

c

The Ribs and Sternum

In addition to protecting the spinal cord, absorbing shocks, and providing flexibility, the vertebral column also serves as an attachment point for twelve pairs of **ribs**, which in turn function as a scaffold for the body cavity of the upper torso. The upper ribs also attach to the paddle-shaped **sternum** (see Figure 5.6). As you will read in later chapters, this "rib cage" helps protect the lungs, heart, and other internal organs and is vitally important in breathing.

While the axial skeleton provides basic body support and helps protect internal organs, many movements depend on interactions of skeletal muscles with the bones of the appendicular skeleton—the skeletal component we turn to next.

Bones of the axial skeleton make up the body's vertical axis. They include the skull (and facial bones), the vertebral column, and the ribs and sternum.

Intervertebral disks between the vertebrae absorb shocks and serve as flex points.

THE APPENDICULAR SKELETON

"Append" means to hang, and the appendicular skeleton includes the bones of body parts that we sometimes think of as dangling from the main body frame: arms, hands, legs, and feet. It also includes a pectoral girdle at each shoulder and the pelvic girdle at the hips.

The Pectoral Girdle and Upper Limbs

Each **pectoral girdle** (Figure 5.9) has a large, flat shoulder blade—a **scapula**—and a long, slender collarbone, or **clavicle,** that connects to the breastbone (sternum). The rounded shoulder end of the **humerus**, the long bone of the upper arm, fits into an open socket in the scapula. The human arm is capable of remarkably versatile movements; it can swing in wide circles and back and forth, lift objects, or tug on a rope. We owe this freedom of movement to the fact that the pectoral girdles and upper limbs are loosely attached to the rest of the body by muscles. Although this arrangement is sturdy enough under normal conditions, it is vulnerable to strong blows.

Fall on an outstretched arm and you might fracture your clavicle or dislocate your shoulder. The collarbone is the bone most frequently broken.

Each of your upper limbs includes some thirty separate bones. The humerus connects with two bones of the forearm—the **radius** (on the thumb side) and the **ulna** (on the "pinky finger" side). The upper end of the ulna joins the lower end of the humerus to form the elbow joint. The bony bump sometimes (mistakenly) called the "wristbone" is the lower end of the ulna.

The radius and ulna together join the hand at the wrist joint, where they meet eight small, curved *carpal* bones. Ligaments attach these bones to the long bones. Blood vessels, nerves, and tendons pass in sheaths over the wrist; when a blow, constant pressure, or repetitive movement (such as prolonged typing) damages these tendons, the result can be a painful disorder called *carpal tunnel syndrome*. The bones of the hand, the five *metacarpals*, end at the knuckles. *Phalanges* are the bones of the fingers.

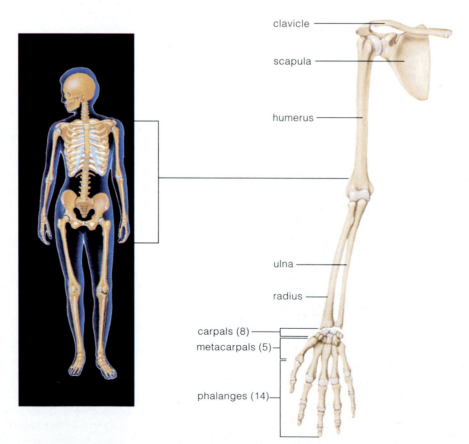

clavicle

scapula

humerus

ulna

radius

carpals (8)

metacarpals (5)

phalanges (14)

Figure 5.9 Bones of the pectoral girdle, the arm, and the hand.

Further reading: Student Guide to InfoTrac on web site →

The Pelvic Girdle and Lower Limbs

For most of us, our shoulders and arms are much more flexible than are our hips and legs. Why? Although there are similarities in the basic "design" of both girdles, this lower part of the appendicular skeleton is adapted to bear the body's entire weight when we are standing. The **pelvic girdle** (Figure 5.10) is much more massive than the combined pectoral girdles, and it is attached to the axial skeleton by extremely strong ligaments. It forms an open basin: A pair of *coxal bones* attach to the lower spine (sacrum) in back, then curve forward and meet at the *pubic arch*. ("Hipbones" are actually the upper *iliac* regions of the coxal bones.) This combined structure is the *pelvis*. In females the pelvis is broader than in males, and it shows other structural differences that are evolutionary adaptations for the function of childbearing. A forensic scientist or paleontologist examining skeletal remains can easily establish the sex of the individual if a pelvis is present.

The legs contain the body's largest bones. In terms of length, the thighbone, or **femur,** ranks number one. It is also extremely strong. When you run or jump, your femurs routinely withstand stresses of several tons per square inch (aided by contracting leg muscles). The femur's ball-like upper end fits snugly into a deep socket in the coxal (hip) bone. The other end connects with one of the bones of the lower leg, the thick, load-bearing *tibia* on the inner (big toe) side. A slender *fibula* parallels the tibia on the outer (little toe) side. The tibia is your shinbone. A triangular kneecap, the patella, helps protect the knee joint. In spite of this protection, however, knees are among the joints most often damaged by athletes, both amateur and professional.

Bones of the ankle and foot correspond closely with those of the wrist and hand. *Tarsal* bones make up the ankle and heel, and the foot contains five long bones, the *metatarsals*. The largest metatarsal, leading to the big toe, supports a great deal of body weight and is thicker and stronger than the others. Like fingers, the toes contain phalanges.

The appendicular skeleton includes bones of the limbs, a pectoral girdle at the shoulders, and a pelvic girdle at the hips.

The thighbone (femur) is the largest bone in the human body and also one of the strongest. The wrists and hands and ankles and feet have closely corresponding sets of bones known respectively as carpals and metacarpals and tarsals and metatarsals.

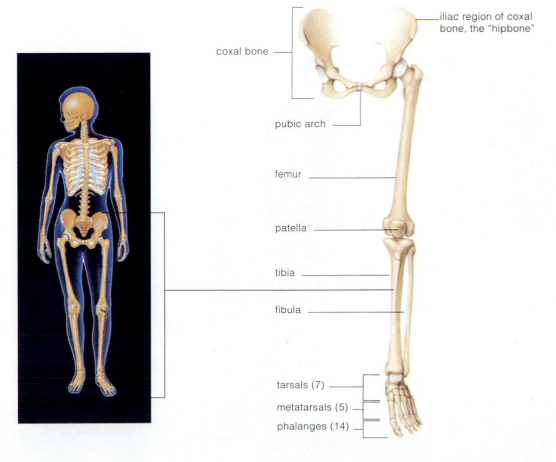

Figure 5.10 Bones of the pelvic girdle, the leg, and the foot.

5.5 JOINTS

Joints are areas of contact or near-contact between bones. There are various types, and each one has some form of connective tissue that bridges the gap between bones. In the most common type of joint, called a **synovial joint**, adjoining bones are separated by a cavity. The articulating ends of the bones are covered with a cushioning layer of cartilage; they are stabilized by ligaments. A capsule of dense connective tissue surrounds the bones of a synovial joint. Cells that line the interior of the capsule secrete a lubricating *synovial fluid* into the joint cavity. (See Figure 4.12.)

Synovial joints are freely movable. Examples are the ball-and-socket joints at the hips. As you know from personal experience, such joints are capable of a wide range of movements: They can rotate and move in different planes—for instance, up-down or side-to-side. Hingelike synovial joints such as the knee and elbow are limited to simple flexing and extending (straightening), like a door hinge. Figure 5.11 shows some of the ways body parts can move at joints.

As a person ages, the cartilage covering the bone ends of freely movable joints may degenerate into *osteoarthritis*. Often, the arthritic joint becomes painfully inflamed. Another condition, *rheumatoid arthritis,* is a degenerative disorder that results when a person's immune system malfunctions and mounts an attack against tissues in the

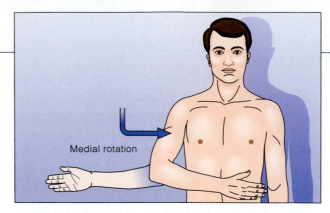

Medial rotation

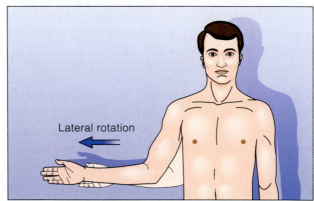

Lateral rotation

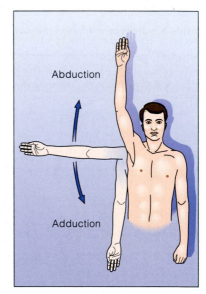

Abduction

Adduction

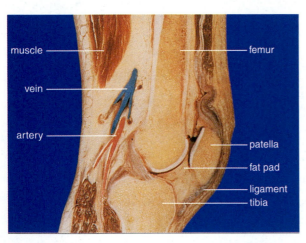

muscle
vein
artery
femur
patella
fat pad
ligament
tibia

Figure 5.11 Ways body parts can move at joints. The synovial joint at the shoulder permits the greatest range of movement. The photograph shows the anatomy of the knee.

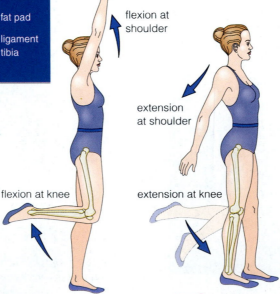

flexion at shoulder

extension at shoulder

flexion at knee

extension at knee

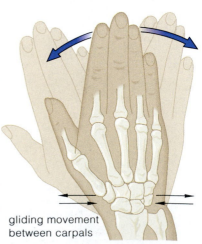

gliding movement between carpals

Further reading: Student Guide to InfoTrac on web site →

affected joint. Then, the synovial membrane becomes inflamed and thickens, cartilage is eroded away, and the bones fall out of proper alignment. The bone ends may eventually fuse together. As described in *Science Comes to Life* at the right, surgeons now routinely replace joints such as hips and knees when they become seriously damaged (Figure 5.12).

In **cartilaginous joints**, cartilage fills the space between bones. Then, only slight movement is possible. Such joints occur between vertebrae and between the breastbone and some of the ribs.

A **fibrous joint** has no cavity, and fibrous connective tissue unites the bones. For instance, fibrous joints hold your teeth in their sockets, and they loosely connect the flat skull bones of a fetus. During childbirth, the loose connections allow the bones to slide over each other and so prevent skull fractures. A newborn baby's skull still has fibrous joints and soft areas called fontanels. During childhood, the joints harden into *sutures*. Much later in life the skull bones may become completely fused.

Joints are areas of contact or near-contact between bones.

The major types of joints include freely movable synovial joints, such as the ball-and-socket joints at the hips, cartilaginous joints, and fibrous joints.

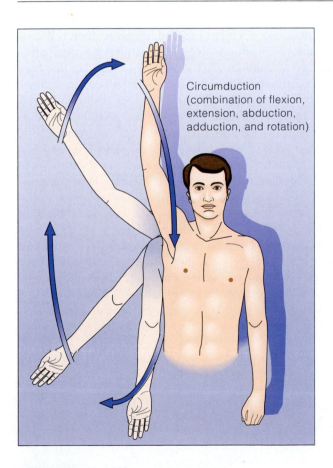

Circumduction
(combination of flexion, extension, abduction, adduction, and rotation)

Knees, hips, fingers, shoulders—physicians are replacing all these joints in procedures that might have seemed daring only a few decades ago. Some hospitals have entire wings devoted to joint replacement, with highly skilled staff wise in the ways of helping patients adjust to their "replacement parts." Improved techniques and postoperative care are one reason for the blossoming of joint replacements. An experienced surgeon can "install" a new hip joint in 45 minutes, a new knee in about two hours. And many patients today are encouraged to begin using their new joint within a day or two after surgery. For instance, under the watchful eye of a physical therapist, a patient with a new hip may be walking and climbing stairs 48 hours after the procedure.

Another factor is improved materials, which have dramatically increased the number of years a replacement joint—called a *prosthesis*—can be expected to last. The early method of securing a new joint with an adhesive has gone by the boards. Now, the surgeon implants a prosthesis of metal (such as titanium) or a combination of metal and plastic. The surface of the new joint has small pits. As soon as a patient begins using the joint, including using it to support body weight, this puts mechanical stress on the bone ends. The stress in turn "kick starts" remodeling by the patient's osteoblasts (Section 5.2) and the bone ends grow into the pits in the replacement joint, making a strong bond between joint and bone. With proper care and normal use, a new knee or hip can be expected to last twenty years or more.

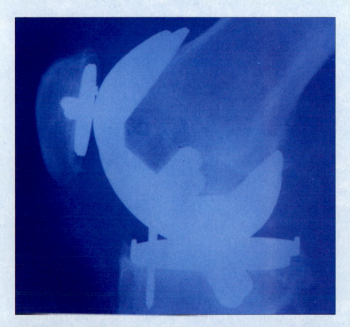

Figure 5.12 An artificial knee joint. Notice how the end of the patient's femur has been fitted around a projection of the new joint (center), while another projection has been fitted into the tibia below. The hatlike projection at the upper left attaches to the patella—the kneecap.

Figure 4.6 on page 72 introduced the three types of muscle tissue: skeletal, cardiac, and smooth. You may have noticed that skeletal and cardiac muscle look very different from smooth muscle, and each type of muscle does have its specialized function. However, all muscle cells are specialized to generate force—that is, to shorten—by contracting. After a muscle contracts, it can relax and lengthen. These muscle activities require energy from ATP. Later, in Section 5.8, we'll look at the mechanisms involved.

When we speak of the body's "muscular system," we're talking only about skeletal muscle. Only **skeletal muscle** interacts with the skeleton to move the body, its limbs, or other parts. The movements range from delicate adjustments that help us keep our balance to the "cool moves" you might execute on a dance floor. Our skeletal muscles also help stabilize joints between bones, as you'll read shortly. And while we won't focus on it in this chapter, our muscles also generate body heat.

How Skeletal Muscles and Bones Interact

You have more than 600 skeletal muscles, and each one helps produce some kind of body movement. In general, one end of a muscle, the *origin*, is attached to a bone that stays relatively motionless during a movement. The other end of the muscle, called the *insertion*, is attached to the bone that moves most. When a skeletal muscle contracts, it pulls on the bones to which it is attached. Said another way, the skeleton and the muscles attached to it are like a system of levers in which bones (rigid rods) move about joints (fixed points). Most joints have nearby muscle attachments. This means that a muscle only needs to contract a short distance to produce a major movement of some body part.

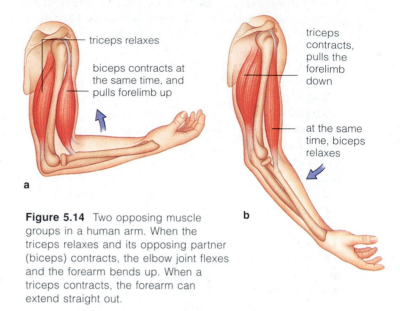

Figure 5.14 Two opposing muscle groups in a human arm. When the triceps relaxes and its opposing partner (biceps) contracts, the elbow joint flexes and the forearm bends up. When a triceps contracts, the forearm can extend straight out.

triceps relaxes

biceps contracts at the same time, and pulls forelimb up

a

triceps contracts, pulls the forelimb down

at the same time, biceps relaxes

b

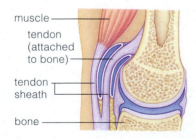

Figure 5.13 A tendon sheath. This is not the same as a bursa, a small sac also filled with synovial fluid. Bursae are cushions *between* bone and skin or tendons.

muscle

tendon (attached to bone)

tendon sheath

bone

A skeletal muscle contains bundles of hundreds to thousands of muscle cells, which look like long, striped fibers. (The term "muscle fiber" is sometimes used as another name for a muscle cell.) Connective tissue bundles the muscle cells together and extends past them to form tendons. Each tendon—which, remember, is a cord or strap of dense connective tissue—attaches some muscle to bone. Tendons are a bit like "duct tape" in that they increase the stability of joints by helping keep the adjoining bones properly aligned. Tendons commonly rub directly against bones. However, they slide inside fluid-filled sheaths that help reduce the friction. Your knees, wrists, and finger joints have such sheaths (Figure 5.13).

Many muscles are arranged as pairs or groups. Some work together (that is, *synergistically*) to promote the same movement. Others work in opposition (*antagonistically*) so that the action of one opposes or reverses the action of the other. Figure 5.14 shows an antagonistic pair of muscles, the biceps and triceps of the arm. Try extending your right arm forward, then place your left hand over the biceps in the upper right arm and slowly "bend the elbow." Can you feel the biceps contract? When the biceps relaxes and its partner (the triceps) contracts, your arm extends and straightens. Such coordinated action comes partly from *reciprocal innervation* by nerves from the spinal cord. When one muscle group is stimulated, no signals are sent to the opposing group, so it does *not* contract.

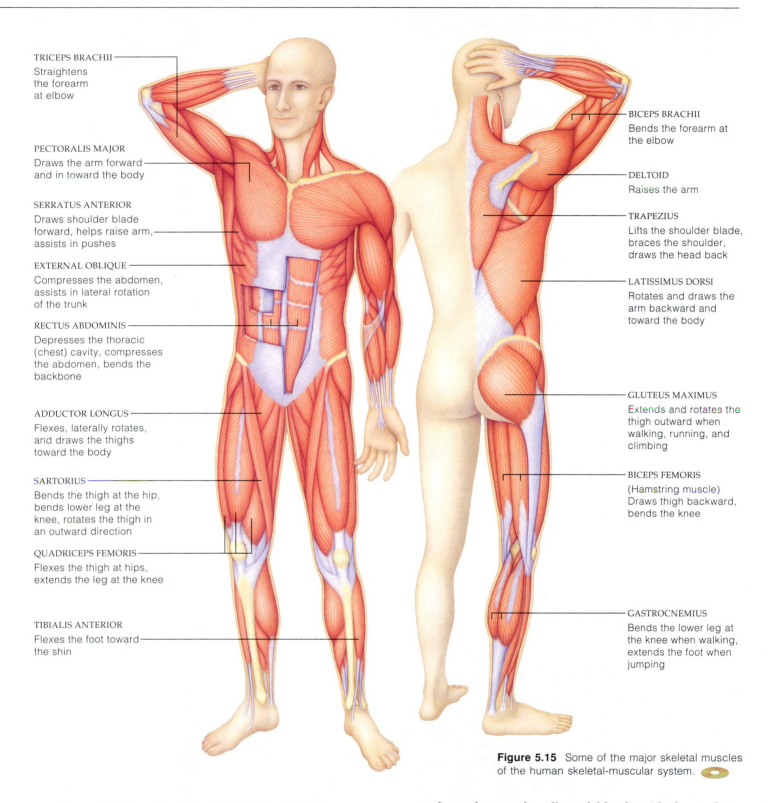

TRICEPS BRACHII
Straightens the forearm at elbow

PECTORALIS MAJOR
Draws the arm forward and in toward the body

SERRATUS ANTERIOR
Draws shoulder blade forward, helps raise arm, assists in pushes

EXTERNAL OBLIQUE
Compresses the abdomen, assists in lateral rotation of the trunk

RECTUS ABDOMINIS
Depresses the thoracic (chest) cavity, compresses the abdomen, bends the backbone

ADDUCTOR LONGUS
Flexes, laterally rotates, and draws the thighs toward the body

SARTORIUS
Bends the thigh at the hip, bends lower leg at the knee, rotates the thigh in an outward direction

QUADRICEPS FEMORIS
Flexes the thigh at hips, extends the leg at the knee

TIBIALIS ANTERIOR
Flexes the foot toward the shin

BICEPS BRACHII
Bends the forearm at the elbow

DELTOID
Raises the arm

TRAPEZIUS
Lifts the shoulder blade, braces the shoulder, draws the head back

LATISSIMUS DORSI
Rotates and draws the arm backward and toward the body

GLUTEUS MAXIMUS
Extends and rotates the thigh outward when walking, running, and climbing

BICEPS FEMORIS
(Hamstring muscle) Draws thigh backward, bends the knee

GASTROCNEMIUS
Bends the lower leg at the knee when walking, extends the foot when jumping

Figure 5.15 Some of the major skeletal muscles of the human skeletal-muscular system.

Figure 5.15 shows the body's major skeletal muscles. Some are superficial, others deep in the body wall. Some, such as facial muscles, attach to the skin. The trunk has muscles of the thorax (chest), backbone, abdominal wall, and pelvic cavity. Other groups of muscles attach to upper and lower limbs. Fibrous connective tissue encloses the muscle cells and blends with the tendons that attach the muscle to bone.

Skeletal muscles transmit contractile force to bones and make them move. Tendons strap skeletal muscles to bone.

5.8 A CLOSER LOOK AT MUSCLES

Bones move—they are pulled in some direction—when the skeletal muscles attached to them shorten. When a skeletal muscle shortens, the individual muscle cells in it shorten. And when a muscle cell shortens, many units of contraction inside that cell are shortening. Each of these basic units of contraction is a **sarcomere**.

Figure 5.16 shows how bundles of cells in a skeletal muscle run parallel with the muscle. In each muscle cell are **myofibrils,** threadlike structures bundled together in parallel. Myofibrils contain myoglobin, a reddish pigment similar to hemoglobin. Like hemoglobin, myoglobin contains iron groups that attract and bind oxygen.

Each myofibril is divided into many sarcomeres, which are arranged one after the other along the length of the myofibril. Dark bands called Z lines mark the two ends of each sarcomere. A sarcomere contains many filaments, oriented parallel with its long axis like pencils in a box. Certain differences in length and positioning make skeletal (and cardiac) muscle look striped ("striated") under the microscope. Some filaments are thin, others are thick. Each thin filament is like two strands of pearls, twisted together. The "pearls" are molecules of **actin**, a globular protein with the ability to contract:

one actin molecule   portion of one thin filament

Other proteins (coded green) are near actin's surface grooves. Each thick filament is made of molecules of **myosin**, another contractile protein. Each myosin molecule has a long tail and a double head that projects from the filament's surface:

one myosin molecule   part of one thick filament

Thus myofibrils, muscle cells, and muscle bundles all run in the same direction. What is the function of this consistent, parallel orientation? It focuses the force of muscle contraction onto a bone in a particular direction.

How do sarcomeres shorten to contract a muscle? The answer starts in each sarcomere, with sliding and pulling interactions between Z lines. Actin filaments extend from each Z line to a sarcomere's center. A set of myosin filaments partly overlaps them, but does not extend all the way to the Z lines (Figure 5.16). When a muscle contracts, actin filaments from opposite sides of the sarcomere slide over myosin filaments, which don't move. And so the sarcomere shortens.

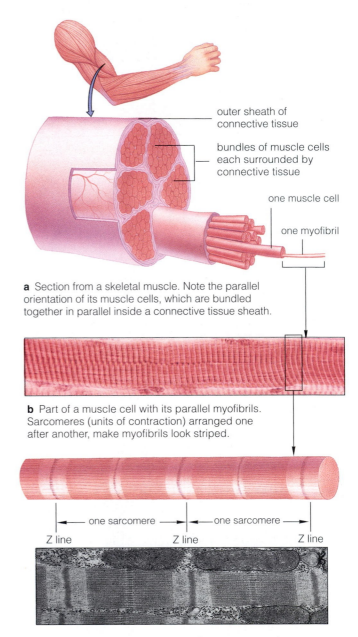

a Section from a skeletal muscle. Note the parallel orientation of its muscle cells, which are bundled together in parallel inside a connective tissue sheath.

b Part of a muscle cell with its parallel myofibrils. Sarcomeres (units of contraction) arranged one after another, make myofibrils look striped.

c Diagram and electron micrograph of two sarcomeres from one myofibril. This view shows how two portions called Z lines define the two ends of each sarcomere. Mitochondria (the oval organelles next to the sarcomeres) provide ATP energy for muscle action.

Figure 5.16 Components of a skeletal muscle.

Further reading: Student Guide to InfoTrac on web site

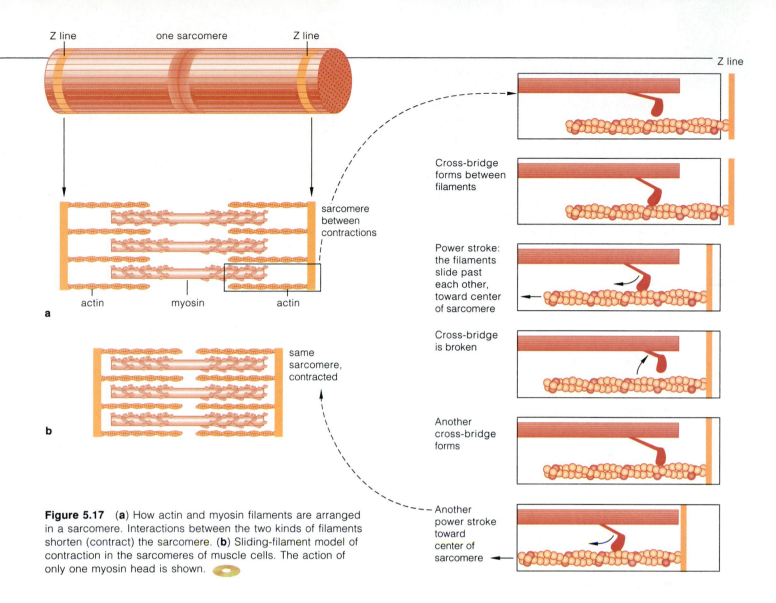

Figure 5.17 (**a**) How actin and myosin filaments are arranged in a sarcomere. Interactions between the two kinds of filaments shorten (contract) the sarcomere. (**b**) Sliding-filament model of contraction in the sarcomeres of muscle cells. The action of only one myosin head is shown.

Labels in figure (a): Z line, one sarcomere, Z line, sarcomere between contractions, actin, myosin, actin. (b) same sarcomere, contracted.

Labels in figure at right: Z line; Cross-bridge forms between filaments; Power stroke: the filaments slide past each other, toward center of sarcomere; Cross-bridge is broken; Another cross-bridge forms; Another power stroke toward center of sarcomere.

Biologists studying this interplay between myosin and actin named the mechanism the **sliding-filament model** of contraction. An important feature of this model is that myosin and actin interact through the formation of cross-bridges. A **cross-bridge** is an attachment between a myosin head and a binding site on actin. A cross-bridge forms when a myosin head attaches to an adjacent actin filament.

Interactions between myosin and actin filaments are cyclic and they are powered by ATP energy. In each cycle, a cross-bridge forms between a myosin head and a binding site on actin. The linked myosin and actin filaments move; then the myosin heads detach from the actin (Figure 5.17*a*, right). We can analyze the cycle beginning when enzyme action splits an ATP molecule into ADP and phosphate, briefly creating a "high-energy" form of myosin. In this state, the myosin binds actin. (As you will read, calcium ions released in a muscle cell provide the actual "start signal" for the attachment of myosin to actin.) After myosin binds actin, the energy released from

splitting ATP drives a short "power stroke" in which the myosin head tilts toward the center of the sarcomere. This stroke also pulls the attached actin filaments toward the center of the sarcomere, which shortens. Then the cycle repeats. Energy from ATP causes each myosin head to detach, the attachment steps occur with the next actin binding site in line, and the actin filaments move a bit more. A single contraction takes a series of power strokes.

In the absence of ATP, myosin cross-bridges cannot detach. When a person dies, ATP production stops, cross-bridges stay locked in place, and skeletal muscles become rigid. This *rigor mortis* lasts up to 60 hours after death.

A skeletal muscle shortens through the combined decreases in length of its sarcomeres, the basic units of contraction.

ATP-driven interactions between myosin and actin filaments shorten the many sarcomeres of a muscle cell. Together, these interactions account for contraction of the muscle cell.

CONTROL OF MUSCLE CONTRACTION

Skeletal muscle cells contract under commands from the nervous system. Motor neurons of the nervous system deliver signals that stimulate contraction, and you will read how that happens in Chapter 11. For the moment, we are interested in how commands from the nervous system stimulate a muscle cell to contract. Figure 5.18 summarizes much of the following discussion.

When stimulation arrives at a muscle cell, signals spread. Moving rapidly, the signals reach small, tubelike extensions of the cell's plasma membrane. The small tubes, called *T tubules*, connect with a system of membrane-bound chambers that lace around the cell's myofibrils;

section from spinal cord

motor neuron

a Signals from the nervous system travel along spinal cord, down motor neuron.

b Endings of motor neuron terminate next to a muscle cell.

section from a skeletal muscle

part of one muscle cell

c Signals travel along muscle cell's plasma membrane to sarcoplasmic reticulum around cell's myofibrils.

Figure 5.18 Pathway for signals from the nervous system that stimulate contraction of skeletal muscle.

you can see an artist's rendition of this arrangement in part *d* of Figure 5.18. The membrane system, called the **sarcoplasmic reticulum** (SR), is a modified version of ER, the endoplasmic reticulum described in Chapter 3. SR takes up and releases calcium ions. When a nerve impulse arrives, it triggers the release of calcium ions from the SR. The ions diffuse into myofibrils and reach actin filaments. That sets the stage for contraction.

Molecules of two proteins, *troponin* and *tropomyosin*, nestle along the surface of actin filaments (Figure 5.19*a* and *b*). When incoming calcium binds to troponin, the binding site on the actin filament is uncovered. This allows myosin cross-bridges to attach to the site, and the cycle described earlier continues. When new nervous system signals shut off, the SR membranes take up calcium, an active process that is powered by ATP. Now the binding site on actin once more is covered up, myosin can't bind to actin, and the muscle cell relaxes.

Junctions between Nerves and Muscles

The nerve impulses that stimulate a skeletal muscle cell to contract arrive at sites called **neuromuscular junctions**. At these sites, the branched endings of certain motor neuron extensions (its *axons*) come very close to the muscle cell membranes, as in Figure 5.18*b* and Figure 5.21. The neuron endings don't quite touch a muscle cell; there is a narrow gap between them (a type of *synapse*). The neurons "send" their messages across the gap to muscle cells by way of a chemical messenger, a *neurotransmitter*,

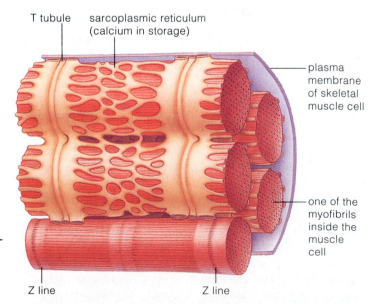

T tubule   sarcoplasmic reticulum (calcium in storage)

plasma membrane of skeletal muscle cell

one of the myofibrils inside the muscle cell

Z line   Z line

d Signals trigger the release of calcium ions from sarcoplasmic reticulum threading among the myofibrils. The arrival of calcium allows actin and myosin filaments in the myofibrils to interact and bring about contraction.

Further reading: Student Guide to InfoTrac on web site ➔

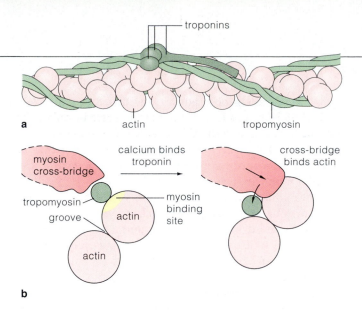

a

b

Figure 5.19 Arrangement of troponin, tropomyosin, and actin filaments in skeletal muscle cells. When calcium binds with a troponin, tropomyosin moves away from the actin, exposing the cross-bridge binding sites.

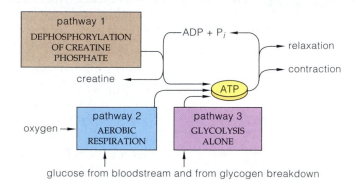

Figure 5.20 Three possible metabolic pathways by which ATP can form in muscles in response to physical exercise.

called acetylcholine (ACh). ACh can excite or inhibit muscle and gland cells throughout the body, as well as some cells in the brain and spinal cord. When the neuron is stimulated, calcium channels open in the plasma membrane of the endings, and calcium ions (Ca^{++}) from the extracellular fluid flow inside. This causes vesicles in the axon ending to release ACh. When ACh binds to receptors on the muscle cell membrane, it has an excitatory effect that may set in motion the events that cause the muscle cell to contract. Curare, a poison extracted from a South American shrub, blocks the binding of ACh by muscle cells. In so doing, it prevents contraction in muscles—including those required for breathing.

Like skeletal muscle, cardiac (heart) muscle also can respond to nerve impulses, but it also beats (contracts) spontaneously—that is, without outside stimulation, by way of a mechanism that is a property of the muscle itself (Chapter 7). Smooth muscle, the muscle in the walls of blood vessels and internal organs, responds to several control systems, including nerves, hormones, and spontaneous mechanisms.

Sources of Energy for Contraction

All cells require ATP, but only in muscle cells does the demand skyrocket so quickly. When a resting muscle cell is ordered to contract, phosphate donations from ATP must occur twenty to one hundred times faster. But a cell has only a small supply of ATP at the start of contractile activity. At such times, the cell forms ATP by a fast reaction. An enzyme simply transfers phosphate from **creatine phosphate**, an organic compound, to ADP. A cell has about five times as much creatine phosphate as ATP,

so this reaction is good for a few contractions. It also buys time for a relatively slower ATP-forming pathway to begin (Figure 5.20).

During prolonged, moderate exercise, the oxygen-requiring reactions of aerobic respiration provide most of the ATP for contraction. For about ten minutes, an active muscle cell taps its store of glycogen for glucose (the starting substrate). For the next half hour or so of sustained activity, it depends on glucose and fatty acids delivered by the bloodstream. For contractile activity longer than this, fatty acids are the main fuel source (for a refresher on this pathway, see Section 3.16).

What happens if you exercise so hard that your respiratory and circulatory systems can't deliver enough oxygen for aerobic respiration in some muscles? Then, glycolysis (which, recall, is *anaerobic*) will contribute more of the total ATP being formed. Since glycolysis doesn't fully break down a glucose molecule, the net ATP yield is small. But muscle cells use this metabolic route as long as glycogen stores continue to provide glucose—or at least until **muscle fatigue** sets in. This is a state in which a muscle cannot contract, even if it is being stimulated. Fatigue is probably due to the **oxygen debt** that results from prolonged muscle activity, when muscles use more ATP than aerobic respiration can deliver. Then they switch to glycolysis, as just described, which produces lactic acid as a by-product. Together with the already low ATP supply, the rising acidity in muscle tissue disrupts the ability of muscle cells to contract. Deep, rapid breathing helps repay the oxygen debt.

Commands from the nervous system initiate contractile activity in muscle cells. These signals trigger the formation of myosin cross-bridges, the first step in contraction. The availability of ATP in muscle cells affects whether and for how long they can contract.

PROPERTIES OF WHOLE MUSCLES

Muscle Tension

Whether a muscle cell actually shortens as cross-bridges form in its sarcomeres depends on the external forces acting on it. Collectively, the cross-bridges exert **muscle tension**. This is the name for a mechanical force that a contracting muscle exerts on an object, such as a bone. Opposing it is a load, either the weight of an object or gravity's pull on the muscle. Only when muscle tension exceeds the load does a stimulated muscle shorten.

An *isometrically* contracting muscle develops tension but does not shorten. It supports a load in a constant position, as when you hold a glass of lemonade in front of you. An *isotonically* contracting muscle shortens and moves a load. With *lengthening* contraction, though, an external load is greater than the muscle tension, so the muscle lengthens during the period of contraction. This happens to your leg muscles when you walk downstairs.

A muscle's tension relates to the formation of cross-bridges in its cells and the number of cells recruited into action. A skeletal muscle contains a large number of cells, but not all of them contract at the same time. Together, a motor neuron and the muscle cells under its control are called a **motor unit** (Figure 5.21).

The number of cells in motor units varies from perhaps three or four in the motor unit of an eye muscle up to several hundred in some motor units of a leg muscle. Small motor units provide the "fine-tuning" required for precise control of your muscles.

How Motor Units Function

By stimulating a motor unit with an electrical impulse, we can induce the sort of isometric contraction that would result from stimulation by a motor neuron. It takes a few

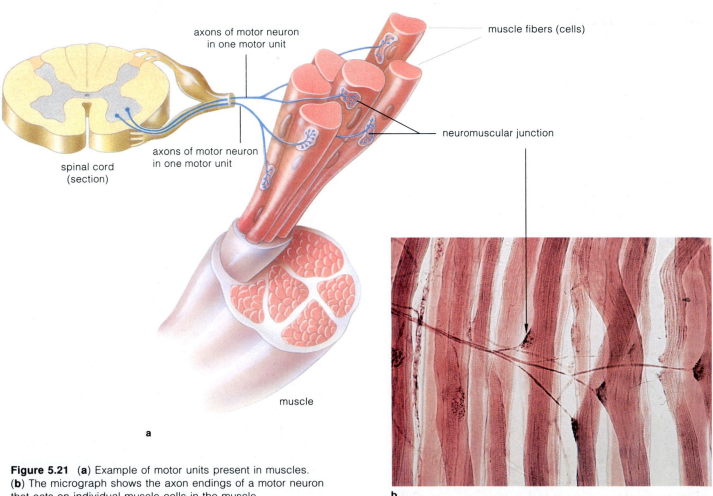

axons of motor neuron in one motor unit

muscle fibers (cells)

neuromuscular junction

axons of motor neuron in one motor unit

spinal cord (section)

muscle

a

b

Figure 5.21 (**a**) Example of motor units present in muscles. (**b**) The micrograph shows the axon endings of a motor neuron that acts on individual muscle cells in the muscle.

Further reading: Student Guide to InfoTrac on web site ➡

seconds for tension to increase, then it peaks and declines. This response is a **muscle twitch** (Figure 5.22). The duration of a twitch depends on the load and cell type. For example, fast-acting muscle cells rely on glycolysis (not efficient but fast) and use up their supply of ATP energy faster than slow-acting cells do. If a motor unit is stimulated again before a twitch response is completed, it twitches once more. The strength of the contraction depends on how far the twitch response has proceeded by the time the second signal arrives. The effect of the new contraction is added to that of the contraction already under way, a phenomenon called **temporal summation** (Figure 5.22c). As a result, with additional stimuli the strength of contraction increases.

Our muscles normally operate near or at maximum temporal summation, a condition called **tetany**. (In the disease *tetanus*, muscles remain contracted, possibly fatally, due to the effects of a bacterial toxin.) In order to postpone *muscle fatigue*, in which hardworking muscle cells outstrip their supply of ATP and temporarily become unable to respond to stimulation, sets of motor units alternate sustaining the contraction of a muscle.

Individual cells in a motor unit always contract according to an **all-or-none principle**: They either contract fully in response to stimulation, or they do not respond at all. If a muscle is contracting only weakly—as might happen in your forearm muscles when you pick up a pencil—it is because the nervous system is activating only a few motor units. In a stronger contraction—when you heft a stack of books, say—a larger number of motor units are activated. Even when muscles are relaxed, some of their motor units are contracted. This steady, low-level contracted state, called *muscle tone*, helps stabilize joints and maintain general muscle health.

"Fast" and "Slow" Muscle

Your body has two general types of skeletal muscle. "Slow" or "red" muscle appears crimson because its cells are packed with myoglobin—the red, oxygen-binding protein—and are served by larger numbers of the tiny blood vessels called capillaries. (Red muscle is the "dark meat" in chicken or turkey.) Red muscle contracts fairly slowly, but because its cells are well equipped to generate lots of ATP aerobically, the contractions can be sustained for a long time. For example, some muscles of the back and legs—called postural muscles because they aid body support—must contract for long periods when a person is standing. They have a high proportion of red muscle cells. By contrast, muscles of your hand have fewer capillaries and relatively more "fast" or "white" muscle cells, which have fewer mitochondria and less myoglobin. Fast muscle cannot sustain contractions for

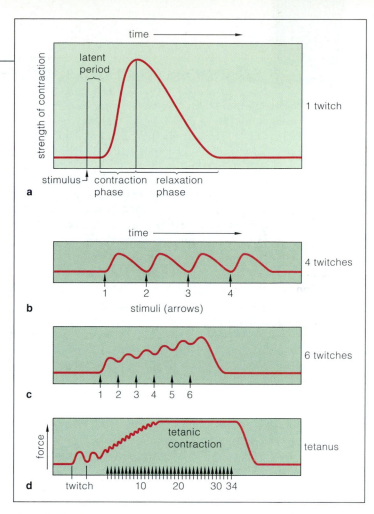

Figure 5.22 Recordings of twitches in artificially stimulated muscles. (**a**) A single twitch. (**b**) Two stimulations per second cause a series of twitches. (**c**) Six per second cause a summation of twitches. (**d**) About 20 per second cause a tetanic contraction.

long, but it can contract rapidly and powerfully for short periods.

When athletes train rigorously, one goal is to increase the relative size and contractile strength of fast or slow fibers in their muscles. A sprinter will benefit from larger, stronger fast muscle fibers in the thighs, while a distance swimmer will follow a regimen designed to increase the number of mitochondria in shoulder muscle cells. Even in the face of many negative research findings, some athletes resort to the synthetic hormones called anabolic steroids to rapidly build muscle mass, a practice this chapter's *Focus on Your Health* essay examines.

A motor neuron and the muscle cells under its control are called a motor unit.

A muscle twitch is a brief muscle contraction that occurs when a single, brief stimulus of a given strength activates a certain number of motor units.

A motor unit and its cells always contract according to an all-or-none principle.

Muscles normally operate at or near tetany, a state near or at maximum temporal summation.

5.11 MUSCLE MATTERS

Muscle makes up more than 40 percent of the human body by weight, and most of that is skeletal muscle. Muscle cells adapt to the activity demanded of them. When severe nerve damage or prolonged bed rest prevents a muscle from being used at all, the muscle will rapidly begin to waste away, or *atrophy*. Over time, affected muscles can lose up to three-fourths of their mass, with a corresponding loss of strength. More commonly, the skeletal muscles of a sedentary person stay basically healthy but cannot respond to physical demands in the same way that well-worked muscles can.

Exercise: Making the Most of Your Muscles

One of the best ways to maintain or improve the work capacity of your muscles is to **exercise** them—that is, to increase the level of contractile activity. To increase muscle endurance, nothing beats regular *aerobic exercise*—activities such as walking, biking, organized aerobics classes, jogging, and swimming (Figure 5.23). Aerobic exercise works muscles at a rate at which the body can keep them supplied with oxygen, and it has the following effects on muscle cells:

1. There is an increase in the number and the size of mitochondria, the organelles that make ATP.

2. The number of blood capillaries supplying muscle tissue increases. This increased blood supply brings more oxygen and nutrients to the muscle tissue and removes metabolic wastes more efficiently.

3. There is more of the oxygen-binding pigment myoglobin in muscle tissues.

Together, these changes produce muscles that are more efficient metabolically and can work longer without becoming fatigued. By contrast, *strength training* involves intense, short-duration exercise, such as weight lifting. It affects fast muscle cells, which form more myofibrils and more enzymes of glycolysis. These changes translate into whole muscles that are larger and stronger (Figure 5.24), but such bulging muscles don't have much endurance. They fatigue rapidly.

Starting at about age 30, a person's muscle tension gradually begins to decrease. As a practical matter, this means that, once you enter your third decade of life, you may exercise just as long and intensely as a younger person, but your muscles cannot adapt in response to the same extent. Even so, some adaptation is possible and it is highly beneficial. Aerobic exercise improves your blood circulation and endurance, and even modest strength training slows the loss of muscle tissue that is an inevitable part of aging.

Figure 5.23 Swimming—an excellent form of aerobic exercise.

Uses—and Abuses—of Muscle-Building Substances

Aerobic exercise may replace fat with muscle, but it does not build significantly larger skeletal muscles. The only way to do that is through dedicated strength training. Some people, especially competitive athletes, want larger (and stronger) muscles, and they want them *now*. In spite of legal sanctions and persistent warnings from health professionals, they choose to use anabolic steroids and other substances that can produce a dramatic increase in muscle mass.

An anabolic steroid is a muscle-building hormone. Most of the substances in this category are synthetic hormones that were developed in the 1930s as therapeutic drugs that could mimic the sex hormone testosterone. Synthetic testosterone has numerous legitimate medical uses. For example, it is used to help treat impotence in men and some menopausal symptoms in women. Boys who are deficient in growth hormone may receive testosterone as part of hormone-replacement therapy.

Researchers also have looked at the potential use of synthetic testosterone supplements to counteract losses of lean muscle mass and bone due to aging. The results of these preliminary studies have been encouraging, and research is ongoing into possible benefits of synthetic testosterone for preventing or minimizing age-related ailments that arise from changes in sex hormone levels.

As with other potent drugs, only prescription use is legal. The roughly twenty varieties of synthetic anabolic steroids stimulate the synthesis of protein molecules, including muscle proteins. Using anabolic steroids while engaged in a weight-training exercise program can lead to rapid gains in lean muscle mass and strength. For this reason, anabolic steroids are popular among weight lifters, football players, and other athletes who specialize

Further reading: Student Guide to InfoTrac on web site →

in events that call for "brute power" or explosive muscle responses. Users commonly combine daily oral doses with a single hefty injection each month.

THE "ANDRO" AND CREATINE CONTROVERSIES   Perhaps because synthetic anabolic steroids are banned in sport, many athletes have turned to natural substances that also can be used to rapidly increase muscle mass and strength. One of these, androstenedione—"andro," for short—made headlines when baseball superstar Mark McGuire admitted using it. In the liver, enzymes convert androstenedione to testosterone, which promotes muscle growth. Androstenedione is legal, as is another popular muscle-building supplement, the amino acid creatine. As described in Section 5.9, creatine helps muscles replenish their supply of ATP. We normally take in most of what the body needs in fish or red meat, and the rest is synthesized in the liver, kidneys, and pancreas. Athletes who take massive doses of creatine can gain twenty pounds of muscle, or more, in a few months.

A DARKER SIDE OF STEROID USE   On the other hand, physicians, researchers, and athletes themselves report a long list of minor and major side effects of using steroids and supplements like creatine. In men, acne, baldness,

shrinking testes, and infertility are the first signs of steroid toxicity. The drugs also may be linked to early onset of a cardiovascular disease, atherosclerosis. And there is some evidence that consistent use contributes to kidney damage and to various types of cancer—including cancer of the testicles. Megadoses of creatine are thought to severely stress, and possibly permanently damage, the kidneys.

Women athletes also use steroids to gain a competitive advantage, and they can suffer ill effects, too. For example, steroids can trigger the development of a deep voice, pronounced facial hair, and irregular menstrual periods. A woman's breasts may shrink and her clitoris may become abnormally enlarged.

'ROID RAGE   Not all steroid users develop such severe physical side effects. In fact, more common are mental difficulties, called *'roid rage* or *bodybuilder's psychosis*. Some steroid-using men experience irritability and increased aggressiveness. Many competitive athletes look upon the added aggressiveness as a plus. Other men, however, experience uncontrollable aggression, delusions, and wildly manic behavior. In a famous case, one steroid user purposely drove his car at high speed into a tree.

Figure 5.24 The oversized muscles of a bodybuilder.

SUMMARY

1. Bones are the structural elements of the human skeleton. They function in movement by interacting with skeletal muscles to which they are attached. Bones also help protect and support other body parts and store minerals. Blood cells form in red bone marrow.

2. Bones are organs, and as such they include more than one type of tissue. Bone tissue is a connective tissue with both living and nonliving components. The living cells are osteocytes. In addition to bone tissue, bones incorporate other types of connective tissue, nerve tissue, and epithelium (in the walls of blood vessels).

3. A bone develops as osteoblasts secrete collagen fibers and a ground substance of protein and carbohydrate. The secretions eventually surround each osteoblast; the ground substance becomes hardened (mineralized) as calcium salts are deposited within it. The mature living bone cells, osteocytes, reside within spaces (lacunae) in the bone tissue.

4. The two kinds of bone tissue are compact bone and spongy bone. In spongy bone, needlelike struts are fused together in a latticework. In some bones, red marrow fills the spaces between struts. In the long bones of adults, most of the red marrow is replaced by yellow marrow. Denser compact bone is organized as thin, concentric layers (osteons) around small canals, which are channels for nerves and blood vessels that serve the bone tissue.

5. Bones develop following a cartilage model. The growth of long bones occurs at the bone ends, called epiphyses. Until long bone growth stops in late adolescence or early adulthood, each epiphysis is separated from the bone shaft by an epiphyseal plate of cartilage. The plates are replaced by bone when growth ends.

6. Bone tissue constantly "turns over" as minerals are deposited and withdrawn from it. Osteoblasts deposit bone and osteoclasts break it down. This process, called remodeling, is largely controlled by hormones and is a key element in maintaining calcium homeostasis in the body. Before adulthood, bones grow via remodeling.

7. The human skeleton has 206 bones (Table 5.2). It has an axial portion (skull, backbone, ribs, and breastbone) and an appendicular portion (limb bones, pelvic girdle, and pectoral girdles). The axial skeleton forms the body's vertical axis and is a central support structure. The appendicular skeleton provides support for upright posture and interacts with skeletal muscles during most movements. Intervertebral disks are shock pads and flex points in the backbone.

8. Together with skeletal muscles, the skeleton works like a system of levers in which rigid rods (bones) move about at fixed points (joints). In a synovial joint, a fluid-filled cavity separates adjoining bones. Such joints are freely movable. In cartilaginous joints, cartilage fills the space between bones and only slight movement is possible. In fibrous joints, there is no cavity, and fibrous connective tissue unites the bones. A limb can be moved and rotated around a joint because of the way pairs or groups of muscles are arranged relative to joints.

9. The body has more than 600 muscles, arranged as pairs or as muscle groups. Smooth, cardiac, and skeletal muscle tissue all contract (shorten) when stimulated. Only skeletal muscle interacts with the skeleton to bring about movement of the body or its parts (Table 5.3). The origin end of a skeletal muscle is attached to a bone that stays relatively motionless during a movement. The insertion end is attached to the muscle that moves most. Some muscles work together to promote the same movement. Others work antagonistically; the action of one opposes or reverses the action of the other.

10. Skeletal and cardiac muscle cells contain many threadlike myofibrils, which contain actin (thin) and myosin (thick) filaments. The filaments are organized in orderly arrays in sarcomeres, which are the basic units of muscle contraction.

11. Sarcomeres contract when nerve stimulation triggers the release of calcium ions from a membrane system (sarcoplasmic reticulum) in the muscle cell. Calcium binding alters the actin filaments so that the heads of adjacent myosin molecules (embedded in thick filaments) can bind to them. ATP provides the energy to drive the crossbridge power strokes that cause actin filaments to slide past the myosin filaments, shortening the sarcomere. This is the sliding-filament model of muscle contraction.

12. A motor neuron and the muscle cells it controls make up a motor unit. The fewer cells in the motor units in a muscle, the more finely the muscle's contractions can be controlled.

13. Neuromuscular junctions are synapses between a motor neuron and muscle cells. Neural stimulation triggers the release of a neurotransmitter (ACh) that sets in motion the biochemical events that lead to contraction. Individual cells in a motor unit always contract in an all-or-none fashion.

14. Human muscles normally operate near or at maximum temporal summation, a condition called tetany.

Review Questions

1. What are the functions of bones? What is a joint? 5.1, 5.5

2. How does skeletal muscle contribute to homeostasis? 5.7

3. Sketch and label the fine structure of a muscle, down to one of its myofibrils. Identify the basic unit of contraction in a myofibril. 5.7

4. How do actin and myosin interact in a sarcomere to bring about muscle contraction? What roles do ATP and calcium play? 5.7, 5.8

5. What is a motor unit? Why does a rapid series of muscle twitches yield a stronger overall contraction than a single twitch? 5.10

Table 5.2 Review of the Skeleton

FUNCTIONS OF BONE:

1. *Movement.* Interact with skeletal muscles to maintain or change the position of body parts.

2. *Support.* Support and anchor muscles.

3. *Protection.* Many bones form hard compartments that enclose and protect soft internal organs.

4. *Mineral storage.* Reservoir for mineral ions, which are deposited or withdrawn and so help maintain ion concentrations in body fluids.

5. *Blood cell formation.* Some contain red marrow, where blood cells are produced.

BONES OF THE SKELETON:

Appendicular portion

Pectoral girdles: clavicle and scapula
Arm: humerus, radius, ulna
Wrist and hand: carpals, metacarpals, phalanges (of fingers)
Pelvic girdle (six fused bones at the hip)
Leg: femur (thighbone), patella, tibia, fibula
Ankle and foot: tarsals, metatarsals, phalanges (of toes)

Axial portion

Skull: cranial bones and facial bones
Rib cage: sternum (breastbone) and ribs (12 pairs)
Vertebral column: vertebrae (26)

Table 5.3 Review of Skeletal Muscle

FUNCTION OF SKELETAL MUSCLE:
Contraction (shortening) that moves the body and its parts.

MAJOR COMPONENTS OF SKELETAL MUSCLE CELLS:

Myofibril: Strands containing filaments of the contractile proteins actin and myosin.
Sarcomere: The basic units of muscle contraction.

Other:

Motor unit: A motor neuron and the muscle cells it controls.
Neuromuscular junction: Synapse between a motor neuron and muscle cells.

What muscle characteristics would your training regimen aim to develop? How would you alter it to train marathoners?

2. Growth hormone, or GH, is used clinically to spur growth in children who are unusually short because they have a GH deficiency. However, it is useless for a short but otherwise normal 25-year-old to request GH treatment from a physician. Why?

3. If bleached human bones found lying in the desert were carefully examined, which of the following would not be present? Haversian canals, a marrow cavity, osteocytes, calcium.

4. For young women, the recommended daily allowance (RDA) of calcium is 800 milligrams. During Hilde's pregnancy, the RDA is 1,200 milligrams a day. Why, and what might happen to a pregnant woman's bones without the larger amount?

Self-Quiz *(Answers in Appendix V)*

1. _____ and _____ systems work together to move the body and specific body parts.

2. The three types of muscle tissue are _____, _____, and _____.

3. _____ are shock pads and flex points.
   a. vertebrae   c. lumbar bones
   b. cervical bones   d. intervertebral disks

4. Haversian canals are characteristic of what tissue?
   a. adipose   d. epithelial
   b. bone   e. muscle
   c. cartilage

5. The _____ is the basic unit of muscle contraction.
   a. myofibril   c. muscle fiber
   b. sarcomere   d. myosin filament

6. Muscle contraction requires _____.
   a. calcium ions   c. arrival of a nerve impulse
   b. ATP   d. all of the above

7. Match the M words with their defining feature.
   ____ muscle   a. actin's partner
   ____ muscle twitch   b. all in the hands
   ____ muscle tension   c. blood cell production
   ____ myosin   d. a muscle cannot contract
   ____ marrow   e. motor unit response
   ____ metacarpals   f. force exerted by cross-bridges
   ____ myofibrils   g. muscle cells bundled in
   ____ muscle fatigue   connective tissue
       h. threadlike parts in a muscle cell

Critical Thinking: You Decide *(Key in Appendix VI)*

1. You are training young athletes for the 100-meter dash. They need muscles specialized for speed and strength, *not* endurance.

Selected Key Terms

actin *5.8*
all-or-none principle *5.10*
appendicular skeleton *5.2*
axial skeleton *5.2*
bone marrow *5.1*
brain case *5.3*
cartilaginous joint *5.5*
clavicle *5.4*
cross-bridge *5.8*
epiphyseal plate *5.1*
femur *5.4*
fibrous joint *5.5*
humerus *5.4*
intervertebral disk *5.3*
ligament *5.2*
mandible *5.3*
motor unit *5.10*
muscle twitch *5.10*
myofibril *5.8*
myosin *5.8*

neuromuscular junction *5.9*
osteocyte *5.1*
osteon *5.1*
pectoral girdle *5.4*
pelvic girdle *5.4*
radius *5.4*
rib *5.3*
sarcomere *5.8*
sarcoplasmic reticulum *5.9*
scapula *5.4*
sinus *5.3*
skeletal muscle *5.7*
skull *5.3*
sliding-filament model *5.8*
sternum *5.3*
synovial joint *5.5*
temporal summation *5.10*
tendon *5.2*
tetany *5.10*
ulna *5.4*

Readings

Hoberman, J. M., and C. E. Yesalis. February 1995. "The History of Synthetic Testosterone." *Scientific American.*

Kearney, J. T. June 1996. "Training the Olympic Athlete." *Scientific American.* This article by a renowned sports physiologist describes the specialized training demands for elite athletes in several sports.

Sherwood, L. 1997. *Human Physiology.* Third Edition. Pacific Grove, California: Brooks/Cole.

6

DIGESTION AND NUTRITION

Weighty Worries

Have you ever worried about being "too fat" or eating the "wrong" foods? If not, you might consider yourself a rare bird. Many Americans have poor eating habits. They eat lots of snack foods, skip meals, or eat too much and too fast. In general, the typical American diet tends to be high in fat, and that is a problem on several counts. A growing number of people, both adults and children, are classified as obese—having an excess of fat in adipose tissues. By current nutritional standards, the proportion of body fat relative to total tissue should be 18 to 24 percent for a female under thirty years old. For males in the same age category, it should be in the range of 12 to 18 percent. Older people are "allowed" to stretch the upper limit a bit, but not much. While the exact standards vary, most experts agree that anyone is obese who carries more than about a quarter of their body weight as fat—and that probably includes at least 36 million Americans. Section 6.10 delves into some other aspects of this topic.

The world distribution of some illnesses suggests that the body may pay a high price for certain dietary habits. For example, in the United States a diet laden with fat and cholesterol is a major contributor to both obesity and heart disease. Obesity also increases an adult's risk of developing *diabetes mellitus*, a metabolic disease that is becoming disturbingly common in the U.S. In places where salt-cured and smoke-cured foods are staples, the rates of stomach and esophageal cancer are disturbingly high. Certain colon disorders, such as the development of small sacs (called *diverticula*) that can become severely inflamed, are far more prevalent among people whose diets lack sufficient fiber, as is common in the United States.

The flip side of overeating often is dieting to lose weight. Keeping lost weight off is another matter. Like most other mammals, we humans are well supplied with fat-storing cells, in our adipose tissue. This is a result of our evolutionary heritage. Fat-storing cells are an adaptation for survival, an energy warehouse that opens up when food is not available. Once those cells have formed, they are in the body to stay. Variations in food intake only influence how full or how empty each fat-storing cell gets.

When we diet and the fat warehouse opens, the brain interprets this as "starvation" and issues commands for a metabolic slowdown. The body uses energy far more efficiently even for the basics, such as breathing and digesting food. It now takes less food to do the same things! As you have probably heard, dieting does no good without a long-term commitment to exercise (Figure 6.1). This is because your skeletal muscles also adapt to "starvation," and they burn less energy than before. If and when you stop dieting, depleted fat cells quickly refill. For most people, the only way to keep off extra weight is through consistently moderate food intake and regular exercise.

Our emotions can influence weight gain and loss, sometimes to extremes. People who suffer from the eating disorder *anorexia nervosa* make a seriously

Figure 6.1 People taking a positive step toward maintaining a healthy body weight.

flawed assessment of their body weight. They typically are terrified of being fat, yet dread being hungry. An anorexic person purposefully starves and may over-exercise as well. She or he may also fear growing up and becoming sexually mature. Most common among younger women, anorexia nervosa is a potentially fatal condition.

Another extreme case is the "binge-purge" disorder called *bulimia*—literally, having an oxlike appetite. During an hour-long eating binge, a bulimic might consume 50,000 Calories worth of food, then vomit or take laxatives to get rid of it. Some bulimics simply want to lose weight, but for other people the disorder is a symptom of deep emotional problems.

With these thoughts to reflect on, we start a tour of **nutrition**. This word refers to all of the processes by which we ingest food, digest it, and absorb the released nutrients. Later, by way of the metabolic pathways described in Chapter 3, cells convert those nutrients to the body's own carbohydrates, lipids, proteins, and nucleic acids.

The central player in nutrition is our **digestive system.** By mechanical and chemical means, this organ system reduces food to particles, then to molecules that are small enough to be absorbed into the internal environment. It also eliminates unabsorbed residues. Other organ systems, especially those shown in Figure 6.2, contribute to nutrition. With their help, the digestive system makes its ongoing contribution to homeostasis.

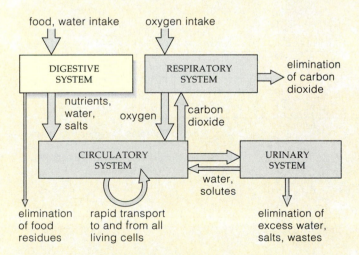

Figure 6.2 Functional links between the digestive system and other organ systems that work together to supply cells with raw materials and eliminate wastes.

KEY CONCEPTS

1. Nutrition encompasses all of the processes by which we take in and digest food, then absorb nutrients that are converted to the body's own carbohydrates, proteins, lipids, and nucleic acids.

2. The digestive system has specialized parts for food transport, processing, and storage. These include the mouth, stomach, and small and large intestines.

3. As food moves through the digestive system, it is mechanically broken apart and chemically broken down, nutrients are absorbed, and unabsorbed residues are eliminated. Accessory structures produce enzymes and other substances that aid the digestive process.

4. To maintain an acceptable body weight and overall health, energy intake must balance energy output by way of metabolic activity and physical exertion. Dietary glucose from complex carbohydrates usually is the body's main source of immediately usable energy.

5. The digestion and absorption of food make a vital contribution to homeostasis. Interactions among the digestive, circulatory, respiratory, and urinary systems supply the body's cells with raw materials, dispose of wastes, and maintain the volume and composition of extracellular fluid.

6. Adequate nutrition also requires that the body be supplied with vitamins, minerals, and certain amino acids and fatty acids that the body itself cannot produce.

CHAPTER AT A GLANCE

OVERVIEW OF THE DIGESTIVE SYSTEM

Our **digestive system** is a tube with two openings and many specialized organs. It extends from the mouth to the anus and is also called the gastrointestinal (GI) tract.

One of the first interesting things to learn about the GI tract is that while food or food residues are in it, technically the material is still *outside* the body. Nutrients don't "officially" enter the body until they move from the space inside the digestive tube—its *lumen*—into the bloodstream. How that happens is a major topic of this chapter.

Stretched out, the gastrointestinal tract would be 6.5 to 9 meters (21 to 30 feet) long in an adult. From beginning to end, mucus-coated epithelium lines all surfaces facing the lumen. The thick, moist mucus protects the wall of the tube and enhances diffusion across its inner lining. When we eat, substances advance in one direction, from the mouth (also called the oral cavity) through the pharynx, esophagus, stomach, small intestine, and large intestine. The large intestine ends in the rectum, anal canal, and anus. Figure 6.3 shows the digestive system of an adult and summarizes the functions of its parts.

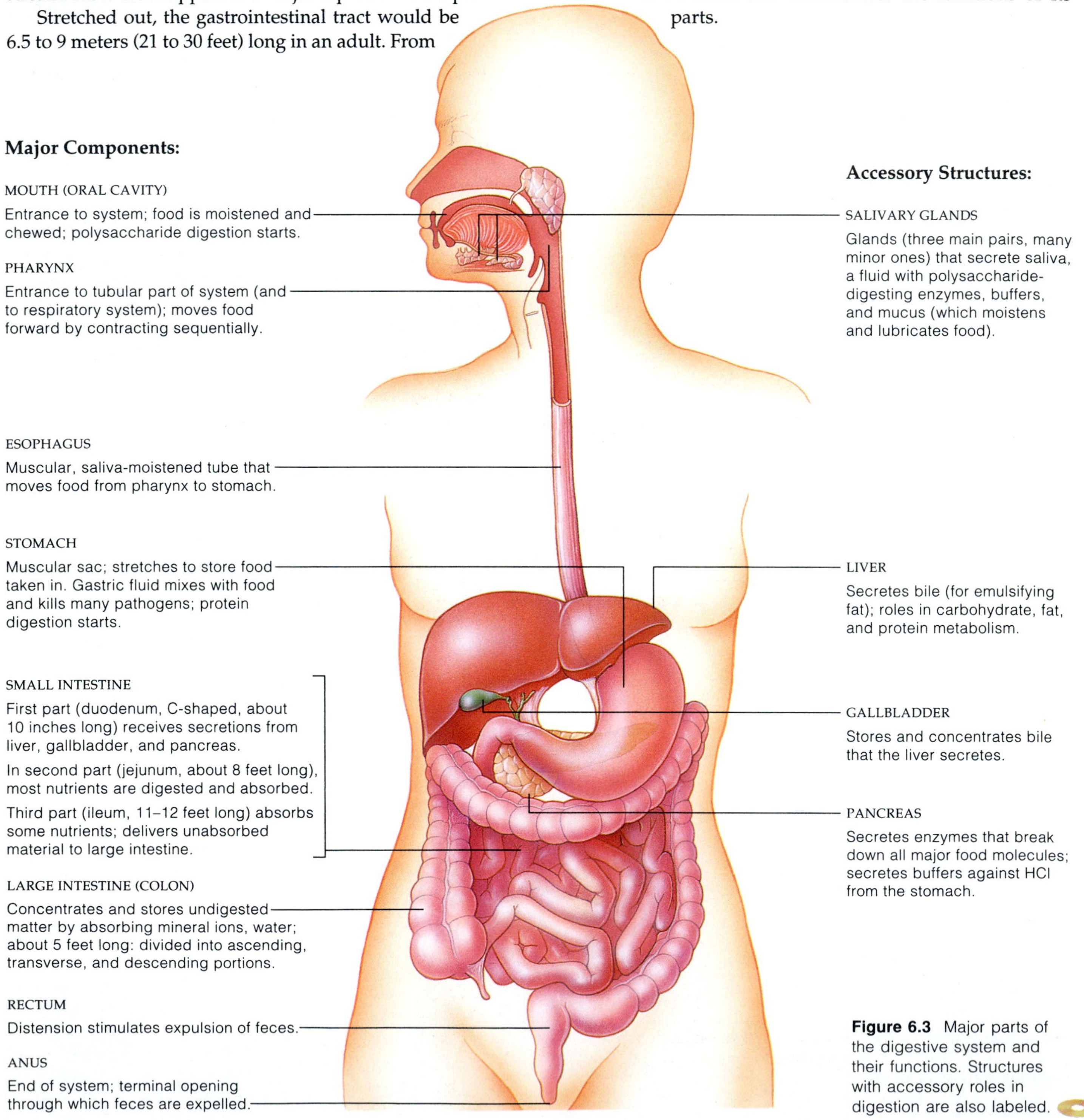

Major Components:

MOUTH (ORAL CAVITY)

Entrance to system; food is moistened and chewed; polysaccharide digestion starts.

PHARYNX

Entrance to tubular part of system (and to respiratory system); moves food forward by contracting sequentially.

ESOPHAGUS

Muscular, saliva-moistened tube that moves food from pharynx to stomach.

STOMACH

Muscular sac; stretches to store food taken in. Gastric fluid mixes with food and kills many pathogens; protein digestion starts.

SMALL INTESTINE

First part (duodenum, C-shaped, about 10 inches long) receives secretions from liver, gallbladder, and pancreas.

In second part (jejunum, about 8 feet long), most nutrients are digested and absorbed.

Third part (ileum, 11–12 feet long) absorbs some nutrients; delivers unabsorbed material to large intestine.

LARGE INTESTINE (COLON)

Concentrates and stores undigested matter by absorbing mineral ions, water; about 5 feet long: divided into ascending, transverse, and descending portions.

RECTUM

Distension stimulates expulsion of feces.

ANUS

End of system; terminal opening through which feces are expelled.

Accessory Structures:

SALIVARY GLANDS

Glands (three main pairs, many minor ones) that secrete saliva, a fluid with polysaccharide-digesting enzymes, buffers, and mucus (which moistens and lubricates food).

LIVER

Secretes bile (for emulsifying fat); roles in carbohydrate, fat, and protein metabolism.

GALLBLADDER

Stores and concentrates bile that the liver secretes.

PANCREAS

Secretes enzymes that break down all major food molecules; secretes buffers against HCl from the stomach.

Figure 6.3 Major parts of the digestive system and their functions. Structures with accessory roles in digestion are also labeled.

The various parts of the digestive system work in a coordinated way, carrying out the following basic tasks:

1. **Mechanical processing and motility.** Movements that break up, mix, and propel food material.

2. **Secretion.** Release of digestive enzymes and other substances into the digestive tube lumen.

3. **Digestion.** Chemical breakdown of food into particles, then into nutrient molecules small enough to be absorbed.

4. **Absorption.** Passing of digested nutrients and fluid across the tube wall and into blood or lymph.

5. **Elimination.** Expulsion of undigested and unabsorbed residues from the end of the GI tract.

Various accessory structures secrete enzymes and other substances that have essential roles in different aspects of digestion and absorption. These parts include glands in the wall of the GI tract, the salivary glands, and the liver, gallbladder, and pancreas.

Layers of the Digestive Tube

From the esophagus onward, the wall of the digestive tube has four layers (Figure 6.4). The *mucosa* (the innermost layer of epithelium) faces the lumen—the space through which food passes. The mucosa is surrounded by the *submucosa*. This is a layer of connective tissue with blood and lymph vessels and local networks of nerve cells. The next layer is *smooth muscle*—usually two sublayers, one in a circular orientation and the other oriented lengthwise. An outer layer, the *serosa*, is a very thin serous membrane (Section 4.7). Circular arrays of smooth muscle in sections of the GI tract are *sphincters*. Contractions of the sphincter muscles can close off a passageway. In your stomach, they help pace the forward movement of food and prevent backflow.

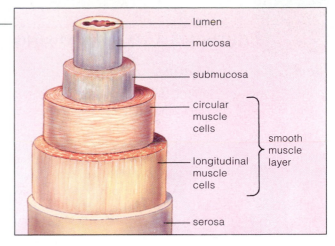

Figure 6.4 The four-layered wall of the gastrointestinal tract. The layers are not drawn to scale. Compare what you see here to the diagram in Figure 6.5a below.

Motility

The muscle layers of the digestive tube mix the tube's contents and propel them from one region to the next by wavelike contractions called **peristalsis** (Figure 6.5a). In peristalsis, rings of circular smooth muscle contract behind food and relax in front of it. The food distends the tube wall, peristalsis forces the food onward and expands the next wall region, and so on. In the intestines, **segmentation** also occurs (Figure 6.5b). Rings of smooth muscle in the wall repeatedly contract and relax, creating an oscillating (back-and-forth) movement. This movement constantly mixes the contents of the lumen and forces the material in it against the wall's absorptive surface.

The digestive tube extends from the mouth to the anus. For most of the length the tube wall consists of four layers, including smooth muscle.

Muscular movements move swallowed food from one tube region to the next, mix the lumen contents, and increase their contact with the tube's absorptive surface.

Figure 6.5 (**a**) A peristaltic wave down the stomach, produced by alternating contraction and relaxation of muscles in the stomach wall. (**b**) Segmentation, or oscillating movement, in the intestines.

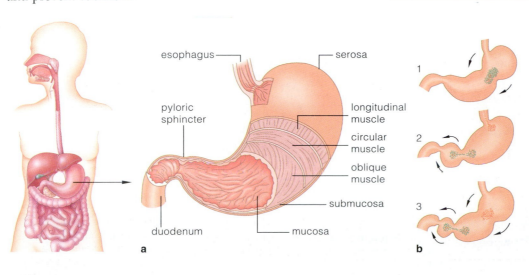

CHEWING AND SWALLOWING

Mouth, Teeth, and Salivary Glands

In the **oral cavity,** or mouth, the food you eat begins to be broken apart by chewing, and enzymes begin chemical digestion of polysaccharides (starch). Adults usually have thirty-two teeth (Figure 6.6a); young children have just twenty "primary teeth." A tooth's main regions are the crown and the root (Figure 6.6b). The crown is coated with hardened calcium deposits, the tooth enamel, which is the hardest substance in the body. It covers a thick bonelike layer of living material called dentin. Dentin and an inner pulp extend into the root. The pulp cavity contains blood vessels and nerves—something you know about if you have ever had a root canal.

Our teeth are real engineering marvels. They can take decades of mechanical stress and exposure to acids in food, enzymes, and other chemicals. Chisel-shaped incisors bite off chunks of food, and cone-shaped cuspids (canines) tear it. Premolars and molars, with broad crowns and rounded cusps, grind it. Bacteria that cause tooth decay (caries) flourish on food residues in the mouth, especially the sugar sucrose. Daily flossing and gentle brushing are a must to avoid a bacterial infection of the gums, which can lead to gingivitis. This inflammation can spread to the periodontal membrane that helps anchor a tooth in the jaw. In advanced periodontal disease, bacteria slowly destroy a tooth's bony socket. Loosened, the tooth may fall out, and other complications may arise.

Chewing mixes food with saliva, a fluid secreted from several strategically located **salivary glands** (Figure 6.6c). A large parotid gland nestles just in front of each ear. Submandibular glands lie just below the lower jaw in the floor of the mouth, and sublingual glands are under your tongue. Saliva flows through ducts that open into the mouth.

Saliva is mostly water, but it includes other important substances. One, the enzyme **salivary amylase**, breaks down starch; chew a bit of soda cracker and you can feel it turning to "mush" as salivary amylase goes to work. A buffer, bicarbonate (HCO_3^-), keeps the pH of your mouth between 6.5 and 7.5, a range within which salivary amylase can function, even when you eat acidic foods. Saliva also contains mucins, modified proteins that help bind bits of food into a softened, lubricated ball called a **bolus**. After you swallow food, starch digestion continues in the stomach until acidic secretions penetrate the bolus and inactivate salivary amylase.

The roof of the mouth, a bone-reinforced section of the **palate**, provides a hard surface against which the tongue can press food as it mixes it with saliva. Contractions of the tongue muscle force the food bolus into the **pharynx** (throat). This passageway connects with both the windpipe, or trachea (Figure 6.7), which leads to the lungs, and the **esophagus**, which leads to the stomach. Mucus secreted by the membrane lining the pharynx and esophagus lubricates the food, and in so doing it helps move the bolus on its way to the stomach.

molars
(12)

premolars
(8)

canines (4)

incisors
(8)

lower jaw

upper jaw

a

enamel

dentin

pulp cavity
(contains
nerves and
blood vessels)

root canal

periodontal
membrane

bone

crown

gingiva
(gum)

root

b

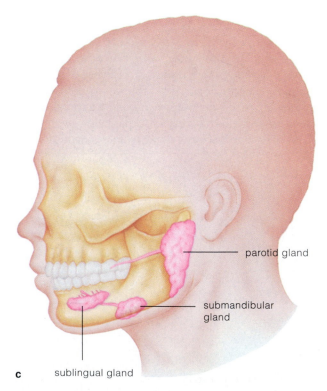

parotid gland

submandibular
gland

c

sublingual gland

Figure 6.6 (**a**) Locations of the different types of teeth. (**b**) Anatomy of a human tooth. (**c**) Locations of the salivary glands.

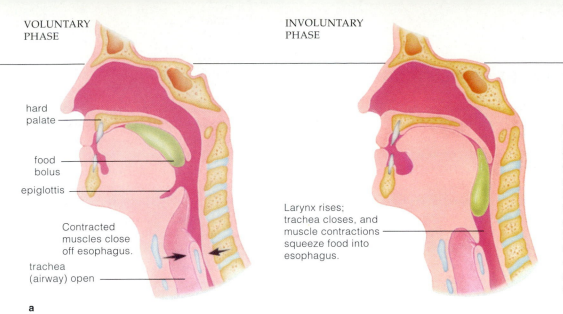

VOLUNTARY PHASE

INVOLUNTARY PHASE

hard palate

food bolus

epiglottis

Contracted muscles close off esophagus.

trachea (airway) open

Larynx rises; trachea closes, and muscle contractions squeeze food into esophagus.

a

Figure 6.7 Swallowing and peristalsis. (**a**) Contractions of the tongue push the food bolus into the pharynx. Next, the vocal cords seal off the larynx and the epiglottis bends downward, helping to keep the trachea closed. Contractions of throat muscles then squeeze the food bolus into the esophagus. (**b,c**) Finally, peristalsis in the esophagus moves the bolus through a sphincter, and food enters the stomach.

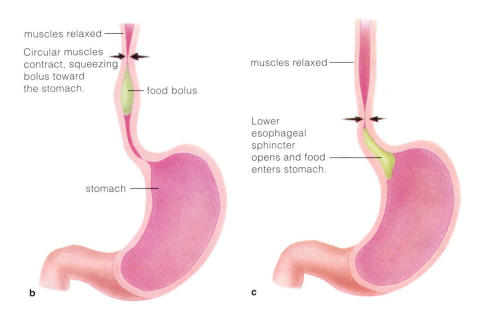

muscles relaxed

Circular muscles contract, squeezing bolus toward the stomach.

food bolus

stomach

muscles relaxed

Lower esophageal sphincter opens and food enters stomach.

b

c

Swallowing

Swallowing a mouthful of food might seem straightforward, but it actually involves an elaborate sequence of events. Swallowing begins when voluntary skeletal muscle contractions push a bolus into the pharynx, stimulating sensory receptors in the pharynx wall. The receptors trigger your "swallowing reflex," in which involuntary muscle contractions prevent food from moving up into your nose and help keep it from entering the trachea. As this reflex gets under way, the vocal cords are stretched tightly across the entrance to the larynx (your "voicebox"). Then, a flaplike valve, the *epiglottis*, is pressed down over the vocal cords as a secondary seal. For a moment, breathing is prevented as food moves into the esophagus—hence, you normally don't choke when you swallow. When swallowed food reaches the lower end of the esophagus, it enters the stomach through a sphincter (Figure 6.7b and c).

When swallowed food "goes down the wrong way," entering the trachea instead of the esophagus, a person can choke to death in as little as four minutes. An emergency procedure called the Heimlich maneuver (Figure 6.8) can dislodge food from the trachea by elevating the diaphragm muscle, which separates the chest cavity and abdominal cavity. This reduces the volume of the chest cavity, forcing air up the trachea—sometimes with enough force to eject an obstruction.

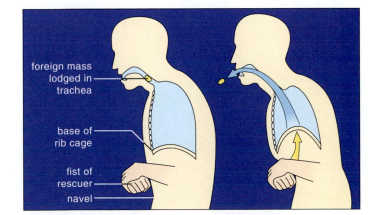

foreign mass lodged in trachea

base of rib cage

fist of rescuer

navel

Figure 6.8 The Heimlich maneuver. To perform the maneuver, stand behind the victim, make a fist, then position the fist, thumb side in, against the victim's abdomen. The fist must be slightly above the navel and well below the rib cage. Next, press the fist into the abdomen with a sudden upward thrust. Repeat the thrust several times if needed. Once the blockage is dislodged, a physician must examine the person at once.

The teeth and tongue begin the mechanical breakup of food. Enzymes in saliva begin the chemical digestion of food.

Swallowing has both voluntary and involuntary phases. Peristalsis moves a swallowed bolus down the esophagus to the lower esophageal sphincter, the gateway to the stomach.

DIGESTION IN THE STOMACH AND THE SMALL INTESTINE

Your stomach and small intestine are your body's premier food-processing organs. Both have layers of smooth muscles, the contractions of which break apart, mix, and move food onward. Digestive enzymes and other secretions enter the lumen of both organs (Table 6.1). Carbohydrates begin being broken down in the mouth, as you've just read, and protein breakdown starts in the stomach. But digestion of nearly all of the carbohydrates, lipids, proteins, and nucleic acids in food is completed in the small intestine.

The Stomach

The **stomach**, a muscular, expandable sac, has three main functions. First, it mixes and stores ingested food—in an adult, as much as several liters. Second, its secretions help dissolve and start the chemical breakdown of food particles, especially proteins. Third, it helps control passage of food into the small intestine. When the stomach is empty, its walls crumple into folds called *rugae* (Figure 6.9a).

The stomach wall surface facing the lumen is lined with glandular epithelium. Each day, some of the lining's gland cells secrete about two liters of hydrochloric acid (HCl), mucus, and other substances. These include pepsinogens—precursors of digestive enzymes called **pepsins**. Other gland cells secrete *intrinsic factor*, a protein required for vitamin B_{12} to be absorbed in the small intestine. Along with water, these secretions make up the stomach's highly acidic **gastric fluid**. Combined with stomach contractions, the acidity converts swallowed boluses into a thick mixture called **chyme**. The acidity kills most microbes in food. It also can cause "heartburn" when gastric fluid backs up into the esophagus.

Protein digestion starts when the high acidity denatures proteins and exposes their peptide bonds. It also converts pepsinogens to active pepsins, which break the bonds. Protein fragments accumulate in the stomach's lumen. Meanwhile, gland cells secrete *gastrin*. This hormone stimulates the cells that secrete HCl and pepsinogen.

Why don't HCl and pepsin break down the stomach lining? Control mechanisms usually assure that enough mucus and bicarbonate are secreted to protect the lining. However, these protective mechanisms, collectively called the "gastric mucosal barrier," can be disrupted. A common cause is infection by a bacterium, *Helicobacter pylori*. Bacterial toxins cause inflammation that, among other effects, breaks down tight junctions between cells of the stomach lining that normally prevent HCl from passing between cells. Hydrogen ions and pepsins diffuse into the lining, triggering events that further damage tissues. The resulting open sore is a *peptic ulcer*. Not everyone who is infected with *H. pylori* develops peptic ulcers. Heredity, chronic emotional stress, smoking, and excessive use of aspirin and alcohol are contributing factors. Antibiotic therapy cures peptic ulcers in patients who test positive for *H. pylori*.

Waves of peristalsis empty the stomach. The waves mix chyme and build force as they approach the pyloric sphincter at the stomach's base. The arrival of a strong contraction closes the sphincter, so most of the chyme is squeezed back. Only a small amount moves into the small intestine at a given time. In fact, depending on the volume and composition of chyme—mainly its fat content and acidity—it can take from two to six hours for a full stomach to empty. Regardless, food is not moved along faster than it can be processed.

Alcohol and a few other substances begin to be absorbed across the stomach wall. Liquids imbibed on an empty stomach pass rapidly from the stomach to the

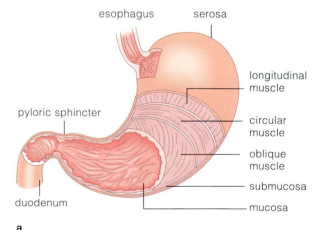

esophagus   serosa

longitudinal muscle

pyloric sphincter

circular muscle

oblique muscle

submucosa

duodenum

mucosa

a

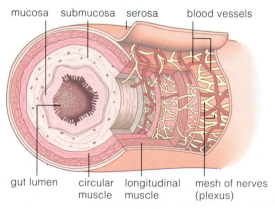

mucosa   submucosa   serosa   blood vessels

gut lumen   circular muscle   longitudinal muscle   mesh of nerves (plexus)

b

Figure 6.9 (**a**) Cutaway view of the stomach. (**b**) Structure of the small intestine. Notice that the intestinal wall has smooth muscle layers that are oriented in different directions. The serosa is an outer layer of connective tissue.

Table 6.1 Major Digestive Enzymes, Their Substrates, and Their Breakdown Products

Enzyme	Source	Where Active	Substrate	Main Breakdown Products
CARBOHYDRATE DIGESTION:				
Salivary amylase	Salivary glands	Mouth, stomach	Polysaccharides	Disaccharides, oligosaccharides
Pancreatic amylase	Pancreas	Small intestine	Polysaccharides	Disaccharides, monosaccharides
Disaccharidases	Intestinal lining	Small intestine	Disaccharides	MONOSACCHARIDES* (e.g., glucose)
PROTEIN DIGESTION:				
Pepsins	Stomach lining	Stomach	Proteins	Protein fragments
Trypsin and chymotrypsin	Pancreas	Small intestine	Proteins	Protein fragments
Carboxypeptidase	Pancreas	Small intestine	Peptides	AMINO ACIDS*
Aminopeptidase	Intestinal lining	Small intestine	Peptides	AMINO ACIDS*
FAT DIGESTION:				
Lipase	Pancreas	Small intestine	Triglycerides	FREE FATTY ACIDS, MONOGLYCERIDES*
NUCLEIC ACID DIGESTION:				
Pancreatic nucleases	Pancreas	Small intestine	DNA, RNA	NUCLEOTIDES*
Intestinal nucleases	Intestinal lining	Small intestine	Nucleotides	NUCLEOTIDE BASES, MONOSACCHARIDES*

*Breakdown products small enough to be absorbed into the internal environment.

small intestine, where further absorption occurs. Gastric emptying slows when food is in the stomach. That is why a person feels the effects of alcohol more slowly when drinking accompanies a meal.

The Small Intestine

Your **small intestine** is only about an inch and a half in diameter, but it is 6 meters long—some 20 feet! Figure 6.3 shows its three sections—the duodenum, the jejunum, and the ileum. The stomach and the ducts from the **pancreas**, **liver**, and **gallbladder** deliver about 9 liters of fluid to the duodenum every day. Nearly all of the fluid is absorbed across the small intestine's epithelial lining.

Chyme entering the duodenum triggers hormonal signals that stimulate a brief flood of digestive enzymes from the pancreas. As part of "pancreatic juice," these enzymes act on carbohydrates, fats, proteins, and nucleic acids. For example, like pepsin in the stomach, the pancreatic enzymes trypsin and chymotrypsin digest the polypeptide chains of proteins into peptide fragments. The fragments are then broken down to amino acids by *carboxypeptidase* from the pancreas and by *aminopeptidase* (present on the surface of the intestinal mucosa). In addition, the pancreas secretes bicarbonate, which buffers HCl from the stomach. This buffering action maintains a chemical environment in which pancreatic enzymes can function. As Chapter 12 will describe, hormones secreted by the pancreas have vital roles in nutrition.

Fat digestion requires enzyme action. It also requires *bile*, which the liver continually secretes. Bile contains bile salts, bile pigments, cholesterol, and lecithin (a phospholipid). When no food is moving through the GI tract, a sphincter closes off the main bile duct and bile backs up into the saclike gallbladder. As a meal is digested, the gallbladder contracts and empties bile into the small intestine.

Bile salts speed up fat digestion by emulsification. Recall that most fats in the human diet are triglycerides, which are insoluble in water. For this reason, they tend to clump into fat globules in chyme. When intestinal wall movements mix chyme, fat globules break up into droplets that become coated with bile salts. Bile salts carry negative charges, so the coated droplets repel each other and form an emulsion—that is, they stay separated. Emulsion droplets give fat-digesting enzymes a much greater surface area upon which to act. Because they are emulsified, triglycerides can be broken down much more rapidly to fatty acids and monoglycerides.

Liver cells use cholesterol to synthesize bile salts, and they also secrete cholesterol into bile. This is how the body excretes cholesterol. When there is chronically more cholesterol than available bile salts can dissolve, the excess may separate out (precipitate). *Gallstones,* which are mostly hard clumps of cholesterol, can develop in the gallbladder. They cause severe pain if they become lodged in bile ducts.

Carbohydrate breakdown starts in the mouth, and protein breakdown starts in the stomach.

In the small intestine, most large organic molecules are digested to molecules small enough to be absorbed into the internal environment.

Enzymes secreted into the small intestine by the pancreas act on carbohydrates , fats, proteins, and nucleic acids in chyme. Bile salts synthesized in the liver speed up the digestion of fats (triglycerides) by emulsifying them. Bile salts are stored (in bile) in the gallbladder.

ABSORPTION OF NUTRIENTS IN THE SMALL INTESTINE

Structure Is the Key to Function

Unlike the stomach, which can only absorb alcohol and a few other substances, the small intestine is where the vast majority of nutrients are absorbed.

Figure 6.10 shows the structure of the wall of the small intestine. Notice first how profusely folded the mucosa is. And notice that the folds all project into the lumen. Look closer (a microscope helps!), and you can see that each one of the large folds has amazing numbers of ever tinier projections. An even closer view would show that the epithelial cells at the surface of these tiny projections have a brushlike crown of still *smaller* projections, all exposed to the intestinal lumen.

What is the benefit of piling projections on folds and projections on projections? It produces a highly favorable surface-to-volume ratio (Section 2.1). Together, all the projections from the intestinal mucosa enormously increase the surface area available for interacting with chyme and absorbing nutrients from it. Without that huge surface area, absorption would take place hundreds of times more slowly—which of course would not be fast enough to sustain human life.

Figure 6.10c, d shows some **villi** (singular: villus), the absorptive "fingers" on each fold of the intestinal mucosa. Each villus is only about a millimeter long, but the mucosa has millions of them. Their density gives the mucosa a velvety appearance. Inside each villus are small blood vessels (an arteriole and a venule) and a lymph vessel, all of which function in moving substances to and from the general circulation (Figure 6.10e).

Most cells in the epithelium covering each villus bear microvilli (singular: microvillus). Each **microvillus** is an ultrafine, threadlike projection of the epithelial cell's plasma membrane, and each epithelial cell has about 1,700 of them. This dense array is what gives the epithelium of villi its common name, the "brush border." Gland cells in the lining make and secrete digestive enzymes (Table 6.1). Also, some phagocytic cells patrol and help protect the lining.

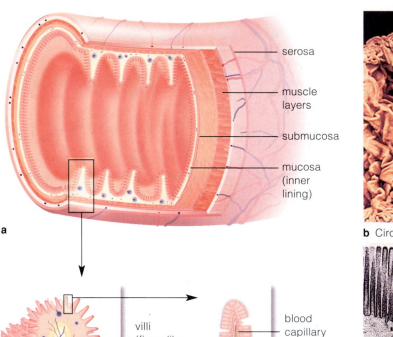

a

serosa

muscle layers

submucosa

mucosa (inner lining)

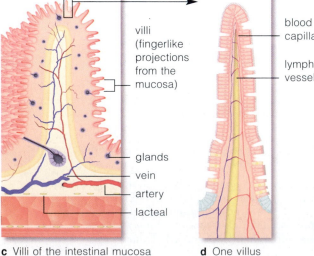

villi (fingerlike projections from the mucosa)

glands

vein

artery

lacteal

c Villi of the intestinal mucosa

blood capillary

lymph vessel

d One villus

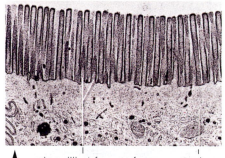

b Circular folds of the mucosa

microvilli at free surface of absorptive cells

cytoplasm

absorption

mucus secretion

hormone secretion

phagocytosis, lysozyme secretion

e Functions of epithelial cells of villus

Figure 6.10 (**a**,**b**) Surface structures of the small intestine. Notice the permanent circular folds of the intestinal mucosa. (**c**) Each fold has a profusion of villi. A villus is a fingerlike absorptive structure.

(**d**) Fine structure of a single villus. Monosaccharides and most amino acids that cross the intestinal lining enter small blood vessels (capillaries) in the villus. Fats enter vessels that carry the tissue fluid called lymph.

(**e**) Sketches of the types of epithelial cells at the free surface of a villus. Each villus has a crown of microvilli facing the intestinal lumen.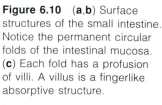

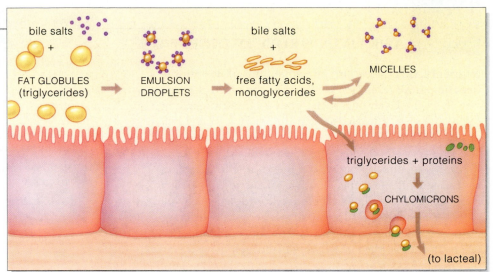

1 Digestion of carbohydrates to monosaccharides, and proteins to amino acids, by pancreatic enzymes.

2 Active transport of monosaccharides and amino acids across plasma membrane of cells; then nutrients move by facilitated diffusion into blood.

1 Emulsification. Wall movements break up fat globules into small droplets. Bile salts keep globules from re-forming. Pancreatic enzymes digest droplets to fatty acids and monoglycerides.

2 Formation of micelles as bile salts combine with digested products and phospholipids. Products readily slip into and out of the micelles.

3 Concentration of monosaccharides and fatty acids in micelles enhances gradients that lead to their diffusion across lipid bilayer of cells' plasma membranes.

4 Reassembly of products into triglycerides inside cells. These join with proteins, then are expelled (by exocytosis) into the internal environment.

Figure 6.11 Digestion and absorption in the small intestine.

What Are the Absorption Mechanisms?

Absorption, recall, is the passage of nutrients, water, salts, and vitamins into the internal environment. The vast absorptive surface area of the intestinal wall facilitates the process, but so does the action of smooth muscle in the intestinal wall. Remember from Section 6.1 that this segmentation helps mix the contents of the lumen and force it against the wall's absorptive surface:

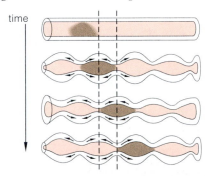

By the time food is halfway through your small intestine, most of it has been broken apart and digested. Water moves across the intestinal lining by osmosis. Mineral ions also are selectively absorbed. Some nutrients are absorbed directly. For instance, transport proteins in the plasma membrane of brush border cells actively shunt monosaccharides and amino acids across the lining. By contrast, fatty acids and monoglycerides must diffuse across the plasma membrane's lipid bilayer of the plasma membrane, with the help of bile salts (Figure 6.11).

Recall that fatty acid and monoglyceride molecules—the products of the digestion of lipids—have hydrophobic regions. They do not dissolve in watery chyme. In fact, left alone they would float like a film of oil on a puddle and not come near the surface of absorptive cells. Instead, however, the molecules clump together with bile salts, along with cholesterol and other substances, and form tiny droplets. Each droplet is a **micelle.**

If concentration gradients are favorable when a micelle comes into contact with the surface of an absorptive cell, nutrient molecules diffuse out of the micelle, into the solution bathing the microvilli. They continue diffusing on into epithelial cells. Inside an epithelial cell, fatty acids and monoglycerides quickly reunite into triglycerides. The triglycerides combine with proteins into microscopic particles (chylomicrons) that leave the cells by exocytosis and enter tissue fluid. The triglycerides enter lymph vessels called **lacteals,** which eventually drain into blood vessels. Once glucose and amino acids are absorbed, they enter blood vessels directly.

With its richly folded intestinal mucosa, millions of villi, and hundreds of millions of microvilli, the small intestine provides a vast surface area for absorbing nutrients.

Substances pass through brush border cells that line the free surface of each villus by active transport, osmosis, or diffusion across the lipid bilayer of plasma membranes.

THE LIVER'S ROLE IN DIGESTION

Nutrient-laden blood in intestinal villi flows to the **hepatic portal vein**. This vein carries the blood through vessels in the **liver**, one of the body's largest organs. In the liver, excess glucose is taken up before a *hepatic vein* returns the blood to the general circulation (Figure 6.12). The liver converts much of this glucose to a storage form, glycogen. As already noted, the liver's role in digestion is to secrete bile—as much as 1,500 ml every day—which is stored in the gallbladder and released when chyme enters the small intestine. It also processes incoming nutrient molecules into substances the body requires (such as blood plasma proteins) and removes toxins ingested in food or already circulating in the bloodstream.

Besides its digestive functions, the liver helps maintain homeostasis in other ways. For instance, it adjusts the concentrations of blood's organic substances (Table 6.2). It also inactivates many hormone molecules and sends them to the kidneys for excretion (in urine). Ammonia (NH_3) that is produced when cells break down amino acids can be toxic to cells, especially in the nervous system. The circulatory system carries ammonia to the liver, where it is converted to a much less toxic waste product, urea, which also is excreted in urine.

Several major liver diseases can have a terrible impact on a person's health. Long-term, heavy alcohol use is a common cause of the liver scarring called *cirrhosis*. While liver transplantation is becoming more successful, many people who suffer from advanced cirrhosis still die of liver failure. At least five types of *hepatitis* are caused by viruses that infect and damage the liver. Types A and E are transmitted by way of contaminated food or water.

Table 6.2 Some Activities That Depend on Liver Functioning
1. Carbohydrate metabolism
2. Partial control of synthesis of proteins in blood; assembly and disassembly of certain other proteins
3. Urea formation from nitrogen-containing wastes
4. Assembly and storage of some fats; fat digestion (bile is formed by the liver)
5. Inactivation of many chemicals (such as hormones and some drugs)
6. Detoxification of many poisons
7. Degradation of worn-out red blood cells
8. Immune response (removal of some foreign particles)
9. Red blood cell formation (liver absorbs, stores factors needed for red blood cell maturation)

Symptoms include nausea, abdominal pain, loss of appetite, and jaundice (yellowing of the skin). Types B and C hepatitis are transmitted sexually or in contaminated blood (Chapter 16). Type B can cause cirrhosis, and it dramatically increases the chance of developing liver cancer. Type C hepatitis usually produces mild symptoms, though it can become chronic and cause serious damage. Hepatitis type D infects only people who already have Type B, but it nearly always becomes chronic, with devastating long-term effects.

Simple sugars and amino acids directly enter the epithelial cells of villi. Both bile produced by the liver and pancreatic enzymes help in the digestion of fats.

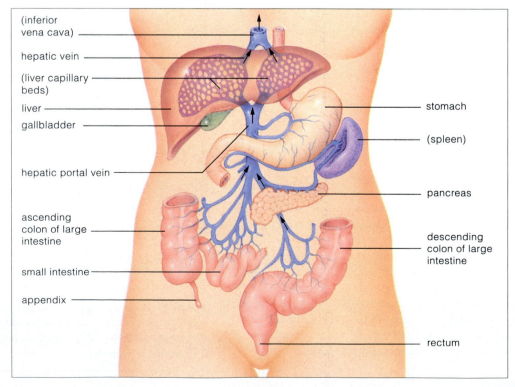

(inferior vena cava)
hepatic vein
(liver capillary beds)
liver
gallbladder
hepatic portal vein
ascending colon of large intestine
small intestine
appendix
stomach
(spleen)
pancreas
descending colon of large intestine
rectum

Figure 6.12 Hepatic portal system. Arrows show the direction of blood flow.

THE LARGE INTESTINE

Anything not absorbed in the small intestine moves into the **large intestine**, where "leftovers" are concentrated as water and salts are reabsorbed. Cells in the lining actively transport sodium ions out of the lumen; and as the ion concentration in the lumen drops, water moves out by osmosis (Section 3.6). The concentrated material is stored and then eliminated as *feces*, a mixture of undigested and unabsorbed food, water, and bacteria. Its typical brown color comes mainly from bile pigments in it.

Bacteria make up about 30 percent of the dry weight of feces. Such microorganisms, including *Escherichia coli*, normally inhabit our intestines and are nourished by the food residues there. Their metabolic activity produces useful fatty acids and some vitamins (such as vitamin K), which are absorbed into the bloodstream. However, feces of humans and other animals also can contain disease-causing organisms. Health authorities use the presence of "coliform bacteria," including *E. coli*, in water and food supplies as a measure of fecal contamination.

The large intestine is about 1.2 meters (5 feet) long. It begins as a blind pouch, the **cecum** (Figure 6.13). The cecum merges with the **colon**, which is divided into four regions in an inverted U-shape. The ascending colon travels up the right side of the abdomen, the transverse colon continues across to the left side, and the descending colon then turns downward. The sigmoid colon makes an S-curve and connects with the **rectum**.

Short, longitudinal bands of smooth muscle in the colon wall are gathered at their ends, like a series of full skirts nipped in at elastic waistbands. As they contract and relax, the contents of the lumen move back and forth against the wall's absorptive surface. Nerves largely control the motion, which is similar to segmentation in the small intestine but much slower.

With each new meal, hormonal signals and nervous system commands induce large portions of the ascending and transverse colon to contract at the same time. Within a few seconds, the lumen's existing contents move as much as three-fourths of the colon's length and make way for incoming food. When feces distend the rectal wall, the stretching triggers *defecation*—expulsion of feces from the body. From the rectum feces move into the **anal canal**. The nervous system also controls defecation. It can stimulate or inhibit contractions of sphincter muscles at the **anus**, the terminal opening of the GI tract.

The normal frequency of defecation ranges from three times a day to once a week. In *constipation*, food residues remain in the colon for too long, too much water is reabsorbed, and the feces become dry and hard. This not only can make defecation difficult, but can also cause abdominal discomfort, headache, even nausea. Constipation is a common cause of the enlarged rectal blood vessels known as *hemorrhoids*. Hard feces also may become lodged in the slender **appendix**, which projects from the cecum like the

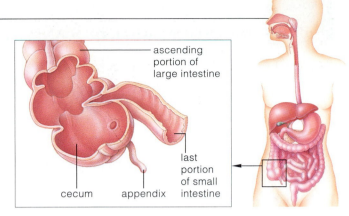

little finger of a glove. No one has ever discovered a digestive function for the appendix, but it is colonized by white blood cells that defend the body against ingested bacteria. Feces that obstruct the normal blood flow and mucus secretion to the appendix can cause *appendicitis*—an inflamed appendix. Symptoms include pain near the navel or right side of the abdomen, nausea, and vomiting. Any suspected case of appendicitis is a medical emergency. If an infected appendix bursts, bacteria are spewed into the abdominal cavity. The victim can develop *peritonitis*, a potentially life-threatening infection. Stress, lack of exercise, and other factors can cause constipation, but a major cause is a lack of bulk in the diet.

Bulk is the volume of fiber (mainly cellulose from plants) and other undigested food material that cannot be decreased by absorption in the colon. Much of it consists of *insoluble fiber* such as cellulose and other plant compounds that humans cannot digest (we lack the required enzymes) and that does not easily dissolve in water. Wheat bran is just one example. (*Soluble* fiber consists of plant carbohydrates such as fruit pectins that swell or dissolve in water.) People who eat too little fiber are more likely to form diverticula—knoblike sacs where the inner colon lining protrudes through the intestinal wall. Inflammation of a diverticulum is called *diverticulitis*. If one bursts, as sometimes happens, peritonitis may develop.

Diarrhea can occur when an irritant (such as a bacterial toxin) causes the mucosal lining of the small intestine to secrete more water and salts than the large intestine can absorb. It can also result when infections, stress, or other factors speed up peristalsis in the small intestine, so that there is not time for enough water to be absorbed. Diarrheal diseases are dangerous in part because they deplete the body of the water and salts needed for proper functioning of nerve and muscle cells.

Figure 6.13 Cecum and appendix of the large intestine.

In the large intestine, water and salts are reabsorbed from food residues entering from the small intestine. The remaining concentrated residues are stored and later eliminated as feces.

DIGESTION CONTROLS AND NUTRIENT TURNOVER

Controls over Digestion

As you probably know by now, homeostatic feedback loops operate when physical and chemical conditions change in the internal environment. By contrast, controls over the digestive system act *before* food is absorbed into the internal environment. They respond to the volume and chemical makeup of material in the lumen of the GI tract. The nervous system, local nerve networks in the digestive tube wall, and the endocrine system interact to exert control (Figure 6.14).

For instance, in the first few hours after a meal, food distends the walls of the stomach and later those of the small intestine. This stretching triggers signals from sensory receptors in the wall of these digestive tube regions. Some of the signals give you (by way of signal processing in your brain) that "full" feeling after you eat. Others can lead to muscle contractions in the tube wall or secretion of digestive enzymes and other substances into the lumen. Various parts of the brain monitor such activities and coordinate them with events such as blood flow to the small intestine, where nutrients are being absorbed.

There are a number of gastrointestinal hormones. *Gastrin* is secreted into the bloodstream by endocrine cells in the stomach's lining when amino acids and peptides are in the stomach. It mainly stimulates the secretion of acid into the stomach. Acid secretion can be inhibited by *somatostatin,* which is produced by another type of endocrine cell in the stomach lining. Endocrine cells in the lining of the small intestine secrete other hormones. As acid arrives in the small intestine, *secretin,* a peptide hormone, stimulates the pancreas to secrete bicarbonate. *Cholecystokinin* (CCK) is released in response to fat in the small intestine. It enhances the actions of secretin, stimulates pancreatic enzyme secretion, and stimulates gallbladder contractions. Secretin and CCK also slow the rate of stomach emptying. *Glucose insulinotropic peptide* (GIP) is released in response to the presence of glucose and fat in the small intestine. It slows both the secretions of gastric glands and the rate of gastric emptying. GIP also stimulates the release of insulin, which is necessary for glucose uptake by cells or storage (as glycogen) by the liver.

Nutrient Turnover and Organic Metabolism

Once nutrient molecules have entered the body, they are shuffled and reshuffled according to body needs at any given time. The nervous system and endocrine system interact to control these and other aspects of organic metabolism; Figure 6.15 summarizes the main pathways.

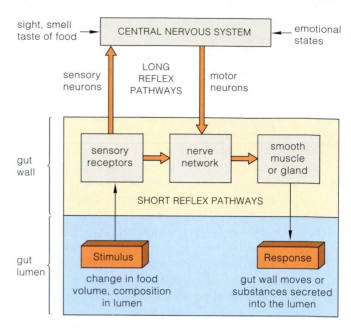

Figure 6.14 Local and long-distance reflex pathways called into action when food is in the digestive tract.

Most carbohydrates, lipids, and proteins are broken down continually. Their component parts are picked up and reused as building blocks or as energy sources.

When you eat, your body builds up pools of organic compounds. Excess carbohydrates and other dietary molecules are transformed mostly into fats, which are stored in adipose tissue. Some carbohydrates are converted to glycogen in the liver and in muscle tissue. When organic molecules are being absorbed and stored, most cells use glucose as their main energy source.

As already described, during a meal and for several hours thereafter, glucose moves into cells, where it will be used for energy. However, if a person fasts or engages in rigorous physical activity, his or her glucose supplies become depleted. Then the body taps into its fat stores for energy. Fat stored mainly in adipose tissue is broken down to glycerol and fatty acids, which are released into blood. Glycerol is converted to glucose in the liver. Cells can take up the circulating fatty acids and use them to make ATP.

Disorders That Disrupt Absorption

A variety of mild to serious disorders result from maladies that disrupt the production of needed enzymes or harm the absorptive intestinal lining. Their ill effects can range from flatulence (gas) and diarrhea to life-threatening malnutrition (Section 6.11). Anything that interferes with the uptake of nutrients across the lining of the small intestine can lead to a **malabsorption disorder**.

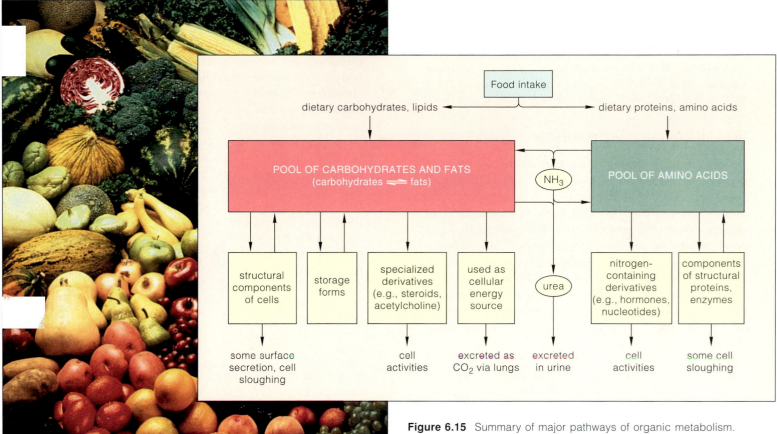

Figure 6.15 Summary of major pathways of organic metabolism. Carbohydrates, fats, and proteins are continually being broken down and resynthesized. Urea forms mainly in the liver.

For example, *lactose intolerance* results from a deficiency of the enzyme lactase. It occurs most often in adults, and prevents normal breakdown and absorption of lactose, the sugar found in milk, ice cream, and many other milk products. People who have certain diseases affecting the pancreas, including the genetic disorder *cystic fibrosis*, do not produce the pancreatic enzymes needed for normal digestion and absorption of fats and other nutrients. Cystic fibrosis also has devastating effects on the lungs, which we consider in later chapters.

Disorders such as *Crohn's disease* and intolerance to the gluten in wheat can so severely damage the intestinal lining that much of the intestine must be removed. In some cases a patient must be fed artificially, receiving "meals" of a nutrient-rich liquid through a feeding tube.

Food allergies also disrupt digestion and absorption, but for a different reason. Like other allergies (described in Chapter 8), they are skewed responses of the immune system in which a particular food is interpreted as an "invader." The most common food culprits are shellfish, eggs, and wheat. Reactions can be almost instant or occur over a period of hours. Depending on the person and the food involved, symptoms typically include diarrhea,

vomiting, and sometimes swelling or tingling of mucous membranes. Some food allergies can be lethal. Peanuts are notorious offenders, because in affected people even a tiny amount can trigger *anaphylaxis*—a body-wide allergic response in which blood pressure plummets, among other frightening symptoms (Section 8.11).

A food allergy or any other disorder that produces prolonged, severe diarrhea or vomiting can lead to electrolyte imbalances that disrupt the functioning of the nervous system and threaten the body's ability to maintain homeostasis in other ways as well. Several bacteria, viruses, and protozoa that infect the digestive system cause severe diarrhea. Worldwide, such infections are a leading cause of death. Chapter 16 takes a closer look at these and some other infectious diseases.

The nervous system and the endocrine system exert control over conditions in the digestive system in response to the volume and composition of material in the digestive tract.

When glucose supplies run low, the body metabolizes fat molecules as the main energy source.

THE BODY'S NUTRITIONAL REQUIREMENTS

You nourish yourself by eating foods that are suitable sources of energy and raw materials. We measure food energy in **kilocalories** (kcal). Each kilocalorie is 1,000 calories of heat energy, the amount needed to raise the temperature of 1 kilogram of water by 1°C. The word Calorie, with a capital "C," is simply shorthand for a kilocalorie. Now, we consider the raw materials in food.

New, Improved Food Pyramids

A few million years ago, our prehuman ancestors ate mostly fresh fruits and other fibrous plant material. From that nutritional beginning, humans in many parts of the world came to prefer low-fiber, high-fat foods. They still do, even when they start in elementary school to study **food pyramids**—charts of well-balanced diets. Nutritionists regularly revise the charts to reflect additional research. Figure 6.16 shows a recent version. It gives daily portions of foods that will supply an adult male of average size with 55–60 percent complex carbohydrates, 15–20 percent proteins (less for females), and 20–25 percent fats. Not everyone agrees with these proportions. Some popular diet regimes suggest a mix of 40 percent complex carbohydrates, 30 percent proteins, and 30 percent fats for peak body performance.

Carbohydrates

Starch and, to a lesser extent, glycogen should be the main carbohydrates in the human diet. These complex carbohydrates are easily broken down to glucose, the body's main energy source (Chapter 3). Good sources of starch include fleshy fruits, cereal grains, and legumes, including peas and beans.

Foods rich in complex carbohydrates have another advantage: They typically are high in fiber, including the insoluble fiber that adds needed bulk to feces. In a contrast that is well worth noting, simple sugars do not have the fiber of complex carbohydrates. Neither do they have the vitamins and minerals of whole foods. In one week—and every week—the average American eats up to two pounds of refined sugar (sucrose). In the ingredients lists on packaged foods, these sugars often are disguised as corn syrup, corn sweeteners, dextrose, and so on.

MILK, YOGURT, CHEESE GROUP
Choose 2–3 servings from these alternatives:

1 cup nonfat or lowfat milk
1 cup nonfat or lowfat yogurt
1-1/2 ounces lowfat, unprocessed cheese
2 ounces lowfat, processed cheese

FRUIT GROUP
Choose 2–4 servings from these alternatives:

1 medium-size fruit, such as apple, mango
1 wedge cantaloupe or pineapple
1 cup fresh berries
3/4 cup unsweetened fruit juice
1/2 cup canned unsweetened fruit
1/4 cup dried fruit

Figure 6.16 Food guide pyramid. The daily portions of the various foods shown give an idea of what makes up a well-balanced diet. Alternatives listed are a general guide to the kinds of foods in each group. Select foods from the "red-hot" tip of the pyramid only sparingly. (Extra fats and simple sugars add calories but provide few vitamins and minerals.) Where there is a range of serving sizes, choosing the smallest one will help limit total caloric intake to about 1,600 kcal. Choosing the largest serving size will raise the total to about 2,800 kcal.

ADDED FATS AND SIMPLE SUGARS
Restrict intake (other foods provide ample amounts)

Limit salad dressings, butter, cream, margarine, syrups, table sugar, soft drinks, candies, sweet desserts

LEGUME, NUT, POULTRY, FISH, MEAT GROUP
Choose 2–3 servings from these alternatives:

1-1/4 to 1-1/2 cups cooked legumes
2-1/2 to 3 ounces cooked poultry (no skin)
2-1/2 to 3 ounces cooked fish
2-1/2 to 3 ounces trimmed, cooked, lean red meat
1 egg (or two egg whites)
5 tablespoons peanut butter

VEGETABLE GROUP
Choose 3–5 servings from these alternatives:

1/2 cup chopped raw or cooked vegetables
1 cup raw, leafy vegetables

BREAD, CEREAL, RICE, PASTA GROUP
Choose 6–11 servings from these alternatives:

1 slice whole-grain bread
1/2 cup cooked rice (brown rice has more nutrients)
1/2 cup cooked pasta
1/2 cup cooked unsweetened cereal, such as oatmeal
1 ounce dry unsweetened cereal, such as
shredded wheat

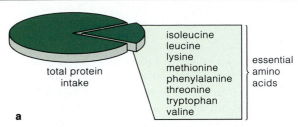

Figure 6.17 (**a**) The eight essential amino acids, a small portion of the total protein intake. All eight must be available at the same time, in certain amounts, if cells are to build their own proteins. Milk and eggs have high amounts of all eight in proportions that humans require; they are among the *complete* proteins.

Nearly all plant proteins are incomplete, so vegetarians should construct their diet carefully to avoid protein deficiency. For example, they can combine different foods from the three columns of incomplete proteins shown in (**b**). Also, vegetarians who avoid dairy products and eggs should take vitamin B$_{12}$ and B$_2$ (riboflavin) supplements. Animal protein is a luxury in most societies. Yet their cuisines include good combinations of plant proteins, including rice/beans, chili/cornbread, tofu/rice, and lentils/wheat bread.

No Limiting Amino Acid	Low in Lysine	Low in Methionine, Other Sulfur-Containing Amino Acids	Low in Tryptophan
legumes: soybean tofu soy milk	legumes: peanuts	legumes: beans (dried) black-eyed peas garbanzos lentils lima beans mung beans peanuts	legumes: beans (dried) garbanzos lima beans mung beans peanuts
cereal grains: wheat germ	cereal grains: barley buckwheat cornmeal oats rice rye wheat		cereal grains: cornmeal
nuts		nuts: hazelnuts	nuts: almonds English walnuts
milk cheeses (except cream cheese) yogurt eggs meats	nuts, seeds: almonds cashews coconut English walnuts hazelnuts pecans pumpkin seeds sunflower seeds	fresh vegetables: asparagus broccoli green peas mushrooms parsley potatoes soybeans Swiss chard	fresh vegetables: corn green peas mushrooms Swiss chard

b

Lipids

The body can't function without fats and other lipids. For instance, the phospholipid lecithin is a necessary component of cell membranes. Fats serve as energy reserves, cushion many organs (such as the eyes and kidneys), and provide insulation beneath the skin. The body also stores fat-soluble vitamins in adipose tissues. The brain of a young child does not develop properly unless it is supplied with cholesterol and saturated fat.

The liver can synthesize most of the fats the body needs, including cholesterol, from protein and carbohydrates. The ones it cannot produce are called **essential fatty acids**, but whole foods provide plenty of them. Linoleic acid is an example. Just one teaspoon a day of corn oil, olive oil, or some other polyunsaturated fat in food provides enough of it.

Currently butter and other fats make up 40 percent of the kilocalories in the average diet in the United States. Most of the medical community agrees the proportion should be 30 percent or less.

Like other animal fats, butter is a saturated fat that tends to raise the level of cholesterol in the blood. As described in earlier chapters, the body needs cholesterol for the synthesis of bile salts, steroid hormones, and cell membranes. For many people, though, too much cholesterol may interfere seriously with blood circulation (Section 7.14). Some brands of "diet" frozen yogurt, potato chips, and other foods contain "fake fats"—edible but nondigestible oils, such as sucrose polyester. These products can be useful for people who have trouble cutting back on their fat intake. On the downside, though, fat substitutes often produce intestinal gas and other unwanted side effects.

Proteins

When the digestive system digests and absorbs proteins, their amino acids become available for the body's own protein synthesis. Of the twenty common amino acids, eight are **essential amino acids**. Our cells cannot synthesize them, so we must obtain them from food. The eight are isoleucine, leucine, lysine, methionine, phenylalanine, threonine, tryptophan, and valine (Figure 6.17).

Most animal proteins are *complete*, meaning their ratios of amino acids match human nutritional needs. Plant proteins are *incomplete*, meaning they lack one or more of the essential amino acids. To get adequate amounts of needed amino acids, vegetarians must eat certain combinations of different plants.

Because enzymes and other proteins are vital for the body's structure and function, a protein-deficient diet may have severe consequences. Protein deficiency is most damaging among the young, for the brain grows and develops rapidly during early life. Even mild protein starvation during the mother's pregnancy or for some months after a child is born can retard the child's growth and affect mental and physical performance. The final section in this chapter describes some of the severe impacts of undernutrition, which is a critical problem in the poorest nations.

In terms of total caloric intake, the bulk of a well-balanced diet consists of complex carbohydrates, which are the body's preferred energy sources.

Lipids—which are essential components of cell membranes, serve as energy reserves, and have other roles—should make up no more than about 30 percent of calories.

Proteins, which provide the body's essential amino acids, should make up the remainder of dietary calories.

VITAMINS AND MINERALS

Vitamins are organic substances that are essential for growth and survival, for no other substances can play their metabolic roles. In the course of evolution, animal cells have lost the ability to synthesize these substances, so we must obtain vitamins from food.

At a minimum, our cells need the 13 vitamins listed in Table 6.3. Each vitamin has specific metabolic functions. Many chemical reactions use several types, and the absence of one affects the functions of others.

Minerals are *inorganic* substances that are essential for growth and survival; no other substances can play their metabolic roles. For instance, all of your cells need iron for their electron transport chains. Your red blood cells can't function without iron in hemoglobin, the oxygen-carrying pigment in blood. And neurons stop functioning without sodium and potassium (Table 6.4).

People who are in good health get all the vitamins and minerals they need from a balanced diet of whole foods.

Table 6.3 Vitamins: Sources, Functions, and Effects of Deficiencies or Excesses*

Vitamin	Common Sources	Main Functions	Signs of Severe Long-Term Deficiency	Signs of Extreme Excess
FAT-SOLUBLE VITAMINS:				
A	Its precursor comes from beta carotene in yellow fruits, yellow or green leafy vegetables; also in fortified milk, egg yolk, fish liver	Used in synthesis of visual pigments, bone, teeth; maintains epithelia	Dry, scaly skin; lowered resistance to infections; night blindness; permanent blindness	Malformed fetuses; hair loss; changes in skin; liver and bone damage; bone pain
D	D_3 formed in skin and in fish liver oils, egg yolk, fortified milk; converted to active form elsewhere	Promotes bone growth and mineralization; enhances calcium absorption	Bone deformities (rickets) in children; bone softening in adults	Retarded growth; kidney damage; calcium deposits in soft tissues
E	Whole grains, dark green vegetables, vegetable oils	Possibly inhibits effects of free radicals; helps maintain cell membranes; blocks breakdown of vitamins A and C in gut	Lysis of red blood cells; nerve damage	Muscle weakness, fatigue, headaches, nausea
K	Colon bacteria form most of it; also in green leafy vegetables, cabbage	Blood clotting; ATP formation via electron transport	Abnormal blood clotting; severe bleeding (hemorrhaging)	Anemia; liver damage and jaundice
WATER-SOLUBLE VITAMINS:				
B_1 (thiamine)	Whole grains, green leafy vegetables, legumes, lean meats, eggs	Connective tissue formation; folate utilization; coenzyme action	Water retention in tissues; tingling sensations; heart changes; poor coordination	None reported from food; possible shock reaction from repeated injections
B_2 (riboflavin)	Whole grains, poultry, fish, egg white, milk	Coenzyme action	Skin lesions	None reported
Niacin	Green leafy vegetables, potatoes, peanuts, poultry, fish, pork, beef	Coenzyme action	Contributes to pellagra (damage to skin, gut, nervous system, etc.)	Skin flushing; possible liver damage
B_6	Spinach, tomatoes, potatoes, meats	Coenzyme in amino acid metabolism	Skin, muscle, and nerve damage; anemia	Impaired coordination; numbness in feet
Pantothenic acid	In many foods (meats, yeast, egg yolk especially)	Coenzyme in glucose metabolism, fatty acid and steroid synthesis	Fatigue, tingling in hands, headaches, nausea	None reported; may cause diarrhea occasionally
Folate (folic acid)	Dark green vegetables, whole grains, yeast, lean meats; colon bacteria produce some folate	Coenzyme in nucleic acid and amino acid metabolism	A type of anemia; inflamed tongue; diarrhea; impaired growth; mental disorders	Masks vitamin B_{12} deficiency
B_{12}	Poultry, fish, red meat, dairy foods (not butter)	Coenzyme in nucleic acid metabolism	A type of anemia; impaired nerve function	None reported
Biotin	Legumes, egg yolk; colon bacteria produce some	Coenzyme in fat, glycogen formation, and amino acid metabolism	Scaly skin (dermatitis), sore tongue, depression, anemia	None reported
C (ascorbic acid)	Fruits and vegetables, especially citrus, berries, cantaloupe, cabbage, broccoli, green pepper	Collagen synthesis; possibly inhibits effects of free radicals; structural role in bone, cartilage, and teeth; role in carbohydrate metabolism	Scurvy, poor wound healing, impaired immunity	Diarrhea, other digestive upsets; may alter results of some diagnostic tests

*The guidelines for appropriate daily intakes are being worked out by the Food and Drug Administration.

Table 6.4 Major Minerals: Sources, Functions, and Effects of Deficiencies or Excesses*

Mineral	Common Sources	Main Functions	Signs of Severe Long-Term Deficiency	Signs of Extreme Excess
Calcium	Dairy products, dark green vegetables, dried legumes	Bone, tooth formation; blood clotting; neural and muscle action	Stunted growth; possibly diminished bone mass (osteoporosis)	Impaired absorption of other minerals; kidney stones in susceptible people
Chloride	Table salt (usually too much in diet)	HCl formation in stomach; contributes to body's acid–base balance; neural action	Muscle cramps; impaired growth; poor appetite	Contributes to high blood pressure in susceptible people
Copper	Nuts, legumes, seafood, drinking water	Used in synthesis of melanin, hemoglobin, and some electron transport chain components	Anemia, changes in bone and blood vessels	Nausea, liver damage
Fluorine	Fluoridated water, tea, seafood	Bone, tooth maintenance	Tooth decay	Digestive upsets; mottled teeth and deformed skeleton in chronic cases
Iodine	Marine fish, shellfish, iodized salt, dairy products	Thyroid hormone formation	Enlarged thyroid (goiter), with metabolic disorders	Goiter
Iron	Whole grains, green leafy vegetables, legumes, nuts, eggs, lean meat, molasses, dried fruit, shellfish	Formation of hemoglobin and cytochrome (electron transport chain component)	Iron-deficiency anemia, impaired immune function	Liver damage, shock, heart failure
Magnesium	Whole grains, legumes, nuts, dairy products	Coenzyme role in ATP-ADP cycle; roles in muscle, nerve function	Weak, sore muscles; impaired neural function	Impaired neural function
Phosphorus	Whole grains, poultry, red meat	Component of bone, teeth, nucleic acids, ATP, phospholipids	Muscular weakness; loss of minerals from bone	Impaired absorption of minerals into bone
Potassium	Diet provides ample amounts	Muscle and neural function; roles in protein synthesis and body's acid–base balance	Muscular weakness	Muscular weakness, paralysis, heart failure
Sodium	Table salt; diet provides ample to excessive amounts	Key role in body's acid–base balance; roles in muscle and neural function	Muscle cramps	High blood pressure in susceptible people
Sulfur	Proteins in diet	Component of body proteins	None reported	None likely
Zinc	Whole grains, legumes, nuts, meats, seafood	Component of digestive enzymes; roles in normal growth, wound healing, sperm formation, and taste and smell	Impaired growth, scaly skin, impaired immune function	Nausea, vomiting, diarrhea; impaired immune function and anemia

*The guidelines for appropriate daily intakes are being worked out by the Food and Drug Administration.

Generally, vitamin and mineral supplements are necessary only for strict vegetarians, the elderly, and people suffering a chronic illness or taking medication that affects the body's use of specific nutrients. For example, vitamin K supplements help older women retain calcium and diminish bone loss (osteoporosis). Vitamins E, C, and the beta-carotene precursor for vitamin A lessen some aging effects and improve immune function by inactivating free radicals. (A free radical, remember, is an atom or group of atoms that is highly reactive because it has an unpaired electron.)

However, metabolism varies in its details from one person to the next, so no one should take massive doses of any vitamin or mineral supplement except under medical supervision. Also, excessive amounts of many vitamins and minerals can harm anyone (Tables 6.3 and 6.4). For example, very large doses of the fat-soluble vitamins A and D can accumulate in tissues, especially in the liver, and interfere with normal metabolic functions. Similarly, sodium is present in plant and animal tissues and is a component of table salt. It has roles in the body's salt–water balance, muscle activity, and nerve function. However, prolonged, excessive intake of sodium may contribute to high blood pressure in some people.

Severe shortages or self-prescribed, massive excesses of vitamins and minerals can disturb the delicate balances in body function that promote health.

Have you ever asked yourself *why* you want to weigh a given amount? Are you simply afraid of being "fat"? This chapter's introduction described **obesity** as an excess of fat in the body's adipose tissues. Usually it is caused by imbalances between caloric intake and energy output. Of course, outside a doctor's office one's perception of obesity can be in the eye of the beholder, partly because cultural standards vary. If a woman who is considered "fat" in the United States goes to Africa, she may well find herself the object of a great deal of male attention. In much of Africa, carrying lots of body fat is considered attractive. However, a lot depends on how healthy you want to be and how long you want to live. In general, thinner people live longer.

In the United States, obesity is the second leading cause of death; roughly 300,000 people die each year from preventable, weight-related cases of type-2 diabetes, heart disease, high blood pressure, breast cancer, colon cancer, gout, gallstones, and osteoarthritis. Ominously, children today are 42 percent fatter than they were in 1980.

In Search of the "Ideal" Weight

Figure 6.18 shows charts for estimating the "ideal" weight for adults. Another indicator of obesity-related health risk is the *body mass index* (BMI). It is determined by the formula:

$$\text{BMI} = \frac{\text{weight (pounds)} \times 700}{\text{height (inches)}^2}$$

If your BMI value is 27 or higher, the health risk increases dramatically. Other factors that influence the risk are smoking, family history of heart disorders, use of sex hormones after menopause, and the distribution of body fat. Studies suggest that fat stored above the waist, as it is in a "beer belly," is an indicator of increased risk.

By itself, dieting cannot lower your BMI value. Eat less, and the body slows its metabolic rate to conserve energy. So how do you keep functioning normally over the long term while maintaining an acceptable weight? *Caloric intake must be balanced with energy output.* In this regard, a value called **basal metabolic rate** (BMR) is a factor. BMR is the amount of energy it takes to sustain the body when a person is resting, awake, and has not eaten for 12–18 hours. BMR varies from person to person; usually, it is higher in males. Other factors that influence a person's energy output include differences in physical activity, age, hormone activity, and emotional state.

To figure out how many kilocalories you should take in daily to maintain a desired weight, multiply that weight (in pounds) by 10 if you are not active physically, by 15 if moderately active, and by 20 if highly active. From the value you get this way, subtract the following amount:

Age	20–34	Subtract	0
	35–44		100
	45–54		200
	55–64		300
	Over 65		400

Figure 6.18 How to estimate the "ideal" weight for an adult. The values given are consistent with a long-term Harvard study into the link between excessive weight and increased risk of heart disorders. Depending on certain factors, such as having a small, medium, or large frame, the "ideal" may vary by plus or minus 10 percent.

Weight Guidelines for Women

Starting with an ideal weight of 100 pounds for a woman who is 5 feet tall, add five additional pounds for each additional inch of height. Examples:

Height (feet)	Weight (pounds)
5' 2"	110
5' 3"	115
5' 4"	120
5' 5"	125
5' 6"	130
5' 7"	135
5' 8"	140
5' 9"	145
5' 10"	150
5' 11"	155
6'	160

Weight Guidelines for Men

Starting with an ideal weight of 106 pounds for a man who is 5 feet tall, add six additional pounds for each additional inch of height. Examples:

Height (feet)	Weight (pounds)
5' 2"	118
5' 3"	124
5' 4"	130
5' 5"	136
5' 6"	142
5' 7"	148
5' 8"	154
5' 9"	160
5' 10"	166
5' 11"	172
6'	178

Table 6.5 Calories Expended in Some Common Activities

Activity	Kcal/hour per pound of body weight	Hours needed to lose 1 lb. fat		
		120 lbs	**155 lbs**	**185 lbs**
Basketball	3.78	7.7	6.0	5.0
Cycling (9 mph)	2.70	10.8	8.4	7.0
Hiking	2.52	11.6	8.9	7.5
Jogging	4.15	7.0	5.4	4.5
Mowing lawn (push mower)	3.06	9.5	7.4	6.2
Racquetball	3.90	7.5	5.8	4.8
Running (9-minute mile)	5.28	5.5	4.3	3.6
Snow skiing (cross-country)	4.43	6.6	5.1	4.3
Swimming (slow crawl)	3.48	8.4	6.5	5.4
Tennis	3.00	9.7	7.5	6.3
Walking (moderate pace)	2.16	13.5	10.4	8.7

To calculate these values for your own body weight, first multiply your weight by the kcal/hour expended for an activity to determine total kcal you use during one hour of the activity. Then divide that number into 3,500 (kcal in a pound of fat) to obtain the number of hours you must perform the activity to burn a pound of body fat.

For instance, if you want to weigh 120 pounds and are very active, $120 \times 20 = 2,400$ kilocalories. If you are 35 years old and moderately active, then you should take in a total of $1,800 - 100$, or 1,700 kcal a day. Along with this rough estimate, factors such as height and gender also must be considered. Males tend to have more muscle and so burn more calories; hence an active woman needs fewer kilocalories than an active man of the same height and weight. Nor does she need as many as another active woman who weighs the same but is several inches taller.

My Genes Made Me Do It

In most adults, energy input balances output, so body weight remains stable over long periods. It is as if the body has a *set point* for its weight and works to counteract deviations from the set point. As you have probably noticed, some people have far more trouble keeping off excess weight than others do. To some extent, a person's age and emotional state have something to do with it. More important, *genes* have a lot to do with it.

In 1995, researchers working with mice isolated a gene that operates only in adipose tissue—fat-storing cells. Named the *ob* gene (guess why), this bit of DNA's biochemical information translates into a hormone called leptin. A nearly identical hormone was isolated in humans. Hormones can be released into the bloodstream, and thereby may hang the tale of obesity.

Leptin is only one of several factors that influence the brain's commands to stimulate or suppress appetite, but it's a crucial one. By assessing the blood concentrations of incoming hormonal signals, an appetite control center in the brain "decides" whether the body has taken in enough food to provide enough fat for the day. If so, commands go out to increase metabolic rates—and to stop eating. However, if a faulty *ob* gene disrupts the normal feedback loop, then proper controls won't operate. Investigators theorize that some obese people may inherit such malfunctioning controls. During one experiment that supports the hypothesis, researchers injected obese mice with leptin. The mice quickly shed their excess weight. One day it may become possible to treat obesity with hormone-based therapy.

Dieting and Exercise

Many overweight people attempt to shed excess fat by dieting, but most eventually regain what they lose. Why? A person's weight set point contributes. As the chapter introduction noted, another reason is that the fat-storing cells of adipose tissue are an adaptation for survival—an energy reserve for times when food isn't available.

In addition, exercise is essential to long-term weight control. Losing a pound of fat requires expending about 3,500 kilocalories. For example, a 120-pound woman can burn a pound of fat by playing about 10 hours of tennis (Table 6.5). Permanent weight loss requires combining a moderate reduction in caloric intake with increased physical activity. This strategy seems to minimize the "starvation" response and increase the rate at which the body burns calories. Exercise also increases muscle mass, and even at rest muscle burns more calories than other types of tissues.

To maintain an acceptable body weight, energy input (caloric intake) must be balanced with energy output in the form of metabolic activity and exercise.

A person's energy output is influenced by basal metabolic rate, physical activity, age, hormone activity, and emotional state.

Permanent weight loss requires a moderate reduction in caloric intake combined with a consistent increase in physical activity.

6.11 MALNUTRITION AND UNDERNUTRITION

If you are reading this book, you almost certainly enjoy the luxury of a varied diet and adequate food supplies. Like millions of affluent people, you may carefully tailor your meals, picking and choosing foods that optimize your intake of certain vitamins, antioxidants, proteins, minerals, fats, and other substances. You may even be in the enviable position of needing to limit your food intake and exercise more because you have access to much more food energy than your body needs.

Many of your fellow humans are less fortunate. Over a lifetime, an average person in a country such as the United States, Italy, or Australia will consume 20 or 30 times as much food as an individual born in poor, undeveloped, and developing nations of Africa, Asia, and South America. Even in affluent countries, people who lack the resources or knowledge to obtain a healthy diet can suffer from malnutrition.

Malnutrition is a state in which body functions or development suffers due to inadequate or unbalanced food intake. A malnourished person may suffer from *undernutrition*—that is, the individual lacks food and thus does not obtain sufficient kilocalories or nutrients to sustain proper growth, body functioning, and development. Another word for undernutrition is *starvation*, and globally it is the most common form of malnutrition. Less often, but probably more common in affluent countries, a malnourished person consumes plenty of food from the standpoint of kilocalories, but has an unbalanced diet that lacks essential vitamins or minerals.

Global Undernutrition

Nutrition researchers estimate that at least half a billion people in the world go hungry every day. Virtually all of them are too poor to obtain adequate food. Over the long term, lack of food energy or essential nutrients disrupts homeostasis and weakens the body's immune defenses, among other effects. Every year up to 20 million people, three-fourths of them children, die of starvation or from diseases that easily ravage a chronically under-nourished body. Many other poorly nourished people, including elderly persons, are chronically weak, lethargic, prone to illness, or mentally impaired. Let's consider some of the most serious effects of severe nutritional deficiencies.

PROTEIN DEFICIENCY Many undernourished people take in not only too few kilocalories but especially too little protein. The result, *protein-energy malnutrition*, causes weakness, weight loss, impaired immunity, and other symptoms in adults, but it is especially devastating to infants and children. The child in Figure 6.19 shows the classic signs of *kwashiorkor*, a condition in which a child who may consume near-normal amounts of calories fails to grow and gain weight and shows other evidence of chronic protein deficiency. The swollen abdomen is due to edema (fluid retention) because there are too few proteins in the child's blood plasma. The osmotic imbalance causes water to leave the bloodstream and collect in tissue spaces. Proteins are also essential for proper functioning of the immune system, so affected children are sickly and susceptible to infection. Impaired brain development often leads to permanent mental retardation. Kwashiorkor typically develops when a child is weaned and its diet shifts from protein-rich breast milk to low-protein starchy foods.

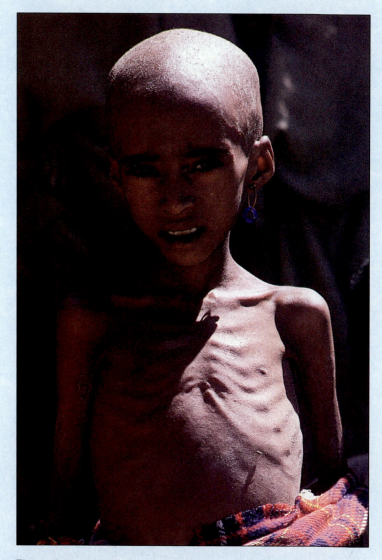

Figure 6.19 A child with kwashiorkor.

Further reading: Student Guide to InfoTrac on web site →

Many of the gruesome photographs of skeletal infants in Africa and elsewhere show the body-wasting disease *marasmus*. Usually marasmus affects very young children who lack both protein and food calories. It is common among bottle-fed infants of poor families in which parents dilute prepared formula, essentially "nourishing" their children mainly with water. Symptoms include low weight and muscle wasting, dry hair and skin, and retarded growth and mental development.

Some effects of kwashiorkor and marasmus can be halted or reversed if the child's diet is improved before the damage is too severe. As a practical matter, however, relief efforts are often too little or too late to save many affected children from death. Those that do survive are likely to function at abnormally low levels both physiologically and intellectually.

OTHER NUTRIENT DEFICIENCIES People in many underdeveloped countries are malnourished because their food sources are extremely limited. A starchy food, such as rice, corn, millet, or cassava, may make up the lion's share of most meals. This lack of variety translates into deficiencies of key vitamin and mineral nutrients. For example, nutritionists estimate that one-sixth of all humans, almost 1 billion people, are anemic because their diet lacks iron. Iron is a component of hemoglobin, the oxygen-transporting protein in blood. Major sources include meat and green, leafy vegetables. Iron deficiency impairs oxygen delivery to cells, and it also limits the ability of cells to extract energy from carbohydrates. In addition, it impairs the functioning of the immune system. Authorities cite a lack of dietary iron as a major factor contributing to high death rates from intestinal and respiratory infections in many countries. It also is the most common nutritional deficiency of children in the United States.

Lack of vitamin A causes *xeropthalmia*, a form of blindness that annually afflicts a quarter of a million children. Globally, xeropthalmia is the leading cause of preventable blindness, and it is irreversible.

Children are also the chief victims of iodide deficiency. This mineral, an ionic form of iodine, is a key component of thyroid hormones, which help govern normal growth and metabolism. In adults, low iodide causes goiter—overgrowth of the thyroid gland, which may be noticeable as a lump at the front or side of the neck. In children an iodide deficiency can permanently retard physical and mental development. Other common ailments associated with undernutrition in poor nations include scurvy (vitamin C deficiency), beriberi (vitamin B_1 deficiency), and rickets (vitamin D deficiency)

Groups at Highest Risk

As you surely have gathered by now, children are the main victims of malnutrition. The damage can begin during fetal development when a pregnant mother does not have access to sufficient protein and other nutrients. In general, babies born to undernourished mothers may be one to three weeks premature; they may have low birth weight and an underdeveloped respiratory system, as well as other potentially life-threatening complications. Elderly individuals also are at greater risk of undernutrition. The causes may be economic, social, or psychological (including clinical depression), and over time the predictable effects are declining immunity and impairment of other physiological functions.

Social and Political Factors in Global Malnutrition

The United Nations and relief organizations estimate that current global food supplies are adequate to meet the minimum nutritional needs of all or most of the earth's human population. Why then do people starve in places like Somalia, Ethiopia, and North Korea? Although the situation is complex, part of the answer is that world food supplies and human populations are not equally distributed. Most of the world's grain and other basic foods are produced in relatively wealthy countries, while 75 percent of the human population lives in developing regions such as Africa. Also, people in developed nations have more access to food because they tend to have more money to pay for it. Efforts to increase food production in poor countries are complicated by the high costs of fuel, fertilizers, and other inputs to agriculture and by the fact that arable land is being converted to other uses as the human population expands (see Chapter 25).

Politics also can get in the way of food distribution. For example, in recent years in various African countries and in embattled regions of central Europe, food earmarked for the general public has been stolen by warring factions and then sold to raise funds or has rotted in storage because warlords want to make sure that it cannot be used to feed their enemies.

Within the next three decades the human population is projected to rise from its current 6 billion to more than 11 billion. Given the recurring food crises we now face, it is difficult to imagine how we can adequately feed so many people. Clearly, it is crucial that the human population stabilize at a number that the planet's resources can sustain over the long term. We will return to this very troubling and challenging issue in Chapter 25.

SUMMARY

1. Nutrition refers to processes by which the body takes in, digests, absorbs, and uses food.

2. Four main activities take place in the digestive system:

   a. Mechanical processing and motility (movements that break up, mix, and propel ingested food through the gastrointestinal tract).

   b. Secretion (release of digestive enzymes and other substances from the salivary glands, pancreas, liver, and glandular epithelium into the lumen of the digestive tube).

   c. Digestion (chemical breakdown of food into particles, then into nutrient molecules small enough to be absorbed).

   d. Absorption (the passage of digested organic compounds, fluid, and ions into the internal environment).

   Undigested and unabsorbed residues are expelled at the end of the system.

3. The human digestive system (the gastrointestinal tract) includes the mouth, pharynx, esophagus, stomach, small intestine, and large intestine (colon). Accessory organs associated with digestion include the salivary glands, liver, gallbladder, and pancreas (see Table 6.6).

4. The GI tract is lined with mucous membrane, and from the esophagus onward the tube wall consists of four basic layers. The innermost mucosa is surrounded by the submucosa, a connective tissue layer. The next layer is smooth muscle, which is surrounded by a thin serous membrane, the serosa. Thickened muscular sphincters at either end of the stomach and at other locations help control the forward movement of ingested material and prevent backflow.

5. Starch digestion begins in the mouth, and protein digestion begins in the stomach. Digestion is completed and most nutrients are absorbed in the small intestine. After being absorbed, monosaccharides and most amino acids go directly to the liver. Fatty acids and triglycerides enter lymph vessels, then the general circulation.

6. Digestion in the small intestine depends on enzymes and other substances secreted by the pancreas, the liver, and the gallbladder. Pancreatic juice includes enzymes that act on all four main categories of organic molecules in food—carbohydrates, fats, proteins, and nucleic acids. Bile, which is secreted by the liver and then stored and released into the small intestine by the gallbladder, contains bile salts that speed up fat digestion via the process of emulsification.

7. Most nutrient absorption occurs across the lining of the small intestine, which is densely folded into villi (projections of the mucosa). Individual epithelial cells of villi bear microvilli on their surface. Microvilli greatly increase the surface area available for absorption.

Table 6.6 Summary of the Digestive System	
MOUTH (oral cavity)	Start of digestive system, where food is chewed and moistened; polysaccharide digestion begins here
PHARYNX	Entrance to tubular part of digestive and respiratory systems
ESOPHAGUS	Muscular tube, moistened by saliva, that moves food from pharynx to stomach
STOMACH	Sac where food mixes with gastric fluid and protein digestion begins; stretches to store food taken in faster than can be processed; gastric fluid destroys many microbes
SMALL INTESTINE	The first part (duodenum) receives secretions from the liver, gallbladder, and pancreas
	Most nutrients are digested and absorbed in the second part (jejunum)
	Some nutrients are absorbed in the last part (ileum), which delivers unabsorbed material to the colon
COLON (large intestine)	Concentrates and stores undigested matter (by absorbing mineral ions and water)
RECTUM	Distension triggers expulsion of feces
ANUS	Terminal opening of digestive system

Accessory Organs:

SALIVARY GLANDS	Glands (three main pairs, many minor ones) that secrete saliva, a fluid with polysaccharide-digesting enzymes, buffers, and mucus (which moistens and lubricates ingested foods)
PANCREAS	Secretes enzymes that digest all major food molecules; secretes buffers against HCl from the stomach
LIVER	Secretes bile (used in fat emulsification); roles in carbohydrate, fat, and protein metabolism
GALLBLADDER	Stores and concentrates bile from the liver

8. Nervous and hormonal controls over the digestive system respond to the volume and composition of food passing through. The response can be a change in muscle activity, in the secretion rate of hormones or enzymes, or in all of these.

9. To maintain acceptable weight and overall health, caloric intake must balance energy output. Complex carbohydrates are the body's preferred energy source. The diet must provide eight essential amino acids, a few essential fatty acids, vitamins, and minerals.

Review Questions

1. What are the main functions of the stomach? The small intestine? The large intestine? 6.1

2. Using the diagram on the next page, list the organs and the accessory organs of the human digestive system. On a separate piece of paper, list the main functions of each. 6.1

3. Name four kinds of molecules that are small enough to be absorbed across the intestinal lining. 6.4

4. What is segmentation? Where does it take place? 6.4, 6.6

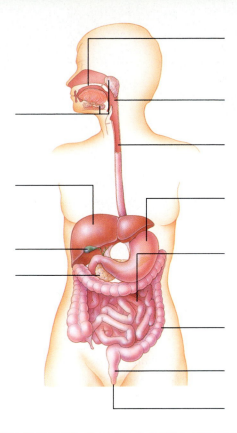

Self-Quiz (Answers in Appendix V)

1. The _____, _____, _____, and _____ systems interact in supplying body cells with raw materials, disposing of wastes, and maintaining the volume and composition of extracellular fluid.

2. Various specialized regions of the digestive system function in _____ and _____ food and in _____ unabsorbed food residues.

3. Maintaining good health and normal body weight requires that _____ intake be balanced by _____ output.

4. The preferred energy sources for the body are complex _____.

5. The human body cannot produce its own vitamins or minerals, and it also cannot produce certain _____ and _____.

6. Which of the following is *not* associated with digestion?
   a. salivary glands   d. gallbladder
   b. thymus gland   e. pancreas
   c. liver

7. Digestion is completed and products are absorbed in the _____.
   a. mouth   c. small intestine
   b. stomach   d. large intestine

8. After absorption, fatty acids and monoglycerides move into the _____.
   a. bloodstream   c. liver
   b. intestinal cells   d. lacteals

9. Excess carbohydrates and proteins are stored as _____
   a. amino acids   c. fats
   b. starches   d. monosaccharides

10. Match each of the following digestive system components with its description.
   ____ liver
   ____ small intestine
   ____ salivary glands
   ____ stomach
   ____ large intestine

   a. secrete substances that moisten food, start polysaccharide breakdown
   b. where protein digestion begins
   c. where water is reabsorbed
   d. where most digestion is completed
   e. receives monosaccharides and amino acids

Critical Thinking: You Decide (Key in Appendix VI)

1. A glass of whole milk contains lactose, protein, triglycerides (in butterfat), vitamins, and minerals. Explain what happens to each component when it passes through your digestive tract.

2. Some nutritionists claim that the secret to long life is to be slightly underweight as an adult. If a person's weight is related partly to diet, partly to activity level, and partly to genetics, what underlying factors could be at work to generate statistics that support this claim?

3. As a person ages, the number of body cells steadily decreases and energy needs decline. If you were planning an older person's diet, what kind(s) of nutrients would you emphasize, and why? Which ones would you recommend less of?

4. Along the lines of question 3, formulate a healthy diet for an actively growing seven-year-old.

5. Raw poultry can carry *Salmonella* or *campylobacter* bacteria, both of which produce toxins that can cause serious diarrhea, among other symptoms. Aside from the discomfort, why does such an infection require immediate medical attention?

Selected Key Terms

absorption 6.1	mineral 6.9
anus 6.6	motility 6.1
basal metabolic rate 6.10	obesity 6.10
digestion 6.1	palate 6.2
digestive system 6.1	pancreas 6.2
elimination 6.1	pepsin 6.2
esophagus 6.2	peristalsis 6.1
essential amino acid 6.8	pharynx 6.2
essential fatty acid 6.8	rectum 6.4
gallbladder 6.3	salivary amylase 6.2
gastrointestinal (GI) tract 6.1	salivary gland 6.2
hepatic portal vein 6.5	secretion 6.1
kilocalorie 6.8	small intestine 6.3
lacteal 6.4	stomach 6.3
large intestine 6.6	villus 6.4
malnutrition 6.11	vitamin 6.9

Readings

Blaser, M. J. February 1996. "The Bacteria behind Ulcers." *Scientific American.*

Jaret, P. January–February 1995. "The Way to Lose Weight." *Health.*

Kassirer, J. 17 September 1998. *New England Journal of Medicine.* Editorial introduces reports on unproven and harmful uses of dietary supplements, some of which were contaminated with heavy metals and with a stimulant that can cause cardiac arrest.

Sherwood, L. 1997. *Human Physiology.* Third edition. Pacific Grove, California: Brooks/Cole.

7

BLOOD AND CIRCULATION

Heartworks

For Dr. Augustus Waller, Jimmie the bulldog was no ordinary pooch. Connected to wires and soaked to his ankles in buckets of salty water, Jimmie was a four-footed explorer of the workings of the heart (Figure 7.1*a*). Press your fingers to your chest a few inches left of the center, between the fifth and sixth ribs, and feel the repetitive thumpings of your heart. The same rhythms intrigued Waller and other nineteenth-century physiologists (Figure 7.1*b*). They wondered: Does each beat of the heart produce a pattern of electrical currents? Could they find out by devising a painless way to record such currents at the body surface?

That's where Jimmie and the buckets of salty water came in. Saltwater happens to be an efficient conductor of electricity. In Waller's experiment, it picked up faint signals from Jimmie's beating heart through the skin of his legs and conducted them to a crude monitoring device. With that device, in the late 1880s Waller made one of the world's first recordings of heart activity. Today we call such a recording an ECG, for *electrocardiogram* (Figure 7.1*c*).

A graph of the normal electrical activity of your own heart would look much the same. The pattern emerged when you were an embryo inside your mother. Patches of newly formed cardiac tissue started to contract. Eventually, one patch took the lead and has been your heart's pacemaker ever since.

If all goes well, that patch of cardiac muscle cells will continue to contract as it should until the day you die. It is a natural pacemaker; it sets the baseline rate at which blood is pumped out of your heart, through the blood vessels, then back to the heart. The rate is moderate, about seventy beats every minute.

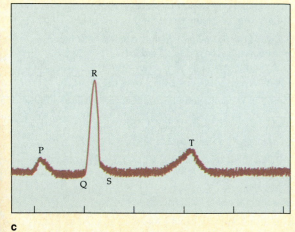

c

Figure 7.1 A bit of history in the making. (**a**) Jimmie the bulldog, taking part in a painless experiment. (**b**) Augustus Waller and his beloved pet bulldog sharing a quiet moment in Waller's study. (**c**) One of the world's first electrocardiograms.

This service of the pacemaker is vital, in part because blood performs such a remarkable array of functions. It brings oxygen and nutrients to cells, transports cell secretions such as hormones, and carries away metabolic wastes. Blood helps stabilize internal pH, and it serves as a highway for phagocytic cells that scavenge tissue debris and fight infections. It also helps equalize body temperature by carrying excess heat from regions of high metabolic activity (such as skeletal muscles) to the skin, where heat can be dissipated.

When your body is at rest, the demands on blood as a transport medium are at their lowest. However, join a fast game of volleyball and the demands by your muscles for blood-borne oxygen and glucose will soon escalate. Within moments, your heart may start pounding 150 times a minute or more, and this helps deliver enough blood to muscles.

We have come a long way from Waller and Jimmie in monitoring the heart. Sensors can now detect the faintest signals of an impending heart attack. Physicians now use computers to analyze a patient's beating heart and build images of it on a video screen. Surgeons substitute battery-powered pacemakers for natural ones that have malfunctioned.

With this chapter we turn to the circulatory system, known also as the **cardiovascular system**. Its physiological role is to move blood, and the substances in blood, rapidly to and from all your living cells. As it performs this function, the cardiovascular system interacts closely with the respiratory and digestive systems. As you probably already suspect, its smooth operation is central to the body's ability to maintain homeostasis.

KEY CONCEPTS

1. Cells survive by exchanging substances with their surroundings. The cardiovascular system rapidly moves substances to and from all living cells. The system consists of the heart, blood, and blood vessels. It is supplemented by a lymphatic system.

2. Blood is confined within the heart and blood vessels and is the transport medium of the cardiovascular system. It transports oxygen, carbon dioxide, plasma proteins, vitamins, hormones, lipids, and other solutes. It also transports heat generated by metabolism.

3. Blood pumped by the heart flows in two circuits. In the pulmonary circuit, the heart pumps deoxygenated blood to the lungs, where it releases carbon dioxide and picks up oxygen. Then blood flows back to the heart. In the systemic circuit, the heart pumps oxygenated blood to all body regions. After giving up oxygen and picking up carbon dioxide in those regions, blood flows back to the heart.

4. In addition to oxygen and carbon dioxide, blood also transports plasma proteins, vitamins, hormones, lipids, and other solutes.

5. The blood vessels called arteries and veins are large-diameter transport tubes. Capillaries are narrow-diameter tubes for diffusion. Venules (small veins) also collect blood from capillaries and channel it to veins. Arterioles (small arteries) have adjustable diameters. They are control points for the distribution of different volumes of blood flow to different regions.

CHAPTER AT A GLANCE

If you are an adult woman of average size, you have about 4 to 5 liters of blood in your body. It amounts to about 6 to 8 percent of your body weight. As in the old saying, human blood *is* thicker than water. It also flows more slowly, and it is rather sticky. But what exactly is this unusual liquid? **Blood** consists of plasma, red blood cells, white blood cells, and cell fragments called platelets. We now consider each of these components.

Plasma

If you fill a test tube with whole blood, treat it to prevent clotting, and whirl it in a centrifuge, the result should resemble the tube in Figure 7.2. About 55 percent of whole blood is **plasma**. The color of homemade lemonade, plasma is mostly water. It transports blood cells and platelets, and more than a hundred other substances. Most of these "substances" are different plasma proteins, which have a variety of functions.

Two-thirds of plasma proteins are molecules of albumin, which the liver manufactures from amino acids and then releases to the bloodstream. Because it is so plentiful in blood plasma—that is, because its concentration is relatively high—albumin has a major influence on the osmotic movement of water between the bloodstream and interstitial fluid. Too little albumin can be one cause of *edema*, swelling that occurs when water leaves the

blood and enters tissues. Said another way, albumin and other plasma proteins determine the fluid volume of a person's blood. Albumin also carries a variety of chemicals through the system, from metabolic wastes to therapeutic drugs.

Other plasma proteins include protein hormones, proteins of the immune system (Section 7.4), and those involved in blood clotting (Section 7.13). Lipoproteins carry cargoes of lipids, and still other plasma proteins transport fat-soluble vitamins.

Plasma also contains ions, glucose and other simple sugars, amino acids, various communication molecules, and dissolved gases—mostly oxygen, carbon dioxide, and nitrogen. The ions (such as Na^+, Cl^-, H^+, and K^+) help maintain extracellular pH and fluid volume.

Red and White Blood Cells

About 45 percent of whole blood—the bottom portion of your centrifuged test tube—consists of erythrocytes, or **red blood cells**. Each erythrocyte is a biconcave disk, rather like a thick pancake with an indentation on each side (Figure 7.3a). The cell's red color comes from the iron-containing protein *hemoglobin*, which transports oxygen used in aerobic respiration. Erythrocytes also carry away some carbon dioxide wastes. You will read about other features of red blood cells in upcoming sections.

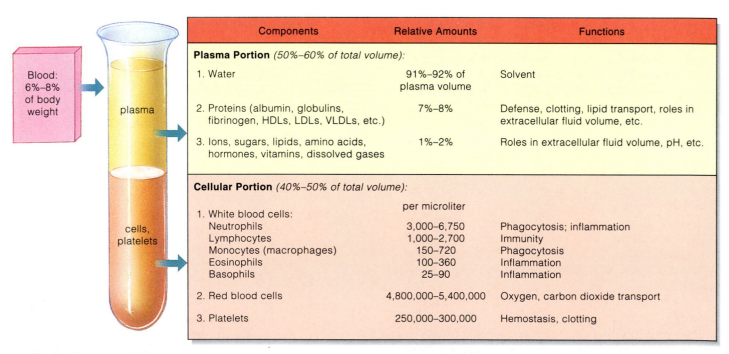

Components	Relative Amounts	Functions
Plasma Portion *(50%–60% of total volume):*		
1. Water	91%–92% of plasma volume	Solvent
2. Proteins (albumin, globulins, fibrinogen, HDLs, LDLs, VLDLs, etc.)	7%–8%	Defense, clotting, lipid transport, roles in extracellular fluid volume, etc.
3. Ions, sugars, lipids, amino acids, hormones, vitamins, dissolved gases	1%–2%	Roles in extracellular fluid volume, pH, etc.
Cellular Portion *(40%–50% of total volume):*		
	per microliter	
1. White blood cells:		
Neutrophils	3,000–6,750	Phagocytosis; inflammation
Lymphocytes	1,000–2,700	Immunity
Monocytes (macrophages)	150–720	Phagocytosis
Eosinophils	100–360	Inflammation
Basophils	25–90	Inflammation
2. Red blood cells	4,800,000–5,400,000	Oxygen, carbon dioxide transport
3. Platelets	250,000–300,000	Hemostasis, clotting

Blood: 6%–8% of body weight

plasma

cells, platelets

Figure 7.2 Components of blood. If a blood sample placed in a test tube is prevented from clotting, it will separate into a layer of straw-colored liquid, the plasma, that floats over the darker-colored cellular portion of blood.

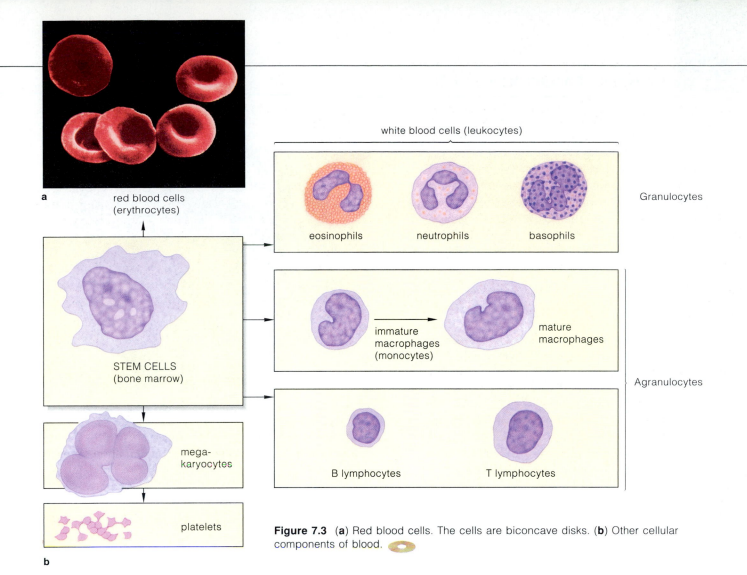

Figure 7.3 (**a**) Red blood cells. The cells are biconcave disks. (**b**) Other cellular components of blood.

Leukocytes, or **white blood cells**, make up a tiny fraction of whole blood. (Along with platelets, they are the thin, off-white, middle layer in your test tube.) Leukocytes have vital functions in day-to-day housekeeping and defense. Some scavenge dead or worn-out cells, as well as any material that is identified as foreign to the body. Others target or destroy specific bacteria, viruses, or other disease agents. Leukocytes arise from **stem cells** in bone marrow. They circulate in blood, but most go to work after they squeeze out of blood capillaries and enter tissues. The number of white blood cells varies, depending on whether a person is sedentary or highly active, healthy or battling an infection.

There are five types of white blood cells, divided into two major classes (Figure 7.3b). The **granulocytes** include *neutrophils*, *eosinophils*, and *basophils*. All have a lobed nucleus. When the cell is stained, various types of granules are visible in its cytoplasm. About two-thirds of all leukocytes are neutrophils, "search and destroy" cells that follow chemical trails to infected, inflamed, or damaged tissues. Eosinophils take part in allergic responses and attacks on invading parasites. Basophils are a bit more mysterious, but they appear to have roles in allergies and inflammation responses.

The leukocytes known as **agranulocytes** have no visible granules in their cytoplasm. One type, *monocytes*, develop into phagocytic macrophages that engulf invading microbes and cellular debris in tissues. Another type, *lymphocytes* (B cells and T cells), have key roles in specific immune responses. Most types of white cells live for only a few days or, during a major infection, a few hours. As described in Chapter 8, others may live for many years.

Platelets

Some stem cells in bone marrow develop into "giant" cells called megakaryocytes. These cells shed fragments of their cytoplasm, which become enclosed in a bit of plasma membrane. The membrane-bound fragments, known as **platelets** (or thrombocytes), are disks about 2 to 4 micrometers across. Each lasts only about a week, but millions are always circulating in our blood. Substances released from platelets initiate blood clotting.

Blood consists of plasma, in which proteins and other substances are dissolved; red blood cells; white blood cells; and platelets.

OXYGEN TRANSPORT IN BLOOD

Hemoglobin, the Oxygen Carrier

If you were to analyze a liter of blood drawn from an artery, you would find only a small amount of oxygen dissolved in the plasma—just 3 milliliters. Yet, like all large, active, warm-blooded animals, humans require a great deal of oxygen to maintain the metabolic activity of their cells. Hemoglobin (Hb) meets this need. In addition to 3 milliliters of dissolved oxygen, a liter of arterial blood generally carries roughly 194 milliliters of oxygen bound to the heme groups of hemoglobin molecules.

Factors That Affect Oxygen Binding

As conditions vary in different tissues and organs, so does the tendency of hemoglobin to bind with and hold onto oxygen. Several factors influence this process, which is vital in helping to maintain homeostasis. The most important factor is the amount of oxygen present relative to the amount of carbon dioxide. Other factors are the temperature and the acidity of tissues. Hemoglobin binds oxygen most readily where blood plasma contains a relatively large amount of oxygen, where the temperature is relatively cool, and where the pH is roughly neutral. This is exactly the environment in our lungs, where the body's blood supply must be oxygenated. By contrast, metabolic activity *requires* oxygen, and it also increases both the temperature and the acidity (lowers the pH) of tissues. Under those conditions, the oxyhemoglobin of red blood cells arriving in tissue capillaries tends to give up oxygen, which then becomes available to metabolically active cells. We can summarize these events this way:

	LUNGS	TISSUES	
more O_2 cooler less acidic	Hb + O_2 → HbO_2	HbO_2 → Hb + O_2	less O_2 warmer more acidic

Hemoglobin also transports some of the carbon dioxide wastes of aerobic metabolism and carries hydrogen ions that help control the pH of body fluids. We will return to hemoglobin's functions in Chapter 9, where we consider the many interacting elements that enable the respiratory system to efficiently transport gases to and from our body cells.

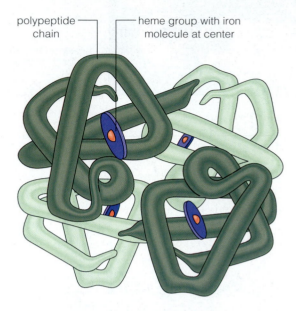

polypeptide chain — heme group with iron molecule at center

Figure 7.4 The structure of hemoglobin. Recall from Chapter 2 (Figure 2.29) that hemoglobin is a globular protein consisting of four polypeptide chains and four iron-containing heme groups. Oxygen binds to the iron in heme groups, which is one reason why humans require iron as a mineral nutrient.

Figure 7.4 shows a hemoglobin molecule's structure. Notice that it has two components: the protein *globin* and nonprotein *heme groups* that contain iron. Globin is built of four linked polypeptide chains, and a heme group is associated with each chain. It is the iron molecule at the center of each heme group that binds oxygen.

Oxygen in the lungs diffuses into the blood plasma, then into individual red blood cells, where it binds with the iron in hemoglobin. This oxygenated hemoglobin, or *oxyhemoglobin*, is deep red. Hemoglobin depleted of oxygen appears scarlet, especially when it is observed through skin and the walls of blood vessels.

Hemoglobin in red blood cells transports oxygen. The oxygen becomes bound to iron molecules in heme groups in each hemoglobin molecule.

Factors that influence hemoglobin binding—and the amount of oxygen available to tissues—include the relative amounts of oxygen and carbon dioxide present in blood and the temperature and the acidity of tissues.

LIFE CYCLE OF RED BLOOD CELLS

As Figure 7.3 shows, red blood cells are derived from stem cells. In general, a **stem cell** is unspecialized—sort of a "blank slate." However, its descendants can differentiate into cells that *are* specialized for a particular function. Red blood cells are the descendants of stem cells in red bone marrow. Recall from Chapter 4 that adults have this kind of marrow in certain bones of the skull, the vertebrae, and the sternum (breastbone). These stem cells begin to crank out red blood cells when they receive a signal from a hormone called *erythropoietin*, which is produced in the kidneys. Roughly 3 million new red blood cells enter your bloodstream each second! As they mature (Figure 7.5), they lose their nucleus, ribosomes, and other structures. Red blood cells do not divide or synthesize new proteins, but they have enough enzymes and other proteins to function for about 120 days.

In "blood doping," some of an athlete's blood is withdrawn and stored. In response to the loss of red blood cells, erythropoietin stimulates the production of replacements. The stored blood is then reinjected several days prior to an athletic event, so that the athlete has more than the normal number of red blood cells to carry oxygen to muscles—and a temporary, though unethical, competitive advantage.

As red blood cells near the end of their useful life, die, or become damaged or abnormal, phagocytes called macrophages ("big eaters") remove them from the blood. Much of this "cleanup" occurs in the spleen, which is located in the upper left abdomen. As a macrophage dismantles a hemoglobin molecule, amino acids from its proteins return to the bloodstream and the iron in its heme groups returns to red bone marrow, where it may be recycled in new red blood cells. The rest of the heme group is converted to the orangish pigment bilirubin, which travels to the liver and is mixed with bile secreted during digestion.

Ongoing replacements from stem cells keep the red blood cell count fairly constant. A **cell count** tallies the number of cells in a microliter of blood. The red blood cell count averages 5.4 million in adult males, 4.8 million in females.

Normally, feedback mechanisms help stabilize the red blood cell count. For example, if you take a ski trip in the Rockies, your body must work harder to obtain enough oxygen because air contains less oxygen per unit volume at high altitudes. First your kidneys secrete erythropoietin, which stimulates stem cells in bone marrow to step up production of red blood cells. New red blood cells enter your bloodstream, increasing your blood's capacity for carrying oxygen. As the oxygen level rises in your blood and in tissues, this information feeds back to the kidneys, erythropoietin production falls, and red blood cell production drops.

Anemia develops when the blood's capacity to carry oxygen is reduced. An anemic person tends to feel tired and cold, and may look pale and be short of breath. In *iron-deficiency anemia*, red blood cells contain less hemoglobin than normal. This disorder generally results from an iron-poor diet and is treatable with a mineral supplement. *Pernicious anemia* is caused by a deficiency of vitamin B_{12}, which is required for red blood cells to mature properly. It usually reflects an underlying disorder in which the intestines are unable to absorb that vitamin from food. Treatment involves regular injections of vitamin B_{12}. Other types of anemia develop when a person has too few red blood cells, as when a hemorrhage causes severe blood loss or when disease or radiation therapy destroys stem cells in bone marrow.

Red blood cells arise from stem cells in red bone marrow and live about 120 days. Ongoing replacements from stem cells keep the red blood cell count fairly constant.

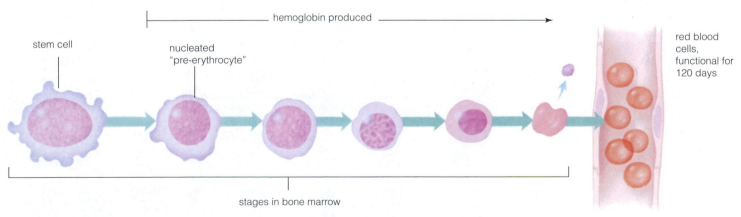

Figure 7.5 Life cycle of a red blood cell. A stem cell in red bone marrow gives rise to a nucleated "pre-erythrocyte," which begins making hemoglobin. Later, when the cell's cytoplasm is full of hemoglobin, its ability to synthesize proteins stops. The nucleus is expelled, the center of the cell caves in, and the cell, now an erythrocyte, is released from the bone marrow into the bloodstream.

BLOOD TYPING

Genetically determined surface proteins mark each of your body cells, including red blood cells, as "self" (see Section 3.3). You probably know that there are different human blood types; the differences are due to variations in the surface markers on red blood cells. Such variations are medically important because immune system proteins called *antibodies* recognize and organize an attack on most entities, including viruses and bacteria, that aren't one of the body's own cells. This happens precisely because the foreign material has "nonself" markers. Any large molecule that prompts such a defensive attack is called an *antigen*. ABO antigens, which are glycoproteins, fill this bill. Chapter 8 delves more fully into antigen–antibody interactions. For the time being, we are interested in what happens when the blood of two people mixes, as it does during a transfusion.

ABO Blood Typing

One kind of marker on human red blood cells has variant forms known as A, B, and O. This group is of particular interest because severe immune responses result when incompatible types are mixed. In type A blood, red blood cells bear A markers. Type B blood has B markers, type AB has both A and B, and type O has neither. If you are type A, your body does not have antibodies against A markers but does have them against B markers. If you are type B, you lack antibodies against B markers, but you do have antibodies against A markers. If you are

type AB, you lack antibodies against either form of the marker. If you are type O, you have antibodies against both forms of the marker, so you are limited to type O donations. In theory, type O people are "universal donors," because they have neither A nor B antigens, and—again, only in theory—type AB people are "universal recipients." In fact, however, there are *many* antigens in blood, any of which can trigger the defensive response known as **agglutination** (Figure 7.6). When commingled blood causes agglutination, antibodies act against the foreign cells and cause them to clump. The clumps can clog small blood vessels, causing severe tissue damage and even death.

Rh Blood Typing

Of the various kinds of surface markers on red blood cells that can cause agglutination, one you may have heard of is the "RH factor." **Rh blood typing** is based on the presence or absence of an Rh marker (so named because it was first identified in the blood of *Rh*esus monkeys). If you are type Rh$^+$, your blood cells bear this marker. If you are type Rh$^-$, they don't. Ordinarily, people do not have antibodies against Rh markers. But a recipient of an Rh$^+$ blood transfusion will produce antibodies against the marker, and these will continue circulating in the blood.

If an Rh$^-$ woman becomes pregnant by an Rh$^+$ man, there is a chance the fetus will be Rh$^+$. During pregnancy or childbirth, some of the fetal red blood cells may leak

Figure 7.6 (**a**) Agglutination responses in blood types O, A, B, and AB when mixed with blood samples of the same and different types. (**b**) Micrographs showing the absence of agglutination in a mixture of two different but compatible types (**top**) and agglutination in a mixture of incompatible blood types (**bottom**).

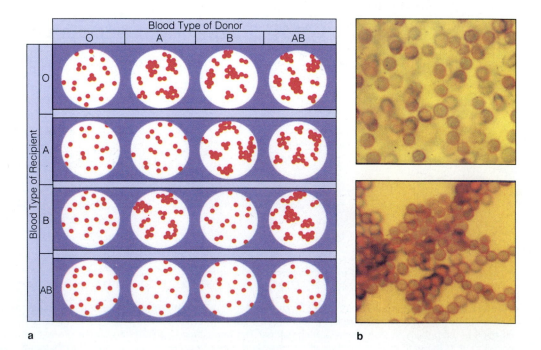

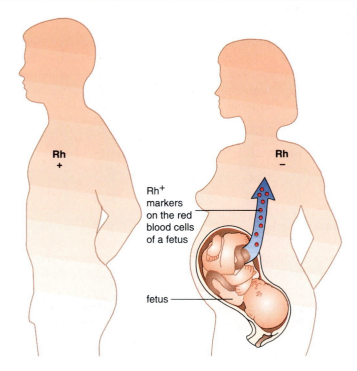

a A forthcoming child of an Rh⁻ woman and Rh⁺ man inherits the genetic instructions for the Rh⁺ marker. Some blood cells from the growing fetus can leak into the mother's bloodstream, passing through the placenta (a complex tissue that forms during pregnancy).

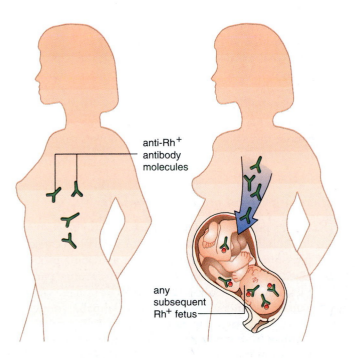

b The foreign markers in the mother's body stimulate production of antibody molecules. Suppose the woman becomes pregnant again. If the new fetus (or any other one) inherits instructions for the Rh⁺ marker, the anti-Rh⁺ antibodies will act against them.

Figure 7.7 Development of antibodies in response to Rh⁺ blood.

into the mother's bloodstream. If they do, her body will produce antibodies against Rh (Figure 7.7). If she becomes pregnant *again*, Rh antibodies will enter the bloodstream of this new fetus. If its blood is type Rh⁺, its mother's antibodies will cause its red blood cells to swell and burst.

In extreme cases, called *hemolytic disease of the newborn*, so many red blood cells are destroyed that the fetus dies. If the condition is diagnosed before or at a live birth, the baby can survive by having its blood replaced with transfusions free of Rh antibodies.

Currently, a known Rh⁻ woman can be treated after her first pregnancy with an anti-Rh gamma globulin (RhoGam) that will protect her next fetus. The drug will inactivate Rh⁺ fetal blood cells circulating in the mother's bloodstream before she can become sensitized and begin producing anti-Rh antibodies. In non-maternity cases, an Rh⁻ person who receives a transfusion of Rh⁺ blood also can have a severe negative reaction if he or she has previously been exposed to the Rh marker.

Besides the ABO and Rh blood markers, hundreds of others are now known to exist. These markers are a bit like needles in a haystack, for they are widely scattered within the human population and usually don't cause problems in transfusions. Reactions do occur, though; and except in cases of extreme emergency, hospitals use a technique called *cross-matching* to exclude the remote possibility that blood to be transfused and that of a patient might be incompatible due to the presence of a rare blood cell marker outside the ABO and Rh groups.

Other Applications of Blood Typing

ABO markers occur not only in blood, but also in semen and saliva. Because blood groups are determined by genes, they are a useful cache of information about a person's genetic heritage. For example, it is common for criminal investigations of murders, rapes, and sometimes other crimes to compare the blood groups of victims and any possible perpetrators. Blood samples also provide DNA for testing, both in criminal cases and for determining the identity of a child's father or mother. We consider this modern application of biotechnology in Chapter 21.

Like all cells, red blood cells bear genetically determined proteins on their surface. These proteins serve as "self" markers and determine a person's ABO (and Rh) blood type.

When incompatible blood types mix, an agglutination response occurs in which antibodies cause potentially fatal clumping of red blood cells.

OVERVIEW OF THE CARDIOVASCULAR SYSTEM

The basic function of the cardiovascular system is the rapid transport of blood. When it works well, it efficiently circulates blood to the immediate neighborhood of every living cell in the body. Blood, in turn, is aptly called the "river of life." It brings cells such essentials as oxygen, nutrients from food, and secretions (such as hormones) from other cells. It also carts away the wastes produced by our metabolism, along with excess heat. In fact, cells depend on circulating blood to make constant pickups and deliveries of an amazingly diverse range of substances, including those that move into or out of the digestive system, as described in Chapter 6, and into and out of the respiratory and urinary systems (Figure 7.8).

Homeostasis is one of our constant themes in this book, so it's good to keep in mind that maintaining it would be impossible were it not for our circulating blood. Cells must exchange substances with blood because that is a key way they adjust to changes in the chemical makeup, volume, and temperature of the interstitial fluid—the "external environment" in which they live.

Cardiovascular System Components

"Cardiovascular" comes from the Greek *kardia* (heart) and the Latin *vasculum* (vessel). Figure 7.9 gives you an overview of the human cardiovascular system, which has three main elements: (1) blood, with its different components; (2) the heart, a muscular pump that generates the pressure required to move blood throughout the body; and (3) blood vessels, which you can think of as tubes having lumens of different diameters.

The heart pumps blood into large-diameter **arteries**. From there blood flows into smaller, muscular **arterioles**, which branch into even smaller-diameter **capillaries**. Blood flows from capillaries into small **venules**, then into large-diameter **veins** that return blood to the heart. The heart pumps constantly. Thus the volume of flow through the entire system each minute is equal to the volume of blood returned to the heart each minute.

As described later on, the rate and volume of blood flow through the cardiovascular system can be adjusted as conditions in the body vary. For example, blood flows rapidly through arteries, but in capillaries it must flow slowly so that there is enough time for substances moving to and from cells to be exchanged with interstitial fluid. This slow flow occurs in *capillary beds*, where blood moves through vast numbers of slender capillaries. By dividing up the blood flow, the small-diameter capillaries handle the same total volume of flow as the large-diameter vessels, but at a slower pace.

Links with the Lymphatic System

The heart's pumping action puts pressure on blood flowing through the cardiovascular system. Partly because of this pressure, small amounts of water and some proteins dissolved in blood are forced out and become part of interstitial fluid. An elaborate network of drainage vessels picks up excess interstitial fluid and reclaimable solutes—including water, proteins, fatty acids, and glycerol—and returns them to the cardiovascular system. This network is part of the lymphatic system, about which you will learn more in Section 7.15.

The cardiovascular system transports substances to and from the interstitial fluid that bathes all living cells.

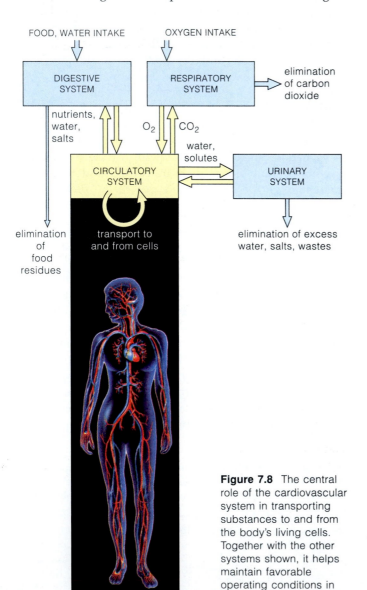

FOOD, WATER INTAKE   OXYGEN INTAKE

DIGESTIVE SYSTEM

RESPIRATORY SYSTEM

elimination of carbon dioxide

nutrients, water, salts

O_2 CO_2

water, solutes

CIRCULATORY SYSTEM

URINARY SYSTEM

elimination of food residues

transport to and from cells

elimination of excess water, salts, wastes

Figure 7.8 The central role of the cardiovascular system in transporting substances to and from the body's living cells. Together with the other systems shown, it helps maintain favorable operating conditions in the internal environment.

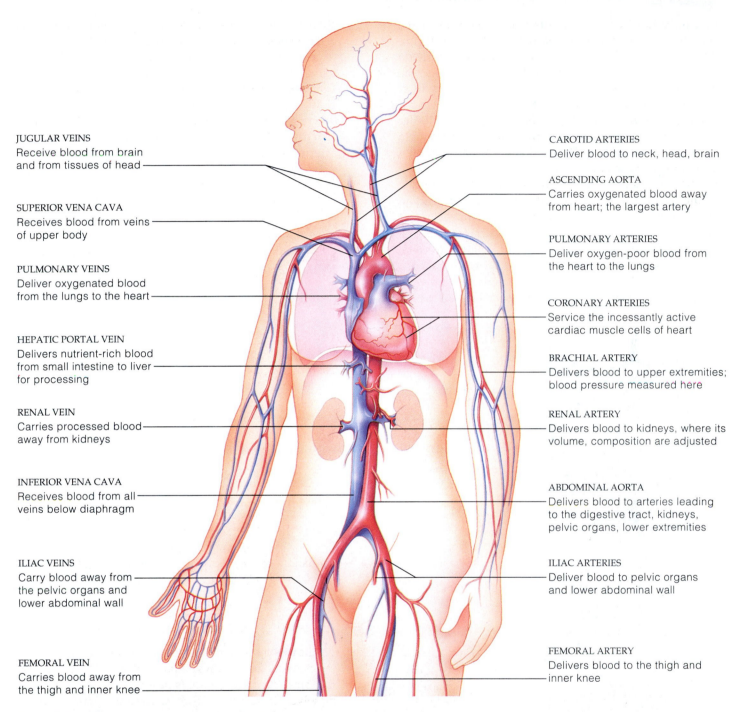

JUGULAR VEINS
Receive blood from brain and from tissues of head

SUPERIOR VENA CAVA
Receives blood from veins of upper body

PULMONARY VEINS
Deliver oxygenated blood from the lungs to the heart

HEPATIC PORTAL VEIN
Delivers nutrient-rich blood from small intestine to liver for processing

RENAL VEIN
Carries processed blood away from kidneys

INFERIOR VENA CAVA
Receives blood from all veins below diaphragm

ILIAC VEINS
Carry blood away from the pelvic organs and lower abdominal wall

FEMORAL VEIN
Carries blood away from the thigh and inner knee

CAROTID ARTERIES
Deliver blood to neck, head, brain

ASCENDING AORTA
Carries oxygenated blood away from heart; the largest artery

PULMONARY ARTERIES
Deliver oxygen-poor blood from the heart to the lungs

CORONARY ARTERIES
Service the incessantly active cardiac muscle cells of heart

BRACHIAL ARTERY
Delivers blood to upper extremities; blood pressure measured here

RENAL ARTERY
Delivers blood to kidneys, where its volume, composition are adjusted

ABDOMINAL AORTA
Delivers blood to arteries leading to the digestive tract, kidneys, pelvic organs, lower extremities

ILIAC ARTERIES
Deliver blood to pelvic organs and lower abdominal wall

FEMORAL ARTERY
Delivers blood to the thigh and inner knee

Figure 7.9 Human cardiovascular system. Arteries, which carry oxygenated blood to tissues, are shaded *red*. Veins, which carry deoxygenated blood away from tissues, are shaded *blue*. Notice, however, that for the pulmonary arteries and veins the roles are reversed.

THE HEART: A DURABLE PUMP

In a lifetime of 70 years, the human **heart** beats some 2.5 billion times. The heart's structure, shown in Figure 7.10, reflects this marvelous organ's role as a durable pump. The heart is mostly cardiac muscle tissue, the **myocardium**. A tough, fibrous sac, the pericardium (*peri* = around), surrounds, protects, and lubricates the heart. The heart's inner chambers have a smooth lining (endocardium) composed of connective tissue and a layer of epithelial cells. The epithelial cell layer, known as endothelium, also lines the inside of blood vessels.

A thick wall, the **septum**, divides the heart into two halves, right and left. Each half has two chambers: an **atrium** (plural: atria) located above a **ventricle**. Flaps of membrane separate the two chambers and serve as a one-way **atrioventricular valve** (AV valve) between them. The AV valve in the right half of the heart is called a *tricuspid valve* because its three flaps come together in pointed cusps (Figure 7.11). In the heart's left half the AV valve consists of just two flaps; it is called the *bicuspid valve* or *mitral valve*. Tough, collagen-reinforced strands (chordae tendineae, or "heartstrings") connect the AV valve flaps to cone-shaped muscles that extend out from the ventricle wall. When a blood-filled ventricle contracts, this arrangement prevents the flaps from opening backward into the atrium. Each half of the heart also has a **semilunar valve** between the ventricle and the arteries leading away from it. During a heartbeat, this valve opens and closes in ways that keep blood moving in one direction through the body.

The heart has its own "coronary circulation." Two **coronary arteries** lead into a capillary bed that services most of its cardiac muscle cells (Figure 7.12). They branch off the **aorta**, the major artery carrying oxygenated blood away from the heart. Coronary arteries become dangerously clogged in some cardiovascular disorders, as described in Section 7.14.

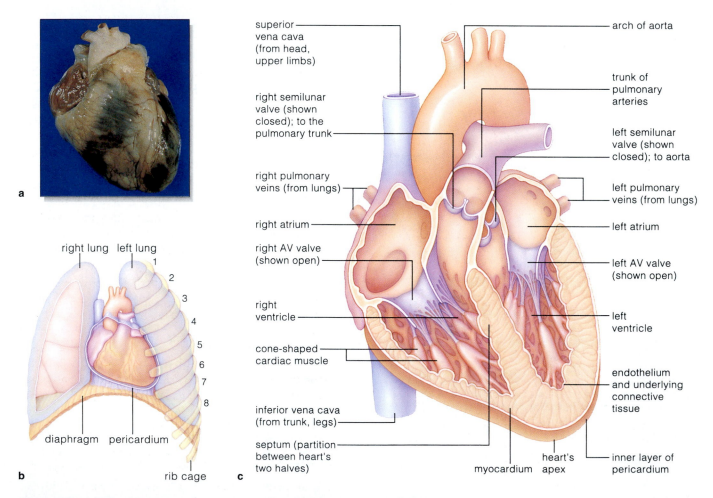

Figure 7.10 (a) The human heart, and (b) its location. (c) Cutaway view showing the heart's internal organization.

Heartbeat: The Cardiac Cycle

Blood is pumped each time the heart beats. A "heartbeat" is one sequence of contraction and relaxation of the heart chambers. The sequence occurs almost simultaneously in both sides of the heart. The contraction phase is called **systole** (SISS-toe-lee), and the relaxation phase is called **diastole** (dye-ASS-toe-lee). This sequence of muscle contraction and relaxation is a **cardiac cycle** (Figure 7.13). Each heartbeat lasts about eight-tenths of a second.

During the cycle, the ventricles relax before the atria contract, and the ventricles contract when the atria relax. When the relaxed atria are filling with blood, the fluid

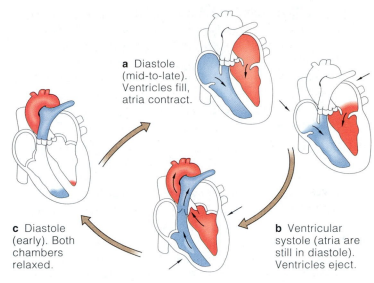

Figure 7.13 Blood flow during part of a cardiac cycle. Blood and heart movements produce a lub-dup sound that can be heard at the chest wall.

a Diastole (mid-to-late). Ventricles fill, atria contract.

b Ventricular systole (atria are still in diastole). Ventricles eject.

c Diastole (early). Both chambers relaxed.

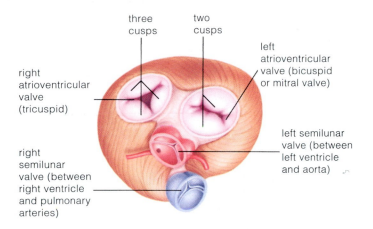

Figure 7.11 The valves of the heart. In this drawing, you are looking down at the heart, and the atria have been removed so that the atrioventricular and semilunar valves are visible.

three cusps

two cusps

left atrioventricular valve (bicuspid or mitral valve)

right atrioventricular valve (tricuspid)

left semilunar valve (between left ventricle and aorta)

right semilunar valve (between right ventricle and pulmonary arteries)

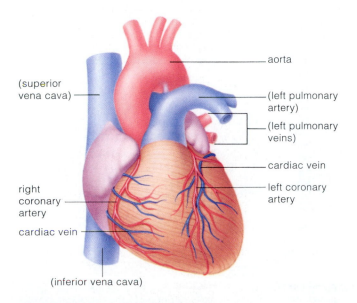

Figure 7.12 Coronary arteries and veins.

(superior vena cava)

aorta

(left pulmonary artery)

(left pulmonary veins)

cardiac vein

left coronary artery

right coronary artery

cardiac vein

(inferior vena cava)

pressure inside them rises and the AV valves open. Blood flows into the ventricles, which are 80 percent filled by the time the atria contract. As the filled ventricles begin to contract, fluid pressure inside them increases, forcing the AV valves shut. Their continuing contraction boosts pressure in the ventricles *above* that in blood vessels leading away from the heart. The pressure forces the semilunar valves open, so blood flows out of the heart and into the aorta and pulmonary artery. After blood has been ejected, the ventricles relax, and the semilunar valves close. For about half a second the atria and ventricles are all in diastole; then the blood-filled atria contract, and the cycle repeats.

The blood and heart movements during the cardiac cycle generate an audible "lub-dup" sound made by the forceful closing of the heart's one-way valves. At each "lub," the AV valves are closing as the two ventricles contract. At each "dup" the semilunar valves are closing as the ventricles relax. With this overview of the heart's structure in mind, we now can look more closely at the vascular routes blood travels each time the heart pumps.

The heart consists of two halves, each divided into an atrium and a ventricle. Valves in each half of the heart help control the direction of blood flow.

During a cardiac cycle, contraction of the atria helps fill the ventricles. Contraction of the ventricles is the driving force for blood circulation away from the heart.

Each half of the heart (atrium and ventricle) pumps blood. The two side-by-side pumps are the basis of two cardiovascular circuits through the body, the pulmonary and systemic circuits (Figure 7.14). Each circuit has its own set of arteries, arterioles, capillaries, venules, and veins.

The Two Circuits

The **pulmonary circuit** receives blood from tissues and circulates it through the lungs for gas exchange. It begins as blood from tissues enters the right atrium, then moves through the AV valve into the right ventricle. As the ventricle fills, the atrium contracts. Blood arriving in the right ventricle is fairly low in oxygen and high in carbon dioxide. When the ventricle contracts, the blood moves through the right semilunar valve into the *main* pulmonary artery, then into the *right* and *left* pulmonary arteries. These arteries carry the blood to the two lungs, where (in capillaries) it picks up oxygen and gives up carbon dioxide that will be exhaled. The freshly oxygenated blood returns through two sets of pulmonary veins to the heart's left atrium, completing the circuit.

The **systemic circuit** carries blood to and from tissues. Oxygenated blood pumped by the left half of the heart moves through the body and returns to the right atrium. The left atrium receives blood from pulmonary veins, and this blood moves through an AV (bicuspid) valve to the left ventricle. This chamber contracts forcefully to send blood coursing through a semilunar valve into the aorta.

As the aorta descends into the torso (see Figure 7.9), major arteries branch off it, funneling blood to organs and tissues where O_2 is used and CO_2 is produced.

Deoxygenated blood returns to the right half of the heart, where it enters the pulmonary circuit. Notice that in both the pulmonary and the systemic circuits, blood travels through arteries, arterioles, capillaries, and venules, and finally returns to the heart in veins. Blood from the head, arms, and chest arrives through the *superior vena cava,* and the *inferior vena cava* collects blood from the lower body.

Arteries branch off the aorta, carrying blood directly to capillary beds in specific tissues and organs. For example, in a resting person, each minute roughly a quarter of the blood pumped into the systemic circulation enters the kidneys via *renal arteries.* Chapter 10 discusses kidney functions, which include removing metabolic wastes.

Substances actually move between blood and tissues in **capillary beds.** A few capillary beds have two types of vessels: "true capillaries" where exchanges between

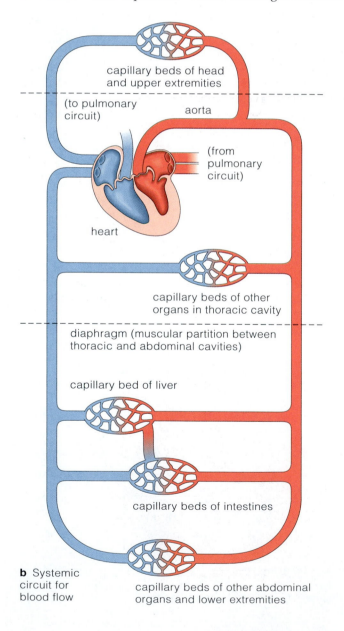

b Systemic circuit for blood flow

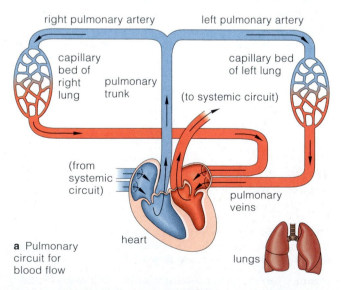

a Pulmonary circuit for blood flow

Figure 7.14 The (**a**) pulmonary and (**b**) systemic circuits for blood flow in the cardiovascular system. (**c**) Distribution of the the heart's output in a person at rest.

blood and tissues take place, and "thoroughfare channels" that connect arterioles and venules. Blood flow into true capillaries is controlled by precapillary sphincters—wispy collars of smooth muscle cells where the capillaries branch from thoroughfare channels. The muscle cells are sensitive to changes in the amount of carbon dioxide in their surroundings. When the CO_2 level rises above a set point—suggesting a homeostatic "need" for more oxygen-bearing blood in the area—the sphincter relaxes and more blood flows through the capillary. When the CO_2 level falls, the sphincter contracts and less blood flows through.

A Detour through the Liver

After a meal the blood passing through capillary beds in the GI tract (taking up nutrients) detours through the *hepatic portal vein* to another capillary bed in the liver. As blood seeps through this second bed, the liver can remove impurities and process absorbed substances. In part of this processing, the liver synthesizes cholesterol. This chapter's *Science Comes to Life* explains how, in people who have an unhealthy excess of cholesterol in their blood, potent drugs called statins can dramatically reduce the liver's cholesterol output.

Blood leaving the liver's capillary bed enters the general circulation through a *hepatic vein*. (The liver also receives its own supply of oxygenated blood via the *hepatic artery*.)

A short pulmonary circuit carries blood through the lungs for gas exchange. A long systemic circuit transports blood to and from tissues.

After you eat, blood in capillary beds in the GI tract is diverted to the liver, which processes absorbed nutrients and removes impurities before the blood re-enters the general circulation.

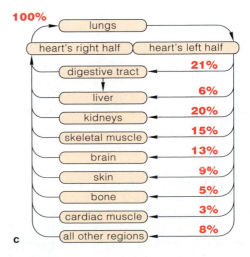

c

In the 1980s, researchers looking for chemical weapons against cancer stumbled onto a revolution in treating heart disease. They discovered that one class of experimental drugs, called *statins*, dramatically reduced the amount of "bad" (LDL) cholesterol carried in the blood (see Section 7.14). Further studies revealed at least four reasons why. To begin with, statins—with names such as pravastatin, lovastatin, atorvastatin, and others—interrupt the metabolic pathway in the liver that creates cholesterol. They also increase the liver's output of receptor proteins that bind with and remove LDL cholesterol from the bloodstream. Other experiments showed that statins *raise* the blood level of "good" cholesterol (HDLs), and they lower blood levels of triglycerides (Chapter 2). Like high LDL cholesterol, having a chronically high level of triglycerides in the blood seems to be a risk factor in heart disease.

Clinical trials of statins began in the mid-1990s, and the results started a revolution in clinical care of heart patients. Depending on the dose prescribed, a statin can lower a person's LDL by up to 60 percent, and triglycerides by about 25 percent. These effects translate into dramatically reduced risks of heart attacks and brain strokes, both of which can be caused when blood flow is blocked by the build-up of fatty, cholesterol-rich deposits, called plaques, in blood vessels (Figure 7.15). In study after study, heart-related deaths have plummeted as much as 45 percent in patients known to be at high risk of heart attack.

One of the most encouraging recent findings about statins is that they also benefit people who don't have "problem" cholesterol, but still seem to develop plaques, possibly due to heredity. In these patients, using a statin apparently can reduce both the overall death rate due to a heart malfunction and the chance of having a stroke.

Every year in the United States, cardiovascular diseases claim about 750,000 lives. It's no wonder, then, that today at least half a dozen statins—the new "wonder drugs" for heart care—are in common use.

Figure 7.15 Plaques (the whitish blobs) in blood vessels that service the heart.

HOW THE HEART CONTRACTS

The Cardiac Conduction System

We think of our heart as an organ that reliably thumps away, but a crucial 1 percent of the cardiac muscle cells do not contract. Instead, their job is to function as the heart's **cardiac conduction system**. As noted in the chapter introduction, this system includes self-excitatory pacemaker cells. These cells spontaneously produce and conduct the electrical impulses (called action potentials) that stimulate contractions in the heart's contractile cells. Because of the cardiac conduction system, even if all nerves leading to the heart are severed, the heart will keep on beating.

In our skeletal muscle tissue, the ends of the fiberlike cells are attached to bones or tendons. But cardiac muscle cells are "built" differently. They branch, then connect with one another at their endings (Figure 7.16). Communication junctions called *intercalated discs* bridge the plasma membranes on the ends of abutting cells. With each heartbeat, signals calling for contraction spread so rapidly across the junctions that cardiac muscle cells contract together, almost as if they were a single unit.

Normally, excitation begins with a small mass of cells in the upper wall of the right atrium. This **sinoatrial (SA) node** generates wave after wave of excitation, usually 70 or 80 times a minute. Each wave spreads over both atria and causes them to contract. It then rapidly reaches the **atrioventricular node** (AV node) in the septum dividing the two atria. Notice in Figure 7.17 that bundles of conducting fibers extend from the AV node to each ventricle. At intervals along each bundle, conducting cells called *Purkinje fibers* branch off and make contact with muscle cells in the ventricles. When a stimulus reaches the AV

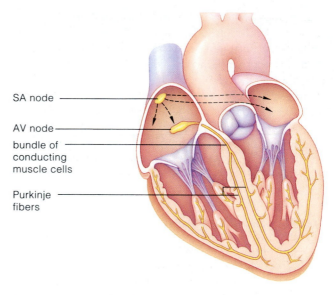

Figure 7.17 Location of specialized cardiac muscle cells that conduct signals for contraction through the heart.

node, it slows, almost pausing, before it quickly passes along the bundles to Purkinje fibers and on to contractile muscle fibers in each ventricle. The slow conduction in the AV node gives the atria time to finish contracting before the wave of excitation spreads to the ventricles.

Of all cells of the cardiac conduction system, the SA node fires off impulses at the highest frequency and is the first region to respond in each cardiac cycle. It is called the **cardiac pacemaker** because its rhythmic firing is the basis for the normal rate of heartbeat. The ECG in Figure 7.26, Section 7.14, traces a normal heartbeat sequence. People whose SA node chronically malfunctions may have an artificial pacemaker implanted to provide a regular stimulus for their heart contractions.

Neural Controls over Heart Rate

The nervous system triggers the contraction of skeletal muscle, but it can only *adjust* the rate and strength of cardiac muscle contraction. Stimulation by one set of nerves increases the force and rate of heart contractions, while stimulation by another set of nerves can slow heart activity. The centers for neural control of heart functions are in the spinal cord and parts of the brain. They are discussed more fully in Chapter 11.

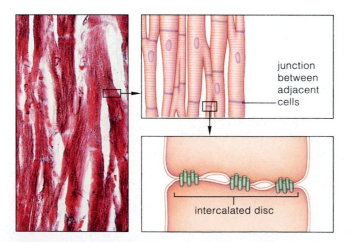

Figure 7.16 Intercalated discs containing communication junctions at the ends of adjacent cardiac muscle cells. Signals travel rapidly across the junctions and cause cells to contract nearly in unison.

The SA node is the cardiac pacemaker—it establishes a regular heartbeat. Its spontaneous, repeated excitation signals spread along a system of muscle cells that stimulate contractile tissue in the atria, then the ventricles, in a rhythmic cycle.

BLOOD PRESSURE AND VELOCITY IN THE CARDIOVASCULAR SYSTEM

Blood Pressure

Heart contractions generate **blood pressure**—the fluid pressure blood exerts against vessel walls. Blood pressure is highest in the aorta; then it drops along the systemic circuit. The pressure typically is measured when a person is at rest (Figure 7.18). For an adult, 120/80 is in the normal range. The first number is *systolic pressure*, which is the peak of pressure in the aorta while the left ventricle contracts and pushes blood into the aorta. The second number, *diastolic pressure*, measures the lowest blood pressure in the aorta, when the heart is relaxed and blood is flowing out of the aorta.

Values for systolic and diastolic pressure provide vital health information. Elevated blood pressure can be associated with various ailments, such as *atherosclerosis* (Section 7.14) and kidney disease.

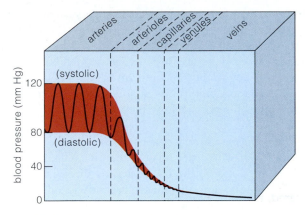

Figure 7.19 Blood pressure. This diagram plots measurements of the drop in fluid pressure for a given volume of blood moving through the systemic circuit.

Friction and other factors combine to create resistance to the movement of blood in vessels. When friction occurs, energy is lost (as heat). In the systemic circulation, resistance increases dramatically as flowing blood moves from arteries to arterioles. The increasing resistance causes circulating blood to lose energy, and so blood pressure drops, as you can see in the diagram in Figure 7.19.

Resting blood pressure depends mainly on centers in a part of the brain called the medulla oblongata. As described in Chapter 11, the centers integrate information from sensory receptors in cardiac muscle tissue and in certain arteries, such as the aorta and the **carotid arteries** in the neck. They use this information to coordinate the rate and strength of heartbeats with changes in the diameter of arterioles and, to some extent, of veins. The next section explains how these controls work.

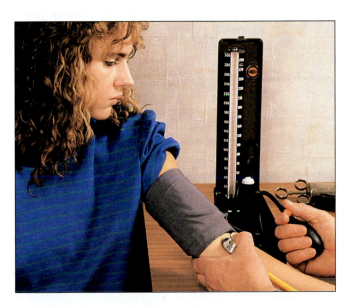

Figure 7.18 Measuring blood pressure with a device called a sphygmomanometer. A hollow cuff attached to a pressure gauge is wrapped around the upper arm. Then it is inflated with air to a pressure above the highest pressure of the cardiac cycle (at systole, when the ventricles contract). Above the systolic pressure, no sounds are heard through a stethoscope positioned below the cuff, because no blood is flowing through the vessel.

As air in the cuff is slowly released, some blood flows into the artery. The turbulent flow causes soft tapping sounds, and when this first occurs, the value on the gauge is the systolic pressure—about 120 mm mercury (Hg) in young adults at rest. (This means the measured pressure would force mercury to move upward 120 millimeters in a narrow glass column.)

Next, more air is released from the cuff. Just after the sounds become dull and muffled, blood flows continuously, and so the turbulence and tapping sounds stop. The silence corresponds to the diastolic pressure at the end of a cardiac cycle, just before the heart pumps out blood. Generally the reading is about 80 mm Hg. In this example, the *pulse* pressure (the difference between the highest and lowest pressure readings) is 120 − 80, or 40 mm Hg.

Blood Velocity

Pumped by your heart's muscular left ventricle, blood entering the systemic circulation is moving quite rapidly when it leaves the heart in the aorta. Just as a river's flow slows when it becomes divided into many small channels, the flow of arterial blood slows as the total cross-sectional area of vessels increases. Its velocity is greatest in the aorta, drops in the more numerous arterioles, and slows to a relative crawl in countless narrow capillaries. This slowdown allows time for materials to diffuse between capillaries and tissues. The velocity increases again as blood moves into veins for the return trip to the heart.

Blood pressure is highest in the aorta and drops along the rest of the systemic circuit. The velocity of blood flow varies with the total cross-sectional area of the vascular network.

The "vascular" part of the cardiovascular system consists of arteries, arterioles, capillaries, venules, and finally veins. Figure 7.20 diagrams the structure of these vessels. The reason we show it is because, as with all body parts, structure is key to the functions of blood vessels. All our vessels transport blood, but as you'll now see, there are important differences in how they "manage" blood flow and blood pressure.

The wall of an artery consists of several tissue layers (Figure 7.20*a*). The outer layer is mainly collagen, which anchors the vessel to the tissue it runs through. A thick middle layer of smooth muscle is sandwiched between thinner layers containing elastin. The innermost layers include a basement membrane and a thin sheet of endothelium. Collectively, these layers form a thick, muscular, and elastic wall. In a large artery the wall bulges

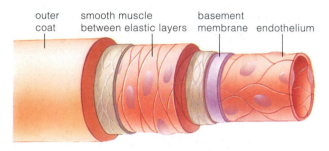

a ARTERY

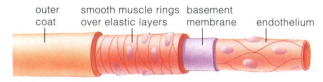

b ARTERIOLE

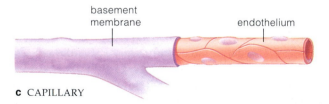

c CAPILLARY

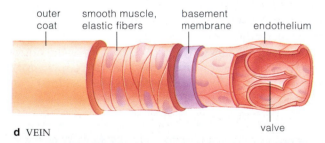

d VEIN

Figure 7.20 Structure of blood vessels. The basement membrane around the endothelium contains a form of collagen especially rich in proteins and polysaccharides.

slightly under the pressure surge caused when a ventricle contracts. In arteries near the body surface, as in the wrist, you can feel the surges as your **pulse**.

The bulging of artery walls helps keep blood flowing on through the system. How? For a fleeting moment, some of the blood pumped during the systole phase of each cardiac cycle is stored in the "bulge"; the elastic recoil of the artery then forces that stored blood onward during diastole, when heart chambers are relaxed. In addition to stretchable walls, arteries also have large diameters. For this reason, they present little resistance to blood flow, so blood pressure does not drop much in the large arteries of the systemic and pulmonary circuits.

Resistance at Arterioles

Arteries branch into arterioles, which have a smaller diameter. The wall of an arteriole also has fewer layers than that of an artery. It has rings of smooth muscle over a single layer of elastic fibers (Figure 7.20*b*). This structure enables arterioles to dilate (enlarge in diameter) when the smooth muscle relaxes or to constrict (shrink in diameter) when the smooth muscle contracts. As shown in Figure 7.19, arterioles offer more resistance to blood flow than other vessels do. As the blood flow slows, it can be controlled in ways that divert greater or lesser amounts of the total volume to different body regions. For example, you get drowsy after a large meal in part because control signals divert blood away from your brain in favor of your digestive system. As described shortly, the diameter of arterioles can dilate or constrict in response to signals from the nervous system and endocrine system, and even in response to a change in local chemical conditions.

Capillaries: Specialized for Diffusion

Your body has about 2 miles of arteries and veins but a whopping 62,000 miles of capillaries. Each capillary bed is a site where substances can diffuse between blood and interstitial fluid. This is truly where "the rubber meets the road" when it comes to exchanges of gases (oxygen and carbon dioxide), nutrients, and wastes. As befits its functional role in diffusion, a capillary has the thinnest wall of any blood vessel—a single layer of flat endothelial cells, separated from one another by narrow gaps (Figure 7.20*c*). Most capillaries have such a small lumen that red blood cells must squeeze through them single file. Thus a single capillary presents high resistance to blood flow. Yet so many of them are present in a capillary bed that their combined diameters are greater than the combined diameters of arterioles leading into them (Figure 7.21). As a result, a capillary bed presents *less total resistance* to flow than do the arterioles leading into it, and the total drop in blood pressure is less steep in this region.

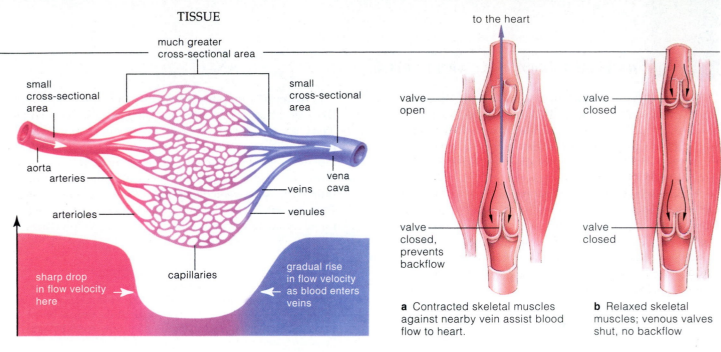

Figure 7.21 Comparison of cross-sectional areas represented by arteries, arterioles, and capillaries in the circulatory system.

a Contracted skeletal muscles against nearby vein assist blood flow to heart.

b Relaxed skeletal muscles; venous valves shut, no backflow

Figure 7.22 How contracting skeletal muscles can increase fluid pressure in a vein. Notice the structure of the vein valve.

Venules and Veins

Capillaries merge into venules. In terms of their function, venules are a bit like capillaries. Some solutes diffuse across their wall, which is only a little thicker than that of a capillary. Venules merge into veins, the large-diameter vessels leading back to the heart.

Veins are blood reservoirs. Collectively they contain 50 to 60 percent of the body's total blood volume. A vein wall is thin enough to bulge under pressure, more so than an arterial wall. The wall also contains some smooth muscle (Figure 7.20*d*). When blood must circulate faster (for instance, during exercise), the smooth muscle in veins contracts. The wall stiffens, the vein bulges less, and venous pressure rises. This drives more blood to the heart. Venous pressure also rises when contracting skeletal muscle—especially in the legs and abdomen—bulges against adjacent veins (Figure 7.22). This muscle activity is an important factor in returning blood through the venous system.

Some veins, mainly in the limbs, have valves. When blood starts moving backward due to gravity, it pushes the valves closed, preventing backflow. Weakened venous valves can result from inherited defects, obesity, pregnancy, and other factors. The walls of a *varicose vein* have become overstretched because, over time, weak valves have allowed blood to pool there.

How Vessels Help Control Blood Pressure

Certain arteries, all arterioles, and even veins have key roles in homeostatic mechanisms that work to maintain adequate blood pressure over time. Centers in the brain's medulla control resting blood pressure. When the centers detect an abnormal *increase* in blood pressure, they order

the heart to beat more slowly and contract less forcefully. They also order smooth muscle cells in arteriole walls to relax. The result is **vasodilation**—an enlargement (dilation) of the vessel diameter. When the centers detect an abnormal *decrease* in blood pressure, they command the heart to beat faster and contract more forcefully. Neural signals also cause the smooth muscle cells of arterioles to contract. The resulting decrease in the vessel diameter is called **vasoconstriction**. In some body regions arterioles have receptors for hormones that help maintain blood pressure due to their ability to trigger vasoconstriction or vasodilation.

As already noted, the nervous and endocrine systems also control how blood is allocated to different body regions at different times. In addition, the conditions in a particular part of the body can exert local control over blood flow. For example, when you run, the tissue level of oxygen in your hardworking skeletal muscles falls, and levels of carbon dioxide, H^+, potassium ions, and other substances rise. The changes in local chemistry cause the smooth muscle in arterioles to relax. With the vasodilation, more blood flows past the active muscles, delivering more raw materials and carrying away cell wastes. At the same time, arterioles in your digestive tract and kidneys constrict.

Smokers are using a drug—nicotine—to alter their blood pressure, though not for the better. The nicotine in tobacco is a potent vasoconstrictor, and thus leads to abnormally high blood pressure—and increased stress on the whole cardiovascular system.

Overall, the heart's output and the total resistance to blood flow through the vascular system determine blood pressure. Much of the control is exerted at arterioles.

Capillaries thread to within 0.01 millimeter of almost every living cell in your body. Most solutes, including oxygen and carbon dioxide, diffuse across the capillary wall. Certain proteins enter or leave by endocytosis or exocytosis, and certain ions probably pass through pores in capillary walls and spaces between endothelial cells.

Fluid also enters and leaves capillaries in response to various types of pressure (Figure 7.23). You may recall from Chapter 2 that the force a fluid exerts against a surface is called hydrostatic pressure. In capillaries, this is the same as blood pressure, and it forces some water out of capillaries, especially at the arterial end. This process is called **ultrafiltration**. Normally, water also moves *into* capillaries in response to osmotic pressure. That is, the water follows its concentration gradient into capillaries as the concentration of solutes there rises. This process is called **reabsorption**.

Hydrostatic pressure exerted by interstitial fluid and the concentration of solutes in that fluid also have some effect on water movements. On balance, more water tends to leave capillaries than to enter them. As water moves in either direction, some solutes follow. Such fluid and solute movements help maintain the proper fluid balance between blood and the surrounding tissues. They are vital to maintaining the blood volume needed for adequate blood pressure.

Capillary beds are diffusion zones where exchanges occur between the blood and interstitial fluid. Such exchanges help maintain the blood volume required for proper blood pressure. They also help maintain the proper fluid balance between blood and tissues.

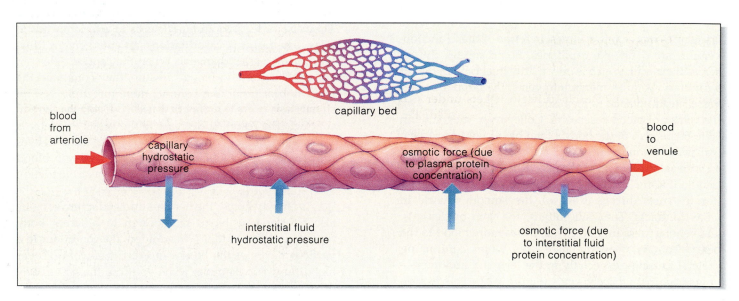

Figure 7.23 Fluid movements in a capillary bed. The movements help maintain the distribution of extracellular fluid between the bloodstream and interstitial fluid.

At the arteriole end of a capillary, the difference between capillary blood pressure and interstitial fluid pressure causes some water to leave the capillary. Part of this water will be picked up by the lymphatic system and eventually return to the bloodstream.

Water enters a capillary by osmosis, following its concentration gradient. (Plasma has a greater solute concentration and therefore a lower water concentration.) Fluid loss at the arteriole end of a capillary bed tends to be balanced by fluid intake at the venule end.

Edema is a condition in which excess fluid accumulates in interstitial spaces. This happens to some extent during exercise. As arterioles dilate in local tissues, capillary blood pressure increases and more fluid is forced out of capillaries into tissues. Edema also can result from an obstructed vein or from heart failure.

Overview of Hemostasis

Small blood vessels are delicate. They can easily tear or be damaged by a cut or blow. **Hemostasis** is the name of a process that stops the bleeding and so prevents excessive blood loss. This process includes spasms in affected blood vessels, the formation of platelet plugs, and the coagulation, or clotting, of blood.

When a blood vessel first ruptures, smooth muscle in a damaged vessel wall contracts in an automatic response called a spasm. The blood vessel constricts, so blood flow through it slows or stops. This response can last for up to half an hour, and it is extremely important in stemming the immediate loss of blood. Then, while vessel spasms reduce blood flow, platelets aggregate and create a temporary plug in the damaged wall. They also release the hormone serotonin and other chemicals that help prolong the spasm and attract more platelets. Lastly, blood *coagulates*—that is, it converts to a gel—and forms a clot.

Two Clotting Mechanisms

A blood clot forms in one of two ways. The first is called an *intrinsic clotting mechanism* because it involves substances that are in the blood itself. It gets under way when a protein in the blood plasma is activated. This triggers reactions that lead to the formation of an enzyme (thrombin), which acts on a large, rod-shaped plasma protein called fibrinogen. The fibrinogen rods stick to one another, forming long, insoluble threads (fibrin) that also stick to one another. The result is a net in which blood cells and platelets become entangled (Figure 7.24). The entire mass is a blood clot. Given a little time, the clot retracts into a compact mass, drawing the torn walls of the vessel back together.

Blood also can coagulate through an *extrinsic clotting mechanism*. "Extrinsic" means that the series of reactions leading to blood clotting is triggered by the release of enzymes and other substances *outside* the blood itself. These substances come from damaged blood vessels or from the surrounding tissues. The substances lead to the formation of thrombin, and the remaining steps parallel those of the intrinsic pathway.

Aspirin reduces the aggregation of platelets, which is why it is sometimes prescribed (in small doses) to help prevent blood clots. A clot that forms in an unbroken blood vessel can be a serious threat. A clot that stays where it forms is called a *thrombus*, and the condition is spoken of as a *thrombosis*. Even scarier is an *embolus*, a clot that breaks free and circulates through the bloodstream. A person who suffers an *embolism* in the heart, lungs, brain, or some other organ may suddenly die when the roving clot shuts down the organ's blood supply.

Hemostasis refers to processes that slow or stop the flow of blood from a ruptured vessel. The mechanisms include spasms that constrict blood vessel walls, the formation of platelet plugs, and blood clotting.

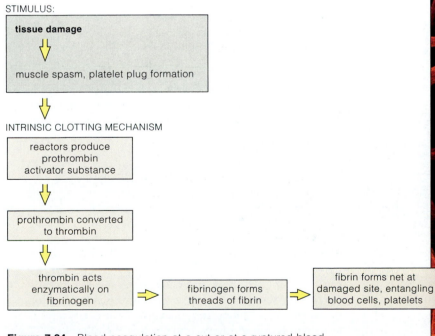

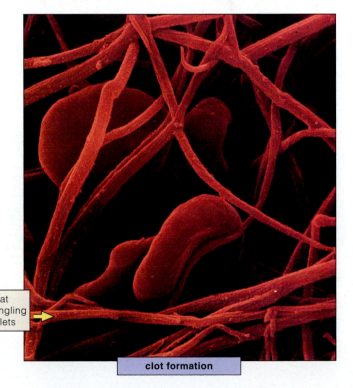

clot formation

Figure 7.24 Blood coagulation at a cut or at a ruptured blood vessel. The micrograph shows red blood cells trapped in a fibrin net.

7.14 CARDIOVASCULAR DISORDERS

In the United States, more than 40 million people have cardiovascular disorders. Most common are *hypertension*, which is sustained high blood pressure, and *atherosclerosis*, which is a progressive thickening of the arterial wall and progressive narrowing of the arterial lumen. Both affect blood circulation and so cause most *heart attacks* (damage to or death of heart muscle) and *strokes* (brain damage). In most heart attacks, one of the first signs is a "crushing" pain behind the breastbone that lasts a half hour or more. Frequently, the pain radiates into the left arm, shoulder, or neck. The pain can be mild but usually is excruciating. Often the person also sweats, vomits or feels nauseated and dizzy, or loses consciousness.

RISK FACTORS   Lifestyle choices and other factors increase the risk of cardiovascular disorders. Smoking is a major risk factor, as is lack of regular exercise. Other common risk factors include obesity (Chapter 6), having high blood cholesterol, an inherited predisposition to heart failure, and hypertension. Aging contributes, in that the older you get, the greater the risk. Gender also is a factor. Until age 55, females are at lower risk than are males, because the female hormone estrogen provides some protection. Estrogen declines rapidly after a woman reaches menopause, usually in her early fifties.

HYPERTENSION   Hypertension results from gradual increases in the flow resistance through small arteries. In time, blood pressure stays elevated, even when a person is resting. Heredity may be a factor; the disorder tends to run in families. Diet is also a factor. For instance, high salt intake can raise blood pressure in susceptible people. Regardless of the cause, high blood pressure boosts the heart's workload. The heart may eventually enlarge and fail to pump blood effectively. High blood pressure also can cause artery walls to "harden" and so reduces the delivery of oxygen to the brain, heart, and other vital organs.

Hypertension kills about 180,000 Americans every year. It is a "silent killer" because affected people may have no outward symptoms. Even when they know their blood pressure is high, some resist helpful medication, diet changes, and getting regular exercise. Of the roughly 23 million Americans who are hypertensive, most do not seek treatment.

ATHEROSCLEROSIS   In *arteriosclerosis,* arteries thicken and lose elasticity. In **atherosclerosis**, this condition gets worse as cholesterol and other lipids build up in the arterial wall. When this **atherosclerotic plaqu**e enlarges enough to protrude into the artery lumen, there is less room for flowing blood (Figure 7.25).

With their narrow diameters, coronary arteries and their branches are highly susceptible to clogging by plaque or occlusion by a clot. When such an artery is narrowed to one-quarter of its former diameter, the

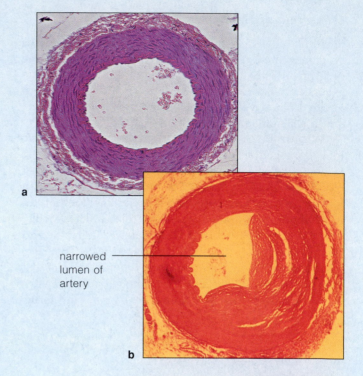

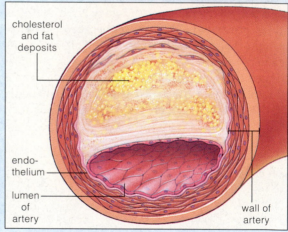

narrowed lumen of artery

cholesterol and fat deposits

endo-thelium

lumen of artery

wall of artery

Figure 7.25 Sections from (**a**) a normal artery. and (**b**) one with its lumen narrowed by a plaque. (**c**) Sketch of an atherosclerotic plaque.

resulting symptoms can range from mild chest pain (*angina pectoris*) to a full-scale heart attack.

Atherosclerosis involving coronary arteries can be diagnosed through a stress electrocardiogram. This is a recording of the electrical activity of the cardiac cycle while a person is exercising on a treadmill. It can also be diagnosed by *angiography,* in which a dye is injected that causes plaques to show up on X rays. A severe blockage may require surgery. In *coronary bypass surgery,* a section of a large vessel taken from the chest is stitched to the aorta and to the coronary artery below the affected region.

In *laser angioplasty,* laser beams vaporize the plaques. In *balloon angioplasty,* a small balloon is inflated inside a blocked artery to flatten a plaque and increase the artery's diameter. In combination with statin drug therapy such procedures can reduce the immediate threat of a heart attack, but do not cure the underlying problem.

Recall from Chapter 6 that the liver normally produces enough cholesterol to meet body needs. Regularly eating cholesterol-rich foods typically increases the blood's load of cholesterol *beyond* what is required. In the bloodstream, proteins bind cholesterol and triglycerides, forming *low-density lipoproteins,* or **LDLs**. LDLs can bind to receptors on body cells, which is how cells take up the cholesterol used to build cell membranes and for other activities. Excess cholesterol becomes attached to proteins as *high-density lipoproteins,* or **HDLs,** and is carried back to the liver where it is metabolized, mixed into bile, and excreted.

In some people, for a variety of reasons, cells remove too little LDL from the blood. The blood level of LDL rises—and so does the risk of atherosclerosis. LDLs, with their bound cholesterol and triglycerides, infiltrate the walls of arteries. There, abnormal smooth muscle cells multiply, other harmful changes occur, and cholesterol accumulates in endothelial cells and the gaps between them. Although they are not shown in Figure 7.25*c,* calcium deposits form microscopic slivers of bone on top of the lipids. A fibrous net forms over the entire mass, the atherosclerotic plaque bulging into the artery lumen.

The plaque's bony slivers shred the endothelium. Platelets gather at the damaged site, and as described in Section 7.13, they secrete chemicals that set in motion the formation of a blood clot. The situation gets worse as fatty globules in the plaque become oxidized (by free radicals). These changes trigger inflammation, which in turn leads to further damage. Meanwhile, as plaques and clots grow, they can narrow or block an artery. A clot that breaks loose can cause a life-threatening embolism (Section 7.13).

Implicated in many cases of artery damage is the amino acid *homocysteine,* which forms when the body breaks down another amino acid, methionine (found in meats and dairy foods). Methionine is an important nutrient, but a chronically high level of homocysteine in the blood can encourage atherosclerotic plaques to form. Taking B vitamins (including folic acid) may help remove homocysteine before it can trigger damaging changes in arteries.

ARRYTHMIAS ECGs reveal **arrhythmias,** irregular heart rhythms (Figure 7.26). Not all arrhythmias are abnormal. For example, endurance athletes may have a below-average resting cardiac rate, or *bradycardia.* In an adaptation to regular strenuous exercise, the athlete's nervous system has adjusted the cardiac pacemaker's contraction rate downward. There is more time for the ventricles to fill, so each contraction pumps blood more efficiently.

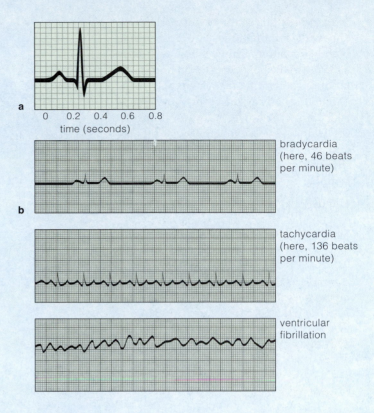

Figure 7.26 (**a**) ECG of a single, normal heartbeat. The P wave is generated by electrical signals from the SA node that stimulate contraction of the atria. As the stimulus moves over the cardiac muscle of the ventricles by way of Purkinje fibers, it is recorded as the QRS wave complex. After the ventricles contract, they go through a brief period of recovery. Electrical activity during this period is marked by the T wave. (There is also an atrial recovery period "hidden" in the QRS complex.) (**b**) ECG readings for bradycardia, tachycardia, and ventricular fibrillation.

A cardiac rate above 100 beats per minute, called *tachycardia,* occurs normally during exercise or stressful situations. Serious tachycardia can be triggered by drugs (including caffeine, nicotine, alcohol, and cocaine), excessive hormone output by the thyroid gland (hyperthyroidism), and other factors.

Coronary artery disease or some other disorders may cause abnormal rhythms that can degenerate rapidly into an extreme medical emergency called *ventricular fibrillation.* In parts of the ventricles, cardiac muscle contracts haphazardly, and blood pumping suffers. Within seconds, the person loses consciousness and death may be near. With luck, a strong electrical shock to the patient's heart, or the use of defibrillating drugs, can restore a normal rhythm before the damage becomes too serious.

THE LYMPHATIC SYSTEM

We conclude this chapter with a brief look at how the lymphatic system supplements blood circulation. Think of this section as a bridge to the next chapter, on immunity, because the lymphatic system also helps defend the body against injury and attack. As Figure 7.27 shows, the system consists of drainage vessels, lymphoid organs, and lymphoid tissues. The tissue fluid that has moved into the vessels is called **lymph**.

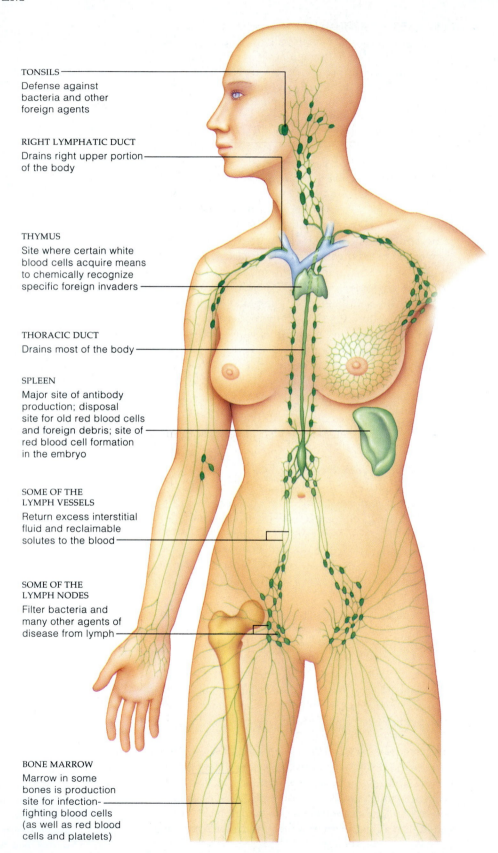

TONSILS
Defense against bacteria and other foreign agents

RIGHT LYMPHATIC DUCT
Drains right upper portion of the body

THYMUS
Site where certain white blood cells acquire means to chemically recognize specific foreign invaders

THORACIC DUCT
Drains most of the body

SPLEEN
Major site of antibody production; disposal site for old red blood cells and foreign debris; site of red blood cell formation in the embryo

SOME OF THE LYMPH VESSELS
Return excess interstitial fluid and reclaimable solutes to the blood

SOME OF THE LYMPH NODES
Filter bacteria and many other agents of disease from lymph

BONE MARROW
Marrow in some bones is production site for infection-fighting blood cells (as well as red blood cells and platelets)

Figure 7.27 Components of the human lymphatic system and their functions. The small *green* ovals show some of the major lymph nodes. Patches of lymphoid tissue in the small intestine and in the appendix also are part of the lymphatic system.

The Lymph Vascular System

Part of the lymphatic system, called the lymph vascular system, consists of many tubes that collect water and solutes from interstitial fluid and transport them to ducts of the cardiovascular system. These "tubes" include **lymph capillaries** and **lymph vessels.** The lymph vascular system has three functions, which we could call the "three Ds"—drainage, delivery, and disposal.

To begin with, the system's vessels are drainage channels. They collect water and solutes that have leaked out of the blood in capillary beds (due to the fluid pressure there) and return those needed substances to the bloodstream. Second, the system also picks up fats that the body has absorbed from the small intestine and delivers them to the bloodstream, in the way described in Section 6.4. Third, it transports foreign cells and material, and cellular debris, from the body's tissues to the lymph vascular system's efficiently organized disposal centers, the lymph nodes.

The lymph vascular system starts at capillary beds, where fluid enters the lymph capillaries. There is no obvious entrance to these capillaries. Instead, water and solutes move into their tips at flaplike "valves." These are areas where endothelial cells overlap (Figure 7.28a).

Lymph capillaries merge into lymph vessels, which have a larger diameter. They also have smooth muscle in their wall and valves in their lumen that prevent backflow. They converge into collecting ducts that drain into veins in the lower neck. This is how the lymph fluid is returned to circulating blood. Movements of the skeletal muscles and of the rib cage (during breathing) help move fluid through our lymph vessels, just as they do for veins.

Lymphoid Organs and Tissues

Other components of the lymphatic system are players in the body's defenses against injury or attack. These parts are the lymph nodes, the spleen, and the thymus, as well as the tonsils and patches of tissue in the small intestine, in the appendix, and in airways leading to the lungs.

The **lymph nodes** are strategically located at intervals along lymph vessels (Figures 7.27 and 7.28b). Before lymph enters the bloodstream, it trickles through at least one of these nodes. A lymph node has several inner chambers where many white blood cells take up residence after they have been produced in bone marrow. During an infection, lymph nodes become battlegrounds where armies of lymphocytes form and where foreign agents are destroyed. Macrophages in the nodes help clear the lymph of bacteria, cellular debris, and other substances.

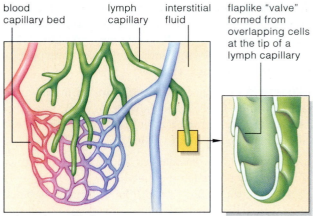

a Lymph capillaries

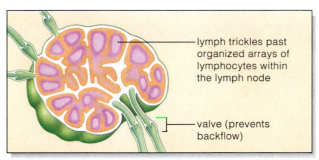

b A lymph node, cross section

Figure 7.28 (**a**) Some of the lymph capillaries at the start of the drainage network called the lymph vascular system. (**b**) Cutaway diagram of a lymph node. Its inner chambers are packed with highly organized arrays of infection-fighting white blood cells.

The largest lymphoid organ, the **spleen**, is a filtering station for blood and a holding station for lymphocytes. The spleen has inner chambers filled with red and white "pulp." The red pulp is a reservoir of red blood cells and macrophages. (In a developing embryo, the spleen produces red blood cells.) The white pulp has masses of lymphocytes in close proximity to blood vessels. If a specific invader reaches the spleen during an infection, the lymphocytes are mobilized to destroy it just as they are in lymph nodes.

In the **thymus**, lymphocytes multiply, differentiate, and mature into fighters of specific disease agents. The thymus produces hormones that influence these events. It is central to immunity, our focus in Chapter 8.

The lymph vascular system returns tissue fluid to the blood, transports fats, and carries debris and foreign material to lymph nodes. Lymph nodes and other lymphoid organs function in the body's systems of defense.

SUMMARY

1. The human cardiovascular system consists of the heart, a muscular pump; blood vessels, including arteries, arterioles, capillaries, venules, and veins; and blood. The system's function is the rapid internal transport of substances to and from cells.

2. Blood, a fluid connective tissue, helps maintain favorable conditions for cells. It delivers oxygen and other substances to the interstitial fluid around cells. It also picks up cell products and wastes from that fluid.

3. Blood consists of plasma, red and white blood cells, and cell fragments called platelets.

   a. Plasma, the liquid part of blood, transports blood cells and platelets. Plasma water is a solvent for plasma proteins, simple sugars, amino acids, mineral ions, vitamins, hormones, and several gases.

   b. Red blood cells carry oxygen from the lungs to all body tissues. They are packed with hemoglobin, an iron-containing pigment molecule that binds reversibly with oxygen. Red blood cells also transport some carbon dioxide (also bound to hemoglobin) from interstitial fluid back to the lungs (to be exhaled).

   c. Certain phagocytic white blood cells scavenge dead or worn-out cells and other debris and cleanse tissues of anything detected as not belonging to the body. Other white blood cells (lymphocytes) form armies that destroy specific bacteria, viruses, and other disease agents.

   d. Platelets function in blood clotting.

4. An internal partition divides the human heart into two halves, each with two chambers, an atrium and a ventricle. The partition separates the blood flow into two circuits, one pulmonary and the other systemic.

   a. In the pulmonary circuit, deoxygenated blood in the heart's right half is pumped to capillary beds in the lungs. The blood picks up oxygen, then flows to the heart's left atrium.

   b. In the systemic circuit, the left half of the heart pumps oxygenated blood to body tissues. There, cells take up oxygen and release carbon dioxide. The blood, now deoxygenated, flows to the heart's right atrium.

5. Contraction of the heart's ventricles drives blood through both circuits of the cardiovascular system. Blood pressure is highest in contracting ventricles, then progressively drops in arteries, arterioles, capillaries, then veins. It is lowest in relaxed atria.

   a. Arteries are an elastic pressure reservoir. They smooth out pressure changes resulting from heartbeats and so smooth out blood flow through capillaries.

   b. Arterioles are control points for distributing different volumes of blood to different regions.

   c. Capillary beds are diffusion zones where blood and interstitial fluid exchange substances.

   d. Venules overlap capillaries and veins somewhat in function.

   e. Veins are a blood-volume reservoir that can be tapped to adjust the volume of flow back to the heart.

6. The vascular portion of the lymphatic system takes up water and plasma proteins that seep out of blood capillaries, then returns them to the blood circulation. It also transports absorbed fats. Other lymphatic system components, including lymph nodes, have major roles in general body defense and in immune responses.

Review Questions

1. What is blood plasma, and what is its function? What are the cellular components of blood? *7.1*

2. Define the functions of the cardiovascular system and the lymphatic system. *7.5`*

3. Label the heart's components: *7.6*

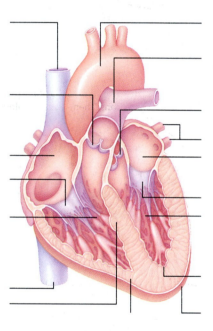

4. Define a "heartbeat," giving the sequence of events that make it up. *7.6*

5. Distinguish between the systemic and pulmonary circuits. *7.7*

6. State the main function of blood capillaries. What drives water and solutes out of and into capillaries in capillary beds? *7.11, 7.12*

7. State the main functions of venules and veins. What forces work together in returning venous blood to the heart? *7.11*

8. What is the function of hemostasis? What are the two ways a blood clot can form? *7.13*

Self-Quiz (Answers in Appendix V)

1. Cells directly exchange substances with _____ .
   a. blood vessels c. interstitial fluid
   b. lymph vessels d. both a and b

2. Which are *not* components of blood?
   a. plasma
   b. blood cells and platelets
   c. gases and other dissolved substances
   d. all of the above are components of blood

3. The _____ produces red blood cells, which transport _____ and some _____ .
   a. liver; oxygen; mineral ions
   b. liver; oxygen; carbon dioxide
   c. bone marrow; oxygen; hormones
   d. bone marrow; oxygen; carbon dioxide

4. The _____ produces white blood cells, which function in _____ and _____ .
   a. liver; oxygen transport; defense
   b. lymph glands; oxygen transport; stabilizing pH
   c. bone marrow; day-to-day housekeeping; defense
   d. bone marrow; stabilizing pH; defense

5. In the pulmonary circuit, the heart's _____ half pumps _____ blood to capillary beds inside the lungs; then _____ blood flows to the heart.
   a. left; deoxygenated; oxygenated
   b. right; deoxygenated; oxygenated
   c. left; oxygenated; deoxygenated
   d. right; oxygenated; deoxygenated

6. In the systemic circuit, the heart's _____ half pumps _____ blood to all body regions; then _____ blood flows to the heart.
   a. left; deoxygenated; oxygenated
   b. right; deoxygenated; oxygenated
   c. left; oxygenated; deoxygenated
   d. right; oxygenated; deoxygenated

7. Blood pressure is high in _____ and lowest in _____ .
   a. arteries; veins c. arteries; ventricles
   b. arteries; relaxed atria d. arterioles; veins

8. _____ contraction drives blood through the pulmonary circuit and the systemic circuit; blood pressure is highest in contracting _____ .
   a. Atrial; ventricles   c. Ventricular; arteries
   b. Atrial; atria   d. Ventricular; ventricles

9. Which is not a function of the lymphatic system?
   a. delivers disease agents to disposal centers
   b. produces lymphocytes
   c. delivers oxygen to cells
   d. returns water and plasma proteins to blood

10. Match the type of blood vessel with its major function.
    _____ arteries   a. diffusion
    _____ arterioles   b. control of blood distribution
    _____ capillaries   c. transport, blood volume reservoirs
    _____ veins   d. blood transport and pressure regulators

11. Match the following circulation components with their descriptions
    _____ capillary beds   a. two atria, two ventricles
    _____ lymph vascular system b. driving force for blood
    _____ heart chambers   c. zones of diffusion
    _____ heart contractions   d. starts at capillary beds

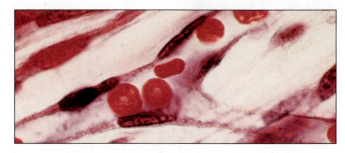

Figure 7.29 Light micrograph of branching blood vessel.

Critical Thinking: You Decide (Key in Appendix VI)

1. A patient suffering from hypertension may receive drugs that decrease the heart's output, dilate arterioles, or increase urine production. In each case, how would the drug treatment help relieve hypertension?

2. *Aplastic anemia* develops when certain drugs or radiation destroy red bone marrow, including stem cells that give rise to red and white blood cells and platelets. Predict some likely symptoms of aplastic anemia. Include at least one symptom related to each type of formed element in blood.

3. Shirelle, who is using a light microscope to examine a human tissue specimen, sees red blood cells moving single file through thin-walled tubes. She makes a photomicrograph (Figure 7.29). What type of blood vessel has she captured on film?

Selected Key Terms

agglutination 7.4
agranulocyte 7.1
aorta 7.6
arrhythmia 7.14
arteriole 7.5
artery 7.5
atrioventricular node 7.9
atrioventricular valve 7.6
atrium 7.6
blood 7.1
blood pressure 7.10
capillary 7.5
capillary bed 7.7
cardiac conduction system 7.9
cardiac cycle 7.6
cardiovascular system CI
coronary artery 7.6
diastole 7.6
granulocyte 7.1
heart 7.6
hemostasis 7.13
lymph 7.15

lymphatic system 7.15
lymph node 7.15
lymphoid organ 7.15
lymph vascular system 7.15
myocardium 7.6
platelet 7.1
pulmonary circuit 7.7
pulse 7.11
red blood cell (erythrocyte) 7.1
semilunar valve 7.6
sinoatrial node 7.9
spleen 7.15
stem cell 7.1
systemic circuit 7.7
systole 7.6
thymus 7.15
vasoconstriction 7.11
vasodilation 7.11
vein 7.5
ventricle 7.6
venule 7.5
white blood cell (leukocyte) 7.1

Readings

Radetsky, P. March 1995. "The Mother of All Blood Cells." *Discover.*

Zamir, M. September–October 1996. "Secrets of the Heart." *The Sciences.* Effects of exercise on heart function.

IMMUNITY

Desperate Measures for Desperate Times

You've probably had several vaccinations in your life, each to protect you from a serious disease like tetanus. Have you ever wondered how the concept of vaccination came about? It's a story of some daring experiments by both scientists and nonscientists. The plague they hoped to conquer was smallpox—a disease everyone wanted desperately to avoid. Armed only with keen observations, bold hypotheses, and the guts to test them, a few brave men and women changed the world. Here is how it all happened.

Until about a century ago, smallpox epidemics swept repeatedly through the world's cities. Some outbreaks were so severe that only half of those stricken survived. Survivors had permanent scars on the face, neck, shoulders, and arms—but they rarely contracted the disease again. They were "immune" to smallpox.

No one knew what caused smallpox. But the idea of acquiring immunity had powerful appeal. In 12th-century China, healthy people gambled with deliberate infections. They sought out survivors of

mild cases of smallpox (who were only mildly scarred), then removed crusts from the scars, ground them up, and inhaled the powder. By the late 17th century, Mary Montagu, wife of the English ambassador to Turkey, was championing inoculation. She even injected bits of smallpox scabs into her children. So did the Prince of Wales. Others soaked threads in the fluid from smallpox sores, then poked the threads into the skin.

People who survived these practices acquired immunity to smallpox—but many also developed raging infections. To make matters even more dicey, the crude inoculation procedures also opened the door to other infectious diseases.

While all this was going on, Edward Jenner was growing up in the English countryside. At the time, it was common knowledge that smallpox never afflicted someone who had had a much milder disease called cowpox (because it was transmitted from cattle). In 1796, Jenner, then a physician, injected material from a cowpox sore into the arm of an uninfected boy.

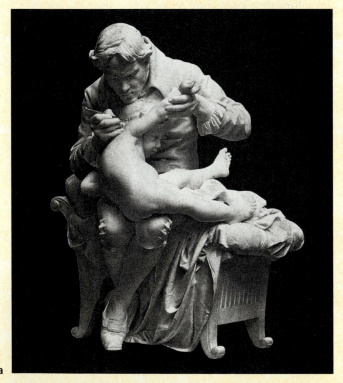

a

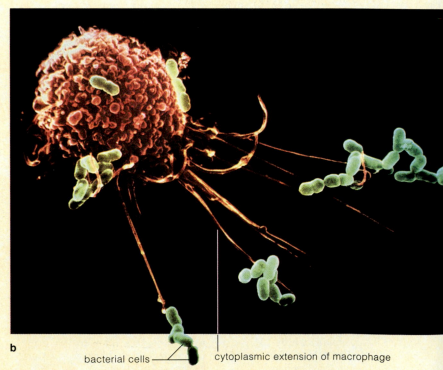

b

Figure 8.1 (**a**) Statue honoring Edward Jenner's development of vaccination against smallpox, one of the most dreaded diseases in human history. (**b**) False-color scanning electron micrograph of a defense cell (a macrophage). Threadlike extensions of its cytoplasm make contact with bacteria that the macrophage will engulf and destroy.

bacterial cells —   cytoplasmic extension of macrophage

Six weeks later, after the reaction subsided, Jenner injected fluid from *smallpox* sores into the boy (Figure 8.1). He hypothesized that the earlier injection might produce immunity to smallpox, and he was right. The boy did not get smallpox.

When news of Jenner's procedure reached France, the French initially treated it as a joke. They called it "vaccination," which translates as "encowment." Much later a French chemist, Louis Pasteur, devised similar procedures for other diseases. Pasteur also called his procedures vaccinations, and only then did the term become respectable.

By Pasteur's time, improved microscopes were revealing diverse bacteria, fungal spores, and other previously invisible forms of life. As Pasteur himself discovered, even ordinary air is full of microorganisms.

Did some cause diseases? He guessed they probably could. Could they settle into food or drink and cause it to spoil? He proved that they did. Pasteur also found a way to kill most of the suspect disease agents in food or beverages. He and others knew that boiling killed these agents. Being a wine connoisseur, he also knew you cannot boil wine—or beer or milk, for that matter—and end up with the same beverage. He devised a way to heat the fluids at a temperature low enough not to ruin them but high enough to kill most of the microorganisms. We still depend on his pioneering methods, which were named *pasteurization* in his honor.

In the late 1870s Robert Koch, a German physician, linked a specific microorganism to a specific disease—namely, anthrax. In one experiment, Koch injected blood from infected animals into uninfected ones. The recipients of the injection ended up with blood that teemed with cells of a bacterium (*Bacillus anthracis*)—and they developed anthrax. Even more convincing, injections of bacterial cells cultured outside the body also caused the disease!

Thus, by the beginning of the 20th century, the promise of understanding the basis of immunity loomed large—and the battles against infectious diseases were about to begin in earnest. Since that time, advances in microscopy, biochemistry, and molecular biology have enormously increased our understanding of the body's defenses. We now know much more about its responses to tissue damage in general, and have greater insights into its immune responses to specific pathogens or cancerous cells. These responses are the focus of this chapter.

KEY CONCEPTS

1. The human body has physical, chemical, and cellular defenses against harmful microorganisms, cancer cells, and other agents that can destroy tissues and sometimes even cause death.

2. In the early stages when a tissue is being invaded and damaged, white blood cells and certain proteins in blood plasma escape from capillaries. They execute a rapid, general counterattack in response to a general alarm. They do not react to a particular pathogen, so we call their response a nonspecific inflammatory response. Phagocytic white blood cells ingest invading agents and clean up tissue debris. Plasma proteins promote phagocytosis, and some also destroy invaders directly.

3. If the invasion does not stop, certain white blood cells make immune responses. Those cells can chemically recognize features on molecules that are abnormal or foreign to the body, such as those on bacteria and viruses. If the foreign or abnormal molecule triggers an immune response, it is called an antigen.

4. In one type of immune response, some of the white blood cells produce enormous quantities of antibodies. Antibodies are molecules that bind to a specific antigen and tag it for destruction.

5. In another type of immune response, cells specialized for killing directly destroy other body cells that have become abnormal.

CHAPTER AT A GLANCE

THREE LINES OF DEFENSE

Every day we encounter a huge assortment of pathogens. **Pathogen** is the general name for viruses, bacteria, fungi, protozoa, and parasitic worms that cause disease. The body surface bars most pathogens. If pathogens do make it across the barriers, cells and chemical weapons mount specific and nonspecific attacks against them.

Surface Barriers to Invasion

Usually, pathogens can't get past our skin or the linings of other body surfaces. Skin, remember, is relatively dry, and there is a thick layer of dead cells at its surface—conditions that are fine for harmless bacteria. In fact, few pathogens can compete with the dense populations of harmless bacteria that normally teem on our skin surface. If conditions change, however, so does the bacterial balance. For example, the skin between the toes of sweaty feet encased in sneakers is moist and warm. These conditions favor the growth of locker room fungi that cause *athlete's foot*.

Similarly, normally "friendly" bacteria in the mucosal lining of the GI tract help protect you. In females, lactate produced by *Lactobacillus* bacteria in the vaginal mucosa helps maintain a low vaginal pH that most bacteria and fungi cannot tolerate. Some antibiotics commonly pre-scribed to cure bacterial infections can trigger a vaginal yeast infection because the drug also kills *Lactobacillus*.

Patches of tissue in mucous membranes also contain white blood cells (lymphocytes) that function in specific immune responses, as Section 8.4 describes.

The inner walls of the branching, tubular respiratory airways leading to your lungs are coated with a sticky mucus. That mucus contains protective substances such as **lysozyme**, an enzyme that chemically attacks and helps destroy many bacteria. Broomlike cilia in the airways sweep out the pathogens.

Lysozyme and some other chemicals in tears, saliva, and gastric fluid offer more protection (for example, when tears give your eyes a sterile washing). Urine's low pH and flushing action help bar pathogens from invading the urinary tract. In adults, mild diarrhea can rid the lower GI tract of irritating pathogens; blocking it can prolong infection. In children, however, diarrhea must be controlled to prevent dangerous dehydration.

Nonspecific and Specific Responses

All animals react to tissue damage, and we humans are no exception. In our bodies, specialized types of white blood cells and plasma proteins serve as another line of defense if a pathogen breaches the surface barriers. When a tissue becomes damaged, these defenders take part in a *nonspecific response*—they react to tissue damage in

general, not to specific pathogens. A familiar example is inflammation, the subject of Section 8.3. The white blood cells known as lymphocytes—the T cells and B cells described shortly—may mount a *specific* response against a particular pathogen, not against general tissue damage. This specific kind of defense is called an **immune response**, and it occurs when lymphocytes recognize a unique molecular feature on an invading pathogen. An immune response is the body's third line of defense against invasion and tissue damage. Table 8.1 summarizes the three lines of defense. Much of the rest of this chapter is devoted to nonspecific and specific responses, and we begin in the next section with proteins in blood plasma that play important roles in both types of responses.

Table 8.1 The Human Body's Three Lines of Defense against Pathogens
BARRIERS AT BODY SURFACES (*nonspecific* targets)
Intact skin; mucous membranes at other body surfaces
Infection-fighting chemicals in tears, saliva, and so on
Normally harmless bacteria on body surfaces, which outcompete pathogens
Flushing effect of tears, saliva, urination, diarrhea; sneezing, and coughing
NONSPECIFIC RESPONSES (*nonspecific* targets)
Inflammation
1. Fast-acting white blood cells (neutrophils, eosinophils, and basophils)
2. Macrophages (also take part in immune responses)
3. Complement proteins, blood-clotting proteins, and other infection-fighting substances
Organs with pathogen-killing functions (such as lymph nodes)
Some cytotoxic cells (e.g., NK cells) with a range of targets
IMMUNE RESPONSES (*specific* targets only)
1. White blood cells (T cells, B cells, and macrophages that interact with them)
2. Communication signals (e.g., interleukins) and chemical weapons (e.g., antibodies, perforins)

Intact skin, mucous membranes, antimicrobial secretions, and other barriers at the body surface are the body's first line of defense against invasion and tissue damage.

Generalized responses to invasion (such as inflammation) are the second line of defense.

Immune responses against specific invaders, executed by lymphocytes and their chemical weapons, are the third line of body defense.

8.2 COMPLEMENT PROTEINS

A set of plasma proteins has roles in nonspecific *and* in specific defenses. Collectively, they are the **complement system** (because they "complement" other defenses). About twenty kinds of complement proteins circulate in the blood in an inactive form. The system can be activated two ways. A type of complement protein called C1 can bind to a complex that consists of a defender protein (an antibody) that is bound to part of an invader (an antigen). Alternatively, a different complement protein can interact with carbohydrate molecules on the surfaces of various microorganisms and viruses. If even a few molecules of a complement protein are activated, they trigger a cascade

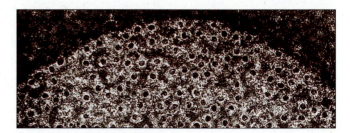

Figure 8.2 Micrograph of a cell surface, showing pores formed by membrane attack complexes.

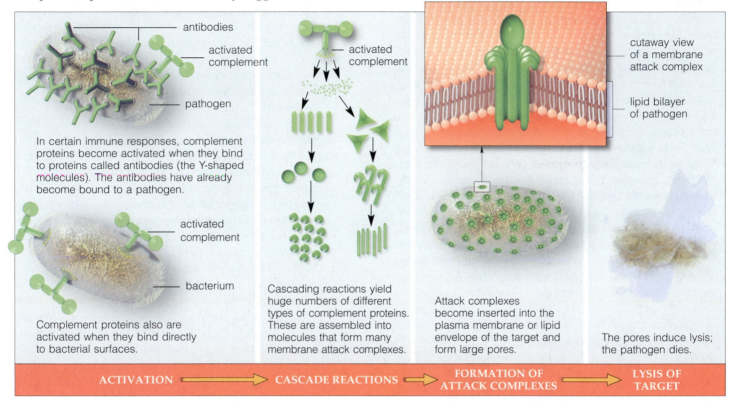

In certain immune responses, complement proteins become activated when they bind to proteins called antibodies (the Y-shaped molecules). The antibodies have already become bound to a pathogen.

Complement proteins also are activated when they bind directly to bacterial surfaces.

Cascading reactions yield huge numbers of different types of complement proteins. These are assembled into molecules that form many membrane attack complexes.

Attack complexes become inserted into the plasma membrane or lipid envelope of the target and form large pores.

The pores induce lysis; the pathogen dies.

ACTIVATION ⟹ **CASCADE REACTIONS** ⟹ **FORMATION OF ATTACK COMPLEXES** ⟹ **LYSIS OF TARGET**

of "snowballing" reactions. Molecules of one kind of complement protein activate huge numbers of another kind, and so on. The deployment of so many molecules has the following effects.

Some complement proteins unite to form **membrane attack complexes**. These structures have an interior channel (Figures 8.2 and 8.3). When inserted into the plasma membrane of many pathogens, they form pores— essentially, holes—in the membrane. The result is **lysis**, or disintegration, of the cell when its plasma membrane is disrupted. A lysed cell dies almost instantly. Membrane attack complexes also are inserted into the lipid coats of some bacteria. Lysozyme molecules diffuse through the resulting pores and digest a structural element of the bacterial cell it needs to survive.

Some activated complement proteins promote inflammation, a nonspecific defense response described next.

Figure 8.3 How a membrane attack complex forms. One reaction pathway starts when complement binds to bacterial surfaces. Another operates in immune responses to specific invaders. Both produce membrane pore complexes that induce lysis in pathogens.

Their cascades of reactions create concentration gradients that attract phagocytic white blood cells to an irritated or damaged tissue. Activated complement proteins also can bind to many invaders. Phagocytes, in turn, have receptors that bind to the complement proteins. As a result, the "complement-coated" invader sticks to the phagocyte, which ingests and kills it.

The complement system is a set of about twenty kinds of plasma proteins circulating in blood. It takes part in cascades of reactions that help defend against many bacteria, some parasites, and enveloped viruses.

INFLAMMATION

The Roles of Phagocytes and Their Kin

Certain types of white blood cells take part in an initial response to tissue damage. White blood cells, remember, develop from stem cells in bone marrow. Many of them circulate in blood and lymph. A great many take up stations in lymph nodes as well as in the spleen, liver, kidneys, lungs, and brain. Still others are in connective tissue beneath the skin and in mucous membranes. To review the lymphatic system and the various kinds of blood cells, see Sections 7.15 and 7.1, respectively.

Like SWAT teams, three kinds of white blood cells (all granulocytes) react swiftly to danger in general. They live for only a few hours or days and so are not adapted for sustained battles. **Neutrophils**, the most abundant kind, phagocytize bacteria. They ingest and kill bacteria and digest the dead cells to molecular bits. **Eosinophils** secrete enzymes that make holes in parasitic worms; they also phagocytize foreign proteins and help control allergic responses (see Section 7.11). **Basophils** secrete substances (such as histamine) that may help keep inflammation going after it starts. The names of these different granulocytes relate to their staining properties in the laboratory. The "neutral" granules of neutrophils take up various dyes well, eosinophils "prefer" the red dye eosin, and basophils take up a basic blue dye best.

Although slower to act, the white blood cells called **macrophages** are the "big eaters." They can live for months, engulfing and digesting nearly any foreign agent (Figure 8.4a). They also help clean up damaged tissue. Immature macrophages in blood are called monocytes.

The Inflammatory Response

An inflammatory response develops when something damages or kills cells of a tissue. Infections, punctures, burns, and other insults are the triggers. In the course

Table 8.2	Local Signs and Causes of Inflammation
Redness	Vasodilation, increased blood flow to site
Warmth	Vasodilation, greater flow of blood carrying more metabolic heat to site
Swelling	Capillaries made more permeable, plasma and plasma proteins move into interstitial spaces
Pain	Increased fluid pressure and local chemical signals stimulate pain receptors

of **acute inflammation**, the fast-acting phagocytes, complement proteins, and some other plasma proteins escape from the bloodstream at capillary beds in the damaged tissue (Figure 8.4b). Signs that they have entered interstitial fluid in a region of tissue include redness, swelling, heat, and pain (Table 8.2).

Mast cells, which dwell in tissues and function like basophils, act during an inflammatory response. They release **histamine** and other chemicals into interstitial fluid. Their secretions are local chemical signals that trigger vasodilation of arterioles that thread through the damaged tissue. Vasodilation, remember, is an increase in a vessel's diameter after smooth muscle in its wall has relaxed. When arterioles become engorged with blood, the affected tissue reddens and gets warmer, owing to blood-borne metabolic heat.

Released histamine also increases the permeability of the thin-walled capillaries in the tissue. It induces the endothelial cells making up the capillary wall to pull apart farther at the narrow clefts between them. Thus the capillaries become "leaky" to water and solutes (plasma) that normally do not leave the blood. When some proteins leak out, osmotic pressure increases in the surrounding interstitial fluid. In the process called ultrafiltration (Section 7.12) a small amount of protein-free plasma is

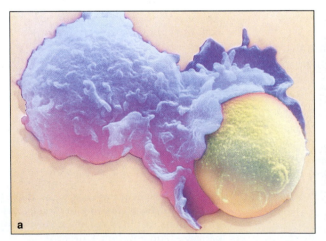

Figure 8.4 (**a**) A macrophage about to engulf a yeast cell, which looks a little like a lemon in this image. (**b**) A white blood cell squeezing out of a blood capillary, at a narrow gap between endothelial cells.

Further reading: Student Guide to InfoTrac on web site

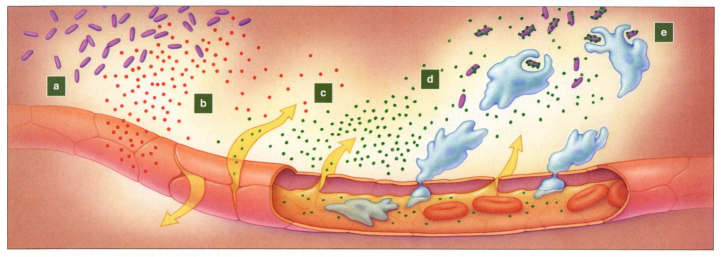

a Bacteria invade a tissue and directly kill cells or release metabolic products that damage tissue.

b Mast cells in tissue release histamine, which then triggers arteriolar vasodilation (hence redness and warmth) as well as increased capillary permeability.

c Fluid and plasma proteins leak out of capillaries; localized edema (tissue swelling) and pain result.

d Plasma proteins attack bacteria. Clotting factors wall off inflamed area.

e Neutrophils, macrophages, and other phagocytes engulf invaders and debris. Activated complement attracts phagocytes and directly kills invaders.

Figure 8.5 Acute inflammation in response to a bacterial invasion. The response involves delivering phagocytes and plasma proteins to the tissue. Together, these components of blood inactivate, destroy, or isolate the invaders, remove chemicals and cellular debris, and prepare the tissue for subsequent repair. These are their functions in all inflammatory responses.

pushed out through crevices in the capillary wall. At the same time, less tissue fluid is reabsorbed across the wall. This shift in the fluid balance across the capillary wall causes edema in the affected tissue. The tissue swells with fluid, and the swelling and inflammatory chemicals cause pain. A person typically avoids voluntary movements that might aggravate the pain. This behavior favors tissue repair processes.

Within hours of the first physiological responses to tissue damage, neutrophils are squeezing across capillary walls. They swiftly go to work. Macrophages arrive later, and engage in sustained action (Figure 8.5). While macrophages are engulfing pathogens and debris, they secrete chemical mediators—substances that are communication signals. Mediators called *chemotaxins* attract more phagocytes. One type of **interleukin** ("between lymphocytes"), a protein called IL-1, carries signals between B and T lymphocytes, as described shortly. *Lactoferrin* directly kills bacteria. *Endogenous pyrogen* (literally, an "internal firemaker") might trigger the release of prostaglandins (discussed in Chapter 12), which in turn trigger an increase in the "set point" on the body's thermostat. What we call a **fever** is a body temperature that has climbed to the higher set point.

A fever of about 39°C (100°F) is actually a helpful mechanism. It increases body temperature to a level that is too hot for the functioning of most pathogens. It also promotes an increase in defense activities. During the

fever, IL-1 induces drowsiness, which reduces the body's demands for energy, so more energy can be diverted to the tasks of defense and tissue repair. Macrophages take part in the cleanup and repair operations.

Among the plasma proteins that leak into the tissue are complement proteins and clotting factors (Section 7.13). The resulting blood clots wall off the inflamed area and typically prevent or delay the spread of invaders and toxic chemicals into the surrounding tissues. After the inflammation subsides, anticlotting factors that had also escaped from the capillaries dissolve the clots.

An inflammatory response develops in a region of tissue when cells there are damaged or killed, as by infection. Inflammation occurs during both nonspecific and specific defenses of tissues.

Mast cells in damaged or invaded tissues secrete histamine, which causes arterioles to dilate and increases the permeability of capillaries to fluid and plasma proteins. The vasodilation warms and reddens the tissue. Edema results from the fluid imbalance across capillary walls. The tissue swelling causes pain.

During inflammation, phagocytes such as macrophages engulf invaders and debris and secrete chemical mediators. The response requires plasma proteins, such as complement proteins that target invaders for destruction, as well as clotting factors that wall off the inflamed tissue.

THE IMMUNE SYSTEM

Defining Features

Sometimes physical barriers and inflammation are not enough to overwhelm an invader, and an infection becomes well established. Then, white blood cells known as **B** and **T lymphocytes** form armies that join the battle.

B and T cells are central to the body's third line of defense—the immune system. We define the **immune system** by two key features. The first is *immunological specificity*. This is the ability of certain kinds of lymphocytes to zero in on specific pathogens and eliminate them. The second feature is *immunological memory*—that is, some lymphocytes that form during an initial confrontation are set aside for a future battle with the same pathogen.

A straightforward operating principle governs the immune system: it can tell the difference between cells of the body itself and foreign cells or other substances. The immune system can accomplish this feat because each kind of cell, virus, or substance has features on its surface that give it a unique identity. Examples are the glycoproteins that are self markers on human red blood cells (Section 7.4). Lymphocytes recognize the body's own self markers and normally ignore them because "self" is not a threat.

There are two basic points about lymphocytes that can be introduced at this place in our story. First, there are two groups of T lymphocytes, *helper T cells* and *cytotoxic T cells*. Their functions are different, as described shortly. Second, any T or B cell that has not yet been exposed to its "own" specific nonself opponent is called a "virgin" T or B cell.

When a virgin lymphocyte first recognizes a specific nonself marker, the encounter stimulates the lymphocyte to divide. Then its descendants divide, and so on and on until a huge population of lymphocytes forms. The population is a *clone*; that is, each descendant is genetically identical to the first, "parent" lymphocyte.

Next, subgroups of the newly formed cells become specialized to respond to the foreign agent in different ways. Some groups become *effector cells*—lymphocytes that are immediately available to engage and destroy the enemy. Other subgroups become **memory cells**. These cells enter a resting phase. Instead of attacking the specific agent that triggered the initial response, memory cells "remember" it. If the same kind of agent appears again, they will join a larger, more rapid response to it.

Any molecular feature that triggers the formation of lymphocyte armies and is their target is an **antigen**. The most important antigens are certain proteins at the surface of pathogens or tumor cells, or ones that are unbound but toxic. As you will see, lymphocytes can recognize nonself because they have receptors that bind to such targeted "foreign" features.

To recap, immunological specificity and memory involve three events: *recognition* of an antigen, *repeated cell divisions* that form huge populations of lymphocytes, and then specialization (*differentiation*) of the lymphocytes into subpopulations of effector and memory cells, each with receptors for a particular antigen.

Antigen-Presenting Cells— The Triggers for Immune Responses

The plasma membrane of every nucleated cell in the body incorporates various proteins. These include **MHC markers**, proteins named after the *m*ajor *h*istocompatibility *c*omplex genes that encode the instructions for making them. Some MHC markers are common at the surface of all nucleated body cells. Others are found only on the body's macrophages and lymphocytes.

To get an idea how MHC proteins operate, suppose bacteria enter a cut on your finger. Lymph vessels pick up some of the inflamed tissue's interstitial fluid and deliver it, along with some invading cells, to nearby lymph nodes. There, macrophages engulf the "nonself" foreign cells, which become enclosed in vesicles with digestive enzymes that break the antigen molecules into fragments. These fragments bind to MHC molecules and form **antigen-MHC complexes.** Then the vesicles move to the cell's plasma membrane and fuse with it. Now, the complexes are displayed at the surface of the macrophage.

Similar events take place with cancer cells or cells that are infected by an intracellular ("inside the cell")

Figure 8.6
Molecular cues that stimulate lymphocytes to make immune responses.

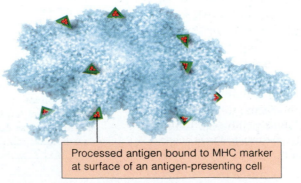

MHC marker designating "self" (present only at the surface of the body's own cells)

T cells and B cells ignore this.

Processed antigen bound to MHC marker at surface of an antigen-presenting cell

T cell recognition initiates immune response.

Antigen (any unprocessed foreign or abnormal molecular configuration that lymphocytes recognize as "nonself")

B cell recognition initiates immune response.

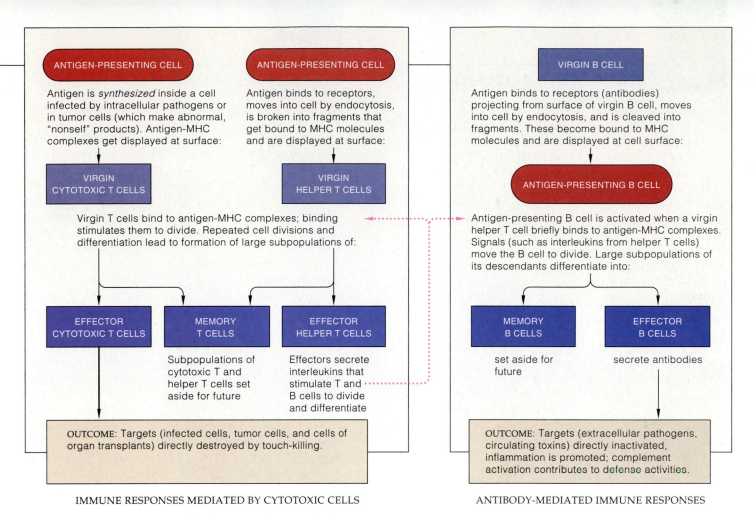

IMMUNE RESPONSES MEDIATED BY CYTOTOXIC CELLS

ANTIGEN-PRESENTING CELL

Antigen is *synthesized* inside a cell infected by intracellular pathogens or in tumor cells (which make abnormal, "nonself" products). Antigen-MHC complexes get displayed at surface:

VIRGIN CYTOTOXIC T CELLS

ANTIGEN-PRESENTING CELL

Antigen binds to receptors, moves into cell by endocytosis, is broken into fragments that get bound to MHC molecules and are displayed at surface:

VIRGIN HELPER T CELLS

Virgin T cells bind to antigen-MHC complexes; binding stimulates them to divide. Repeated cell divisions and differentiation lead to formation of large subpopulations of:

EFFECTOR CYTOTOXIC T CELLS

MEMORY T CELLS

Subpopulations of cytotoxic T and helper T cells set aside for future

EFFECTOR HELPER T CELLS

Effectors secrete interleukins that stimulate T and B cells to divide and differentiate

OUTCOME: Targets (infected cells, tumor cells, and cells of organ transplants) directly destroyed by touch-killing.

ANTIBODY-MEDIATED IMMUNE RESPONSES

VIRGIN B CELL

Antigen binds to receptors (antibodies) projecting from surface of virgin B cell, moves into cell by endocytosis, and is cleaved into fragments. These become bound to MHC molecules and are displayed at cell surface:

ANTIGEN-PRESENTING B CELL

Antigen-presenting B cell is activated when a virgin helper T cell briefly binds to antigen-MHC complexes. Signals (such as interleukins from helper T cells) move the B cell to divide. Large subpopulations of its descendants differentiate into:

MEMORY B CELLS

set aside for future

EFFECTOR B CELLS

secrete antibodies

OUTCOME: Targets (extracellular pathogens, circulating toxins) directly inactivated, inflammation is promoted; complement activation contributes to defense activities.

Figure 8.7 Overview of the key interactions among B and T lymphocytes during an immune response. Usually, both types of lymphocytes are activated when an antigen has been detected.

pathogen. Such abnormal body cells synthesize "nonself" substances. Then, as with the macrophage example, the antigen is incorporated into antigen-MHC complexes that are displayed at the cell surface. Any cell that processes and displays an antigen that is bound with a suitable MHC molecule is called an **antigen-presenting cell,** or **APC**. When lymphocytes recognize such a cell, they respond with the rounds of cell division that form great armies of effector and memory cells (Figure 8.6).

Key Players in Immune Responses

In any immune response, the same kinds of white blood cells are called into action. Figure 8.7 is an overview of how the cells interact. In brief, the recognition of antigen-MHC complexes activates **helper T cells**. These T lymphocytes produce and secrete chemical signals that cause other responsive T or B lymphocytes to divide. Large populations of effector cells and memory cells will be the result. Recognition also activates **cytotoxic T cells**. These lymphocytes can eliminate infected body cells or tumor cells by "touch-killing." When they touch a target, they deliver lethal chemicals into it. Immune responses by T cells are called *cell-mediated* responses. By contrast, B cells produce the antigen-binding receptor

molecules we know as **antibodies**. During an immune response, effector B cells secrete huge numbers of antibody molecules. For this reason, we say that B cells provide *antibody-mediated* responses.

Control of Immune Responses

Removal of an antigen stops an immune response. In fact, by the time the tide of battle turns, the effector cells and their secretions have already destroyed most antigen-bearing agents in the body. With fewer antigen molecules around to stimulate the cells, the response declines, then stops. In addition, cells with suppressor roles produce chemical signals that help shut down immune responses.

Nonself markers called antigens trigger immune responses when they encounter lymphocytes that can recognize them.

Armies of T and B cells are produced in immune responses. Attacks on nonself substances are carried out by helper T cells, cytotoxic T cells, and B cells (and the antibodies they make). In addition to these effector cells, an immune response produces memory cells that can help mount a larger, more rapid response if the same invader returns.

The antigen-presenting cells and lymphocytes interact inside lymphoid tissue that is strategically located all over the body (Figure 8.8). For example, there are nodules of lymphoid tissue just under the mucous membranes of your respiratory, digestive, and reproductive systems. There, antigen-presenting cells and lymphocytes can intercept invaders that penetrate surface barriers. The simple lymphoid organs we call **tonsils** ring the throat (pharynx)—a major point of entry for pathogens.

Consider also antigen in tissue fluid that is entering the lymph vascular system. Because lymph vessels eventually drain into the bloodstream, antigen *could* become distributed to every body region. However, before antigen can reach the blood, it must trickle through lymph nodes—which are packed with defending cells. Even in the few cases where antigen manages to enter the blood, defending cells in the spleen intercept it.

In lymph nodes, cells are organized for maximum effectiveness. Antigen-presenting cells make up the front line, engulfing invaders. They process and display an antigen, thus calling lymphocyte comrades into action.

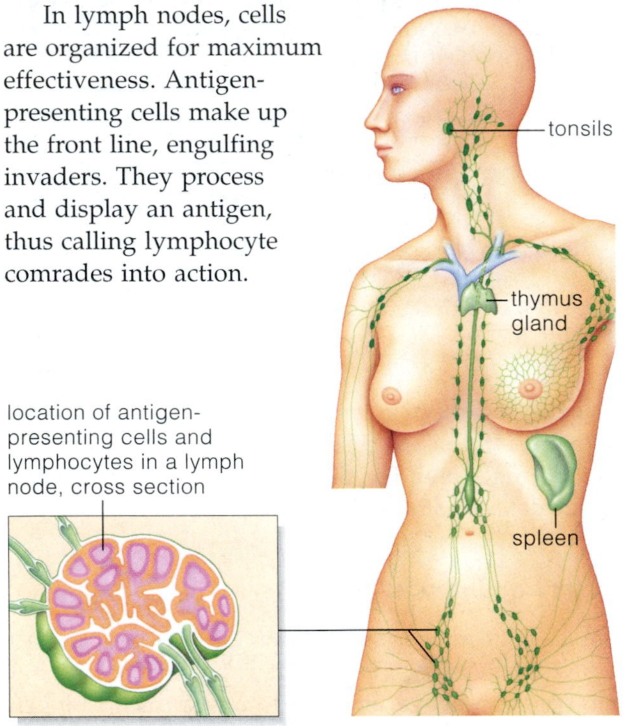

location of antigen-presenting cells and lymphocytes in a lymph node, cross section

tonsils

thymus gland

spleen

Figure 8.8 Organized arrays of antigen-presenting cells and lymphocytes in lymph nodes.

The cell divisions that produce subgroups of effector and memory cells take place in lymph nodes. As lymph drains through, it moves effector activities to the back of the organ and beyond. Meanwhile, virgin and memory cells circulate through the lymph node, patrolling at the front line.

Antigen-presenting cells and lymphocytes intercept and battle pathogens inside lymphoid organs and tissues, especially the lymph nodes.

How T Cells Form and Become Active

Let's begin our close-up look at immune responses with T cells. Recall from Chapter 7 that T cells arise from stem cells in bone marrow. However, they do not finish developing until they travel to the thymus. There they become differentiated—specialized as helper T cells and cytotoxic T cells. Each T cell acquires many identical T cell receptor molecules, or TCRs, on its surface. It is the TCR that can bind to a specific antigen. At least a billion different TCRs are made in the thymus, which is why there are enough T cells to react with billions of different antigens. Bristling with their TCRs, T cells leave the thymus, circulate in lymph and blood, and move into lymph nodes and the spleen as virgin T cells—T cells that have not yet been stimulated by an antigen.

The TCRs of virgin T cells ignore MHC markers on body cells if there is no antigen bound to them. They also ignore antigens by themselves. TCRs *do* recognize and bind with antigen-MHC complexes on antigen-presenting cells. This binding, and chemicals released by the APC, stimulate T cells to divide (Figure 8.9). Repeated divisions produce large clones (populations of genetically identical cells). These clonal descendants differentiate into subgroups of effector cells and memory cells. Each descendant has the same TCR for a particular antigen-MHC complex.

What Effector T Cells Do

What do subpopulations of effector T cells do? Effector helper T cells secrete interleukins, the chemical mediators that promote cell divisions and then differentiation of responsive virgin T and B cells. Effector cytotoxic T cells are killers; they respond to antigen-MHC complexes at the surface of tumor cells and of body cells already attacked by intracellular pathogens (such as viruses). The complex is a "double signal" that tells the killers to attack any cell bearing it (Figure 8.9c).

Effector cytotoxic T cells destroy infected cells with a touch-kill mechanism. They secrete *perforins*, protein molecules that form doughnut-shaped pores in a target cell's plasma membrane. (The pores look similar to the ones in Figure 8.3.) These effectors also secrete chemicals that cause the genetically programmed death of a target cell. This programmed cell death is called **apoptosis.** The affected cell's cytoplasm dribbles out, its organelles break down, and its DNA breaks apart. Having made its lethal hit, the cytotoxic T cell moves on.

Cytotoxic T cells also contribute to the rejection of tissue and organ transplants. Parts of MHC markers on donor cells are different enough from the recipient's to be recognized as antigens. Other parts are similar enough to complete the double signal. MHC typing minimizes the risk.

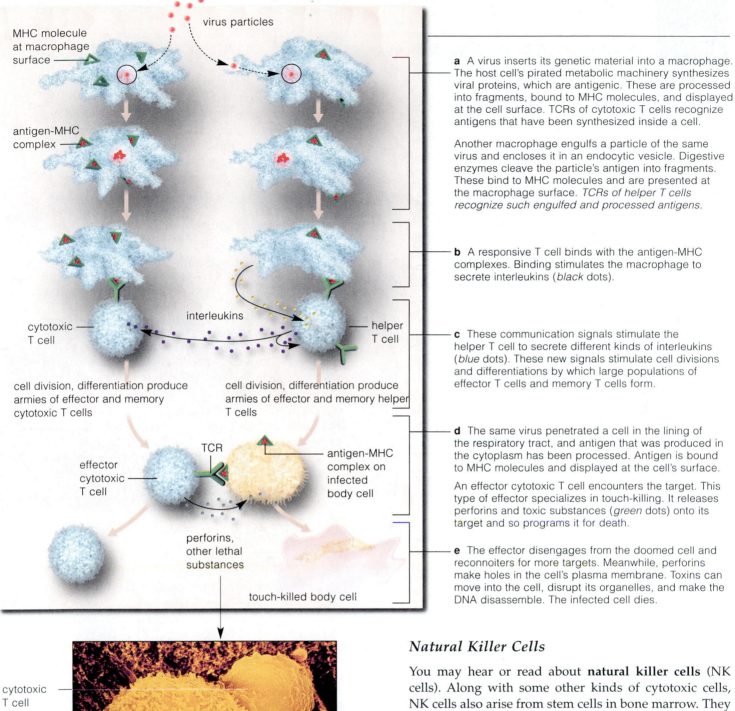

MHC molecule at macrophage surface

antigen-MHC complex

virus particles

a A virus inserts its genetic material into a macrophage. The host cell's pirated metabolic machinery synthesizes viral proteins, which are antigenic. These are processed into fragments, bound to MHC molecules, and displayed at the cell surface. TCRs of cytotoxic T cells recognize antigens that have been synthesized inside a cell.

Another macrophage engulfs a particle of the same virus and encloses it in an endocytic vesicle. Digestive enzymes cleave the particle's antigen into fragments. These bind to MHC molecules and are presented at the macrophage surface. *TCRs of helper T cells recognize such engulfed and processed antigens.*

b A responsive T cell binds with the antigen-MHC complexes. Binding stimulates the macrophage to secrete interleukins (*black* dots).

cytotoxic T cell

interleukins

helper T cell

c These communication signals stimulate the helper T cell to secrete different kinds of interleukins (*blue* dots). These new signals stimulate cell divisions and differentiations by which large populations of effector T cells and memory T cells form.

cell division, differentiation produce armies of effector and memory cytotoxic T cells

cell division, differentiation produce armies of effector and memory helper T cells

d The same virus penetrated a cell in the lining of the respiratory tract, and antigen that was produced in the cytoplasm has been processed. Antigen is bound to MHC molecules and displayed at the cell's surface.

effector cytotoxic T cell

TCR

antigen-MHC complex on infected body cell

An effector cytotoxic T cell encounters the target. This type of effector specializes in touch-killing. It releases perforins and toxic substances (*green* dots) onto its target and so programs it for death.

perforins, other lethal substances

e The effector disengages from the doomed cell and reconnoiters for more targets. Meanwhile, perforins make holes in the cell's plasma membrane. Toxins can move into the cell, disrupt its organelles, and make the DNA disassemble. The infected cell dies.

touch-killed body cell

Natural Killer Cells

You may hear or read about **natural killer cells** (NK cells). Along with some other kinds of cytotoxic cells, NK cells also arise from stem cells in bone marrow. They appear to be a type of lymphocyte, but not T or B cells, and we don't really know much about them. They do not require an antigen-MHC complex signal to be aroused; they may recognize odd molecular features at the surface of their targets. NK cells patrol for tumor cells and virus-infected cells, which they then touch-kill.

cytotoxic T cell

tumor cell

T cells arise in bone marrow. Later on, in the thymus, they acquire TCRs (receptors for self markers and bound antigen).

Effector helper T cells secrete interleukins that trigger the cell divisions and differentiation into armies against specific antigens. Effector cytotoxic cells touch-kill infected cells or tumor cells, even foreign cells of transplants.

Figure 8.9 (**a**) Example of a T-cell-mediated immune response. In this case, the response involves an antigen-MHC complex that activates T cells. (**b**) Scanning electron micrograph of a cytotoxic T cell caught in the act of touch-killing a tumor cell.

ANTIBODY-MEDIATED RESPONSES

B Cells and the Targets of Antibodies

Like T cells, B cells also arise from stem cells in bone marrow and start down a pathway that will culminate in full differentiation—that is, they will be fully specialized for their function. Along *their* pathway, however, B cells start manufacturing many, many copies of a single kind of antibody molecule.

All antibodies are proteins, but the antigen binding site of each kind matches only one kind of antigen. Antibody molecules are more or less Y-shaped, with a tail and two arms that bear identical antigen receptors. Section 8.9 provides a closer look at these molecules. For now, you can simply think of them as Y-shaped structures, like the ones sketched in Figure 8.10.

When a B cell is maturing, each freshly synthesized antibody molecule moves to the plasma membrane. Its tail becomes embedded in the membrane's lipid bilayer and the two arms stick out above it. Soon the cell bristles with antigen receptors (the bound antibodies), and it is ready to join the body's defenses as a virgin B cell.

When its antigen receptors lock onto a target, the B cell does something you might not expect. *It becomes an antigen-presenting cell.* First, an endocytic vesicle forms around the bound antigen and moves it into the cell where it will be digested into fragments. The fragments bind to MHC molecules and are presented at the B cell surface. When TCRs of a responsive helper T cell bind to the antigen-MHC complex, signals are transferred

Figure 8.10 Antibody-mediated immune response. This example is a response to a bacterial invasion. The inset is a computer model of an antigen fragment (*pink*) bound to the cleft of an MHC protein.

a A virgin B cell encounters unbound antigen in tissue fluid. Antigen receptors (in this case, membrane-bound antibodies) bind the antigen. An endocytic vesicle moves bound antigen into the cell for processing. Antigen-MHC complexes are displayed at the cell surface; the B cell has become an antigen-presenting cell.

b TCRs of a helper T cell bind to antigen-MHC complexes on the B cell. Binding activates the T cell and stimulates the B cell to prepare to divide. Then the cells disengage.

c Unprocessed antigen binds to the B cell. Meanwhile, the helper T cell secretes interleukins (*purple* dots). Both events trigger repeated cell divisions and differentiation that yield large armies of antibody-secreting effector B and memory B cells.

d Antibody molecules released from the effector B cells enter extracellular fluid. When they contact a bacterial cell that is the target, they bind to antigen on its surface. Binding tags the cell for destruction.

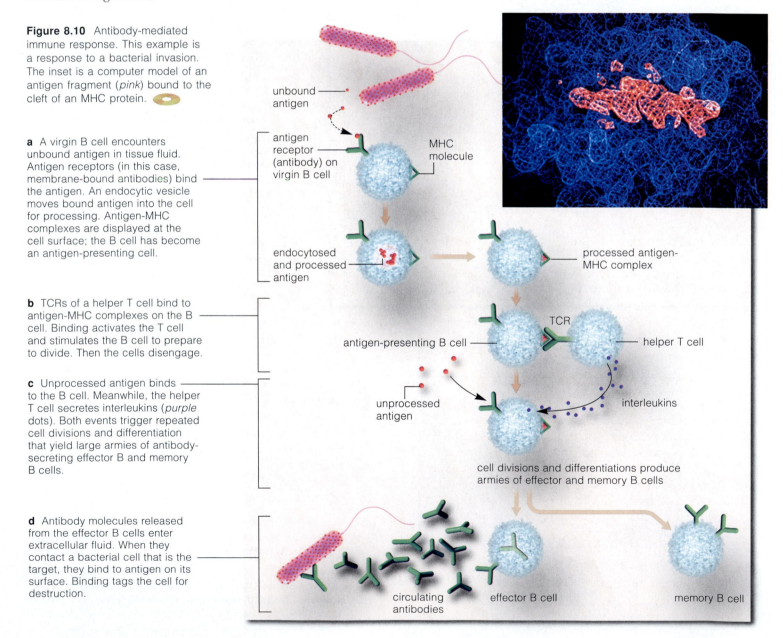

unbound antigen

antigen receptor (antibody) on virgin B cell

MHC molecule

endocytosed and processed antigen

processed antigen-MHC complex

antigen-presenting B cell

TCR

helper T cell

unprocessed antigen

interleukins

cell divisions and differentiations produce armies of effector and memory B cells

circulating antibodies

effector B cell

memory B cell

Further reading: Student Guide to InfoTrac on web site →

between the T and B cell. The cells soon disengage. Then the B cell encounters unprocessed antigen, and its surface antibodies bind to it. The binding, along with interleukins secreted from nearby helper T cells, prompts the B cell to divide. Its clonal descendants differentiate into effector B and memory B cells. The effectors (also called plasma cells) make and secrete huge numbers of antibody molecules. When a circulating antibody binds to antigen, it tags an invader for destruction, as by phagocytes and the activation of complement.

The main targets of antibody-mediated responses are extracellular pathogens and toxins that are freely circulating in tissues or body fluids. Antibodies *cannot* enter cells and bind to pathogens or toxins hidden there.

The Immunoglobulins

During immune responses, B cells produce several classes of antibody molecules. Collectively antibodies are called **immunoglobulins**, or Igs. As described in Section 8.9, they are the proteins that result from genetic events that proceed while B cells mature and while an immune response is under way. The different immunoglobulins have antigen-binding sites, as well as other sites with specialized functions.

IgM antibodies are the first to be secreted during immune responses. They trigger the cascade reactions that produce complement proteins. They also bind targets in clumps that phagocytes can readily handle. (Recall the agglutination responses described in Section 7.4?) The *IgD* antibodies associate with IgM on virgin B cells, but their function is not yet understood.

About three-fourths of our antibodies are *IgG* antibodies. These antibodies are long-lasting. They activate complement proteins and block many toxins. They also are the only antibodies that can cross the placenta, protecting both the fetus and newborn with the mother's acquired immunities. IgGs are secreted into the early milk produced by mammary glands and then are absorbed into a suckling newborn's bloodstream.

IgA antibodies enter mucus-coated surfaces of the respiratory, digestive, and reproductive tracts, where they may neutralize infectious agents. Mother's milk delivers them to the mucous lining of a newborn's GI tract.

IgE triggers inflammation after attacks by parasitic worms and other pathogens. As described later, it also figures in allergies. The tails of IgE antibodies bind to basophils and mast cells, and the antigen receptors face outward. Antigen-binding induces basophils and mast cells to release substances that promote inflammation and allergic reactions.

Antibodies that are secreted by B cells bind to antigens of extracellular pathogens or toxins and tag them for disposal, as by phagocytes and complement proteins.

ORGAN TRANSPLANTS: BEATING THE (IMMUNE) SYSTEM

Every year, thousands of people with severe heart, lung, kidney, and liver disease receive donated replacement organs. Organ transplants are risky, in part because, except in the case of identical twins, the organ recipient's immune system will perceive donated tissues as foreign and attempt to reject them from the body.

STRATEGIES FOR PREVENTING REJECTION   To help prevent rejection, before an organ is to be transplanted the MHC markers of a potential donor are analyzed to determine how closely they match those of the patient. Because such tissue grafts generally succeed only when the donor and recipient share at least 75 percent of their MHC markers, the favored donor is a close relative of the recipient, such as a parent or sibling, who is likely to have a similar genetic makeup. More commonly, the donated organ comes from a fresh cadaver. In addition to having well-matched MHC markers, donor and recipient also must have compatible blood types, as discussed in Section 7.4.

After surgery, the organ recipient receives drugs, such as cyclosporin, that suppress the immune system. The treatment also may include other therapies designed to fend off an attack by B and T cells. Suppression of the immune system means that the patient must take large doses of antibiotics to control infections. In spite of the difficulties, however, many organ recipients survive for years beyond the surgery.

Some researchers are actively trying to genetically alter pigs to create varieties with organs bearing common human MHC markers. Why pigs? A major reason is that, anatomically and physiologically, pig organs are surprisingly similar to ours. In theory, such transgenic animals could be sources of readily transplantable organs. Transplantation of organs from one species of animal to another is called *xenotransplantation*. It is a fascinating topic about which you can read more in Chapter 21.

IMMUNE-PRIVILEGED TISSUES   There are some intriguing exceptions to the "rule" that transplanted foreign tissues provoke a recipient's immune defenses. Two examples are tissues of the eye, and the testicles. In simple terms, the plasma membrane of cells of these organs apparently bears receptor proteins that can detect activated lymphocytes in the surroundings. Before such a lymphocyte can launch an attack, the protein signals the soon-to-be-besieged cell to secrete a chemical that triggers apoptosis ("cell suicide") in the approaching lymphocytes—averting an attack. Our ability to readily transplant the cornea—the outermost layer of the eye that is vital to clear vision—is a practical benefit of this mechanism.

How Do Antigen-Specific Receptors Form?

The variety of antigens in our surroundings is mind-boggling. Pathogens are everywhere in and on food, water, air, soil, and people. Collectively, however, T and B cell populations in the body have an equally staggering number of different antigen receptors—enough, in fact, to recognize about a billion different antigens. To get an idea of how this diversity can arise, we'll briefly examine how the body comes by its vast array of different antibodies—the antigen receptors of B cells.

As you may remember, all of the antigen receptors of a single B (or T) cell are identical, and all are proteins. Every antibody molecule on a B cell consists of four polypeptide chains bonded together, in a configuration that gives the antibody its Y-shape (Figure 8.11). Certain parts of each chain, the *variable* regions, fold in ways that produce grooves and bumps with a certain charge distribution. This arrangement is what gives an antibody its specificity, because the only antigen that can bind with it is one that has complementary grooves, bumps, and charge distribution.

All cells carry genetic instructions in the form of DNA in chromosomes. Each *gene* is a defined stretch of the DNA. In a newly formed B cell, the full genetic instructions for making a particular antibody are not "set"—the necessary parts are there, but they haven't yet been put together in their final form. An analogy is a rail yard that has many different kinds of railroad cars that can be joined in many different combinations to form a train. Similarly, an immature B cell's DNA has regions containing many chemically different segments that can be combined in many different ways to form the instructions for making the variable regions of an antibody molecule.

As the B cell matures, one kind of segment contacts another kind and binds with it. *Which* segment is tapped from each DNA region is pure chance. Next, a third kind of segment, carrying information for the antibody's constant region, is attached to the first two. After several more steps, the three segments form a gene with the instructions for the constant and variable regions of a particular antibody. Because there are so many genetic options in the immature lymphocyte, the final DNA sequence codes for only one of at least billions of possible antigen receptors. The same kinds of DNA rearrangements help produce the variable regions of the TCR molecules of maturing T cells.

Some years ago, Macfarlane Burnet developed a **clonal selection hypothesis** that helped point the way to our current view of receptor diversity. He proposed that antigen "selects" (binds to) one lymphocyte from all the various types in the body, because that lymphocyte has the receptor specific for it. Repeated cell divisions then give rise to a clone of cells that carry out the response (Figure 8.12).

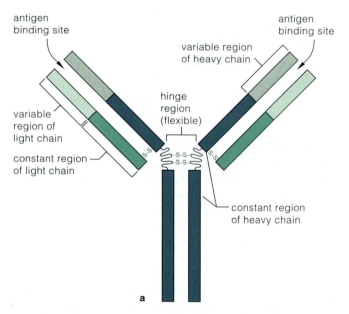

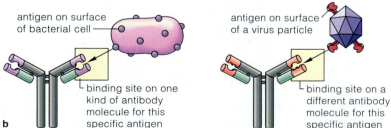

Figure 8.11 Antibody structure. (**a**) An antibody molecule has four polypeptide chains, often joined in a Y shape. (**b**) Some regions are almost the same in all antibodies. But the molecular configuration in one region is unique for each kind of antibody; it is the binding site for one kind of antigen only. The antigen fits into the site's grooves and onto its protrusions, as shown in Figure 8.9.

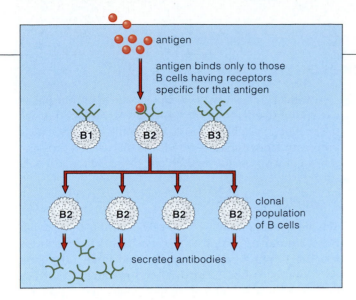

Figure 8.12 Clonal selection of a B cell that made the specific antibody that can combine with a specific antigen. Only B cells (and T cells) that are selected by an antigen are activated and give rise to a clonal population of immunologically identical cells.

Immunological Memory

The clonal selection theory explains how a person can have *immunological memory* of a first encounter with antigen. The term refers to the body's capacity to make a *secondary* immune response to any subsequent encounter with the same type of antigen that provoked the primary response (Figure 8.13).

Memory cells that form during a primary immune response do not engage in battle. They circulate for years or for decades. Compared to the virgin cells that initiate a primary response, these patrolling battalions have far more cells, so they intercept antigen far sooner. Effector cells form sooner, in greater numbers, so the infection is terminated before the host (you) gets sick. Even more memory T and B cells form during a secondary response (Figure 8.14). In terms of a host's long-term survival, these preparations for subsequent meeting with a pathogen bestow a significant advantage.

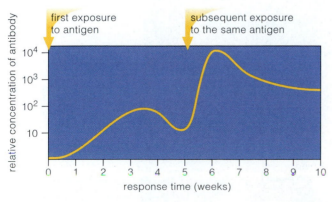

Figure 8.14 An example of differences in magnitude and duration between a primary and a secondary immune response to the same antigen. (Here, the secondary response starts at week 5.)

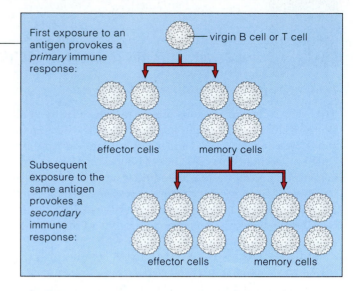

Figure 8.13 Immunological memory. Not all B and T cells are used in a primary immune response to an antigen. A large number continue to circulate as memory cells, which become activated during a secondary immune response.

A skin test for tuberculosis can provide visible evidence of immunological memory. A health care worker scratches a bit of TB antigen into a small patch of the patient's skin. In people who have a positive reaction to the skin test, a hard, red swelling develops at the test site, usually within forty-eight hours. Even in someone who has no medical history of the disease, this response is visible evidence of immunological memory. It indicates that there has been an immune response against the tuberculosis bacterium, which the person's immune system must have encountered in the past.

As this example suggests, over time, the composition of memory T and B cell populations is shaped by the particular kinds of antigens to which a person has been exposed. For instance, if you grew up in North America in the late 20th century, you almost certainly have hosts of memory cells that can fight off pathogens common in that particular environment. You may *not* have ready contingents of memory cells primed to battle, say, bacteria present in the water supply of Mongolia or Paris, France. That is one reason why wise tourists travel prepared to cope, at least temporarily, with intestinal problems caused by exposure to unfamiliar waterborne bacteria.

Recombination of segments drawn at random from receptor-encoding regions of DNA gives each new T or B cell a gene sequence for one of billions of possible antigen receptors.

Immunological specificity means that the clonal descendants of an antigen-selected cell will react only with the selecting antigen.

Immunological memory means that a person can make a secondary (faster, greater) immune response to the pathogen that caused the primary response.

PRACTICAL APPLICATIONS OF IMMUNOLOGY

Immunization: "Borrowing" Immunity

The word **immunization** refers to procedures that promote increased immunity against specific diseases. In *active immunization,* an antigen-containing preparation called a **vaccine** is injected into the body or taken orally, sometimes according to a schedule (Figure 8.15). A first injection elicits a primary immune response. A "booster shot" given later elicits a secondary response. As a result, more effector cells and memory cells form and can provide long-lasting protection against the disease.

Many vaccines are manufactured from killed or extremely weakened ("attenuated") pathogens. For example, weakened poliovirus particles are used for the Sabin polio vaccine. Other vaccines are based on inactivated forms of natural toxins, such as the bacterial toxin that causes tetanus. Today many vaccines are made with harmless genetically engineered viruses (Chapter 21). These "transgenic" viruses incorporate genes from three or more different viruses in their genetic material. After a person is vaccinated with an engineered virus, body cells use the new genes to produce antigens, and immunity is established.

Passive immunization often helps people who are already infected with pathogens, including the ones that cause diphtheria, tetanus, measles, hepatitis B, rabies, and some other diseases. A person receives injections of antibody molecules purified from some other source. The best source is someone who already has produced a large amount of the required antibody. The effects are not lasting, because the recipient's own B cells are not producing antibodies. However, injected antibodies may counter the immediate attack.

Vaccines are powerful weapons, but they *can* fail or have adverse effects. In rare cases, a vaccine can damage the nervous system or result in chronic immunological problems. Hence it is a good idea to discuss the pros and cons of any vaccination with your physician.

Monoclonal Antibodies

The antibodies produced by B cells contribute so powerfully to immune responses because they are amazingly specific. Commercially prepared **monoclonal antibodies** harness this power for medical and research uses. They can be obtained by laboratory techniques that yield large quantities of a desired antibody derived from B cells. The term "monoclonal antibody" refers to the fact that the antibody molecules are made by a population of cells cloned from a single antibody-producing cell.

At one time laboratory mice were the "factories" for making monoclonal antibodies. Typically, a mouse was immunized with a specific antigen, and then B cells were extracted from its spleen. The B cells were fused with a line of cancer cells, producing what are called hybridoma cells (because they are "hybrids" of the two cell types). Like cancer cells, hybridoma cells multiply incessantly, and some of them produce the antibody of interest. There are drawbacks to using mice to make monoclonal antibodies, though, and the advent of recombinant DNA technology is changing the way many monoclonal antibodies are produced. Genetic instructions for making an antibody are inserted into bacteria, or into cultured mammalian cells, and the cells are grown in a laboratory or commercial operation. As Chapter 21 describes, some

RECOMMENDED VACCINES	RECOMMENDED AGES
Hepatitis B	Birth–2 months
Hepatitis B booster	1–4 months
Hepatitis B booster	6–18 months
Hepatitis B assessment	11–12 years
DTP (Diphtheria; Tetanus; and Pertussin, or whooping cough)	2, 4, and 6 months
DTP booster	15–18 months
DTP booster	4–6 years
DT	11–16 years
HiB (*Hemophilus influenzae*)	4 and 6 months
HiB booster	12–15 months
Polio	2 and 4 months
Polio booster	6–18 months
Polio booster	4–6 years
MMR (Measles, Mumps, Rubella)	12–15 months
MMR booster	4–6 years
MMR assessment	11–12 years
Varicella	12–18 months
Varicella assessment	11–12 years

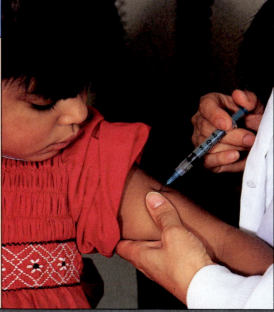

Figure 8.15 From the Centers for Disease Control and Prevention, the 1998 immunization guidelines for children in the United States. Physicians routinely immunize infants and children. Low-cost or free vaccinations are available at many community clinics and health departments.

researchers are engineering plants, such as corn, to make "plantibodies" that will be safe and effective for humans.

Monoclonal antibodies have become extremely important diagnostic tools. Their ability to recognize and bind to a particular kind of antigen is the key that allows them to detect substances in the body—be it a bacterium, another antibody, or a chemical—even if only a very small amount of the substance is present. Uses include home pregnancy tests and screening for prostate cancer and some sexually transmitted diseases. Monoclonal antibodies also have a range of potential uses in cancer treatment, including acting as highly specific "magic bullets" to deliver lethal drugs to cancer cells only, sparing the body's healthy tissues.

Interferons and Lymphokines

Various body cells, including ones infected by a virus, can make and secrete **interferons**. When they reach an uninfected cell, these defensive proteins trigger a chemical attack that inhibits viral replication. *Gamma* interferon, which is produced by T cells, has other functions, too. It calls natural killer cells into action and also enhances the activity of macrophages. Genetically engineered gamma interferon is used to treat hepatitis C, the chronic, serious viral disease mentioned earlier. Some kinds of cells (not lymphocytes) produce *beta* interferon. This protein has recently been approved for the treatment of one type of *multiple sclerosis*, a disease in which the immune system attacks parts of the nervous system.

Lymphoctes make communication chemicals called *lymphokines* ("lymphocyte movers"). They include the interleukins which, recall, help spur B and T cells to action (Section 8.3). Dozens of interleukins have been identified, and there is hope some can be harnessed to strengthen immune responses. Lymphocytes often infiltrate cancer tumors. In studies where researchers removed such cells from a patient and exposed them to an interleukin, the "parent" lymphocytes produced huge populations of tumor-infiltrating lymphocytes that were highly efficient at killing tumor cells. Called LAK cells (for Lymphokine-Activated Killers), they appear to be somewhat effective when they are injected back into a patient.

Immunization promotes enhanced immunity to specific diseases by stimulating the production of both effector and memory lymphocytes.

Monoclonal antibodies and cytokines (such as interleukins and gamma interferon) have become important tools in medical research, testing, and the treatment of disease.

Every autumn millions of people get a "flu shot"—a vaccination they hope will protect them against that season's most threatening *influenza* virus. If your Italian is not too rusty, you might recognize that "influenza" is Italian for "influence." In the Middle Ages some Italians credited a widespread outbreak of flu to the influence of the stars. Then as now, everybody dreads the flu, and for good reason. Even a mild case can cause fever, aching muscles, and coldlike symptoms. In people with a weakened immune system—a group that includes the elderly, young children, and those with immunity disorders—a case of the flu can quickly develop into pneumonia and become life-threatening.

Modern flu vaccines are an effort to stop influenza in its tracks. However, even after many research advances in recent years, we still are struggling to keep up with our microscopic foes (Figure 8.16). Why? Because influenza viruses are genetically clever.

There are three broad groups of influenza viruses, which are known simply as A, B, and C. Each group also has many strains, and new ones are constantly popping up. Once you suffer your way through an infection by one strain, you are immune to it, but the various strains of the different influenza groups are different enough that each one is a brand-new threat to the body.

The genetics can be complicated, but basically different flu strains have—surprise—different, unique surface antigens. In addition, a given influenza strain can rapidly change its genetic makeup—that is, it can mutate. There are various ways the change can occur. A common one, which could be called "gene swapping," has been observed in the laboratory. It wasn't a pretty sight! Two slightly different influenza viruses infect the same cell, then they exchange bits of their genetic material. When the "dirty deed" is done, the result is new virus particles that can be genetically different

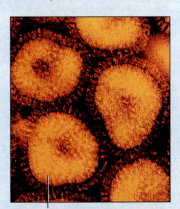

virus particle

Figure 8.16 Particles of a virus that causes influenza in humans.

from any influenza virus that has ever existed. As you might guess, when this happens, any vaccine that has been developed to combat the parent virus strain might as well be thrown out the window. Your personal immunity to the former strain is useless, too. With its slightly different antigen, the virus will go unrecognized by your existing memory cells and can attack your body as a completely new threat.

Allergies

In at least 15 percent of the people in the United States, normally harmless substances can provoke immune responses. These substances are known as *allergens*, and the response to them is an **allergy**. Common allergens are pollen (Figure 8.17*a*), a variety of foods and drugs, dust mites, fungal spores, insect venom, and cosmetics. Some responses start within minutes; others are delayed. Either way, the allergens trigger mild to severe inflammation of mucous membranes and sometimes other tissues.

Some people are genetically predisposed to develop allergies. Infections, emotional stress, or changes in air temperature also may cause reactions that otherwise might not occur. When an allergic person first is exposed to certain antigens, IgE antibodies are secreted and bind to mast cells (Figure 8.17*b*). When the IgE binds antigen, mast cells secrete chemical mediators—prostaglandins, histamine, and other substances—that fan inflammation. They also cause airways to constrict. In *hay fever*, the

allergic response is miserably evident in stuffed sinuses, a drippy nose, and sneezing.

In some cases, the allergen spreads and promotes wide-ranging inflammation in body tissues, triggering a life-threatening condition called *anaphylactic shock*. For example, a person who is allergic to wasp or bee venom can die within minutes of a single sting. Air passages to the lungs rapidly constrict, closing almost completely. Fluid escapes swiftly from grossly dilated, permeable (and hence extremely leaky) blood vessels all over the body. The person's blood pressure plummets, which can lead to the complete collapse of the cardiovascular system. The emergency treatment for anaphylactic shock is an injection of the hormone epinephrine. People who know they are at risk (usually because they've already had a bad reaction to an allergen) can carry the necessary medication with them, just in case.

Antihistamines are anti-inflammatory drugs that are often used to relieve the short-term symptoms of allergies. Over the long term, however, a person may try a desensitization program. First, skin tests are used to identify offending allergens. Inflammatory responses to some of them can be blocked if the patient's body can be stimulated to make IgG instead of IgE. Gradually, larger and larger doses of specific allergens are administered. Each time, the person's body produces more circulating IgG molecules and IgG memory cells. The IgG will bind with allergen it encounters and block its attachment to IgE—and thereby block inflammation.

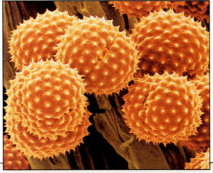

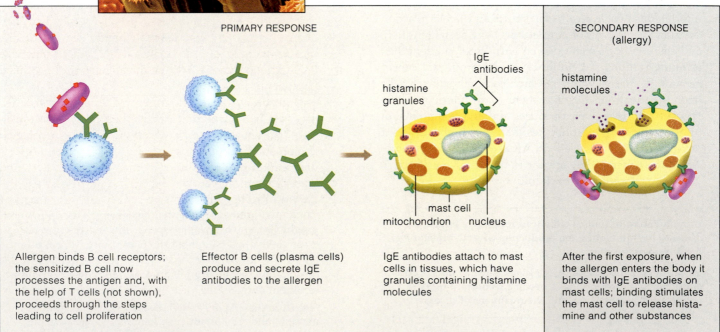

Allergen (antigen) enters the body

PRIMARY RESPONSE

SECONDARY RESPONSE
(allergy)

IgE antibodies

histamine granules

histamine molecules

mast cell

mitochondrion   nucleus

Allergen binds B cell receptors; the sensitized B cell now processes the antigen and, with the help of T cells (not shown), proceeds through the steps leading to cell proliferation

Effector B cells (plasma cells) produce and secrete IgE antibodies to the allergen

IgE antibodies attach to mast cells in tissues, which have granules containing histamine molecules

After the first exposure, when the allergen enters the body it binds with IgE antibodies on mast cells; binding stimulates the mast cell to release histamine and other substances

Figure 8.17 (**a**) Micrograph of ragweed pollen. (**b**) The basic steps leading to an allergic response.

Table 8.3 Some Disorders of the Immune System

Cause/Risk Factors	Major Effects	Symptoms
AIDS Human immunodeficiency virus (HIV), transmitted in body fluids such as blood, semen, breast milk	Destroys CD4 (helper T) cells, thereby disrupting immune responses	Unexplained weight loss, fatigue, enlarged lymph glands, fever, diarrhea, infections and tumors that are uncommon in the general population
Immunodeficiency (non-HIV) Genetic disorder; radiation exposure; various drugs used during cancer therapy, organ transplants, and other medical procedures; protein-deficient diet	Destruction of lymphoid tissues (radiation) or impairment of growth and division of lymphocytes; immunosuppressive effects of stress are strongly suspected but not confirmed	Extreme vulnerability to infection and development of tumors
Lymphoma Any of a group of cancers of lymphocytes; sometimes associated with immune system suppression by drugs administered during an organ transplant. Also may be associated with HIV infection; *Burkitt's lymphoma* is caused by the Epstein-Barr virus.	Unchecked growth of lymphocytes in the spleen, lymph nodes, bone marrow, and many other locations, leading to severe impairment or destruction of those tissues	Enlargement of lymph nodes in the spleen and in any other organ where lymphocyte growth is unchecked

Autoimmune Disorders

In an **autoimmune response**, the immune system's powerful weapons are unleashed against normal body cells or proteins. An example is *rheumatoid arthritis*. Sufferers are genetically predisposed to this disorder. Their macrophages, T cells, and B cells get activated by antigens associated with the joints. Immune responses are made against their body's collagen molecules and apparently against antibody that has bound to an as-yet unknown antigen. More damage to joint tissues is inflicted by activation of the complement system and inflammation. Skewed repair mechanisms compound the problem. Eventually the affected joints fill with synovial membrane cells and become immobile.

Another common autoimmune disease is type 1 diabetes. This is a type of *diabetes mellitus*, in which the pancreas does not secrete enough of the hormone insulin for proper absorption of glucose from the blood. For reasons that are still being investigated, the immune system attacks and destroys the insulin-secreting cells. A viral infection may trigger the response. Chapter 12 looks at the various forms of diabetes in more detail.

The autoimmune disease *systemic lupus erythematosus* (SLE) primarily affects younger women, but males can develop it as well. A characteristic symptom is a rash on the face that extends from cheek to cheek, roughly in the shape of a butterfly. The rash is just one sign that the affected person has developed antibodies to her or his own DNA and other self components. Antigen-antibody complexes accumulate in joints, blood vessel walls, the skin, and the kidneys. In addition to the "butterfly" rash, symptoms include fatigue, painful arthritis, and in some cases a near-total breakdown of kidney function. Drug treatments can help relieve many symptoms, but there is no cure. Table 8.3 lists a few other immune system disorders.

Deficient Immune Responses

When the body does not have enough functioning lymphocytes, its immune responses are not effective. Both T cells and B cells are in short supply in *severe combined immune deficiency* (SCID). This disorder usually is inherited, and infants born with it may die early in life. The lack of adequate immune responses makes them highly vulnerable to infections that are not life-threatening to the general population.

Infection by the human immunodeficiency virus (HIV) causes **AIDS** (acquired immunodeficiency syndrome). HIV is transmitted when body fluids of an infected person enter another person's tissues. The virus cripples the immune system by attacking helper T cells and macrophages. This leaves the body dangerously susceptible to infections and to some otherwise rare forms of cancer. We return to this topic in Chapter 16, which considers the growing global threat of infectious diseases.

An allergic reaction is an immune response to a substance in the external environment that usually is harmless.

Autoimmune responses are attacks by lymphocytes on normal body cells or proteins. Immunodeficiency is the inability to mount a normal immune response.

SUMMARY

1. The body protects itself from pathogens with physical and chemical barriers at body surfaces. It also is protected by nonspecific and specific responses of the white blood cells listed in Table 8.4.

2. Intact skin and mucous membranes lining body surfaces are physical barriers to infection. The chemical barriers include glandular secretions (such as tears, saliva, and gastric juice) and metabolic products of bacteria that normally inhabit body surfaces.

3. An inflammatory response develops in body tissues that have become damaged (for example, by infection).

   a. The response starts with vasodilation of arterioles that increases the blood flow to the tissue, which reddens and becomes warmer as a result. Blood capillary permeability increases and the resulting local edema causes swelling and pain.

   b. Pathogens and dead or damaged body cells release the substances that trigger increased permeability of capillaries. White blood cells leave the bloodstream and enter the tissue, where they release various chemical mediators and engulf invaders. Plasma proteins also enter the tissue. Complement proteins bind pathogens and induce their lysis. They also attract phagocytes. Blood-clotting proteins help repair damaged blood vessels.

4. An immune response has three characteristics:

   a. It has specificity, meaning it is directed against one antigen alone. An antigen is a unique molecular feature that lymphocytes recognize as foreign (nonself).

   b. Each response shows memory, which means that a subsequent encounter with the same antigen triggers a more rapid, secondary response of greater magnitude.

   c. Normally, an immune response is not made against the body's own self marker proteins.

5. Antigen-presenting cells process and bind fragments of antigen to their own MHC markers. Lymphocyte receptors can bind to displayed antigen-MHC complexes. Binding is a start signal for an immune response.

6. After recognition of antigen, an immune response advances by repeated cell divisions that form clones of B and T cells. These differentiate (become specialized) into subpopulations of effector and memory cells. Chemical mediators such as interleukins secreted by white blood cells drive the responses. The effector helper T cells, cytotoxic T cells, effector B cells, and antibodies act at once. The memory cells are set aside for secondary responses.

7. T cells arise in bone marrow but continue to develop in the thymus, where they acquire TCRs. These T-cell receptors recognize and bind antigen-MHC complexes on antigen-presenting cells. B cells arise in bone marrow. As they mature, they synthesize antigen receptors (antibodies) that become positioned at their surface.

8. Effector cytotoxic T cells directly destroy virus-infected cells, tumor cells, and cells of tissue or organ transplants. Effector B cells (plasma cells) produce and secrete great numbers of antibodies that freely circulate.

9. Antibodies are protein molecules, often Y-shaped, and each has binding sites for one kind of antigen. Only B cells make them. When antibody binds antigen, toxins are neutralized, pathogens are tagged for destruction, or pathogens are prevented from attaching to body cells.

10. In active immunization, vaccines provoke immune responses, with production of effector and memory cells. In passive immunization, injections of purified antibodies help patients combat an infection.

11. An allergic reaction is an immune response to a generally harmless substance. An autoimmune response is an attack by lymphocytes on normal body cells. Immunodeficiency is a weakened or nonexistent capacity to mount an immune response.

Table 8.4 Summary of Major White Blood Cells and Their Roles in Defense	
Cell Type	**Main Characteristics**
MACROPHAGE	Phagocyte; has roles in nonspecific defense responses; presents antigen to T cells; cleans up and helps repair tissue damage
NEUTROPHIL	Fast-acting phagocyte; takes part in inflammation, not in sustained responses; most effective against bacteria
EOSINOPHIL	Secretes enzymes that attack certain parasitic worms
BASOPHIL AND MAST CELL	Secrete histamines, other substances that act on small blood vessels to produce inflammation; contribute to allergies
LYMPHOCYTES	All take part in most immune responses; after antigen recognition, form clonal populations of effector and memory cells.
B cell	Effectors secrete antibodies that protect the host in specialized ways
Helper T cell	Effectors secrete interleukins that stimulate rapid divisions and specialization of both B cells and T cells
Cytotoxic T cell	Effectors kill infected cells, tumor cells, and foreign cells by a touch-kill mechanism
NATURAL KILLER (NK) CELLS	Cytotoxic cell of undetermined affiliation; kills virus-infected cells and tumor cells

Review Questions

1. While you're jogging in the surf, your toes land on a jellyfish. Soon the bottoms of your toes are swollen, red, and warm to the touch. What events caused these signs of inflammation? *8.3*

2. Distinguish between:
   a. neutrophil and macrophage *8.3*
   b. cytotoxic T cell and natural killer cell *8.4, 8.6*
   c. effector cell and memory cell *8.4*
   d. antigen and antibody *8.4*

3. Tell how a macrophage becomes an antigen-presenting cell. *8.4*

4. What is the difference between an allergy and an autoimmune response? *8.12*

Self-Quiz *(Answers in Appendix V)*

1. _____ are barriers to pathogens at body surfaces.
   a. Intact skin and mucous membranes
   b. Tears, saliva, and gastric fluid
   c. Resident bacteria
   d. all are correct

2. Macrophages are derived from _____.
   a. basophils b. neutrophils c. monocytes d. eosinophils

3. Complement functions in defense by _____.
   a. neutralizing toxins
   b. enhancing resident bacteria
   c. promoting inflammation
   d. forming pore complexes that cause lysis of pathogens
   e. both a and b are correct
   f. both c and d are correct

4. _____ are certain molecules that lymphocytes recognize as foreign and that elicit an immune response.
   a. Interleukins   c. Immunoglobulins   e. Histamines
   b. Antibodies   d. Antigens

5. Immunoglobulins designated _____ increase antimicrobial activity in mucus-coated surfaces of some organ systems.
   a. IgA   c. IgG   e. IgD
   b. IgE   d. IgM

6. Antibody-mediated responses work best against _____.
   a. intracellular pathogens
   b. extracellular pathogens
   c. toxins
   d. both b and c
   e. all are correct

7. The most important antigens are _____.
   a. nucleotides   c. steroids
   b. triglycerides   d. proteins

8. _____ would be a target of an effector cytotoxic T cell.
   a. Extracellular virus particles in blood
   b. A virus-infected body cell or tumor cell
   c. Parasitic flukes in the liver
   d. Bacterial cells in pus
   e. Pollen grains in nasal mucus

9. Development of a secondary immune response is based on populations of _____.
   a. memory cells
   b. circulating antibodies
   c. effector B cells
   d. effector cytotoxic T cells
   e. mast cells

10. Match the immunity concepts:
    ____ inflammation
    ____ antibody secretion
    ____ phagocyte
    ____ immunological memory
    ____ vaccination
    ____ allergy
    a. neutrophil
    b. effector B cell
    c. nonspecific response
    d. deliberately provoking memory cell production
    e. basis of secondary immune response
    f. nonprotective immune response

Critical Thinking: You Decide *(Key in Appendix VI)*

1. Rob's bumper sticker reads "Have you thanked your resident bacteria today?" Explain why he appreciates the bacteria that normally reside on the skin and mucous membranes.

2. Given what you now know about how foreign invaders trigger immune responses, explain why mutated forms of viruses, which have altered surface proteins, pose a monitoring problem for a person's memory cells.

3. Researchers have been trying to develop a way to get the immune system to accept foreign tissue as "self." Speculate on some of the clinical applications of such a development.

4. Ellen developed chicken pox when she was in kindergarten. Later in life, when her children developed chicken pox, she remained healthy even though she was exposed to countless virus particles each day. Explain why.

5. Quickly review Section 4.11 on homeostasis. Then write a short essay on how the immune response contributes to stability in the internal environment.

Selected Key Terms

acute inflammation *8.3*	immunoglobulin *8.7*
allergy *8.12*	interferon *8.10*
antibody *8.4*	interleukin *8.3*
antigen *8.4*	lymphoid organ *8.5*
antigen-MHC complex *8.4*	lysis *8.2*
antigen-presenting cell *8.4*	lysozyme *8.1*
apoptosis *8.6*	macrophage *8.3*
autoimmune response *8.12*	memory cell *8.4*
B lymphocyte (B cell) *8.4*	MHC marker *8.4*
clonal selection hypothesis *8.9*	monoclonal antibodies *8.10*
complement system *8.2*	natural killer cell *8.6*
cytotoxic T cell *8.4*	pathogen *8.1*
effector cell *8.4*	perforin *8.6*
fever *8.3*	TCR *8.6*
helper T cell *8.4*	T lymphocyte (T cell) *8.4*
immune system *8.4*	vaccine *8.10*
immunization *8.10*	

Readings

Engelhard, V. August 1994. "How Cells Process Antigens." *Scientific American.* Includes an excellent summary of how the immune system functions.

Johnson, H., F. Bazer, B. Szente, and M. Jarpe. May 1994. "How Interferons Fight Disease." *Scientific American.*

Laver, W., N. Bischofberger, and R. Webster. January 1999. "Disarming Flu Viruses." *Scientific American.*

THE RESPIRATORY SYSTEM

Conquering Chomolungma

To experienced climbers, Chomolungma may be the ultimate challenge (Figure 9.1). The summit of this Himalayan mountain, also known as Everest, is 9,700 meters (29,108 feet) above sea level. It is the highest place on earth.

Chomolungma's challenge goes beyond iced-over vertical rock, driving winds, blinding blizzards, and heart-stopping avalanches. It is the extreme danger that oxygen-poor air poses to the brain. In 1996, nine climbers, including two of the world's best professional mountaineers, perished during a blinding storm near the mountain's summit. While the severe weather contributed to the tragedy, so did the bodily effects of extreme altitude.

Most of us live at low elevations. Of the air we breathe, one molecule in five is oxygen. In mountains higher than 3,300 meters (10,000 feet), the breathing game changes. The Earth's gravitational pull is not as great, and gas molecules spread out—so breathing the way we do at sea level will not deliver enough oxygen to our lungs. The oxygen deficit can cause headaches, shortness of breath, heart palpitations, even loss of appetite, nausea, and vomiting. These are the telling symptoms of "altitude sickness."

At Chomolungma's base camp, climbers are 6,300 meters (19,000 feet) above sea level. More than half of the atmospheric oxygen is below them. At 7,000 meters (23,000 feet), oxygen and other gaseous molecules are extremely diffuse. The exceedingly low air pressure and scarcity of oxygen combine to make smaller blood vessels leak. Plasma fluid trickles out through gaps between the endothelial cells lining the blood vessels. In a climber's brain and lungs, tissues swell with plasma fluid, a condition called edema. Unless the edema is reversed, a climber may become comatose and die.

Figure 9.1 A climber approaching the summit of Chomolungma, where oxygen is dangerously scarce.

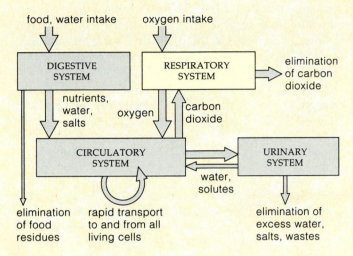

food, water intake oxygen intake

DIGESTIVE SYSTEM RESPIRATORY SYSTEM elimination of carbon dioxide

nutrients, water, salts oxygen carbon dioxide

CIRCULATORY SYSTEM URINARY SYSTEM

water, solutes

elimination of food residues rapid transport to and from all living cells elimination of excess water, salts, wastes

Figure 9.2 Interactions between the respiratory system and other organ systems in the body.

Given the risks, experienced climbers keep their bodies in top physical condition and live for a time at high elevations before their assault on the summit. They know that breathing "thinner" air containing less oxygen than air at sea level triggers some important physiological responses—including the formation of billions of additional red blood cells and more blood capillaries. Cells also make more mitochondria— the organelles where aerobic respiration takes place (Section 3.9). As you know from your reading in Chapter 3, aerobic respiration in cells uses oxygen and produces carbon dioxide wastes that must be removed from the body. The body's adaptations for meeting these requirements include structures (such as lungs) and functions (such as breathing) that facilitate gas exchange. The organs and mechanisms for gas exchange are what we mean in this chapter when we talk about "respiration" and the respiratory system. Along with other organ systems, it contributes to homeostasis—that is, maintaining proper internal operating conditions for all living cells (Figure 9.2).

Few of us will ever find ourselves near the peak of Chomolungma, pushing to the limit our ability to obtain oxygen. We won't need the bottled oxygen that elite climbers use. (Although at extreme altitude extra oxygen is no guarantee; most of the climbers who died on Everest in 1996 were using it.) Here in the lowlands, however, disease, smoking, and other environmental insults may affect gas exchange in more ordinary ways. And the risks can be just as great, as you will clearly understand by this chapter's end.

KEY CONCEPTS

1. As highly active animals, humans require a great deal of energy to power their metabolism. At the cellular level, the energy comes mainly from aerobic respiration in mitochondria. Aerobic respiration is an ATP-producing metabolic pathway that requires oxygen and produces carbon dioxide wastes.

2. By a physiological process called respiration, the body moves oxygen into the internal environment and releases carbon dioxide to the external environment.

3. Oxygen diffuses into the body as a result of a pressure gradient. The pressure of this gas is higher in air than it is in metabolically active tissues, where cells rapidly use oxygen. Carbon dioxide follows its own gradient, in the opposite direction. Its pressure is higher in tissues, where it is a by-product of metabolism, than it is in air.

4. The respiratory system includes a pair of lungs, which provide a respiratory surface—an extensive, thin, moist layer of epithelium. Blood flowing in the cardiovascular system picks up oxygen and gives up carbon dioxide at this respiratory surface.

5. Gas exchange is most efficient when the rate of air flow matches the rate of blood flow. The nervous system brings the rates into balance by controlling the rhythmic pattern and the rate and depth of breathing.

CHAPTER AT A GLANCE

Airways to the Lungs

Getting oxygen from air and expelling carbon dioxide are the key functions of the **respiratory system** (Figure 9.3). **Respiration** is the name for the process by which oxygen moves inward and CO_2 is released.

During quiet breathing, air typically enters (and leaves) the respiratory system by way of the nose. Hairs at the entrance to the *nasal cavity* and in the cavity's ciliated epithelial lining filter out dust and other large particles. Incoming air also is warmed in the nose and picks up moisture from mucus. The nasal cavity's two chambers are separated by a septum (wall) of cartilage and bone. Paranasal sinuses above and behind the nasal cavity are linked with it by channels. (This is why nasal sprays prescribed for colds or allergies can help unclog mucus-filled sinuses.) Tear glands constantly produce moisture that drains into the nasal cavity. When you cry, the flow increases and you get "sniffles." From the nasal cavity, air moves into the **pharynx**. This is the entrance to both the **larynx** (an airway) and the esophagus (the tube leading to the stomach). Nine pieces of cartilage form the larynx. Viewed from above, it resembles a triangular doughnut. One point of the triangle, the thyroid cartilage, is the "Adam's apple."

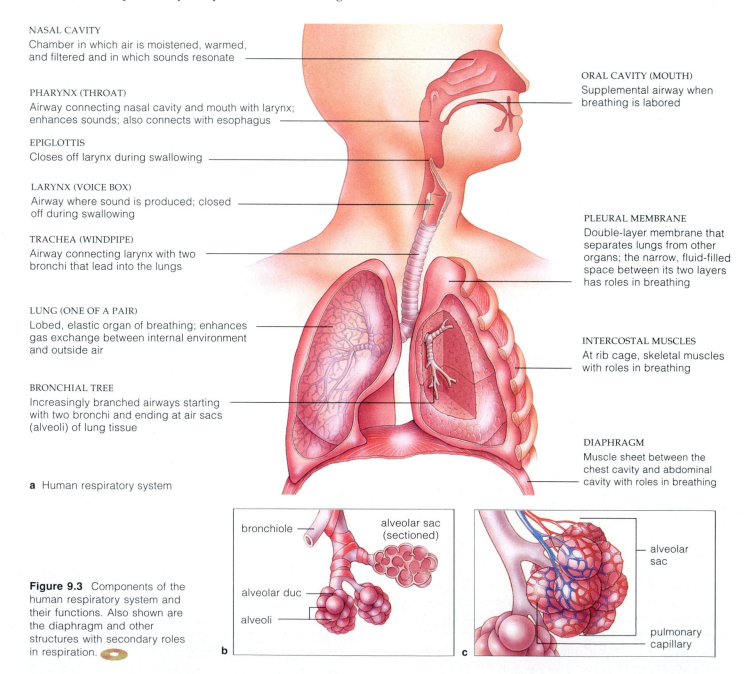

NASAL CAVITY
Chamber in which air is moistened, warmed, and filtered and in which sounds resonate

PHARYNX (THROAT)
Airway connecting nasal cavity and mouth with larynx; enhances sounds; also connects with esophagus

EPIGLOTTIS
Closes off larynx during swallowing

LARYNX (VOICE BOX)
Airway where sound is produced; closed off during swallowing

TRACHEA (WINDPIPE)
Airway connecting larynx with two bronchi that lead into the lungs

LUNG (ONE OF A PAIR)
Lobed, elastic organ of breathing; enhances gas exchange between internal environment and outside air

BRONCHIAL TREE
Increasingly branched airways starting with two bronchi and ending at air sacs (alveoli) of lung tissue

a Human respiratory system

ORAL CAVITY (MOUTH)
Supplemental airway when breathing is labored

PLEURAL MEMBRANE
Double-layer membrane that separates lungs from other organs; the narrow, fluid-filled space between its two layers has roles in breathing

INTERCOSTAL MUSCLES
At rib cage, skeletal muscles with roles in breathing

DIAPHRAGM
Muscle sheet between the chest cavity and abdominal cavity with roles in breathing

bronchiole

alveolar sac (sectioned)

alveolar duc

alveoli

alveolar sac

pulmonary capillary

Figure 9.3 Components of the human respiratory system and their functions. Also shown are the diaphragm and other structures with secondary roles in respiration.

b

c

Figure 9.4 Color-enhanced scanning electron micrograph of cilia (*gold*) in the respiratory tract. The orange-colored cells secrete mucus.

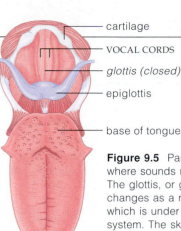

Figure 9.5 Paired human vocal cords, where sounds required for speech originate. The glottis, or gap between the vocal cords, changes as a result of skeletal muscle action, which is under the control of the nervous system. The sketches show what the glottis looks like when it is closed and opened.

Fortunately, our airways are nearly always open. The flaplike *epiglottis*, attached to the larynx, points up during breathing. However, recall from Chapter 6 that when you swallow, the larynx moves up so that the epiglottis partly covers the opening of the larynx. This helps prevent food from entering the respiratory tract and causing choking.

From the larynx, air moves into the **trachea,** or windpipe. Press gently at the lower front of your neck, and you can feel some of the bands of cartilage that ring the tube, adding strength and helping to keep it open. The trachea branches into two airways, one leading to each lung. Each airway is a **bronchus** (plural: bronchi). Its epithelial lining includes mucus-secreting cells and is carpeted with cilia (Figure 9.4). Bacteria and airborne particles stick in the mucus; then the upward-beating cilia sweep the debris-laden mucus up toward the mouth.

Near the entrance to the larynx, part of a mucous membrane forms two pairs of horizontal folds. The lower pair are the **vocal cords** (Figure 9.5). Every time you exhale, air is forced through the *glottis*, a gap between the vocal cords that is the opening to the larynx. Air moving through it makes the cords vibrate. By controlling the vibrations we can make different kinds of sounds. Using our lips, teeth, tongue, and the soft roof over the tongue (the soft palate), we can modify the different sounds into patterns of vocalization, such as speech and song.

Sometimes an infection causes the mucous lining of the vocal cords to become irritated and inflamed. The tissues become swollen, making it difficult for the vocal cords to vibrate. *Laryngitis*, with its characteristic hoarseness, is the result.

Sites of Gas Exchange in the Lungs

The **lungs** are elastic, cone-shaped organs separated from each other by the heart. The left lung has two lobes, the right lung three. Our lungs are located within the rib cage above the **diaphragm**, which is a sheet of muscle between the thoracic (chest) and abdominal cavities. The lungs are soft and spongy, and they don't attach directly to the wall of the chest cavity. Instead, each lung is enclosed by a pair of thin membranes called **pleurae**. You can visualize this arrangement if you think of pushing your closed fist into a fluid-filled balloon. A lung occupies the same kind of position as your fist, and the pleural membrane folds back on itself (like the balloon does) to form a closed *pleural sac*. An extremely narrow *intrapleural space* (*intra* means between) separates the membrane's two facing surfaces. A thin film of lubricating *intrapleural fluid* in the space reduces chafing between the membranes. Pneumonia and some other ailments can cause one or both pleurae to become inflamed and swollen. In an ailment called *pleurisy*, the membranes become inflamed. They can dry out and rub against each other—an extremely painful situation—or they can oversecrete fluid, which hampers breathing movements.

Inside each lung, the bronchi narrow as they branch repeatedly, forming "bronchial trees." These narrowing airways are **bronchioles**. Their endings, the **respiratory bronchioles**, have outpouchings from their walls. Each cup-shaped pouch is an **alveolus** (plural: alveoli), and each lung has about 150 million of them. Usually alveoli are clustered as a larger pouch, an *alveolar sac*. Alveoli are the sites where gases diffuse between the lungs and lung capillaries (Figure 9.3*b* and *c*). The respiratory system's role in respiration ends at alveoli, where the cardiovascular system takes over the task of moving gases.

Collectively the alveoli provide a huge surface area for the diffusion of gases. If they were stretched out as a single layer, they would cover the body several times over—or the floor of a racquetball court!

Taking up oxygen and removing carbon dioxide are the major functions of the respiratory system.

Deep inside the lungs, in alveoli at the ends of respiratory bronchioles, gases are exchanged with lung capillaries.

Alveoli are the end points of the respiratory system's role in gas exchange. From there, the cardiovascular system takes over the remaining tasks of respiration.

A CLOSER LOOK AT GAS EXCHANGE

Pressure Gradients and Transport Pigments

Gas exchange in the body relies on the tendency of oxygen and carbon dioxide to diffuse down their respective concentration gradients—or, as we say for gases, their *pressure gradients*. Said another way, when molecules of either gas are more concentrated outside the body, they tend to move into the body—and vice versa.

At sea level, a given volume of dry air is about 78 percent nitrogen, 21 percent oxygen, 0.04 percent carbon dioxide, and 0.96 percent other gases. Each gas exerts only *part* of the total pressure exerted by the whole mix of gases. That is, each exerts a "partial pressure"(Figure 9.6). At sea level, atmospheric pressure is about 760 mm Hg. (Atmospheric pressure is measured with a mercury barometer, and Hg is the chemical symbol for the element mercury.) Thus, at sea level the partial pressure of oxygen is 21 percent of 760—about 160 mm Hg. The partial pressure of carbon dioxide is about 0.3 mm Hg.

To meet the metabolic needs of a large, active animal such as a human, gas exchange must be efficient. Various factors influence the process. To begin with, gases enter and leave the body by crossing a thin **respiratory surface** of epithelium. This respiratory surface must be moist, because gases cannot diffuse across it unless they are dissolved in fluid. Two factors affect how many gas molecules can move across the respiratory surface in a given period of time. The first is surface area, and the second is the partial pressure gradient across it. The larger the surface area and the steeper the partial pressure gradient, the faster diffusion takes place. The millions of thin-walled alveoli in healthy lungs provide a huge surface area for gas exchange.

Gas exchange also gets a boost from respiratory pigments, mainly the hemoglobin in red blood cells. Each hemoglobin molecule binds with up to four oxygen molecules in the lungs, where the oxygen concentration is high. When circulating blood carries red blood cells into tissues where the oxygen concentration is low, hemoglobin *releases* oxygen. Thus, by carrying oxygen away from the respiratory surface, hemoglobin helps maintain the pressure gradient that helps draw oxygen into the lungs—and into blood in lung capillaries. Gas exchange is enhanced even more by ventilation, the mechanism of breathing described in Section 9.4.

Gas Exchange in Unusual Environments

At high altitude or deep under water the rules of gas exchange change. For instance, recall from this chapter's introduction that the partial pressure of oxygen decreases as altitude increases. People who aren't acclimatized to

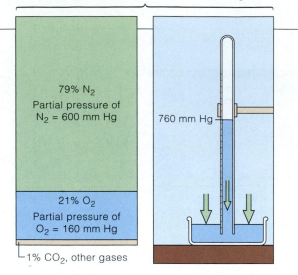

Figure 9.6 Gas partial pressures.

Total atmospheric pressure = 760 mm Hg

79% N_2
Partial pressure of N_2 = 600 mm Hg

760 mm Hg

21% O_2
Partial pressure of O_2 = 160 mm Hg

1% CO_2, other gases

the thinner air at high altitudes can become *hypoxic*—their tissues are chronically short of oxygen. Higher than about 2,400 meters (about 8,000 feet), brain respiratory centers compensate for the oxygen deficiency by triggering *hyperventilation* (breathing faster and more deeply).

Underwater, pressure greatly increases with depth. To prevent the lungs from collapsing under the increased pressure, a diver may inhale pressurized air (from tanks), which increases the total pressure of gases in the diver's lungs. As a diver ascends, however, the pressure of the surrounding water *decreases* and gaseous nitrogen, N_2, tends to move from tissues into the bloodstream much more rapidly than it would at sea level. If the ascent is too rapid, the N_2 comes out of solution faster than the diver can exhale it—and bubbles of N_2 form. Too many bubbles cause pain in joints and elsewhere. You may have heard the expression *the bends* for what is otherwise called *decompression sickness*. Bubbles that block blood flow to the brain can damage hearing, vision, and even lead to paralysis. Deeper than about 150 meters, the high partial pressure of N_2 can produce extreme euphoria (*nitrogen narcosis*). Affected divers have reportedly offered their airtank mouthpieces to fish! For deep descents knowledgeable divers use a nitrogen-free gas mixture.

Hypoxia also occurs in *carbon monoxide poisoning*. Carbon monoxide, a colorless, odorless gas, is present in automobile exhaust fumes and in smoke from burning tobacco, coal, charcoal, or wood. It binds to hemoglobin at least 200 times more tightly than oxygen does. Even tiny amounts can tie up half of the body's hemoglobin and seriously hamper oxygen delivery to tissues.

Gas exchange depends on steep partial pressure gradients between the outside and inside of the body. The greater the area of the respiratory surface and the larger the partial pressure gradient, the faster diffusion takes place.

9.3 BREATHING AS A HEALTH HAZARD

In cities, in certain occupations, and anyplace near a smoker, airborne particles and irritating gases put extra workloads on the lungs. Breathing polluted air over a long period of time—and, for some people, even breathing "clean" air that contains allergens—can result in any of several disorders.

Bronchitis can be brought on when air pollution increases mucus secretions and interferes with ciliary action in the lungs. Ciliated epithelium in the bronchioles is especially sensitive to cigarette smoke. Mucus and the particles it traps—including bacteria—accumulate in airways, coughing starts, and the bronchial walls become inflamed. Bacteria or chemical agents start destroying the wall tissue. Cilia are lost from the lining, and mucus-secreting cells multiply as the body attempts to get rid of the accumulating debris. Eventually scar tissue forms and can block parts of the respiratory tract.

In an otherwise healthy person, even acute bronchitis is easily treated. When inflammation continues, however, scar tissue builds up and the bronchi become clogged with more and more mucus. Also, the walls of alveoli break down, and stiffer fibrous tissue comes to surround them. Remaining alveoli enlarge, and the balance between air flow and blood flow is skewed. The result is *emphysema*, a disorder in which the lungs are so distended and inelastic that gases cannot be exchanged efficiently (see Figure 9.7). Running, walking, even exhaling can be difficult. About 1.3 million people in the United States have emphysema.

Smoking, chronic colds, and other respiratory ailments sometimes make a person susceptible to emphysema later in life. "Secondhand smoke" inhaled by a nonsmoker also may contribute to emphysema and other ills, including lung cancer. Many emphysema sufferers lack a functional gene coding for antitrypsin, a protein that inhibits tissue-destroying enzymes produced by bacteria. Emphysema

Figure 9.8 An asthma sufferer using an aerosol inhaler.

can develop over 20 or 30 years. By the time the disease is detected, lung tissue is permanently damaged.

Millions of people suffer from *asthma*, a chronic, incurable respiratory condition in which breathing can become so difficult so quickly that the victim literally feels in imminent danger of suffocating. Among other triggers, allergens can set off an asthma attack. Pollen, dairy products, shellfish, flavorings and other foods, pet hairs or dandruff—even the dung of tiny mites in house dust—all can be culprits. In susceptible people, attacks also can result from noxious fumes, stress, strenuous exercise, or a respiratory infection.

Asthma's scary symptoms are related to a basic feature of respiratory anatomy. Stiff rings and plates of cartilage in the walls of bronchi hold the tubes open (as they do the trachea). However, this cartilage gives way to smooth muscle as the bronchi branch into bronchioles in the lungs. In an asthma attack the smooth muscle contracts in strong spasms that cause the bronchioles to suddenly narrow. Bronchioles even may close off completely. At the same time, mucus pours from bronchial epithelium, clogging the constricted passages even more.

Various strategies and medications can help asthma sufferers. One tactic is to identify the sources of allergens and limit exposure to those substances. Preventive drugs also can help. Aerosol inhalants squirt a fine mist into the airways (Figure 9.8). A drug in the mist dilates bronchial passages and helps restore free breathing. Aerosol inhalants must be used only with medical supervision.

Figure 9.7 (**a**) Normal human lungs (this lung tissue looks darker than normal because it has been chemically preserved). (**b**) Lungs from a person with emphysema.

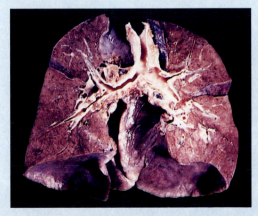

a

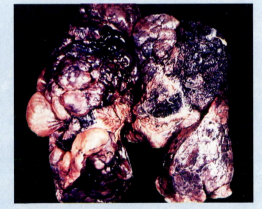

b

BREATHING—CYCLIC REVERSALS IN AIR PRESSURE GRADIENTS

The Respiratory Cycle

At twelve breaths per minute, you will take about 500 million breaths by age 75—even more if you consider that young children breathe faster than adults do. If breathing stops for as little as five minutes, the brain can suffer irreversible damage.

There is a cyclic pattern to breathing, which ventilates the lungs. This pattern is known as a **respiratory cycle**. Ventilation consists of **inspiration** (a single breath of air drawn into the airways) and **expiration** (a single breath out). Inspiration always is an active, energy-requiring action. When someone is breathing quietly, inspiration begins with the contraction of the diaphragm and, to a lesser extent, the external intercostal muscles (see Figure

9.3). These muscle movements increase the volume of the thoracic cavity (Figure 9.9). If you take a deep breath, the volume increases further because neck muscles contract and raise the sternum and the first two ribs.

During each respiratory cycle, the volume of the thoracic cavity increases, then decreases. When the volume changes, pressure gradients between the lungs and the air outside the body are *reversed*. To understand how this reversal affects breathing, begin by remembering that the atmospheric pressure—760 mm Hg at sea level—is exerted by the combined weight of all the atmospheric gases on all the airways. Before inspiration, the pressure inside all alveoli (called *intrapulmonary pressure*) is also 760 mm Hg (Figure 9.9c).

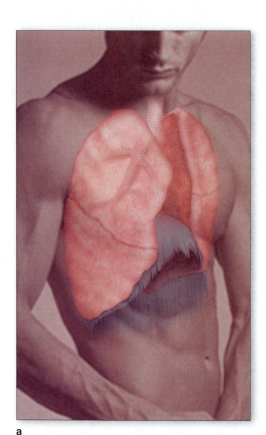

a

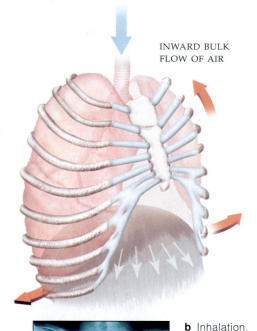

INWARD BULK
FLOW OF AIR

OUTWARD BULK
FLOW OF AIR

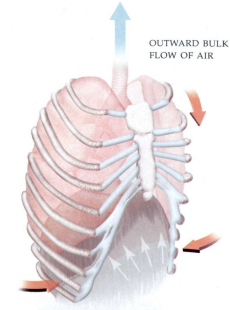

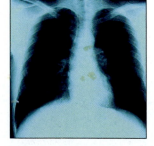

b Inhalation. Diaphragm contracts and moves down. The external intercostal muscles contract and lift rib cage upward and outward. The lung volume expands.

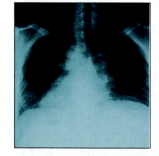

c Exhalation. Diaphragm and external intercostal muscles return to the resting positions. Rib cage moves down. Lungs recoil passively.

Figure 9.9 (**a**) Location of the lungs relative to the diaphragm. (**b,c**) Changes in the size of the thoracic cavity during a respiratory cycle. The two x-ray images show how the maximum possible inhalation changes the volume of the thoracic cavity. (**d–f**) Changes in lung volume and intrapulmonary pressure during a respiratory cycle.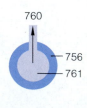

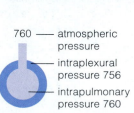

760 —— atmospheric pressure
—— intraplexural pressure 756
—— intrapulmonary pressure 760

d BEFORE INHALATION

760
lungs expanded
—— 764
—— 759

e DURING INHALATION

760
—— 756
—— 761

f DURING EXHALATION

Another pressure gradient helps keep the lungs close to the thoracic cavity wall all during the respiratory cycle, even during expiration, when lungs have a far smaller volume than the thoracic cavity (Figure 9.9*b*). When the cavity expands, so do the lungs, because there is a *negative pressure gradient* across the lung wall. Here is how the mechanism works in a resting person.

To start with, pressure inside the pleural sac is about 756 mm Hg—4 mm Hg *less* than atmospheric pressure. In addition, pleural sac pressure is exerted within the thoracic cavity but *outside* the lungs. As pleural sac pressure pushes in on the lung's wall, the pressure inside the lungs pushes outward. The pressure difference—a bit lower in the pleural sac than inside the lungs—is enough to make the lungs stretch, filling the expanded thoracic cavity.

Recall from Chapter 2 that a biologically important property of water is cohesiveness—the many hydrogen bonds linking water molecules prevent them from being easily pulled apart. This property comes into play every time you take a breath. The cohesiveness of water molecules in the fluid inside the pleural sac helps keep your lungs close to the thoracic wall, in much the same way that two wet panes of glass resist being pulled apart. The panes easily slide back and forth, however. Similarly, intrapleural fluid "glues" the lungs to the thoracic wall. So, when the thoracic cavity expands as you inhale, so do your lungs.

Figure 9.9*a* shows what happens as you start to inhale. The dome-shaped diaphragm flattens, and the rib cage is lifted up and outward. As the thoracic cavity expands, the lungs expand with it. At that time, the air pressure in alveolar sacs is lower than the atmospheric pressure. Fresh air follows the gradient and flows down the airways, then into the alveoli.

During normal quiet breathing, the second action of the respiratory cycle is passive—it doesn't require an energy outlay. The muscles that brought about inspiration simply relax and the lungs recoil, much the way a rubber band does if you stretch it and then let it relax. As the lung volume shrinks, this change compresses the air in the alveolar sacs. Now, pressure in the sacs *exceeds* atmospheric pressure. Air follows the gradient, out of the lungs (Figure 9.9*b*).

Expiration becomes active and requires energy only when your lungs must rapidly expel more air—for instance, during aerobic exercise. Then, muscles in the abdominal wall contract, increasing the pressure in the abdomen and pushing your diaphragm upward. When this happens, the volume of the thoracic cavity decreases, the internal intercostal muscles contract, and they pull the thoracic wall down and inward. The chest wall flattens, further reducing the dimensions of the thoracic

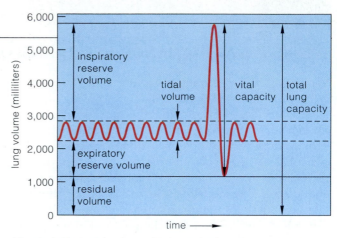

Figure 9.10 Lung volume. During quiet breathing, a tidal volume of air enters and leaves the lungs. Forced inspiration delivers more air to them; forced expiration releases some air that normally stays in the lungs. A residual volume is trapped in partially filled alveoli even during the strongest expiration.

cavity. The lung volume decreases even more when the elastic tissue of the lungs recoils.

Lung Volumes

About 500 milliliters (two cupfuls) of air enter or leave your lungs in a normal breath. This volume of air is called **tidal volume**. You can increase the amount of air you inhale or exhale, however. In addition to air taken in as part of the tidal volume, a person can forcibly inhale roughly 3,100 milliliters of air, called the *inspiratory reserve volume*. By forcibly exhaling, you can expel an additional *expiratory reserve volume* of about 1,200 milliliters of air. **Vital capacity** is the maximum volume of air that can move out of the lungs after you inhale as deeply as possible. It is about 4,800 milliliters for a healthy young man and about 3,800 milliliters for a healthy young woman. As a practical matter, people rarely take in more than half their vital capacity, even when they breathe deeply during strenuous exercise. At the end of your deepest exhalation, your lungs still are not completely emptied of air; roughly another 1,200 milliliters of *residual volume* remain (Figure 9.10).

How much of the 500 milliliters of inspired air is actually available for gas exchange? Between breaths, about 150 milliliters of exhaled "dead" air remain in the airways and never reach the alveoli. Thus only about 350 (500 − 150) milliliters of fresh air reach the alveoli with each inhalation. An adult typically breathes at least twelve times per minute. This rate of ventilation supplies the alveoli with (350 × 12) or 4,200 milliliters of fresh air—a little more than the volume of soda pop in four 1-liter bottles—every 60 seconds.

In the respiratory cycle, the air movements of breathing occur because of cyclic increases and decreases in the volume of the chest cavity and the corresponding changes in pressure gradients between the lungs and outside air.

GAS EXCHANGE AND TRANSPORT

Ventilation draws oxygen into your lungs and moves carbon dioxide out. But ventilation is not the same as respiration. *Respiration* provides the body as a whole with oxygen for aerobic respiration in cells and disposes of carbon dioxide. Physiologists divide respiration into two phases, "external" and "internal." During *external* respiration, oxygen moves from alveoli into the blood, and carbon dioxide moves in the opposite direction. During *internal* respiration, oxygen moves from the blood into tissues, and carbon dioxide moves from tissues into the blood. We now examine these two respiratory phases.

Gas Exchange in Alveoli

The alveoli in your lungs are ideally constructed for their function of gas exchange. The wall of each alveolus is a single layer of epithelial cells, supported by a gossamer-thin basement membrane. Hugging the alveoli are lung capillaries (Figure 9.11a). They, too, have a gossamer-thin basement membrane around their wall. In between the two basement membranes is a thin film of interstitial fluid. It may seem like a lot of layers, but together they form a *respiratory membrane* that is far narrower than even a fine baby hair. This is why oxygen and carbon dioxide can diffuse rapidly across it (Figure 9.11b, c).

Figure 9.12 shows the partial pressure gradients for oxygen and carbon dioxide throughout the respiratory system of a resting person. Oxygen diffuses across the respiratory membrane and into the bloodstream, and carbon dioxide diffuses outward.

Certain cells in the epithelium of alveoli secrete *pulmonary surfactant*. This substance reduces the surface tension of the watery fluid film between alveoli. Without it, the force of surface tension can collapse the delicate alveoli.

This happens to premature babies whose incompletely developed lungs do not yet have working surfactant-secreting cells. The result is a potentially lethal disorder called *infant respiratory distress syndrome*.

Gas Transport between Blood and Metabolically Active Tissues

Blood plasma can carry only so much dissolved oxygen and carbon dioxide. To meet the body's requirements, the gas transport must be improved. The hemoglobin in red blood cells binds and transports both O_2 and CO_2. This pigment enables blood to carry some 70 times more oxygen than it otherwise would and to transport 17 times more carbon dioxide away from tissues.

OXYGEN TRANSPORT Air inhaled into your alveoli contains plenty of oxygen and relatively little carbon dioxide. Just the opposite is true of blood arriving from tissues—which, recall, enters lung capillaries at the "end" of the pulmonary circuit (Section 7.8). Thus, in the lungs, oxygen diffuses down its pressure gradient into the blood plasma, then into red blood cells, where up to four oxygen molecules rapidly form a weak, reversible bond with each hemoglobin molecule. Hemoglobin with oxygen bound to it is called **oxyhemoglobin**, or HbO_2.

The amount of HbO_2 that forms depends on the interplay of several factors. One is the partial pressure of oxygen. In general, the higher its partial pressure, the more oxygen will be picked up, until all hemoglobin binding sites have oxygen attached to them. HbO_2 holds onto oxygen rather weakly and will give it up in tissues where the partial pressure of oxygen is lower than in the blood.

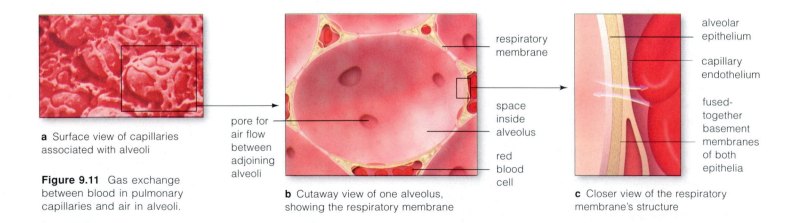

a Surface view of capillaries associated with alveoli

Figure 9.11 Gas exchange between blood in pulmonary capillaries and air in alveoli.

pore for air flow between adjoining alveoli

respiratory membrane

space inside alveolus

red blood cell

b Cutaway view of one alveolus, showing the respiratory membrane

alveolar epithelium

capillary endothelium

fused-together basement membranes of both epithelia

c Closer view of the respiratory membrane's structure

Certain conditions reduce hemoglobin's affinity for binding oxygen. They occur in tissues with high metabolic activity—and, therefore, with a greater demand for oxygen. For example, the binding of oxygen weakens as temperature rises or as acidity increases and pH falls. Several events contribute to a falling pH: the reaction that forms HbO_2 releases hydrogen ions (H^+), making the blood more acidic. Blood pH also falls as the level of carbon dioxide given off by active cells increases.

When tissues chronically receive less oxygen than proper functioning requires, red blood cells increase their production of a compound called 2,3-diphosphoglycerate, or DPG for short. DPG reversibly binds hemoglobin. The more of it that is bound to hemoglobin, the *less* tightly hemoglobin binds oxygen—thus the more oxygen is available to tissues.

CARBON DIOXIDE TRANSPORT Aerobic respiration in cells produces carbon dioxide as a waste. For this reason, there is more carbon dioxide in metabolically active tissues than in blood flowing through the nearby capillaries. Following its pressure gradient, carbon dioxide diffuses into these capillaries, and it will be transported toward the lungs in one of three ways. About 7 percent of the carbon dioxide stays dissolved in plasma. About another 23 percent binds with hemoglobin in red blood cells, forming the compound *carbaminohemoglobin* ($HbCO_2$). Most of the carbon dioxide, about 70 percent, is converted to ions of bicarbonate (HCO_3^-) and transported in the plasma. Bicarbonate forms when carbon dioxide combines with water. The reaction has two steps. First carbonic acid (H_2CO_3) forms; then it dissociates (separates) into bicarbonate ions and hydrogen ions:

$$CO_2 + H_2O \rightleftharpoons \underset{\text{carbonic acid}}{H_2CO_3} \rightleftharpoons \underset{\text{bicarbonate}}{HCO_3^-} + H^+$$

This reaction takes place in the blood plasma and in red blood cells. However, it is much faster in red blood cells, which contain the enzyme **carbonic anhydrase**. This enzyme increases the reaction rate by at least 250 times. Freshly minted bicarbonate ions in red blood cells rapidly diffuse into the plasma, which will carry them to the lungs. The enzyme-mediated conversion makes the blood level of carbon dioxide drop rapidly. This in turn helps maintain the gradient that keeps carbon dioxide diffusing from interstitial fluid into the bloodstream.

The reactions just described are reversed in alveoli, where the partial pressure of carbon dioxide is *lower* than it is in surrounding capillaries. The carbon dioxide that forms as a result of the reactions diffuses into the alveolar sacs. From there it is exhaled.

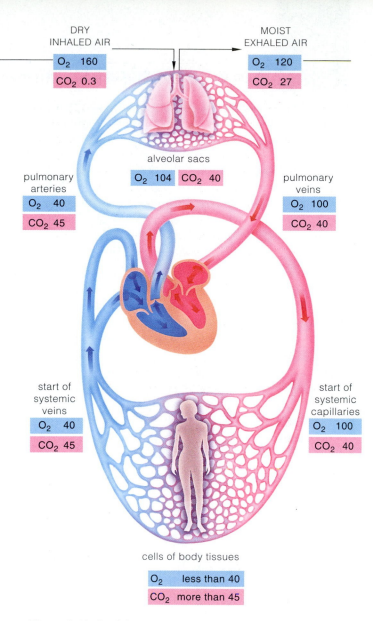

Figure 9.12 Partial pressure gradients for oxygen and carbon dioxide through the respiratory tract. Remember that *each gas moves from regions of higher to lower partial pressure.*

What happens to the hydrogen ions that are formed in red blood cells along with bicarbonate? Hemoglobin binds some of them and thus acts as a buffer (Chapter 2). In blood plasma, certain plasma proteins also bind H^+. Such buffering mechanisms are extremely important in homeostasis, for they help prevent an abnormal decline in blood pH.

Driven by its partial pressure gradient, oxygen diffuses from alveoli, through interstitial fluid, and into lung capillaries. Carbon dioxide diffuses in the opposite direction, driven by its partial pressure gradient.

Hemoglobin in red blood cells greatly enhances the oxygen-carrying capacity of the blood.

CONTROLS OVER GAS EXCHANGE

Homeostasis depends on efficient gas exchange in our bodies. Day in and day out, the respiratory system must deliver a steady stream of oxygen to our respiring cells and carry away carbon dioxide wastes. At least two levels of controls accomplish this regulation. The first we will consider is control exerted by the nervous system. The second level is local, in the lungs.

Matching Air Flow with Blood Flow

Gas exchange is most efficient when a person's breathing is regulated so that the rate of air flow is matched with the rate of blood flow. The nervous system acts to balance the flow rates by controlling the breathing *rhythm* and also its *magnitude*—the rate and depth of breathing.

The nervous system governs both inhalation and exhalation. It performs this function by way of controls that respond to the levels of both oxygen and carbon dioxide in arterial blood. In these events, respiratory centers in the brain play a central role.

BREATHING RHYTHM The diaphragm muscle and the muscles that move the rib cage are under the control of neurons in a system of nerve cells (neurons) running through the brain stem, at the lower rear of the brain. This system of nerve cells is called the reticular formation; Chapter 11 describes its functions in some detail. For the moment, we are concerned with two small clusters of cells in the reticular formation (in the medulla region of the brain stem). One cluster of cells coordinates signals calling for inspiration; the other coordinates the signals calling for expiration. The resulting rhythmic contractions are fine-tuned by centers in a nearby part of the brain stem (the pons) that can stimulate or inhibit both clusters.

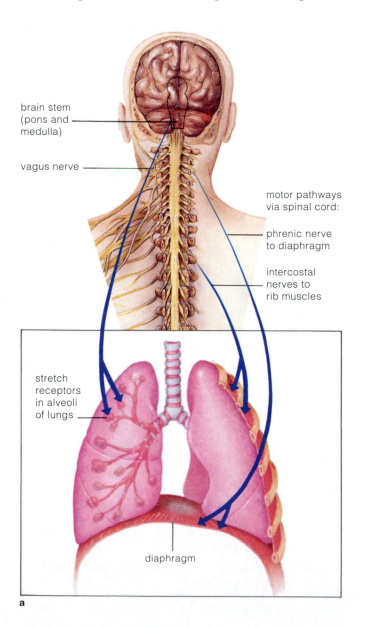

brain stem
(pons and
medulla)

vagus nerve

motor pathways
via spinal cord:

phrenic nerve
to diaphragm

intercostal
nerves to
rib muscles

stretch
receptors
in alveoli
of lungs

diaphragm

a

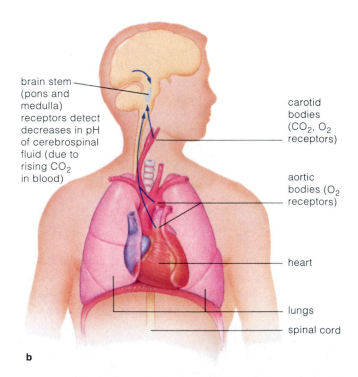

brain stem
(pons and
medulla)
receptors detect
decreases in pH
of cerebrospinal
fluid (due to
rising CO₂
in blood)

carotid
bodies
(CO₂, O₂
receptors)

aortic
bodies (O₂
receptors)

heart

lungs

spinal cord

b

Figure 9.13 Controls over breathing. (**a**) In normal quiet breathing, cell clusters (reticular formation) in the brain stem coordinate signals to the diaphragm and muscles that move the rib cage, triggering inhalation. When a person breathes deeply or rapidly, another cell cluster receives signals from stretch receptors in the lungs and coordinates signals for exhalation. (**b**) Different sensory receptors detect changes in the concentrations of carbon dioxide and oxygen in the blood.

As Figure 9.13a suggests, when you inhale, signals from the reticular formation travel nerve pathways to the diaphragm and chest. These signals stimulate the rib muscles and diaphragm to contract. As described in Section 8.4, this causes the rib cage to expand, and air moves into the lungs. When the diaphragm and chest muscles relax, elastic recoil returns the rib cage to its unexpanded state, and air in the lungs moves out. When you breathe rapidly or deeply, stretch receptors in airways send signals to the brain control centers, which respond by inhibiting contraction of the diaphragm and rib cage muscles—so you exhale.

BREATHING MAGNITUDE   You might assume that nervous system controls over breathing mainly involve monitoring the level of oxygen in blood. However, the nervous system is *more* sensitive to changes in the level of carbon dioxide. Both gases are monitored in blood flowing through arteries. When the conditions warrant, nervous system signals adjust contractions of the diaphragm and muscles in the chest wall and so adjust the rate and depth of your breathing.

Sensory receptors in the medulla of the brain can detect rising carbon dioxide levels. How? The mechanism is indirect, but (fortunately!) extremely sensitive. The receptors detect hydrogen ions that are produced when dissolved CO_2 leaves the bloodstream and enters fluid— called *cerebrospinal fluid*—that bathes the medulla. In cerebrospinal fluid, the drop in pH that goes along with increasing H^+ stimulates receptors that signal the change to the brain's respiratory centers (Figure 9.13b). In response, both the rate and the depth of breathing increase—and soon the blood level of CO_2 falls. This is another example of a negative feedback loop that helps maintain homeostasis.

Our brain also receives input from other sensory receptors, including **carotid bodies**, where the carotid arteries branch to the brain, and **aortic bodies** in arterial walls near the heart. Both types of receptors detect changes in levels of CO_2 and of oxygen in arterial blood. They also detect changes in blood pH. The brain responds by increasing the ventilation rate, so more oxygen can be delivered to tissues.

Local Chemical Controls

Local controls over air flow operate in the lungs, in the millions of alveoli. For example, if your heart begins pumping hard and fast but your lungs aren't ventilating at a corresponding pace, blood flows too fast and air moves too sluggishly for the efficient disposal of carbon dioxide. An increase in the blood level of carbon dioxide affects smooth muscle in the walls of bronchioles. Then,

the bronchioles dilate, increasing the air flow. Similarly, a decrease in carbon dioxide levels causes the bronchiole walls to constrict, so the air flow decreases.

Local controls also work on lung capillaries. When too much air is flowing into the lungs relative to blood flow, the oxygen level rises in some parts of the lungs. The increase affects smooth muscle in blood vessel walls. The vessels dilate, so more blood flows there. On the other hand, if the volume of air flow relative to blood flow is too small, the vessels constrict and blood flow decreases.

Disrupted Controls over Gas Exchange

You can voluntarily hold your breath, but not for long. As CO_2 builds up in your blood, nervous system "orders" force you to take a breath. The mechanisms by which the nervous system regulates the respiratory cycle normally operate under involuntary control, as you've just read. However, in some situations, a person can fail to breathe when the arterial CO_2 level falls below a set point. Breathing that stops briefly and then resumes spontaneously is called *apnea*. During certain times in the normal sleep cycle (see Section 11.10), breathing may stop for one or two seconds or even minutes—in extreme cases, as often as 500 times a night.

Aging takes its toll on the respiratory system. *Sleep apnea* is a common problem in the elderly, because the mechanisms for sensing changing oxygen and carbon dioxide levels gradually become less effective over the years. Also, as we age our lungs lose some of their elasticity, and this along with other changes reduces the efficiency of ventilation. Even so, staying physically fit, and maintaining a "lung-healthy" lifestyle in other ways, can go a long way toward keeping the respiratory system functioning well throughout life.

When the respiratory system is in good condition, gas exchange mechanisms in the body are like parts of a fine Swiss watch—intricately coordinated and smoothly operating. The *Focus* essay in Section 9.7 outlines some ailments in which this seamless functioning falters, including those closely linked to tobacco use.

The nervous system and local chemical controls (such as those in the lungs) regulate the rates of air flow and blood flow in the respiratory system, adjusting them in ways that enhance gas exchange.

Respiratory centers in the brain stem control the rhythmic pattern of breathing and the rate and depth of breathing. These controls contribute to homeostasis by helping to maintain appropriate levels of carbon dioxide, oxygen, and hydrogen ions in arterial blood.

TOBACCO AND OTHER THREATS TO THE RESPIRATORY SYSTEM

People who start smoking tobacco begin wreaking havoc with their lungs. Smoke from just one cigarette can prevent cilia in bronchioles from beating for hours. Noxious particles it contains can stimulate mucus secretion and kill the infection-fighting phagocytes that normally patrol the respiratory epithelium. The ensuing "smoker's cough" can pave the way for bronchitis and emphysema.

Today we know that cigarette smoke—including "secondhand smoke" inhaled by a nonsmoker—causes lung cancer and contributes to other ills. In the body, some compounds in coal tar and cigarette smoke are converted to highly reactive forms. These are the real carcinogens (cancer-causing substances; Chapter 22); they provoke genetic damage leading to lung cancer. Susceptibility to lung cancer is related to the number of cigarettes smoked per day and how often and how deeply the smoke is inhaled (Figure 9.14). Cigarette smoking causes at least 80 percent of all lung cancer deaths. The American Cancer Society estimates that tobacco use costs the U.S. economy more that $100 billion a year in health care costs and lost productivity due to illness. The chart below lists the health risks known to be associated with tobacco smoking, as well as the benefits of quitting. Table 9.1 summarizes some health problems, from lung cancer and pneumonia to various injuries, that can impair the functioning of the respiratory system.

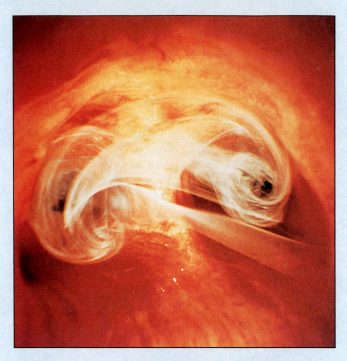

Figure 9.14 Cigarette smoke swirling through the human windpipe and down the two bronchial tubes to the lungs.

Risks Associated with Smoking	Benefits of Quitting
SHORTENED LIFE EXPECTANCY: Nonsmokers live 8.3 years longer on average than those who smoke two packs daily from the mid-20s on	Cumulative risk reduction; after about 15 years, life expectancy of ex-smokers approaches that of nonsmokers
CHRONIC BRONCHITIS, EMPHYSEMA: Smokers have 4–25 times more risk of dying from these diseases than do nonsmokers	Greater chance of improving lung function and slowing down rate of deterioration
LUNG CANCER: Cigarette smoking is the major cause of lung cancer	After 15 years, risk approaches that of nonsmokers
CANCER OF MOUTH: 3–10 times greater risk among smokers	After 15 years, risk is reduced to that of nonsmokers
CANCER OF LARYNX: 2.9–17.7 times more frequent among smokers	After 15 years, risk is reduced to that of nonsmokers
CANCER OF ESOPHAGUS: 2–9 times greater risk of dying from this	Risk proportional to amount smoked; quitting should reduce it
CANCER OF PANCREAS: 2–5 times greater risk of dying from this	Risk proportional to amount smoked; quitting should reduce it
CANCER OF BLADDER: 7–10 times greater risk for smokers	Risk decreases gradually over 7–10 years to that of nonsmokers
CORONARY HEART DISEASE: Cigarette smoking is a major contributing factor	Risk drops sharply after a year; after 10–15 years, risk reduced to that of nonsmokers
EFFECTS ON OFFSPRING: Women who smoke during pregnancy have more stillbirths, and weight of liveborns averages less (hence, babies are more vulnerable to disease, death)	When smoking stops before fourth month of pregnancy, risk of stillbirth and lower birth weight eliminated
IMPAIRED IMMUNE SYSTEM FUNCTION: Increase in allergic responses, destruction of macrophages in respiratory tract	Avoidable by not smoking
BONE HEALING: Evidence suggests that surgically cut or broken bones require up to 30 percent longer to heal in smokers, possibly because smoking depletes the body of vitamin C and reduces the amount of oxygen reaching body tissues. Reduced vitamin C and reduced oxygen interfere with production of collagen fibers, a key component of bone. Research in this area is continuing.	Avoidable by not smoking

Information provided by the American Cancer Society.

Table 9.1 Some Common Disorders of the Respiratory System

Cause/Risk Factors	Major Effects	Symptoms
LUNG CANCER		
Most cases associated with tobacco smoking or regular exposure to smoke (*passive smoking*). Other risk factors include inhaling asbestos fibers, exposure to radioactivity, and living where air pollution is high	Tumors develop in lung tissue, with resulting loss of respiratory surface for gas exchange. Tumor spread may trigger pneumonia, lung collapse, or accumulation of fluid between the pleural membranes	Persistent cough or chronic bronchitis, chest pain, shortness of breath, coughing up blood. Weight loss, especially if cancer has spread to other tissues
INFECTIONS		
Bronchitis Often associated with colds or flu. Can be *acute* or *chronic;* may be a symptom of lung cancer	Increased mucus secretion in bronchi and trachea, which leads to coughing. Chronic form can result in inflammation of bronchial walls and scarring	Cough, breathlessness, chest pain. Yellow or greenish phlegm; sometimes, fever
Pneumonia Infection by viruses or bacteria; inhaling toxic fumes	Inflammation of lung tissue; fluid buildup in alveoli	Dry cough, fever; shortness of breath. Symptoms sometimes include chills, blood in mucus, chest pain, blue tinge to skin
Histoplasmosis Fungal infection, most common in southeastern United States	Inflammation and increased mucus secretion in respiratory tract; can spread to retina of the eye and cause blindness	Flulike symptoms, including cough and fever
Influenza Viral infection, usually spread from person to person by contact with infectious particles (coughs, sneezes, discarded tissues, unwashed hands)	Initial infection of mucous membranes in nose and throat, spreading into the trachea and lungs. May be followed by bacterial infection, which leads to bronchitis or pneumonia	Chills, fever, headache, muscle pain, sore throat, usually followed by cough, chest pain, and runny nose due to increased mucus secretion
Tuberculosis Infection by the bacterium *Mycobacterium tuberculosis*. TB had become rare in the United States until the advent of HIV and AIDS. Now this disease is becoming common once again among AIDS sufferers	Destruction of patches of lung tissue; possible spread of infection to other body regions	Flulike illness; if infection progresses, mild fever, night sweats, fatigue, cough, and chest pain. Mucus containing blood and/or pus. Can be fatal if not treated
Allergic rhinitis (hay fever) Airborne pollen, animal hair, feathers, or other irritants	Release of histamine, which triggers inflammation and fluid production in nasal passages, sinuses, eyelids, and surface layer of eyes	Sneezing, runny nose, watery, itchy eyes; dry throat, wheezing
INJURIES		
Pneumothorax Injury, such as knife or bullet wound or broken rib, that permits air to enter pleural space	Air enters space between pleural membranes, eliminating pressure difference between lungs and external environment	Lung collapses
Asbestosis and *silicosis* Inhalation of particles or fibers of asbestos and silica, respectively	Irritation of lung tissue followed by development of scar tissue, so lungs become stiff and inflexible. Occasionally, lung cancer	Breathlessness; chronic cough with heavy phlegm. Increased susceptibility to other lung disorders
EMPHYSEMA		
Inflammation resulting from chronic bronchitis or severe asthma	Enzymatic destruction of walls of alveoli and deterioration of gas exchange	Difficulty breathing, usually becoming progressively worse. Can lead to heart failure and death
PULMONARY EMBOLISM		
Blood clot in artery leading to the lungs	Blockage of blood flow to lungs	Heart failure; general collapse of the circulatory system
RESPIRATORY DISTRESS SYNDROME		
Lack of *surfactant* in the lungs of a premature newborn. Normally, surfactant acts to keep lung alveoli open so gas exchange can proceed	Alveoli begin to close up within hours of birth	Labored, rapid breathing. Can cause death in premature infants. Also affects infants born to diabetic mothers

SUMMARY

1. Body cells rely mainly on aerobic respiration, a metabolic pathway that requires oxygen and produces carbon dioxide. By way of gas exchange, a function of the respiratory system, the human body as a whole acquires oxygen and disposes of carbon dioxide.

2. Air is a mixture of oxygen, carbon dioxide, and other gases. Each gas exerts a partial pressure, and each tends to move (diffuse) from areas of higher to lower partial pressure. This tendency plays an essential role in the operation of the respiratory system.

3. In the lungs, oxygen and carbon dioxide diffuse across a respiratory surface—a moist, thin layer of epithelium. Airways carry gases to and from one side of the respiratory surface, and blood vessels carry gases to and away from the other side.

4. Airways of the respiratory system include the nasal cavity, pharynx, larynx, trachea, bronchi, and bronchioles. Gas exchange occurs in millions of saclike alveoli, which are located at the end of the terminal bronchioles.

5. During inhalation, the chest cavity expands, pressure in the lungs falls below atmospheric pressure, and air flows into the lungs. During normal exhalation, these steps are reversed.

6. Driven by its partial pressure gradient, oxygen in the lungs diffuses from alveolar air spaces into the pulmonary capillaries. Then it diffuses into red blood cells and binds weakly with hemoglobin. In tissues where cells are metabolically active, hemoglobin gives up oxygen, which diffuses out of the capillaries, across interstitial fluid, and into cells.

7. Hemoglobin binds with or releases oxygen in response to shifts in oxygen levels, carbon dioxide levels, pH, and temperature.

8. Driven by its partial pressure gradient, carbon dioxide diffuses from cells, across interstitial fluid, and into the bloodstream. Most CO_2 reacts with water to form bicarbonate; the reactions are reversed in the lungs. There, carbon dioxide diffuses from lung capillaries into the air spaces of the alveoli, then is exhaled.

9. Gas exchange is governed both by the nervous system and by local chemical controls in the lungs. The nervous system governs oxygen and carbon dioxide levels, monitoring those gases in arterial blood by way of sensory receptors. These include carotid bodies (at branches of carotid arteries leading to the brain), aortic bodies (in an arterial wall near the heart), and receptors in the medulla of the brain. Blood levels of carbon dioxide are most important in triggering neural responses that regulate breathing. The responses are generated by respiratory centers in the brain, which coordinate the neural signals that regulate inspiration and exhalation.

Review Questions

1. Label the components of the human respiratory system and the structures that enclose some of its parts:

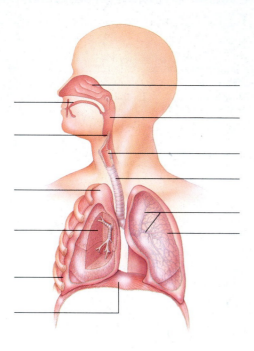

2. How do partial pressure gradients figure in gas exchange? *9.2*

3. What is oxyhemoglobin? Where does it form? *9.5*

4. What drives oxygen from alveolar air spaces, through interstitial fluid, and across capillary epithelium? What drives carbon dioxide in the reverse direction? *9.4, 9.5*

5. How does hemoglobin help maintain the oxygen partial pressure gradient during gas transport in the body? What reactions enhance the transport of carbon dioxide throughout the body? *9.5*

6. Gas exchange is most efficient when the rates of air flow and blood flow are balanced. Give an example of a local control that comes into play in the lungs when the two rates are imbalanced. Do the same for a neural control. *9.6*

7. Why does your breathing rate increase as you exercise? What happens to your heart rate at the same time—and why? *9.6*

Self-Quiz *(Answers in Appendix V)*

1. A partial pressure gradient of oxygen exists between _____.
   a. air and lungs
   b. lungs and metabolically active tissues
   c. air at sea level and air at high altitudes
   d. all of the above

2. The _____ is an airway that connects the nose and mouth with the _____.
   a. oral cavity; larynx
   b. pharynx; trachea
   c. trachea; pharynx
   d. pharynx; larynx

3. Oxygen air must diffuse across _____ to enter the blood.
   a. pleural sacs   c. a moist respiratory surface
   b. alveolar sacs   d. both b and c

4. Each lung encloses a _____.
   a. diaphragm
   b. bronchial tree
   c. pleural sac
   d. both b and c

5. Gas exchange occurs at the _____.
   a. two bronchi
   b. pleural sacs
   c. alveolar sacs
   d. both b and c

6. Breathing _____.
   a. ventilates the lungs
   b. draws air into airways
   c. expels air from airways
   d. causes reversals in pressure gradients
   e. all of the above

7. After oxygen diffuses into lung capillaries it also diffuses into
   _____ and binds with _____ .
   a. interstitial fluid; red blood cells
   b. interstitial fluid; carbon dioxide
   c. red blood cells; hemoglobin
   d. red blood cells; carbon dioxide

8. Due to its partial pressure gradient, carbon dioxide diffuses
   from cells, into interstitial fluid, and into the _____; in the
   lungs, carbon dioxide diffuses into the _____ .
   a. alveoli; bronchioles
   b. bloodstream; bronchioles
   c. alveoli; bloodstream
   d. bloodstream; alveoli

9. Hemoglobin performs which of the following respiratory
   functions:
   a. transports oxygen
   b. transports some carbon dioxide
   c. acts as a buffer to help maintain blood pH
   d. all of the above

10. Most carbon dioxide in the blood is in the form of _____ .
    a. carbon dioxide
    b. carbon monoxide
    c. carbonic acid
    d. bicarbonate

11. Match the following respiratory components with their
    descriptions.
    ____ bronchus   a. site of gas exchange
    ____ alveolus   b. fine bronchial tree branching
    ____ hemoglobin   c. throat
    ____ trachea   d. air-conducting tube; windpipe
    ____ bronchiole   e. respiratory pigment
    ____ pharynx   f. airway leading into a lung

Critical Thinking: You Decide (Key in Appendix VI)

1. People occasionally poison themselves with carbon monoxide
by building a charcoal fire in an enclosed area. Assuming help
arrives in time, what would be the *most* effective treatment: placing
the victim outdoors in fresh air, or administering pure oxygen?
Explain your answer.

2. Skin divers and swimmers
sometimes purposely hyperven-
tilate. Doing so doesn't increase
the oxygen available to tissues. It
does increase blood pH (making
it more alkaline), and it decreases
the blood level of carbon dioxide.
Based on your reading in this
chapter, what effect is hyperventilation likely to have on the
neural controls over breathing?

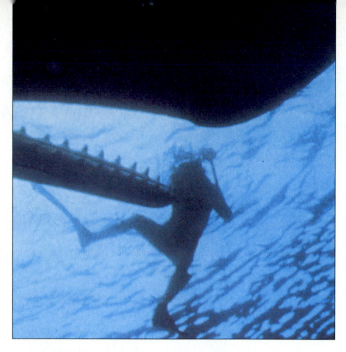

Figure 9.15 A diver inspecting a whale.

3. Underwater, we humans can't compete with whales and other
air-breathing marine mammals, which can stay submerged for
extended periods (Figure 9.15). At the beach one day you meet a
diver who tells you that special training could allow her to swim
underwater without breathing for an hour. From what you know
of respiratory physiology, explain why she is mistaken.

4. When you sneeze or cough, abdominal muscles contract
suddenly, pushing your diaphragm upward. After reviewing the
discussion of the respiratory cycle in Section 9.4, explain why this
change expels air out your nose and mouth with explosive force.

Selected Key Terms

alveolar sac *9.1*	lung *9.1*
alveolus *9.1*	oxyhemoglobin *9.5*
aortic body *9.6*	pharynx *9.1*
bronchiole *9.1*	pleura *9.1*
bronchus *9.1*	respiration *9.1*
carbonic anhydrase *9.5*	respiratory bronchiole *9.1*
carotid body *9.6*	respiratory surface *9.2*
diaphragm *9.1*	tidal volume *9.4*
expiration *9.4*	trachea *9.1*
inspiration *9.4*	vital capacity *9.4*
larynx *9.1*	vocal cords *9.1*

Readings

American Cancer Society. *Dangers of Smoking; Benefits of Quitting
and Relative Risks of Reduced Exposure*, rev. ed. New York: American
Cancer Society.

Mortality from Smoking in Developed Countries 1950–2000. A 1994
publication by scientists of Britain's Imperial Cancer Research
Fund, the World Health Organization, and the American Cancer
Research Fund. Research finds that worldwide, smoking now
kills 3 million people every year. If current patterns do not change,
by the time today's young smokers reach middle age, 10 million
may be dying annually because of their habit. That is one person
every three seconds.

Sherwood, L. 1997. *Human Physiology*. Third edition. Pacific Grove,
California: Brooks/Cole.

10

WATER–SALT BALANCE AND EXCRETION

Survival at Stovepipe Wells

In Death Valley, the sand dunes at Stovepipe Wells cover some 14 square miles. Just south of the dunes are 200 square miles of waterless salt flats. The whole parched region has claimed more than a few human lives.

Without adequate water, a person can't survive for long anywhere, let alone in Death Valley (Figure 10.1). Homeostasis in the internal environment is a major theme in our study of human anatomy and physiology, and maintaining water balance is crucial to this internal stability. Among other considerations, our living cells are mostly water, and water is required for many critical chemical reactions inside them. Water in blood and interstitial fluid is a solvent for important salts and other molecules. In addition, most of the potentially toxic nitrogen-containing wastes of protein metabolism are moved out of the body as part of the watery fluid called urine.

Although individuals differ, an adult routinely loses about 2,400 milliliters (two and a half quarts) of water daily. Much of this loss is in urine, which the body is obligated to produce in order to rid itself of certain metabolic wastes. More water exits in feces, as vapor in exhaled air, and in sweat. In the sweltering Death Valley summer, there is no halt to the body's need to remove wastes, but sweating increases dramatically as a mechanism to help dissipate heat. Whereas a person might normally lose about 100 milliliters of water in sweat a day, in extreme heat—or during extremely heavy exercise—the loss can rise to 900 milliliters or

Figure 10.1 Hikers in Death Valley National Monument must carry ample water to replenish the body's supply. Otherwise, they risk potentially serious disruption of homeostasis in the body's finely tuned balance of water and solutes.

more *per hour*. The body cannot tolerate such rapid water loss for long, unless the water is replaced. So, within the body of a hiker at Stovepipe Wells, a remarkable homeostatic juggling act goes on. In response to changing internal conditions, the brain issues commands you experience as thirst—telling you, in effect, to drink water. Water-conserving mechanisms are also set in motion while other mechanisms operate simultaneously to cool the body, dispose of metabolic wastes, and make other required exchanges with the external environment. These activities coordinate with the operations of other organ systems, as sketched in Figure 10.2.

Depending on circumstances, we also may take in too much water, eat foods that contain large amounts of sodium, or do something else that upsets the optimal balance of water and solutes in blood and body fluids. Fortunately, two fist-sized organs, our kidneys, come to the rescue. Water and solutes from the blood flow continuously through the kidneys, where adjustments occur in how much water and which solutes are reabsorbed into body fluids or disposed of as urine. Those adjustments help the body cope with everything from water imbalances to antibiotic drugs, and they will be our primary focus in this chapter. We will also examine the homeostatic mechanisms that operate to maintain body temperature within limits that support life.

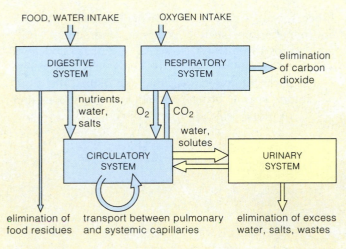

Figure 10.2 Links between the urinary system and other organ systems that maintain stable operating conditions in the internal environment.

KEY CONCEPTS

1. The body continually takes in and loses water and dissolved substances (solutes). It continually produces metabolic wastes. Even with all the inputs and outputs, the overall volume and composition of the body's extracellular fluid remain relatively constant.

2. The urinary system is crucial to balancing the intake and output of water and solutes. This system continually filters water and solutes from the blood. It reclaims some amount of both and eliminates the rest. Different amounts are reclaimed at different times, depending on what is necessary to maintain homeostasis of the extracellular fluid.

3. The urinary system includes two kidneys. The kidneys are blood-filtering organs. Packed inside each one are about 1 million filtering structures called nephrons.

4. Water and solutes from blood enter kidney nephrons, where filtration (under pressure) and selective reabsorption take place. After this processing, most of the water and necessary solutes are returned to the bloodstream.

5. Water and solutes that are not returned to the blood leave the body as urine. At any given time, control mechanisms influence whether the urine is concentrated or dilute. Hormones play key roles in these adjustments.

6. Internal (core) body temperature must be maintained in a range suitable for metabolism. Metabolic activity produces most body heat. Overall, core temperature depends on the balance between heat gained and heat lost to the environment.

7. Internal body temperature is maintained within a favorable range through controls over a person's metabolic activity, as well as through adaptations in physiology and behavior.

CHAPTER AT A GLANCE

THE CHALLENGE: SHIFTS IN EXTRACELLULAR FLUID

If you are an adult female in good health, by weight your body is about 50 percent fluid. If you are an adult male, the same number is about 60 percent. Clearly, fluid is vital both in our anatomy (body structures) and in our physiology (body functions). This fluid occurs both outside our cells and inside them. Recall that interstitial fluid fills the spaces between living cells and other components of our body tissues. The "fluid tissue" blood—which, recall, is mostly watery plasma—circulates in blood vessels. Together, interstitial fluid, blood plasma, and the relatively small amounts of other fluids (such as lymph) outside cells are the body's **extracellular fluid**, or ECF.

The fluid *inside* our cells is **intracellular fluid**. From previous chapters you know that there is a constant exchange of gases and other substances between intracellular and extracellular fluid. Those exchanges are crucial for keeping cells functioning smoothly, and they cannot occur properly unless the volume and composition of the ECF are stable.

Yet the ECF constantly changes, because gases, cell secretions, ions, and other materials enter or leave it. To maintain stable conditions in the ECF, especially the concentrations of water and vital ions such as sodium (Na^+) and potassium (K^+), there must be mechanisms that remove substances as they enter the extracellular fluid or add needed ones as they leave it. This task is the job of the **urinary system**. Before examining the parts of this system, however, we will look more closely at the traffic of substances into and out of extracellular fluid.

Water Gains and Losses

Ordinarily, each day you take in about as much water as your body loses (Table 10.1). Two processes account for these water gains:

1. Absorption from ingested liquids and solid foods

2. Metabolic reactions that produce water as a by-product

Thirst influences how much water we take in. When a water deficit occurs in body tissues, our brain compels us to seek out water—for example, from a water fountain or a cold drink from the refrigerator. We will be looking at the thirst mechanism later in the chapter.

The body *loses* water in the following ways:

1. Excretion in urine
2. Evaporation from the lungs and skin
3. Sweating
4. Elimination in feces

Of these four routes of water loss, **urinary excretion** is the one over which the body can exert the most control. Urinary excretion eliminates excess water, as well as excess or harmful solutes, in the form of urine. Some water also evaporates from the respiratory surfaces of the lungs and from our skin. These are sometimes called "insensible" water losses because a person is not always aware they are taking place. As noted in Chapter 6, in a healthy person very little water that enters the GI tract is lost; most is absorbed, not eliminated in feces.

Solute Gains and Losses

Solutes enter the body's extracellular fluid mainly by four routes:

1. Absorption from ingested liquids and solid food
2. Secretion from cells
3. Respiration
4. Metabolism

When we eat, a variety of nutrients (including glucose) and mineral ions (such as potassium and sodium ions) are absorbed from the GI tract. So are drugs and food additives. Throughout the body, living cells also secrete substances into interstitial fluid and blood. The respiratory system brings oxygen into the blood, and respiring cells add carbon dioxide to it.

Table 10.1	Normal Daily Balance between Water Gain and Water Loss in Adult Humans		
Water Gain (milliliters)		**Water Loss (milliliters)**	
Ingested in solids:	850	Urine:	1,500
Ingested as liquids:	1,400	Feces:	200
Metabolically derived:	350	Evaporation:	900
	2,600		2,600

Further reading: Student Guide to InfoTrac on web site →

Extracellular fluid *loses* mineral ions and metabolic wastes in these ways:

1. Urinary excretion
2. Respiration
3. Sweating

Carbon dioxide is the most abundant metabolic waste, and we get rid of it by exhaling it from our lungs. All other major wastes—and there are more than 200 of them—leave in urine. One, *uric acid*, is formed when cells break down nucleic acids. If it builds up in the ECF, it can crystallize and collect in the joints, causing the painful condition called *gout*.

Other major metabolic wastes include by-products of protein metabolism. One of these, *ammonia*, is formed in "deamination" reactions—reactions in which nitrogen-containing amino groups are removed from amino acids. Ammonia can be highly toxic if it accumulates in the body. *Urea* is produced in the liver when ammonia combines with carbon dioxide. It is the main waste product when cells break down proteins. About 40 to 60 percent of the urea filtered from blood in the kidneys is reabsorbed. The rest is excreted. Protein breakdown also produces phosphoric acid, sulfuric acid, creatine, and small amounts of other, nitrogen-containing compounds, some of which are toxic. These also are excreted.

Although sweat carries away a small percentage of nitrogen-containing wastes (urea and uric acid), by far the most are removed by our kidneys while they are filtering other substances from the blood. In addition to filtering water and wastes, the kidneys also work to maintain the balance of important ions such as sodium, potassium, and calcium. (These ions are also called **electrolytes** because a solution in which they are dissolved will carry an electric current.)

Normally only a little of the water and solutes that enter the kidneys leaves as urine. In fact, except when you take in lots of fluid (without exercise), all but about 1 percent of the water is returned to the blood. However, the chemical composition of the fluid that is returned has been adjusted in vital ways. Just how this happens will be our focus as we turn our attention to the urinary system.

Each day the body gains water ingested in liquids and solid foods and from metabolism. It loses an approximately equal amount of water through urinary excretion, evaporation, sweating, and elimination in feces.

The body gains solutes by way of ingestion, secretion, metabolism, and respiration. Solutes exit via urinary excretion, respiration, and sweating.

By adjusting blood's volume and composition, the kidneys help maintain tolerable conditions in the extracellular fluid.

Focus on Our Environment

10.2 IS YOUR DRINKING WATER POLLUTED?

You might want to investigate the answer to the above question. Polluted drinking water is a fact of life for more and more people, and sometimes the contaminants can come from surprising sources.

In many countries, including the United States, aquifers and other groundwater sources of drinking water now contain not only pure water, but pesticides, fertilizers, and hazardous organic chemicals (Figure 10.3). Trace amounts of roughly 800 synthetic organic chemicals appear in different water supplies, and some are suspect in kidney disorders, birth defects, and certain other ills. Some cause cancer in laboratory animals. While the test doses are high, such findings can be troubling.

Our drinking water can even be contaminated with chemicals aimed at *cleaning up* pollution. That is the case with MTBE—short for methyl tertiary butyl ether. MTBE is an additive that oxygenates gasoline. It increases the "octane" (a measure of energy content) and also makes gas burn cleaner.

Gasoline refiners have been adding MTBE to certain grades of gasoline for decades. Under the gun to produce cleaner-burning gas (to reduce air pollution from vehicle emissions), many companies began adding MTBE to *all* their gasoline in the mid-1990s. By 2000, MTBE was turning up in wells and public water supplies from Maine to California. How did it get there? One known source is leakage from underground storage tanks—the kind that store the gas at gas stations. How much should consumers worry about drinking MBTE-laced water? At this writing there is little in the way of published research findings on MBTE's possible or actual health effects, if any. Meanwhile, the U.S. Environmental Protection Agency has classified MTBE as a "potential human carcinogen"—that is, a potential cancer-causing chemical.

Figure 10.3 Leaking barrels in a hazardous waste dump near Washington, D.C. Such dumps are now illegal.

THE URINARY SYSTEM

The kidneys are the central components of the urinary system (Figure 10.4). Each **kidney** is a bean-shaped organ about the size of a rolled-up pair of socks. Internally, it has several lobes. In each lobe, an outer *cortex* wraps around a central region, the *medulla,* as you can see sketched in Figure 10.5. The whole kidney is enclosed within a tough coat of connective tissue, the *renal capsule* (from the Latin *renes,* meaning kidneys).

Each kidney lobe contains blood vessels and slender tubes called **nephrons**. Nephrons are the functional heart of the kidneys, for they are the structures that filter water and solutes from blood. Most of the filtered material— the filtrate—is reabsorbed back into the bloodstream. The rest continues on through *collecting ducts,* where more reabsorption occurs, and into the kidney's central cavity, the *renal pelvis.*

In addition to the two kidneys, the urinary system includes "plumbing" that transports or stores urine. Once urine has formed, it flows from a kidney into a tubelike **ureter**, then on into a storage organ, the **urinary bladder**. It leaves the bladder through the **urethra**, a muscular tube that opens at the body surface.

More than a million nephrons are packed inside each of your kidneys. Each nephron is a tiny tubular structure shaped a little like the piping under a sink (see Figure 10.6). Its wall is a single layer of epithelial cells, but the cells and junctions between them vary in different parts of the tube. Some tube regions are extremely permeable to water and solutes. Others *prevent* the passage of solutes

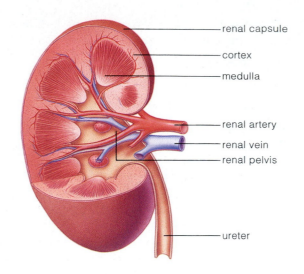

Figure 10.5 A kidney and the major blood vessels leading into and out of it.

Figure 10.4 Organs of the urinary system and their functions. The two kidneys, two ureters, and urinary bladder are located outside the peritoneum, the lining of the abdominal cavity.

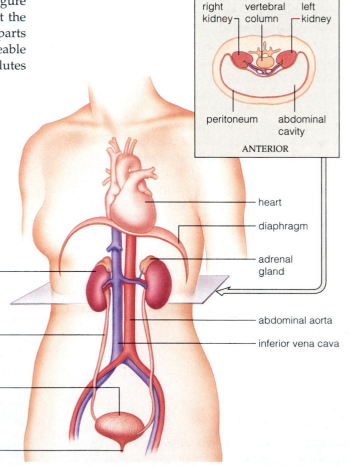

KIDNEY (one of a pair)

Constantly filters water and all solutes except proteins from blood; reclaims water and solutes as the body requires and excretes the remainder, as urine

URETER (one of a pair)

Channel for urine flow from a kidney to the urinary bladder

URINARY BLADDER

Stretchable container for temporarily storing urine

URETHRA

Channel for urine flow between the urinary bladder and body surface

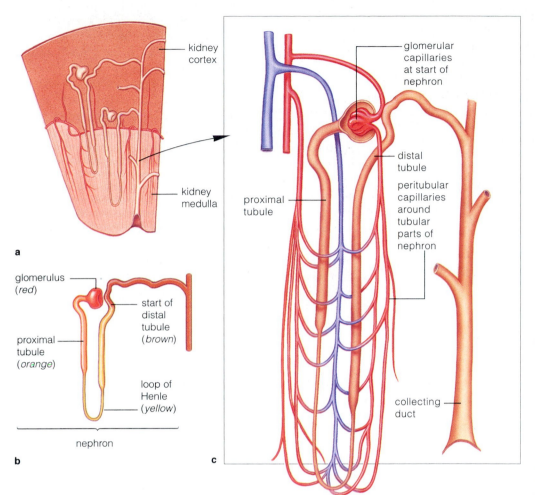

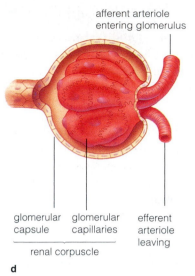

kidney cortex

glomerular capillaries at start of nephron

distal tubule

peritubular capillaries around tubular parts of nephron

proximal tubule

kidney medulla

a

glomerulus (*red*)

start of distal tubule (*brown*)

proximal tubule (*orange*)

loop of Henle (*yellow*)

nephron

b

collecting duct

c

afferent arteriole entering glomerulus

glomerular capsule

glomerular capillaries

efferent arteriole leaving

renal corpuscle

d

Figure 10.6 Simplified diagrams of a nephron and its association with two sets of blood capillaries.

(**a**) Orientation of a nephron relative to the cortex and medulla of a kidney lobe. (**b**) Functional regions of a nephron. (**c**) The two interconnected sets of blood capillaries associated with the nephron. A capsule at the start of the nephron (the glomerular capsule) houses the first set, the glomerular capillaries. The second set, the peritubular capillaries, thread around all tubular parts of the nephron. (**d**) The glomerulus, the nephron's unit that filters blood.

except via active transport systems built into the plasma membrane (Section 3.6). As you will see, such differences influence the movement of water and solutes across different parts of the nephron wall.

An *afferent arteriole* delivers blood to each nephron (afferent means toward). Filtration starts at the **renal corpuscle**, where the nephron wall balloons around a tiny cluster of blood capillaries called the **glomerulus** (Figure 10.6). The ballooned, cuplike wall region, the **glomerular (Bowman's) capsule**, receives water and solutes filtered from blood. The rest of the nephron is tubular. Filtrate flows from the cup into the **proximal tubule** (closest to the glomerular capsule), then through a hairpin-shaped **loop of Henle** and into the **distal tubule** (most distant from the glomerular capsule). This distal tubule empties into a collecting duct.

Unlike capillaries in most other parts of the body, the glomeruli—capillaries inside a glomerular capsule—do not link arterioles and venules. Accordingly, they do not channel blood immediately back to the general circulation. Instead, they converge to form an *efferent*

arteriole (efferent means "away from"). This arteriole branches into yet another set of capillaries, which are called **peritubular** ("around the tubule") **capillaries** (Figure 10.6c). The peritubular capillaries weave around the nephron's tubular parts. Water and essential solutes reabsorbed from the tubule enter them and in this way return to the bloodstream. Peritubular capillaries merge into venules, which carry filtered blood out of the kidneys.

The urinary system consists of two kidneys, two ureters, the urinary bladder, and the urethra.

Kidney nephrons receive water and solutes filtered from blood. A tiny cluster of capillaries called a glomerulus is the nephron's blood-filtering unit.

From the glomerulus, filtrate enters the nephron's tubular portion, which ultimately delivers the fluid that enters it to a collecting duct. Peritubular capillaries surrounding tubular nephron regions recapture water and essential solutes.

Filtration, Reabsorption, and Secretion

Urine is a fluid that rids the body of water and solutes over and above the amounts needed to maintain homeostatic stability in the extracellular fluid (Table 10.2). Urine forms by a sequence of three processes: filtration, reabsorption, and secretion.

Blood pressure generated by the heart's contractions drives **filtration**, which takes place at the glomerulus (Figure 10.7). It forces water and various small solutes (such as glucose, sodium, and urea) out of the blood in glomerular capillaries and into the cuplike space inside the glomerular capsule. Blood cells, proteins, and other large solutes are left behind. The filtrate moves from the glomerular capsule into the proximal tubule.

Reabsorption takes place along the tubular parts of the nephron. Most of the filtrate's water and solutes—including sodium ions, vitamins, amino acids, and glucose—move *out* of the nephron (by diffusion or active transport) across the tubule wall. Then they move into adjacent capillaries.

Secretion also occurs across parts of the tubule wall, but in the *opposite* direction. Substances diffuse out of the capillaries and *into* cells of the wall. They then move across the cells, which secrete them into the forming urine. Among other functions, this highly controlled process rids the body of excess hydrogen ions (H^+) and potassium ions. It also prevents some metabolic wastes (such as uric acid and some breakdown products of hemoglobin) and water-soluble foreign substances (such as penicillin and certain pesticides) from accumulating in blood. Drug testing of athletes and employees in certain professions relies on the use of urinalysis to detect drug residues secreted into the urine. Section 10.6 gives a few examples of how urinalysis can help a physician evaluate a patient's health.

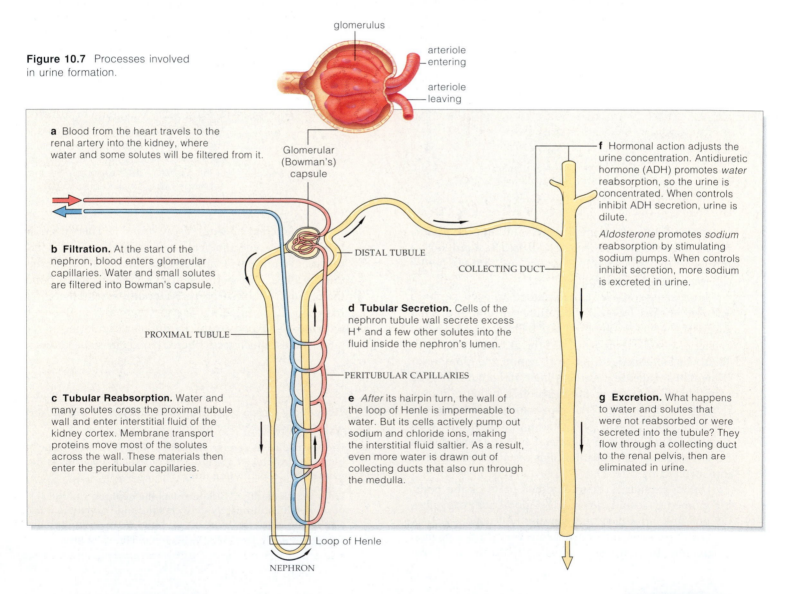

Figure 10.7 Processes involved in urine formation.

glomerulus

arteriole entering

arteriole leaving

a Blood from the heart travels to the renal artery into the kidney, where water and some solutes will be filtered from it.

Glomerular (Bowman's) capsule

f Hormonal action adjusts the urine concentration. Antidiuretic hormone (ADH) promotes *water* reabsorption, so the urine is concentrated. When controls inhibit ADH secretion, urine is dilute.

Aldosterone promotes *sodium* reabsorption by stimulating sodium pumps. When controls inhibit secretion, more sodium is excreted in urine.

b Filtration. At the start of the nephron, blood enters glomerular capillaries. Water and small solutes are filtered into Bowman's capsule.

DISTAL TUBULE

COLLECTING DUCT

d Tubular Secretion. Cells of the nephron tubule wall secrete excess H^+ and a few other solutes into the fluid inside the nephron's lumen.

PROXIMAL TUBULE

PERITUBULAR CAPILLARIES

c Tubular Reabsorption. Water and many solutes cross the proximal tubule wall and enter interstitial fluid of the kidney cortex. Membrane transport proteins move most of the solutes across the wall. These materials then enter the peritubular capillaries.

e *After* its hairpin turn, the wall of the loop of Henle is impermeable to water. But its cells actively pump out sodium and chloride ions, making the interstitial fluid saltier. As a result, even more water is drawn out of collecting ducts that also run through the medulla.

g Excretion. What happens to water and solutes that were not reabsorbed or were secreted into the tubule? They flow through a collecting duct to the renal pelvis, then are eliminated in urine.

Loop of Henle

NEPHRON

Table 10.2	Typical Kinds and Daily Amounts of Solutes in Normal Urine
Solute	Amount of Solute / Day
Urea	20.0–35.0 grams
Sodium	4.0–6.0 grams
Chloride	6.0–9.0 grams
Potassium	2.5–3.5 grams
Creatinine	1.0–1.5 grams
Calcium	0.01–0.30 grams

Urination is urine flow from the body (a technical term for it is "micturition"). Urination is a reflex response. As the urinary bladder fills, tension increases in the smooth muscle of its strong walls. Where the bladder joins the urethra, smooth muscle acts as an *internal urethral sphincter* that helps prevent urine from flowing into the urethra. As tension in the bladder wall increases, though, the sphincter relaxes; at the same time, the bladder walls contract and force urine through the urethra.

The internal sphincter is not under our voluntary control, but a person *can* exert control over the *external urethral sphincter* formed by skeletal muscle closer to the urethral opening. Learning to control this external sphincter is the basis of urinary "toilet training" in young children. Adults sometimes experience *stress incontinence* when they cough, laugh, or strain during exercise. Then, an abrupt increase in muscle-generated pressure inside the abdomen forces urine through the external sphincter.

Kidney stones are deposits of uric acid, calcium salts, and other substances that have settled out of urine and collected in the renal pelvis. Smaller kidney stones usually pass naturally from the body during urination. Larger ones can become lodged in the renal pelvis or ureter or, on rare occasions, in the bladder or urethra. The blockage can partially dam urine flow and cause intense pain and kidney damage. Large kidney stones must be removed medically or surgically. In *lithotripsy*, high-energy sound waves blast the stone to fragments that are small enough to pass out in the urine.

Factors That Influence Blood Filtration

Each day, more blood flows through your kidneys than through any other organ except the lungs. Each minute, about 1.5 quarts of blood course through them! That is nearly one-fourth of a resting person's cardiac output. How can the kidneys handle blood flowing through on such a massive scale? There are two mechanisms.

To begin with, the afferent arterioles delivering blood to a glomerulus have a wider diameter—and therefore less resistance to flow—than do most arterioles. This means that the hydrostatic pressure caused by heart contractions does not drop much when blood enters them. By contrast, the efferent arteriole that receives blood from glomeruli offers *high* resistance to blood flow. Because of these properties of the afferent and efferent arterioles, blood pressure in the glomerular capillaries is higher than in other capillaries.

Second, glomerular capillaries are highly permeable. They do not allow blood cells or protein molecules to escape, but compared to other capillaries they are 10 to 100 times more permeable to water and small solutes. Because of the high hydrostatic pressure and greater capillary permeability, an adult's kidneys can filter an average of 45 gallons (180 liters) per day.

At any time, the volume of blood flowing to the kidneys affects the filtration rate. Neural, endocrine, and local controls keep that volume fairly constant even when a person's blood pressure changes. For example, when you run a race or dance until dawn, the nervous system diverts an above-normal volume of blood away from the kidneys toward your skeletal muscles. When this happens, neural signals direct arterioles in different parts of your body to constrict or dilate in coordinated ways, so that less of the flow volume reaches the kidneys.

As another example, cells in the walls of arterioles leading to glomeruli respond to arterial blood pressure. When blood pressure decreases, they secrete chemicals that induce vasoconstriction, so the kidneys receive less of the body's total blood volume. But when blood pressure rises, they vasodilate, so more blood flows in.

Lastly, the filtration rate depends on how fast the kidney tubules are reabsorbing water. As you will see in the next section, reabsorption is partly under the control of hormones.

Urine forms through the sequential processes of filtration, reabsorption, and secretion. It includes water and solutes not needed to maintain the extracellular fluid, as well as water-soluble wastes.

Two mechanisms permit the kidneys to filter a large amount of blood at a rapid rate. First, blood pressure in glomerular capillaries is higher than in other capillaries. Second, glomerular capillaries are highly permeable.

Neural, hormonal, and local controls ensure that the volume of blood flowing to kidney glomeruli remains relatively constant, even when blood pressure varies.

REABSORPTION OF WATER AND SODIUM

Your kidneys precisely adjust how much water and sodium your body excretes or conserves. It makes no difference whether you drink too much or too little water at lunch, eat a bag of salty peanuts, or follow a low-sodium diet—your kidneys will make the necessary adjustments to keep the water and salt content of your body relatively constant. In this section we consider how some of this regulation is accomplished.

Reabsorption in the Proximal Tubule

Of all the water and sodium the kidneys filter, about two-thirds is reabsorbed right away. This happens at the **proximal tubule**—the part of the nephron closest to the glomerulus (see Figure 10.7c). As Figure 10.8 shows, epithelial cells of the proximal tubule wall have transport proteins at their outer surface. Nearly all cells have such proteins, which function as sodium "pumps." In this case, the proteins actively transport sodium ions (which, remember, are important electrolytes) from the filtrate inside the tubule into the interstitial fluid. Sodium ions (Na^+) are positively charged; negatively charged ions, including chloride (Cl^-), follow the sodium. Glucose and amino acids are reabsorbed by active transport that is linked to sodium reabsorption. Table 10.3 gives the average daily reabsorption values for a few substances.

This outward movement of solutes has several effects. It *reduces* the solute concentration inside the tubule, and it *increases* the concentration of solutes in the interstitial fluid outside the tubule. The wall of the proximal tubule

is quite permeable to water, so water follows the osmotic gradient and moves passively out of the tubule. From there, the water—and solutes—move into peritubular capillaries, returning to the bloodstream.

As water leaves the proximal tubule, the volume of fluid remaining inside it is reduced a great deal. Even so, at this point the total concentration of solutes in the filtrate—sodium, especially—has changed only a little.

Reabsorption in Other Parts of the Nephron

The situation changes after filtrate moves on through the proximal tubule and enters the loop of Henle. This hairpin-shaped structure descends into the kidney medulla. Review Figure 10.7 to see the arrangement. In the interstitial fluid surrounding the loop, the solute concentration increases steadily the deeper the loop plunges into the medulla. Several factors contribute to this and other fluctuations in chemical conditions that are about to occur.

The descending limb of the loop is permeable to water. Because the solute concentration outside it is greater than that inside, some more water moves out by osmosis. It, too, is reabsorbed. As the water leaves, the solute concentration in the fluid that remains *inside* the descending limb increases until it matches that in the interstitial fluid.

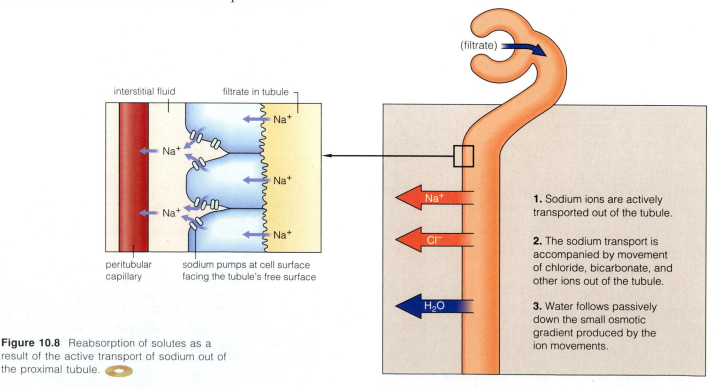

Figure 10.8 Reabsorption of solutes as a result of the active transport of sodium out of the proximal tubule.

interstitial fluid   filtrate in tubule

peritubular capillary   sodium pumps at cell surface facing the tubule's free surface

(filtrate)

Na^+

Cl^-

H_2O

1. Sodium ions are actively transported out of the tubule.

2. The sodium transport is accompanied by movement of chloride, bicarbonate, and other ions out of the tubule.

3. Water follows passively down the small osmotic gradient produced by the ion movements.

Table 10.3 Average Daily Reabsorption Values for a Few Substances

	Amount Filtered	Amount Excreted	Proportion Reabsorbed
Water	180 liters	1.8 liters	99%
Glucose	180 grams	None, normally	100%
Sodium ions	630 grams	3.2 grams	99.5%
Urea	54 grams	30 grams	44%

Sodium is actively transported out of the nephron in the ascending limb of the loop of Henle. In this part of the tubule, water cannot cross the tubule wall. Naturally, as sodium (and chloride) ions leave the filtrate, the solute concentration rises outside the tubule and falls inside it. (This rise in the concentration of solutes outside the tubule favors the osmotic movement of water out of the descending limb.)

Another factor also is playing into the increase in solute concentration outside the tubule. Remember, as described a few paragraphs earlier, that the amount of solutes in filtrate leaving the proximal tubule is not much different than it was when the filtrate entered. However, a very high solute concentration develops anyway in the deeper portions of the medulla. Why? Urea contributes to the steep gradient. As water is reabsorbed, the urea left behind in the filtrate becomes concentrated. Some of it will be excreted in urine, but when filtrate eventually moves into the final portion of the collecting duct, some urea also will diffuse out. A portion of it enters the interstitial fluid in the inner medulla—and this further increases the concentration of solutes there. For this reason, solute concentration is always highest in the very deepest parts of the inner medulla.

With so many solutes—but no water—having left the fluid in the ascending limb of the tubule, solutes are not very concentrated there. Hence, the filtrate that finally reaches the distal tubule in the kidney cortex is quite dilute, with a low sodium concentration. As you will read in the next section, the stage is set for the excretion of urine that is either highly dilute or highly concentrated—or anywhere in between.

A reabsorption mechanism operates in the tubular portion of the nephron. By way of this mechanism, which is adjustable by hormones, water and solutes can be retained or excreted as required to maintain the extracellular fluid.

Most water is reabsorbed across the permeable walls of the proximal tubule and the descending limb of the loop of Henle. Sodium is pumped outward in the ascending limb.

10.6

KEEPING YOUR URINARY TRACT HEALTHY

Urinary tract infections routinely plague millions of people. Women especially are susceptible to bladder infections because of their urinary anatomy: The female urethra is short, just a little over an inch long. (An adult male's urethra is about 9 inches long.) Its external opening also is close to the anus. Therefore it is relatively easy for bacteria from outside the body to make their way to a female's bladder and trigger the inflammation called *cystitis*—or even all the way to the kidneys to cause *pyelonephritis*. In both sexes, urinary tract infections sometimes result from sexually transmitted microbes, including the microorganisms that cause *chlamydia*. Chapter 16 provides more information on this topic.

In males, the prostate gland wraps around the urethra. As a man ages, his prostate may swell either occasionally or chronically, narrowing the urethra and preventing urine from draining effectively. When this happens, the growth of bacteria can trigger an infection. Urinary problems can also be an early warning sign of prostate cancer (Chapter 22).

URINALYSIS The procedure called *urinalysis* analyzes the composition of urine and is used to help diagnose illness. For example, the presence of glucose in urine may be a sign of diabetes, whereas white blood cells (pus) frequently indicate a urinary tract infection. Red blood cells can reveal bleeding due to infection, kidney stones, cancer, or an injury. High levels of albumin and other proteins in urine may indicate severe hypertension, kidney disease, and some other disorders. Bile pigments enter the urine when liver functions are impaired by cirrhosis and hepatitis.

There are simple measures everyone can take to help keep their urinary tract healthy. Drink plenty of fluids (Figure 10.9), and practice careful hygiene to minimize the opportunity for bacteria to enter the urethra. People who are susceptible to bladder infections may also want to limit their intake of alcohol, caffeine, and spicy foods, all of which can irritate the bladder.

Figure 10.9 Drinking plenty of water helps maintain a healthy urinary system.

Because so much water and sodium are reabsorbed from the nephron's proximal tubule and loop of Henle, the volume of dilute urine reaching the start of the distal tubule (farthest from the glomerulus) has been greatly reduced. Yet if even that reduced volume were excreted "as is," the body would rapidly become depleted of both water and sodium. Fortunately, controlled adjustments in reabsorption are made at cells located in the walls of distal tubules and collecting ducts. Two hormones are the agents of control. One of them, **ADH** (antidiuretic hormone) influences water reabsorption. Another, called **aldosterone**, influences the reabsorption of sodium.

How ADH Influences Water Reabsorption

Let's first consider how ADH helps adjust the rate at which water is reabsorbed. The hypothalamus in the brain controls the release of ADH from the posterior pituitary gland. It triggers ADH secretion either when the solute concentration of extracellular fluid rises above a set point—say, after you've eaten that bag of salty chips—or when blood pressure falls. The solute concentration also can increase when a person takes in too little water or becomes dehydrated (for example, by exercising hard outdoors on a hot day); severe bleeding (hemorrhage) might cause a rapid decline in blood pressure. ADH acts on distal tubules and collecting ducts—which, recall, are situated in the kidney cortex. ADH makes their walls more permeable to water (Figure 10.10). Thus, additional

water is reabsorbed from the dilute urine inside the distal tubules of nephrons. With its volume now reduced somewhat, the urine passes down through the collecting ducts, which plunge down into the medulla. Remember from the previous section that solute concentrations in the surrounding interstitial fluid are high in the inner medulla. This detour encourages the reabsorption of even more water. When all is said and done, the secretion of ADH means that only a small volume of very concentrated urine is excreted.

Conversely, when you take in more water than the body needs during a given period, the solute concentration in your extracellular fluid falls. This state of affairs inhibits ADH secretion—and without ADH, the walls of the distal tubules and collecting ducts are less permeable to water. Less water is reabsorbed, and a large volume of dilute urine can be excreted. This is how the body rids itself of the excess water.

A *diuretic* is any substance that promotes the loss of water in urine. Caffeine is a mild diuretic; it reduces the reabsorption of sodium along nephron tubules, so less water is retained (and hence more is excreted). Alcohol also is a diuretic; it acts by suppressing the release of ADH. (Among other lessons we can draw from this fact is that drinking beer to replenish body fluids after exercise is not the way to rehydrate the body. Cold water does the job a great deal more effectively.) A diuretic may be prescribed as part of medical treatment for hypertension (Section 7.14), because the resulting water loss helps reduce blood pressure.

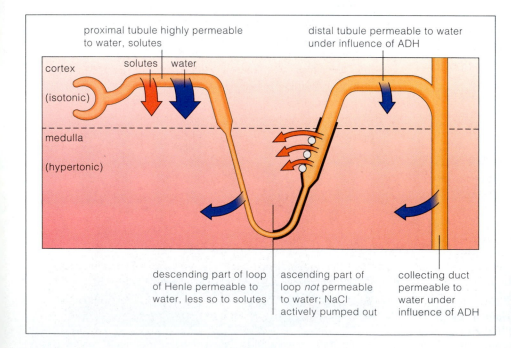

proximal tubule highly permeable to water, solutes

distal tubule permeable to water under influence of ADH

cortex
(isotonic)

solutes   water

medulla
(hypertonic)

descending part of loop of Henle permeable to water, less so to solutes

ascending part of loop *not* permeable to water; NaCl actively pumped out

collecting duct permeable to water under influence of ADH

How Aldosterone Influences Sodium Reabsorption

Now let's examine how aldosterone helps adjust the rate at which sodium is reabsorbed. When the body loses more sodium than it takes in (e.g, in sweat), the volume of extracellular fluid falls. This is because, as just described, where sodium goes, water follows. Sensory receptors in the walls of blood vessels in the kidneys and elsewhere detect the decrease and call into action kidney cells that secrete an

Figure 10.10 How ADH affects permeability characteristics of different parts of the nephron. (Most of the solute movements in the dital tubule and collecting duct are related directly or indirectly to active transport mechanisms.)

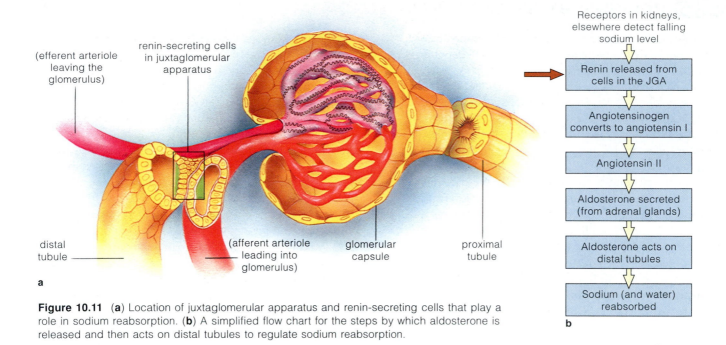

Figure 10.11 (**a**) Location of juxtaglomerular apparatus and renin-secreting cells that play a role in sodium reabsorption. (**b**) A simplified flow chart for the steps by which aldosterone is released and then acts on distal tubules to regulate sodium reabsorption.

Labels in figure (a): (efferent arteriole leaving the glomerulus); renin-secreting cells in juxtaglomerular apparatus; distal tubule; (afferent arteriole leading into glomerulus); glomerular capsule; proximal tubule; a

Flow chart in figure (b): Receptors in kidneys, elsewhere detect falling sodium level → Renin released from cells in the JGA → Angiotensinogen converts to angiotensin I → Angiotensin II → Aldosterone secreted (from adrenal glands) → Aldosterone acts on distal tubules → Sodium (and water) reabsorbed; b

enzyme called *renin*. Those cells are part of the **juxtaglomerular apparatus** (JGA). *Juxta-* means next to, and this "apparatus" is an area where arterioles of the glomerulus come into contact with a nephron's distal tubule (Figure 10.11*a*).

Renin acts on molecules of an inactive protein (called angiotensinogen) that circulates in the bloodstream. Then, enzyme action removes part of the molecule, leaving angiotensin I. A follow-up reaction converts angiotensin I to angiotensin II. Among other effects, angiotensin II stimulates cells of the adrenal cortex, the outer portion of a gland perched on top of each kidney, to secrete aldosterone (Figure 10.11*b*). Aldosterone, recall, causes cells of the distal tubules and collecting ducts to reabsorb sodium faster, so less sodium is excreted in urine. On the other hand, when the extracellular fluid contains *too much* sodium, aldosterone secretion is inhibited. Less sodium is reabsorbed, and more is excreted.

In both instances, water "follows the salt." When less sodium is excreted, so is less water, and when more sodium is excreted, more water leaves the body as well.

For various reasons, sodium-regulating mechanisms do not operate properly in some people, and their bodies cannot fully dispose of excess sodium. Their tissues retain excess water, which leads to a rise in blood pressure. Abnormally high blood pressure (hypertension) can adversely affect the kidneys as well as the vascular system and brain (Chapter 7). Some hypertensive people can help control their blood pressure by restricting their intake of sodium chloride. This means limiting their intake of salt-laden processed foods and table salt, which is virtually 100 percent NaCl.

Salt–Water Balance and Thirst

The body does not rely entirely on events in the kidneys when it needs water. The same stimuli that lead to ADH secretion and increased reabsorption of water in the kidneys also stimulate thirst. Suppose you eat a box of salty popcorn at the movies. Soon the salt is absorbed into your bloodstream. The solute concentration in your extracellular fluid rises, and your hypothalamus detects the increase and prompts the secretion of ADH. In addition, a **thirst center** is stimulated. Signals from this cluster of nerve cells in the hypothalamus can inhibit the production of saliva. Your brain interprets the resulting mouth dryness as "thirst" and "tells" you to seek fluids.

In fact, a "cottony mouth" is one early sign that the body is becoming dehydrated. Dehydration is common after severe blood loss, burns, or diarrhea. It also results after profuse sweating, after the body has been deprived of water for a long time, or as a side effect of some medications. Then, thirst can become especially intense.

ADH enhances water reabsorption at distal tubules and collecting ducts when the body must conserve water. As a result, the urine is concentrated.

When excess water must be excreted, ADH secretion is inhibited. Less water is reabsorbed, and urine is dilute.

When the body must conserve sodium, aldosterone enhances its reabsorption at distal tubules and collecting ducts.

Under appropriate conditions, signals from a thirst center in the brain trigger the sensation of thirst.

ACID–BASE BALANCE

In addition to maintaining the volume and composition of extracellular fluid, the kidneys also help keep the extracellular fluid from becoming too acidic or too basic (alkaline). This overall **acid–base balance** is maintained through control over the concentrations of hydrogen ions (H^+) and other dissolved ions. Buffer systems, respiration, and urinary excretion provide the control.

Normal extracellular pH in the human body must be maintained between 7.37 and 7.43. As you know, acids lower the pH and bases raise it. A variety of acidic and basic substances enter the blood by absorption from the GI tract. They're also produced during normal metabolism. Typically, cell activities produce excess acids, which dissociate into H^+ and other fragments. This lowers the pH. The effect is minimized when excess hydrogen ions react with buffer molecules. One example is the **bicarbonate–carbon dioxide buffer system**:

$$H^+ + HCO_3^- \rightleftharpoons H_2CO_3 \rightleftharpoons H_2O + CO_2$$

bicarbonate   carbonic acid

In this case, the buffer system neutralizes excess H^+, and the carbon dioxide that forms during the reactions is exhaled from the lungs (Section 9.5). Like other buffer systems in the body, this one has only a temporary effect; it does not *eliminate* excess H^+. Only the urinary system can do that. In so doing it restores the buffer.

The reactions just described for the lungs proceed in reverse in the cells of the nephron tubule's walls, as diagrammed in Figure 10.12. The diagram may look a bit daunting, but overall the steps aren't difficult to understand. Bicarbonate (HCO_3^-) produced by the reverse reactions moves into interstitial fluid, and from there into peritubular capillaries. They deliver it into the general circulation, where it buffers excess acid. What happens to the H^+ formed in cells? It is secreted into the nephron. There, it may combine with bicarbonate ions in the filtrate (those ions can't cross the tubule wall) to form CO_2—which can be returned to the blood and exhaled. Or it may combine with phosphate ions or ammonia (NH_3) and be excreted in urine. In such ways, hydrogen ions are permanently removed from extracellular fluid.

Adjustments in acid–base balance and other kidney functions are essential to maintaining homeostasis. *Science Comes to Life* discusses health problems that develop when the kidneys cannot effectively add or remove solutes from blood, and describes some clinical methods for dealing with them.

Along with buffering systems and the respiratory system, kidneys help keep the extracellular fluid from becoming too acidic or too basic (alkaline).

Buffering systems prevent large fluctuations in pH. The urinary system eliminates excess hydrogen ions and also restores bicarbonate used up in buffering reactions.

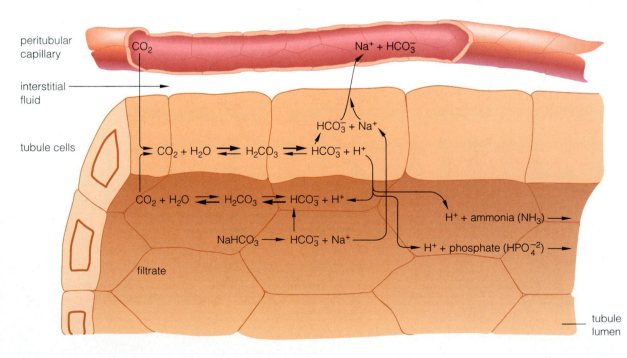

Figure 10.12 How the bicarbonate–carbon dioxide buffer system in the kidneys helps regulate pH.

10.9 COPING WITH KIDNEY DISORDERS

From the preceding sections, you can sense that good health depends on normal kidney function. Disorders or injuries that interfere with it can have mild to severe effects. As you've already read, for example, uric acid, calcium salts, and other wastes can settle out of urine and collect in the renal pelvis as kidney stones.

Nephritis is an inflammation of the kidneys. It can be caused by a range of factors, including bacterial infections. As you may recall from Chapter 8, an inflamed tissue tends to swell as fluid accumulates within it. However, because a kidney is "trapped" inside the tough renal capsule, it cannot increase in size. As a result, hydrostatic pressure builds up in or around glomerular capillaries, blocking them and hampering or preventing the passage of blood. Then, of course, blood filtering becomes difficult to impossible.

Glomerulonephritis is an umbrella term for a large number of disorders (often involving faulty immune responses) that can severely damage the kidneys. Hypertension and diabetes can disrupt blood circulation to and within the kidneys, sometimes virtually blocking the flow of blood through the glomeruli. An estimated 13 million people in the United States have kidneys in which the nephrons have been so damaged that the filtering of blood *and* formation of urine are seriously impaired. Control of the volume and composition of the extracellular fluid is disturbed, and toxic by-products of protein breakdown can accumulate in the bloodstream. Patients can suffer nausea, fatigue, and memory loss. In advanced cases, death may result. A kidney dialysis machine can restore the proper solute balances. Like the kidney itself, the machine helps maintain extracellular fluid by selectively removing and adding solutes to the patient's bloodstream.

"Dialysis" refers to the exchange of substances across a membrane between solutions of differing compositions. In *hemodialysis*, the dialysis machine is connected to an artery or a vein, and then blood is pumped through tubes made of a material similar to cellophane. The tubes are submerged in a warm-water bath. The precise mix of salts, glucose, and other substances in the bath sets up the correct gradients with the blood. Dialyzed blood is returned to the body (Figure 10.13).

Hemodialysis generally takes about four hours; blood must circulate repeatedly before solute concentrations in the body are improved. The procedure must be performed three times a week and is used as a temporary measure in patients with reversible kidney disorders. In chronic cases, the procedure must be used for the rest of the patient's life or until a functional kidney is transplanted.

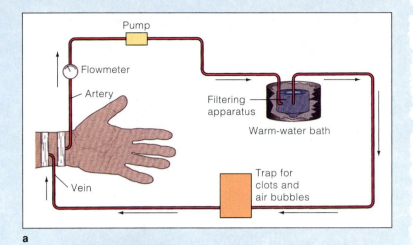

a

b

Figure 10.13 (**a**) Schematic drawing of steps in hemodialysis, one type of kidney dialysis. (**b**) Patient undergoing hemodialysis.

Hemodialysis invariably does some damage to the patient's red blood cells as they move through the tubing. As an alternative, in *peritoneal dialysis* two liters of fluid of the proper composition are put into the abdominal cavity, left in place for a period of time, and then drained out. Here, the lining of the cavity (the peritoneum) serves as the dialysis membrane. The procedure usually must take place several times every day.

Although chronic kidney disease can impose some inconveniences, with proper treatment and a controlled diet, many people are able to pursue a surprisingly active, close-to-normal lifestyle.

MAINTAINING THE BODY'S CORE TEMPERATURE

When you walk into a heated shop from a frigid winter street—or into a frigid air-conditioned shop from a sweltering summer street, you purposely change the temperature of the environment just outside your body. Such changes are a part of everyday life, and they trigger slight increases or decreases in the body's normal **core temperature**. "Core" refers to the temperature of the torso and head. Normal human core temperature is about 37°C (98.6°F).

Heat is an inevitable by-product of metabolic activity. (Even as you sit reading this book, you are producing roughly one kilocalorie of heat per hour per kilogram of your body weight.) If that heat were to accumulate internally, your core body temperature would steadily rise. Above 41°C (105.8°F), some protein enzymes become denatured and virtually shut down. Likewise, the rate of enzyme activity generally *decreases* by at least half when body temperature drops by 10°F. If it drops below 35°C (95°F), you are courting danger. Reduced functioning of enzymes causes the heart rate to fall, and heat-generating mechanisms such as shivering stop. At this core temperature a person's breathing slows, and he or she may lose consciousness. Below 80°F the heart may well stop beating entirely. Given these physiological facts, humans require mechanisms that help maintain core body temperature within narrow limits (Figure 10.14).

We humans are **endotherms**, which means "heat from within." Our body temperature is controlled mainly by metabolic activity and by homeostatic controls—negative feedback loops that adjust physiological responses for conserving or dissipating heat. We also can change our behavior in ways (such as changing clothes) that supplement the physiological controls.

Responses to Cold Stress

Table 10.4 summarizes the major responses to cold stress. They are are governed in the brain, by the hypothalamus—a centrally located structure that includes both neurons and endocrine cells (see Section 11.8). When the outside temperature drops, thermoreceptors (*thermo* means heat) at the body surface detect the decrease. When their signals reach the hypothalamus, neurons command smooth muscle in the walls of arterioles in the skin to contract. The resulting **peripheral vasoconstriction** reduces blood flow to capillaries near the body surface, so body heat is retained. This response to cold stress can be quite effective. For example, when your fingers or toes get cold, up to 99 percent of the blood that would otherwise flow to their skin is diverted.

In the **pilomotor response** to falling outside temperature, your body hair can "stand on end." This happens because smooth muscle controlling the erection of body hair is stimulated to contract. This creates a layer of still air close to the skin that reduces heat losses from the body. (This response is most effective in mammals with more body hair than humans!) Heat loss can be restricted even more by behaviors that reduce the amount of body surface exposed for heat exchange—as when you put on a sweater or hold your arms tightly against your body.

When other responses can't counter cold stress, the hypothalamus calls for increased activity in skeletal muscle, similar to the low-level contractions that produce muscle tone. The result? You start **shivering**. Rhythmic tremors begin as your skeletal muscles contract about ten to twenty times per second, and heat production throughout the body increases several times over.

Prolonged or severe exposure to cold can lead to a hormonal response that elevates the rate of metabolism in cells. This *nonshivering heat production* is especially notable in a specialized type of adipose tissue called "brown fat." Heat is generated as the lipid molecules are broken down. Babies (who are unable to shiver) have this tissue in the neck and armpits and near their kidneys; adults have little unless they are cold-adapted.

Hypothermia is a condition in which, despite all physiological defenses, the body core temperature falls below the normal range. A drop of only a few degrees leads to mental confusion; as described earlier, further cooling can lead to coma and death. Some victims of extreme hypothermia, mainly children, have survived prolonged immersion in ice-cold water. One reason is that mammals, including humans, have a **dive reflex**. When a mammal is submerged, the heart rate slows and blood is shunted to the brain and other vital organs.

Freezing often destroys tissues, a condition we call *frostbite*. Cells that have frozen may be salvageable if thawing is precisely controlled. This sometimes can be done in a hospital.

Table 10.4	Summary of Human Responses to Cold Stress and to Heat Stress	
Environmental Stimulus	Main Responses	Outcome
Drop in temperature	Vasoconstriction of blood vessels in skin; pilomotor response; behavior changes (e.g., putting on a sweater)	Heat is conserved
	Increased muscle activity; shivering; nonshivering heat production	More heat produced
Rise in temperature	Vasodilation of blood vessels in skin; sweating; changes in behavior; heavy breathing	Heat is dissipated from body
	Reduced muscle activity	Less heat produced

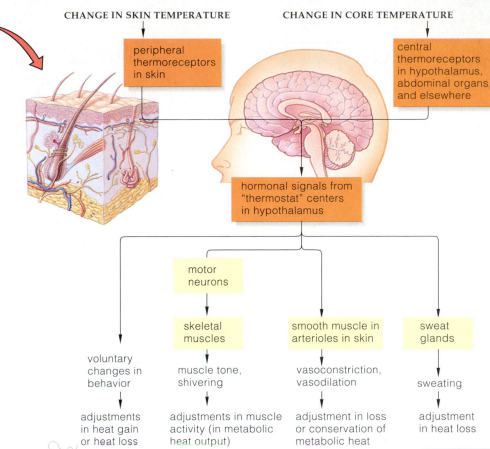

peripheral thermoreceptors in skin

central thermoreceptors in hypothalamus, abdominal organs, and elsewhere

hormonal signals from "thermostat" centers in hypothalamus

motor neurons

skeletal muscles

smooth muscle in arterioles in skin

sweat glands

voluntary changes in behavior

muscle tone, shivering

vasoconstriction, vasodilation

sweating

adjustments in heat gain or heat loss

adjustments in muscle activity (in metabolic heat output)

adjustment in loss or conservation of metabolic heat

adjustment in heat loss

Figure 10.14 Homeostatic controls over internal body temperature.

Responses to Heat Stress

Table 10.4 also summarizes the main responses to heat stress. When core temperature rises above a set point, the hypothalamus once again plays a central role in ordering responses. In **peripheral vasodilation**, hypothalamic signals cause blood vessels in the skin to dilate. More blood flows to the skin, where the excess heat the blood carries is dissipated.

Evaporative heat loss is another response that can be influenced by the hypothalamus, which can activate sweat glands. Your skin has 2.5 million or more sweat glands, and considerable heat is dissipated when the water they give up to the skin surface evaporates. With prolonged heavy sweating the body loses important salts—especially sodium chloride—as well as a great deal of water (Section 10.5). Such electrolyte losses may so alter the character of the internal environment that the affected person faints.

Sometimes peripheral blood flow and evaporative heat loss can't adequately counter heat stress. The result is *hyperthermia*, a condition in which the core temperature increases above normal. If the increase isn't too great, a person can suffer *heat exhaustion*, in which vasodilation and water losses from heavy sweating cause a drop in blood pressure. The skin feels cold and clammy, and the affected person may collapse.

When heat stress is great enough to completely break down the body's temperature controls, *heat stroke* occurs. Sweating stops, the skin becomes dry, and body temperature rapidly increases to a level that can be lethal.

Recall from Chapter 8 that when someone has a fever, the hypothalamus has reset the "thermostat" that dictates what the body's core temperature will be. The normal response mechanisms are brought into play, but they are carried out to maintain a higher temperature.

At the onset of fever, heat loss decreases, heat production increases, and the person feels chilled. When a fever "breaks," peripheral vasodilation and sweating increase as the body attempts to restore the normal core temperature; then, the person feels warm. The controlled increase in body temperature during a fever seems to enhance the body's immune response. For that reason, using fever-reducing drugs such as aspirin or ibuprofen may actually interfere with fever's beneficial effects. A severe fever can be quite dangerous, however, and requires medical supervision.

Physiological responses to cold stress include peripheral vasoconstriction, the pilomotor response, shivering, and sometimes nonshivering heat production.

Responses to heat stress include peripheral vasodilation and evaporative heat loss. The hypothalamus governs temperature regulation.

SUMMARY

Control of Extracellular Fluid

1. Inside the body, the cellular environment consists of certain types and amounts of substances dissolved in water. This extracellular fluid fills tissue spaces and (in the form of blood plasma) blood vessels. Its volume and composition are maintained only when the daily intake of water and solutes is in balance. The following processes maintain the balance:

   a. The body gains water by absorption from the GI tract and by metabolism. Water is lost by urinary excretion, evaporation from the lungs and skin, sweating, and elimination in feces.

   b. Solutes are gained by absorption from the GI tract, secretion, respiration, and metabolism. They are lost by excretion, respiration, and sweating.

   c. Losses of water and solutes are controlled mainly by adjusting the volume and composition of urine.

2. The human urinary system consists of two kidneys, two ureters, a urinary bladder, and a urethra.

3. Blood is filtered and urine forms in kidney nephrons. Each nephron interacts closely with two sets of blood capillaries (glomerular and peritubular).

   a. A nephron has a cup-shaped beginning (glomerular capsule), then three tubelike regions (proximal tubule, loop of Henle, and distal tubule, which empties into a collecting duct).

   b. The glomerular (Bowman's) capsule surrounds a set of highly permeable capillaries. Together, they are a blood-filtering unit, the glomerulus.

   c. Blood pressure forces water and small solutes out of the capillaries, into the cup. Most of the filtrate is reabsorbed by the tubules and returned to the blood. A portion is excreted as urine.

4. Urine forms in the nephron by three processes:

   a. Filtration of blood at the glomerulus of a nephron, which puts water and small solutes into the nephron.

   b. Reabsorption. Water and solutes to be retained leave the nephron's tubular parts and enter the peritubular capillaries that thread around them. A small volume of water and solutes remains in the nephron.

   c. Secretion. Some ions and a few other substances leave the peritubular capillaries and enter the nephron, for disposal in urine.

5. Many solutes are reabsorbed passively, following their concentration gradients back into the bloodstream. In other instances, active transport is required. Sodium is reabsorbed by active transport. Water reabsorption is always passive, occurring along its osmotic gradient.

6. Urine becomes more or less concentrated by the action of two hormones, ADH and aldosterone. These act on cells of distal tubules and collecting ducts as follows:

   a. ADH is secreted when the body must conserve water; it enhances reabsorption from the distal nephron tubule and collecting ducts. Inhibition of ADH allows more water to be excreted.

   b. Aldosterone conserves sodium by enhancing its reabsorption in the distal tubule. Inhibition of aldosterone allows more sodium to be excreted. Because "water follows salt," aldosterone indirectly influences water reabsorption.

7. Together with the respiratory system and other mechanisms, the kidneys also help maintain the body's overall acid–base balance.

Control of Body Temperature

1. Core body temperature is determined by the balance between metabolically produced heat and the heat absorbed from and lost to the environment.

2. Humans are endotherms. Their core body temperature is controlled largely by metabolic activity and by precise controls over heat produced and heat lost. Humans also can make behavioral responses that help maintain core body temperature within a normal range.

Review Questions

1. Label the component parts of this kidney and nephron: *10.3*

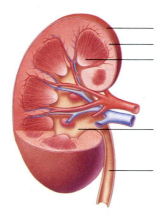 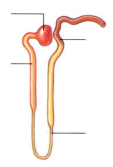

2. How does urine formation help maintain the body's internal environment? *10.1*

3. Define filtration, tubular reabsorption, and secretion. Explain how urine formation helps maintain the internal environment. *10.4*

4. Which hormone or hormones promote (a) water conservation, (b) sodium conservation, and (c) thirst behavior? *10.7*

5. Explain the role of the kidneys in helping to maintain the balance of acids and bases in the internal environment. *10.8*

6. Name and define the physical processes by which the human body gains and loses heat. What are the main physiological responses to cold stress and heat stress? *10.10*

Self-Quiz (Answers in Appendix V)

1. The body gains water by _____ .
   a. gastrointestinal absorption   c. a thirst mechanism
   b. metabolism   d. all of the above

2. The body loses water by way of the _____ .
   a. skin   d. urinary system
   b. lungs   e. c and d
   c. digestive system   f. a through d

3. Water and small solutes enter nephrons during _____ .
   a. filtration   c. secretion
   b. reabsorption   d. both a and b

4. Kidneys return water and small solutes to blood by _____ .
   a. filtration   c. secretion
   b. reabsorption   d. both a and c

5. A few substances move out of the peritubular capillaries and are moved into the nephron during _____ .
   a. filtration   c. secretion
   b. reabsorption   d. both a and c

6. A nephron's reabsorption mechanism depends on _____ .
   a. osmosis across the nephron wall
   b. active transport of sodium across the nephron wall
   c. a steep solute concentration gradient
   d. all of the above

7. _____ promotes water conservation.
   a. ADH   c. aldosterone
   b. renin   d. both b and c

8. _____ enhances sodium reabsorption.
   a. ADH   c. aldosterone
   b. renin   d. both b and c

9. Match the following salt–water balance concepts:
   ____ aldosterone   a. blood filter of a nephron
   ____ nephron   b. controls sodium reabsorption
   ____ thirst mechanism   c. occurs at nephron tubules
   ____ reabsorption   d. site of urine formation
   ____ glomerulus   e. controls water gain

Critical Thinking: You Decide (Key in Appendix VI)

1. A urinalysis reveals that the patient's urine contains glucose, hemoglobin, and white blood cells (pus). Are any of these substances abnormal in urine? Explain.

2. As a person ages, nephron tubules lose some of their ability to concentrate urine. What is the effect of this change?

3. In 1912, the ocean liner *Titanic* left Europe on her maiden voyage across the Atlantic to America. Late at night several days into the voyage, the vessel collided with an iceberg off the coast of Newfoundland. The *Titanic* began to sink (Figure 10.15), and there were not enough lifeboats to hold even half of the 2,200 passengers and crew. The surrounding sea was calm; its temperature was around 29°F. When rescue ships arrived on the scene several hours later, more than 1,500 bodies were recovered from the water. Many wore life jackets, and none had drowned. Probably they had died of _____ . If so, how did their blood flow, metabolism, and skeletal muscle action change prior to death?

Figure 10.15 Sinking of the *Titanic*, based on eyewitness accounts.

4. Fatty tissue holds the kidneys in place. Extremely rapid weight loss may cause this tissue to shrink so that the kidneys slip from their normal position. On rare occasions, the slippage may put a kink in one or both ureters and block urine flow. Speculate on what might then happen to the kidneys.

Selected Key Terms

ADH *10.7*
aldosterone *10.7*
core temperature *10.10*
distal tubule *10.3*
endotherm *10.10*
extracellular fluid *10.1*
filtration *10.4*
glomerular (Bowman's) capsule *10.3*
glomerulus *10.3*
juxtaglomerular apparatus *10.7*
kidney *10.3*
loop of Henle *10.3*
nephron *10.3*
peripheral vasoconstriction *10.10*

peripheral vasodilation *10.10*
peritubular capillary *10.3*
pilomotor response *10.10*
proximal tubule *10.5*
reabsorption *10.5*
renal corpuscle *10.3*
secretion *10.4*
thirst center *10.7*
ureter *10.3*
urethra *10.3*
urinary bladder *10.3*
urinary excretion *10.1*
urinary system *10.1*
urine *10.4*

Readings

Raloff, J. March 21, 1998. "Drugged Waters." Science News. A nontechnical article that looks at the problem of water pollution by powerful drugs used in health care.

Sherwood, L. 1997. *Human Physiology*. Third edition. Pacific Grove, California: Brooks/Cole.

Smith, H. 1961. *From Fish to Philosopher*. New York: Doubleday. Paperback. This entertaining, classic book traces the evolutionary path that produced the human kidney.

11

THE NERVOUS SYSTEM

Dr. Parkinson's Discovery

In 1817, a curious English physician named James Parkinson carefully recorded his observations of certain symptoms he had noticed in people navigating the streets of London. They walked slowly, taking short, shuffling steps. And their limbs trembled, sometimes violently. Dr. Parkinson did not understand the specific cause of these symptoms, but he did rightly suspect a serious problem in the body's ability to control its skeletal muscles—control that comes from the nervous system. Later generations of researchers took up the search for the cause of what came to be known as *Parkinson's disease*, or PD.

In Dr. Parkinson's day, people with PD probably paid a heavy social price for their disease. In our own time, some public figures with PD (Figure 11.1) have discussed their medical condition openly and lent their support to Parkinson's disease research. Their courage and candor have helped increase public awareness of a disease that has defied attempts to find a cure.

Today efforts to understand and treat PD are going strong. We know that neurons in two areas in the lower rear of the brain—the *substantia nigra* ("black substance") and the *locus ceruleus* ("blue place")—degenerate and die in PD, though we do not know just why. Healthy substantia nigra neurons secrete the neurotransmitter dopamine, and the locus ceruleus neurons secrete one called norepinephrine. As more of the neurons die, the body's supply of these communication chemicals dries up. As a result, other neurons in clusters called *basal nuclei* do not receive the chemical orders they require to operate properly. Slowly and surely, a range of body functions break down.

It takes years for the first PD symptoms to show up. There are four main ones—slow body movements (called *bradykinesia*), muscle tremors, rigidity at one or more major joints, and an inability to keep or regain one's balance when the body's position suddenly changes (as might happen when you go down steps). In people with advanced PD, the long-term deficit of dopamine and norepinephrine apparently damages neurons involved in thinking and related "higher" brain functions. About a third of PD patients develop the physical brain changes and symptoms of *Alzheimer's disease* (see Sections 11.3 and 11.4).

We tend to associate PD with adults over 50, but it can strike much earlier in life. In about 15 percent of cases heredity seems to play a role. Studies of such patients uncovered suspect changes (mutations) in two genes. Possibly, in PD cases that are not hereditary, an external factor could trigger similar changes. Again, however, the scientific sleuthing has been difficult. Genes provide instructions for building specific proteins, but no one has yet found out which proteins are coded by the "PD genes," or what their jobs in the nervous system might be.

What about treatments? In the 1960s physicians began prescribing a dopamine precursor called L-dopa (for levodopa). After the body converts it to dopamine, this drug can relieve major symptoms. However, it has worrisome long-term effects, including jerky body movements. Newer drugs, called dopamine agonists

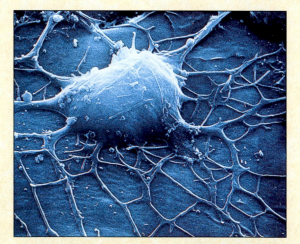

Figure 11.1 (**a**) Muhammad Ali during his prime as a champion athlete. After being diagnosed with Parkinson's disease, Mr. Ali established the Muhammad Ali Parkinson's Research Center in Phoenix, Arizona. (**b**) A motor neuron, one of the communication cells of the nervous system.

("movers"), do not have to be converted to dopamine and also can give significant symptom relief with fewer side effects. Some researchers believe that if treatment with a dopamine agonist starts early enough, the drug may even delay the progression of PD.

There are surgical treatments for PD, although none of them is a cure and all are expensive—and hence only available to a few. Some patients with advanced Parkinson's have an operation that destroys specific areas of brain tissue. This reduces many symptoms and also can relieve side effects from using L-dopa. In one experimental procedure, called "deep brain stimulation," a battery and one or more electrodes are implanted in the brain. A constant barrage of mild electric shocks from the electrode stimulates healthy neurons in ways that reduce uncontrolled muscle contractions and body movements.

At one time, biologists believed that once we reach adulthood, no new nerve cells could arise in the brain. Recent research has demonstrated that new nerve cells *can* form in the brain, although slowly. This discovery has stimulated new research on PD, aimed at finding ways to promote the formation of neurons to replace those PD has killed.

With this brief look at a disease that disrupts the nervous system, we have an introduction to just how important that system is in each moment of our daily lives. It governs nearly every body movement. Directly or indirectly, it is also responsible for our ability to perceive joy and pain, to read a book, and to remember what we've read. Constant, coordinated signals travel rapidly through its communication lines—**nerves** composed of the cells called neurons (Figure 11.1*b*). These signals control the muscles we use for walking, scratching an itch, and making facial expressions—not to mention food digestion, breathing, and many other vital functions that help maintain homeostasis.

The nervous system has two main parts. Our brain and spinal cord make up the *central nervous system* (CNS). Its task is to receive and integrate signals from inside and outside the body world and coordinate responses. Nerves make up the *peripheral nervous system*. Its task in the body is to carry messages between the central nervous system and other body regions. You may remember from Section 4.4 that the nervous system consists of two general types of cells. **Neurons** are specialized for communication, while **neuroglia** ("nerve glue") provide structural and functional support. The rest of this chapter fleshes out this general picture, beginning with the structure and functioning of neurons.

KEY CONCEPTS

1. Neurons are the basic units of communication in the nervous system. Collectively, they detect and integrate information about external and internal conditions. Then they act on muscles and glands in ways that produce appropriate responses.

2. The inside of a neuron has a negative electrical charge, relative to the outside. When a neuron is stimulated to action—that is, to send a message—this charge difference across the membrane may briefly and abruptly reverse. Such fleeting reversals are called action potentials. They are the means by which messages travel through the nervous system.

3. Action potentials travel along a neuron, but they cannot cross the small gaps *between* neurons or between neurons and some other kinds of cells. Chemical signals bridge the gaps (synapses) and stimulate or inhibit the adjoining neuron, muscle cell, or gland cell.

4. The flow of information through the nervous system depends on the moment-to-moment integration of signals that act upon neurons. Some signals stimulate (excite) receiving neurons, other signals inhibit them.

5. The nervous system includes the brain, spinal cord, and many nerves. The brain and spinal cord make up the central nervous system; paired nerves that thread through the rest of the body make up the peripheral nervous system.

6. Nerves of the peripheral nervous system are divided into somatic nerves, which service skeletal muscles, and autonomic nerves, which service internal organs.

7. The brain is complex in both structure and function. In its three main regions—the forebrain, the midbrain, and the hindbrain—are centers that receive, integrate, store, and respond to information.

CHAPTER AT A GLANCE

There are three classes of neurons in the nervous system. **Sensory neurons** respond to a specific type of stimulus (light, pressure, or another form of energy). A sensory neuron then relays information about the stimulus to the spinal cord and brain. There, **interneurons** receive the sensory input, integrate it with other information, then send signals that influence the activity of other neurons. **Motor neurons** relay information *away from* the brain and spinal cord to muscles or glands—the body's effectors, which carry out responses.

Neurons are less than half of the volume of your nervous system, and at most 20 percent of its cells. The rest are **neuroglia**, or "glia." Glia physically support and protect neurons; they also help maintain proper concentrations of vital ions in the fluid around neurons. The glia called astrocytes (see Figure 4.4) give structure to the brain, as connective tissues do elsewhere in the body. Other glia separate groups of neurons; still others, called *Schwann cells*, insulate the long extensions (axons) of sensory and motor neurons, so they can conduct signals rapidly. In the CNS *microglia* are macrophages that engulf and dispose of dead cells and microbes.

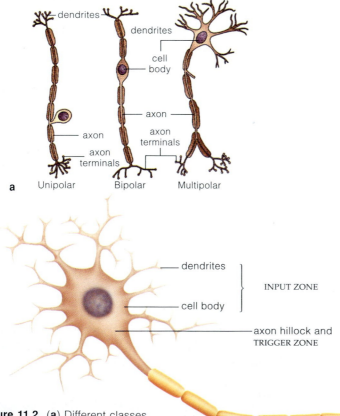

a Unipolar   Bipolar   Multipolar

dendrites
dendrites
cell body
axon
axon
axon terminals
axon terminals

b

dendrites
INPUT ZONE
cell body
axon hillock and TRIGGER ZONE

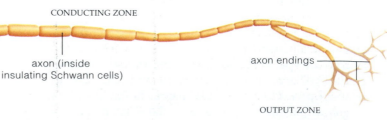

CONDUCTING ZONE

axon (inside insulating Schwann cells)

axon endings

OUTPUT ZONE

Figure 11.2 (**a**) Different classes of neurons based on the number of their cytoplasmic extensions. Unipolar cells have one axon and branching dendrites. Bipolar cells have one dendrite and one axon. Multipolar cells, like the motor neuron shown in the chapter introduction, have one axon and many dendrites. (**b**) A motor neuron.

Parts of a Neuron and Their Functions

A neuron is an *excitable* cell—it can respond to certain stimuli by producing electrical signals. It is by way of such signals that neurons perform their vital function of communication. That function is mirrored in neuron structure. Every neuron has a cell body, where its nucleus and various organelles are. Slender extensions of its cytoplasm project out from the cell body. There are two kinds, called **axons** and **dendrites**. Typically, the cell body and dendrites are "input zones" where a neuron receives incoming signals. As you can see in Figure 11.2, dendrites tend to be shorter than axons and their number and length vary, depending on the type of neuron. The axon is the neuron's "conducting zone," carrying outgoing signals. Except in sensory neurons, new signals are triggered at a plump patch of plasma membrane where the cell body and axon meet. In motor neurons and interneurons this "trigger zone" is called the *axon hillock* ("little hill"). The branched endings of an axon are "output zones," where messages are sent to other cells.

We will look closely at the signal-sending function of neurons in Section 11.2. For the moment, our focus is on the electrical and chemical conditions that set the stage for a signal to come about.

Ready for Action: A Neuron at Rest

When a neuron is resting there is a steady difference in electric charge—in *voltage*—across its plasma membrane. The cytoplasmic fluid next to the membrane is negatively charged, compared to the interstitial fluid right outside the membrane. We measure the charges in units called millivolts. For many neurons, the steady charge difference across the plasma membrane is about –70 millivolts. (The minus sign indicates that the cytoplasm side of the membrane is more negative.) This charge difference is called the **resting membrane potential**. Its name indicates that it has the potential to do physiological work. As described shortly, various kinds of signals occur in the nervous system. If a strong enough signal arrives at a patch of membrane in a resting neuron's input zone, it may cause the voltage difference across the plasma membrane in the trigger zone to *reverse*, just for an instant. That reversal, called an **action potential**, is a neuron's communication signal, or a "nerve impulse."

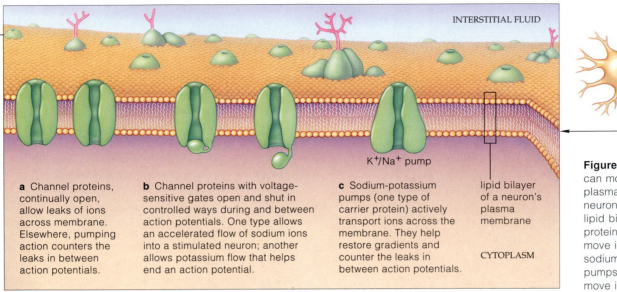

INTERSTITIAL FLUID

K⁺/Na⁺ pump

CYTOPLASM

a Channel proteins, continually open, allow leaks of ions across membrane. Elsewhere, pumping action counters the leaks in between action potentials.

b Channel proteins with voltage-sensitive gates open and shut in controlled ways during and between action potentials. One type allows an accelerated flow of sodium ions into a stimulated neuron; another allows potassium flow that helps end an action potential.

c Sodium-potassium pumps (one type of carrier protein) actively transport ions across the membrane. They help restore gradients and counter the leaks in between action potentials.

lipid bilayer of a neuron's plasma membrane

Figure 11.3 How ions can move across the plasma membrane of a neuron. Spanning the lipid bilayer are channel proteins that passively move ions across and sodium-potassium pumps that actively move ions across.

trigger zone

Restoring the Potential for Action

When an incoming signal reverses the charge difference across a patch of membrane in a neuron's trigger zone, that bit of membrane can't receive another signal until its resting membrane potential is restored. Two factors aid the restoration process. First, recall from Chapter 3 that the membrane's lipid bilayer prevents charged substances—including potassium ions (K^+) and sodium ions (Na^+) from passing through it. Thus differences in ion concentrations can build up across the membrane. However, ions can flow from one side to the other in controlled ways, through the interior of channel proteins that span the bilayer (Figure 11.3). Some channels stay open, so that ions can diffuse ("leak") through them all the time. Other channels have "gates"; they open only when the neuron is adequately stimulated.

In a resting motor neuron the gated sodium channels are shut. The neuron's plasma membrane also is much more permeable to K^+ than it is to Na^+. Hence, each ion has a concentration gradient across the membrane:

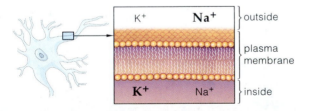

K⁺   **Na⁺** } outside

plasma membrane

K⁺   Na⁺ } inside

The concentration gradients determine the direction in which Na^+ and K^+ tend to diffuse across the membrane (from the larger letter toward the smaller letter) through the interior of channel proteins. Sodium tends to diffuse in, and potassium tends to diffuse out. There also is an *electric* gradient across the neuron's plasma membrane. That is, the inside of the cell is a bit more negative than the outside, partly because the cytoplasm contains many negatively charged proteins. Also, K^+ can readily diffuse out of the neuron, down its concentration gradient.

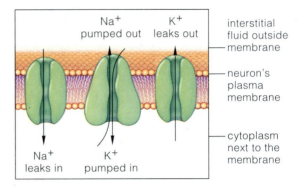

Na⁺ pumped out   K⁺ leaks out   — interstitial fluid outside membrane

— neuron's plasma membrane

Na⁺ leaks in   K⁺ pumped in   — cytoplasm next to the membrane

Figure 11.4 Pumping and leaking processes that maintain the distribution of sodium and potassium ions across a resting neuron's plasma membrane. The inward and outward movements for each kind of ion are balanced.

Together, these factors mean that in a resting motor neuron, some Na^+ is constantly leaking *into* the cell, down its electrochemical gradient, and K^+ is leaking *out* down *its* concentration gradient.

A neuron can't respond to an incoming signal unless the concentration and electric gradients across its plasma membrane are in place. Yet, as you've just read, the Na^+ and K^+ leaks never stop—and one might think that the net amount of K^+ in the cell would continue to fall while the amount of Na^+ would slowly and surely increase. This potentially disastrous imbalance doesn't develop because a resting neuron expends energy on an active transport mechanism that maintains the gradients (Figure 11.4). Spanning the membrane are carrier proteins called **sodium-potassium pumps**. With energy from the cell's supply of ATP, they actively transport potassium *in* and sodium *out*.

A resting neuron maintains a resting membrane potential—a difference in electric charge across the plasma membrane. An action potential (nerve impulse) is a brief, abrupt reversal of the charge difference. Ion pumps restore and maintain the resting membrane potential.

A CLOSER LOOK AT ACTION POTENTIALS

As you've just read, an action potential is a brief, sudden reversal of the "resting" ion balance across a neuron's plasma membrane. It only results, however, if an incoming stimulus is strong enough. For instance, suppose you gently stroke your arm, putting a bit of pressure on the skin. In tissues beneath the skin surface are receptor endings—input zones of sensory neurons. Like a pillow that "pooches in" where you lay your head on it, patches of plasma membrane around the receptor endings deform under the pressure of your fingertips. This causes certain types of ion channels to open. Some ions flow across, so the voltage difference across the membrane changes. In this case, although you may barely touch your arm, the fact that you can feel it means that action potentials must have been triggered.

In physiological terms, the pressure of your touch produced a graded, local signal at the input zones of sensory neurons. *Graded* means that signals at an input zone can vary in magnitude. They can be large or small, depending on the intensity of the stimulus (in this instance, the touch pressure). *Local* means the signals don't spread far from the point of stimulation.

When a stimulus is intense enough or long-lasting, graded signals can spread out of the neuron's input zone and into a trigger zone. There, if the voltage difference across the plasma membrane changes by a certain minimum amount, an action potential will result. The minimum voltage change needed to produce an action potential is the **threshold level of stimulation**. It can be reached at any patch of membrane that has voltage-sensitive, gated channels for sodium ions.

An appropriate stimulus causes sodium ions to flow into the neuron (Figure 11.5). As these positively charged ions flow in, the cytoplasm side of the membrane becomes less negative. Then, more gates open, more sodium enters, and so on. The "snowballing" influx of sodium is an example of positive feedback, in which an event intensifies as a result of its own occurrence:

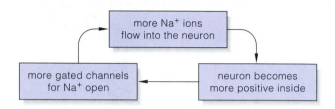

When the threshold level is reached, the opening of more sodium gates no longer depends on the strength of the original stimulus. Because the positive-feedback cycle is under way, the inward-rushing sodium itself is enough to cause more sodium gates to open.

An All-or-Nothing Spike

Figure 11.6 shows a recording of the voltage difference across a neuron's plasma membrane before, during, and after an action potential. Notice how the membrane potential spikes once threshold is reached. Every single action potential in the neuron spikes to the same level above threshold as an *all-or-nothing* event. That is, once the positive-feedback cycle starts, nothing will stop the full spiking. If threshold is not reached, the disturbance to the plasma membrane will subside just as soon as the stimulus is removed (Figure 11.6*b*).

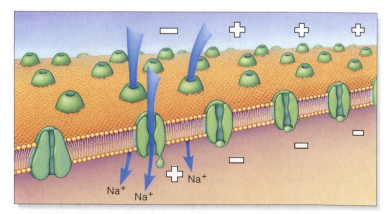

a Membrane at rest (inside negative with respect to the outside). An electrical disturbance (*red* arrow) spreads from an input zone to an adjacent trigger region of the membrane, which has many gated sodium channels.

b A strong disturbance initiates an action potential. Sodium gates open, the sodium inflow decreases the negativity inside; this causes more gates to open, and so on, until threshold is reached and the voltage difference across the membrane reverses.

Figure 11.5 Propagation of an action potential along the axon of a motor neuron.

Each spike lasts for about a millisecond. At the place on the membrane where the charge reversed, the gated sodium channels close and the influx of sodium stops. About halfway through the action potential, potassium channels open, so potassium ions flow out and restore the original voltage difference across the membrane. And sodium–potassium pumps restored ion gradients, as you've read. Later, after the resting membrane potential has been restored, most potassium gates are closed and sodium gates are in their initial state, ready to be opened again when a suitable stimulus arrives.

Propagation of Action Potentials

The changes in membrane potential that lead up to each action potential are self-propagating—that is, they spread by themselves—and they don't lose strength. When the change spreads from one membrane patch to another, roughly the same number of gated channels open. For a brief period after an action potential, a membrane patch can't be restimulated because there is a time lag before its sodium gates can open again. That is why action potentials do not spread back into the trigger zone, but propagate *away* from it.

The input zone of a neuron receives graded, local signals. If the signals are long-lasting or intense enough to reach the threshold level of stimulation, they can trigger an action potential.

An action potential is an all-or-nothing event, and it always propagates away from a neuron's trigger zone.

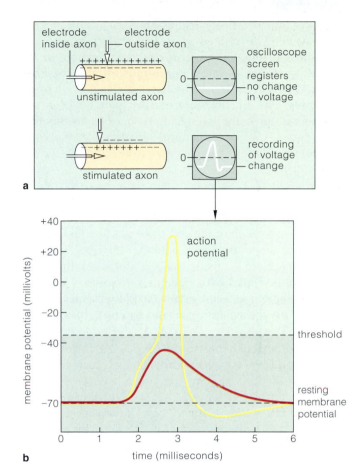

b

Figure 11.6 Action potentials. (**a**) Electrodes placed inside and outside an axon can be used to measure voltage across the membrane and changes when the axon is stimulated. Changes show up as deflections in a beam of light on an oscilloscope screen. The solid *white* line is a recording of an action potential. (**b**) The *yellow* line is a typical waveform for an action potential. The *red* line represents a local signal that did not reach the threshold level, so no spiking occurred.

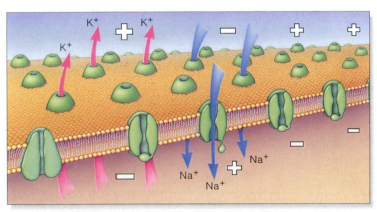

c The reversal causes sodium gates to shut and potassium gates to open (at *pink* arrows). Potassium follows its gradient (out of the neuron). Voltage is restored. The disturbance produced by the action potential triggers an action potential at the adjacent membrane site, and so on, away from the point of stimulation.

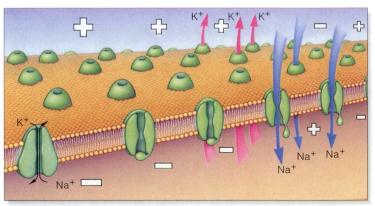

d The inside of the membrane becomes negative again following each action potential, but the sodium and potassium concentration gradients are not yet fully restored. Active transport at sodium–potassium pumps restores the gradients.

CHEMICAL SYNAPSES

When action potentials reach a neuron's output zone, they usually do not go any farther. But their arrival may prompt the neuron to release one or more of the signaling molecules called **neurotransmitters**. These are substances that diffuse across junctions called **chemical synapses**. The junctions are narrow gaps between one neuron's output zone and the input zone of a neighboring cell (Figure 11.7). Some chemical synapses occur between two neurons, others between a neuron and a muscle cell or gland cell.

At a chemical synapse, one of the two cells stores neurotransmitter molecules in synaptic vesicles in its cytoplasm. This is the *pre*synaptic cell. Gated channels for calcium ions span the cell's plasma membrane, and they open when an action potential arrives. There are more calcium ions outside the cell, and when they flow in (down their gradient), synaptic vesicles fuse with the plasma membrane. Then, neurotransmitter molecules are released into the synaptic cleft and diffuse across it. They

bind with specific receptor proteins on the membrane of the *post*synaptic cell. Binding changes the shape of these proteins, so that a channel opens up through their interior. Ions then cross the plasma membrane by diffusing through the channels (Figure 11.7c).

How a postsynaptic cell responds depends on the type and concentration of neurotransmitter molecules in the cleft, what kinds of receptors the cell bears, and the number and responsiveness of gated channels in its membrane. Such factors help determine whether a neurotransmitter will have an *excitatory* effect and help drive the postsynaptic cell's membrane toward the threshold of an action potential. Or they may influence whether it has an *inhibitory* effect, driving the membrane away from the threshold.

A Smorgasbord of Signals

Consider the neurotransmitter **acetylcholine** (ACh). Depending on the circumstances, ACh can excite *or* inhibit responses in cells of muscles, glands, the brain, and the spinal cord. Figure 11.8 shows a chemical synapse between a motor neuron and a muscle cell. ACh released from the motor neuron diffuses across the gap and binds to receptors on the muscle cell membrane. It has excitatory effects on this kind of cell, triggering action potentials that cause muscle contraction. ACh receptors at some neuromuscular junctions are destroyed in the disease *myasthenia gravis*. Typical symptoms are drooping eyelids, muscle weakness, and fatigue.

Serotonin is another neurotransmitter. It acts on brain cells that govern sleeping, sensory perception, regulation of body temperature, and emotional states. *Norepinephrine* affects brain regions concerned with emotional states, dreaming, and awaking. Other neurotransmitters that act on different parts of the brain are *dopamine* and *GABA* (gamma aminobutyric acid). When they are stimulated in the necessary way, sensory neurons release *substance P*, a neurotransmitter that conveys information about pain.

Two debilitating degenerative diseases underscore how important neurotransmitters are to normal life. The chapter introduction described Parkinson's disease, with its progressive, severe loss of voluntary muscle control due to a lack of dopamine in certain parts of the brain. A loss or degeneration of neurons that manufacture and release acetylcholine may play a key role in *Alzheimer's disease* (AD), with its attacks on personality and intellect.

Signaling molecules known as **neuromodulators** can magnify or reduce the effects of a neurotransmitter

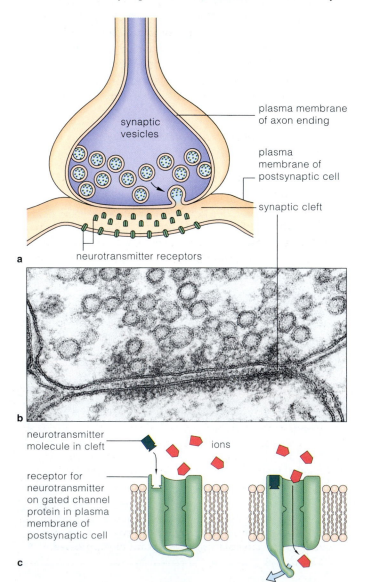

a

b

c

neurotransmitter molecule in cleft

ions

receptor for neurotransmitter on gated channel protein in plasma membrane of postsynaptic cell

synaptic vesicles

plasma membrane of axon ending

plasma membrane of postsynaptic cell

synaptic cleft

neurotransmitter receptors

Figure 11.7 Chemical synapses. (**a**) A chemical synapse between two neurons. (**b,c**) A neurotransmitter carries signals from the presynaptic cell to the postsynaptic cell.

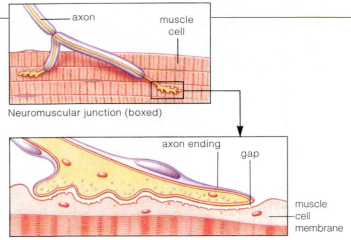

Neuromuscular junction (boxed)

Motor end plate (troughs in muscle cell membrane)

Figure 11.8 Example of a neuromuscular junction. The axon's myelin sheath stops at the junction, so the membranes of the two cells are exposed. There are troughs in the muscle cell membrane where the axon endings are positioned.

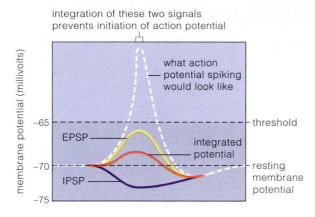

Figure 11.9 Synaptic integration. The *yellow* line shows how an EPSP of a certain magnitude would register on an oscilloscope screen *if it were acting alone*. The *purple* line shows the effect for a single IPSP. The *red* line shows their effect on a postsynaptic cell membrane when they arrive at the same time. When these two signals are integrated (the *red* line), threshold is not reached, so no action potential is initiated in the target cell. ⊙

on neighboring or distant neurons. Neuromodulators include *endorphins*—natural painkillers that inhibit the release of substance P from nerves. Athletes who push themselves beyond normal fatigue can experience a euphoric "high" as endorphin release increases. Endorphins also may have roles in memory, learning, and some mental disorders, and in control of body temperature and sexual behavior.

Synaptic Integration

Between 1,000 and 10,000 communication lines form synapses with a typical neuron in your brain. And your brain contains at least *100 billion* neurons, humming with messages about doing what it takes to be a human.

At any moment, many excitatory and inhibitory signals are washing over the input zones of a postsynaptic cell. Some signals drive its membrane voltage closer to threshold; others maintain the resting level or drive it away from threshold. In effect, *the signals are competing for control of the membrane potential at the trigger zone.*

All synaptic signals are graded potentials. The ones called EPSPs (for excitatory postsynaptic potentials) have a *depolarizing* effect. This simply means they bring the membrane closer to threshold. IPSPs (inhibitory postsynaptic potentials) may have a *hyperpolarizing* effect (driving the membrane away from threshold) or may help maintain the membrane at its resting level.

In the process of **synaptic integration**, competing signals that reach an input zone of a neuron at the same time are summed. Summation is the means by which signals arriving at a neuron are reinforced, suppressed, or sent onward to other cells in the body. Figure 11.9 shows what recordings of an EPSP, an IPSP, and their summation might look like.

Integration involves both spatial and temporal summation. *Spatial summation* occurs when neurotransmitter molecules from several presynaptic cells reach a neuron's input zone at the same time. On the other hand, if the cell is excited by a rapid series of action potentials so that a neurotransmitter is released repeatedly, over a short time period, the response is a *temporal summation*.

Removing Neurotransmitter Molecules from the Synaptic Cleft

The flow of signals through the nervous system depends on the rapid, controlled removal of neurotransmitter molecules from synapses. Some neurotransmitter molecules diffuse out of the gap. Enzymes cleave others right in the synapse, as when acetylcholinesterase breaks down ACh. Also, membrane transport proteins actively pump the neurotransmitter molecules back into presynaptic cells or into neighboring neuroglia.

Some drugs block the reuptake of certain neurotransmitters. For example, cocaine blocks the uptake of dopamine, which then lingers in synapses and keeps on stimulating target cells—with disastrous effects described in Section 11.13. Some antidepressant drugs (such as Prozac) alter a person's mood by blocking the reuptake of serotonin.

Neurotransmitters cross the synapse between two neurons or between a neuron and a muscle cell or gland cell. These signaling molecules may excite or inhibit activity of different kinds of receiving cells.

In synaptic integration, excitatory and inhibitory signals acting on a postsynaptic cell are combined (summed). In this way messages traveling through the nervous system can be reinforced or downplayed, sent onward or suppressed.

INFORMATION PATHWAYS

Synaptic integration allows signals arriving at any neuron in the body to be reinforced or dampened, sent on or suppressed. However, the *direction* in which a given signal will travel depends on the organization of neurons in different body regions.

For example, many of the billions of neurons in your brain are grouped in blocks. Regional blocks of hundreds or thousands of neurons receive excitatory and inhibitory signals. They integrate signals entering the block, then send out new ones in response. Some regions have neurons organized as *divergent* circuits, with neuron processes fanning out from one block to form connections with others. Some have neurons arranged in *convergent* circuits, with signals from many neurons funneled to just a few. In still other regions, neurons synapse back on themselves, repeating signals among themselves over and over like echoes. These *reverberating* circuits include the ones that make your eye muscles rhythmically twitch as you sleep.

In each cablelike **nerve**, long axons of many sensory neurons, motor neurons, or both, permit long-distance communication between the brain or spinal cord and the rest of the body. Connective tissue bundles most of the axons in parallel (Figure 11.10). Each axon has a **myelin sheath**, which speeds the rate at which action potentials propagate. The sheath consists of the glia called **Schwann cells,** wrapped around the long axons like jelly rolls. An exposed node, or gap, separates each cell from the next one. There, voltage-sensitive, gated sodium channels pepper the plasma membrane (Figure 11.11).

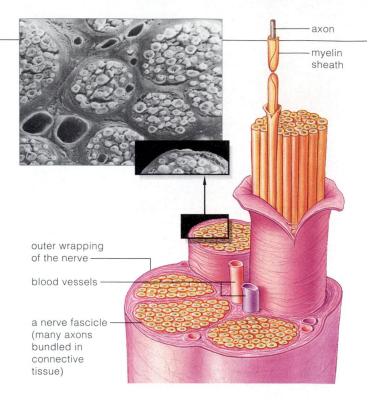

Figure 11.10 Structure of a nerve. Axons in the nerve are bundled together inside wrappings of connective tissue.

In a manner of speaking, action potentials jump from node to node. (Hence biologists sometimes use the term saltatory conduction, after a Latin word meaning "to jump.") The sheathed regions between nodes hamper the movement of ions across the plasma membrane, so ion disturbances tend to flow along the membrane until the next node in line. At each node, the flow of ions can produce a new action potential. In large sheathed axons, action potentials propagate at a remarkable 120 meters (nearly 400 feet) per second.

There are no Schwann cells in the central nervous system. There, processes from glia called oligodendrocytes form the sheaths of myelinated axons. In *multiple sclerosis*, myelin sheaths around axons in the spinal cord slowly degenerate. A mutant gene may predispose a person to the disease, but the actual trigger may be a viral infection. Symptoms include serious progressive weakening of muscles, fatigue, and numbness.

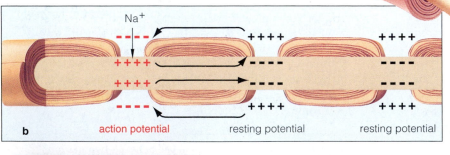

a "Jellyrolled" Schwann cells of an axon's myelin sheath

Figure 11.11 Propagation of an action potential along a motor neuron having a myelin sheath. (**a**) The sheath is a series of Schwann cells, each wrapped like a jelly roll around the axon. Each "jelly roll" blocks ion movements across the membrane. But ions can cross it at nodes between Schwann cells. (**b**) The unsheathed nodes have dense arrays of gated sodium channels. When an action potential propagating down the axon reaches a node, sodium gates open, Na+ rushes in, and another action potential results. This new ion disturbance spreads rapidly to the next node and triggers another action potential, and so on down the line. (**c**) An axon wrapped in a myelin sheath (cross section).

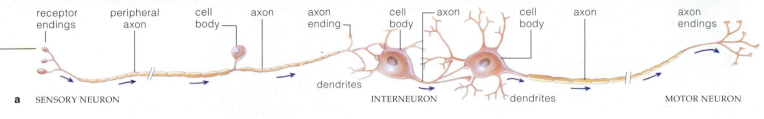

receptor endings — peripheral axon — cell body — axon — axon ending — cell body — axon — cell body — axon — axon endings

dendrites

dendrites

a SENSORY NEURON   INTERNEURON   MOTOR NEURON

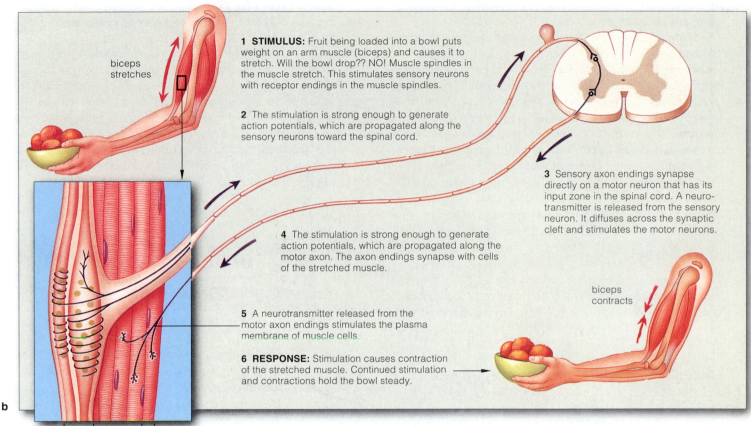

biceps stretches

1 STIMULUS: Fruit being loaded into a bowl puts weight on an arm muscle (biceps) and causes it to stretch. Will the bowl drop?? NO! Muscle spindles in the muscle stretch. This stimulates sensory neurons with receptor endings in the muscle spindles.

2 The stimulation is strong enough to generate action potentials, which are propagated along the sensory neurons toward the spinal cord.

3 Sensory axon endings synapse directly on a motor neuron that has its input zone in the spinal cord. A neurotransmitter is released from the sensory neuron. It diffuses across the synaptic cleft and stimulates the motor neurons.

4 The stimulation is strong enough to generate action potentials, which are propagated along the motor axon. The axon endings synapse with cells of the stretched muscle.

biceps contracts

5 A neurotransmitter released from the motor axon endings stimulates the plasma membrane of muscle cells.

6 RESPONSE: Stimulation causes contraction of the stretched muscle. Continued stimulation and contractions hold the bowl steady.

b

muscle spindle   muscle cell

Figure 11.12 (**a**) General direction of information flow in the nervous system. Sensory nerves relay information *into* the spinal cord and brain, where their neurons synapse with interneurons. Interneurons in the spinal cord and brain integrate the signals. Many synapse with motor neurons, which carry signals *away* from the spinal cord and brain. (**b**) Organization of nerves in a reflex arc that deals with muscle stretching. In a skeletal muscle, stretch-sensitive receptors of a sensory neuron are located in muscle spindles. The stretching generates action potentials, which reach axon endings in the spinal cord. These synapse with a motor neuron that carries signals to contract, from the spinal cord back to the stretching muscle.

Reflex Arcs

Figure 11.12 is a specific example of information flow through the nervous system. It shows how sensory and motor neurons of certain nerves take part in a path known as the stretch reflex. A **reflex** is a simple, stereotyped movement (one that is always the same) in response to a stimulus. In the simplest **reflex arcs**, sensory neurons synapse directly on motor neurons.

The stretch reflex works to contract a muscle after gravity or some other load has caused the muscle to stretch. Suppose you hold out a large bowl and keep it stationary as someone puts peaches into it. The peaches add weight to the bowl, and when your hand starts to drop, a muscle in your arm (the biceps) is stretched.

In the muscle, stretching activates receptor endings that are a part of muscle spindles—sensory organs in which specialized cells are enclosed in a sheath that runs parallel with the muscle. The receptor endings are the input zones of sensory neurons whose axons synapse with motor neurons in the spinal cord (Figure 11.12). Axons of the motor neurons lead back to the stretched

muscle. Action potentials that reach the axon endings trigger the release of ACh, which initiates contraction. As long as receptor activity continues, the motor neurons are excited even more, and this allows them to maintain your hand's position.

In the vast majority of reflex pathways, the sensory neurons also interact with a number of interneurons, which then activate or suppress all the motor neurons necessary for a coordinated response.

Interneurons are organized in information-processing blocks. Cablelike nerves that have long axons of sensory neurons, motor neurons, or both, connect the brain and spinal cord with the rest of the body.

Reflex arcs, in which sensory neurons synapse directly on motor neurons, are the simplest paths of information flow.

THE NERVOUS SYSTEM: AN OVERVIEW

Thus far our discussion has focused on the signals that travel through the nervous system. Figure 11.13 now gives you an overview of how the system is organized into its two major divisions.

Humans have the most intricately wired nervous system in the animal world. Investigators typically approach its complexity by dividing it into two main regions, according to function. The brain and spinal cord make up the **central nervous system** (CNS). All of the nervous system's interneurons are in the CNS. The **peripheral nervous system** (PNS) consists mainly of nerves that thread through the rest of the body and carry signals into and out of the central nervous system.

As you can see in Figure 11.13a, the peripheral nervous system is further subdivided into *somatic* and *autonomic* subdivisions, and the autonomic nerves are subdivided yet again. The roles of those nerves are one topic in the next section.

The peripheral nervous system consists of thirty-one pairs of spinal nerves that connect with the spinal cord, and twelve pairs of cranial nerves that connect directly with the brain (Figure 11.13b). At some places in the peripheral nervous system, cell bodies of several neurons occur in clusters called *ganglia* (singular: ganglion). Both the CNS and the PNS also have glia, such as the oligodendrocytes (CNS) and Schwann cells (PNS) mentioned in previous sections.

Figure 11.13 (**a**) Divisions of the nervous system. (**b**) View of the nervous system showing the brain, spinal cord, and some major peripheral nerves. Twelve pairs of cranial nerves extend from different regions of the brain stem.

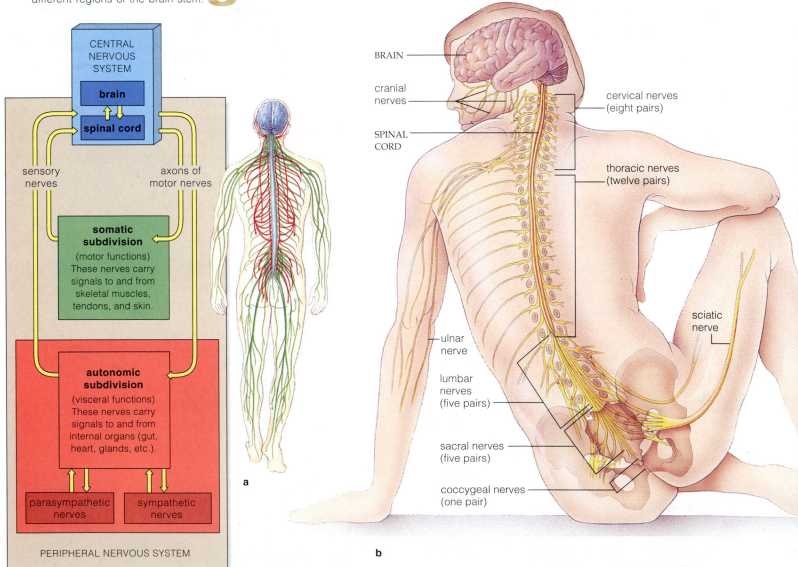

Table 11.1 The Nervous System under Attack

Cause/Risk Factors	Major Effects
Meningitis Viral or bacterial infection of meninges; often fatal	Inflammation of the membrane covering the brain and/or spinal cord. Symptoms include headache, stiff neck, vomiting.
Encephalitis Usually, infection by a virus, such as herpes simplex; also, HIV infection	Inflammation of the brain. Common symptoms are fever, confusion, seizures.
Epilepsy Brain injury or infection, drug overdose, metabolic imbalance. Often, no specific cause can be identified	Abnormal electrical activity that temporarily alters one or more brain functions. Attacks are sometimes triggered by fatigue, stress, or flashing lights. Symptoms include jerking body movements, brief (petit mal) or longer (grand mal) loss of consciousness.
Multiple sclerosis Cause unknown. Possibly an autoimmune disease with viral origins. Some people may have genetic susceptibility	Progressive destruction of myelin sheaths of neurons in brain and spinal cord, causing increasing muscle weakness, fatigue, and other symptoms.
Alzheimer's disease Genetic origin in some cases. Victims have lower than normal levels of acetylcholine in brain tissues	Progressive degeneration of neurons and shrinking of the brain. Abnormal buildup of masses of amyloid protein, leading to progressive loss of memory and intellectual functions.
Concussion Violent blow to the head or neck	Disruption of electrical activity of brain neurons. Symptoms include brief unconsciousness, blurred vision.
Neuralgia Physical injury or viral infection; specific cause often unknown	Irritation of or damage to a nerve leading to intermittent bouts of severe pain, especially in the face and head.

Throughout life, our remarkable nervous system integrates the array of body functions in ways that help maintain homeostasis. Operations of the CNS also give us much of our "humanness," in ways later sections will discuss. Table 11.1 gives a short list of common disorders that can hamper these activities or bring them to a halt.

The nervous system is divided into the central nervous system and the peripheral nervous system. The CNS consists of the brain and spinal cord, and the PNS consists of nerves that thread through the rest of the body and carry signals to and from the central region.

AN ENVIRONMENTAL ASSAULT ON THE NERVOUS SYSTEM

Have you ever heard someone called "mad as a hatter"? The phrase originated a century ago in Great Britain, when hat makers used the heavy metal mercury to process felt. Traces of that substance entered the bloodstream, crossed the blood-brain barrier, and caused irreparable neurological damage. In an adult, symptoms of mercury poisoning include muscle tremors, impaired vision, difficulty walking and talking, and emotional problems. Victims were often believed to be mentally deranged.

Industrial mercury compounds may be dumped or washed from the atmosphere into waterways, lakes, and marshlands. There, metabolic processes in bacteria convert such compounds to methyl mercury, which dissolves in fatty tissues in fish and other animals. Consumed in food (especially fish) by a pregnant woman, methyl mercury becomes concentrated in adipose (fat) cells and breast milk—and also wreaks havoc on the developing fetus. In Minimata, Japan, in the 1960s, more than 200 women living near a mercury-discharging factory gave birth to mentally retarded, physically deformed babies. In the late 1980s, authorities in Michigan, Minnesota, and Wisconsin warned nursing mothers and women who were even *considering* pregnancy not to eat *any* fish from many lakes in the region. Those advisories are still in force. In fact, the U.S. Environmental Protection Agency estimates that mercury contamination in fish is rising by 3 to 5 percent per year. Ocean fish have also been found to carry large amounts of mercury in their tissues.

From a health standpoint, consumers are well advised to limit their intake of fish taken from waters where mercury pollution may be a problem. Pregnant women or those considering pregnancy should be especially careful. Citizens also can support action on the part of government, industry, and individuals to minimize the entry of mercury-containing pollutants into the environment. Recent studies show that emissions from garbage incinerators and coal-fired power plants may now be even more worrisome sources of mercury than are wastewater discharges. In Sweden, where stringent controls over waste-burning were instituted almost two decades ago, the amount of mercury released into the air by human activities has declined by 90 percent.

THE MAJOR EXPRESSWAYS

Let's now take a look at the peripheral nervous system and the spinal cord. The two interconnect as the major expressways for information flow through the body.

Peripheral Nervous System

SOMATIC AND AUTONOMIC SUBDIVISIONS   In humans, the peripheral nervous system includes thirty-one pairs of *spinal* nerves, which connect with the spinal cord. The system also includes twelve pairs of *cranial* nerves, which connect directly with the brain.

Cranial and spinal nerves are also classified by their function. The ones that carry signals about moving your head, trunk, and limbs are **somatic nerves**. The sensory axons inside them deliver information from receptors in skin, skeletal muscles, and tendons to the central nervous system. Their motor axons deliver commands from the brain and spinal cord to the body's skeletal muscles.

By contrast, spinal and cranial nerves dealing with smooth muscle, cardiac (heart) muscle, and glands are the **autonomic nerves**. They carry signals to and from visceral parts of the body—internal organs and structures.

Unlike somatic neurons, single autonomic neurons don't extend the entire distance between muscles or glands and the central nervous system. Instead, *preganglionic* ("before a ganglion") *neurons* have cell bodies inside the spinal cord or brain stem, but their axons travel through nerves to autonomic system ganglia *outside* the CNS. There, the axons synapse with *postganglionic* ("after a ganglion") *neurons*, which make the actual connection with the body's muscles and glands (effectors).

SYMPATHETIC AND PARASYMPATHETIC NERVES   Figure 11.14 shows the two categories of autonomic nerves. We call them *parasympathetic* and *sympathetic* nerves Normally they work antagonistically, with the signals from one opposing those of the other. However, both these groups of nerves carry excitatory and inhibitory signals to internal organs. Often their signals arrive at the same time at muscle or gland cells and compete for control. In such cases, synaptic integration at the cellular level leads to minor adjustments in an organ's activity.

Parasympathetic nerves dominate when the body is not receiving much outside stimulation. They tend to

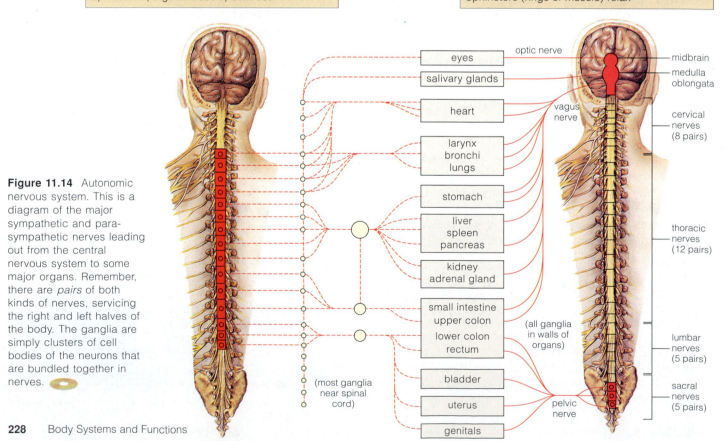

SYMPATHETIC OUTFLOW FROM THE SPINAL CORD

Examples of Responses
Heart rate increases
Pupils of eyes dilate (widen, let in more light)
Glandular secretions in airways to lungs decrease
Salivary gland secretions thicken
Stomach and intestinal movements slow down
Sphincters (rings of muscle) contract

PARASYMPATHETIC OUTFLOW FROM THE SPINAL CORD AND BRAIN

Examples of Responses
Heart rate decreases
Pupils of eyes constrict (let in less light)
Glandular secretions in airways to lungs increase
Salivary gland secretions become more watery
Stomach and intestinal movements increase
Sphincters (rings of muscle) relax

Figure 11.14 Autonomic nervous system. This is a diagram of the major sympathetic and parasympathetic nerves leading out from the central nervous system to some major organs. Remember, there are *pairs* of both kinds of nerves, servicing the right and left halves of the body. The ganglia are simply clusters of cell bodies of the neurons that are bundled together in nerves.

eyes
salivary glands
heart
larynx bronchi lungs
stomach
liver spleen pancreas
kidney adrenal gland
small intestine upper colon lower colon rectum
bladder
uterus
genitals

optic nerve
midbrain
medulla oblongata
cervical nerves (8 pairs)
vagus nerve
thoracic nerves (12 pairs)
(all ganglia in walls of organs)
lumbar nerves (5 pairs)
sacral nerves (5 pairs)
pelvic nerve

(most ganglia near spinal cord)

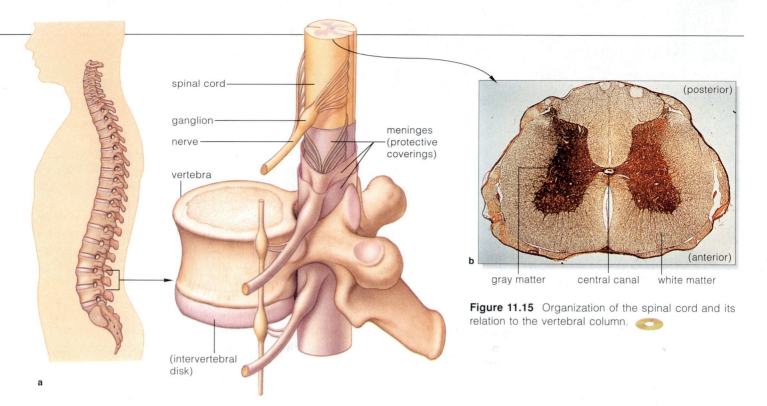

spinal cord

ganglion

nerve

vertebra

meninges
(protective
coverings)

(intervertebral
disk)

a

(posterior)

(anterior)

b

gray matter   central canal   white matter

Figure 11.15 Organization of the spinal cord and its relation to the vertebral column.

slow down the body overall and divert energy to basic "housekeeping" tasks, such as digestion.

Sympathetic nerves dominate during heightened awareness, excitement, or danger. They tend to shelve housekeeping tasks. For example, as you read this, sympathetic nerves are prompting your heart to beat a bit faster, and parasympathetic nerves are commanding it to beat a bit slower. Integration of these opposing signals influences the heart rate. If something scares or excites you, parasympathetic input to your heart drops. At the same time, sympathetic nerves release norepinephrine, a signaling molecule that makes your heart beat faster and makes you breathe faster and sweat. In this state of intense arousal, you are primed to fight (or play) hard or to get away fast. This is called the **fight-flight response**.

When the stimulus for the fight-flight response stops, sympathetic activity may fall and parasympathetic activity may rise. This "rebound effect" can occur after someone has been mobilized, say, to rush onto a street to save a child from an oncoming car. The person may well collapse as soon as the child has been swept out of danger.

The Spinal Cord

The **spinal cord** is a vital expressway for signals between the peripheral nervous system and the brain. It threads through a canal made of bones of the vertebral column (Figure 11.15). Most of the cord consists of nerve tracts (bundles of myelinated axons). Because the myelin sheaths of these axons are white, the tracts are called *white matter*. The cord also contains dendrites, cell bodies of neurons, interneurons, and neuroglial cells. These form

its *gray matter*. In cross section, the cord's gray matter looks a little bit like a butterfly. The cord lies inside a closed channel formed by the bones of your vertebral column. Those bones, and ligaments attached to them, protect the soft nervous tissue of the cord. So do three coverings of connective tissue called *meninges* layered around the spinal cord; these meninges also cover the brain, and we will discuss them in the next section.

In addition to carrying signals between the peripheral nervous system and the brain, the spinal cord is a control center for reflexes. Sensory and motor neurons involved in many reflex movements of skeletal muscle make direct connections in the cord; such *spinal reflexes* do not require input from the brain. When you jerk your hand away from a hot stove burner, you are experiencing a spinal reflex in action. Higher brain centers also receive information about the sensory stimulus, and so you become aware of "hot burner!" even as your hand is moving away from it. In addition to all of the above, the spinal cord also contributes to some *autonomic reflexes*, which deal with internal organ functions such as bladder emptying.

The peripheral nervous system consists of the nerves to and from the brain and spinal cord.

Its somatic nerves deal with skeletal muscle movements. Its autonomic nerves govern functions of internal organs, such as the heart and glands. Autonomic nerves are divided into parasympathetic nerves (for housekeeping functions) and sympathetic nerves (for aroused states).

The spinal cord carries signals between peripheral nerves and the brain. It also is a control center for certain reflexes.

The spinal cord merges with the **brain**, a master control center that receives, integrates, stores, and retrieves sensory information. The brain also coordinates responses to information by adjusting activities throughout the body.

Like the spinal cord, the brain is protected by bones (of the cranium) and by the three **meninges**. These are membranes of connective tissue layered between the skull bones and the brain tissue itself. Meninges cover and protect the fragile CNS neurons and blood vessels that service the tissue. Folds in the tough, outermost one (the *dura mater*) separate the brain into left and right hemispheres. The meninges also enclose spaces that are filled with **cerebrospinal fluid**, which cushions and helps nourish the brain. We will return to its functions shortly.

The brain has three divisions, the hindbrain, midbrain, and forebrain. Table 11.2 summarizes the functions of each division. In the hindbrain and midbrain, the most ancient nervous tissue still houses centers that control many simple, basic reflexes; we call this tissue the **brain stem**. Over evolutionary time, expanded layers of gray matter developed over the brain stem. The most recent additions have been correlated with our species' increasing reliance on three major sensory organs: the nose, ears, and eyes. They are topics in Chapter 12.

Hindbrain

The medulla oblongata, cerebellum, and pons are all components of the hindbrain. The **medulla oblongata** contains reflex centers for a number of vital tasks, such as respiration and blood circulation. It also coordinates motor responses with certain complex reflexes, such as coughing. In addition, the medulla influences brain centers that help you sleep or wake up.

The **cerebellum** integrates signals from your eyes, inner ears, and muscle spindles with motor signals from the forebrain to coordinate movement and balance. It helps control motor dexterity. Some of its activities may also be crucial in human language and other forms of mental "dexterity."

Bands of many axons extend from the cerebellum into the **pons** ("bridge"). Although lodged in the brain stem, the pons directs the signal traffic between the cerebellum and the higher integrating centers of the forebrain.

Midbrain

The midbrain coordinates reflex responses to sights and sounds. It has a roof of gray matter, the *tectum* (Latin for roof), where visual and auditory sensory input converges before being sent on to higher brain centers.

Table 11.2	Regions of the Brain	
FOREBRAIN	Cerebrum	In two cerebral hemispheres, centers for coordinating sensory and motor functions, for memory, and for abstract thought. Most complex coordinating center; intersensory association, memory circuits
	Olfactory lobes	Relaying of sensory input from the nose to olfactory structures of cerebrum
	Thalamus	Major coordinating center for sensory signals; relay station for most sensory impulses to cerebrum
	Hypothalamus	Neural-endocrine coordination of visceral activities (e.g., water-solute balance, temperature control, carbohydrate metabolism)
	Limbic system	Via interacting scattered brain centers (including hypothalamus), coordination of skeletal muscle and internal organ activity underlying emotional expression
	Pituitary gland	"Master" endocrine gland (controlled by hypothalamus). Control of growth, metabolism, etc.
	Pineal gland	Control of some daily (circadian) rhythms
MIDBRAIN	Tectum	Largely reflex coordination of visual, tactile, auditory input; contains nerve tracts ascending to thalamus, descending from cerebrum
HINDBRAIN	Pons	"Bridge" of nerve tracts from cerebrum to both sides of cerebellum. Also contains longitudinal tracts connecting forebrain and spinal cord
	Cerebellum	Unconscious coordination of motor activity underlying limb movements, maintaining posture, spatial orientation
	Medulla oblongata	Contains tracts extending between pons and spinal cord; reflex centers involved in respiration, cardiovascular function, etc.

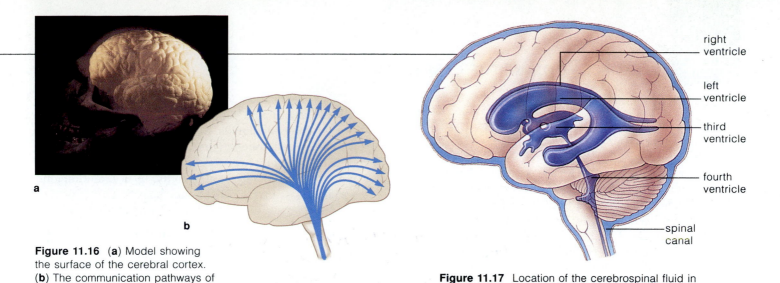

Figure 11.16 (**a**) Model showing the surface of the cerebral cortex. (**b**) The communication pathways of the reticular formation.

Figure 11.17 Location of the cerebrospinal fluid in the brain.

right ventricle
left ventricle
third ventricle
fourth ventricle
spinal canal

Forebrain

The forebrain is the most highly developed region of your brain. Its parts include the cerebrum, olfactory bulbs, and the thalamus and hypothalamus. In the **cerebrum**, information processing occurs and sensory input and motor responses are integrated. The two *olfactory bulbs* deal with sensory information about smell. The **thalamus** is mainly a relay switchboard for sensory information. In it, signals arriving via sensory nerve tracts are relayed to small clusters of neuron cell bodies (called *nuclei*), then relayed onward to the cerebrum or elsewhere. The nuclei also process some outgoing motor information; recall from the chapter introduction that the neurotransmitter deficits of Parkinson's disease disrupt the functioning of *basal nuclei* in the thalamus.

Below the thalamus, the **hypothalamus** evolved into the body's "supercenter" for controlling homeostatic adjustments in the activities of internal organs. As noted in previous chapters, for example, it helps govern states such as thirst and hunger; the hypothalamus also has roles in sexual behavior and emotional expression, such as sweating with fear.

The Reticular Formation

By now, you may be thinking that the brain is neatly subdivided into three main regions. This isn't the case, however. An evolutionarily ancient mesh of interneurons still extends from the uppermost spinal cord, on through the brain stem, and into higher integrative centers of the cerebral cortex (Figure 11.16). This major network of interneurons is the **reticular formation**. It operates as a low-level motor pathway to the medulla oblongata and spinal cord. Through these links, the reticular formation helps govern muscle activity associated with maintaining balance, posture, and muscle tone. It also can activate centers in the cerebral cortex and so help govern activities of the whole nervous system.

Brain Cavities and Canals

Our brain and spinal cord would both be highly vulnerable to damage if they were not protected by bones and meninges. In addition, both of them are surrounded by a transparent **cerebrospinal fluid.** This extracellular fluid is secreted from specialized capillaries inside a system of fluid-filled cavities and canals in the brain. The cavities in the brain (called ventricles) connect with each other and with the central canal of the spinal cord. All are filled with cerebrospinal fluid. The fluid also fills the space between the innermost layer of the meninges and the brain itself (Figure 11.17). Because the enclosed cerebrospinal fluid can't be compressed, it helps cushion the brain and spinal cord from sudden, jarring movements.

The bloodstream exchanges substances with the extracellular fluid, which in turn exchanges substances with neurons. However, the structure of brain capillaries makes them relatively impermeable to certain substances. Continuous tight junctions fuse endothelial cells of the capillary walls (Section 4.6). As a result substances must pass *through* the cells, rather than between them, to reach the brain. This **blood-brain barrier** helps control *which* blood-borne substances are allowed to reach the brain's neurons. Transport proteins embedded in the plasma membrane of those cells selectively transport glucose and other water-soluble substances across the barrier. However, lipid-soluble substances are another matter. They quickly diffuse through the lipid bilayer of the plasma membrane. This "lipid loophole" in the blood-brain barrier is one reason why caffeine, nicotine, alcohol, barbiturates, heroin, and anesthetics can rapidly affect brain function.

The brain's main divisions are the hindbrain, midbrain, and forebrain. Their functions range from reflex controls over basic survival functions (as in the brain stem) to the complex integration of sensory information and motor responses.

The Cerebral Hemispheres

If you are an average-sized adult, your brain weighs about 1,300 grams (three pounds). It contains at least 100 billion neurons!

The human cerebrum looks somewhat like the much-folded nut in a walnut shell (Figure 11.18). A deep fissure divides the cerebrum into left and right **cerebral hemispheres**. Each has a thin, outer layer of gray matter, the **cerebral cortex**. (The cerebral cortex weighs about a pound, and if you were to stretch it flat, it would cover a surface area of two and a half square feet.) Below this is the white matter (axons), and the basal nuclei—patches of gray matter in the thalamus.

Each cerebral hemisphere receives, processes, and coordinates responses to sensory input mainly from the opposite side of the body. (For instance, "cold" signals from an ice cube in your left hand travel to the right hemisphere.) The left hemisphere deals mainly with speech, analytical skills, and mathematics. In most people it dominates the right hemisphere, which deals more with visual-spatial relationships, music, and other creative enterprises. A transverse band of nerve tracts, the corpus callosum, carrries signals back and forth between the hemispheres and coordinates their functioning.

Each hemisphere is divided into four tissue regions: the frontal, occipital, temporal, and parietal lobes, which process different signals. EEG and PET scans can reveal activity in each lobe. Figure 11.19 gives a few examples. (EEG, short for electroencephalogram, is simply a recording of summed electrical activity in a brain region.) Section 2.2 describes PET scans.

Functions of Cerebral Cortex Areas

Everything people comprehend, communicate, remember, and voluntarily act upon arises in the cerebral cortex; it governs our conscious behavior. Functionally, the cortex is divided into *motor* areas (control of voluntary motor activity), *sensory* areas (perception of the meaning of sensations), and *association* areas (the integration of information that precedes a conscious action). None of these areas functions alone; consciousness arises by way of interactions throughout the cortex. The following paragraphs give some examples.

MOTOR AREAS In the frontal lobe of each hemisphere, the entire body is spatially mapped out in the primary motor cortex. This area controls coordinated movements of skeletal muscles. Thumb, finger, and tongue muscles get much of the area's attention. This gives you an idea of how much control is required for voluntary hand movements and verbal expression (Figure 11.20).

Also in the frontal lobe are the premotor cortex, Broca's area, and the frontal eye field. The premotor cortex deals with learned patterns or motor skills. Use a computer keyboard, play a piano concerto, dribble a basketball— such repetitive movements are evidence that your motor cortex is coordinating the simultaneous and sequential movements of a number of muscle groups. Broca's area (usually in the left hemisphere) and a corresponding area in the right hemisphere control tongue, throat, and lip muscles used in speech. It kicks in when we are about to speak, even when we plan voluntary motor activities other than speaking. Hence, you can talk on the telephone and write down a message, all at the same time. Above Broca's area is the frontal eye field, which controls voluntary eye movements.

SENSORY AREAS Sensory areas occur in different parts of the cortex. In the parietal lobe, the body is spatially mapped out in the primary somatosensory cortex. This area is the main receiving center for sensory input from the skin and joints (Section 12.2). The parietal lobe also has a primary cortical area dealing with perception of taste. At the back of the occipital lobe is the primary visual cortex, which receives sensory inputs from your eyes (Section 12.9). Perception of sounds and of odors arises in primary cortical areas in each temporal lobe.

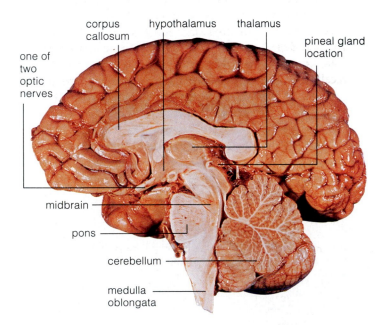

corpus callosum   hypothalamus   thalamus

pineal gland location

one of two optic nerves

midbrain

pons

cerebellum

medulla oblongata

Figure 11.18 Right hemisphere of the human brain, sagittal view. Not visible is the reticular formation, which extends between the upper spinal cord and the cerebrum. Each hemisphere has a cerebral cortex, a layer of gray matter about 2–4 millimeters (1/8 inch) thick. Interneuron cell bodies and dendrites, unmyelated axons, glia, and blood vessels make up the gray matter.

Further reading: Student Guide to InfoTrac on web site →

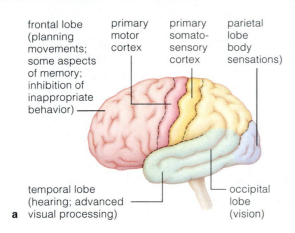

frontal lobe (planning movements; some aspects of memory; inhibition of inappropriate behavior)

primary motor cortex

primary somato-sensory cortex

parietal lobe body sensations)

temporal lobe (hearing; advanced visual processing)

occipital lobe (vision)

a

b Motor cortex activity when speaking

Prefrontal cortex activity when generating words

Visual cortex activity when observing words

Figure 11.19 (**a**) Primary receiving and integrating centers for the human cerebral cortex. Primary cortical areas receive signals from receptors on the body's periphery. Association areas coordinate and process sensory input from different receptors. The PET scans (**b**) show which brain regions were active when a person performed three specific tasks: speaking, generating words, and observing words.

ASSOCIATION AREAS Association areas occupy all parts of the cortex except primary motor and sensory regions. Each integrates, analyzes, and responds to many inputs. For instance, the visual association area surrounds the primary visual cortex. It helps us recognize something we see by comparing it with visual memories. Neural activity in the most complex association area, the prefrontal cortex, is the basis for complex learning, intellect, and personality. Without it, we would be incapable of abstract thought, judgment, planning, and concern for others.

Connections with the Limbic System

The prefrontal cortex interacts intimately with the **limbic system**, which is located inside the cerebral hemispheres.

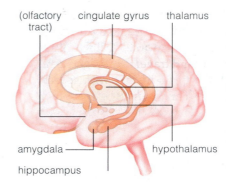

(olfactory tract)   cingulate gyrus   thalamus

amygdala   hypothalamus

hippocampus

Figure 11.21 Key structures in the limbic system, which encircles the upper brain stem. The amygdala and cingulate gyrus are especially important in emotions. The hypothalamus is a clearinghouse for emotions and visceral activity. Both the hippocampus and amygdala help convert stimuli into long-term memory (Section 11.11).

It governs our emotions and has roles in memory (Figure 11.21). The limbic system includes parts of the thalamus along with the hypothalamus, the amygdala, and the hippocampus. It is distantly related to olfactory centers and still deals with the sense of smell. That's one reason why you may feel warm and fuzzy when your brain recalls the cologne of a special person who wore it.

The limbic system is called our emotional-visceral brain. It's connections with other brain centers allow it to correlate organ activities with self-gratifying behavior, such as eating and sex. Signals based on reasoning in the cerebral cortex often can override or dampen rage, hatred, or other "gut reactions."

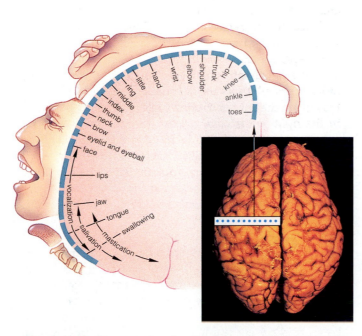

Figure 11.20 Diagram of a slice through the primary motor cortex of the left cerebral hemisphere. The distortions to the human body draped over the diagram indicate which body parts receive the most precise control. The photograph is a "top down" view of both cerebral hemispheres.

The cerebrum is divided into two hemispheres. Each hemisphere receives, processes, and coordinates responses to sensory input mainly from the opposite side of the body. The corpus callosum carries signals between them.

The left hemisphere deals mainly with speech, analytical skills, and mathematics. It usually dominates the right hemisphere, which deals more with creative activity, such as visual-spatial relationships and music.

The cerebral cortex, the outermost layer of gray matter of each hemisphere, contains motor, sensory, and association areas. Communication among these areas governs conscious behavior. The cerebral cortex also interacts with the limbic system, which governs emotions and memory.

Science Comes to Life

EXPERIMENTS WITH A "SPLIT BRAIN"

Some decades ago Roger Sperry and his coworkers demonstrated intriguing differences in perception between the two halves of the cerebrum of epileptics. Severe *epilepsy* is characterized by seizures, sometimes as often as every half hour. The seizures are analogous to an electrical storm in the brain. Sperry's patients were so debilitated by their condition that it was impossible for them to lead normal lives. Conventional treatments had been ineffective, so Sperry asked: Would *cutting* the corpus callosum of epileptics confine the electrical storm to one hemisphere, leaving at least the other hemisphere to function normally? Earlier studies of laboratory animals and of humans whose corpus callosum had been damaged suggested this might be so.

He performed the surgery, and the electrical storms did subside in frequency and intensity. Cutting the neural bridge ended what must have been positive feedback of ever intensifying electrical disturbances between the two hemispheres. The "split-brain" patients were able to lead what seemed, on the surface, entirely normal lives. But then Sperry devised some elegant experiments to test whether their conscious experience was indeed "normal." Given that the corpus callosum contains 200 million axons, surely *something* was different. Something was. "The surgery," he later reported, "left these people with two separate minds, that is, two spheres of consciousness. What is experienced in the right hemisphere seems to be entirely outside the realm of awareness of the left."

Sperry presented the two hemispheres of split-brain patients with two different portions of the same visual stimulus. It was known at the time that visual connections to and from one hemisphere are mainly concerned with the opposite half of the visual field (Figure 11.22*a*). Sperry projected words—say, COWBOY—onto a screen so that COW

fell in the left half of the visual field, and BOY fell in the right (Figure 11.22*b*).

The subjects of this experiment reported *seeing* the word BOY. The left hemisphere, which controls language, perceived only the letters BOY. However, when asked to write the perceived word with the left hand—a hand that was deliberately blocked from a subject's view—the subject wrote COW. The right hemisphere "knew" the other half of the word (COW) and had directed the left hand's motor response. But it couldn't tell the left hemisphere what was going on because of the severed corpus callosum. The subject knew that a word was being written, but could not say what it was!

Thus Sperry showed that signals across the corpus callosum coordinate the functioning of the two cerebral hemispheres, each of which had responded to visual signals from the opposite side of the body.

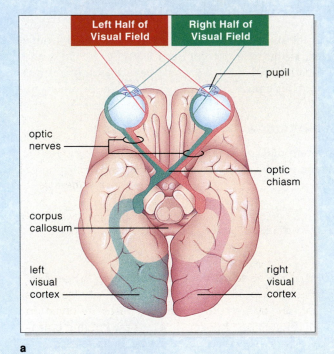

a

Figure 11.22 (**a**) The human eye gathers visual information at the retina, a layer of densely packed light receptors. Light from the *left* half of the visual field strikes receptors on the right side of both retinas. Parts of two optic nerves carry signals from the receptors to the right cerebral hemisphere. Light from the *right* half of the visual field strikes receptors on the left side of both retinas. Parts of the optic nerves carry signals from them to the left hemisphere.

(**b**) Response of a split-brain patient to different portions of a visual stimulus.

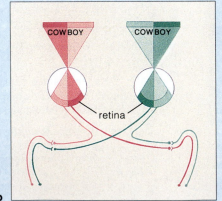

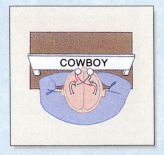

b

11.11 MEMORY

Even before you were born, your brain began to build your **memory**—to store and retrieve information about an individual's unique experiences. Learning and adaptive modifications of our behavior would be impossible without it. Information is stored in stages. *Short-term* storage is a stage of neural excitation that lasts a few seconds to a few hours. It is limited to bits of sensory information—numbers, words of a sentence, and so on. In *long-term* storage, seemingly unlimited amounts of information get tucked away more or less permanently, as shown in Figure 11.23.

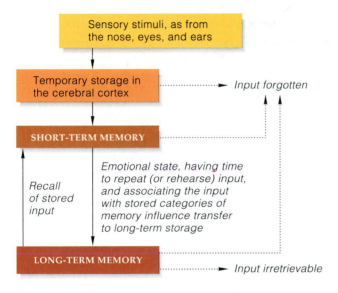

Figure 11.23 Stages of memory processing, starting with the temporary storage of sensory inputs in the cerebral cortex.

Not all of the sensory input bombarding the cerebral cortex ends up in memory storage. Only some is selected for transfer to brain structures involved in short-term memory. Information in these holding bins is processed for relevance, so to speak. If irrelevant, it is forgotten; otherwise it is consolidated with the banks of information in long-term storage structures.

The human brain processes facts separately from skills. Dates, names, faces, words, odors, and other bits of explicit information are *facts*, soon forgotten or filed away in long-term storage, along with the circumstance in which they were learned. Hence you might associate, say, the smell of baking bread with your grandmother's kitchen. By contrast, *skills* are gained by practicing specific motor activities. A skill such as slam dunking a basketball or playing a piano concerto is best recalled by actually performing it, rather than by recalling the circumstances in which the skill was first learned.

Separate memory circuits handle different kinds of input. A circuit leading to fact memory (Figure 11.24*a*)

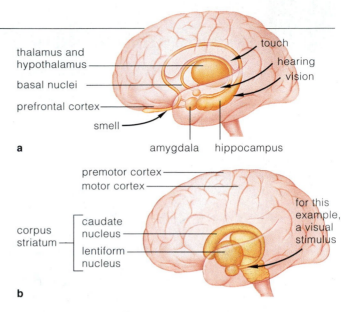

Figure 11.24 Possible circuits involved in (**a**) fact memory and (**b**) skill memory.

starts with inputs at the sensory cortex that flow to the amygdala and hippocampus in the limbic system. The amygdala is the gatekeeper, connecting the sensory cortex with parts of the thalamus and parts of the hippocampus that govern emotional states. Information flows on to the prefrontal cortex, where multiple banks of fact memories are retrieved and used to stimulate or inhibit other parts of the brain. The new input also flows to basal nuclei, which send it back to the cortex in a feedback loop that reinforces the input until it can be consolidated in long-term storage.

Skill memory also starts at the sensory cortex, but this circuit routes sensory input to the corpus striatum ("layered body"), which promotes motor responses (Figure 11.24*b*). Motor skills entail muscle conditioning, and as you might suspect, the circuit extends to the cerebellum, the brain region that coordinates motor activity.

Amnesia is a loss of fact memory. How severe the loss is depends on whether the hippocampus, amygdala, or both are damaged, as by a head blow. Amnesia does not affect a person's capacity to learn new skills. By contrast, Parkinson's disease destroys basal nuclei and learning ability, although skill *memory* remains. Alzheimer's disease, which usually starts late in life, is linked to physical changes in the cerebral cortex and hippocampus. Affected people often can remember long-standing information, such as their Social Security number, but they have trouble recalling what just happened to them. In time they grow confused, depressed, and incoherent.

Memory, the storage and retrieval of sensory information, results from circuits between the cerebral cortex and parts of the limbic system, thalamus, and hypothalamus. Sensory input is processed through short-term and long-term storage.

The spectrum of consciousness includes sleeping and aroused states, during which neural chattering shows up as wavelike patterns in EEGs. As mentioned earlier, EEGs are electrical recordings of the summed frequency and strength of membrane potentials at the surface of the brain (Figure 11.25). PET scans also can show the precise location of brain activity as it takes place.

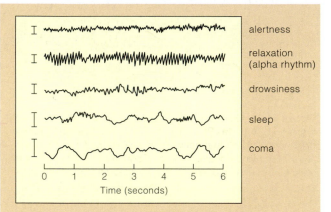

Figure 11.25   EEG patterns. Vertical bars mark a fifty-microvolt range of electrical responses (EEG waves). Irregular horizontal graph lines show waves recorded in "trains" of one after the other, about ten per second.

The *alpha rhythm* is the prominent wave pattern for someone meditating (relaxed, with eyes closed). Wave trains become larger, slower, and more erratic during the transition to sleep. This *slow-wave sleep* pattern occurs when sensory input is low and the mind is more or less idling. Awaken subjects from slow-wave sleep, and they usually say they weren't dreaming; they often seemed to be mulling over recent, ordinary events.

REM sleep punctuates slow-wave sleep. *Rapid Eye Movements* accompany this pattern (eyes jerk under closed lids), as do irregular breathing, faster heartbeat, and twitching fingers. Most people awakened from REM sleep say they were experiencing vivid dreams. A shift to low-amplitude, higher frequency wave trains marks the transition from sleep or deep relaxation into alert wakefulness, *EEG arousal*. Then, someone consciously focuses on external stimuli or on one's own thoughts.

Part of the reticular formation promotes chemical changes that influence whether you stay awake or fall asleep. Serotonin, a neurotransmitter released from one of its sleep centers, inhibits other neurons that arouse the brain and maintain wakefulness. At high levels, serotonin triggers drowsiness and sleep. Substances released from another brain center inhibit serotonin's effects and bring about wakefulness.

The spectrum of consciousness, which includes sleeping and states of arousal, is influenced by the reticular formation.

A **drug** is a substance introduced into the body to provoke a specific physiological response. For thousands of years people in nearly every culture have used drugs as medicines, to alter mental states as part of religious or social rituals, or simply for personal effects. **Drug abuse** is use of a drug in a way that harms health or interferes with a person's ability to function in society.

Psychoactive drugs act on the central nervous system by binding to receptors in the neuron plasma membrane. Such receptors normally bind to neurotransmitter molecules, which transmit chemical messages among neurons. When drug molecules bind such receptors, the result is a change in the chemical messages neurons send or receive—and in associated mental and physical states.

Drug Effects

Psychoactive drugs exert their effects on brain regions that govern states of consciousness and behavior. Various drugs also influence physiological events, such as heart rate, respiration, sensory processing, and muscle coordination. Many affect a pleasure center in the hypothalamus and artificially fan the sense of pleasure we associate with eating, sexual activity, or other behaviors. Psychoactive drugs fall into one of five categories: stimulants, drugs that reduce brain activity in some way, pain relievers (such as morphine and heroin), psychedelics and hallucinogens (such as LSD), and deliriants (such as inhalants).

Caffeine, nicotine, cocaine, and amphetamines (including "speed") are all **stimulants**—they increase alertness and physical activity at first, then lead to depression. Nicotine mimics the neurotransmitter acetylcholine, triggering a variety of metabolic effects (such as increased heart rate and blood pressure). Chapter 9 detailed some of the devastating health effects of long-term nicotine use.

Amphetamines are synthetic chemicals that chemically resemble dopamine and norepinephrine, natural signaling molecules that stimulate the brain's pleasure center, in the hypothalamus. Over time, however, the brain responds to amphetamines by producing less of its own signaling molecules, and an unhealthy *tolerance* develops, as described shortly. Chronic users may become malnourished and suffer cardiovascular problems.

Cocaine stimulates the pleasure center by *blocking* the reabsorption of dopamine and other signaling molecules. Cocaine use harms not only the cardiovascular system (a single episode can cause heart failure), but weakens the immune system as well. Many authorities rank the social ills associated with crack cocaine use as the most serious of all illegal drug problems.

Alcohol is a prime example of a drug that dampens brain activity. Depending on circumstances, as little as an ounce or two can produce disorientation, uncoordinated motor functions, and diminished judgment. *Blood alcohol concentration* (BAC) measures the percentage of alcohol in the blood. In most states, someone with a BAC of 0.08 per milliliter is considered legally drunk. When the BAC reaches 0.15 to 0.4, a person is obviously intoxicated and cannot function normally physically or mentally. A BAC greater than 0.4 can be lethal.

Tolerance and Dependence

Often the body develops *tolerance* to a drug, meaning that it takes larger or more frequent doses of the drug to produce the same effect. Tolerance reflects *physical drug dependence*. As you may recall from Chapter 6, the liver produces enzymes that detoxify drugs circulating in the bloodstream. Tolerance develops when the level of detoxifying liver enzymes increases in response to the consistent presence of the drug in the blood. In effect, a user must increase his or her intake to keep one step ahead of the liver's increasing ability (up to a point) to break down the drug.

Psychological drug dependence, or *habituation*, develops when a user begins to crave the feelings associated with using a particular drug. The inability to "feel good" or function normally without a regular supply of a drug—whether it is caffeine, alcohol, nicotine, or crack cocaine (Figure 11.26)—is a clear sign of habituation. Table 11.3 lists some warning signs of potentially serious drug dependence. Both habituation and tolerance are evidence that the user has become addicted to a substance.

Drug Action and Interactions

Not surprisingly, a drug's effects depend on the dose, and typically the effects of a higher dose are more intense

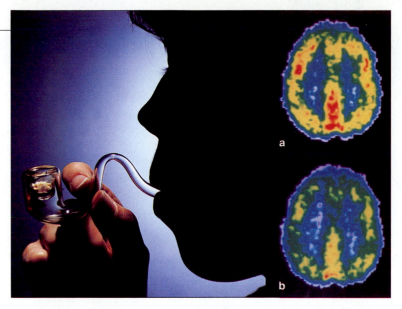

Figure 11.26 Smoking crack puts cocaine in the brain in less than 8 seconds. (**a**) PET scan of a horizontal section of the brain, showing normal activity. (**b**) PET scan of a comparable section showing cocaine's effect. *Red* indicates greatest activity; *yellow*, *green*, and *blue* indicate successively inhibited activity.

than those of a smaller dose. For example, smoking just a little marijuana may pleasurably intensify some sensory perceptions (such as listening to music), but a higher dose may provoke exaggerated, even paranoid, perceptions.

When different psychoactive drugs are used together, they can interact, sometimes in ways that are extremely dangerous. In a *synergistic* interaction, two drugs used together have a much more powerful effect than they would have separately. Alcohol and barbiturates (such as Seconal and Nembutal) both depress the central nervous system, for example; when they are used at the same time they can lethally depress respiratory centers in the brain. Some drug combinations are *antagonistic*—one drug blocks the effects of another. Others are *potentiating*—one drug *enhances* the effects of another. For instance, you may have noticed that labels on allergy medications often warn against drinking alcohol while using the product. The warnings are there because even a little alcohol may deepen the drowsiness antihistamines can cause, seriously impairing the user's ability to function normally.

Table 11.3 Warning Signs of Drug Dependence*
1. Tolerance—it takes increasing amounts of the drug to produce the same effect.
2. Habituation—it takes continued drug use over time to maintain self-perception of functioning normally.
3. Inability to stop or curtail use of the drug, even if there is persistent desire to do so.
4. Concealment—not wanting others to know of the drug use.
5. Extreme or dangerous behavior to get and use a drug, as by stealing, asking more than one doctor for prescriptions, or jeopardizing employment by drug use at work.
6. Deterioration of professional and personal relationships.
7. Anger and defensive behavior when someone suggests there may be a problem.
8. Preference of drug use over previously customary activities.

*Three or more of these signs may be cause for concern.

Psychoactive drugs exert their effects on brain regions that govern states of consciousness and behavior.

Some psychoactive drugs also influence physiological events, such as heart rate, respiration, sensory processing, and muscle coordination.

Habituation and tolerance to a drug are evidence of addiction.

SUMMARY

1. The nervous system detects, interprets, and responds directly to sensory stimuli.

2. A neuron can receive and respond to stimuli because of its membrane properties. An unstimulated neuron shows a steady voltage difference across its plasma membrane (the inside is more negative than the outside). The resting neuron maintains concentration gradients of potassium ions, sodium ions, and other ions across the membrane.

3. With adequate stimulation, the voltage difference across the membrane changes, exceeding a certain threshold level. Then, gated sodium channels across the membrane open and close rapidly and suddenly reverse the voltage difference, which recording devices register as a spike (action potential).

4. Action potentials propagate themselves along the neuron membrane until they reach an output zone, where axon endings form a chemical synapse with another neuron or a muscle or gland cell. The presynaptic cell releases a neurotransmitter into the synaptic cleft. The neurotransmitter excites or inhibits the postsynaptic cell. Integration is the moment-by-moment combining of all signals—excitatory and inhibitory—acting on a neuron.

5. The direction of information flow through the body is established by the organization of neurons into circuits and pathways. Local circuits are sets of interacting neurons confined to a single region in the brain or spinal cord. Nerve pathways extend from neurons in one body region to neurons or effectors in different regions. Reflex arcs, in which sensory neurons directly signal motor neurons that act on muscle cells, are the simplest pathways. In more complex reflexes, interneurons coordinate and refine the responses.

6. The central nervous system consists of the brain and spinal cord. The peripheral nervous system consists of nerves and ganglia in other body regions. Nerves are pathways for signals between the central nervous system and the peripheral nervous system.

7. The peripheral nervous system's somatic subdivision deals with skeletal muscles concerned with voluntary body movements and sensations arising from skin, muscles, and joints. Its autonomic subdivision deals with the functions of the heart, lungs, glands, and other internal organs.

8. The spinal cord has nerve tracts that carry signals between the brain and the peripheral nervous system. It also is a center for some direct reflex connections that underlie limb movements and internal organ activity.

9. The brain has three divisions:
    a. The hindbrain includes the medulla oblongata, pons, and cerebellum and contains reflex centers for vital functions and muscle coordination.
    b. Midbrain centers coordinate and relay visual and auditory information. The midbrain, medulla oblongata, and pons make up the brain stem. The reticular formation, a network of interneurons, extends the length of the brain stem. It helps govern activities of the entire nervous system.
    c. The forebrain includes the cerebrum, thalamus, hypothalamus, and limbic system. The thalamus relays sensory information and helps coordinate motor responses. The hypothalamus monitors internal organs and influences behaviors related to their functioning (such as thirst and sexual activity). The limbic system, which has roles in learning, memory, and emotional behavior, includes pathways that link parts of the thalamus and hypothalamus with other forebrain regions.

10. The cerebral cortex has regions devoted to specific functions, such as receiving and integrating information from sense organs and coordinating motor responses. States of consciousness vary between total alertness and deep coma. The levels are governed by the reticular activating system.

11. Memory occurs in short-term and long-term stages. Long-term storage depends on chemical or structural changes in the brain.

Review Questions

1. Define sensory neuron, interneuron, and motor neuron. *11.1*

2. What are the functional zones of a motor neuron? *11.1*

3. Distinguish between a local signal at the input zone of a neuron and an action potential. *11.2*

4. What is a synapse? Explain the difference between an excitatory and an inhibitory synapse. Define synaptic integration. *11.3*

5. What is a reflex? Describe the events of a stretch reflex. *11.4*

6. What are the parts of the central nervous system? Of the peripheral nervous system? *11.5*

7. Label the parts of the brain in the diagram below. *11.8*

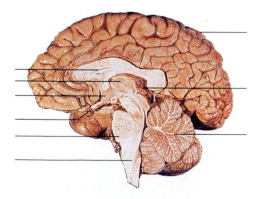

8. Distinguish between the following:
   a. neurons and nerves *CI, 11.4*
   b. somatic system and autonomic system *11.7*
   c. parasympathetic and sympathetic nerves *11.7*

Self-Quiz *(Answers in Appendix V)*

1. The nervous system senses, interprets, and issues commands for responses to _____. Its communication lines are organized gridworks of nerve cells, or _____.

2. A neuron responds to adequate stimulation with _____, a type of self-propagating signal.

3. When action potentials arrive at a synapse between a neuron and another cell, they stimulate the release of molecules of a _____ that diffuse over to that cell.

4. The moment-by-moment combining of all signals acting on all the different synapses on a neuron is called _____.

5. Interactions among neurons in your body _____.
   a. involve neurotransmitters
   b. are mediated by memory
   c. are mediated by learning and reasoning
   d. all of the above are correct

6. In the simplest kind of reflex, _____ directly signal _____, which act on muscle cells.
   a. sensory neurons; interneurons
   b. interneurons; motor neurons
   c. sensory neurons; motor neurons
   d. motor neurons; sensory neurons

7. The accelerating flow of _____ ions through gated channels across the membrane is the actual trigger for an action potential.
   a. potassium
   b. sodium
   c. hydrogen
   d. a and b are correct

8. _____ nerves slow down the body overall and divert energy to basic housekeeping tasks; _____ nerves slow down housekeeping tasks and increase overall activity during times of heightened awareness, excitement, or danger.
   a. Autonomic; somatic
   b. Sympathetic; parasympathetic
   c. Parasympathetic; sympathetic
   d. Peripheral; central

9. Match each of the following central nervous system regions with some of its functions.
   ____ spinal cord
   ____ medulla oblongata
   ____ hypothalamus
   ____ limbic system
   ____ cerebral cortex

   a. receives sensory input, integrates it with stored information, coordinates motor responses
   b. monitors internal organs and related behavior (e.g., thirst, hunger, sex)
   c. governs emotions
   d. coordinates reflexes (e.g., for respiration, blood circulation)
   e. makes reflex connections for limb movements, internal organ activity

Critical Thinking: You Decide *(Key in Appendix VI)*

1. In some cases of ADD (attention deficit disorder) the impulsive, erratic behavior typical of so-called hyperactive youngsters can be normalized with drugs that *stimulate* the central nervous system. Explain this finding in terms of neurotransmitter activity in the brain.

2. *Meningitis* is an inflammation of the meninges that cover the brain and spinal cord. Diagnosis involves making a "spinal tap" (lumbar puncture) and analyzing a sample of cerebrospinal fluid for signs of infection. Why analyze this fluid and not blood?

3. In newborns and premature babies, the blood-brain barrier is not fully developed. Explain why this might be reason enough to pay careful attention to their diet.

4. At one time people deeply feared contracting the disease called *tetanus* (caused by the bacterium *Clostridium tetani*) because it generally meant an agonizing death. The bacterial toxin blocks the release of neurotransmitters released by interneurons that help control motor neurons. The result is continuous contraction of skeletal muscles and, eventually, the heart muscle also. Victims are fully aware of their plight, for the brain is not affected. What neurons does the toxin affect?

Selected Key Terms

action potential *11.1*
autonomic nerve *11.7*
axon *11.1*
blood-brain barrier *11.6*
brain *11.8*
central nervous system *11.5*
cerebral cortex *11.9*
cerebrospinal fluid *11.8*
cerebrum *11.8*
chemical synapse *11.3*
dendrite *11.1*
hypothalamus *11.8*
interneuron *11.1*
limbic system *11.9*
memory *11.11*
meninges *11.8*
motor neuron *11.1*
myelin sheath *11.4*
nerve *11.4*
nerve tract *11.5*
neuroglia *11.1*
neuromodulator *11.3*
neurotransmitter *11.3*
parasympathetic nerve *11.7*
peripheral nervous system *11.5*
reflex *11.4*
reflex arc *11.4*
resting membrane potential *11.1*
reticular formation *11.8*
Schwann cell *11.4*
sensory neuron *11.1*
sodium-potassium pump *11.1*
somatic nerve *11.7*
spinal cord *11.7*
sympathetic nerve *11.7*
synaptic integration *11.3*
thalamus *11.8*

Readings

Beardsley, T. August 1997. "The Machinery of Thought." *Scientific American.*

Gazzaniga, M. July 1998. "The Split Brain Revisited." *Scientific American.*

Kempermann, G. and F. Gage. May 1999. "New Nerve Cells for the Adult Brain." *Scientific American.*

SENSORY RECEPTION

Nosing Around

That charming protuberance called your nose knows more than you might think. In the upper reaches of the nasal cavity, patches of yellowish epithelium are crammed with some 10 million specialized neurons. The neurons have hairlike cilia with receptor proteins jutting out from the surface. They are receptors for odor molecules. Molecules in inhaled air are sensory calling cards of a rose's perfume, garlic's pungency, and the human body's secretions, just to mention a few of the 10,000 different scents the average person can distinguish (Figure 12.1).

As you will see in the following pages, sensory receptors are gateways into the nervous system. A recent exciting discovery indicates that there are several hundred *different* types of protein receptors in the olfactory epithelium, each one built according to a genetic blueprint in a particular segment of DNA (a gene). There may even be as many as 1,000 different receptor types coded by a large family of "smell genes." Each receptor can latch onto odor molecules of a specific size and shape. Most likely, many "smells" result from stimulation of a combination of receptors.

a

Figure 12.1 (**a**) The sense of smell—an ancient sensory capacity, and just one of the ways humans gain information about the world. (**b**) Without the ability to sense the position of body parts, these children could not even stay upright— let alone frolic joyfully in a meadow.

By contrast, your taste buds distinguish only four basic types of flavors (sweet, sour, salty, and bitter), and your ability to visually distinguish colors relies on just three general types of photoreceptors (for red, green, or blue light wavelengths). In short, your ability to sniff the difference between pizza pie and apple pie may be the product of one of the more complex elements of the human nervous system.

In this chapter we turn to the means by which the body receives signals from the external and internal environments. As you will read, once those signals are received they are decoded in ways that give rise to awareness of sounds, sights, odors, and other sensations—including those that enable us to feel physical pain and keep our balance. Sensory neurons, nerve pathways, and brain regions are required for these tasks. Together they represent the portions of the nervous system that are called *sensory systems*.

KEY CONCEPTS

1. Sensory systems are portions of the nervous system. Each one consists of specific types of sensory receptors, nerve pathways from receptors to the brain, and brain regions that receive and process sensory information.

2. A stimulus is a form of energy that activates a specific type of sensory receptor, which is either a sensory neuron or a specialized cell next to it. Photoreceptors detect light energy, thermoreceptors detect heat energy, and so on.

3. A sensation is conscious awareness of change in some aspect of the external or internal environment. It begins when sensory receptors detect a specific stimulus. The stimulus energy is converted to a graded, local signal that may help initiate an action potential.

4. Information about the stimulus becomes encoded in the number and frequency of action potentials sent to the brain along particular nerve pathways. Then specific brain regions translate the information into a sensation.

5. The somatic sensations include touch, pressure, temperature, pain, and muscle sense.

6. Taste, smell, hearing, and vision are special senses.

CHAPTER AT A GLANCE

b

Sensory systems, the front doors of the nervous system, receive information about specific changes inside and outside the body and notify the spinal cord and brain of what is going on. The systems all have sensory receptors, nerve pathways from the receptors to the brain, and brain regions where sensory information is processed and translated into sensations. A **sensation** is conscious awareness of a stimulus. It is not the same as **perception**—understanding what the sensation means. *Compound* sensations arise when information about different stimuli is integrated at the same time. For example, our perception of "wetness" comes from simultaneous inputs about pressure, touch, and temperature.

Types of Sensory Receptors

Sensory receptors are the receptionists at the front door of the nervous system. As listed in Table 12.1, there are six major categories of sensory receptors, based on the type of stimulus energy that each type detects:

Mechanoreceptors detect forms of mechanical energy (changes in pressure, position, or acceleration).

Thermoreceptors detect infrared energy (heat).

Nociceptors (pain receptors) detect tissue damage.

Chemoreceptors detect chemical energy of specific substances dissolved in the fluid surrounding them.

Osmoreceptors detect changes in water volume (solute concentration) in the surrounding fluid.

Photoreceptors detect visible light.

Sensory Pathways

Recall, from Chapter 11, that sensory axons carry signals from receptors to the brain. Before they can do so, the stimulus energy must be converted to action potentials, signals that can travel through the nervous system. When a stimulus disturbs the ion balance across the plasma membrane of a receptor ending, some ions flow through protein channels in a patch of the membrane. In other words, the disturbance has triggered a local, graded potential (Section 11.2). Such signals don't spread far from the point of stimulation, and they vary in magnitude.

When a stimulus is intense or when it is repeated fast enough for a summation of local signals, action potentials may be the result. Action potentials propagate themselves from the receptor endings of a sensory neuron to its axon endings. There, neurotransmitter molecules are released from the presynaptic cell and affect the activity of the postsynaptic cell next to it—either an interneuron or a motor neuron. If the effect on the postsynaptic cell is to trigger action potentials there, those signals flow along an information pathway (a nerve tract) leading to the brain. Figure 12.2 depicts this sequence using a light-hearted example of a NASA "planetary rover" running over the foot of a "Martian."

The action potentials that are propagated along sensory neurons are not like a wailing ambulance siren. Their amplitude doesn't vary. How, then, does the brain assess the nature of a given stimulus? It depends on three factors: (1) which sensory area receives signals

Category	Examples	Stimulus
Table 12.1 Major Categories of Sensory Receptors		
MECHANORECEPTORS		
Touch, pressure	Certain free nerve endings and Pacinian corpuscles in skin	Mechanical pressure against body surface
Baroreceptor	Carotid sinus (artery).	Pressure changes in fluid (blood) that bathes them
Stretch	Muscle spindle in skeletal muscle	Stretching
Auditory	Hair cells in organ inside ear	Vibrations (sound or ultrasound waves)
Balance	Hair cells in organ inside ear	Fluid movement
THERMORECEPTORS	Certain free nerve endings	Change in temperature (heating, cooling)
NOCICEPTORS (PAIN RECEPTORS)*	Certain free nerve endings	Tissue damage (e.g., distortions, burns)
CHEMORECEPTORS		
Internal chemical sense	Carotid bodies in blood vessel wall	Substances (O_2, CO_2, etc.) dissolved in extracellular fluid
Taste	Taste receptors of tongue	Substances dissolved in saliva, etc.
Smell	Olfactory receptors of nose	Odors in air, water
OSMORECEPTORS	Hypothalamic osmoreceptors	Change in water volume (solute concentration) of fluid that bathes them
PHOTORECEPTORS		
Visual	Rods, cones of eye	Wavelengths of light

*Extremely intense stimulation of any sensory receptor also may be perceived as pain.

from nerve pathways, (2) the *frequency* of signals that are traveling along each axon that is part of the pathway, and (3) the *number* of axons that responded.

First, specific sensory areas of the brain can interpret action potentials only in certain ways. For example, have you ever wondered why you "see stars" when your eye is poked, even in the dark? This happens because the mechanical pressure on the eye's photoreceptors generated signals that the associated optic nerve carried to a specific sensory area in your brain. That area always interprets incoming signals as "light." (If the signals were somehow rerouted to the sensory area for sound, you would "hear" the poke in the eye.) As described in Section 12.2, the brain has what amounts to a detailed map of the sources of different sensory stimuli.

Second, when a stimulus is strong, receptors fire action potentials more frequently than they do if the stimulus is weaker. The same receptor can detect the sounds of a throaty whisper and a wild screech. The brain senses the difference through frequency variations in the signals that the receptor sends to it.

Third, a stronger stimulus can recruit more sensory receptors than a weaker stimulus can. Gently tap a spot of skin on your arm and you activate a few receptors. Press hard on the same spot and you activate more receptors in a larger area. The increased disturbance translates into action potentials in many sensory axons at the same time. The brain interprets the combined activity as an increase in stimulus intensity. Figure 12.3 is an example of this effect.

In some cases the frequency of action potentials decreases or stops even when a stimulus is being maintained at constant strength. A decrease in the response to a stimulus is called **sensory adaptation**. After you have put on clothing, for example, you no longer are aware of its pressure against your skin. Some mechanoreceptors in your skin adapt rapidly to the sustained stimulation; they are of the type that can only signal a change in the stimulus (when it starts and stops). By contrast, other receptors adapt slowly or not at all; they help the brain monitor particular stimuli all the time. For instance, if you pick up a full sack of sugar, slowly adapting stretch receptors

The interneuron sends the message on to the brain.

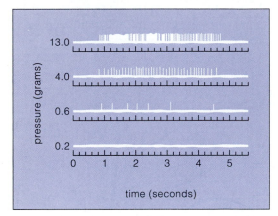

The message travels from the stimulated sensory neuron to interneurons inside the spinal cord.

Figure 12.3 Action potentials recorded from a single pressure receptor of the human hand. The recordings correspond to variations in stimulus strength. A thin rod was pressed against the skin with the pressure indicated on the vertical axis of this diagram. Vertical bars above each thick horizontal line represent individual action potentials. Notice the increases in frequency, which correspond to increases in stimulus strength.

in your arm's skeletal muscle will fire off action potentials and continue to do so for as long as you hold the sack.

The following sections explore specific examples of the sensory receptors we have been discussing. Receptors that are present at more than one location in the body contribute to **somatic** ("of the body") **sensations**. Other receptors are restricted to sense organs, such as the eyes or ears. They contribute to the **special senses**.

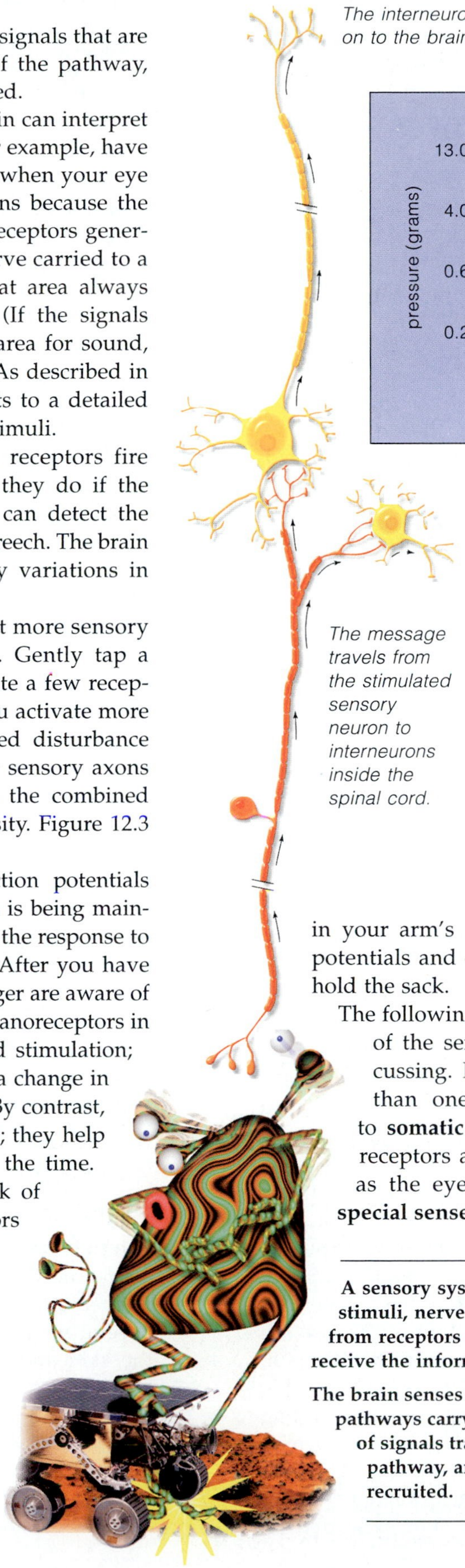

A sensory system has sensory receptors for specific stimuli, nerve pathways that conduct information from receptors to the brain, and brain regions that receive the information.

The brain senses a stimulus based on which nerve pathways carry the incoming signals, the frequency of signals traveling along each axon of that pathway, and the number of axons that have been recruited.

Figure 12.2 NASA Rover running over a Martian's foot and stimulating the receptor endings of a sensory neuron. This is an example of a sensory nerve pathway leading away from a sensory receptor to the brain. The sensory neuron is coded *red*; interneurons are coded *yellow*.

SOMATIC SENSATIONS

Somatic sensations begin with receptors in the body's surface tissues, skeletal muscles, and walls of internal organs. Information from receptors travels into the spinal cord and on into the **somatosensory cortex**, part of the gray matter of the cerebral hemispheres (Section 11.8). The interneurons of this part of the brain are organized in a way that maps the body's surface. The largest areas of the map correspond to body parts where the density of sensory receptors is greatest. Such body parts have the most sensory acuity and require the most intricate control. They include the fingers, thumbs, and lips (Figure 12.4).

Receptors Near the Body Surface

You can discern sensations of touch, pressure, cold, warmth, and pain near your body surface. Regions with the greatest number of sensory receptors, such as the fingertips and the tip of the tongue, are most sensitive to stimulation. Other regions, such as the back of the hand and neck, do not have nearly as many receptors per unit of area and are far less sensitive.

    Free nerve endings are the simplest receptors. They are thinly myelinated or unmyelinated ("naked") branched endings of sensory neurons in the epidermis or in the underlying dermis of skin. They also occur in internal tissues. Different types are mechanoreceptors, thermoreceptors, and pain receptors (nociceptors). All adapt slowly to stimulation. One subgroup of free nerve endings gives rise to a sense of prickling pain—like the "jab" you feel when you stick your finger with a pin. Another subgroup contributes to the sensations of itching or warming that are provoked by chemicals, including histamine. Two thermoreceptive types of free nerve endings are most sensitive to temperatures that are higher or lower than body core temperature, respectively. One mechanoreceptive type coils around hair follicles and detects the movement of the hair inside (Figure 12.5). That might be how, for instance, you become aware that a spider is gingerly feeling its way across your arm.

    Encapsulated receptors, which are surrounded by a capsule of epithelial or connective tissue, are common near the body surface. One type, Meissner's corpuscle, adapts very slowly to low-frequency vibrations. There are many of these receptors in your lips, fingertips, eyelids, nipples, and genitals. The bulb of Krause, an encapsulated thermoreceptor, is activated at 20°C (68°F) or lower. Below 10°C it contributes to painful freezing sensations. Ruffini endings are other slowly adapting encapsulated receptors. They respond to steady touching and pressure, and to temperatures above 45°C (113°F).

    The Pacinian corpuscle is an encapsulated receptor that is widely distributed deep in the skin, in the dermis.

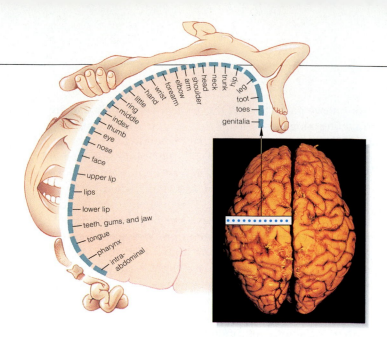

Figure 12.4 Differences in how various body parts are represented in the somatosensory cortex. This region is a strip of cerebral cortex, a little wider than 2.5 centimeters (an inch), from the top of the head to just above the ear.

There, it may have a role in our ability to perceive fine textures, like the delicate fuzz on an apricot. It also is located near freely movable joints (like shoulder and hip joints) and in some internal organs. Onionlike layers of membrane alternating with fluid-filled spaces enclose this sensory ending. This arrangement enhances the receptor's ability to detect rapid pressure changes associated with touch and vibrations.

Muscle Sense

Sensing limb motions and the body's position in space requires mechanoreceptors in skeletal muscles, joints, tendons, ligaments, and skin. Examples include stretch receptors of muscle spindles. As you may recall from Section 11.4, muscle spindles are embedded in skeletal muscle tissue and run parallel with the muscle cells. Their responses to stimulation depend on how much and how fast the muscle stretches.

A Closer Look at the Sensation of Pain

Pain is a perception of injury to some body region. The most important pain receptors are subpopulations of free nerve endings. The technical name for them is nociceptor (from the Latin word *nocere*, "to do harm"). Several million of them are distributed throughout the skin and in internal tissues, excluding the brain. Sensations of *somatic pain* start with nociceptors in skin, skeletal muscles, joints, and tendons. Sensations of *visceral* pain, which is associated with the internal organs, are related to excessive chemical stimulation, muscle spasms, muscle fatigue, inadequate blood flow to organs, and other abnormal conditions.

Further reading: Student Guide to InfoTrac on web site →

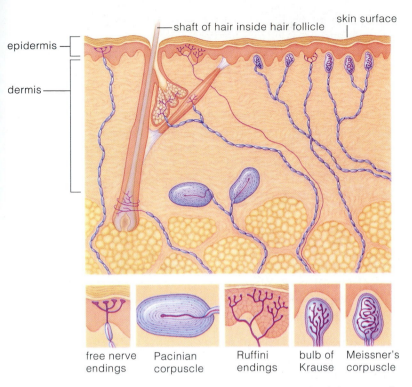

free nerve endings | Pacinian corpuscle | Ruffini endings | bulb of Krause | Meissner's corpuscle

Figure 12.5 Sensory receptors in human skin. Subgroups of free nerve endings function as mechanoreceptors (such as those around hair follicles, which detect hair movements), thermoreceptors, and pain receptors. Pacinian corpuscles detect rapid pressure changes associated with vibrations and touch. Ruffini endings detect steady touching and pressure. The bulb of Krause is most sensitive to cold. Meissner's corpuscle detects vibrations of lower frequency (slower pressure changes) than those detected by Pacinian corpuscles.

When cells are damaged, they release chemicals that activate neighboring pain receptors. The most potent, the bradykinins, open the floodgates for an outpouring of histamine, prostaglandins, and other substances that take part in the inflammatory response (Section 8.3).

Signals from pain receptors reach interneurons in the spinal cord and prompt them to release Substance P. This neurotransmitter causes signals to be relayed along two pathways. Some signals reach the thalamus and the sensory cortex, which rapidly analyzes the type and intensity of pain. Other signals alert the brain stem and limbic system, which mediates arousal and emotional responses to pain. Then, signals from the hypothalamus and midbrain stimulate the release of endorphins and enkephalins—natural opiates (morphinelike substances) that, like the morphine derived from opium poppies, reduce pain perception. Morphine, hypnosis, and natural childbirth techniques may also stimulate the release of these natural opiates.

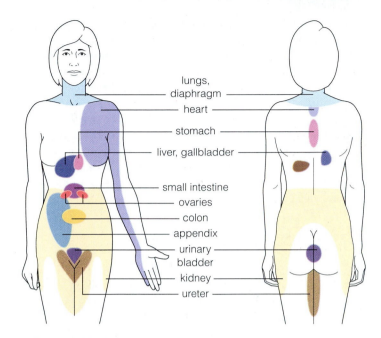

Figure 12.6 Referred pain. Receptors in some internal organs detect painful stimuli. Instead of localizing the pain, at the organs, the brain projects the sensation to the skin areas indicated.

Referred Pain

A person's perception of pain often depends on the brain's ability to identify the affected tissue. Get hit in the face with a snowball and you "feel" the contact on facial skin. However, sensations of pain from some internal organs may be wrongly projected to part of the skin surface. This response, called "referred pain," is related to the way the nervous system is constructed. Possibly, sensory inputs from the skin and from certain internal organs enter the spinal cord along common nerve pathways, so the brain can't accurately identify their source. For example, a heart attack can be felt as pain in skin above the heart and along the left shoulder and arm (Figure 12.6).

Referred pain is not the same as the *phantom pain* reported by amputees. Often they sense the presence of a missing body part, as if it were still there. In some undetermined way, severed sensory nerves continue to respond to the amputation. The brain projects the pain back to the missing part, past the healed region.

Diverse mechanoreceptors detect touch, pressure, heat and cold, pain, limb motions, and changes in the body's position in space. Responses to signals from these receptors give rise to somatic sensations.

We turn now to examples of the special senses, starting with those of taste and smell. Both are *chemical* senses; their sensory pathways start at chemoreceptors, which are activated when they bind a chemical substance that is dissolved in the fluid bathing them. These receptors wear out, and new ones replace them. In both pathways, sensory information travels from the receptors through the thalamus and on to the cerebral cortex, where perceptions of the stimulus take shape and undergo fine-tuning. The input also travels to the limbic system, which can integrate it with emotional states and stored memories (Sections 11.9 and 11.11).

researchers are beginning to discover that our taste sense actually encompasses a richly complex set of molecular mechanisms. *Science Comes to Life* examines the findings of some of these intriguing studies.

The olfactory element of taste is extremely important. In addition to odor molecules in inhaled air, molecules of volatile chemicals are released as you chew food. These waft up into the nasal passages. There, as described next, the "smell" inputs contribute to the perception of a smorgasbord of complex flavors. This is why anything that dulls your sense of smell—such as a head cold—also seems to diminish food's flavor.

Gustation: The Sense of Taste

In the human body, **taste receptors** are part of sense organs called taste buds (Figure 12.7). You have about 10,000 taste buds, which are scattered over your tongue, the roof of your mouth (the palate), and your throat.

A taste bud has a pore through which fluids in the mouth (including, of course, saliva) contact the surface of receptor cells. Of thousands of perceived tastes, all are some combination of four primary tastes: sweet, sour, salty, and bitter. The stimulated receptor in turn stimulates a sensory neuron, which conveys the message to centers in the brain, where the stimulus is interpreted.

Strictly speaking, the flavors of most foods are some combination of the four basic "tastes," plus sensory input from olfactory receptors in the nose. Simple as this sounds,

Olfaction: The Sense of Smell

Olfactory receptors (Figure 12.8) detect water-soluble or volatile (easily vaporized) substances. Recent studies suggest that when odor molecules bind to receptors on olfactory neurons in cells of the nose's olfactory epithelium, the resulting action potential travels directly to olfactory bulbs in the frontal area of the brain. There, other neurons forward the message to a center in the cerebral cortex, which interprets it as "fresh bread," "pine tree," or some other substance.

From an evolutionary perspective, olfaction is one of the most ancient senses—and for good reason. Food, potential mates, and predators give off chemical substances that can diffuse through air (or water) and so give clues or advance warning of their whereabouts.

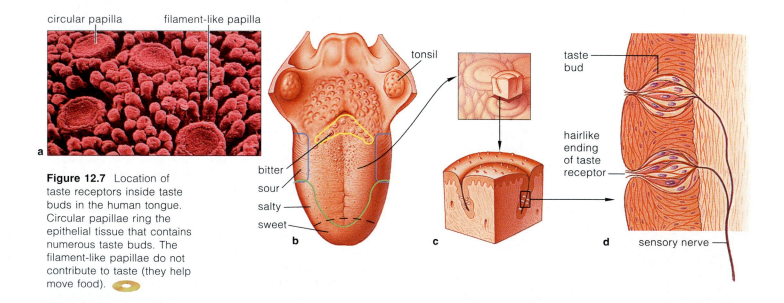

Figure 12.7 Location of taste receptors inside taste buds in the human tongue. Circular papillae ring the epithelial tissue that contains numerous taste buds. The filament-like papillae do not contribute to taste (they help move food).

circular papilla filament-like papilla

a

tonsil

bitter
sour
salty
sweet

b

c

taste bud

hairlike ending of taste receptor

d

sensory nerve

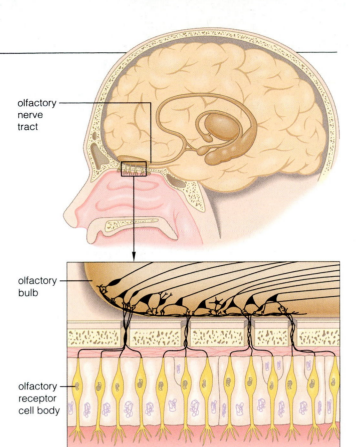

Figure 12.8 Sensory nerve pathway leading from olfactory receptors in the nose to primary receiving centers in the brain.

Even humans, with their rather insensitive sense of smell, have about 10 million olfactory receptors in patches of olfactory epithelium in the upper nasal passages. (The nose of a bloodhound has more than 200 million.)

About half an inch inside the nose, next to the vomer bone (Section 5.3), humans have a vomeronasal organ, or "sexual nose." (Some other mammals also have one.) Its receptors detect pheromones, which are signaling molecules with roles in animals' social interactions. Pheromones affect the behavior—and perhaps the physiology—of other individuals. For instance, one or more pheromones in the sweat of females may account for the common observation that women of reproductive age who are in regular, close contact with one another often come to have their menstrual periods on a similar schedule.

Taste (gustation) relies on receptors in sensory organs called taste buds in the tongue. The receptors bind molecules dissolved in fluid. Associated sensory neurons relay the message to the brain. The four basic tastes are sweet, sour, salty, and bitter.

Olfaction (smell) relies on receptors in patches of epithelium in the upper nasal passages. Neural signals along olfactory neurons travel directly to the olfactory bulbs in the brain.

A TASTE OF SOME SENSORY SCIENCE

Without your taste buds, every meal would be an essential but dull exercise in food intake. With these sensory organs, eating is not only a means of securing nutrients, but also one of life's pleasures. In ways that have been poorly understood until the last decade, the sensory receptors in taste buds distinguish sweet, sour, salty, and bitter—which then become mingled (together with olfactory inputs) into our conscious perceptions of countless flavors. Thanks to methods of molecular biology, we have begun to understand the cellular and molecular details underlying this powerful chemical sense.

Essentially, a taste bud is an organized array of 50 to 150 taste receptor cells. Each receptor cell functions for at most two weeks before being replaced. In that time, it can respond to "tastants" (molecules we can perceive as having a taste) in at least two—and in some cases as many as all four—of the taste classes.

Different kinds of molecules and mechanisms are responsible for each of the four basic tastes. Which taste class (or combination of classes) is ultimately perceived depends on the chemical nature of the "signal" and also on how it is processed by the receptor. In each case, some event causes the receptor cell to release a neurotransmitter (Chapter 10), which triggers action potentials in a nearby sensory neuron.

Many chemical salts, including sodium ions, are perceived as salty. The receptor cell's response is due to the influx of Na^+ through sodium ion channels in its plasma membrane. A receptor responds with a "sour" message when tastant molecules release hydrogen ions that block membrane potassium ion channels. Caffeine and quinine taste bitter, while many "sweet" tastants are sugars (of course!), alcohols, and amino acids. Experiments indicate that both the bitter and the sweet taste messages sent on to the nervous system are mediated by the activity of specific proteins within the receptor cell.

There also are differences in taste receptor sensitivity to various tastants. The receptors tend to be exquisitely sensitive to bitter tastants, which can be detected in extremely low concentrations. Some researchers speculate that this may be an evolutionary adaptation to the fact that many naturally occurring substances that are poisonous to mammals (strychnine is one) are bitter-tasting. Taste receptors must be exposed to somewhat higher concentrations of sour tastants before the stimulus registers; they are least sensitive to (and hence require the highest concentrations of) sweet and salty substances. Relatively small quantities of artificial sweeteners can significantly sweeten foods because their molecular characteristics make them 150 times (aspartame) to more than 600 times (saccharin) as potent as plain sucrose.

SENSE OF HEARING

Sounds are traveling vibrations, or wavelike forms of mechanical energy. For example, clapping produces waves of compressed air. Each time hands clap together, molecules are forced outward, so a low-pressure state is created in the region they vacated. The pressure variations can be depicted as a wave form, and the *amplitude* of its peaks corresponds to loudness. The *frequency* of a sound is the number of wave cycles per second. Each cycle extends from the start of one wave to the start of the next:

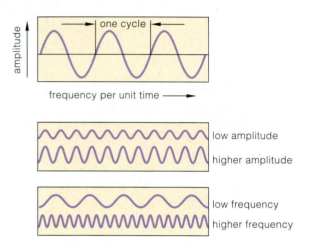

The sense of hearing starts with vibration-sensitive mechanoreceptors deep in the ear. When sound waves travel into the ear's auditory canal, they eventually reach a membrane and make it vibrate. The vibrations then cause a fluid inside the ear to move, the way water in a waterbed sloshes if you push back and forth on its side wall. In your ear, the movement of fluid bends the tips of hairs on mechanoreceptors. With enough bending, the end result will be action potentials that ultimately reach the brain, where they are interpreted as a "sound."

The Ear

A human ear consists of three regions (Figure 12.9a), each with its own role in hearing. The *outer ear* provides a pathway by which sound waves can enter the ear, setting up vibrations. As described shortly, the vibrations are amplified in the *middle ear.* Inside the *inner ear,* vibrations of different sound frequencies are "sorted out" as they stimulate different patches of receptors. The inner ear houses several structures, including *semicircular canals* which are involved in balance—the topic of Section 12.6. It also contains the coiled **cochlea,** where key events in hearing take place. As you'll now read, a coordinated sequence of events in the ear's various regions provides the brain with the auditory input it can interpret to give us a hearing sense.

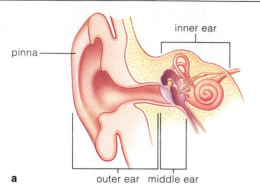

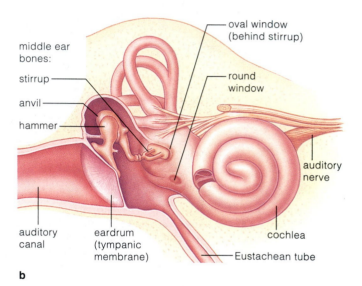

How We Hear

Hearing begins when the external flaps of the outer ear collect sound waves and channel them inward through the auditory canal to the **tympanic membrane** (the eardrum). Sound waves cause the tympanic membrane to vibrate. This vibration in turn causes vibrations in a leverlike array of three tiny bones of the middle ear: the *malleus* ("hammer"), *incus* ("anvil"), and stirrup-shaped *stapes*. The vibrating bones transmit their motion to the *oval window,* an elastic membrane over the entrance to the cochlea. The oval window is much smaller than the tympanic membrane, and as the middle-ear bones vibrate against its small surface with the full energy that struck the tympanic membrane, the force of the original vibrations is amplified.

Now the action, so to speak, shifts to the cochlea. If we could uncoil the cochlea, we would see that a fluid-filled chamber folds around an inner *cochlear duct* (Figure 12.9c). Each long "arm" of the outer chamber functions as a separate compartment (the *scala vestibuli* and *scala tympani*, respectively). The amplified vibrations of the oval window are strong enough to produce pressure waves in the fluid within the chambers. In turn, these waves are transmitted to the fluid in the cochlear duct. On the floor

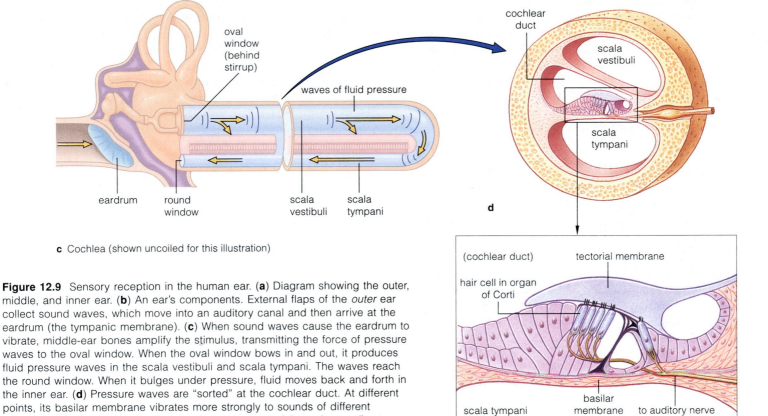

c Cochlea (shown uncoiled for this illustration)

Figure 12.9 Sensory reception in the human ear. (**a**) Diagram showing the outer, middle, and inner ear. (**b**) An ear's components. External flaps of the *outer* ear collect sound waves, which move into an auditory canal and then arrive at the eardrum (the tympanic membrane). (**c**) When sound waves cause the eardrum to vibrate, middle-ear bones amplify the stimulus, transmitting the force of pressure waves to the oval window. When the oval window bows in and out, it produces fluid pressure waves in the scala vestibuli and scala tympani. The waves reach the round window. When it bulges under pressure, fluid moves back and forth in the inner ear. (**d**) Pressure waves are "sorted" at the cochlear duct. At different points, its basilar membrane vibrates more strongly to sounds of different frequencies. (**e**) When the basilar membrane and tectorial membrane vibrate, hair cells sandwiched between them bend. With enough bending, they fire off action potentials that travel via the auditory nerve to the brain.

of the cochlear duct is a *basilar membrane*, and resting on the basilar membrane is a specialized organ, the **organ of Corti**, which includes sensory **hair cells**. Slender projections at the tips of the cells abut against an overhanging **tectorial** ("rooflike") **membrane**, which is not a membrane at all but a firm, jellylike noncellular structure. Pressure waves in the cochlear fluid vibrate the basilar membrane, and its movements can press hair cell projections against the tectorial membrane so that the projections bend. Affected hair cells release a neurotransmitter, which triggers action potentials in neurons of the auditory nerve—the route by which the action potentials reach the brain.

Different sound frequencies cause different parts of the basilar membrane to vibrate—and, accordingly, to bend different patches of hair cells. Pressure waves that vibrate at a higher frequency (more waves per second) set up vibrations in regions of the membrane nearer the entrance to the coil, and waves of lower frequency cause vibrations nearer the tip of the coil. Apparently, the total number of hair cells that are stimulated in a given region determines the loudness of a sound. The perceived tone or "pitch" of a sound depends on the frequency of the vibrations that excite different groups of hair cells—the higher the frequency, the higher the pitch.

Eventually, pressure waves en route through the cochlea push against the *round window*, a membrane at the far end of the cochlea. As the round window bulges outward toward the air-filled middle ear, it serves as a "release valve" for the force of the waves. Air also moves through an opening in the middle ear into the *Eustachian tube*. This tube runs from the middle ear to the throat (pharynx) and permits air pressure in the middle ear to be equalized with the pressure of outside air. When you change altitude (say, riding in aircraft), this equalizing process makes your ears "pop."

Sounds such as amplified music and the thundering of jet engines are so intense that prolonged exposure to them can cause permanent damage to the inner ear (see *Focus on Our Environment*, Section 12.7). Sounds produced by these modern technologies exceed the functional range of the evolutionarily ancient hair cells in the ear.

The sense of hearing relies on mechanoreceptors called hair cells, which are attached to membranes within the cochlea of the inner ear. Pressure waves generated by sound cause membrane vibrations that bend hair cells. The bending results in action potentials in neurons of the auditory nerve.

SENSE OF BALANCE

Like most other animals, we humans must have a sense of the "natural" position for the body (and its parts), given the predictable way we return to it after being tilted or turned upside down. The baseline against which our brain assesses the body's displacement from its natural position is called the *equilibrium position*.

The sense of balance relies partly on input from receptors in our eyes, skin, and joints. In addition, there are organs of equilibrium located in a part of the inner ear called the **vestibular apparatus**. This "apparatus" is a closed system of fluid-filled canals and sacs. Look closely at Figure 12.10a and you can see that there are three canals. They are called **semicircular canals**, and they are positioned at right angles to one another, corresponding to the three planes of space. Inside them, some sensory receptors monitor dynamic equilibrium—that is, rotating head movements. Elsewhere in the vestibular apparatus are receptors that monitor straight-line movements—acceleration and deceleration.

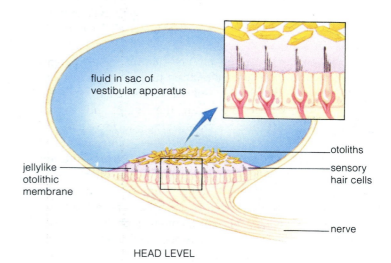

vestibular apparatus, a system of fluid-filled sacs and canals inside the ear

posterior canal

horizontal canal

superior canal

utricle

saccule

nerve

a vestibular apparatus (part of each inner ear) consists of a utricle, a saccule, and the three canals labeled here.

fluid pressure (*blue* arrows)

a gelatinous cupula that stimulates hair cells inside when it bends under pressure

hair cells, which synapse with sensory endings leading to nerve

a

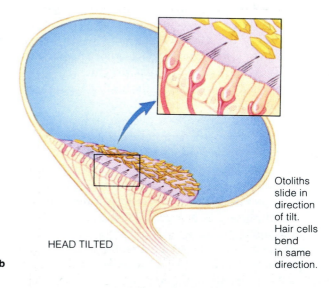

fluid in sac of vestibular apparatus

jellylike otolithic membrane

otoliths

sensory hair cells

nerve

HEAD LEVEL

HEAD TILTED

Otoliths slide in direction of tilt. Hair cells bend in same direction.

b

Figure 12.10 (**a**) Vestibular apparatus, an organ of equilibrium. Its three semicircular canals, positioned at angles corresponding to the three planes of space, detect changes in angular (rotational) movements. (**b**) Moving in response to gravity, the otoliths in otolith organs bend projections of hair cells and so stimulate the endings of sensory neurons that are part of a nerve.

The receptors attuned to *dynamic* equilibrium are clustered in a *crista*, which is present on a ridge of the swollen base of each semicircular canal. As in the cochlea, these receptors are sensory hair cells; their delicate hairs project up into a jellylike *cupula* ("little cap"). When your head rotates horizontally or vertically or tilts diagonally, fluid in a canal corresponding to that direction is displaced in the opposite direction. As the fluid presses against the cupula, the hairs bend. This bending is the first step in a sequence of events leading to action potentials that travel to the brain—in this case, along the vestibular nerve.

The vestibular receptors attuned to *static* equilibrium are in two fluid-filled sacs, the *utricle* and *saccule*. Each sac contains an *otolith organ,* which has hair cells embedded in a jellylike "membrane." The material also contains hard bits of calcium carbonate called *otoliths* ("ear stones"). Movements of the membrane and otoliths signal changes in the head's orientation relative to gravity, as well as straight-line acceleration and deceleration. For example, if you tilt your head to one side, the otoliths slide in that direction, the membrane mass shifts, and projections of the hair cells bend (Figure 12.10b). The otoliths also press on hair cells if your head accelerates, as happens if you set off in an automobile or take a plunge in a roller-coaster.

Action potentials from the different parts of the vestibular apparatus travel to reflex centers in the brain stem. The brain integrates the incoming signals with information from the eyes and muscles, then orders compensatory movements that help you keep your balance when you stand, walk, or move your body in other ways.

Motion sickness can occur when extreme or continuous linear, angular, or vertical motion overstimulates hair cells in the canals of the balance organs. It may also result from conflicting signals from the ears and eyes about motion or the head's position. People who are carsick, airsick, or seasick suffer nausea and may vomit; action potentials triggered by the sensory input reach a brain center that governs the vomiting reflex.

Balance involves a sense of the natural position for the body or its parts.

The sense of balance relies in large measure on input from the vestibular apparatus, a closed system of fluid-filled canals and sacs in the inner ear.

The semicircular canals are positioned at angles corresponding to the three planes of space. Sensory receptors within them detect rotational head movement, acceleration, and deceleration.

The fluid-filled utricles and saccules contain otolith organs, in which sensory hair cells are embedded in a jellylike membrane. Movements of the membrane and otoliths signal changes in the head's orientation relative to gravity, as well as straight-line acceleration and deceleration.

12.7

NOISE POLLUTION: AN ATTACK ON THE EARS

In the United States, roughly 10 million people, including a growing number of musicians, have lost some hearing due to damage caused by loud noise. The National Institutes of Health has estimated that high noise levels in the home, on the job, or in recreational pursuits put another 20 million Americans at risk of hearing loss. One-third of adults in the United States will suffer significant damage to their hearing by the time they are 65. Researchers believe that most cases are due to the long-term effects of living in a noisy world.

The loudness of a sound is measured in *decibels*. A quiet conversation occurs at about 50 decibels, whereas rustling papers make noise at a mere 20 decibels. The delicate sensory hair cells in the inner ear (Figure 12.11) begin to be damaged when a person is exposed to sounds louder than about 75–85 decibels over long periods. Unfortunately, our society is permeated with sounds above that threshold.

For example, a snowmobile or "boom box" stereo system cranks out sound at well over 100 decibels, as can a personal "Walkman"-type stereo if the volume is turned up too high. At 130 decibels—typical of a rock concert or shotgun blast—permanent damage can occur much more quickly. Many health professionals recommend protective earwear for anyone who regularly operates a power mower or other noisy lawn-care equipment. Such protection is (or should be) mandatory for workers in some noisy professions, such as airline service personnel or construction workers who operate jackhammers or chain saws.

For many people it is difficult to avoid noise pollution completely. However, individuals *can* use ear protection when it seems prudent and avoid frequent, prolonged exposure to very loud noise. Following these simple guidelines can help ensure that your hearing lasts as long as you do.

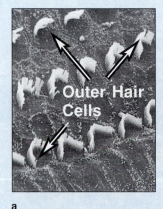
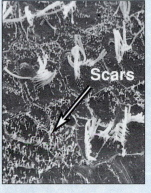

a   b

Figure 12.11 (**a**) Healthy sensory hair cells of the inner ear. (**b**) Hair cells damaged by exposure to loud noise.

VISION: AN OVERVIEW

All organisms may be sensitive to light. **Vision**, however, requires (1) a system of photoreceptors and (2) brain centers that can receive and interpret the patterns of action potentials from different portions of the photoreceptor system. The sense of vision is an awareness of the position, shape, brightness, distance, and movement of visual stimuli. Our **eyes** are sensory organs that contain a tissue with a dense array of photoreceptors.

Eye Structure

The eye has three layers (Table 12.2), sometimes called "tunics." The outer layer consists of a sclera and a transparent **cornea**. The middle layer consists mainly of a choroid, ciliary body, and iris. The key feature of the inner layer is the retina (Figure 12.12).

The *sclera*—the dense, fibrous "white" of the eye—protects most of the eyeball, except for a "front" region formed by the cornea. Moving inward, the thin, dark-pigmented *choroid* underlies the sclera. It prevents light from scattering inside the eyeball and contains most of the eye's blood vessels.

Behind the transparent cornea is a round, pigmented **iris** (after *irid*, which means "colored circle"). The iris has more than 250 measurable features (such as pigments and fibrous tissues). For this reason some advanced security systems use "iris scans" as a form of identification. Look closely at someone's eye, and you will see a "hole" in the center of the iris. This *pupil* is the entrance for light. When bright light hits the eye, circular muscles in the iris contract and shrink the pupil. In dim light, radial muscles contract and enlarge the pupil.

Behind the iris is a saucer-shaped **lens**, with onionlike layers of transparent proteins. Ligaments attach the lens to smooth muscle of the *ciliary body*; this muscle functions in focusing light, as we will see shortly. The lens focuses incoming light onto a dense layer of photoreceptor cells behind it, in the retina. A clear fluid, *aqueous humor* (body fluids were once called "humors"), bathes both sides of the lens. A jellylike substance (*vitreous humor*) fills the chamber behind the lens.

The **retina** is a thin layer of neural tissue at the back of the eyeball. It has a basement layer, a pigmented epithelium that covers the choroid. Resting on the basement layer are densely packed photoreceptors that are functionally linked with a variety of neurons. Axons from some of these neurons converge to form the optic nerve at the back of the eyeball. The optic nerve is the trunk line to the thalamus—which, recall, sends signals on to the **visual cortex.** The place where the optic nerve exits the eye is a "blind spot" because no photoreceptors are present there.

Table 12.2   Components of the Eye

THREE LAYERS FORMING THE WALL OF THE EYEBALL

Fibrous Tunic	Sclera. *Protects eyeball*
	Cornea. *Focuses light*
Vascular Tunic	Choroid. *Blood vessels nutritionally support wall cells; pigments prevent light scattering*
	Ciliary body. *Its muscles control lens shape; its fine fibers hold lens in upright position*
	Iris. *Adjustments here control incoming light*
	Pupil. *Serves as entrance for light*
Sensory Tunic	Retina. *Absorbs and transduces light energy*
	Fovea. *Increases visual acuity*
	Start of optic nerve. *Carries signals to brain*

INTERIOR OF THE EYEBALL

Lens	*Focuses light on photoreceptors*
Aqueous humor	*Transmits light, maintains pressure*
Vitreous body	*Transmits light, supports lens and eyeball*

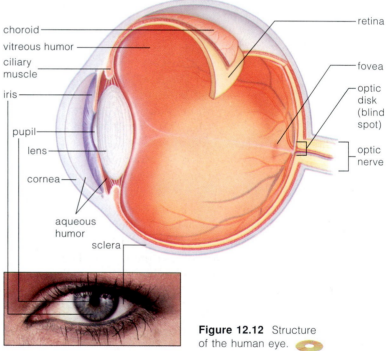

Figure 12.12 Structure of the human eye.

Focusing Mechanisms

The surface of the cornea is curved. This means that incoming light rays hit it at different angles and, as they pass through the cornea, their trajectories (paths) bend. There, because of the way the rays were bent at the curved cornea, the rays converge at the back of the eyeball. They stimulate the retina in a pattern that is upside-down and

Figure 12.13 The pattern of retinal stimulation in the eye. Being curved, the cornea alters the trajectories of light rays as they enter the eye. The pattern is upside-down and reversed left to right, compared to the stimulus.

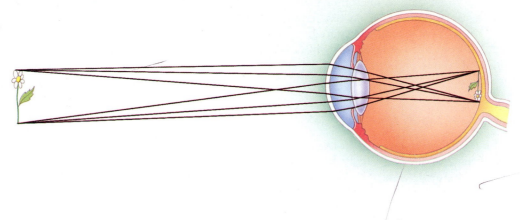

Figure 12.14 Focusing light on the retina by adjusting the lens (visual accommodation). A muscle encircling the lens attaches to it by fiberlike ligaments. (**a**) Close objects are brought into focus when the muscle contracts and makes the lens bulge, so the focal point moves closer. (**b**) Distant objects are brought into focus when the muscle relaxes and makes the lens flatten, so the focal point moves farther back.

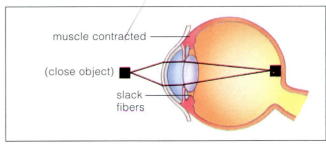

muscle contracted

(close object)

slack fibers

a Accommodation for close objects (lens bulges)

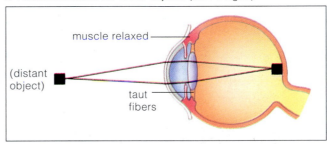

muscle relaxed

(distant object)

taut fibers

b Accommodation for distant objects (lens flattens)

reversed left to right relative to the original source of the light rays. Figure 12.13 gives a simplified diagram of this outcome. The "upside-down and backwards" orientation is corrected in the brain.

Visual Accommodation

Light rays from sources at different distances from the eye strike the cornea at different angles. As a result, they will be focused at different distances behind it. Therefore, adjustments must be made so that the light will be focused precisely on the retina. Normally, the lens can be adjusted so that the focal point coincides exactly with the retina. Ciliary muscle adjusts the shape of the lens. The muscle encircles the lens and attaches to it by fiber-like ligaments (Figure 12.12). When the muscle contracts, the lens bulges, so the focal point moves closer. When the muscle relaxes, the lens flattens, so the focal point moves farther back (Figure 12.14). Such adjustments are called **accommodation**. If they are not made, rays from very distant objects will be in focus at a point slightly in front of the retina, and rays from very close objects will be focused behind it.

Sometimes the lens cannot be adjusted enough to place the focal point on the retina. Sometimes also, the eyeball is not shaped quite right. The lens is too close or too far away from the retina, so accommodation alone cannot produce a precise match. Eyeglasses or contact lenses can correct these problems, which are called far-sightedness and nearsightedness, respectively. Section 12.10 examines these and some other, more serious eye disorders.

Eyes are sensory organs specialized for photoreception.

In the outer eye layer, the sclera protects the eyeball and the cornea focuses light.

In the middle layer, the choroid prevents light scattering, the iris controls incoming light, and the ciliary body and lens aid in focusing light on photoreceptors. Photoreception occurs in the retina of the inner layer.

Adjustments in the position or shape of the lens focus incoming visual stimuli onto the retina.

FROM NEURON SIGNALING TO VISUAL PERCEPTION

We continue now with a superb example of how different neurons can interrelate to one other in the nervous system—the sensory pathway from the retina into the brain. This is how raw visual information is received, transmitted, and combined; it leads to awareness of light and shadows, of colors, and of near and distant objects in the world around us.

Organization of the Retina

The flow of information begins as light reaches the retina, at the back of the eyeball. The retina's basement layer, a pigmented epithelium, covers the choroid. Resting on this layer are densely packed photoreceptors called **rod cells** and **cone cells** (Figure 12.15). Rod cells detect very dim light. During the night or in darkened places, they contribute to coarse perception of movement by detecting changes in light intensity across the field of vision. Cone cells detect bright light. They contribute to daytime vision and color perception.

Neurons in the eye are organized in distinct layers above the rods and cones. As Figure 12.16 shows, information flows from these photoreceptors to the *bipolar* interneurons, then to the interneurons called *ganglion*

cells. The axons of ganglion cells form the two optic nerves to the brain.

Before visual signals depart from the retina, they converge dramatically. Input from 125 million photoreceptors converges on a mere 1 million ganglion cells. Signals also flow laterally among *horizontal* cells and *amacrine* cells. Both types of neurons act in concert to dampen or enhance the signals before they reach ganglion cells. Thus, a great deal of synaptic integration and processing goes on even before visual information is sent to the brain.

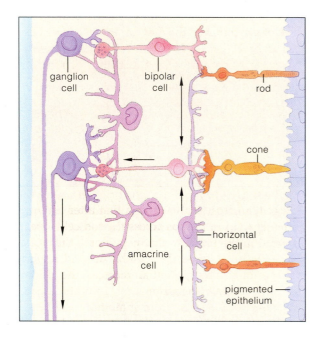

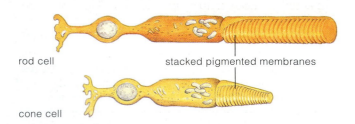

Figure 12.15 Photoreceptors: rods and cones.

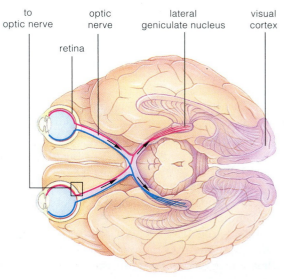

Figure 12.16 Sensory pathway from the retina to the brain.

Neuron Responses to Light

ROD CELLS A rod cell's outer segment contains stacks of several hundred membrane disks, each peppered with about 10^8 molecules of a visual pigment called **rhodopsin**. The membrane stacking and the extremely high density of pigments greatly increase the likelihood of intercepting light energy, which comes packaged as *photons*. (Each photon is a given amount of light energy.) The action potentials that result from the absorption of even a few photons can lead to conscious awareness of objects in dimly lit surroundings, as in dark rooms or late at night.

Each rhodopsin molecule consists of a protein (opsin) to which a signal molecule, *cis*-retinal, is bound. The retinal is derived from vitamin A, which is one reason why lack of vitamin A in the diet can impair vision, especially at night. When the cis-retinal absorbs light energy, it is temporarily converted to a slightly different form, called *trans*-retinal:

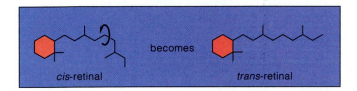

cis-retinal   becomes   *trans*-retinal

This triggers a cascade of reactions that alter activity at ion channels and ion pumps across the rod cell's plasma membrane. Gated sodium channels close, the voltage difference changes across the membrane, and a graded potential results. This potential reduces the ongoing release of a neurotransmitter that inhibits adjacent sensory neurons. Released from inhibition, the sensory neurons start sending signals about the visual stimulus on toward the brain.

CONE CELLS Daytime vision and the sense of color start with photon absorption by red, green, and blue cone cells, each with a different kind of visual pigment. Here again, the absorption of photons reduces the release of a neurotransmitter that otherwise inhibits the neurons that are next to the photoreceptors.

Near the center of the retina is a funnel-shaped depression called the **fovea**. There, photoreceptors are arrayed in the greatest density, and the visual acuity (ability to discriminate) also is greatest. The fovea's dense array of cone cells enables you to discriminate between adjacent points in space with precision.

RESPONSES AT RECEPTIVE FIELDS Collectively, the neurons in the eye respond to light in organized ways. The retina's surface is organized into "receptive fields," which are restricted areas that influence the activity of individual sensory neurons. For example, for each

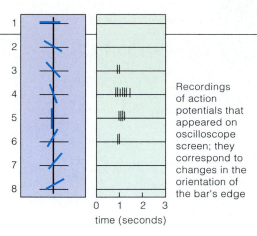

Recordings of action potentials that appeared on oscilloscope screen; they correspond to changes in the orientation of the bar's edge

time (seconds)

Figure 12.17 Example of experiments into the nature of receptive fields for visual stimuli. David Hubel and Torsten Wiesel implanted an electrode in the brain of an anesthetized cat. They positioned the cat in front of a small screen upon which different patterns of light were projected—in this case, a hard-edged bar. Light or shadow falling on a restricted portion of the screen excited or inhibited signals by a single neuron in the visual cortex.

Tilting the bar at different angles produced changes in the neuron's activity. A nearly vertical bar image produced the strongest signal (*numbered 4 in the sketch*). When the bar image tilted slightly, signals were less frequent. When the image tilted past a certain angle, signals stopped.

ganglion cell, the field is a tiny circle. Some cells respond best to a tiny spot of light, ringed by dark, in the field's center. Others respond to a rapid change in light intensity, to a spot of one color, or to motion. In one experiment, cells generated signals about a suitably oriented bar (Figure 12.17). Such cells do not respond to diffuse, uniform illumination, which could produce a confusing array of signals to the brain.

ON TO THE VISUAL CORTEX The part of the outside world that a person actually sees is his or her "visual field." The right side of both retinas intercepts light from the left half of the visual field; the left side intercepts light from the right half. The optic nerve leading out of each eye delivers the signals about a stimulus from the left visual field to the right cerebral hemisphere, and signals from the right to the left hemisphere (see Figure 12.16).

Axons of the optic nerves end in the lateral geniculate nucleus, an island of gray matter in the cerebrum. Its layers each have a map corresponding to receptive fields of the retina. Each map's interneurons deal with one aspect of a visual stimulus—form, movement, depth, color, texture, and so on. After initial processing all the visual signals travel rapidly, at the same time, to different parts of the visual cortex. There, final integration produces organized electrical activity and the sensation of sight

Each eye is an outpost of the brain, collecting and analyzing visual information, then sending it on for further processing in the brain through a highly organized sensory pathway.

12.10 DISORDERS OF THE EYE

Two-thirds of all the sensory receptors in your body are photoreceptors in your eyes. And those photoreceptors do more than detect light. They also allow you to see the world in a rainbow of colors. In fact, for humans, eyes are the single most important source of information about the outside world.

Given their essential role, it is no surprise that attention is paid to eye disorders. Problems that disrupt normal eye functions range from injuries and infectious diseases to inherited abnormalities and changes that come naturally. The outcomes range from some relatively harmless conditions, such as nearsightedness, to total blindness. Each year, many millions of people must deal with such consequences.

COLOR BLINDNESS Occasionally, some or all of the cone cells that selectively respond to light of red, green, or blue are missing. The rare people who have only one of the three kinds of cones are totally color-blind. They see the world only in shades of gray.

Consider a common inherited abnormality, *red-green color blindness*. It shows up most often in males, for reasons described in Chapter 19. The retina lacks some or all of the cone cells with pigments that normally respond to light of red or green wavelengths. Most of the time, color-blind persons merely have trouble distinguishing

red from green in dim light. However, some cannot distinguish between the two even in bright light.

FOCUSING PROBLEMS Other inherited abnormalities arise from misshapen eye structures that affect the eye's ability to focus light. A person with *astigmatism*, for example, has one or both corneas with an uneven curvature; they cannot bend incoming light rays to the same focal point.

Nearsightedness (myopia) commonly occurs when the horizontal axis of the eyeball is longer than the vertical axis. It also occurs when the ciliary muscle responsible for adjusting the lens contracts too strongly. Then, images of distant objects are focused in front of the retina instead of on it (Figure 12.18a). *Farsightedness* (hyperopia) is the opposite problem. The vertical axis of the eyeball is longer than the horizontal axis (or the lens is "lazy"), so close images are focused behind the retina (Figure 12.18b).

Even a normal lens loses some of its natural flexibility as we grow older. This normal stiffening is why people over 40 years old often must start wearing eyeglasses.

EYE DISEASES As with other body parts, the eye and its functions are vulnerable to infection and disease. In some parts of the United States, for example, a fungal infection of the lungs, *histoplasmosis*, can damage the retina if the pathogen moves to the eye. This complication can cause a partial or total loss of vision. As another example, *herpes simplex*, a virus that causes skin sores, also can infect the cornea and cause ulcers there.

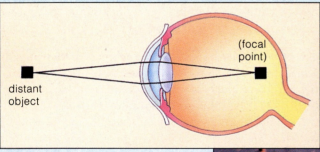

a Focal point in nearsighted vision. The example shows flamingos in Tanzania, East Africa. In a nearsighted person, the birds in flight in the background would be out of focus.

Figure 12.18 Examples of nearsighted and farsighted vision.

Trachoma is a highly contagious disease that has blinded millions, mostly in North Africa and the Middle East. The culprit is a bacterium that also is responsible for the sexually transmitted disease chlamydia (Chapter 16). The eyeball and the lining of the eyelids (the conjunctiva) become damaged. Then, other bacteria can enter the damaged tissues and cause secondary infections. In time the cornea can become so scarred that blindness follows.

AGE-RELATED PROBLEMS *Cataracts*, the gradual clouding of the eye's lens, are associated with aging, although an injury or diabetes can also cause them to develop. The underlying change may be an alteration in the structure of transparent proteins that make up the lens. This change in turn may skew the trajectory of incoming light rays. If the lens becomes totally opaque, no light can enter the eye.

Glaucoma results when too much aqueous humor builds up inside the eyeball. Blood vessels that service the retina collapse under the increased fluid pressure. An affected person's vision deteriorates as neurons of the retina and optic nerve die off. Although chronic glaucoma often is associated with advanced age, the problem actually starts in a person's middle years. If detected early, the fluid pressure can be relieved by drugs or surgery before the damage becomes severe.

EYE INJURIES *Retinal detachment* is the eye injury we read about most often. It may follow a physical blow to the head or an illness that tears the retina. As the jellylike vitreous body oozes through the torn region, the retina is lifted from the underlying choroid. In time it may peel away entirely, leaving its blood supply behind. Early symptoms of the damage include blurred vision, flashes of light that occur in the absence of outside stimulation, and loss of peripheral vision. Without medical help, the person may become totally blind in the damaged eye.

CORRECTIVE TECHNOLOGIES Today many different procedures are used to correct eye disorders. In *corneal transplant surgery*, the defective cornea is removed; then an artificial cornea (made of clear plastic) or a natural cornea from a donor is stitched in place. Within a year, the patient is fitted with eyeglasses or contact lenses. Similarly, cataracts can be surgically corrected by removing the lens and replacing it with an artificial one, although the operation is not always successful.

Severely nearsighted people may opt for *radial keratotomy*, a surgical procedure in which small, spokelike incisions are made around the edge of the cornea to flatten it more. When all goes well, the adjustment eliminates the need for corrective lenses. Several newer procedures use a laser to reshape the cornea.

Retinal detachment may be treatable with *laser coagulation*, a painless technique in which a laser beam seals off leaky blood vessels and "spot welds" the retina to the underlying choroid.

b Focal point in farsighted vision. In a farsighted person, the birds in the foreground would be out of focus, as in this photograph.

SUMMARY

1. A stimulus is a specific form of energy that the body detects by means of sensory receptors. A sensation is a conscious awareness that stimulation has occurred. Perception is understanding what the sensation means.

2. Sensory receptors are endings of sensory neurons or specialized cells next to them. They respond to stimuli, which are specific forms of energy, such as mchanical pressure and light.

  a. Mechanoreceptors detect mechanical energy associated with changes in pressure (e.g., sound waves), changes in position, or acceleration.

  b. Thermoreceptors detect the presence of or changes in radiant energy from heat sources.

  c. Nociceptors (pain receptors) detect tissue damage. Their signals are perceived as pain.

  d. Chemoreceptors, such as taste receptors, detect chemical substances dissolved in the body fluids that are bathing them.

  e. Osmoreceptors detect changes in water volume (hence solute concentrations) in the surrounding fluid.

  f. Photoreceptors, such as rods and cones of the retina, detect light.

3. A sensory system has sensory receptors for specific stimuli and nerve pathways from those receptors to receiving and processing centers in the brain. The brain assesses a particular stimulus based on which nerve pathway is delivering the signals, the frequency of signals traveling along each axon of the pathway, and the number of axons that were recruited into action.

4. Somatic sensations include touch, pressure, pain, temperature, and muscle sense. The receptors associated with these sensations are not localized in a single organ or tissue.

5. The special senses include taste, smell, hearing, balance, and vision. The receptors associated with these senses typically are in sense organs or another specific body region.

6. The senses of taste and smell both involve sensory pathways from chemoreceptors to processing regions in the cerebral cortex and limbic system.

7. The sense of hearing requires components of the outer, middle, and inner ear that respectively collect, amplify, and respond to sound waves. Organs of equilibrium in the inner ear detect gravity, velocity, acceleration, and other factors that affect body positions and movements.

8. Vision requires eyes (sensory organs with a dense array of photoreceptors) and a capacity for image formation in the brain, based upon incoming patterns of visual stimulation.

Review Questions

1. What is a stimulus? When a receptor cell detects a specific kind of stimulus, what happens to the stimulus energy? *12.1*

2. Name six categories of sensory receptors and the type of stimulus energy that each type detects. *12.1*

3. How do somatic sensations differ from special senses? *12.1*

4. Where in the body are you likely to find free nerve endings? What are some functions of the various kinds? *12.2*

5. What is pain? Describe one type of pain receptor. *12.2*

6. What are the stimuli for taste receptors? *12.3*

7. What is the pathway of information flow from an olfactory receptor to the brain? *12.3*

8. List the components of the ear. *12.5*

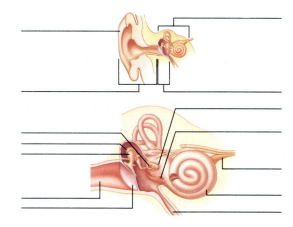

9. In the ear, sound waves cause the tympanic membrane to vibrate. What happens next in the middle ear? In the inner ear? *12.5*

10. Label the components of the eye: *12.8*

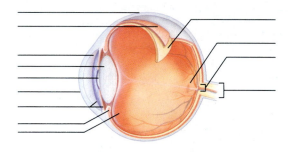

11. How does vision differ from photoreception? What sensory apparatus does vision require? *12.8*

12. How does the eye focus the light rays of an image? What do nearsighted and farsighted mean? *12.8*

Self-Quiz *(Answers in Appendix V)*

1. A _____ is a specific form of energy that can elicit a response from a sensory receptor.

2. Awareness of a stimulus is called a _____ .

3. _____ is understanding what particular sensations mean.

4. A sensory system is composed of _____ .
   a. nerve pathways from specific receptors to the brain
   b. sensory receptors
   c. brain regions that deal with sensory information
   d. all of the above

5. _____ detect energy associated with changes in pressure, in body position, or acceleration.
   a. Chemoreceptors   c. Photoreceptors
   b. Mechanoreceptors   d. Thermoreceptors

6. Detecting substances present in the body fluids that bathe them is the function of _____ .
   a. thermoreceptors   c. mechanoreceptors
   b. photoreceptors   d. chemoreceptors

7. Which of the special senses is based on the following events: Membrane vibrations cause fluid movements, which lead to bending of mechanoreceptors and initiation of action potentials.
   a. taste   c. hearing
   b. smell   d. vision

8. Detecting light energy is the function of _____ .
   a. Pacinian corpuscles   c. rods and cones
   b. vestibular hair cells   d. free nerve endings

9. Vision requires _____ .
   a. a tissue with dense arrays of photoreceptors
   b. eyes
   c. image-forming centers in the brain
   d. all of the above
   e. b and c only

10. The outer layer of the eye includes the _____ .
    a. lens and choroid   c. retina
    b. sclera and cornea   d. both a and c are correct

11. The inner layer of the eye includes the _____ .
    a. lens and choroid   c. retina
    b. sclera and cornea   d. start of optic nerve

12. Match each of the following terms with the appropriate description.
    _____ somatic senses   a. produced by strong
          (general senses)   stimulation
    _____ special senses   b. endings of sensory
    _____ variations in   neurons or specialized
          stimulus intensity   cells next to them
    _____ action potential   c. taste, smell, hearing,
    _____ sensory receptor   balance, and vision
                              d. frequency and number of
                                 action potentials
                              e. touch, pressure,
                                 temperature, pain, and
                                 muscle sense

Critical Thinking: You Decide (Key in Appendix VI)

1. Astronauts living in a space vehicle eat much of their food directly from sealed pouches. They have often complained that such food tastes bland. Given what you know about the senses of taste and smell and the link between them, can you explain why?

2. Juanita consulted her doctor because she was experiencing recurring dizziness. Her doctor asked her whether "dizziness" meant she had sensations of lightheadedness, as if she were going to faint, or whether it meant she had sensations of *vertigo*—that is, a feeling that she herself or objects near her were spinning around. Why was this clarification important for the diagnosis?

3. Michael, a 3-year old, experiences chronic *middle-ear infection*, which is becoming quite common among youngsters, in part due to an increase in antibiotic-resistant bacteria. This year, despite antibiotic treatment, an infection became so advanced that he had trouble hearing. Then his left eardrum ruptured and a jellylike substance dribbled out. The pediatrician told Michael's parents not to worry, that if the eardrum had not ruptured on its own she would have had to insert a drainage tube into it. Speculate on why the pediatrician concluded that this would have been a necessary procedure.

4. Jill is diagnosed with sensorineural deafness, a disorder in which sound waves are transmitted normally to the inner ear but they are not translated into neural signals that travel to the brain. Sometimes the cause is a problem with the auditory nerve, but in Jill's case it has to do with a problem in the inner ear itself. What part of the inner ear is the most likely location of the disruption?

5. Larry goes to the doctor complaining that he can't see the right side of the visual field with either eye. Where in the visual signal-processing pathway is the problem?

Selected Key Terms

accommodation *12.8*
chemoreceptor *12.1*
cochlea *12.5*
cone cell *12.9*
cornea *12.8*
eye *12.8*
fovea *12.9*
hair cell *12.5*
iris *12.8*
lens *12.8*
mechanoreceptor *12.1*
nociceptor *12.1*
organ of Corti *12.5*
perception *12.1*
photoreceptor *12.1*
retina *12.8*

rhodopsin *12.9*
rod cell *12.9*
semicircular canal *12.6*
sensation *12.1*
sensory receptor *12.1*
sensory system *12.1*
somatic sensation *12.1*
somatosensory cortex *12.2*
stimulus *12.1*
tectorial membrane *12.5*
thermoreceptor *12.1*
tympanic membrane *12.5*
vestibular apparatus *12.6*
vision *12.8*
visual cortex *12.9*

Readings

McLaughlin, S., and R. Margolskee. November–December 1994. "The Sense of Taste." *American Scientist*.

Sacks, O. 1998. *The Man Who Mistook His Wife for a Hat*. Real life stories from physician-writer Oliver Sacks, revealing what can be learned about sensory processing in the brain from patients who have suffered brain injuries or other disorders. New York: Simon & Schuster.

Sherwood, L. 1997. *Human Physiology*. Third edition. Pacific Grove, Calif.: Brooks/Cole.

Travis, J. April 10, 1999. "Making Sense of Scents." *Science News*.

Wright, K. April 1994. "The Sniff of Legend." *Discover*. Speculation on the existence of a sensory pathway activated by human pheromones.

Wu, C. April 4, 1998. "Private Eyes." *Science News*. A survey of the emerging technologies of biometrics, and the use of the eye's iris as an identifying character.

13 THE ENDOCRINE SYSTEM

Rhythms and Blues

The pineal gland, a lump of tissue in your brain, secretes the hormone melatonin. When the brain receives sensory signals from your eyes about the waning light at sunset, for example, the pineal gland steps up its melatonin secretion. Molecules of melatonin are picked up by the bloodstream and transported to target cells—in this case, to certain brain neurons. Those neurons are involved in sleep behavior, a lowering of body temperature, and possibly other physiological events. At sunrise, when the eye detects the light of a new day, melatonin production slows down. Your body temperature increases, and you wake up and become active.

The cycle of sleep and arousal is evidence of an internal *biological clock* that seems to tick in synchrony with day length. The clock apparently is influenced by melatonin, and it can be disturbed by a change in circumstances. Jet lag is an example. A traveler from the United States to Paris starts her vacation with four days of disorientation. Two hours past midnight she is sitting up in bed, wondering where the coffee and croissants are. Two hours past noon she is ready for bed. Her body will gradually shift to a new routine as melatonin secretion becomes adjusted so that the hormone signals begin arriving at their target neurons on Paris time.

Figure 13.1 Too little light and too much melatonin, a hormone, may trigger seasonal affective disorder (SAD), or "winter blues," in some people.

Some people are affected by severe *winter blues*, or seasonal affective disorder (SAD). Symptoms typically include depression and an overwhelming desire to sleep (Figure 13.1). Their discomfort may result from a biological clock that is out of synchrony with the changes in day length during winter (when days are shorter and nights longer). Their symptoms worsen when they are given doses of melatonin. And they improve dramatically when they are exposed to intense light, which shuts down pineal activity.

Hormones, our main topic in this chapter, are part and parcel of life's tempos. Some hormones are crucial to basic biological events, such as sexual development and reproduction. They and other signaling molecules are secreted by various organs and tissues and, as we will see, hormones have a range of targets throughout the human body. Indeed, they ultimately control many body functions and influence many aspects of behavior.

KEY CONCEPTS

1. Hormones and other signaling molecules have a central role in integrating the activities of individual cells in ways that benefit the whole body.

2. Only the cells with receptors for a specific hormone are its targets. Hormones operate by serving as signals for change in the activity of target cells.

3. Many types of hormones affect the activity of a target cell's genes and its machinery for synthesizing proteins. Other types are signals that call for changes in existing cell molecules and structures. Some hormones exert their effects by altering membrane characteristics, such as changing the permeability of the plasma membrane to a particular solute.

4. Some hormones help the body adjust to short-term changes in diet and in levels of activity. Other hormones help spur long-term adjustments in cell activities that bring about body growth, development, and reproduction.

5. The hypothalamus and pituitary gland interact in ways that coordinate the activities of a number of endocrine glands. Together, they exert control over many body functions.

6. Particular hormones are secreted in response to signals from other hormones or from the nervous system, or in response to changes in local chemical conditions or environmental cues.

CHAPTER AT A GLANCE

Hormones and Other Signaling Molecules

Our cells must continually respond to changing conditions by taking up and releasing various chemical substances. As in other vertebrates, the responses of millions to many billions of a person's cells must be integrated in ways that benefit the whole body. This crucial integration depends upon signaling molecules—hormones, neurotransmitters, local signaling molecules, and pheromones. Each type of signaling molecule acts on *target cells*. A target cell is any cell that has receptors for the molecule and that may alter its activities in response. A target may or may not be next to the cell that sends the signal.

Hormones are secreted by endocrine glands, endocrine cells, and some neurons, and they travel the bloodstream to target cells some distance away. They are this chapter's focus.

As Chapter 11 described, neurotransmitters operate differently. They are released from the axon endings of neurons, then act swiftly on abutting target cells by diffusing across the tiny gap (synapse) that separates the cells (Section 13.3). In addition, many types of body cells release **local signaling molecules**, which alter conditions within localized regions of tissues. For instance, targets of prostaglandins include smooth muscle cells in the walls of bronchioles, which then constrict or dilate and so alter air flow in lungs (Section 9.6). Certain exocrine glands secrete **pheromones**, which we touched on briefly in Chapter 12. Unlike other signaling molecules, a pheromone acts on targets *outside* the body. It diffuses through air (or water) to the cells of other animals of the same species. Pheromones help integrate social behaviors, such as those related to reproduction. Recall from Section 12.3 that the vomeronasal organ in the human nose is a pheromone detector. Do pheromones act at the subconscious level to trigger impressions, such as spontaneous good or bad "feelings" about someone you just met? We do not know the answer, but studies are under way.

Discovery of Hormones and Their Sources

The word hormone dates back to the early 1900s. W. Bayliss and E. Starling were trying to find out what triggers the secretion of pancreatic juices when food travels through the canine gut. As they knew already, acids mix with food in the stomach, and the pancreas secretes an alkaline solution after the acidic mixture moves into the small intestine. Was the nervous system or something else stimulating the pancreatic response?

To find the answer, Bayliss and Starling blocked nerves—but not blood vessels—leading to a laboratory animal's upper small intestine. Later, when acidic food entered the intestine, the pancreas was still able to respond. Extracts of cells from the intestinal lining—a glandular epithelium—also provoked the response. Therefore, gland cells had to be the source of the pancreas-stimulating substance, which came to be called secretin.

Proof that secretin existed and evidence on how it acts confirmed the centuries-old idea that *internal secretions picked up by the bloodstream influence the activities of the body's organs*. Starling coined the word "hormone" for such secretions (after the Greek *hormon*, "to set in motion"). Later, researchers identified other hormones and their sources. Figure 13.2 shows the locations of the major sources of hormones in the human body.

Collectively, these sources of hormones came to be viewed as the **endocrine system**. The name implies that there is a separate control system for the body, apart from the nervous system. (*Endon* means within; *krinein* means separate.) However, biochemical studies and electron microscopy have revealed that endocrine sources and the nervous system function in intricately connected ways, as you will see shortly.

Today we know that two or more hormones also often affect the same process, in three common ways:

1. *Opposing interaction.* The effect of one hormone opposes the effect of another. Insulin, for example, promotes a decrease in the glucose level in the blood, and glucagon promotes an increase.

2. *Synergistic interaction.* The combined action of two or more hormones is required to produce a particular effect on target cells. For instance, a woman's mammary glands cannot produce and secrete milk without the synergistic interaction of the hormones prolactin, oxytocin, and estrogen.

3. *Permissive interaction.* One hormone exerts its effect on a target cell only when a different hormone first "primes" the target cell. For example, even if one of her eggs is fertilized, a woman can only become pregnant if the lining of her uterus has been exposed to estrogens, then to progesterone.

Hormones are signaling molecules secreted by endocrine glands, endocrine cells, and some neurons. The bloodstream distributes hormones to distant target cells.

Together, the various glands and cells that secrete hormones make up the endocrine system.

Hormones may interact in opposition, as when the effect of one blocks the effect of another; synergistically, as when the actions of two or more hormones are necessary to produce the effect on target cells; or permissively, as when a target cell must be primed by exposure to one hormone before it can respond to a second one.

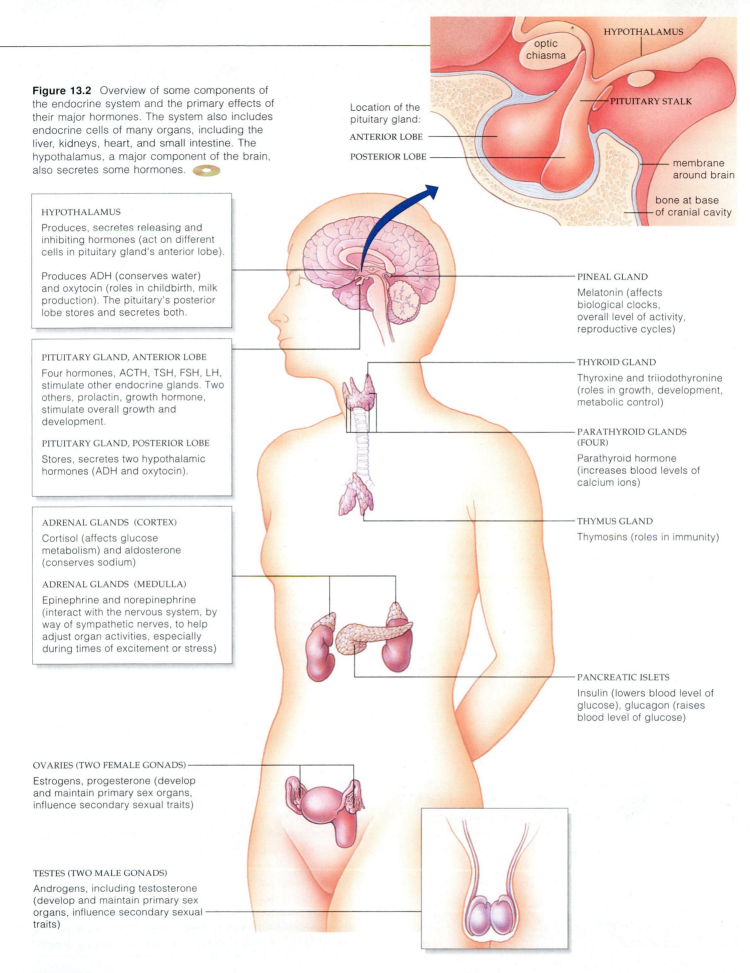

Figure 13.2 Overview of some components of the endocrine system and the primary effects of their major hormones. The system also includes endocrine cells of many organs, including the liver, kidneys, heart, and small intestine. The hypothalamus, a major component of the brain, also secretes some hormones.

Location of the pituitary gland:

ANTERIOR LOBE

POSTERIOR LOBE

optic chiasma

HYPOTHALAMUS

PITUITARY STALK

membrane around brain

bone at base of cranial cavity

HYPOTHALAMUS

Produces, secretes releasing and inhibiting hormones (act on different cells in pituitary gland's anterior lobe).

Produces ADH (conserves water) and oxytocin (roles in childbirth, milk production). The pituitary's posterior lobe stores and secretes both.

PITUITARY GLAND, ANTERIOR LOBE

Four hormones, ACTH, TSH, FSH, LH, stimulate other endocrine glands. Two others, prolactin, growth hormone, stimulate overall growth and development.

PITUITARY GLAND, POSTERIOR LOBE

Stores, secretes two hypothalamic hormones (ADH and oxytocin).

ADRENAL GLANDS (CORTEX)

Cortisol (affects glucose metabolism) and aldosterone (conserves sodium)

ADRENAL GLANDS (MEDULLA)

Epinephrine and norepinephrine (interact with the nervous system, by way of sympathetic nerves, to help adjust organ activities, especially during times of excitement or stress)

OVARIES (TWO FEMALE GONADS)

Estrogens, progesterone (develop and maintain primary sex organs, influence secondary sexual traits)

TESTES (TWO MALE GONADS)

Androgens, including testosterone (develop and maintain primary sex organs, influence secondary sexual traits)

PINEAL GLAND

Melatonin (affects biological clocks, overall level of activity, reproductive cycles)

THYROID GLAND

Thyroxine and triiodothyronine (roles in growth, development, metabolic control)

PARATHYROID GLANDS (FOUR)

Parathyroid hormone (increases blood levels of calcium ions)

THYMUS GLAND

Thymosins (roles in immunity)

PANCREATIC ISLETS

Insulin (lowers blood level of glucose), glucagon (raises blood level of glucose)

SIGNALING MECHANISMS

Overview of Hormone Action

Hormones and other signaling molecules interact with protein receptors of target cells. Those interactions affect physiological processes in various ways. Some kinds of hormones induce a target cell to increase its uptake of glucose or some other substance. Others stimulate or inhibit the target in ways that alter its rates of protein synthesis, modify existing proteins or other structures in the cytoplasm, or even change cell shape.

Two factors have a major influence on responses to hormone signals. First of all, different hormones activate different kinds of mechanisms in target cells. Second, not all types of cells can respond to a given signal. For instance, many types of cells have receptors for cortisol, so that hormone's effects are widespread in the body. If only a few cell types have receptors for a hormone, its effects in the body are limited to the receptive areas.

We assign hormones to chemical families, depending on their "parent" molecule. As indicated in Table 13.1, steroid hormones are derived from cholesterol. A second group, the amines, are synthesized from the amino acid tyrosine. A third group, including peptide and protein hormones, consists of nonsteroid hormones that are derived from amino acids *other than* tyrosine.

Steroid Hormones

Steroid hormones, recall, are lipid-soluble molecules synthesized from cholesterol (Section 2.10). The synthesis process takes places in the endoplasmic reticulum and mitochondria of cells that secrete steroid hormones. The adrenal glands and primary reproductive organs (ovaries and testes) have such cells. Testosterone, one of the sex hormones, is one such product.

Consider how testosterone affects the development of secondary sexual traits associated with maleness. The developmental steps will advance normally only if target cells have working receptors for testosterone. In a genetic disorder called *testicular feminization syndrome*, these receptors are defective. The affected person has a male's genetic makeup (XY) and has functional testes that can secrete testosterone. Target cells cannot respond to the

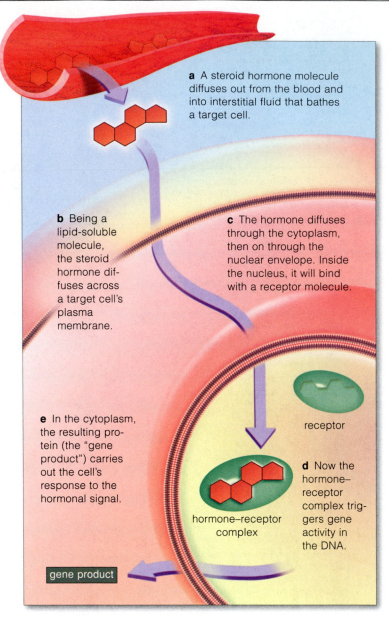

a A steroid hormone molecule diffuses out from the blood and into interstitial fluid that bathes a target cell.

b Being a lipid-soluble molecule, the steroid hormone diffuses across a target cell's plasma membrane.

c The hormone diffuses through the cytoplasm, then on through the nuclear envelope. Inside the nucleus, it will bind with a receptor molecule.

e In the cytoplasm, the resulting protein (the "gene product") carries out the cell's response to the hormonal signal.

receptor

d Now the hormone–receptor complex triggers gene activity in the DNA.

hormone–receptor complex

gene product

Figure 13.3 Example of a mechanism by which a steroid hormone initiates changes in a target cell's activities.

hormone, however, so the person's secondary sexual traits, including breasts, are female. Some researchers speculate that the French heroine Joan of Arc had this disorder; as evidence they cite reliable historical accounts that she had no monthly menstrual periods and showed other symptoms of testicular feminization.

How does a steroid hormone such as testosterone exert effects on cells? Being lipid soluble, it may diffuse directly across the lipid bilayer of a target cell's plasma membrane (Figure 13.3). Once it is inside the cytoplasm, the hormone molecule usually moves into the nucleus and binds to some type of receptor. Or, in some cases, it binds to a receptor molecule in the cytoplasm, then the hormone-receptor complex enters the nucleus. The shape of the complex allows it to interact with a specific

Table 13.1	Two Main Categories of Hormones
STEROIDS	Estrogens, testosterone, progesterone, aldosterone, cortisol
NONSTEROIDS:	
Amines	Norepinephrine, epinephrine
Peptides	ADH, oxytocin, TRH
Proteins	Insulin, growth hormone, prolactin
Glycoproteins	FSH, LH, TSH

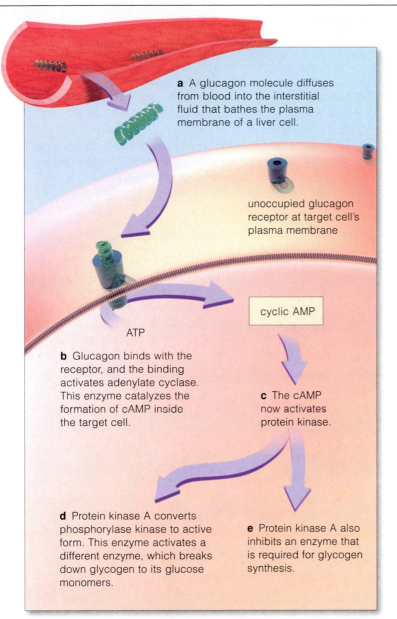

a A glucagon molecule diffuses from blood into the interstitial fluid that bathes the plasma membrane of a liver cell.

unoccupied glucagon receptor at target cell's plasma membrane

cyclic AMP

ATP

b Glucagon binds with the receptor, and the binding activates adenylate cyclase. This enzyme catalyzes the formation of cAMP inside the target cell.

c The cAMP now activates protein kinase.

d Protein kinase A converts phosphorylase kinase to active form. This enzyme activates a different enzyme, which breaks down glycogen to its glucose monomers.

e Protein kinase A also inhibits an enzyme that is required for glycogen synthesis.

Figure 13.4 Example of a mechanism by which a peptide hormone initiates changes in the activity of a target cell. When the hormone glucagon binds at a receptor, it spurs reactions inside the cell. In this case, cyclic AMP, a type of second messenger, relays the signal to the cell's interior.

region of the cell's DNA. Regions of DNA are genes that contain instructions for making proteins. Different complexes stimulate (or inhibit) protein synthesis by switching certain genes on (or off). The resulting proteins carry out the target cell's response to the hormonal signal.

Steroid hormones also may exert effects in another way. Research suggests that some may have receptors on cell membranes and may alter the membrane properties in ways that modify the functions of the target cell.

One other point: Thyroid hormones and vitamin D are not steroid hormones, but they behave like them. The genes that code for their receptors are part of the same group that codes for steroid hormones.

Amine, Peptide, and Protein Hormones

Many hormones are derived from amino acids. In this group are *amine* hormones, *peptide* hormones, and *protein* (polypeptide) hormones. Due to their chemical makeup, such hormones are soluble in water, but not in lipid. Thus, like other water-soluble signaling molecules, they cannot cross the lipid-rich plasma membrane of a target cell. Instead, they require the aid of membrane proteins in order to exert their effect on target cells.

For an example, let's focus on a liver cell that has receptors for glucagon, a peptide hormone. This type of receptor spans the plasma membrane. Part of it extends into the cytoplasm. When a receptor binds glucagon, the cell produces a **second messenger**. (The hormone itself is the "first messenger.") Second messengers are small molecules in the cytoplasm that relay signals from hormone-receptor complexes into a cell. In this case, the second messenger is cAMP (cyclic adenosine monophosphate). Glucagon binding triggers activity at a membrane-bound enzyme system (Figure 13.4).

An activated enzyme (adenyl cyclase) starts a cascade of reactions by converting ATP to cyclic AMP. Many molecules of cyclic AMP form. They act as signals to convert many molecules of a protein kinase, another type of enzyme, to active form. These act on other enzymes, and so on until a final reaction converts stored glycogen in the cell to glucose. In short order, the number of molecules involved in the final response to the glucagon-receptor complex is enormous.

For a slightly different example, consider a muscle cell with receptors for insulin, which is a protein hormone. Among other things, when insulin binds to the receptor, the complex induces molecules of proteins called glucose transporters to move through the cytoplasm and insert themelves into the plasma membrane. They help cells take up glucose faster. The signal also activates enzymes that catalyze reactions by which glucose is stored.

Amine hormones include melatonin and epinephrine, among others. Like glucagon, epinephrine combines with specific receptors at the target cell surface. Binding triggers the release of cyclic AMP as a second messenger that assists in the target cell response.

Steroid hormones stimulate or inhibit protein synthesis by entering the nucleus of target cells and switching certain genes on or off.

Amine and peptide hormones enter cells either by receptor-mediated endocytosis or by activating membrane proteins. They bind to receptors and so activate second messengers in the cell.

Second messengers have various functions. Often they trigger an amplified response to the hormone.

Recall from Chapter 11 that the **hypothalamus**, located deep in the forebrain, monitors internal organs and states related to their functioning, such as eating. It influences certain behaviors, such as sexual behavior. It also secretes some hormones. Suspended from its base by a slender stalk is a lobed gland about the size of a pea. This **pituitary gland** and the hypothalamus interact closely as a major neural-endocrine control center.

The pituitary's *posterior* lobe stores and secretes two hormones that are synthesized by the hypothalamus. Its *anterior* lobe produces and secretes a half dozen of its own hormones, most of which govern the release of hormones from other endocrine glands (Table 13.2).

Posterior Lobe Hormones

Figure 13.5 shows the cell bodies of certain neurons in the hypothalamus. Notice that their axons extend down into the posterior lobe, then end next to a capillary bed. The neurons produce antidiuretic hormone (ADH) and oxytocin, then store them in the axon endings. When either hormone is released, it diffuses through interstitial fluid and enters capillaries, then travels the bloodstream to its targets. ADH acts on cells of nephrons and collecting ducts in the kidneys. Kidneys, remember, filter blood and rid the body of excess water and salts (in urine). ADH promotes water reabsorption when the body must conserve water.

Through the gamut of events that can cause blood pressure to drop—from the loss of body water through sweating in a warm classroom to severe blood loss due to

injury—the hypothalamus keeps tabs on the fluctuations, so to speak, and releases ADH into the bloodstream when blood pressure falls below a set point. ADH causes the arterioles in some tissues to constrict, and so systemic blood pressure rises. (For this reason, ADH is sometimes called vasopressin.)

Oxytocin has roles in reproduction in both males and females. For example, it triggers muscle contractions in the uterus during labor and causes milk to be released when a mother nurses her infant.

Anterior Lobe Hormones

Inside the pituitary stalk, a capillary bed picks up hormones secreted by the hypothalamus and delivers them to a *second* capillary bed in the anterior lobe. There the hormones leave the bloodstream and then act on target cells. As Figure 13.6 shows, different cells of the anterior pituitary secrete six hormones that they themselves produce:

Corticotropin	ACTH
Thyrotropin	TSH
Follicle-stimulating hormone	FSH
Luteinizing hormone	LH
Prolactin	PRL
Growth hormone (somatotropin)	GH (or STH)

All these hormones have widespread effects in the body. ACTH and TSH orchestrate secretions from the adrenal glands and thyroid gland, respectively, as described shortly. FSH and LH act through the gonads to influence reproduction, the main topic of Chapter 14. Prolactin acts

Table 13.2 Hormones Released from the Pituitary Gland				
Pituitary Lobe	Secretions	Designation	Main Targets	Primary Actions
POSTERIOR Nervous tissue (extension of hypothalamus)	Antidiuretic hormone	ADH	Kidneys	Induces water conservation required in control of extracellular fluid volume (and, indirectly, solute concentrations)
	Oxytocin		Mammary glands	Induces milk movement into secretory ducts
			Uterus	Induces uterine contractions
ANTERIOR Mostly glandular tissue	Corticotropin	ACTH	Adrenal cortex	Stimulates release of adrenal steroid hormones
	Thyrotropin	TSH	Thyroid gland	Stimulates release of thyroid hormones
	Gonadotropins: Follicle-stimulating hormone	FSH	Ovaries, testes	In females, stimulates egg formation; in males, helps stimulate sperm formation
	Luteinizing hormone	LH	Ovaries, testes	In females, stimulates ovulation, corpus luteum formation; in males, promotes testosterone secretion, sperm release
	Prolactin	PRL	Mammary glands	Stimulates and sustains milk production
	Growth hormone (also called somatotropin)	GH (STH)	Most cells	Promotes growth in young; induces protein synthesis, cell division; roles in glucose, protein metabolism in adults

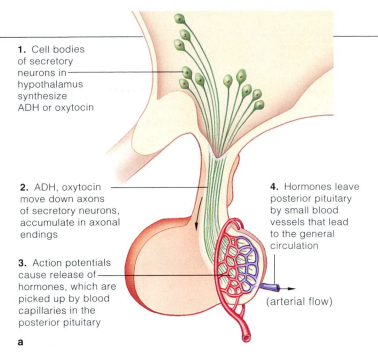

1. Cell bodies of secretory neurons in hypothalamus synthesize ADH or oxytocin

2. ADH, oxytocin move down axons of secretory neurons, accumulate in axonal endings

3. Action potentials cause release of hormones, which are picked up by blood capillaries in the posterior pituitary

4. Hormones leave posterior pituitary by small blood vessels that lead to the general circulation

(arterial flow)

a

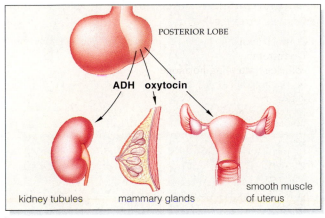

POSTERIOR LOBE

ADH   oxytocin

kidney tubules   mammary glands   smooth muscle of uterus

b

Figure 13.5 (**a**) Functional links between the hypothalamus and the posterior lobe of the pituitary. (**b**) Main targets of the posterior lobe secretions.

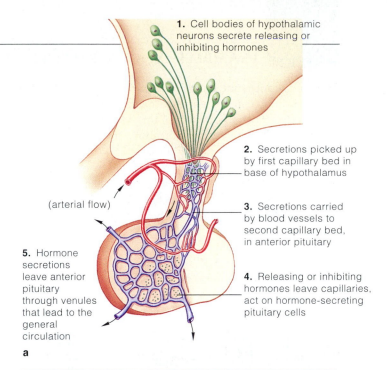

1. Cell bodies of hypothalamic neurons secrete releasing or inhibiting hormones

2. Secretions picked up by first capillary bed in base of hypothalamus

3. Secretions carried by blood vessels to second capillary bed, in anterior pituitary

(arterial flow)

5. Hormone secretions leave anterior pituitary through venules that lead to the general circulation

4. Releasing or inhibiting hormones leave capillaries, act on hormone-secreting pituitary cells

a

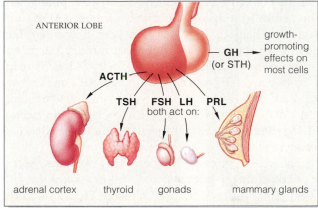

ANTERIOR LOBE

ACTH

TSH   FSH LH   PRL
both act on:

GH (or STH) → growth-promoting effects on most cells

adrenal cortex   thyroid   gonads   mammary glands

b

Figure 13.6 (**a**) Functional links between the hypothalamus and the anterior lobe of the pituitary. (**b**) Main targets of the anterior lobe secretions.

on different cell types. However, it is best known as the hormone that stimulates and sustains milk production in mammary glands, after other hormones have primed the tissues.

Growth hormone (GH) affects most body tissues. It stimulates protein synthesis and cell division in target cells, and it profoundly influences growth, especially of cartilage and bone. GH is equally important as a "metabolic hormone." Throughout life it stimulates cells to take up amino acids and promotes the breakdown and release of fat stored in adipose tissues, thereby increasing the amount of fatty acids available to cells. GH also moderates the rate at which cells take up glucose, which helps maintain appropriate blood levels of that cellular fuel.

Most hypothalamic hormones that act in the anterior lobe of the pituitary are **releasers**—they stimulate target cells to secrete other hormones. For example, GnRH (gonadotropin-releasing hormone) brings about the

secretion of FSH and LH—both of which are classified as gonadotropins. Similarly, TRH (thyrotropin-releasing hormone) stimulates the secretion of TSH. Other hypothalamic hormones are **inhibitors**—they block secretions from their targets in the anterior pituitary. One example, somatostatin, inhibits secretion of growth hormone and thyrotropin.

The hypothalamus and pituitary gland produce eight hormones and interact to control their secretion.

The pituitary's posterior lobe stores and secretes ADH and oxytocin, both of which target specific cell types.

The anterior lobe of the pituitary produces and secretes six hormones—ACTH, TSH, FSH, LH, PRL, and GH. These trigger the release of other hormones from other endocrine glands, with wide-ranging effects throughout the body.

EXAMPLES OF ABNORMAL PITUITARY OUTPUT

The body does not churn out enormous numbers of hormone molecules. Two researchers, Roger Guilleman and Andrew Schally, realized this when they isolated the first-known releasing hormone. In their four-year attempt to secure TRH, they dissected 500 tons of brains and 7 metric tons of hypothalamic tissue from sheep and ended up with only a single milligram of it.

Yet normal body function depends on the tiny but significant amounts of endocrine gland secretions, which commonly are released in short bursts. Controls over the frequency of secretion prevents hormones from being either overproduced or underproduced. If something interferes with the controls, the body's form and functioning may be altered in abnormal ways.

For instance, *gigantism* results from overproduction of growth hormone during childhood. Affected adults are proportionally like an average-sized person but much larger (Figure 13.7a). *Pituitary dwarfism* results from underproduction of growth hormone. Affected adults are proportionally similar to an average person but much smaller (Figure 13.7b).

What if the output of growth hormone becomes excessive during adulthood, when long bones such as the femur no longer can lengthen? Bone, cartilage, and other connective tissues in the hands, feet, and jaws thicken abnormally. So do epithelia of the skin, nose, eyelids, lips, and tongue. This outcome is called *acromegaly* (Figure 13.7c).

a

b

age 9 · age 16 · age 33 · age 52

c

Figure 13.7 (**a**) Basketball player Manute Bol is 7 feet 6-3/4 inches tall owing to excessive GH production during childhood.

(**b**) Effect of growth hormone (GH) on overall body growth. The person at the center is affected by gigantism, which resulted from excessive GH production during childhood. The person at right displays pituitary dwarfism, which resulted from underproduction of GH during childhood. The person at the left is average in size.

(**c**) Acromegaly, which resulted from excessive production of GH during adulthood. Before this female reached maturity, she was symptom-free.

As another example, ADH secretion may fall or stop if the pituitary's posterior lobe is damaged, as by a blow to the head. This is one cause of *diabetes insipidus*. A person with this disorder excretes a greal deal of dilute urine—so much that serious dehydration can occur. Hormone replacement therapy is an effective treatment.

Generally, endocrine glands release very small amounts of hormones. Control mechanisms govern the frequency of releases. When controls fail, the resulting oversecretion or undersecretion may cause disorders.

Further reading: Student Guide to InfoTrac on web site →

SOURCES AND EFFECTS OF OTHER HORMONES

Table 13.3 lists hormones from endocrine sources other than the pituitary. The remainder of this chapter will provide you with a few examples of their effects and of the controls over their output. The examples will make more sense if you keep the following points in mind.

First, hormones often interact with one another. Second, negative feedback mechanisms usually control the secretions. When a hormone's concentration increases or decreases in some body region, the change triggers events that respectively dampen or stimulate further

secretion. Third, a target cell may react differently to a hormone at different times. Its response depends on the hormone's concentration and on the functional state of the cell's receptors. Fourth, environmental cues, such as light, can be important factors in hormone secretion.

The secretion of a hormone and its effects are influenced by hormone interactions, feedback mechanisms, variations in the state of target cells, and sometimes environmental cues.

Table 13.3   Hormone Sources Other Than the Hypothalamus and Pituitary

Source	Secretion(s)	Main Targets	Primary Actions
PANCREATIC ISLETS	Insulin	Muscle, adipose tissue	Lowers blood sugar level
	Glucagon	Liver	Raises blood sugar level
	Somatostatin	Insulin-secreting cells	Influences carbohydrate metabolism
ADRENAL CORTEX	Glucocorticoids (including cortisol)	Most cells	Promote protein breakdown and conversion to glucose
	Mineralocorticoids (including aldosterone)	Kidney	Promote sodium reabsorption; control salt–water balance
ADRENAL MEDULLA	Epinephrine (adrenalin)	Liver, muscle, adipose tissue	Raises blood level of sugar, fatty acids; increases heart rate, force of contraction
	Norepinephrine	Smooth muscle of blood vessels	Promotes constriction or dilation of blood vessel diameter
THYROID	Triiodothyronine, thyroxine	Most cells	Regulate metabolism; have roles in growth, development
	Calcitonin	Bone	Lowers calcium levels in blood
PARATHYROIDS	Parathyroid hormone	Bone, kidney	Elevates levels of calcium and phosphate ions in blood
THYMUS	Thymosins, etc.	Lymphocytes	Have roles in immune responses
GONADS:			
Testes (in males)	Androgens (including testosterone)	General	Required in sperm formation, development of genitals, maintenance of sexual traits; influence growth, development
Ovaries (in females)	Estrogens	General	Required in egg maturation and release; prepare uterine lining for pregnancy; required in development of genitals, maintenance of sexual traits; influence growth, development
	Progesterone	Uterus, breasts	Prepares, maintains uterine lining for pregnancy; stimulates breast development
PINEAL	Melatonin	Hypothalamus	Influences daily biorhythms
ENDOCRINE CELLS OF STOMACH, GUT	Gastrin, secretin, etc.	Stomach, pancreas, gallbladder	Stimulate activity of stomach, pancreas, liver, gallbladder
LIVER	IGFs (Insulin-like growth factors)	Most cells	Stimulate cell growth and development
KIDNEYS	Erythropoietin	Bone marrow	Stimulates red blood cell production
	Angiotensin*	Adrenal cortex, arterioles	Helps control blood pressure, aldosterone secretion
	Vitamin D$_3$*	Bone, gut	Enhances calcium resorption and uptake
HEART	Atrial natriuretic hormone	Kidney, blood vessels	Increases sodium excretion; lowers blood pressure

*These hormones are not produced in the kidneys but are formed when enzymes produced in kidneys activate specific substances in the blood.

If you look carefully at Table 13.3 on the previous page, you get a sense of how feedback mechanisms control hormone secretions. The hypothalamus, the pituitary, or both, signal other glands to alter their secretory activity. As a result, the concentration of a secreted hormone changes in the blood or elsewhere. With the shift in chemical signals, a feedback mechanism trips into action and blocks or promotes further change.

Recall from Section 4.11 that with *negative feedback*, an increase in the concentration of a secreted hormone triggers activities that put the brakes on secretion. With *positive feedback*, an increase in the concentration of a secreted hormone triggers leads to *additional* secretion.

Negative Feedback from the Adrenal Cortex

We have two adrenal glands, one perched atop each kidney. The outer portion of each gland is the **adrenal cortex**. There, cells secrete two major types of hormones, glucocorticoids and mineralocorticoids.

Glucocorticoids influence metabolism in ways that help raise the blood level of glucose when it falls below a set point. In humans, *cortisol* is the primary glucocorticoid. Among other effects, it promotes protein breakdown in muscle and stimulates the liver to take up amino acids, from which liver cells synthesize glucose in a process called *gluconeogenesis*. Cortisol also reduces how much glucose tissues such as skeletal muscle take up from the bloodstream to use for cellular fuel. This effect is sometimes called "glucose sparing." In addition, cortisol promotes fat breakdown and use of the resulting fatty acids for energy. Glucose sparing is extremely important

in homeostasis: it helps ensure that the concentration of glucose in blood will be adequate to meet the needs of the brain, which generally cannot use other molecules for fuel.

In general, a negative feedback loop to the neuroendocrine control center operates to control the secretion of cortisol. When the blood level of the hormone rises above a set point, the hypothalamus begins to secrete less of the releasing hormone CRH (Figure 13.8). Then the anterior pituitary responds by secreting less ACTH and the adrenal cortex slows its secretion of cortisol. In a healthy person, daily cortisol secretion is highest when the blood glucose level is lowest, usually in the early morning. Chronic severe *hypoglycemia*, a persistent low concentration of glucose in the blood, can develop when a person has a cortisol deficiency. Then, the mechanisms that spare glucose and generate new supplies in the liver do not operate properly.

Glucocorticoids suppress the inflammatory response to tissue injury or infection (Section 8.3). In fact, their secretion increases anytime a person experiences unusual physiological stress. A painful injury, a severe illness, or an allergic reaction may trigger shock, inflammation, or both. Then, increased secretion of cortisol and other signaling molecules is vital to recovery. That is why cortisol-like drugs, such as cortisone, are administered after an asthma attack and episodes of serious inflammatory disorders. (However, prolonged use of large doses of glucocorticoids has serious side effects, including suppressing the immune system.) Cortisone is the active ingredient in many over-the-counter creams and lotions for treating skin irritations.

In the main, **mineralocorticoids** regulate concentrations of mineral salts, such as potassium and sodium, in the extracellular fluid. The most abundant mineralocorticoid is *aldosterone*. Recall that aldosterone acts on the distal tubules of kidney nephrons (Section 10.7), stimulating them to reabsorb sodium ions and excrete potassium ions. Sodium reabsorption in turn promotes reabsorption of water from the tubules as urine is forming. Various circumstances, such as falling blood pressure or falling blood levels of sodium (hence, falling blood volume as "water follows salt"), can trigger aldosterone secretion.

Addison's disease develops when the adrenal cortex does not secrete enough mineralocorticoids and glucocorticoids. Sufferers lose weight, and blood levels of glucose and sodium drop while blood potassium rises. The large sodium losses limit the kidney's ability to conserve water. One result can be dangerously low blood pressure and dehydration as the body loses large amounts of water.

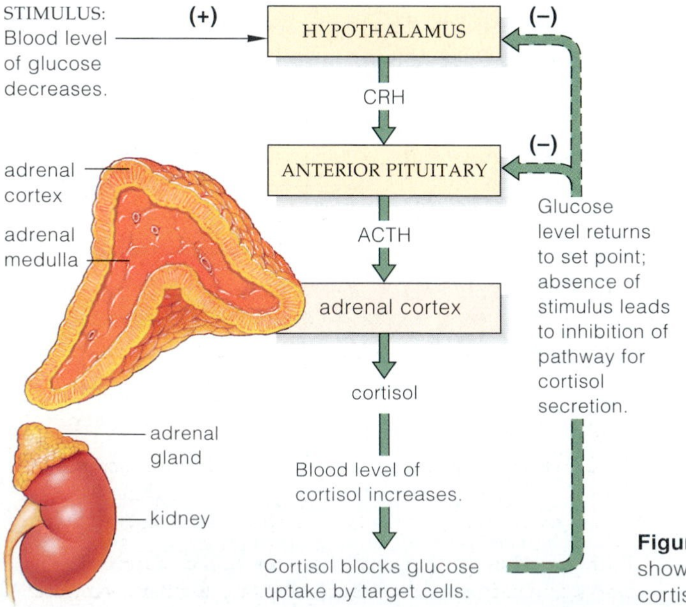

STIMULUS: Blood level of glucose decreases.

(+) HYPOTHALAMUS (−)

CRH

(−) ANTERIOR PITUITARY

ACTH

adrenal cortex

cortisol

Blood level of cortisol increases.

Cortisol blocks glucose uptake by target cells.

Glucose level returns to set point; absence of stimulus leads to inhibition of pathway for cortisol secretion.

adrenal cortex
adrenal medulla
adrenal gland
kidney

Figure 13.8 Location of the adrenal glands. The diagram shows a negative feedback loop that governs secretion of cortisol from the adrenal cortex.

The condition is easily treated with hormone replacement therapy.

In a developing fetus and early in puberty, the adrenal cortex also secretes sizable amounts of sex hormones. The main ones are male sex hormones (androgens), although female sex hormones (estrogens and progesterone) are also produced. In adults, however, most sex hormones are generated by the reproductive organs (Chapter 14).

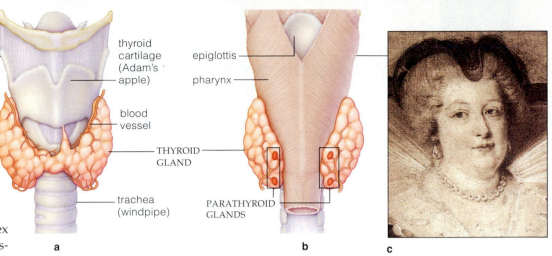

Figure 13.9 The thyroid gland. (**a**) anterior and (**b**) posterior views showing the location of the four parathyroid glands (next to the back of the thyroid). (**c**) A mild case of goiter, displayed by Italian noblewoman Maria de Medici in 1625. At that time, goiter was fairly routine in parts of the world where the typical diet lacked enough iodine for normal thyroid function.

Local Feedback in the Adrenal Medulla

The **adrenal medulla** is the inner part of the adrenal gland (Figure 13.8). It has hormone-secreting neurons that release epinephrine and norepinephrine. (These same substances are considered to be neurotransmitters in some other contexts.) For example, if the hypothalamus sends the appropriate signal via sympathetic nerves to the adrenal medulla, the neuron axons will start to release norepinephrine into the synapse between the axon endings and the target cells. Soon, norepinephrine molecules collect in the synapse, setting the stage for a localized negative feedback mechanism. The accumulating norepinephrine binds to receptors on the axon endings, and in short order the release of norepinephrine shuts down.

Both epinephrine and norepinephrine help regulate blood circulation and carbohydrate metabolism when the body is excited or stressed. For example, they increase heart rate, dilate arterioles in some body regions and constrict them in others, and dilate bronchioles. Thus the heart beats faster and harder, more blood volume is shunted to heart and muscle cells from other regions, and more oxygen flows to energy-demanding cells throughout the body. These are aspects of the "fight-flight" response described in Chapter 11.

Cases of Skewed Feedback from the Thyroid

The **thyroid gland** is located at the base of the neck in front of the trachea, or windpipe (Figure 13.9). Its main secretions, thyroxine (T_4) and triiodothyronine (T_3), affect overall metabolic rate, growth, and development. The thyroid also makes calcitonin, a hormone that helps lower the level of calcium (and phosphate) in blood. Proper feedback control is essential to the functioning of these hormones, a point that is brought into sharp focus by cases of abnormal thyroid output.

For instance, thyroid hormone synthesis requires iodine, a trace element in food. In the absence of iodine, thyroid hormone levels in the blood fall. The anterior pituitary responds to the decrease by secreting the thyroid-stimulating hormone thyrotropin (TSH). Too much TSH overstimulates the thyroid gland, the gland enlarges, and the result is a form of *goiter* (Figure 13.9c). Goiter caused by iodine deficiency is rare in countries where iodized table salt is used. In itself not usually a major health threat, goiter can be a symptom of a more serious disorder.

Too little thyroid output is called *hypothyroidism*. Hypothyroid adults tend to be overweight, sluggish, dry-skinned, intolerant of cold, confused, and depressed. *Cretinism* is a severe hypothyroid condition in children; it can arise from a genetic disorder. Without treatment, an affected child's growth is stunted and the child is mentally retarded. However, with early detection these effects can be prevented.

When blood levels of thyroid hormones are too high, *hyperthyroidism* results. A common hyperthyroid disorder is *Graves' disease*. Affected people have an increased heart rate and sweat profusely; they also are heat-intolerant, have elevated blood pressure, and lose weight even when they eat more food. Apparently, Graves' disease is an autoimmune disorder in which a thyroid-stimulating antibody binds to thyroid cells and causes overproduction of thyroid hormones. Often, affected people are genetically predisposed to the disorder.

Feedback mechanisms control the output of endocrine glands. In many cases, feedback loops to the thalamus, pituitary, or both, govern the secretory activity.

With negative feedback, further secretion of a hormone slows down. Positive feedback steps up further secretion.

HORMONE RESPONSES TO LOCAL CHEMICAL CHANGES

Some endocrine glands or cells don't respond primarily to signals from other hormones or nerves. They respond homeostatically to chemical changes in the immediate surroundings, as in the following examples.

Secretions from Parathyroid Glands

The four **parathyroid glands** respond to chemical change in their immediate surroundings in ways that help maintain homeostatic balance. The parathyroids are located at the back of the thyroid (Figure 13.9b). They secrete parathyroid hormone (PTH) when the concentration of calcium ions in their surroundings falls. Their action affects how much calcium is available to activate enzymes, and for muscle contraction, blood clotting, and many other tasks.

PTH prompts bone cells to secrete enzymes that digest bone tissue and thereby release calcium and phosphate ions that can be used elsewhere in the body. It also stimulates calcium reabsorption from the filtrate flowing through kidney nephrons. In addition, PTH helps to activate vitamin D. The activated form, a hormone, improves calcium absorption from food in the GI tract. In children who have vitamin D deficiency, too little calcium and phosphorus are absorbed, so the rapidly growing bones do not develop properly. The resulting bone disorder, rickets, is characterized by bowed legs, a malformed pelvis, and other skeletal abnormalities, as you can see in Figure 13.10.

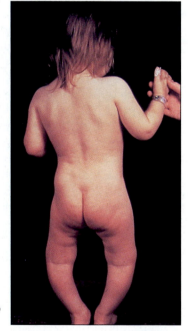

Figure 13.10 A child who is affected by rickets.

Calcium has so many key roles in the body that disorders related to parathyroid functioning can be quite serious. For example, excess PTH (*hyperparathyroidism*) causes so much calcium to be withdrawn from a person's bones that the bone tissue becomes dangerously weakened. Muscle function also deteriorates, and the excess calcium in the bloodstream may cause kidney stones to develop. Operations of the central nervous system may be so disrupted that the affected person eventually dies as a result.

Secretions from Pancreatic Islets

The pancreas is one gland with exocrine *and* endocrine functions. Its *exocrine* cells secrete digetive enzyme into the small intestine (Section 6.4). The *endocrine* cells of the pancreas are located in about 2 million scattered clusters. Each cluster, a **pancreatic islet**, contains three types of hormone-secreting cells:

1. *Alpha cells* secrete the hormone **glucagon**. Between meals, cells use the glucose delivered to them by the bloodstream. The blood glucose level decreases, at which time glucagon secretions cause glycogen (a storage polysaccharide) and amino acids to be converted to glucose in the liver and muscle. Thus, *glucagon raises the glucose level in the blood.*

2. *Beta cells* secrete the hormone **insulin**. After meals, when the blood glucose level is high, insulin stimulates uptake of glucose by muscle and adipose cells. It also promotes synthesis of fats, glycogen, and to a lesser extent, proteins, and inhibits protein conversion to glucose. Thus *insulin lowers the glucose level in the blood.*

3. *Delta cells* secrete *somatostatin*. This hormone acts on beta cells and alpha cells to inhibit secretion of insulin and glucagon, respectively. Somatostatin is part of several hormonal control systems. For example, it is released from the hypothalamus to block secretions of growth hormone; it is also secreted by cells of the GI tract, where it acts locally to inhibit secretion of various substances involved in digestion.

Figure 13.11 shows how interplay among pancreatic hormones helps keep blood glucose levels fairly constant despite great variation in when—and how much—we eat. When the body cannot produce enough insulin or when insulin's target cells cannot respond to it, the body does not store glucose in a normal fashion, and disorders in carbohydrate, protein, and fat metabolism occur.

Insulin deficiency can lead to *diabetes mellitus*. Because target cells cannot take up glucose from blood, that sugar accumulates in blood and is lost in the urine. (*Mellitus* means honey in Greek; early physicians reportedly often tasted their patients' urine to confirm the diagnosis.) As the kidneys shunt excess glucose into the urine, water is lost as well—skewing the body's water–solute balance. Affected people become dehydrated and extremely thirsty. Their insulin-deprived (glucose-starved) cells start breaking down protein and fats for energy, and the person loses weight. Ketones (normal acidic products of fat breakdown) build up in the blood and urine. One result is excess water loss, which can lead to dangerously low blood pressure and heart failure. Another outcome is *metabolic acidosis*, a lower than optimal pH in blood that can seriously disturb brain functioning.

Figure 13.11 A simplified picture of some homeostatic controls over glucose metabolism.

Following a meal, glucose enters the bloodstream faster than cells can use it. The blood glucose level rises, and pancreatic beta cells are stimulated to secrete insulin. Insulin's targets (mainly liver, fat, and muscle cells) use glucose or store excess amounts as glycogen, or use it to synthesize lipids.

Between meals, the blood glucose level decreases. Pancreatic alpha cells are stimulated to secrete glucagon. This hormone's target cells convert glycogen back to glucose, which enters the blood. Also, the hypothalamus prods the pituitary gland to secrete other hormones that slow down conversion of glucose to glycogen in liver, fat, and muscle cells.

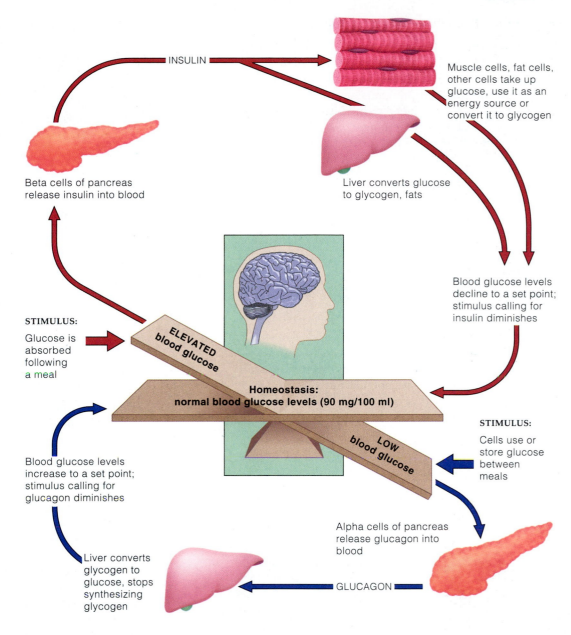

INSULIN

Muscle cells, fat cells, other cells take up glucose, use it as an energy source or convert it to glycogen

Beta cells of pancreas release insulin into blood

Liver converts glucose to glycogen, fats

Blood glucose levels decline to a set point; stimulus calling for insulin diminishes

STIMULUS:
Glucose is absorbed following a meal

ELEVATED blood glucose

Homeostasis:
normal blood glucose levels (90 mg/100 ml)

LOW blood glucose

STIMULUS:
Cells use or store glucose between meals

Blood glucose levels increase to a set point; stimulus calling for glucagon diminishes

Alpha cells of pancreas release glucagon into blood

Liver converts glycogen to glucose, stops synthesizing glycogen

GLUCAGON

In type 1 or "insulin-dependent" diabetes, the body mounts an autoimmune response against its beta cells and destroys them. Only about 10 percent of diabetics are type 1, which is the more immediately dangerous of the two types of diabetes. It may be caused by a combination of genetic susceptibility and viral infection. Because symptoms usually appear during childhood and adolescence, type 1 diabetes also is called "juvenile-onset diabetes." Patients survive with insulin injections, but their life span may be shortened by associated cardiovascular problems.

In type 2 diabetes, insulin levels are close to or above normal, but for various reasons target cells cannot respond properly to the hormone. As affected people grow older, their beta cells deteriorate and they produce less and less insulin. Type 2 diabetes usually occurs in middle age and is sometimes called "maturity-onset diabetes." Obesity increases a person's risk of developing the disorder. Affected people can lead a normal life by controlling their diet and weight and sometimes by taking drugs that improve insulin's action or secretion.

The secretions from some endocrine glands and cells are direct homeostatic responses to a change in the chemical environment of a local area.

Melatonin from the Pineal Gland

Up to this point, we have seen how endocrine glands and endocrine cells respond to other hormones, to signals from the nervous system, and to chemical changes in their surroundings. In humans and other animals, certain hormones are secreted or inhibited in response to cues from the external environment.

Until about 240 million years ago, vertebrates commonly had a light-sensitive "third eye" on top of the head. Some, such as lampreys (a type of parasitic fish), still have one, beneath the skin. Most vertebrates have a modified form of this photosensitive organ, the **pineal gland** described earlier. The human pineal gland secretes the hormone melatonin into the blood and cerebrospinal fluid. Melatonin functions in the development of gonads, in reproductive cycles, and sleep/wake cycles.

As this chapter's introduction explained, melatonin is secreted in the absence of light. This means that melatonin levels in the bloodstream vary from day to night. The levels also change with the seasons, for winter days are shorter than summer days. Decreased melatonin secretion may help trigger the onset of *puberty*, the age at which our reproductive organs and structures start to mature. If disease destroys the pineal gland, a person can enter puberty prematurely.

Hormones from the Heart, GI Tract, and Elsewhere

Beyond the endocrine glands we have been considering, hormones are also made and secreted by specialized cells elsewhere. For example, the two heart atria secrete *atrial natriuretic peptide*, or ANP. This hormone has a variety of effects, including helping to regulate blood pressure. When a person's blood pressure rises, ANP acts to inhibit the reabsorption of sodium ions—and hence water—in the kidneys. As a result, more water is excreted, the blood volume decreases, and blood pressure falls.

Hormones produced in the GI tract include gastrin and secretin. Gastrin, recall, stimulates the secretion of stomach acid when proteins are being digested; secretin prompts the pancreas to secrete bicarbonate (Section 6.7).

The lobed **thymus**, located behind the breastbone (the sternum), between the lungs, is another hormone source. As Chapter 8 noted, T lymphocytes multiply, become specialized (differentiate), and mature in this gland. Those events are influenced by hormones collectively called thymosins. They help guide the development of different categories of mature T cells, such as helper T cells and cytotoxic T cells, that are key elements of immune system functioning (Section 8.4). The thymus is large in children but shrinks to a relatively small size as an adolescent matures into adulthood.

Many of our cells can detect changes in their chemical environment and alter their activity, often in ways that either counteract or amplify the change. The cells secrete *local signaling molecules*, which act on the secreting cell or in its immediate vicinity. Most of the signaling molecules are taken up so quickly that not many are left to enter the general circulation. Prostaglandins and growth factors are examples of such secretions.

Prostaglandins

More than sixteen different kinds of the fatty acids called **prostaglandins** have been identified in tissues throughout the body. In fact, the plasma membranes of most cells contain the precursors of prostaglandins. They are released continually, but the rate at which they are synthesized often increases in response to chemical changes in a local area. The stepped-up secretion can influence neighboring cells as well as the prostaglandin-releasing cells themselves.

There is a long list of known prostaglandin effects. For instance, at least two prostaglandins help adjust blood flow through local tissues. When their secretion is stimulated by epinephrine and norepinephrine, they cause smooth muscle in the walls of blood vessels to constrict or dilate. Similar prostaglandin effects occur in the smooth muscle of lung airways. Prostaglandins may aggravate tissue inflammation and allergic responses to airborne dust and pollen. Certain kidney cells also produce prostaglandins that act as vasodilators or as vasoconstrictors, but exactly how they affect overall nephron function is not well understood.

Prostaglandins have powerful effects on the reproductive system. For example, many women experience painful cramping and excessive bleeding when they menstruate; these effects are due to prostaglandin action on smooth muscle of the uterus. (Antiprostaglandin drugs, such as ibuprofen and aspirin, block prostaglandin synthesis, which is how they alleviate the discomfort.)

Prostaglandins also influence the *corpus luteum*, a glandular structure that develops from cells that earlier surrounded a developing ovum (egg) in the ovary. When pregnancy does not follow the release of an ovum from the ovary (ovulation), a corpus luteum produces a great quantity of prostaglandins. In response, the capillaries that service the corpus luteum constrict, shutting off the blood supply—and the corpus luteum dies. Along with oxytocin, prostaglandins also have roles in stimulating uterine contractions during labor. Prostaglandins in semen may stimulate uterine contractions that help move sperm deeper into the female reproductive tract. We will look much more closely at some of these events in the next chapter.

Growth Factors

Hormonelike proteins called **growth factors** influence growth by regulating the rate at which certain cells divide. *Epidermal growth factor* (EGF) influences the growth of many cell types. So does insulin-like growth factor (IGF) secreted by the liver. *Nerve growth factor* (NGF) is another example. NGF promotes the survival and growth of neurons in the developing embryo. One experiment demonstrated that certain immature neurons survive indefinitely in tissue culture when NGF is present but die within a few days if it is not. NGF apparently defines the direction in which these embryonic neurons grow: target cells lay down a chemical path that leads the elongating processes to them.

In the years since NGF was discovered, several other growth factors with effects on neural cells have been identified. *Science Comes to Life* will give you an inkling of how such growth factors may one day be important tools for treating victims of heart attack, paralyzing spinal cord injuries, and perhaps other conditions.

Pheromones Revisited

Chapter 12 touched on possible human pheromones in its discussion of the sense of smell (Section 12.3). It's not too surprising that we humans may produce and respond to pheromones; after all, some of our close primate relatives (including rhesus monkeys), and bears, coyotes, dogs, various insects, and many other animals produce pheromones that serve as sex attractants, territory markers, and other communication signals to members of their own species. Pheromones are released outside of the individual and then pass through the air or water and so reach another individual.

As noted previously, hypotheses about pheromone activity in humans are still quite tentative, but research on this fascinating subject is continuing.

Endocrine cells in the pineal gland, the thymus, the heart, and the GI tract all produce hormones. Other cells in various tissues produce local signaling molecules, which act on the secreting cell itself or in the immediate vicinity.

Prostaglandins are a family of local signaling molecules that have varied effects throughout the body. Several serve as vasodilators or vasoconstrictors and so help adjust blood flow in affected regions. Other prostaglandins cause smooth muscle (as in airways or the uterus) to contract.

Growth factors help regulate the rate at which certain cells divide. Researchers are exploring their potential clinical use to stimulate regeneration of damaged neurons.

Not too far into the future, a doctor treating someone who has sustained severe injuries, including major damage to the spinal cord, could have a powerful chemical arsenal to work with. Researchers are unraveling how an array of communication molecules, including an "alphabet soup" of growth factors, operate in body tissues. They are also trying to figure out how to convert such knowledge into treatments for specific kinds of tissue damage.

The list of known growth factors is growing rapidly. It includes not only the epidermal growth factor (EGF), insulin-like growth factor (IGF), and nerve growth factor (NGF) described in the text, but also fibroblast growth factor (FGF), platelet-derived growth factor (PDGF), so-called "transforming" growth factors (TGFs), and tumor angiogenesis factors (TAFs).

Studies on rodents have shown that EGF and one type of TGF enhance the healing of some types of wounds. In one experiment, an antibiotic cream that contained TGF-*alpha* accelerated the rate at which severely burned skin regenerated. TGF-*beta* helps regulate the formation of bone and could possibly be marshaled to speed the healing of broken bones. TAFs that stimulate the development of new blood vessels in tumors might be harnessed to perform that function in cardiac muscle damaged by a heart attack. PDGF, which is released by platelets in blood plasma at the site of a wound, apparently stimulates the regeneration of smooth muscle cells in torn vessel walls—and it could perhaps do the same for damaged smooth muscle elsewhere.

Some of the most tragic injuries involve spinal cord damage that quickly results in the death of neurons—and in paralysis. Experiments have already demonstrated that NGF can help severed nerves regrow; there is also evidence that FGF helps to maintain healthy neurons and to repair damaged tissues. (Fibroblasts are cells in loose connective tissue that give rise to various fiberlike molecules with structural roles.) In experimental trials, a substance called GM_1 appears to derail continued deterioration of damaged spinal cord neurons. GM_1 is a ganglioside, a normal component of nerve cell membranes that seems to serve as a communication link between cells. It is already being used experimentally on spinal injury patients.

SUMMARY

1. Cells continually take up and release substances. In the complex human body, these constant withdrawals and secretions must be integrated in ways that ensure cell survival through the whole body.

2. Integration requires the stimulatory or inhibitory effects of signaling molecules.

 a. Signaling molecules are chemical secretions by one cell that adjust the behavior of other, target cells.

 b. Any cell with receptors for the signal is the target. It may or may not be next to the signaling cell.

 c. Hormones, neurotransmitters, local signaling molecules, and pheromones are different kinds of signaling molecules.

3. Steroid and nonsteroid hormones exert their effects on target cells by different mechanisms.

 a. Steroid (and thyroid) hormones have receptors inside their target cells. The hormone-receptor complex binds to the DNA. Binding triggers gene activation and protein synthesis.

 b. Amines and the peptide, polypeptide, and glycoprotein hormones interact with receptors on the plasma membrane of target cells. Responses to them are often mediated by a second messenger, such as cyclic AMP, inside the cell.

 c. Most nonsteroid hormones alter the activity of existing proteins in target cells. The ensuing target cell responses help maintain the internal environment or contribute to normal development or reproductive functioning.

4. The hypothalamus and pituitary gland interact to integrate many body activities.

 a. ADH and oxytocin, two hypothalamic hormones, are stored in and released from the posterior lobe of the pituitary. ADH influences extracellular fluid volume. Oxytocin has roles in reproduction.

 b. Six additional hypothalamic hormones, called releasing and inhibiting hormones, control the secretions by different cells of the pituitary's anterior lobe.

 c. Of the six hormones produced in the anterior lobe, two (prolactin and growth hormone) have general effects on body tissues. Four (ACTH, TSH, FSH, and LH) act on specific endocrine glands.

5. Responses to hormones may be influenced by hormone interactions and homeostatic feedback loops to the hypothalamus and pituitary. They also may be influenced by variations in hormone concentrations, and by the number and kinds of receptors on a target cell.

6. Fast-acting hormones such as parathyroid hormone or insulin generally come into play when the extracellular concentration of a substance must be controlled by homeostatic feedback loops.

7. Hormones such as growth hormone have more prolonged, gradual, and often irreversible effects, such as those on development.

Review Questions

1. Which secretions of the posterior and anterior lobes of the pituitary gland have the targets indicated? (*Fill in the blanks; see Section 13.3.*)

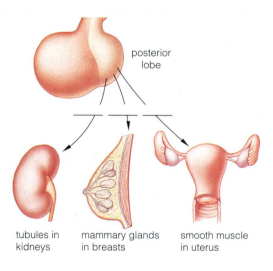

posterior lobe

tubules in kidneys   mammary glands in breasts   smooth muscle in uterus

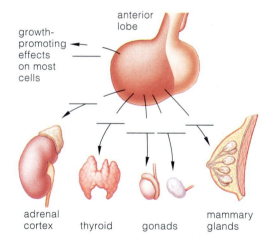

anterior lobe

growth-promoting effects on most cells

adrenal cortex   thyroid   gonads   mammary glands

2. Name the main endocrine glands and state where each is located in the human body. *13.3, 13.5*

3. Distinguish among hormones, neurotransmitters, local signaling molecules, and pheromones. *13.1*

4. A hormone molecule binds to a receptor on a cell membrane. It doesn't enter the cell; rather, the binding activates a second messenger inside the cell that triggers an amplified response to the hormonal signal. Is the signaling molecule a steroid or a nonsteroid hormone? *13.2*

1. _____ are molecules released from a signaling cell that have effects on target cells.
   a. Hormones
   b. Neurotransmitters
   c. Local signaling molecules
   d. Pheromones
   e. a and b
   f. All of the above

2. Hormones are produced by _____.
   a. endocrine glands and cells
   b. some neurons
   c. exocrine cells
   d. a and b
   e. a and c
   f. a, b, and c

3. ADH and oxytocin are hypothalamic hormones secreted from the pituitary's _____ lobe.
   a. anterior
   b. posterior
   c. primary
   d. secondary

4. _____ has effects on body tissues in general.
   a. ACTH
   b. TSH
   c. LH
   d. Growth hormone

5. Which of the following stimulate hormone secretions?
   a. neural signals
   b. local chemical changes
   c. hormonal signals
   d. environment cues
   e. all of the above can stimulate hormone secretion

6. _____ lowers blood sugar levels; _____ raises it.
   a. Glucagon; insulin
   b. Insulin; glucagon
   c. Gastrin; insulin
   d. Gastrin; glucagon

7. The pituitary detects a rising hormone concentration in blood and inhibits the gland secreting the hormone. This is a _____ feedback loop.
   a. positive
   b. negative

8. Second messengers assist _____.
   a. steroid hormones
   b. nonsteroid hormones
   c. only thyroid hormones
   d. both a and b

9. Match the hormone source with the closest description.
   ____ adrenal cortex
   ____ adrenal medulla
   ____ thyroid gland
   ____ parathyroids
   ____ pancreatic islets
   ____ pineal gland
   ____ thymus
   a. affected by day length
   b. cortisol source
   c. roles in immunity
   d. adjust(s) blood calcium level
   e. epinephrine source
   f. insulin, glucagon
   g. hormones require iodine

10. Match the endocrine control concepts.
    ____ oxytocin
    ____ ACTH
    ____ ADH
    ____ estrogen
    ____ growth hormone
    a. released by the anterior pituitary and affects the adrenal gland
    b. influences extracellular fluid volume
    c. has general effects on growth
    d. triggers uterine contractions
    e. a steroid hormone

Critical Thinking: You Decide (Key in Appendix VI)

1. A 20-year-old woman with a malignant brain tumor has her pineal gland removed. What might be some side effects of the loss of the gland?

2. President John F. Kennedy was diagnosed with Addison's disease when he was a young adult. After he began drug treatment with mineralocorticoids, his appearance changed as he

Figure 13.12 President John F. Kennedy, in a photograph taken after he began receiving treatment for Addison's disease.

gained weight (Figure 13.12). Based on your reading of Section 13.6, was the weight gain healthy? What caused it?

3. A physician sees a patient whose symptoms include sluggishness, depression, and intolerance to cold. After eliminating several other possible causes, the doctor diagnoses an endocrine disorder. What endocrine disorder fits the symptoms? Why does the doctor suspect that the underlying cause is a malfunction of the anterior pituitary gland?

4. Marianne is affected by type 1 insulin-dependent diabetes. One day, after injecting herself with too much insulin, she starts to shake and feels confused. Following her doctor's suggestion, she drinks a glass of orange juice—a ready source of glucose—and soon her symptoms subside. What caused her symptoms? How would a glucose-rich snack help?

Selected Key Terms

adrenal cortex *13.6*
adrenal medulla *13.6*
endocrine system *13.1*
hormone *13.1*
hypothalamus *13.3*
local signaling molecule *13.1*
pancreatic islet *13.7*
parathyroid gland *13.7*
pineal gland *13.8*
pituitary gland *13.3*
second messenger *13.2*
thymus *13.8*
thyroid gland *13.6*

Readings

Hadley, M. 2000. *Endocrinology.* 5th ed. Englewood Cliffs, N.J.: Prentice Hall.

Smith, D.F. July/August, 1995. "Steroid Receptors and Molecular Chaperones." *Scientific American Science & Medicine.*

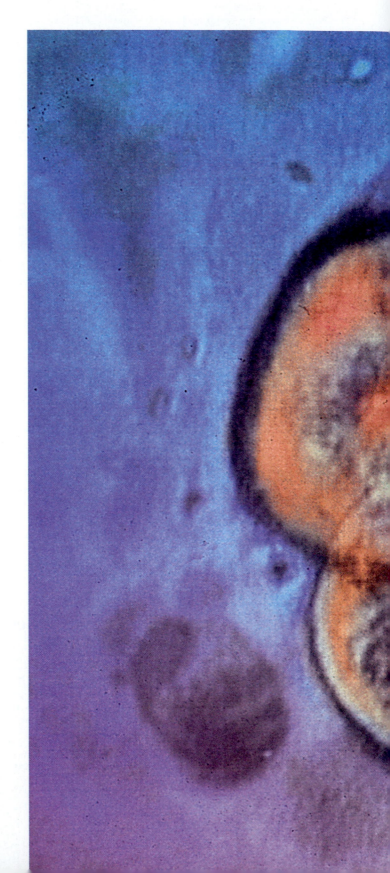

An eight-cell-stage human embryo.

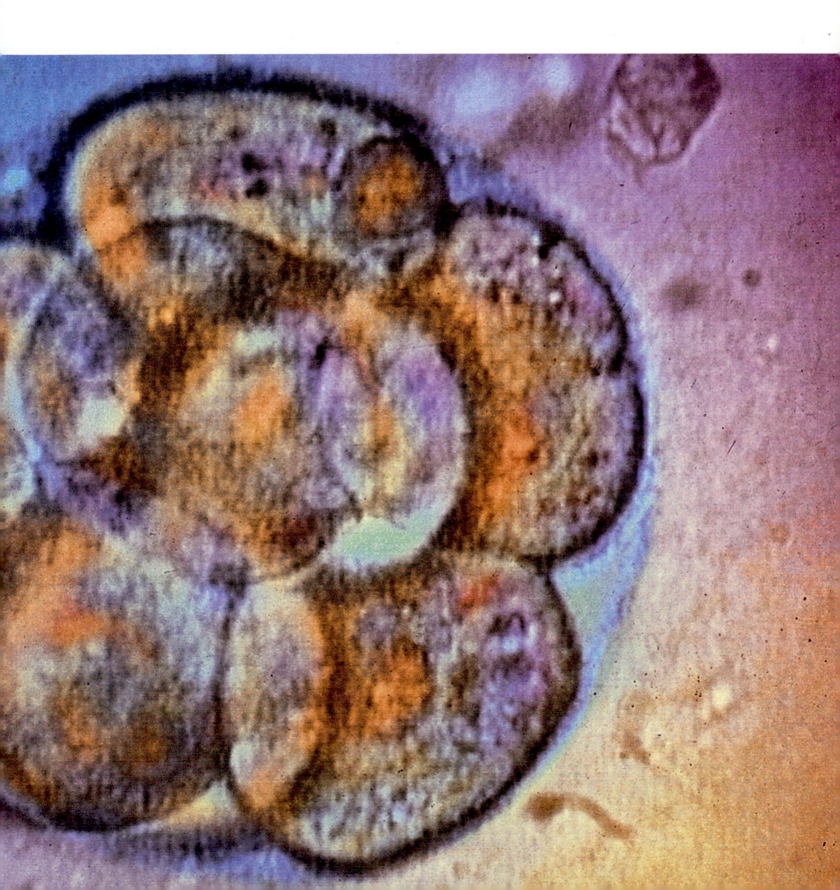

REPRODUCTIVE SYSTEMS

Girls and Boys

An old joke says that you can tell the sex of a baby by the booties it wears—pink or blue. Actually, the way a person's sex is determined is a bit more interesting than that. When you were conceived, a sperm cell from your father fertilized an egg cell in your mother's body. Among its total of twenty-three chromosomes, the egg carried an X chromosome, and among *its* twenty-three chromosomes, the sperm carried either an X or a Y chromosome. Normally, from the time of conception, the embryo that develops from the fertilized egg carries a pair of each of the twenty-three chromosomes, including a combination of the parental sex chromosomes—either XX or XY. But for the first month or so of development, anatomically *you were neither male nor female.*

Apparently, a gene on the Y chromosome governs the fork in the developmental road that determines a person's gender. If the embryo is XX—that is, if no Y chromosome is present—it automatically develops ovaries, the primary female reproductive organs. About eight weeks after a female embryo is conceived, the two ovaries (like the testes) begin to develop from buds in the abdominal wall. After further development, they eventually reside in the pelvic cavity.

In XY embryos, testes—the primary reproductive organs of males—begin to develop during the second month following conception (Figure 14.1). Once they begin to form, the testes start producing testosterone and other sex hormones. These hormones are crucial to the development of a male reproductive system.

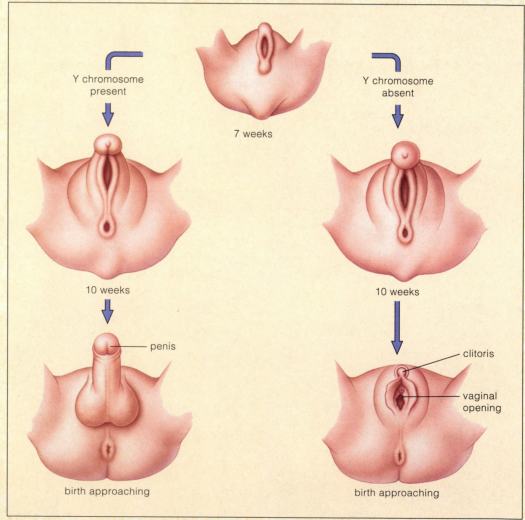

Y chromosome present

7 weeks

Y chromosome absent

10 weeks

10 weeks

penis

clitoris

vaginal opening

birth approaching

birth approaching

Researchers have identified a particular region of the Y chromosome that is the "master gene" for male sex determination. It is called *SRY* (for *s*ex-determining *r*egion of the Y chromosome). The same gene has been identified in DNA from male humans, chimpanzees, rabbits, pigs, horses, cattle, and tigers. As one might guess, the gene was absent in all females tested, because females don't have a Y chromosome. Other tests with mice indicate that the gene becomes active about the time testes start developing. This chapter focuses on the male and female reproductive systems that come into being by way of these fascinating events.

For both men and women, the reproductive system consists of a pair of primary reproductive organs, or *gonads*, plus accessory glands, and ducts. Male gonads are **testes** (singular: testis), and female gonads are **ovaries**. Testes produce sperm; ovaries produce eggs. Both secrete sex hormones that influence reproductive functions and the development of **secondary sexual traits**. Such traits are distinctly associated with maleness and femaleness, although they do not play a direct role in reproduction. Examples are the amount and distribution of body fat, hair, and skeletal muscle.

Gonads look the same in all early human embryos. After seven weeks of development, activation of genes on the sex chromosomes and hormone secretions trigger their development into testes *or* ovaries. The gonads and accessory organs are already formed at birth, but they will not reach full size and become functional until twelve to sixteen years later. In the chapter that follows, the story continues as we trace the step-by-step unfolding of human development from the fertilized egg to ripe old age.

KEY CONCEPTS

1. The human reproductive system consists of a pair of primary reproductive organs (testes in males, ovaries in females), accessory glands, and ducts. Testes produce sperm; ovaries produce eggs. Both kinds of gonads release sex hormones in response to signals from the hypothalamus and pituitary gland.

2. Human males continually produce sperm from puberty onward. The hormones testosterone, LH, and FSH control male reproductive functions.

3. Human females are fertile on a cyclic basis. Each month during their reproductive years, an egg is released from an ovary, and the lining of the uterus is prepared for pregnancy. The hormones estrogen, progesterone, FSH, and LH control this cyclic activity.

4. Sexual intercourse is the natural mechanism that brings together sperm and eggs (the male and female gametes), permitting fertilization and the conception of a new individual.

CHAPTER AT A GLANCE

Figure 14.1 Photograph: The child this woman is expecting has inherited chromosomes from its parents that will determine its sex—and the reproductive structures it will have. **Diagram:** External appearance of developing reproductive organs.

THE MALE REPRODUCTIVE SYSTEM

Where Sperm Form

Figure 14.2 shows the organs of the reproductive system in an adult male, and Table 14.1 lists their functions. In an embryo that is genetically destined to become male, a pair of testes form on the abdominal cavity wall. Before birth, the testes descend into the scrotum, an outpouching of skin that hangs below the pelvic region. By the time of birth, testes are fully formed miniatures of the adult organs.

Figure 14.2a shows the position of the scrotum in an adult male. If sperm cells are to develop properly, the temperature in the interior of the scrotum must remain a few degrees cooler than the core temperature of the rest of the body. A control mechanism works by stimulating or inhibiting the contraction of smooth muscles in the scrotum's wall. It helps assure that the scrotum's internal temperature does not stray far from 95°F. When the air just outside the body becomes too cold, contractions draw the scrotum closer to the body mass, which is warmer. When it is warmer outside, the muscles relax and lower the scrotum.

Packed inside each testis are a large number of small, highly coiled tubes, the **seminiferous tubules**. The formation of sperm begins in these tubules. Section 14.3 looks at that process.

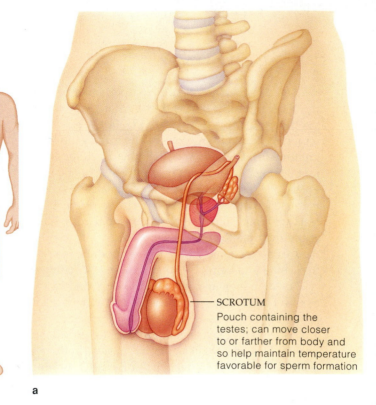

penis

scrotum

SCROTUM
Pouch containing the testes; can move closer to or farther from body and so help maintain temperature favorable for sperm formation

a

Figure 14.2 Above and facing page: (a) Position of the male reproductive system relative to the pelvic girdle and urinary bladder. The diagram in (**b**) shows the components of the system and summarizes their functions.

Where Semen Forms

Human sperm are not quite mature when they leave the testes. First they enter a pair of long, coiled ducts, the epididymides (singular: epididymis). Secretions from

Table 14.1	Organs and Accessory Glands of the Male Reproductive Tract
REPRODUCTIVE ORGANS:	
Testis (2)	Production of sperm, sex hormones
Epididymis (2)	Sperm maturation site and sperm storage
Vas deferens (2)	Rapid transport of sperm
Ejaculatory duct (2)	Conduction of sperm to penis
Penis	Organ of sexual intercourse
ACCESSORY GLANDS:	
Seminal vesicle	Secretion of large part of semen
Prostate gland	Secretion of part of semen
Bulbourethral gland (2)	Production of lubricating mucus

gland cells in the wall of these ducts trigger the finishing touches on sperm. Until sperm leave the body, they are stored in the last stretch of each epididymis.

When a male is sexually aroused, muscle contractions in the walls of his reproductive organs propel mature sperm into and through a pair of thick-walled tubes, the **vas deferentia** (singular: vas deferens). From there, contractions propel sperm through a pair of ejaculatory ducts and then through the urethra. This last tube passes through the penis, the male sex organ, and opens at its tip. The urethra, recall, is a duct that also functions in urine excretion.

Glandular secretions become mixed with sperm as they travel through the urethra. The result is **semen**, a thick fluid that is eventually expelled from the penis during sexual activity. Early in the formation of semen, a pair of seminal vesicles secrete fructose. The sperm use this sugar as an energy source. Seminal vesicles also

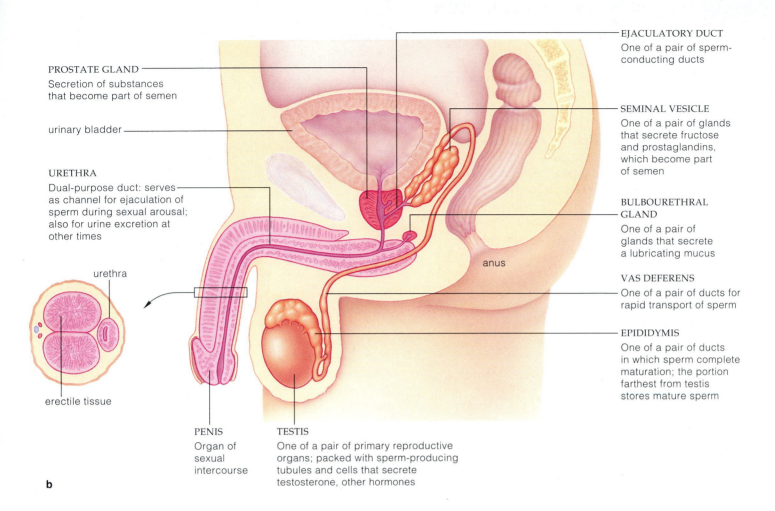

PROSTATE GLAND
Secretion of substances
that become part of semen

urinary bladder

URETHRA
Dual-purpose duct: serves
as channel for ejaculation of
sperm during sexual arousal;
also for urine excretion at
other times

urethra

erectile tissue

b

EJACULATORY DUCT
One of a pair of sperm-
conducting ducts

SEMINAL VESICLE
One of a pair of glands
that secrete fructose
and prostaglandins,
which become part
of semen

BULBOURETHRAL
GLAND
One of a pair of
glands that secrete
a lubricating mucus

anus

VAS DEFERENS
One of a pair of ducts for
rapid transport of sperm

EPIDIDYMIS
One of a pair of ducts
in which sperm complete
maturation; the portion
farthest from testis
stores mature sperm

PENIS
Organ of
sexual
intercourse

TESTIS
One of a pair of primary reproductive
organs; packed with sperm-producing
tubules and cells that secrete
testosterone, other hormones

secrete certain kinds of prostaglandins. These signaling molecules can induce muscle contractions. As noted in Section 13.8, possibly these signaling molecules take effect during sexual activity. At that time, they might induce contractions in the female's reproductive tract and thereby assist sperm movement through it.

Secretions from the **prostate gland** probably help buffer the acidic environment that sperm encounter in the female reproductive tract. Vaginal pH is about 3.5 to 4.0, but sperm motility improves at pH 6. Two **bulbo-urethral glands** secrete mucus-rich fluid into the urethra when the male is sexually aroused. This fluid neutralizes acids in any traces of urine in the urethra, creating a more hospitable environment for the 150 to 350 million sperm that pass through the channel in a typical ejaculation.

The testes and prostate gland both are sites where cancer can develop. At least 5,000 cases of testicular cancer are diagnosed each year in the United States,

mostly among young men, and the cancer kills about half of its victims. Prostate cancer, which is more common among men over 50, kills 40,000 older men annually—almost the same mortality rate recorded for breast cancer in women. As with other cancers, early detection is the key to survival. Causes and treatments of cancers are the subject of Chapter 22.

Testes are the primary reproductive organs (gonads) of an adult male. The male reproductive system also includes accessory glands and ducts.

Sperm develop mainly in the seminiferous tubules of the testes. When sperm are nearly mature, they leave each testis and enter the long, coiled epididymis, where they remain until ejaculated.

Secretions from the seminal vesicles and the prostate gland mix with sperm to form semen.

MALE REPRODUCTIVE FUNCTION

Sperm Formation

Each testis is only about 5 centimeters long, which is smaller than a golfball. Yet packed inside are 125 meters of seminiferous tubules. As many as 30 wedge-shaped lobes divide the interior, and each holds two or three coiled tubules (Figures 14.3 and 14.4).

Inside the walls of the seminiferous tubules are undifferentiated cells called *spermatogonia*. As you can see in Figure 14.4, they are the starting point for ongoing cell divisions, including a type of division called *mitosis* and a type called *meiosis*. Chapter 17 provides the details of mitosis and meiosis; here it is only important to keep in mind that meiosis results in the specialized reproductive cells called *gametes*—sperm and eggs. This chapter's introduction noted that human gametes have twenty-three chromosomes, including one sex chromosome. This is termed a "haploid" number of chromosomes because it is one-half the normal number of chromosomes (a "diploid" number of forty-six) of other human body cells. When two haploid gametes unite at fertilization, the diploid chromosome number is restored.

Spermatogonia develop into *primary spermatocytes*, which become *secondary spermatocytes* after a first round of meiotic division (meiosis I). A second round of cell division (meiosis II) results in *spermatids*. The spermatids gradually develop into *spermatozoa*, or simply **sperm**—the male gametes. The "tail" of each sperm (a flagellum) arises at the end of the process, which takes nine to ten weeks. All the while, the developing cells are nourished by and receive chemical signals from neighboring **Sertoli cells**. They are the cells that line the seminiferous tubule.

The testes produce sperm from puberty onward. Millions are in different stages of development on any given day. A mature sperm has a tail, a midpiece, and a head (Figure 14.4). Inside the head, a nucleus contains DNA organized into chromosomes. An enzyme-containing cap, the **acrosome**, covers most of the head. Its enzymes help the sperm penetrate the extracellular material around an egg at fertilization. In the mid-piece, mitochondria supply energy for the tail's movements.

Hormonal Controls

Male reproductive function depends on testosterone, LH, and FSH (Figure 14.5). **Leydig cells** (also called interstitial cells), located in tissue between the seminiferous tubules in testes, secrete **testosterone**. This hormone

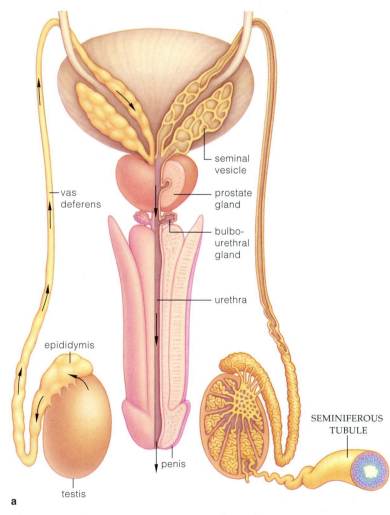

Figure 14.3 (**a**) The male reproductive tract, posterior view. Arrows show the route that sperm take before ejaculation from a sexually aroused male. (**b**) Micrograph of cells in seminiferous tubules.

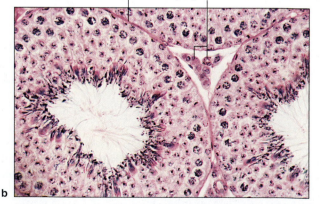

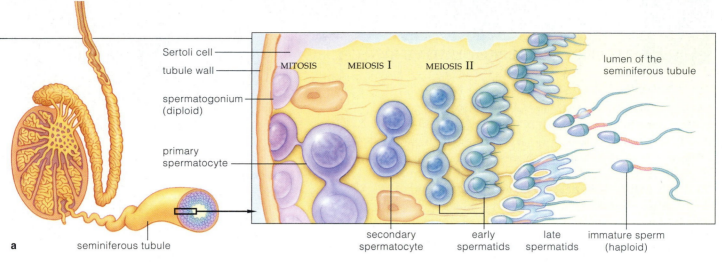

Sertoli cell
tubule wall
MITOSIS   MEIOSIS I   MEIOSIS II
lumen of the
seminiferous tubule
spermatogonium
(diploid)
primary
spermatocyte
secondary
spermatocyte
early
spermatids
late
spermatids
immature sperm
(haploid)

a   seminiferous tubule

Figure 14.4 (**a**) How sperm form, starting with a diploid germ cell. Mitotic cell divisions, then meiosis, produce a cluster of haploid cells that differentiate into sperm. Leydig cells and Sertoli cells interact chemically to produce testosterone, which promotes sperm formation. (**b**) Structure of the mature human sperm.

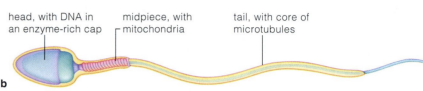

head, with DNA in   midpiece, with   tail, with core of
an enzyme-rich cap   mitochondria   microtubules

b

governs the growth, form, and functions of the male reproductive tract. It stimulates sexual behavior, and promotes development of secondary sexual traits, including facial hair growth and deepening of the voice at puberty.

LH (luteinizing hormone) and **FSH** (follicle-stimulating hormone) are secreted by the anterior lobe of the pituitary gland. They were named for their effects in females, but are chemically the same in males.

The hypothalamus controls secretions of LH, FSH, and testosterone—and thus controls sperm formation (Figure 14.4). When the testosterone level in blood decreases past a set point, the hypothalamus secretes GnRH. This releasing hormone prompts the pituitary's anterior lobe to release LH and FSH, which have targets in the testes. LH stimulates Leydig cells to secrete testosterone, which stimulates diploid germ cells to become sperm. Sertoli cells have FSH receptors. FSH is crucial to establishing spermatogenesis at the time of puberty, but researchers do not know whether it is essential for the normal functioning of mature testes.

A high testosterone level in blood has an inhibitory effect on GnRH release. Also, when the sperm count is high, Sertoli cells release inhibin, a protein hormone that acts on the hypothalamus and pituitary to inhibit the release of GnRH and FSH. Hence, feedback loops to the hypothalamus begin to operate—with a resulting decrease in testosterone secretion and sperm formation.

Sperm formation depends on the hormones testosterone, LH, and FSH. Feedback loops from the testes to the hypothalamus and pituitary gland control their secretion.

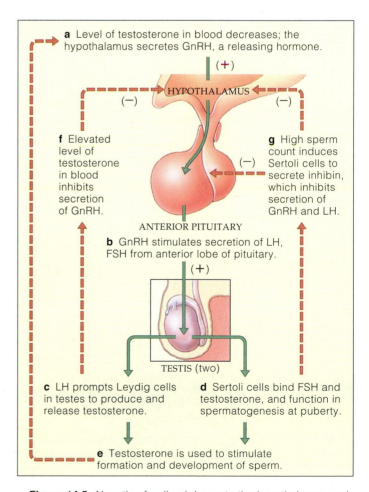

a Level of testosterone in blood decreases; the hypothalamus secretes GnRH, a releasing hormone.

(+)

HYPOTHALAMUS

(−)   (−)

f Elevated level of testosterone in blood inhibits secretion of GnRH.

g High sperm count induces Sertoli cells to secrete inhibin, which inhibits secretion of GnRH and LH.

(−)

ANTERIOR PITUITARY

b GnRH stimulates secretion of LH, FSH from anterior lobe of pituitary.

(+)

TESTIS (two)

c LH prompts Leydig cells in testes to produce and release testosterone.

d Sertoli cells bind FSH and testosterone, and function in spermatogenesis at puberty.

e Testosterone is used to stimulate formation and development of sperm.

Figure 14.5 Negative feedback loops to the hypothalamus and pituitary gland from the testes. Through these loops, excess testosterone production shuts off the mechanisms leading to its production. This helps maintain the testosterone level in amounts required for sperm formation.

THE FEMALE REPRODUCTIVE SYSTEM

The Reproductive Organs

We turn now to the female reproductive system. Figure 14.6 shows its components, and Table 14.2 summarizes their functions. The female's primary reproductive organs, her two ovaries, release sex hormones; during a woman's reproductive years they also produce eggs. The ovarian hormones influence the development of female secondary sexual traits. These traits include the "filling out" of breasts, hips, and buttocks as fat deposits accumulate in those areas.

A female's immature eggs are called **oocytes**. When an oocyte is released from either ovary, it moves into the neighboring **oviduct** (sometimes called a *Fallopian tube*). The oviducts are where fertilization may occur. Fertilized or not, an egg travels down the oviduct into the hollow, pear-shaped **uterus**. In this organ, a new individual can grow and develop. The wall of the uterus consists of a thick layer of smooth muscle (the myometrium) and an interior lining, the **endometrium**. The endometrium includes epithelial tissue, connective tissue, glands, and blood vessels. The lower portion of the uterus is the *cervix*. The *vagina* extends from the cervix to the body surface. It is a muscular tube that receives the penis and sperm and functions as part of the birth canal.

At the body surface are external genitals, collectively called the *vulva*. This region includes organs for sexual stimulation. Outermost are a pair of fat-padded skin folds, the *labia majora*. They enclose a smaller pair of skin folds, the *labia minora*, that are richly endowed with blood vessels. The labia minora partly enclose the **clitoris**, a small organ sensitive to stimulation; if you look back at Figure 14.1, you can see that, from a developmental standpoint, the female's clitoris is analogous to a male's penis.

A female's urethra opens about midway between her clitoris and her vaginal opening. Whereas in males the urethra carries both urine and sperm, in females it is separate and is not involved in reproduction.

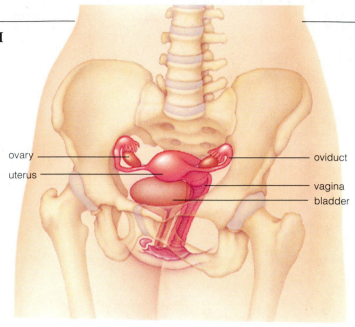

ovary — oviduct
uterus — vagina, bladder

a

Figure 14.6 Above and facing page: (a) Position of the female reproductive system relative to the pelvic girdle and urinary bladder. (**b**) Components of the system and their functions.

Overview of the Menstrual Cycle

Like other female primates, female humans have a **menstrual cycle**. It takes about twenty-eight days to complete one cycle, although this can vary from month to month and from woman to woman. While the cycle advances, an oocyte matures (from a *primary* oocyte to a *secondary* oocyte) and is released from an ovary. All the while, hormones are priming the endometrium to receive and nourish an embryo in case fertilization occurs. If the oocyte is *not* fertilized, a blood-rich fluid starts flowing out through the vaginal canal. This recurring bloody flow is **menstruation**. In effect, the flow means "there is no embryo at this time," and it marks the first day of a new cycle. The uterine "nest" (the disintegrating endometrium) is being sloughed off, only to be reconstructed once again.

The events just sketched out advance through three phases. The cycle starts with a follicular phase. This is the time of menstruation, endometrial disintegration and rebuilding, and oocyte maturation. The next phase is restricted to the release of an oocyte from an ovary. We call it **ovulation**. During the luteal phase of the cycle, an endocrine structure (the corpus luteum, or "yellow body") forms, and the endometrium is primed for pregnancy (Table 14.3).

All three phases are governed by feedback loops to the hypothalamus and pituitary gland from the ovaries. FSH and LH promote cyclic changes in the ovaries. As you will see, FSH and LH also stimulate the ovaries to secrete sex hormones—**estrogens** and **progesterone**—to promote the cyclic changes in the endometrium.

Table 14.2	Female Reproductive Organs
Ovaries	Produce oocytes and sex hormones
Oviducts	Conduct oocytes from ovary to uterus
Uterus	Chamber where new individual develops
Cervix	Secretes mucus that enhances sperm movement into uterus and (after fertilization) reduces the embryo's risk of bacterial infection
Vagina	Organ of sexual intercourse; birth canal

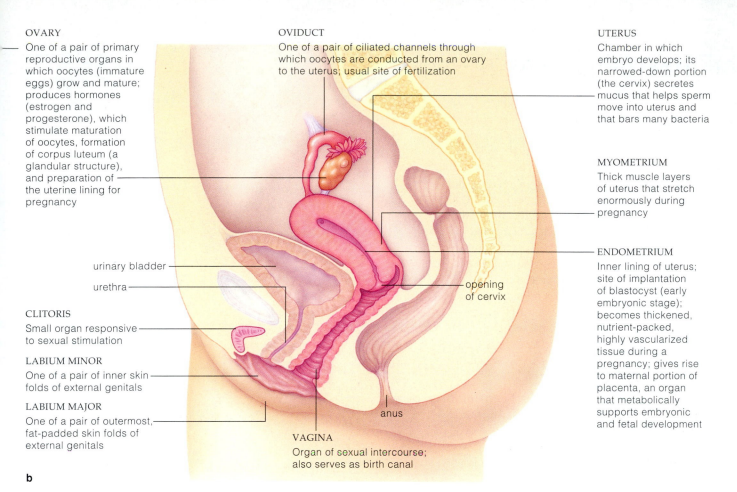

OVARY
One of a pair of primary reproductive organs in which oocytes (immature eggs) grow and mature; produces hormones (estrogen and progesterone), which stimulate maturation of oocytes, formation of corpus luteum (a glandular structure), and preparation of the uterine lining for pregnancy

OVIDUCT
One of a pair of ciliated channels through which oocytes are conducted from an ovary to the uterus; usual site of fertilization

UTERUS
Chamber in which embryo develops; its narrowed-down portion (the cervix) secretes mucus that helps sperm move into uterus and that bars many bacteria

MYOMETRIUM
Thick muscle layers of uterus that stretch enormously during pregnancy

ENDOMETRIUM
Inner lining of uterus; site of implantation of blastocyst (early embryonic stage); becomes thickened, nutrient-packed, highly vascularized tissue during a pregnancy; gives rise to maternal portion of placenta, an organ that metabolically supports embryonic and fetal development

urinary bladder

urethra

opening of cervix

CLITORIS
Small organ responsive to sexual stimulation

LABIUM MINOR
One of a pair of inner skin folds of external genitals

LABIUM MAJOR
One of a pair of outermost, fat-padded skin folds of external genitals

anus

VAGINA
Organ of sexual intercourse; also serves as birth canal

b

Phase	Events	Days of the Cycle*
Table 14.3	**Events of the Menstrual Cycle**	
Follicular phase	Menstruation; endometrium breaks down	1–5
	Follicle matures in ovary; endometrium rebuilds	6–13
Ovulation	Secondary oocyte released from ovary	14
Luteal phase	Corpus luteum forms; endometrium thickens and develops	15–28

*Assumes a 28-day cycle.

A female's first menstruation, or *menarche*, usually occurs between the ages of ten and sixteen. Menstrual cycles continue until *menopause*, in a woman's late 40s or early 50s. By then, her secretion of reproductive hormones has diminished, as has her sensitivity to pituitary reproductive hormones. In some women, the decreasing estrogen levels trigger various temporary symptoms, including moodiness and "hot flashes." These sudden bouts of sweating and feeling uncomfortably warm result from widespread vasodilation of the blood vessels in skin. Section 15.14, which discusses aging in various body systems, describes some other physiological changes associated with menopause. When a woman's menstrual cycles eventually stop, the fertile phase of her life is over.

You may have heard about *endometriosis*, a disorder that arises when endometrial tissue abnormally spreads and grows outside the uterus. Endometrial scar tissue may form on one or both ovaries or oviducts, leading to infertility. In the United States, 10 million women may be affected each year. Possibly the condition develops when menstrual flow backs up through the oviducts and spills into the pelvic cavity. Or perhaps some cells became situated in the wrong place when the female herself was developing before birth, and were stimulated to grow during puberty, when her sex hormones became active. Regardless, the resulting symptoms include pain during menstruation, sex, or urination. Treatment ranges from doing nothing (in mild cases) to surgery to remove the abnormal tissue—sometimes even the whole uterus.

Ovaries, a female's primary reproductive organs, produce oocytes (immature eggs) and sex hormones. Endometrium lines the uterus, a chamber in which embryos develop.

Secretion of sex hormones—estrogens and progesterone—is part of a cyclic menstrual cycle that occurs through a female's reproductive years.

FEMALE REPRODUCTIVE FUNCTION

Cyclic Changes in the Ovary

An average baby girl has about 2 million primary oocytes in her ovaries. By the time she is seven years old, only about 300,000 remain; her body has resorbed the rest. Her primary oocytes entered meiosis I when she was a fetus, but then genetic instructions arrested the division process. Meiosis will resume in one oocyte at a time, starting with her first menstrual cycle. About 400 or 500 will be released during her reproductive years.

In Figure 14.7, sketch *a* is a primary oocyte near an ovary's surface. A layer of **granulosa cells** surrounds and nourishes it. We call the primary oocyte and the cell layer around it a **follicle**. At the start of a menstrual cycle, the hypothalamus is secreting enough GnRH to cause the anterior pituitary to step up *its* secretion of FSH and LH (Figure 14.8). The blood concentration of those two hormones increases, and *that* causes the follicle to grow. (FSH, remember, is short for "follicle-stimulating hormone.")

The oocyte starts to increase in size, and more layers of cells form around it. Glycoprotein deposits accumulate between the oocyte and the layers, widening the space between them. In time the deposits form the **zona pellucida**, a noncellular coating around the oocyte.

FSH and LH stimulate cells outside the zona pellucida to secrete estrogens. An estrogen-containing fluid accumulates in the follicle (now called a secondary or Graafian follicle), and estrogen levels in the blood start to rise. About eight to ten hours before being released from the ovary, the oocyte completes the meiotic cell division (called meiosis I) that was arrested years before. Now, there are two cells. One, a large, **secondary oocyte**, ends up with nearly all the cytoplasm. The other cell is the *first polar body*. (It may divide again.) Chromosomes are allocated to the cells in a way that ensures the secondary oocyte will be haploid—the chromosome number a gamete must have. It now begins a second round of meiosis, which again is arrested—as you will read in Chapter 15, in preparation for fertilization by a sperm.

About halfway through the menstrual cycle, the pituitary gland detects the rising estrogen level. It responds with a brief gush of LH. The LH surge causes rapid vascular changes that make the follicle swell quickly. The surge also induces enzymes to digest the bulging follicle wall. Weakened, it ruptures. Fluid escapes, carrying the secondary oocyte and polar body with it (Figure 14.7e). *The midcycle surge of LH has triggered ovulation—the release of a secondary oocyte from the ovary.*

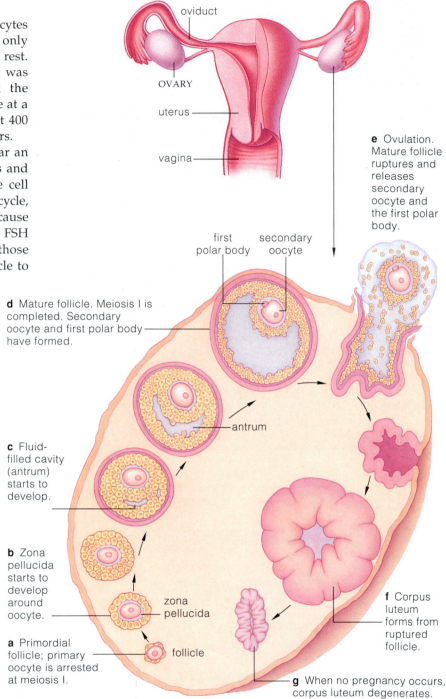

e Ovulation. Mature follicle ruptures and releases secondary oocyte and the first polar body.

first polar body

secondary oocyte

d Mature follicle. Meiosis I is completed. Secondary oocyte and first polar body have formed.

antrum

c Fluid-filled cavity (antrum) starts to develop.

b Zona pellucida starts to develop around oocyte.

zona pellucida

a Primordial follicle; primary oocyte is arrested at meiosis I.

follicle

f Corpus luteum forms from ruptured follicle.

g When no pregnancy occurs, corpus luteum degenerates.

Figure 14.7 Cyclic events in a human ovary (cross section). A folllicle stays in the same place in an ovary all through the menstrual cycle. It does not move around as in this diagram, which only shows the order in which events occur. In the cycle's first phase, the follicle grows and matures. At ovulation, the second phase, the mature follicle ruptures and releases a secondary oocyte. In the third phase, a corpus luteum forms from the follicle's remnants. It self-destructs if pregnancy does not occur.

oviduct

OVARY

uterus

vagina

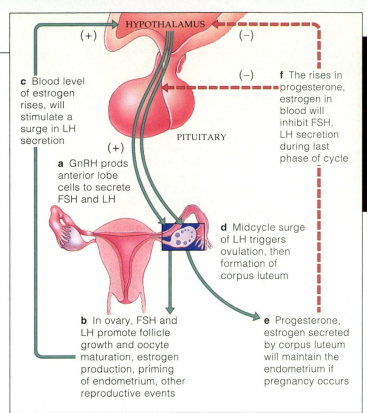

c Blood level of estrogen rises, will stimulate a surge in LH secretion

f The rises in progesterone, estrogen in blood will inhibit FSH, LH secretion during last phase of cycle

HYPOTHALAMUS

(+)   (−)

(−)

PITUITARY

(+)

a GnRH prods anterior lobe cells to secrete FSH and LH

d Midcycle surge of LH triggers ovulation, then formation of corpus luteum

b In ovary, FSH and LH promote follicle growth and oocyte maturation, estrogen production, priming of endometrium, other reproductive events

e Progesterone, estrogen secreted by corpus luteum will maintain the endometrium if pregnancy occurs

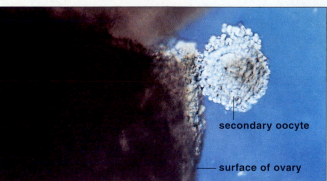

secondary oocyte

surface of ovary

Figure 14.8 Feedback control of hormonal secretion during a menstrual cycle. A positive feedback loop from an ovary to the hypothalamus causes a surge in LH secretion. This surge triggers ovulation. The micrograph above shows a secondary oocyte being released from an ovary at this time. Afterward, negative feedback loops to the hypothalamus and pituitary inhibit FSH secretion. They prevent another follicle from maturing until the cycle is completed.

An ovary releases a secondary oocyte into the abdominal cavity, and from there the oocyte enters an oviduct. Fingerlike projections from the oviduct (called *fimbriae*) extend like an umbrella over part of the ovary. Each "finger" bears beating cilia, and movements of the projections and their cilia sweep the oocyte into the channel. If fertilization takes place, it typically occurs while the oocyte is in the oviduct. At fertilization, the oocyte will finish meiosis II and become a mature ovum, the egg. Chapter 15 looks at the events after fertilization.

Cyclic Changes in the Uterus

The estrogens released early in the menstrual cycle also help pave the way for a possible pregnancy. They stimulate growth of the endometrium and its glands. Then, just before the midcycle LH surge, cells of the follicle wall start secreting progesterone as well as estrogens. Blood vessels grow rapidly in the thickened endometrium. At ovulation, the estrogens act on tissue around the cervical canal, the narrowed portion of the uterus that leads to the vagina. The cervix starts to secrete large amounts of a thin, clear mucus—an ideal medium for sperm to swim through.

After ovulation, another structure dominates events. Granulosa cells left behind in the follicle differentiate into a yellowish glandular structure, the **corpus luteum**. Formation of the corpus luteum results from the midcycle surge of LH. Hence its name, luteinizing hormone.

The corpus luteum secretes some estrogen, along with progesterone. The progesterone prepares the woman's reproductive tract for the arrival of an embryo. For example, it causes mucus in the cervix to become thick and sticky, which may prevent normal vaginal bacteria from entering the uterus. Progesterone also maintains the endometrium during a pregnancy.

A corpus luteum persists for about twelve days. During that time, the hypothalamus signals for a decrease in FSH secretion, which prevents any other follicles from developing. If a developing embryo does not arrive and burrow into the endometrium (a process called *implantation* discussed in Chapter 15), the corpus luteum self-destructs near the end of the menstrual cycle. It does so by secreting prostaglandins, which apparently disrupt the corpus luteum's own functioning.

After the corpus luteum breaks down, progesterone and estrogen levels fall rapidly, so the endometrium also starts to break down. Deprived of oxygen and nutrients, its blood vessels constrict and its tissues die. Blood escapes from the ruptured walls of weakened capillaries. The blood and sloughed endometrial tissues make up the menstrual flow, which continues for three to six days. Then, as the cycle begins anew, rising levels of estrogen stimulate the repair and growth of the endometrium.

Coordinated secretions of estrogen, progesterone, LH, and FSH bring about changes in the ovary and uterus during the menstrual cycle.

A midcycle surge of LH triggers ovulation, the release of the secondary oocyte and the polar body from the ovary.

The hormones released during various phases of the menstrual cycle help pave the way for fertilization and prepare the endometrium and other parts of a female's reproductive tract for pregnancy.

VISUAL SUMMARY OF THE MENSTRUAL CYCLE

By now you may have come to the conclusion that the menstrual cycle is not a simple tune on a biological banjo—it is a full-blown hormonal symphony!

Before continuing your reading, take a moment to review Figure 14.9. It correlates the cyclic events in the ovary and uterus with all the coordinated changes in hormone levels that bring about those events. The illustration may give you a better understanding of what goes on.

Figure 14.9 Changes in the ovary and uterus, correlated with changing hormone levels during the menstrual cycle. *Green* arrows indicate which hormones dominate the cycle's first phase (when the follicle matures), then the second phase (when the corpus luteum forms). (**a**,**b**) FSH and LH cause changes in ovarian structure and function. (**c**,**d**) Estrogen and progesterone from the ovary cause changes in the endometrium.

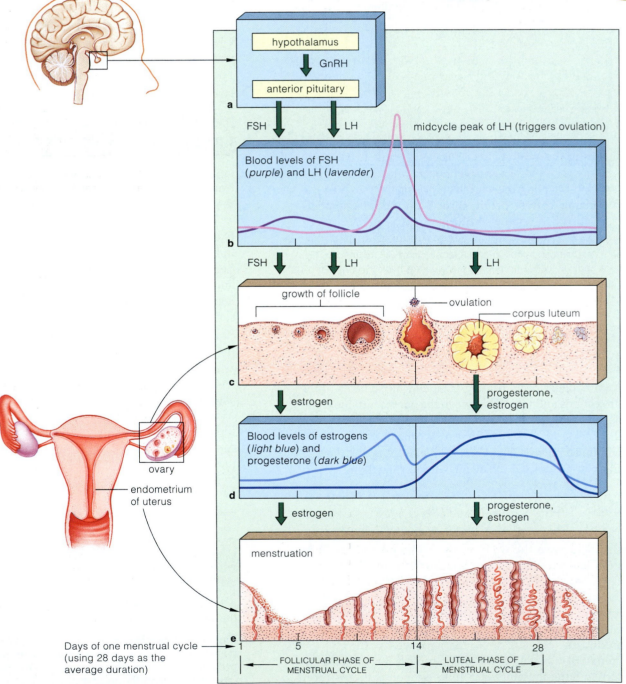

Further reading: Student Guide to InfoTrac on web site →

SEXUAL INTERCOURSE AND FERTILIZATION

Sexual Intercourse

Suppose a secondary oocyte is on its way down an oviduct when a female and male are engaged in sexual intercourse, or **coitus**. The male sex act typically requires erection, in which the limp penis stiffens and lengthens, and ejaculation, forceful expulsion of semen into the urethra and out from the penis. As shown in Figure 14.2, the penis contains cylinders of spongy tissue. One cylinder has a mushroom-shaped tip (the glans penis). Inside it is a dense array of sensory receptors that are activated by friction. In sexually unaroused males, the large blood vessels leading into the cylinders are constricted. In aroused males, these blood vessels vasodilate, so blood flows into the cylinders faster than it flows out. Blood collects in the spongy tissue, and the organ stiffens and lengthens—a mechanism that helps the penis penetrate into the female's vagina.

In females, arousal includes vasodilation of blood vessels in the genital area. This causes vulvar tissues to engorge with blood and swell. Secretions flow from the cervix, lubricating the vagina.

During coitus, pelvic thrusts stimulate the penis as well as the female's clitoris and vaginal wall. The mechanical stimulation triggers rhythmic, involuntary contractions in smooth muscle in the male reproductive tract, especially the vas deferens and the prostate. The contractions rapidly force sperm out of each epididymis. They force the contents of seminal vesicles and the prostate gland into the urethra. The resulting mixture, semen, is ejaculated into the vagina.

During ejaculation, a sphincter closes off the neck of the male's bladder and prevents urine from being excreted. Ejaculation is a reflex response; once it begins, it cannot be halted.

Emotional intensity, hard breathing, and heart pounding, as well as generalized skeletal muscle contractions, accompany the rhythmic throbbing of the pelvic muscles. At orgasm, the culmination of the sex act, strong sensations of release, warmth, and relaxation dominate. Similar events occur during female orgasm.

Some people mistakenly believe that unless a woman experiences orgasm, she cannot become pregnant. Don't believe it. A female can become pregnant from intercourse *regardless* of whether she experiences orgasm, and even if she is not sexually aroused. All that is required is that a sperm meet up with a secondary oocyte that is traveling down one of her oviducts.

Chapter 15 considers the events of conception in some detail. However, it will help round out this chapter's portrait of reproduction to briefly preview the biological sequel when sexual intercourse brings a sperm and an oocyte together.

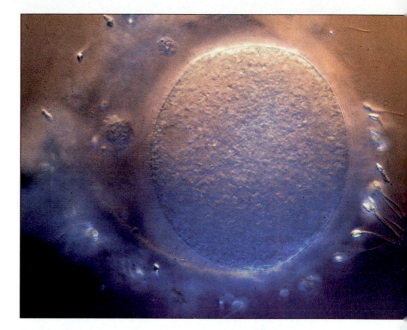

Figure 14.10 A secondary oocyte surrounded by sperm. If fertilization ensues, it will set the stage for the development of a new individual, continuing the human life cycle.

A Preview of Fertilization

If sperm enter the vagina a few days before or after ovulation or anytime between, fertilization may be the outcome. Within thirty minutes after ejaculation, muscle contractions move the sperm deeper into the female reproductive tract. Only a few hundred sperm will actually reach the upper portion of the oviduct, which is where fertilization usually takes place. The remarkable micrograph in Figure 14.10 shows living sperm around a secondary oocyte.

As you will read more fully in the following chapter, the meeting of sperm and secondary oocyte is only the first of several intricately orchestrated events that lead to actual **fertilization**—the fusion of the sperm nucleus with the nucleus of a mature egg.

Sexual intercourse (coitus) typically involves a sequence of physiological changes in both partners.

During arousal, dilation of blood vessels causes increased blood flow to the penis (males) and vulva (females). Orgasm involves muscular contractions (including those leading to ejaculation of semen into the vagina) and sensations of release, warmth, and relaxation.

A female may become pregnant through intercourse even if she is not sexually aroused or does not experience orgasm.

Fertilization can occur when a sperm encounters a secondary oocyte, usually in the oviduct.

CONTROL OF FERTILITY

Many sexually active people choose to exercise control over whether their activity will produce a child. The *Choices* feature (Section 14.9) explores some social and ethical dilemmas that control of fertility presents for some people. Here we consider the biological bases of different forms of birth control.

Natural Birth Control Options

The most effective method of birth control is complete *abstinence*—no sexual intercourse whatsoever. However, the motivation to engage in sex has been evolving for more than 500 million years, and so it is probably unrealistic to expect many people to practice abstinence for long periods.

A modified form of abstinence is the *rhythm method*, also called the "fertility awareness" or *sympto-thermal method*. The idea is to avoid intercourse during the woman's fertile period, beginning a few days before ovulation and ending a few days after. Her fertile period is identified and tracked by keeping records of the length of her menstrual cycles and sometimes by examining her cervical secretions and taking her temperature each morning when she wakes up. (Core body temperature rises by one-half to one degree just after ovulation.) But ovulation can be irregular, and it can be easy to miscalculate. Also, sperm deposited in the vaginal tract a few days before ovulation may survive until ovulation. The method is inexpensive (it requires only a thermometer) and does not require fittings and medical checkups. However, it does come with a high risk of pregnancy (Figure 14.11).

Withdrawal, removing the penis from the vagina before ejaculation, dates back at least two thousand years. It is not very effective. Not only does withdrawal require strong willpower, but fluid released from the penis before ejaculation may contain sperm. *Douching*, or rinsing out the vagina with a chemical right after intercourse, is next to useless. Sperm can move past the cervix and out of reach of the douche within 90 seconds after ejaculation.

Surgical Solutions

Controlling fertility by surgical intervention is less chancy. In vasectomy, a physician makes a tiny incision in a man's scrotum, then severs and ties off each vas deferens. The procedure takes only twenty minutes and requires only a local anesthetic. Afterward, sperm cannot leave the testes and cannot be present in the man's semen. Having a vasectomy apparently does not disrupt hormonal functions and there seems to be no noticeable difference in sexual activity. Vasectomies can be reversed, but scar tissue can prevent sperm from passing through the tubes. About half of men who have

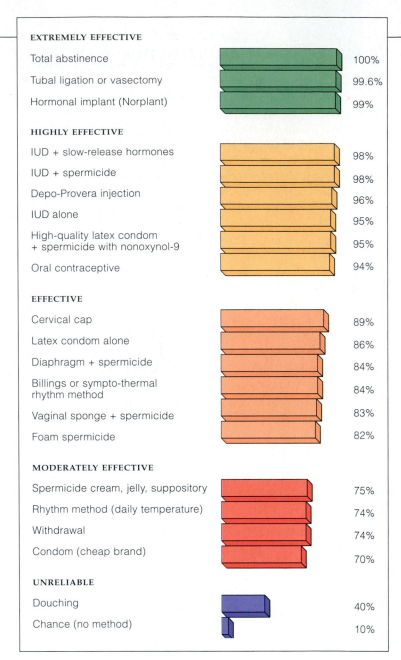

EXTREMELY EFFECTIVE	
Total abstinence	100%
Tubal ligation or vasectomy	99.6%
Hormonal implant (Norplant)	99%
HIGHLY EFFECTIVE	
IUD + slow-release hormones	98%
IUD + spermicide	98%
Depo-Provera injection	96%
IUD alone	95%
High-quality latex condom + spermicide with nonoxynol-9	95%
Oral contraceptive	94%
EFFECTIVE	
Cervical cap	89%
Latex condom alone	86%
Diaphragm + spermicide	84%
Billings or sympto-thermal rhythm method	84%
Vaginal sponge + spermicide	83%
Foam spermicide	82%
MODERATELY EFFECTIVE	
Spermicide cream, jelly, suppository	75%
Rhythm method (daily temperature)	74%
Withdrawal	74%
Condom (cheap brand)	70%
UNRELIABLE	
Douching	40%
Chance (no method)	10%

Figure 14.11 Comparison of the effectiveness of some contraceptive methods in the United States. Percentages shown are based on the number of unplanned pregnancies per 100 couples who used the method as the only form of birth control for one year. For example, "94% effectiveness" for oral contraceptives (the birth control pill) means that, on average, 6 of every 100 women using them will become pregnant.

a vasectomy also develop antibodies to their own sperm, and usually cannot regain their fertility.

In *tubal ligation*, a woman's oviducts are cauterized or cut and tied off. Tubal ligation is usually performed in a hospital. Afterward, some women suffer bouts of pain and inflammation of tissues where the surgery was performed. The operation can be reversed, although major surgery is required and is not always successful.

Physical and Chemical Barriers to Conception

Other methods involve physical or chemical barriers to prevent sperm from entering the uterus and moving to the oviducts. Spermicidal foam and spermicidal jelly are toxic to sperm. They are packaged in an applicator and placed in the vagina just before intercourse. These products are not reliable unless used with another device, such as a diaphragm or condom.

A *diaphragm* is a flexible, dome-shaped device that is inserted into the vagina and positioned over the cervix before intercourse. A diaphragm is fairly effective when fitted initially by a doctor, used with foam or jelly before each sexual contact, and inserted correctly with each use. A variation on the diaphragm, the *cervical cap*, is smaller and can be left in place for as long as three days with just a single dose of spermicide. The *contraceptive sponge* is a soft, disposable disk that contains a spermicide and covers the cervix. It does not require a prescription or special fitting; it is simply wetted and inserted up to 24 hours before intercourse. Because the sponge can slip out of place, the method is about 84 percent reliable.

The *intrauterine device,* or IUD, is a small plastic or metal device that is placed into the uterus for up to two years at a time. It interferes with implantation of a fertilized egg into the uterine wall. Available by prescription, IUDs can cause severe menstrual cramping, and they increase the risk of infection or perforation of the uterus and other problems. Improved types may address such concerns, but any woman considering an IUD should discuss the matter fully with her physician.

Condoms are thin, tight-fitting sheaths of latex or animal skin worn over the penis during intercourse. They are about 85 to 93 percent reliable, and latex condoms help prevent the spread of sexually transmitted diseases. However, condoms can tear and leak, which renders them useless. A pouchlike latex "female condom" that is inserted into the vagina has also been developed.

The most widely used method of fertility control is the *birth control pill*—any of a number of formulations of synthetic estrogens and progesterones. An oral contraceptive suppresses the normal release of anterior pituitary hormones (LH and FSH) required for eggs to mature and be ovulated. Oral contraceptives are prescription drugs. Formulations vary and are selected to match each patient's needs. That is why it is not wise for a woman to borrow oral contraceptives from someone else.

When a woman always takes her daily dose, this is one of the most reliable methods of controlling fertility. It doesn't interrupt sexual activity, and is an easy method to follow. Often, taking an oral contraceptive corrects erratic menstrual cycles and reduces cramping; studies suggest that using birth control pills also reduces the risk of ovarian cancer. Some users experience (usually temporary) side effects, including nausea, weight gain, tissue swelling (edema), and minor headaches. Continued use may lead to blood clots in women predisposed to that disorder (3 out of 10,000). Complications are more likely to develop in women who smoke, and most physicians won't prescribe an oral contraceptive for a smoker.

Injections or implants of progestin (a progesterone) inhibit ovulation. A Depo-Provera injection works for three months and is 96 percent effective. *Norplant* (six rods implanted under the skin) works for five years and is 99 percent effective. The rods contain levonorgestrel, a hormone that prevents implantation of a fertilized egg. Both can cause heavier menstrual periods, and doctors have some trouble removing Norplant rods.

A pregnancy test doesn't register positive until after an embryo implants. According to one view, a woman isn't pregnant until that time, so *morning-after pills* intercept pregnancy. Actually, such pills work up to seventy-two hours after unprotected intercourse by interfering with the hormones that control events between ovulation and implantation. Preven is a kit containing high doses of of birth control pills that suppress ovulation and block the corpus luteum from functioning. RU-486 can block fertilization and implantation. Because it suppresses the action of progesterone (which is needed to maintain pregnancy) it may also induce an abortion when taken within seven weeks of implantation. At this writing, RU-486 is available in Europe, but its use in the United States is still controversial.

Future Options for Fertility Control

Researchers are working to develop new and better methods of fertility control. Examples include implants that biodegrade and so don't require surgical removal. Other efforts are aimed at developing a contraceptive for men that reduces the sperm count, chemicals that "sterilize" the user without surgery, and male and female sterilization procedures that can be reversed more easily.

Many couples cherish the prospect of becoming parents. Hence, controls over fertility also extend in the other direction—to childless couples who want to conceive a child but have problems doing so. This is the topic we turn to next.

There are many options for controlling fertility, some safer or more effective than others.

The most effective methods involve surgery (vasectomy or tubal ligation) or chemical barriers to conception. Each approach has drawbacks; hence, a frank discussion with a physician is the best starting point for fertility control.

COPING WITH INFERTILITY

In the United States, about one in every six couples is *infertile*—unable to conceive a child after a year of trying. Causes run the gamut from hormonal imbalances that prevent ovulation, oviducts blocked by effects of disease, a low sperm count, or sperm that are defective in a way that impairs fertilization.

Artificial Insemination

Artificial insemination was one of the first methods of modern "reproductive technology." In this approach, semen is placed into the vagina or uterus, usually by syringe, around the time the woman is ovulating. The semen may come from a woman's partner, especially if the man has a low sperm count, because his sperm can be concentrated prior to the procedure. In *artificial insemination by donor* (AID), a sperm bank provides sperm from an anonymous donor. AID produces about 20,000 babies in the United States every year.

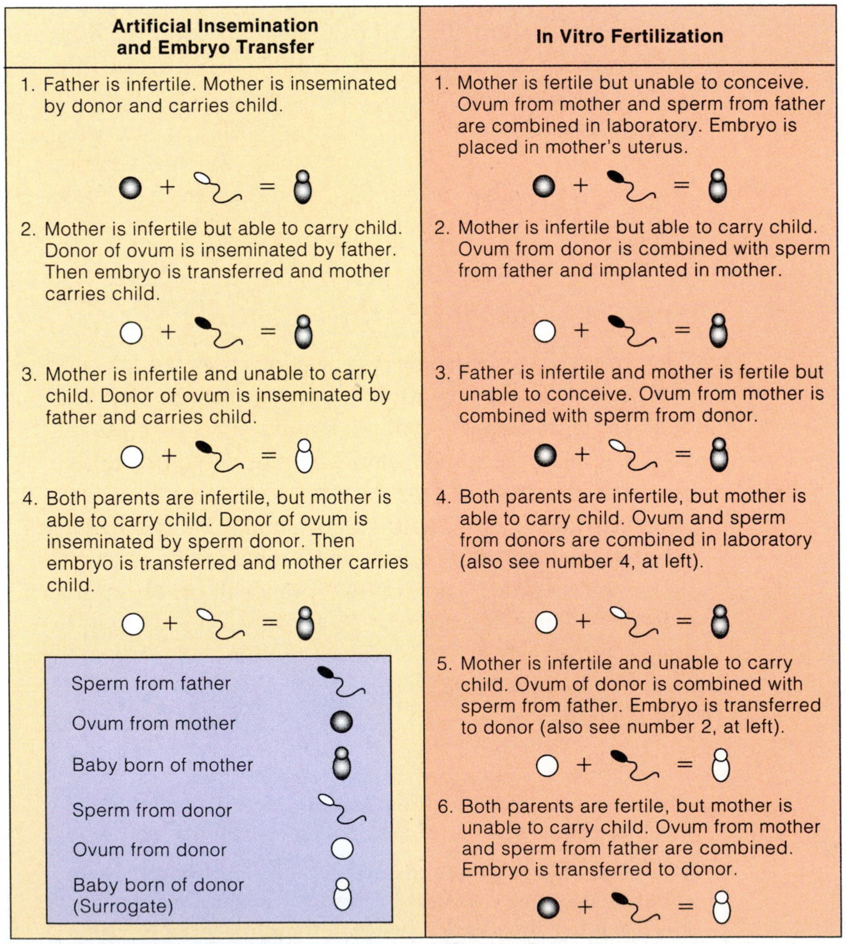

In Vitro Fertilization

With *in vitro fertilization* (IVF)—literally "fertilization in glass"—conception can occur externally. If a couple's sperm and oocytes are normal, they can be used. Otherwise, variations of the technology are available that use sperm, oocytes, or both, from donors (Figure 14.12). Sperm and oocytes are placed in a laboratory dish in a solution that simulates the fluid in oviducts. If fertilization occurs, about 12 hours later *zygotes* (fertilized eggs in their earliest stages of development) are transferred to a solution that will support further development. Two to four days after that, one or more embryos are transferred to the woman's uterus. An embryo implants in about 20 percent of cases. In vitro fertilization can produce more embryos than can be used in a given procedure. The fate of unused embryos (which are stored frozen) has prompted ethical debates and even bitter court battles between divorcing couples.

Several years ago, doctors in Belgium pioneered a variation on IVF in which a single sperm is injected into a single egg using a tiny glass needle. This method appears to have a higher success rate than traditional in vitro fertilization, and it has produced healthy babies for hundreds of couples.

In *IVF with embryo transfer*, a fertile woman "donor" is inseminated with sperm from a male whose female partner is infertile. If a pregnancy results, the developing embryo may be transferred to the infertile woman's uterus. Alternatively, a "surrogate mother" may carry the pregnancy to term. Besides being technically difficult, this approach has serious legal complications. It is not yet a common solution to infertility.

Intrafallopian Transfers

In a technique called GIFT (*g*amete *i*ntra*f*allopian *t*ransfer) sperm and oocytes are collected and placed into an oviduct (fallopian tube). About 20 percent of the time, the oocyte is fertilized and a normal pregnancy ensues. An alternative is ZIFT (*z*ygote *i*ntra*f*allopian *t*ransfer). First, oocytes and sperm are brought together in the laboratory. Then, if fertilization occurs, the zygote is placed in a woman's oviducts. GIFT and ZIFT are about as successful as in vitro fertilization.

In in vitro fertilization, sperm and oocytes are brought together in a laboratory dish, where conception may occur. Other techniques for overcoming infertility include intrafallopian transfers and artificial insemination.

Artificial Insemination and Embryo Transfer	In Vitro Fertilization
1. Father is infertile. Mother is inseminated by donor and carries child.	1. Mother is fertile but unable to conceive. Ovum from mother and sperm from father are combined in laboratory. Embryo is placed in mother's uterus.
2. Mother is infertile but able to carry child. Donor of ovum is inseminated by father. Then embryo is transferred and mother carries child.	2. Mother is infertile but able to carry child. Ovum from donor is combined with sperm from father and implanted in mother.
3. Mother is infertile and unable to carry child. Donor of ovum is inseminated by father and carries child.	3. Father is infertile and mother is fertile but unable to conceive. Ovum from mother is combined with sperm from donor.
4. Both parents are infertile, but mother is able to carry child. Donor of ovum is inseminated by sperm donor. Then embryo is transferred and mother carries child.	4. Both parents are infertile, but mother is able to carry child. Ovum and sperm from donors are combined in laboratory (also see number 4, at left).
Sperm from father / Ovum from mother / Baby born of mother / Sperm from donor / Ovum from donor / Baby born of donor (Surrogate)	5. Mother is infertile and unable to carry child. Ovum of donor is combined with sperm from father. Embryo is transferred to donor (also see number 2, at left).
	6. Both parents are fertile, but mother is unable to carry child. Ovum from mother and sperm from father are combined. Embryo is transferred to donor.

Figure 14.12 New ways to make babies.

DILEMMAS OF FERTILITY CONTROL

About 3 percent of American women of childbearing age have had an abortion; more than 1,500,000 abortions are performed in the United States each year. At one time, abortions were generally forbidden by law in the United States unless the pregnancy endangered the mother's life. In 1973 the Supreme Court ruled (in *Roe v. Wade*) that the government does not have the right to forbid abortions during the early stages of pregnancy (typically up to 5 months). Before this ruling, there were dangerous, traumatic, and often fatal attempts to abort embryos, either by pregnant women themselves or by quacks. During the first trimester (twelve weeks) abortions performed by competent medical personnel are relatively fast, painless, and free of complications. Even with judicial approval, however, abortions in the second and third trimesters are highly controversial unless the mother's life is threatened.

Repeated court actions have reaffirmed the legality of abortion in the United States while permitting states to impose restrictions. Yet court rulings have not done much to quell debate over the issue of legalized abortion. Legal and social conflict rage on. Both sides protest, hold marches, and lobby lawmakers (Figure 14.13). Fights have broken out between the factions, clinics have been bombed and burned, and clinic staff have been harassed. Several doctors have been shot and killed. Abortion foes charge that the physicians themselves are committing murder.

There aren't likely to be easy solutions to the ethical issues associated with fertility control. For some people, a crucial question is, *When does life begin?* During her lifetime, a human female can produce as many as 500 eggs, all of which are alive. During one ejaculation, a human male can release up to half a billion sperm, which also are living cells. Does "life begin" only when a sperm and an egg fuse? Is there a difference between life in general and *a particular life?* Does "meaningful" life begin only when a fetus attains a certain age in the womb? Does a woman have a right to control her reproductive life, even if doing so means terminating a pregnancy? Does the government have the right to intervene in the decision? These are only some of the moral, ethical, and philosophical questions that underlie the public debate on fertility control.

A GLOBAL PERSPECTIVE ON FERTILITY CONTROL

Globally, fertility control is equally controversial. Some cultures and religions strongly discourage efforts to limit births. Meanwhile, the human population growth rate spirals out of control (Chapter 25). For example, in the time it takes you to read this chapter, about 10,700 babies will be born. By the time you go to bed tonight, there will be 257,000 more people on earth than there were last night at that hour. Within a week, the number will reach 1,800,000—about the population of Massachusetts. Worldwide human population growth has so outstripped resources that each year millions of people face the horrors of starvation.

Figure 14.13 Demonstration symbolizing the abortion debate in the United States. It is an understatement to say that emotions run high on both sides of the issue.

Is massive, global birth control a viable solution? Where do individual rights fit into the picture? In China, where the population is expected to top 1.5 billion by 2025 and famine is an ongoing concern, the government has established the most extensive family planning program in the world. Couples who limit themselves to one child receive benefits such as extra food, better housing, free medical care, and salary bonuses. Those who exceed the one-child limit are severely penalized. In this environment, abortion is officially encouraged as an alternative to childbearing.

In some countries, especially poorer ones with rapidly growing populations, some factions are promoting the large-scale use of implantable or injectable contraceptives (Section 14.7). Critics of these proposals fear that the drugs' long-term side effects have not been well studied. Moreover, they worry that in some circumstances (with poor or uneducated women or those having a history of child abuse or drug addiction) authorities might coerce the women into receiving injections or implants.

As diverse societies now move into the 21st century, with burgeoning populations and their associated economic, social, and environmental pressures, one thing seems certain: These and other dilemmas of fertility control will not be easily resolved.

SUMMARY

1. Humans have a pair of primary reproductive organs. These are the sperm-producing testes in males and the egg-producing ovaries in females. Both sexes also have accessory ducts, and glands. Testes and ovaries also produce hormones that influence reproductive functions and secondary sexual traits.

2. The hormones testosterone, luteinizing hormone (LH), and follicle-stimulating hormone (FSH) control the formation of sperm. They are part of feedback loops among the hypothalamus, anterior pituitary, and testes.

3. The hormones estrogen, progesterone, FSH, and LH control the maturation and release of eggs, as well as changes in the lining of the uterus (endometrium). They are part of feedback loops involving the hypothalamus, anterior pituitary, and ovaries.

4. In both males and females, gonadotropin-releasing hormone (GnRH) from the hypothalamus stimulates the anterior pituitary to release LH and FSH.

5. During a menstrual cycle, the following events occur:
   a. The cycle begins when menstrual flow starts, as the endometrial lining of the uterus is shed.
   b. A follicle, which is an oocyte surrounded by a cell layer, matures in an ovary. Under the influence of hormones produced by the follicle, the endometrium of the uterus starts to rebuild.
   c. A midcycle peak of LH triggers the release of a secondary oocyte from the ovary. This event is called ovulation.
   d. A corpus luteum forms from the remainder of the follicle. Progesterone secreted by the corpus luteum prepares the endometrium to receive a fertilized egg. It also helps maintain the endometrium during pregnancy. When fertilization does not occur, the corpus luteum degenerates and the endometrial lining breaks down and is shed through menstruation.

6. Sexual intercourse (coitus) is the usual way in which an egg (a secondary oocyte) becomes fertilized by a sperm. It typically involves a sequence of physiological changes in both partners.

7. Efforts to control human fertility raise important ethical questions. These questions extend to the physical, chemical, surgical, or behavioral interventions used in the control of unwanted pregnancies and in attempts to help infertile couples through reproductive technology.

Review Questions

1. Distinguish between:
   a. seminiferous tubule and vas deferens *14.1*
   b. sperm and semen *14.1, 14.2*
   c. Leydig cells and Sertoli cells *14.1, 14.2*
   d. primary oocyte and secondary oocyte *14.3, 14.4*
   e. follicle and corpus luteum *14.4*
   f. the three phases of the menstrual cycle *14.3*

2. Which hormones influence male reproductive function? *14.2*

3. Which hormones influence the menstrual cycle? *14.4, 14.5*

4. List four events that are triggered by the surge of LH at the midpoint of the menstrual cycle. *14.4*

5. What changes occur in the endometrium during the menstrual cycle? *14.4*

6. Label the components of the human male and female reproductive systems and state their functions. *14.1, 14.3*

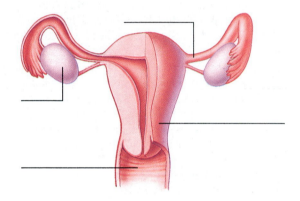

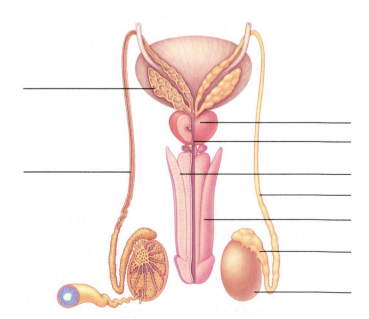

Self-Quiz (Answers in Appendix V)

1. Besides producing gametes (sperm and eggs), the primary male and female reproductive organs also produce sex hormones. The _____ and the pituitary gland control secretion of both.

2. _____ production is continuous from puberty onward in males; _____ production is cyclic and intermittent in females.
   a. Egg; sperm
   c. Testosterone; sperm
   b. Sperm; egg
   d. Estrogen; egg

3. The formation of sperm is controlled through secretion of
   _____ .
   a. testosterone
   c. FSH
   b. LH
   d. all of the above are correct

4. During the menstrual cycle, a midcycle surge of _____ triggers ovulation.
   a. estrogen
   c. LH
   b. progesterone
   d. FSH

5. Which is the correct order for one turn of the menstrual cycle?
   a. corpus luteum forms, ovulation, follicle forms
   b. follicle grows, ovulation, corpus luteum forms

6. In order for sexual intercourse to produce a pregnancy, both partners must experience _____ .
   a. orgasm
   c. an erection
   b. ejaculation
   d. none of the above

Critical Thinking: You Decide (Key in Appendix VI)

1. Counselors sometimes advise a couple who wish to conceive a child to use an alkaline (basic) douche immediately before intercourse. Speculate about what the doctors' reasoning might be.

2. In the "fertility awareness" method of birth control, a woman gauges her fertile period each month by monitoring changes in the consistency of her vaginal mucus. What kind of specific information does such a method provide? How does it relate to the likelihood of getting pregnant?

3. Some women experience *premenstrual syndrome* (PMS), which can include a distressing combination of mood swings, fluid retention (edema), anxiety, backache and joint pain, food cravings, and other symptoms. PMS usually develops after ovulation and lasts until just before or just after menstruation begins. A woman's doctor can recommend strategies for managing PMS, which often include diet changes, regular exercise, and use of diuretics or other drugs. Many women find that taking vitamin B_6 and vitamin E helps reduce pain and other symptoms. Although the precise cause of PMS is unknown, it seems clearly related to the cyclic production of ovarian hormones. After reviewing Figure 14.8, suggest which hormonal changes may trigger PMS in affected females.

4. The absence of menstrual periods, or *amenorrhea,* is normal in pregnant and postmenopausal women and in girls who have not yet reached puberty. However, in females of reproductive age amenorrhea can result from tumors of the pituitary or adrenals. Based on discussion in this chapter and Chapter 13, speculate about why such tumors might disrupt cyclic menstruation.

5. Some infertile couples are willing to go to considerable lengths to have a baby (Figure 14.14). From your reading of Section 14.8, which of the variations of reproductive technologies produces a child that is *least* related (genetically) to the infertile couple. Would you view having a child by that method as preferable to adopting a baby? Why?

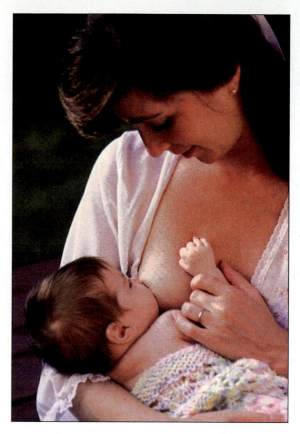

Figure 14.14 An image of parental devotion—and the goal of many infertile couples.

Selected Key Terms

acrosome *14.2*	ovary *CI*
bulbourethral gland *14.1*	oviduct *14.3*
clitoris *14.3*	ovulation *14.3*
coitus *14.6*	progesterone *14.3*
corpus luteum *14.4*	prostate gland *14.1*
endometrium *14.3*	secondary oocyte *14.4*
estrogen *14.3*	secondary sexual trait *CI*
fertilization *14.6*	semen *14.1*
follicle *14.4*	seminiferous tubule *14.1*
FSH *14.2*	Sertoli cell *14.2*
granulosa cell *14.4*	sperm *14.2*
Leydig cell *14.2*	testis *CI*
LH *14.2*	testosterone *14.2*
menstrual cycle *14.3*	uterus *14.3*
oocyte *14.3*	vas deferens *14.1*
orgasm *14.6*	zona pellucida *14.4*

Readings

Fackelmann, K. November 28, 1998. "It's a Girl!" *Science News.* Thought-provoking articles on how reproductive technologies can allow prospective parents to choose their child's sex—and possibly "custom design" some of the child's other traits.

"The Evolution of Sexes." 17 July 1992. *Science.* Fascinating discussion of why (perhaps) there are two sexes of human beings and many other organisms.

Sherwood, L. 1997. *Human Physiology.* Third edition. Pacific Grove, California: Brooks/Cole.

DEVELOPMENT AND AGING

A Life Story, Summarized

The early development of the human body is a momentous journey. Insofar as we understand the sequence of events—using your own beginnings as a handy example—in general the first few weeks of development unfold in the following way.

In the upper end of one of your mother's oviducts, an ovulated oocyte is fertilized. The fertilized egg, now called an embryo, continues down the oviduct toward the uterus (Figure 15.1). On the way, it undergoes several rounds of cell division that transform it into a hollow ball of about sixty-four tiny cells. At this stage, most of the cells make up the wall of the ball, but a small cluster of them—perhaps three to six cells—is situated on the inner wall of the sphere. This cluster of cells, known as the *inner cell mass*, will eventually give

rise to the actual embryo that will continue developing into a fully formed baby. The cells of the outer wall will contribute to vital tissues outside the embryo, about which you will learn much more later in this chapter.

As development proceeds, more rounds of cell division add more and more cells to the inner cell mass. In short order, the mass reorganizes, forming a flattened disk with two tissue layers. Less than a millimeter across, the disk could stretch out across the head of a pin with room to spare. But that speck is an embryo.

A third cell layer forms between the original two. Then, inside those three layers, cells follow genetic commands to change shape and to get moving. Part of one layer curves up and folds over, while other cells are elongating, making their layer thicker. And still

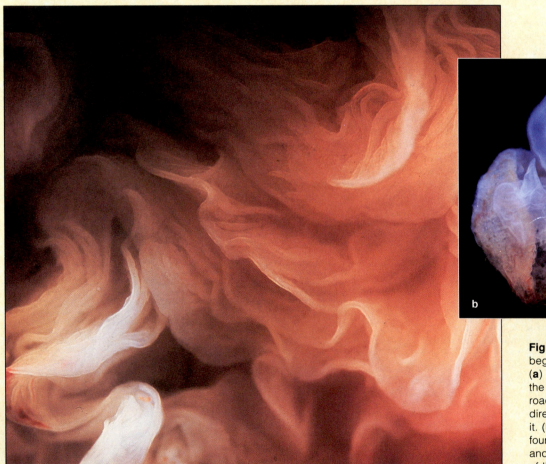

a

b

Figure 15.1 A visual summary of the beginning of human development. (**a**) The billowing entrance to an oviduct, the tubelike road to the uterus. On that road, a sperm traveling from the opposite direction encountered an egg and fertilized it. (**b**) A human embryo, as it appears just four weeks after the moment of fertilization, and (**c**) several years into the journey of life.

other cells are migrating to new destinations. These changes transform the pancakelike embryo into a pale, translucent crescent. Its surface isn't smooth; it has nooks and crannies, bumps and hollows.

By the third week after fertilization, the tubelike forerunners of the embryo's nervous system and circulatory system appear. By the fourth week, its heart starts beating rhythmically in the thickening wall of one tube. On the embryo's sides, budding regions of tissue mark the beginning of arms and legs.

Five weeks into the journey, the limb buds have paddles—the start of hands and feet. Round, dark dots appear under the transparent skin; these are the forming eyes. Around and below them, cells migrate in a coordinated ballet, interacting in ways that sculpt out a nose, cheeks, and a mouth. By now, the embryo is recognizably a human in the making. Later embryonic and fetal events will fill in the details, rounding out contours, adding flesh and fat and hair and nails to your peanut-sized body.

This was your beginning, but not the end of your biological development. As you know from personal experience, body growth and change continue through childhood, adolescence, and the decades of adulthood. From that perspective, human development stops only when we die. This chapter provides what can only be an overview of that journey.

c

STAGES OF DEVELOPMENT

The human body develops in six stages. In the first stage, sperm or eggs—the **gametes**—form and mature in the bodies of the male and female who will become a new individual's parents. The next stage, **fertilization**, begins when a sperm enters a secondary oocyte. After a sequence of steps, its result is a **zygote** ("yolked together"), the first cell of the new individual.

Next comes **cleavage**, when cell divisions convert the zygote to a ball of cells. This is the point in your existence when you first became a multicellular creature. Figure 15.2 is a simple diagram of the first two cleavage divisions. Notice that the first round of cell division in cleavage divides the zygote into two cells; then, in the second round, the two cells each divide but in different planes—one perpendicular to the other. (This "rotational" pattern of cleavage occurs in the embryos of all mammals.) After the third round of cleavage, there are sixteen embryonic cells arranged in a compact ball called a **morula** (from a Latin word for mulberry).

One of the more interesting results of cleavage is that each new cell—called a *blastomere*—ends up with a particular portion of the egg's cytoplasm. And which bit of cytoplasm a blastomere receives helps determine the developmental fate of the blastomere's descendants. For instance, the cytoplasm of one blastomere may have the molecules of a protein that can activate, say, the gene coding for a certain hormone. Therefore, *only* the descendants of that blastomere will make the hormone.

Gastrulation: Primary Tissues Form

Cleavage gives way to **gastrulation**, a stage of wholesale rearrangements of a blastomere's cells. The organizational framework for the whole body is laid out as cells become arranged into three primary tissues, called **germ layers**. The three layers are called **endoderm**, **mesoderm**, and **ectoderm**, respectively. Once the germ layers form, each

Table 15.1	Tissues and Organs Derived from the Three Germ Layers in Human Embryos
Germ Layer	Main Derivatives in the Adult
Endoderm	Various epithelia, as in the gut, respiratory tract, urinary bladder and urethra, and parts of the inner ear; also portions of the tonsils, thyroid and parathyroid glands, thymus, liver, and pancreas
Mesoderm	Cartilage, bone, muscle, and various connective tissues; gives rise to cardio-vascular system (including blood), lymphatic system, spleen, adrenal cortex, excretory and reproductive systems
Ectoderm	Central and peripheral nervous systems; sensory epithelia of the eyes, ears, and nose; epidermis and its derivatives (including hair and nails), mammary glands, pituitary gland, subcutaneous glands, tooth enamel, and adrenal medulla

one splits into subpopulations of cells that will give rise to the body's various tissues and organs (Table 15.1).

Organogenesis, Growth, and Tissue Specialization

Organogenesis is the name for the overall process by which organs form. During this phase, different sets of cells get their basic biological identity—that is, they come to have a specific structure and function. Their identity assigned, cells can give rise to different tissues (nervous tissue, muscle tissue, and so on), which in turn become arranged in organs. The final stage of development, called **growth and tissue specialization**, continues into adulthood. It is the stage in which organs grow in size and take on the specialized properties required for specialized functions—such as pumping blood or filtering wastes.

Three crucial processes accomplish the changes that mold our specialized tissues and organs. The first of these processes is **cell determination.** It establishes which of several possible developmental paths an embryonic cell can follow—for example, whether its fate is to become the forerunner of some kind of nervous tissue, or of epithelium. It is rather like freshman college students being divided into liberal arts majors, business majors, science majors, and so on. In an early embryo, a given cell's fate (and, eventually, the fate of its descendants) depends on where in the embryo the cell originates—the portion of the egg's cytoplasm that a blastomere receives. As an embryo's development progresses, each cell's fate also is influenced by physical interactions and chemical signaling that goes on between groups of cells.

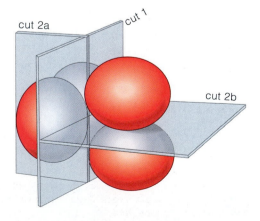

Figure 15.2 Early cleavage planes in a human egg.

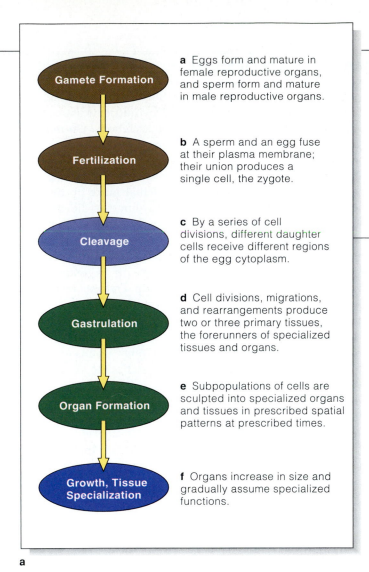

a Eggs form and mature in female reproductive organs, and sperm form and mature in male reproductive organs.

b A sperm and an egg fuse at their plasma membrane; their union produces a single cell, the zygote.

c By a series of cell divisions, different daughter cells receive different regions of the egg cytoplasm.

d Cell divisions, migrations, and rearrangements produce two or three primary tissues, the forerunners of specialized tissues and organs.

e Subpopulations of cells are sculpted into specialized organs and tissues in prescribed spatial patterns at prescribed times.

f Organs increase in size and gradually assume specialized functions.

a

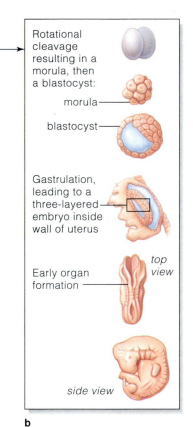

Rotational cleavage resulting in a morula, then a blastocyst:

morula

blastocyst

Gastrulation, leading to a three-layered embryo inside wall of uterus

Early organ formation

top view

side view

b

Figure 15.3 An overview of development. (**a**) The stages of development. (**b**) The developmental journey of a human embryo from fertilization to about 8 weeks. For clarity, the membranes surrounding the embryo are not shown from cleavage onward. Early cleavages result in a morula, which becomes transformed into a blastocyst (Section 15.2). Several stages are shown here in cross section.

Next, a gene-guided process of **cell differentiation** takes place. To continue our analogy, think of a group of college science majors, some of whom go on to specialize in biology while others specialize in physics, still others in chemistry, and so forth. As cells differentiate, they come to have specific structures and the ability to make products that are associated with particular functions. For example, recall from Chapter 8 how various subsets of T cells, with different functions, differentiate in the thymus.

Morphogenesis (quite literally, "the beginning of form") produces the shape and structure of particular body regions. It's a remarkable and complicated process, involving localized cell division and growth and movements of differentiated cells and entire tissues from one site to another. For instance, most of the bones of your face are descended from cells that migrated from the back of your head when you were an early embryo. Morphogenesis also involves the folding of tissue sheets and the programmed death (apoptosis) of certain cells. Thus, morphogenesis at the ends of limb buds first produced paddle-shaped hands at the ends of your arms; then

epithelial cells between lobes in the paddles died on cue, leaving separate fingers.

Figure 15.3 summarizes the six stages of human development. The figure sketches reveal an important concept. By the end of each developmental stage, the embryo has become more complex than it was before. As you will see in sections to come, the structures that develop during one stage serve as the foundation for the stage that follows it. The successful development of an embryo depends on the formation of these structures according to normal patterns, in a prescribed sequence.

Development begins when gametes form in each parent. It proceeds through a sequence of stages: fertilization followed by cleavage of the zygote, followed by gastrulation, organogenesis, and tissue growth and specialization.

Each stage of embryonic development builds on structures that were formed during the stage preceding it. Development cannot proceed properly unless each stage is successfully completed before the next begins.

THE BEGINNINGS OF YOU—EARLY EVENTS IN DEVELOPMENT

Fertilization and Cleavage

Recall from Chapter 14 that if sperm enter a female's vagina at any time during the fertile period of her menstrual cycle, fertilization can occur. As sperm swim through the cervix and uterus and into the oviducts, *capacitation* occurs. In this process, chemical changes structurally weaken the membrane over the sperm's acrosome. Only a sperm that is capacitated ("made able") can fertilize an oocyte (Figure 15.4). Of the millions of sperm in the vagina after an ejaculation, several hundred reach the upper part of an oviduct, which is where fertilization usually occurs. Uterine muscle contractions help move sperm toward the oviducts.

When a capacitated sperm meets up with an oocyte, enzymes are released from the now-fragile region of cell membrane covering the acrosome. Many sperm can reach and bind to the oocyte, and acrosome enzymes clear a path through the zona pellucida. However, usually only one sperm fuses with the oocyte. Why? Almost instantly, chemical changes in the oocyte cell membrane block more sperm from entering.

Fusion with a sperm stimulates an oocyte to finish the cell division process (meiosis II) it began when it was being formed in an ovary (Section 14.4). The result is a mature egg, or **ovum** (plural: ova), plus another polar body. (Remember that one or, often, two polar bodies are produced when meiosis I gives rise to the secondary oocyte; thus there usually are three tiny polar bodies "packaged" with the ovum.) The nuclei of the sperm and ovum swell up, then fuse. Recall that a sperm or oocyte has twenty-three chromosomes, *half* the number present in other body cells. Fertilization combines them into a full diploid set of forty-six chromosomes. Thus a zygote has all the DNA required to guide proper development and functioning of the embryo's body parts.

For three or four days after fertilization, the zygote moves down the oviduct, kept alive by nutrients stored in the ovum and present in maternal secretions. On the way, the embryo undergoes the first three cleavages, which convert the single-celled zygote into a ball of diploid cells (Figure 15.5). The early cleavages don't make the embryo larger. The daughter cells that make up the morula collectively occupy the same volume as did the zygote, although their size, shape, and activities differ.

Sometimes a split separates the two cells produced by the first cleavage, the inner cell mass, or an even later stage. Then, independent embryos develop as *identical twins*, who have the same genetic makeup. *Fraternal twins* result when two different eggs are fertilized at roughly the same time by two different sperm. Fraternal twins need not be the same sex and they do not necessarily resemble each other more than do any other siblings.

When the cluster of dividing cells reaches the uterus, it is a solid ball, the morula. Next, a fluid-filled cavity develops in the ball. This change transforms the morula into a **blastocyst** (*blast-* = bud) having two tissues: a surface epithelium called the trophoblast (*tropho-* = to nourish) and a small clump of cells called the **inner cell mass** (Figure 15.5*e*). The embryo develops from the inner cell mass.

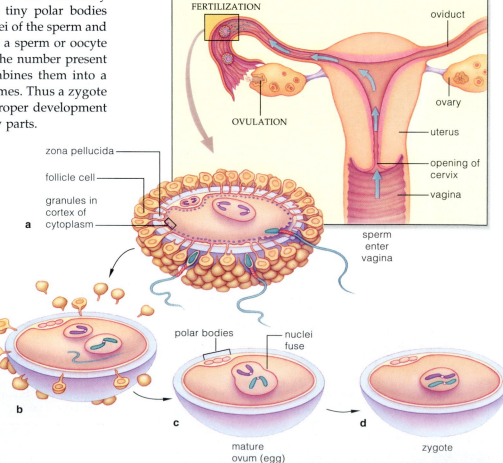

Figure 15.4 Fertilization. (**a**) A number of sperm surround a secondary oocyte. Acrosomal enzymes clear a path through the zona pellucida. (**b**) When a sperm penetrates the secondary oocyte, cortical granules in the egg cytoplasm release substances that make the zona pellucida impenetrable to other sperm. Penetration also stimulates the second meiotic division of the oocyte's nucleus. (**c**) The sperm tail degenerates. The sperm nucleus enlarges and fuses with the egg nucleus. (**d**) With that fusion, fertilization is over. The zygote has formed.

zona pellucida

follicle cell

granules in cortex of cytoplasm

FERTILIZATION

OVULATION

oviduct

ovary

uterus

opening of cervix

vagina

sperm enter vagina

polar bodies

nuclei fuse

mature ovum (egg)

zygote

Identical twins, who started out from the same zygote. The first two blastomeres that formed during cleavage, the inner cell mass, or another early embryonic stage split. The split was the start of two genetically identical, look-alike individuals.

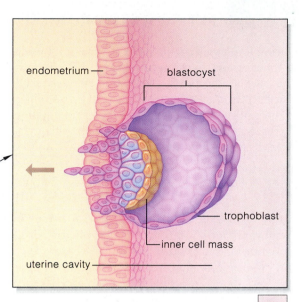

endometrium —

blastocyst

trophoblast

inner cell mass

uterine cavity —

surface layer of cells (trophoblast)

e DAY 5. On day 5, a blastocyst has developed from the morula. It consists of a fluid-filled cavity and an inner cell mass.

inner cell mass

d DAY 4. By 96 hours, divisions have produced a ball of sixteen to thirty-two cells, the morula. The morula will give rise to the embryo and extraembryonic membranes.

f DAYS 6–7. The blastocyst attaches to the endometrium and starts burrowing into it. Implantation is under way.

actual size

c DAY 3. By 72 hours, divisions have produced a ball of six to twelve cells.

Implantation

Implantation takes place about a week after fertilization. It begins as the blastocyst breaks out of the zona pellucida. Then cells of the epithelium invade the endometrium and cross into the underlying connective tissue. This gives the blastocyst a foothold in the uterus, so to speak. As time passes it will sink deep into the connective tissue of the uterus, and the endometrium will close over it.

b DAY 2. The second division, which is completed by about 40 hours, produces the four-cell stage.

Occasionally a fertilized egg implants in the wrong place—in the oviduct or sometimes even in the external surface of the ovary or in the abdominal wall. This condition is called *ectopic (tubal) pregnancy*. Ectopic pregnancy cannot go to full term and must be terminated by surgery. Sometimes it leads to permanent infertility.

a DAY 1. Cleavage begins within 24 hours after fertilization. The first cut is along a plane at right angles to the zygote's equator, in line with the polar bodies.

Implantation is complete a mere fourteen days after the secondary oocyte was ovulated. Menstruation, which would begin at this time if the woman were not pregnant, does not occur because the newly implanted blastocyst secretes HCG (human chorionic gonadotropin). This hormone prods the corpus luteum to keep on secreting estrogen and progesterone, which prevent the lining of the uterus from being shed. By the third week of pregnancy, HCG can be detected in the mother's blood or urine. At-home pregnancy tests use chemicals that change color when HCG is present in urine.

FERTILIZATION

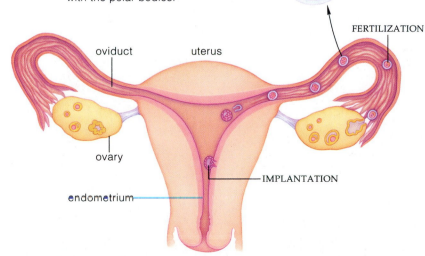

oviduct

uterus

ovary

IMPLANTATION

endometrium

Fertilization of an egg by a sperm produces a zygote, a single cell with a full complement of parental chromosomes. Cleavage and further development produce a multicellular blastocyst that implants in the endometrium of the uterus.

Figure 15.5 Steps from fertilization through implantation.

EXTRAEMBRYONIC MEMBRANES

During implantation, the inner cell mass of the blastocyst is transformed into an **embryonic disk** (Figure 15.6a). This disk is the pancakelike structure described in the chapter introduction. Some cells of the disk will give rise to the embryo during the week after implantation, when all three germ layers will form. Other cells give rise to key extraembryonic membranes, which we now consider.

The Yolk Sac, Amnion, Allantois, and Chorion

A few days after implantation, several **extraembryonic membranes** form. One of them, the **yolk sac**, forms below the embryonic disk. It is the source of early blood cells and of germ cells that will become gametes; parts of the yolk sac give rise to the embryo's digestive tube. Early on, however, the yolk sac stops functioning. Three other extraembryonic membranes, which merit a good deal of our attention, are the amnion, allantois, and chorion.

The embryo develops in a fluid-filled sac enclosed by the **amnion** (Figure 15.6c). Amniotic fluid insulates the embryo, keeps it moist, and absorbs shocks. Just outside the amnion is the **allantois**. As this structure develops, mesoderm on its surface gives rise to several blood vessels that will invade the **umbilical cord**. These vessels are the embryo's contribution to circulatory "plumbing" that will link the embryo with its lifeline, the *placenta* we discuss shortly.

The **chorion** is a protective membrane that encloses the embryo and the other membranes (Figure 15.6b and 15.6c). It develops from the trophoblast and continues the secretion of HCG that began when the embryo implanted. The HCG prevents the uterine lining (the endometrium) from disintegrating during the first three months of pregnancy. After that time, the placenta produces enough progesterone and estrogen to maintain the lining.

The Importance of the Placenta

Three weeks after fertilization, almost a fourth of the inner surface of the uterus has become a spongy tissue, the developing **placenta**. This structure is the structural link through which nutrients and oxygen pass from the mother to the embryo and waste products from the embryo pass back to its mother's bloodstream.

Although the fully developed placenta is considered an organ, it is helpful to think of it as a close association of the embryonic chorion and the superficial cells of the mother's endometrial lining where the embryo implanted. The mother's side of the placenta is a layer of endometrial tissue that contains arterioles and venules. As the chorion develops (from the trophoblast), tiny projections sent out from the blastocyst as it implants in the endometrium develop into many *chorionic villi*. Inside each villus are small blood vessels (Figure 15.7).

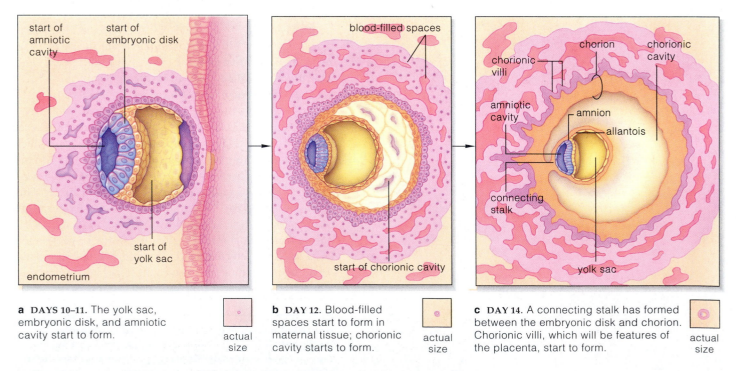

a DAYS 10–11. The yolk sac, embryonic disk, and amniotic cavity start to form.

actual size

b DAY 12. Blood-filled spaces start to form in maternal tissue; chorionic cavity starts to form.

actual size

c DAY 14. A connecting stalk has formed between the embryonic disk and chorion. Chorionic villi, which will be features of the placenta, start to form.

actual size

Figure 15.6 Early formation of extraembryonic membranes (the amnion, chorion, and yolk sac).

While chorionic villi are developing, the erosion of the endometrium that began with implantation continues. As capillaries in the endometrium are broken down, spaces in the eroding endometrial tissue fill with maternal blood. The chorionic villi extend into these spaces. As the embryo develops, very little of its blood ever mixes with that of its mother. Oxygen and nutrients simply diffuse out of the mother's blood vessels, across the blood-filled spaces in the endometrium, then into the embryo's blood vessels. Carbon dioxide and other wastes diffuse in the opposite direction, leaving the embryo. This chapter's *Science Comes to Life* (Section 15.10) describes some methods for screening for birth defects, including using cells removed from chorionic villi in a procedure called *chorionic villus sampling*.

Besides nutrients and oxygen, many other substances taken in by the mother—including alcohol, caffeine, drugs, pesticide residues, and toxins in cigarette smoke—can cross the placenta, as can HIV.

Extraembryonic membranes—the yolk sac, amnion, allantois, and chorion—begin to form shortly after implantation. A region of allantois tissue gives rise to blood vessels that become enclosed inside the umbilical cord.

The umbilical cord links the embryo with the placenta, a spongy tissue in which maternal and embryonic blood vessels are closely associated. By way of the placenta, the embryo's bloodstream can take up nutrients and oxygen from the mother and also discharge wastes that her bloodstream will transport away.

4 weeks

8 weeks

12 weeks

appearance
of the placenta
at full term

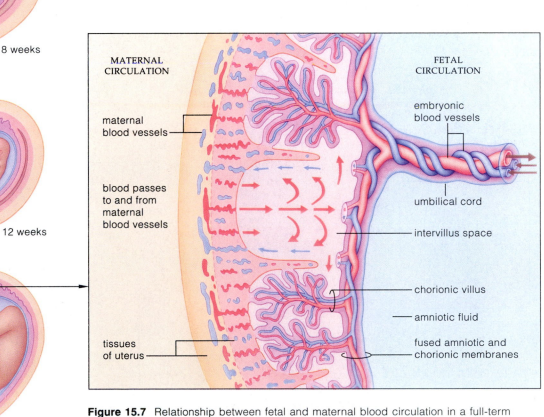

MATERNAL CIRCULATION

FETAL CIRCULATION

embryonic blood vessels

maternal blood vessels

blood passes to and from maternal blood vessels

umbilical cord

intervillus space

chorionic villus

amniotic fluid

tissues of uterus

fused amniotic and chorionic membranes

Figure 15.7 Relationship between fetal and maternal blood circulation in a full-term placenta. Blood vessels extend from the fetus, through the umbilical cord, and into chorionic villi. Maternal blood spurts into spaces between villi. Oxygen, carbon dioxide, and other small solutes diffuse across the placental membrane surface; there is no gross intermingling of the two bloodstreams.

The embryonic period begins shortly after fertilization and lasts for eight weeks. It makes up most of the "first trimester," or three months, of the nine months of human gestation. By about a week after an oocyte is fertilized, the embryonic disk has formed. Soon afterward, gastrulation begins and the three germ layers form. After these layers—the ectoderm, mesoderm, and endoderm—are established, organogenesis begins and the embryo's organs and organ systems start to develop. An embryo can't survive if gastrulation goes awry, and for this reason it is a candidate for the "single most important event" in a person's life.

Gastrulation Revisited

By the time a woman has missed her first menstrual period, gastrulation in the embryo is under way. As the third week of development begins, the two-layered embryonic disk exists. The amnion and chorion also surround the disk, except where a stalk connects the disk to the inner wall of the chorionic cavity.

As a result of some of the cell rearrangements that take place during gastrulation, a faint "primitive streak" appears at the midline of the disk (Figure 15.8). Next,

ectoderm along the midline thickens to establish the forerunner of a **neural tube**. This tube will give rise to the embryo's brain and spinal cord. Some of its cells also give rise to a flexible rod of cells called a *notochord* around which the vertebral column will form.

These events establish the body's long axis and its forthcoming bilateral symmetry. In other words, *the embryonic disk is undergoing morphogenetic events that will provide your body with the basic form that is characteristic of all vertebrates.*

Early in the third week the allantois appears on the yolk sac. As described in Section 15.3, it functions in the formation of early blood vessels. Meanwhile, on the surface of the embryonic disk near the neural tube, the third primary tissue layer—mesoderm—has been forming. Toward the end of the third week, some mesoderm gives

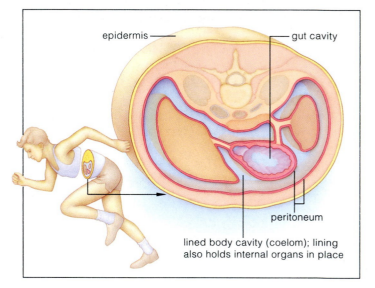

epidermis — — gut cavity

peritoneum

lined body cavity (coelom); lining also holds internal organs in place

Figure 15.8 Hallmarks of the embryonic period of development—the appearance of a primitive streak foreshadowing the brain and spinal cord—and the formation of somites and pharyngeal arches. These are dorsal views (of the embryo's back) except for day 24–25, which is a side view.

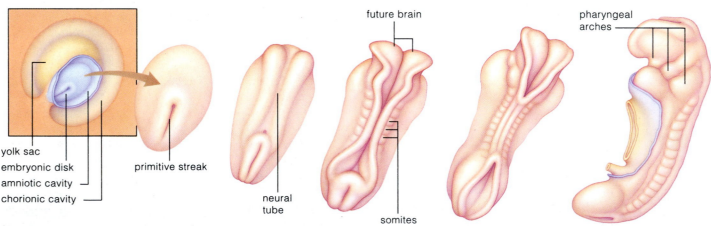

future brain

pharyngeal arches

yolk sac
embryonic disk
amniotic cavity
chorionic cavity

primitive streak

neural tube

somites

DAY 15. A primitive streak appears along the axis of the embryonic disk. This thickened band of cells marks the onset of gastrulation.

DAYS 19–23. Cell migrations, tissue folding, and other morphogenetic events lead to the formation of a hollow neural tube and to somites (bumps of mesoderm). The neural tube gives rise to the brain and spinal cord. Somites give rise to most of the axial skeleton, skeletal muscles, and much of the dermis.

DAYS 24–25. By now, some cells have given rise to pharyngeal arches, which contribute to the face, neck, mouth, nasal cavities, larynx, and pharynx.

Further reading: Student Guide to InfoTrac on web site →

rise to **somites**. From these paired blocks of mesoderm, most bones and the skeletal muscles of the neck and trunk will arise. The dermis overlying these regions comes from somites, also. Pharyngeal arches start to form; with time, they contribute to development of the face, neck, mouth, and other, associated structures. In other mesodermal tissues spaces open up. Eventually, these spaces will coalesce to form the cavity (called the *coelom*) between the body wall and the gastrointestinal tract.

Morphogenesis Revisited

The end of gastrulation marks the beginning of organ formation and morphogenesis—the process, recall, that produces the shape and structure of internal body regions. Consider neurulation, the all-important first stage in the development of the nervous system. Figure 15.9 shows how ectodermal cells at the midline of the embryo elongate and form a neural plate, the first sign that a region of ectoderm is on its way to developing into nervous tissue. Next, cells near the middle become wedge-shaped. Collectively, the changes in cell shape cause the neural plate to fold over and meet at the embryo's midline to form the neural tube.

One form of birth defect, *spina bifida* ("split spine"), occurs when the neural tube fails to develop properly and doesn't close and separate from ectoderm. The infant may be born with a portion of its spinal cord exposed within a cyst and have various kinds of neurological problems, including poor bowel and bladder control. Infection is a constant danger. Neural tube defects afflict two of every 1,000 babies born in the United States. Some cases may be preventable simply through proper maternal nutrition, a topic of Section 15.9. Other neural tube defects can be cured through surgery after a child is born.

"Marching Orders" for Cells

As sketched in Figure 15.9a, one important element in morphogenesis is the folding of sheets of cells. This folding takes place as genetic control mechanisms cause microtubules to lengthen and rings of microfilaments in cells to constrict. The size, shape, and proportions of your body parts emerged through such controlled, localized events. Some embryonic tissues expand more than others, although why is not well understood. Possibly, selective controls over the activity of certain genes are at work.

Morphogenesis also can't occur unless cells "pick up stakes" and migrate from one place to another. Migrating cells send out and use extensions of their cytoplasm called pseudopods (pseudopod means false foot). The pseudopods move migrating cells along prescribed routes. When

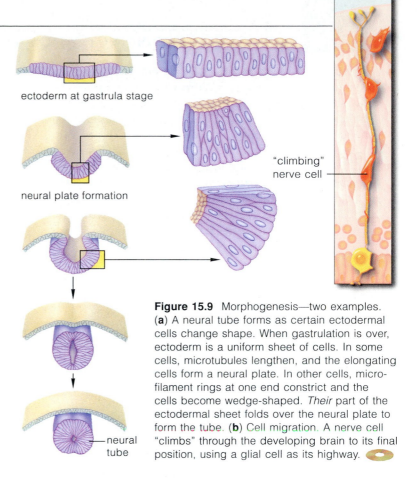

ectoderm at gastrula stage

neural plate formation

"climbing" nerve cell

neural tube

Figure 15.9 Morphogenesis—two examples. (**a**) A neural tube forms as certain ectodermal cells change shape. When gastrulation is over, ectoderm is a uniform sheet of cells. In some cells, microtubules lengthen, and the elongating cells form a neural plate. In other cells, microfilament rings at one end constrict and the cells become wedge-shaped. *Their* part of the ectodermal sheet folds over the neural plate to form the tube. (**b**) Cell migration. A nerve cell "climbs" through the developing brain to its final position, using a glial cell as its highway. ◉

they reach their destination, they establish contact with cells already there. For example, forerunners of neurons interact this way as your nervous system is forming (Figure 15.9b).

How do cells "know" where to go while they are migrating? They respond to adhesive cues, as when migrating Schwann cells stick to adhesion proteins on the surface of axons but not on blood vessels. They also respond to chemical gradients. Their migrations may be coordinated by the synthesis, release, deposition, and removal of specific chemicals in the extracellular matrix; biologists still have much to learn about this process. Adhesive cues also tell the cells when to stop. Cells will migrate to places where the adhesive cues are strongest, but once there, migration is blocked.

Apoptosis, the mechanism of genetically programmed cell death, also helps sculpt body parts. Molecular signals switch on enzymes that are stockpiled inside the cells that are destined to die.

During the third week after fertilization the basic vertebrate body plan emerges in the new individual.

Through morphogenesis, orderly changes take place in an embryo's size, shape, and proportions. Morphogenesis involves cell division, cell migration, the growth and folding of tissues, changes in cell size and shape, and programmed cell death by apoptosis.

EMERGENCE OF DISTINCTLY HUMAN FEATURES

By the end of the fourth week of the embryonic period, the embryo has grown to 500 times its original size. The placenta has been sustaining this remarkable growth spurt. However, now the pace slows. The embryo embarks on a genetically prescribed, intricate program of cell differentiation and morphogenesis. Limbs form, and fingers and toes are sculpted from paddle-shaped structures. The circulatory system becomes more intricate, and the umbilical cord forms. Much emphasis is placed on the all-important head; its growth now surpasses that of any other body region (Figure 15.10*a*). The embryonic period ends as the eighth week draws to a close. The embryo is no longer merely "a vertebrate." As you can see from Figure 15.10*c*, its features now clearly define it as a human fetus.

Gonad Development

Male or female gonads begin to develop by the second half of the first trimester. As described in the introduction to Chapter 14, in an embryo that has inherited both X and Y sex chromosomes, a sex-determining region of the Y chromosome (*SRY*) triggers development of testes at this time. Sex hormones produced by the embryonic testes then influence the development of the entire reproductive system. An embryo with XX sex chromosomes will be female, and ovaries and other female reproductive structures begin to form in her body. Notice that no hormones are required to stimulate development of a female—all that is necessary is the *absence* of testosterone.

After eight weeks the embryo is just over 1 inch long, its organ systems are formed, and it is designated a **fetus**. As the first trimester ends, a heart monitor can detect the fetal heartbeat. The genitals are well formed, and a physician can often determine the baby's sex using ultrasound.

Miscarriage

Miscarriage, the spontaneous expulsion of the contents of the uterus, occurs in more than 20 percent of all conceptions, usually during the first trimester. Often a woman may not even realize she was pregnant. Many factors can trigger a miscarriage (also called spontaneous abortion), but as many as half of all cases occur because the embryo (or the fetus) suffers from one or more genetic disorders that prevent it from developing normally.

Distinctly human features emerge in the first eight weeks of embryonic development. At the end of this period the developing individual is termed a fetus.

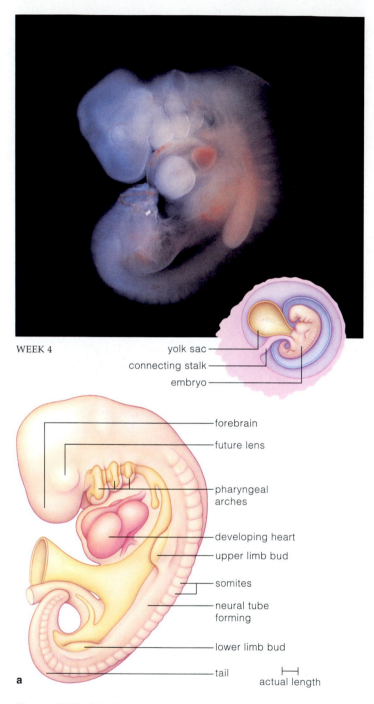

WEEK 4

yolk sac
connecting stalk
embryo

forebrain
future lens
pharyngeal arches
developing heart
upper limb bud
somites
neural tube forming
lower limb bud
tail

a   actual length

Figure 15.10 (**a**) Human embryo at 4 weeks. As is true of all vertebrates, it has a tail and pharyngeal arches. (**b**) The embryo at five to six weeks after fertilization. (**c**) An embryo poised at the boundary between the embryonic and fetal periods. It now has features that are distinctly human. It is floating in fluid within the amniotic sac. The chorion, which normally covers the amniotic sac, has been opened and pulled aside.

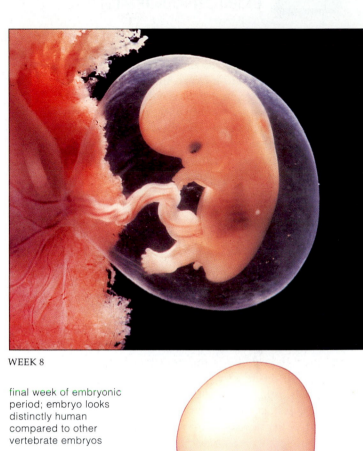

WEEKS 5–6

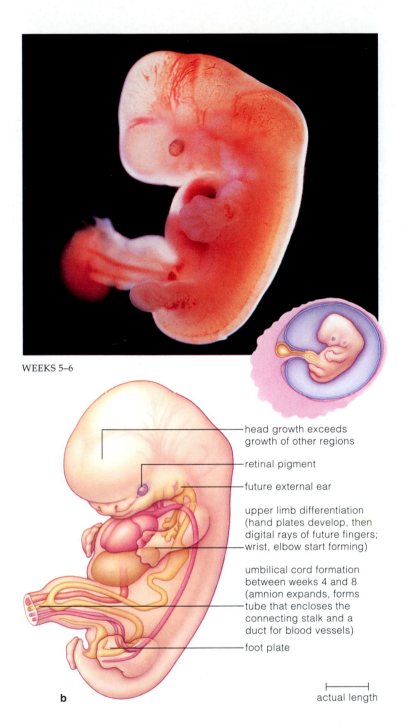

head growth exceeds growth of other regions

retinal pigment

future external ear

upper limb differentiation (hand plates develop, then digital rays of future fingers; wrist, elbow start forming)

umbilical cord formation between weeks 4 and 8 (amnion expands, forms tube that encloses the connecting stalk and a duct for blood vessels)

foot plate

b

actual length

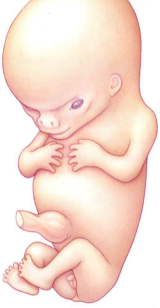

WEEK 8

final week of embryonic period; embryo looks distinctly human compared to other vertebrate embryos

upper and lower limbs well formed; fingers and then toes have separated

primordial tissues of all internal, external structures now developed

tail has become stubby

c

actual length

FETAL DEVELOPMENT

When the fetus is three months old, it is about 4.5 inches long. Soft, fuzzy hair (the lanugo) covers its body. Its reddish skin is wrinkled and protected from abrasion by a thick, cheesy coating called the *vernix caseosa*.

The second trimester of development extends from the start of the fourth month to the end of the sixth. Figure 15.11 shows what the fetus looks like at the sixteenth week of development. Movements of its tiny facial muscles produce frowns and squints, and sucking movements—evidence of a sucking reflex. Before the second trimester ends, the mother can easily sense movements of her fetus's arms and legs. During the sixth month, its eyelids and eyelashes form.

From the Seventh Month to Birth

The third trimester extends from the seventh month until birth. At seven months the fetus is about 11 inches long, and soon its eyes will open. Although the fetus is growing much larger and rapidly becoming "babylike," not until the middle of the third trimester will it be able to survive on its own if born prematurely or removed surgically from its mother's uterus. Although development might seem relatively complete by the seventh month, at that tender age few fetuses can maintain a normal body temperature or breathe normally. However, with intensive medical support, fetuses as young as 23 to 25 weeks have survived early delivery. A baby born before seven months' gestation runs a significant risk of *respiratory distress syndrome*, as described in Chapter 9, because its lungs lack surfactant and so cannot expand adequately. The longer the baby can stay in its mother's uterus, the better its chances. By the ninth month, survival chances increase to about 95 percent.

Fetal Circulation

Over time, the maturation of its organs and organ systems readies the fetus for life outside its mother's body. In the case of the circulatory system, however, the path toward independence requires a detour. Several temporary bypass vessels form and will function until birth. As Figure 15.12 shows, two *umbilical arteries* within the umbilical cord transport deoxygenated blood and metabolic wastes from the fetus to the placenta. There, the fetal blood gives up wastes, takes on nutrients, and exchanges gases with maternal blood. At the molecular level, fetal hemoglobin is slightly different from adult hemoglobin. It binds oxygen more readily, which helps ensure that adequate oxygen will reach developing fetal tissues. The oxygenated blood, enriched with nutrients, returns from the placenta to the fetus in the *umbilical vein*.

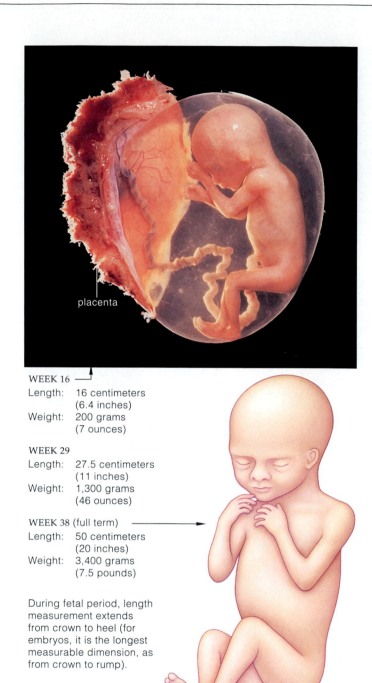

placenta

WEEK 16
Length: 16 centimeters (6.4 inches)
Weight: 200 grams (7 ounces)

WEEK 29
Length: 27.5 centimeters (11 inches)
Weight: 1,300 grams (46 ounces)

WEEK 38 (full term)
Length: 50 centimeters (20 inches)
Weight: 3,400 grams (7.5 pounds)

During fetal period, length measurement extends from crown to heel (for embryos, it is the longest measurable dimension, as from crown to rump).

Figure 15.11 (**top**) The fetus at 16 weeks. During the fetal period, movements begin as soon as nerves establish functional connections with developing muscles. Legs kick, arms wave, fingers grasp, the mouth puckers. These reflex actions will be vital skills in the world outside the uterus. The drawing shows a baby at full term—ready to be born.

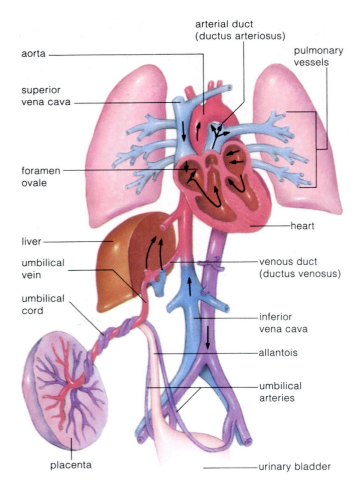

aorta

superior
vena cava

arterial duct
(ductus arteriosus)

foramen
ovale

pulmonary
vessels

liver

umbilical
vein

umbilical
cord

heart

venous duct
(ductus venosus)

inferior
vena cava

allantois

umbilical
arteries

placenta

urinary bladder

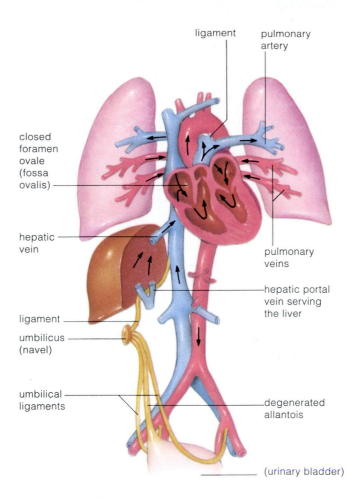

ligament

closed
foramen
ovale
(fossa
ovalis)

hepatic
vein

ligament

umbilicus
(navel)

umbilical
ligaments

pulmonary
artery

pulmonary
veins

hepatic portal
vein serving
the liver

degenerated
allantois

(urinary bladder)

Figure 15.12 The direction of blood flow in fetal circulation (arrows). (**a**) Umbilical arteries carry deoxygenated blood from fetal tissues to the placenta. There blood picks up oxygen and nutrients from the mother's bloodstream and returns to the fetus via the umbilical vein. Blood mainly bypasses the lungs, moving through the foramen ovale and the arterial duct. It bypasses the liver by moving through the venous duct.

(**b**) At birth the foramen ovale closes, and the pulmonary and systemic circuits of blood flow become completely separate. The arterial duct, venous duct, umbilical vein, and portions of the umbilical arteries become ligaments, and the allantois degenerates.

Other temporary vessels divert blood past the lungs and liver. These organs do not develop as rapidly as some others, because (by way of the placenta) the mother's body can perform their functions. The lungs of a fetus are collapsed and will not become functional for gas exchange until the newborn takes its first breaths outside the womb. Hence, its lung tissues receive only enough blood to sustain their development. A little of the blood entering the heart's right atrium flows into the right ventricle and moves on to the lungs. Most of it, however, travels through a gap in the interior heart wall called the *foramen ovale* ("oval opening") or into an arterial duct (*ductus arteriosus*) that bypasses the nonfunctioning lungs entirely.

Likewise, most blood bypasses the fetal liver because the mother's liver performs most liver functions (such as nutrient processing) until birth. Nutrient-laden blood

from the placenta travels through a venous duct (*ductus venosus*) past the liver and on to the heart, which pumps it to body tissues. At birth, blood pressure in the heart's left atrium increases. This causes a valvelike flap of tissue to close off the foramen ovale, which then gradually seals. The closure separates the pulmonary and systemic circuits of blood flow, and the arterial duct collapses (Figure 15.12b). The venous duct gradually closes during the first few weeks after birth.

During the second and third trimesters, the organs and organ systems of a fetus gradually mature. Because the fetus exchanges gases and receives nutrients via its mother's bloodstream, the fetal circulatory system develops temporary vessels that bypass the lungs and liver. At birth, normal circulatory routes begin functioning.

Birth

Birth, or **parturition**, takes place about thirty-nine weeks after fertilization, usually within two weeks of the "due date" (280 days from the start of the woman's last menstrual period). The birth process, "labor," begins when smooth muscle in the uterus starts to contract. Evidently, uterine contractions are the indirect result of a cascade of hormones from the fetus's hypothalamus, pituitary, and adrenal glands—triggered by an as-yet-unknown signal that says, in effect, it's time to be born. The hormonal flood prompts the placenta to produce more estrogen. Rising estrogen in turn calls for a rush of oxytocin and of prostaglandins (also produced by the placenta), which stimulate contractions of the uterus. For the next 2 to 18 hours, the contractions will become stronger, more painful, and more frequent.

Three Stages of Labor

Labor is divided into three stages that we can think of loosely as "before, during, and after." In the first stage, uterine contractions push the fetus against its mother's cervix. Initial contractions are relatively mild and occur about every 15 to 30 minutes. As the cervix gradually dilates to a diameter of about 10 centimeters (4 inches, or "5 fingers"), contractions become more frequent and intense. Usually, the amniotic sac ruptures during this stage, which can last 12 hours or more.

The second stage of labor, actual birth of the fetus, typically occurs less than an hour after the cervix is fully dilated. This stage is usually brief—under 2 hours. Strong contractions of the uterus and abdominal muscles occur every 2 or 3 minutes, and the mother feels an urge to push. Her efforts and the intense contractions move the soon-to-be newborn through the cervix and out through the vaginal canal, usually head first (Figure 15.13). Complications can develop if the baby begins to emerge in a "bottom-first" (*breech*) position, and the attending physician may use hands or forceps to aid the delivery.

After the baby is expelled, the third stage of labor gets under way. Uterine contractions force fluid, blood, and the placenta (now called the afterbirth) from the mother's body. The umbilical cord—the lifeline to the mother—is now severed. Without the placenta to remove wastes, carbon dioxide builds up in the baby's blood. Together with other factors, including handling by medical personnel, this stimulates control centers in the brain, which respond by triggering inhalation—the newborn's crucial first breath.

As the infant's lungs begin to function, the bypass vessels of the fetal circulation begin to close. They soon shut completely. The fetal heart's opening, the foramen

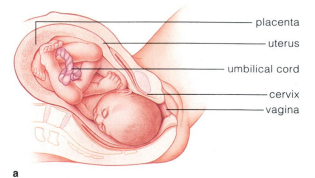

placenta
uterus
umbilical cord
cervix
vagina

a

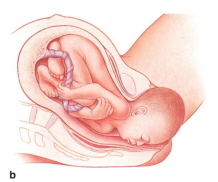

b

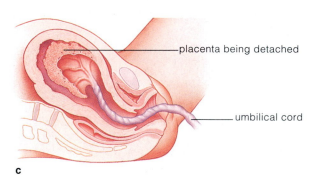

placenta being detached

umbilical cord

c

Figure 15.13 Expulsion of the fetus during birth. The afterbirth—the placenta, fluid, and blood—is expelled shortly afterward.

ovale, normally closes slowly during the first year of life. A lasting reminder of this final stage is the scar we call the navel—the site where the umbilical cord was once attached.

Most full-term pregnancies end in the birth of a healthy infant. Yet babies born prematurely—especially before about eight months of intrauterine life—can suffer complications because their organs have not developed to the point where they can function independently of the mother. Then, attempts to sustain the baby's life under conditions that will permit the necessary additional development may require the most advanced medical

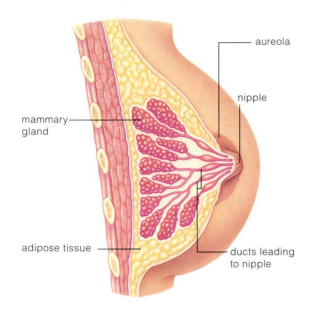

Figure 15.14 Breast of a lactating female. This cutaway view shows the mammary glands and ducts. Chapter 22 describes how to examine breast tissues for cancer.

technology. Even then, the vast majority of extremely premature infants do not survive. The *Choices* essay on this page addresses this and other issues related to uses of advanced technology in caring for desperately ill patients.

Nourishing the Newborn

During pregnancy, estrogen and progesterone stimulate the growth of mammary glands and ducts in the mother's breasts (Figure 15.14). For the first few days after birth, those glands produce colostrum, a colorless fluid low in fat but rich in proteins, antibodies, minerals, and vitamin A. Then prolactin secreted by the pituitary stimulates milk production, or **lactation**.

When the newborn suckles, the pituitary also releases oxytocin, which causes the mother's breast tissues to contract. This forces milk into the ducts. Oxytocin also triggers uterine contractions that "shrink" the uterus back to its normal size.

The mother's cervix dilates during the first stage of labor, and the baby is born during the second stage. In the third stage, uterine contractions expel the placenta.

Lactation (milk production) begins a few days after birth. It is stimulated by the hormone prolactin, which is released from the mother's pituitary and acts on her breast (mammary gland) tissues. Oxytocin released from the pituitary acts to force milk into mammary ducts.

SHOULD THIS BABY BE SAVED?

Our ability to devise ever more sophisticated medical technologies raises serious social issues. One concern is medical treatment for gravely ill newborns, and the high cost of care for extremely premature infants or those born with devastating abnormalities. Nearly 85 percent of early infant deaths occur in premature babies—many still only partially developed fetuses. They cannot be saved no matter how elaborate the technology that is brought to bear. Those that do live often have moderate to severe medical problems. Common ones include blindness and cerebral palsy, a neuromuscular disorder that damages or destroys a person's control over voluntary muscle movements.

Today, there is great societal pressure to control the costs of medical care. To reduce the approximately $5 billion spent annually in the United States on caring for "preemies," not to mention the anguish of parents and affected youngsters alike, some researchers are refining methods for identifying and treating causes of prematurity, including uterine infections in the mother.

Another approach involves efforts to improve our understanding of biochemical indicators—typically, "markers" present in the mother's blood—that can be used to predict when a woman is about to enter premature labor. With such information, her physician could mount preventive countermeasures early on.

More wrenching are cases in which babies are born at full term, but with birth defects so serious that there is little realistic hope for long-term survival. Sometimes an ill infant's death is a foregone conclusion. This is the case with *anencephaly*, in which most of the brain fails to develop. Even so, some parents want high-tech efforts to prolong the child's life. In other instances, only extreme measures, such as organ transplants and complex heart surgeries, can provide a glimmer of hope. Although the ethical issues can be complex and deeply troubling, health insurers and government agencies are increasingly reluctant to pay for expensive or experimental therapies. Should such care be provided only in certain cases? And what if *your* child's life or health were in the balance?

HOW THE MOTHER'S LIFESTYLE AFFECTS EARLY DEVELOPMENT

A woman who decides to become pregnant is committing a large part of her body's resources and functions to the development of a new individual. From fertilization until birth, her future child is at the mercy of her diet, health habits, and lifestyle. This section gives an overview of common concerns in the areas of nutrition, infection risk, and the use of legal and illegal drugs.

Some Basics of Maternal Nutrition

How does a pregnant woman best provide nutrients for her developing child? The same balanced diet that is good for her should provide her future child with all the needed carbohydrates, lipids, and proteins (see Chapter 6). The woman's own demands for vitamins and minerals increase during pregnancy. The placenta absorbs enough for her developing embryo (then fetus) from her bloodstream, except for folate (folic acid). She can reduce the risk that her child will have neural tube defects (Section 15.4) by taking more B-complex vitamins (under her physician's supervision) before conceiving a child and during early pregnancy. There is evidence that smoking reduces the level of vitamin C in a pregnant woman's blood, and in that of her fetus. Smoking may also have a harmful impact on the use of other nutrients in both the mother-to-be and her developing young.

A pregnant woman must eat enough so that her body weight increases between 20 and 35 pounds during her pregnancy, on average. If her weight gain is much less than that, she is stacking the deck against the fetus. Compared to newborns of normal weight, significantly underweight infants have more complications after delivery. As birth approaches, the growing fetus demands more and more nutrients from the mother's body. And during this last phase of pregnancy, the mother's diet profoundly influences the course of her unborn child's development. Poor nutrition especially damages the brain, which grows the most in the weeks just before birth.

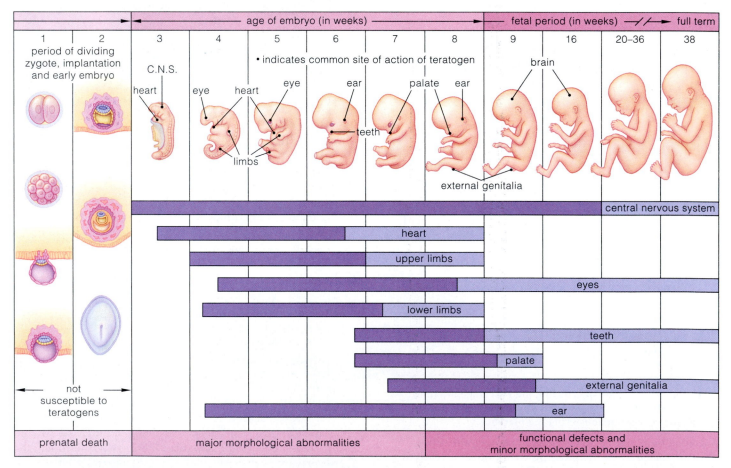

Figure 15.15 Critical periods of embryonic and fetal development. *Light blue* also indicates periods in which organs are most sensitive to damage from alcohol, viral infection, and so on. Numbers signify the week of development.

Risk of Infections

A pregnant woman's antibodies cross the placenta and help protect her developing infant from all but the most severe bacterial infections. Other teratogens—agents that can cause birth defects—are a more serious threat. Some viral diseases can be dangerous during the first six weeks after fertilization, when fetal organs are forming (Figure 15.15). For example, if a pregnant woman contracts German measles during this period, there is a fifty-fifty chance that her embryo will be malformed. If she contracts the virus when the embryo's ears are forming, her newborn may be deaf. The risk of damage to the embryo diminishes with time. For instance, after the fourth month, the infection has no apparent effect on a fetus. Vaccination before pregnancy can eliminate the risk.

Drugs and Alcohol

During its first trimester in the womb, an embryo is extremely sensitive to drugs the mother takes. In the 1960s many women using the drug thalidomide (a tranquilizer) gave birth to infants with missing or severely deformed arms and legs. Although it wasn't known at the time, thalidomide disrupts the steps required for normal limbs to develop. When the connection became clear, thalidomide was withdrawn from the market, but there is a risk that other commonly used tranquilizers—and some sedatives and barbiturates—might cause similar, although less severe, damage. Even certain anti-acne drugs, such as retinoic acid, increase the risk of facial and cranial deformities in a fetus. The widely prescribed antibiotic streptomycin causes hearing problems and may adversely affect the nervous system; a pregnant woman who uses the antibiotic tetracycline may have a child whose teeth are distinctly yellowed.

Like many other drugs, alcohol crosses the placenta and affects the fetus. *Fetal alcohol syndrome* (FAS) is a constellation of defects that can result from alcohol use by a pregnant woman. Tragically, FAS is one of the most common causes of mental retardation in the United States. Babies born with it typically have facial deformities, poor motor coordination, and, sometimes, heart defects (Figure 15.16). Between 60 and 70 percent of alcoholic women give birth to infants with FAS. Some researchers suspect that drinking any alcohol at all during pregnancy may be dangerous for the fetus, and most doctors urge near or total abstinence from alcohol during pregnancy.

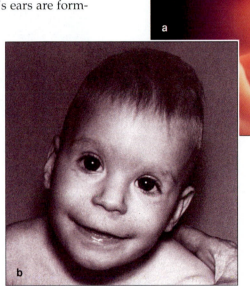

Figure 15.16 (**a**) The fetus at eighteen weeks. (**b**) An infant affected by fetal alcohol syndrome (FAS). Symptoms include a small head, low and prominent ears, poorly developed cheekbones, and a long, smooth upper lip. The child may be likely to encounter growth problems and abnormalities of the nervous system. About 1 newborn in 750 in the United States is affected by this disorder.

Effects of Cigarette Smoke

Beyond its effects on nutrition, cigarette smoke harms the growth and development of a fetus in other ways. Studies have shown that daily smoking during pregnancy leads to an underweight newborn even if the mother's weight, nutritional status, and all other relevant variables match those of pregnant nonsmokers. A pregnant smoker also has a greater risk of miscarriage, stillbirth, and premature delivery. A long-term study at Toronto's Hospital for Sick Children showed that toxic substances in tobacco build up even in the fetuses of nonsmokers who are exposed to *secondhand smoke* at home or work.

The mechanisms by which smoking affects a fetus are not known. However, its demonstrated effects add to the evidence that the placenta, marvelous barrier that it is, cannot protect a developing fetus from every assault.

A woman who is pregnant, or planning to be, can adopt a lifestyle that helps protect her child from harm during early development. Poor maternal nutrition, certain viral infections, unsupervised use of some prescription drugs, alcohol use, and cigarette smoking all pose major risks to a developing embryo and fetus.

15.10   PRENATAL DIAGNOSIS: DETECTING BIRTH DEFECTS

A growing number of options now enable us to detect more than 100 genetic disorders before a child is born.

Amniocentesis samples fluid from within the amnion, the sac that contains the fetus (Figure 15.17). During the fourteenth to sixteenth week of pregnancy, the thin needle of a syringe is inserted through the mother's abdominal wall, into the amnion. The physician must take care that the needle doesn't puncture the fetus and that no infection occurs. Amniotic fluid contains sloughed fetal cells; as the syringe withdraws fluid, some of those cells are included. They are then cultured and tested for genetic abnormalities.

Chorionic villus sampling (CVS) uses tissue from the chorionic villi of the placenta. CVS is tricky. Using ultrasound, the physician guides a tube through the vagina, past the cervix, and along the uterine wall, then removes a small sample of chorionic villus cells by suction. The method can be used by the eighth week of pregnancy; results are available within days. Both CVS and amniocentesis involve a small risk of triggering miscarriage. With CVS there also is a slight chance the future child will have missing or underdeveloped fingers or toes.

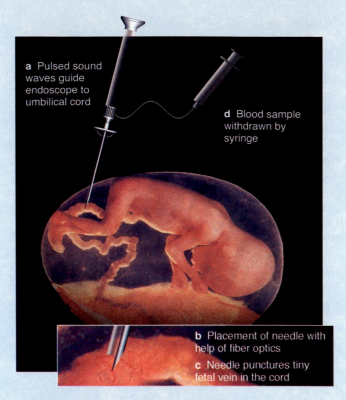

a Pulsed sound waves guide endoscope to umbilical cord

d Blood sample withdrawn by syringe

b Placement of needle with help of fiber optics

c Needle punctures tiny fetal vein in the cord

Figure 15.18 Fetoscopy for prenatal diagnosis.

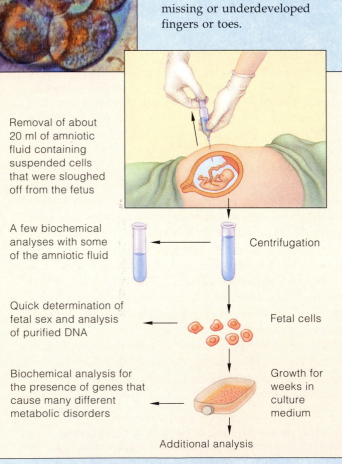

Removal of about 20 ml of amniotic fluid containing suspended cells that were sloughed off from the fetus

A few biochemical analyses with some of the amniotic fluid

Centrifugation

Quick determination of fetal sex and analysis of purified DNA

Fetal cells

Biochemical analysis for the presence of genes that cause many different metabolic disorders

Growth for weeks in culture medium

Additional analysis

Figure 15.17 Procedure for amniocentesis. (*inset, above*) An eight-cell-stage human embryo.

Physicians also have available several methods of embryo screening. In *preimplantation diagnosis*, an embryo conceived by in vitro fertilization (Section 14.8) is analyzed for genetic defects using recombinant DNA technology. The testing occurs at the eight-cell stage (Figure 15.17, *inset*). According to one view, the tiny, free-floating ball is a *pre*-pregnancy stage. Like the unfertilized eggs discarded monthly from a woman (during menstruation), the ball is not attached to the uterus. All cells in the ball have the same genes and are not yet committed to giving rise to specialized cells of a heart, lungs, or other organs. Doctors take one of the undifferentiated cells and analyze its genes for suspected disorders. If the cell has no detectable genetic defects, the ball is inserted into the uterus. Embryo screening is designed to help parents who are at high risk of having children with a genetic birth defect. Even so, it is controversial, raising questions we consider in the *Choices* essay in Chapter 19.

Direct visualization of the developing fetus is possible with *fetoscopy*. A fiberoptic device, an endoscope, uses pulsed sound waves to scan the uterus and visually locate particular parts of the fetus, umbilical cord, or placenta (Figure 15.18). Fetoscopy has been used to diagnose blood cell disorders, such as sickle-cell anemia and hemophilia. Like amniocentesis and CVS, fetoscopy has risks; it increases the risk of a miscarriage by 2 to 10 percent.

15.11 SUMMARY OF DEVELOPMENTAL STAGES

Birth to Adulthood

After a child enters the world, a gene-dictated course of further growth and development leads to adulthood. Table 15.2 summarizes all the prenatal (before birth) and postnatal (after birth) stages of life. A newborn is called a *neonate*. Infancy lasts until about 15 months of age. During this period the child's nervous and sensory systems mature rapidly, and its body becomes longer through a series of "growth spurts." Figure 15.19 shows how our body proportions change as development proceeds through childhood and adolescence. A key feature of adolescence is puberty, the arrival of sexual maturity as the reproductive organs begin to function. Sex hormones trigger the appearance of pubic and underarm hair, other secondary sex characteristics, and behavior changes. A combination of hormones triggers another growth spurt at this time. Boys usually grow most rapidly between the ages of 12 and 15, whereas girls tend to grow most

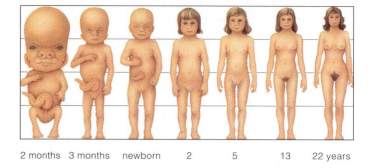

2 months	3 months	newborn	2	5	13	22 years

Figure 15.19 Diagram of changes in the proportions of the human body during prenatal and postnatal growth.

rapidly between the ages of 10 and 13. After several years, the influence of sex hormones causes the cartilaginous plates near the ends of long bones to harden into bone, and growth stops by the early twenties.

Adult Life

Following their initial growth and differentiation, the cells of all complex animals gradually deteriorate. We humans are no different. Although in the United States the average life expectancy is 72 years for males and 79 years for females, we reach the peak of our physical potential in adolescence and early adulthood. A healthy diet, regular exercise, and other beneficial lifestyle habits can go far in keeping a person vigorous for decades of adult life. Even so, after about age 40, body parts begin to undergo structural changes. There is also a gradual loss of efficiency in bodily functions, as well as increased sensitivity to environmentally induced stress. This progressive cellular and bodily deterioration is built into the life cycle of all organisms in which differentiated cells become highly specialized. The process, technically called *senescence*, is what we all know as **aging**.

Aging in humans leads to many structural changes in the body. Beginning around age 40, there is a gradual decline in bone and muscle mass. Our skin develops more wrinkles, and more fat is deposited. Less obvious are gradual physiological changes. This chapter's concluding sections explore the phenomenon of aging, including some ideas about its causes.

Table 15.2	Stages of Human Development: A Summary
PRENATAL PERIOD	
1. Zygote	Single cell resulting from union of sperm and egg at fertilization
2. Morula	Solid ball of cells produced by cleavages
3. Blastocyst	Ball of cells with surface layer and inner cell mass
4. Embryo	All developmental stages from 2 weeks after fertilization until end of eighth week
5. Fetus	All developmental stages from the ninth week until birth (about 39 weeks after fertilization)
POSTNATAL PERIOD	
6. Newborn (neonate)	Individual during the first 2 weeks after birth
7. Infant	Individual from 2 weeks to about 15 months after birth
8. Child	Individual from infancy to about 12 or 13 years
9. Pubescent	Individual at puberty, when secondary sexual traits develop; girls between 10 and 16 years, boys between 13 and 16 years
10. Adolescent	Individual from puberty until about 3 or 4 years later; physical, mental, emotional maturation
11. Adult	Early adulthood (between 18 and 25 years); bone formation and growth completed. Changes proceed very slowly afterward.
12. Old age	Aging culminates in general body deterioration

Following birth, development proceeds through childhood and adolescence, which includes the arrival of sexual maturity at puberty. Puberty is the gateway to the adult phase of life. After about age 30, developmental changes associated with aging become increasingly apparent. The average life span for males currently is about 72 years; it is about 79 years for females.

WHY DO WE AGE?

Most people equate aging with "getting old," a gradual loss of vitality as cells, tissues, and organs function less and less efficiently. Indeed, beginning at about age 40, our skin begins to noticeably wrinkle and sag, body fat tends to accumulate, and injuries to muscles and joints occur more readily and take longer to heal. Stamina declines, and we become increasingly susceptible to disorders ranging from heart disease to arthritis and cancer.

Structural alterations in certain body proteins—including collagen, a key component of many connective tissues—may contribute to many changes we associate with aging (Figure 15.20). A collagen molecule consists of three polypeptide chains twisted into a spiral. This shape is stabilized by molecular bonds that form "cross-links" between segments of the chain. As a person grows older, new cross-links develop, and the protein becomes more and more rigid. As collagen's structure changes, so may the structure and functioning of organs and blood vessels that contain it. Age-related cross-linking is also thought to affect many enzymes and possibly DNA.

No one knows for sure what causes the degenerative changes we associate with aging. Many hypotheses have been proposed; here we look at two that are spurring a great deal of ongoing research.

The Programmed Life Span Hypothesis

Does an internal, biological clock control aging? After all, each species has a maximum life span. For example, we know the maximum is 20 years for dogs and 12 weeks for butterflies. So far as we can document, no human has lived beyond 122 years. The consistency of lifespan within

Figure 15.20 As we age, structural and functional changes affect virtually all body tissues and organs.

species indicates that genes have something to do with aging.

Suppose the body is like a clock shop, with each type of cell, tissue, and organ ticking at its own genetically set pace. Years ago, researchers investigated this possibility. In the laboratory, they grew normal human embryonic cells, all of which divided about 50 times, then died.

In the body, our cells divide eighty or ninety times, at most. As Chapter 17 discusses in some detail, cells duplicate their chromosomes before they divide. Capping the ends of chromosomes are repeated stretches of DNA called *telomeres*. A bit of each telomere is lost during each cycle of cell division. When only a nub remains, cells stop dividing and die.

Cancer cells, and the reproductive cells in our gonads that give rise to sperm and oocytes, are exceptions to this rule. Both these types of cells make telomerase, an enzyme that causes telomeres to *lengthen*. Apparently, that is why such cells can divide over and over, without dying. In recent experiments, somatic (nonreproductive) body cells exposed to this enzyme go on dividing well past the normal life span, with no detectable ill effects.

The Cumulative Assaults Hypothesis

A different hypothesis holds that aging is an outcome of ongoing, cumulative damage to DNA. In this scenario, the damage results from environmental assaults and from a decline in DNA's mechanisms of self-repair.

For example, free radicals, those rogue, short-lived molecules described in Chapter 2, bombard DNA and other biological molecules. This includes DNA inside mitochondria, the cell power plant. Structural changes in DNA endanger the synthesis of proteins, such as enzymes, required for normal life processes. Also, free radicals are implicated in many age-related problems, such as atherosclerosis and Alzheimer's disease.

Problems in the operations by which DNA replicates and repairs itself have been implicated in aging. For instance, researchers have correlated Werner's syndrome, an aging disorder, with a harmful mutation in the gene that carries instructions for an enzyme (called a helicase) that may be essential for DNA self-repair. With a "bad" form of this helicase, accumulating DNA damage may interfere with cell division.

In the final analysis, both hypotheses may have merit. Aging may involve interconnected processes in which genes, hormones, free radical damage, and a decline in DNA repair mechanisms come into play.

Aging may result from a combination of factors, including an internal biological clock that ticks out the life spans of cells, and the accumulation of errors in and damage to DNA.

AGING OF SKIN, MUSCLE, THE SKELETON, AND INTERNAL TRANSPORT SYSTEMS

Skin, Muscles, and the Skeleton

Skin changes often are early obvious signs of aging. The normal replacement of sloughed cells of the epidermis (through cell division) begins to slow. Cells called fibroblasts are a major component of connective tissue, and the number of fibroblasts in the dermis starts to decrease. Also, elastin fibers that give skin its flexibility are slowly replaced with more rigid collagen. As a result of these changes, the skin becomes thinner and less elastic, so it sags and wrinkles develop. Wrinkling increases as fat stores decline in the hypodermis, the subcutaneous layer beneath the dermis. The skin also becomes dryer as sweat and oil glands begin to break down and are not replaced. The loss of sweat glands and subcutaneous fat is one reason why older people tend to have difficulty regulating their body temperature. Hair follicles die or become less active, so there is a general loss of body hair. And as pigment-producing cells die and are not replaced, the remaining body hair begins to appear gray or white.

Fibers in skeletal muscle atrophy, in part because of a corresponding loss of motor neurons that synapse with muscle fibers. In general, aging muscles lose mass and strength, and lost muscle tends to be replaced by fat and, with time, by collagen. Importantly, however, the extent to which this muscle replacement takes place depends on a person's diet and level of exercise. Typically, an 80-year-old has about half the muscle strength he or she had at 30. A program of regular physical activity can help retard all these changes (Figure 15.21).

Our bones become weaker, more porous, and brittle as we age. Over the years bones begin to lose some of their collagen and elastin, but mainly bones weaken from the loss of calcium and other minerals. (Some people may inherit a gene for faulty vitamin D receptors, which reduces the efficiency with which calcium is absorbed and used in bone formation.) After about age 40, minerals begin to be removed. Older people of both sexes may be prone to develop osteoporosis. However, women naturally have less bone calcium than men do, and they begin to lose it earlier. Without medical intervention, some women lose as much as half of their bone mass by age 70. In both women and men, the vertebrae become closer together as the intervertebral disks deteriorate. This is why people may get shorter, often by about a centimeter every ten years from middle age onward.

Nine out of ten people over 40 have some degree of joint breakdown. The cartilage in joints deteriorates from the sheer wear and tear of daily movements, the surfaces of the joined bones begin to wear away, and the joints become more difficult to move. At least 15 percent of adults develop *osteoarthritis*, a chronic inflammation of cartilage, in one or more joints. Although many factors contribute to osteoarthritis, it is most common in older

Figure 15.21 Moderate physical activity, maintained as a person ages, slows age-related deterioration of the musculoskeletal system and promotes cardiovascular health.

people. This suggests that age-related changes in bone are often involved.

Aging of the Cardiovascular and Respiratory Systems

Our heart and lungs also function less efficiently with increasing age. In the lungs, walls of alveoli break down, so there is less total respiratory surface available for gas exchange. If a person does not develop an enlarged heart due to cardiovascular disease, the heart muscle becomes slightly smaller, and its strength and ability to pump blood diminish. As a result, less blood and oxygen are delivered to muscles and other tissues. In fact, decreased blood supply may be a factor in age-related changes throughout the body. Blood transport is also affected by structural changes in aging blood vessels. Elastin fibers in blood vessel walls are replaced with connective tissue containing collagen or become hardened with calcium deposits, and so vessels become stiffer. Cholesterol plaques and fatty deposits often cause further narrowing of arteries and veins (Section 7.14). Thus older people often have higher blood pressure. However, as with the musculoskeletal system, lifestyle choices such as not smoking, eating a low-fat diet, and regular exercise can help a person maintain vigorous respiratory and cardiovascular systems well past middle age.

Changes in connective tissue contribute to age-related declines in the functioning of muscles, bones, joints, and the respiratory and cardiovascular systems.

The Nervous System and Senses

Although some neurons retain the ability to divide, you are born with most of the neurons you will ever have. Chances are, any that are lost will not be replaced. Even in a healthy person, brain neurons die steadily throughout life, and as they do the brain shrinks slightly, losing about 10 percent of its mass after 80 years. On the other hand, the brain has more neurons than it "needs" for various functions. In addition, when certain types of neurons are lost or damaged, other neurons may produce new dendrites and synaptic connections and so take up the slack.

Over time, however, the death of some neurons and structural changes in others apparently do interfere with nervous system functions. For instance, in nearly anyone who lives to old age, tangled clumps of cytoplasmic fibrils develop in the cytoplasm of many neuron cell bodies. These *neurofibrillary tangles* may disrupt normal cell metabolism, although their exact effect is not understood. Clotlike plaques containing protein fragments called *beta amyloid* also develop between neurons.

In people with the dementia called *Alzheimer's disease* (AD), the brain tissue contains masses of neurofibrillary tangles. It also is riddled with beta amyloid plaques (Figure 15.22). At this writing, it is not clear whether the amyloid plaques are a cause of Alzheimer's or simply one effect of another, unknown disease process. For instance, the brains of Alzheimer's patients also have lower-than-normal amounts of the neurotransmitter acetylcholine, and some evidence suggests that this shortage may be related to beta amyloid buildup. Other possibilities under investigation include the hypothesis that AD results from chronic inflammation of brain tissue, similar to the inflammation that triggers arthritis. Certain therapeutic drugs can temporarily help alleviate some symptoms of Alzheimer's disease, which include memory loss and disruptive personality changes.

Some cases of AD are inherited. People who inherit one version of a gene that codes for *apolipoprotein E*, a lipid-binding protein, are at significantly increased risk. Around 16 percent of the U.S. population has one or two copies of this gene, called apoE-4. Those with two copies have a 90 percent chance of developing AD. Of the AD-related genes discovered thus far, most are associated with early-onset Alzheimer's, which develops before the age of 65. However, many thousands of the 4 million Americans who have the disease first showed symptoms in their late 70s or 80s. Does genetics play a role in those cases, also? Possibly. More and more genes are being found to have forms that affect how much amyloid builds up in the brain. However, many investigators believe that AD probably has many contributing factors, including as-yet undetermined environmental ones.

Even otherwise healthy people begin to have some difficulty with short-term memory after about age 60 (the so-called "senior moments") and may find it takes longer to process new information. Perhaps because aging CNS neurons tend to lose some of their insulating myelin sheath, older neurons do not conduct action potentials as efficiently. In addition, neurotransmitters such as acetylcholine may be released more slowly. This may be

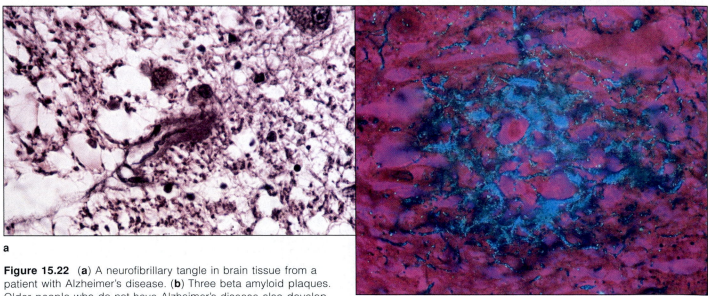

a

b

Figure 15.22 (**a**) A neurofibrillary tangle in brain tissue from a patient with Alzheimer's disease. (**b**) Three beta amyloid plaques. Older people who do not have Alzheimer's disease also develop neurofibrillary tangles and beta amyloid plaques, but not nearly as many.

Further reading: Student Guide to InfoTrac on web site ➝

due to age-related changes in plasma membrane proteins, which cause the membranes of synaptic vesicles to stiffen. As a result of such changes, movements and reflexes become slower, and some coordination is lost.

As we age, our sensory organs become less efficient at detecting or responding to stimuli. For example, people tend to become farsighted as they grow older because the lens of the eye loses its elasticity and is altered in other ways that prevent it from properly flexing to bend incoming light during focusing.

Reproductive Systems and Sexuality

Levels of most hormones remain steady throughout life. However, the sex hormones are exceptions to this rule. Reduced secretion of estrogens and progesterone triggers menopause in women; in older men, falling levels of testosterone reduce fertility. That said, whereas menopause brings a woman's reproductive period to an end, men can and have fathered children into their 80s. Males and females both retain their capacity for sexual response well into old age.

Menopause usually begins in a woman's late 40s or early 50s. Over a period of several years, her menstrual periods become irregular, then stop altogether as her ovaries become less and less sensitive to the hormones FSH and LH (Chapter 13) and gradually stop secreting estrogen in response. Many women report relatively few unpleasant symptoms during menopause. However, declining estrogen levels may trigger "hot flashes" (intense sweating and uncomfortable warmth), thinning of the vaginal walls, and some loss of natural lubrication. Postmenopausal women are also at increased risk of osteoporosis and heart disease. Small doses of estrogen (with progesterone) administered as *hormone replacement therapy* (HRT) can counteract severe side effects of estrogen loss and may even help protect a woman against uterine cancer. However, HRT has its risks and must be carefully discussed with a knowledgeable physician.

After about age 50, men gradually begin to take longer to achieve an erection due to vascular changes that cause the penis to fill with blood more slowly. In the United States, more than half of men over 50 also experience bladder infections, urinary frequency, and other problems caused by age-related enlargement of the prostate gland.

Immunity, Nutrition, and the Urinary System

Other organs and organ systems also change as the years go by. In the immune system, the number of T cells falls and B cells become less active. Older people are more prone to autoimmune diseases, such as rheumatoid

Table 15.3 Some Physiological Changes in Aging

Age	Maximum Heart Rate	Lung Capacity	Muscle Strength	Kidney Function
25	100%	100%	100%	100%
45	94%	82%	90%	88%
65	87%	62%	75%	78%
85	81%	50%	55%	69%

Note: Age 25 is the benchmark for maximal efficiency of physiological functions.

arthritis. Why? One hypothesis holds that, as DNA repair mechanisms become less effective, mutations in genes that code for self-markers are not fixed. If the markers change, this could provoke immune responses against the body's own cells (Section 8.12).

In the aging digestive tract, glands in mucous membranes lining the stomach and small and large intestines gradually degenerate, and the pancreas secretes fewer digestive enzymes. Although it is vital for older people to maintain adequate nutrition, we require fewer food calories as we age. By age 50, the basal metabolic rate (BMR) (Section 6.10) is only 80 to 85 percent of what it was in childhood and will keep falling about 3 percent every decade. Hence people tend to gain weight as they enter middle age, unless they compensate for a falling BMR by consuming fewer calories, increasing their physical activity, or both.

Over time the muscular walls of the large intestine, bladder, and urethra become weaker and less flexible. As the urinary sphincter is affected, many older people experience urine leakage, or *urinary incontinence.* Women who have borne children may have more trouble with urinary incontinence because their pelvic floor muscles are weak. On the other hand, our kidneys may continue to function well, despite the fact that nephrons gradually break down and lose some of their ability to maintain the balance of water and ions in body fluids (Table 15.3). The kidneys generally have more than enough nephrons for satisfactory functioning, which may be why the loss of nephrons in normal aging often has no major effects.

In many body systems, such as the nervous, reproductive, and immune systems, aging correlates with the death of cells that are not replaced and reduced activity of other cells.

Ultimately, aging involves a steady decline in the finely tuned ebb and flow of substances and chemical reactions that maintain homeostasis.

SUMMARY

1. Human development proceeds through six stages:

a. Formation of gametes, during which the egg and sperm mature within the reproductive organs of parents.

b. Fertilization, in which the DNA (on chromosomes) of a sperm and an egg are brought together in a single cell, the zygote.

c. Cleavage, when the fertilized egg (the zygote) undergoes cell divisions that form the early multicelled embryo (the blastocyst). The destiny of cell lineages is established in part by the sector of cytoplasm inherited at this time.

d. Gastrulation, when the organizational framework of the whole body is laid out. Endoderm, ectoderm, and mesoderm form; all the tissues of the adult body will develop from these germ layers.

e. Organ formation (organogenesis), when organs start developing by way of a tightly orchestrated program of cell differentiation and morphogenesis. This is a critical time of early development.

f. Growth and tissue specialization, when organs enlarge and acquire specialized chemical and physical properties. The maturation of tissues and organs continues into the fetal stage and beyond.

2. Cell differentiation and morphogenesis are based on cell interactions that begin during organogenesis. Both are guided by genes. During differentiation, cells come to have specific structures and functions; morphogenesis produces the shape and structure of particular body regions.

3. Embryonic development depends on the formation of four extraembryonic organs:

a. Yolk sac: contributes to the embryo's digestive tube and helps form blood cells and germ cells.

b. Allantois: its blood vessels become umbilical arteries and veins and blood vessels of the placenta; they function in oxygen transport and waste excretion.

c. Amnion: a fluid-filled sac that surrounds and protects the embryo from mechanical shocks and keeps it from drying out.

d. Chorion: a protective membrane around the embryo and the other membranes; a major part of the placenta.

4. The embryo and its mother exchange nutrients, gases, and wastes by way of the placenta (a spongy organ of endometrium and extraembryonic membranes).

5. The placenta provides some protection for the embryo/fetus, but may allow transfer of harmful agents from mother to fetus.

6. Estrogen and progesterone stimulate growth of the mammary glands. At delivery, contractions of the uterus dilate the cervix and expel the fetus and afterbirth. After delivery, nursing causes the secretion of hormones that stimulate milk production and release.

7. As with all complex animals, over time the human body shows changes in structure and a decline in its functional efficiency. Although this aging is part of the life cycle of all animals having extensively specialized cell types, its precise cause is unknown.

Review Questions

1. Define and describe the main features of the following developmental stages: fertilization, cleavage, gastrulation, and organogenesis. *15.1*

2. Label the following stages of early development: *15.1*

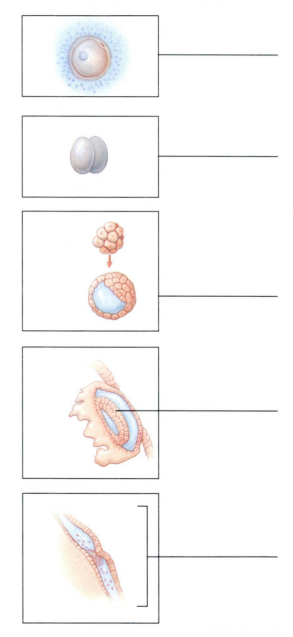

3. Define cell differentiation and morphogenesis, two processes that are critical for development. Which two mechanisms serve as the basic foundation for cell differentiation and morphogenesis? *15.1*

4. Summarize the development of an embryo and fetus. When are body parts such as the heart, nervous system, and skeleton largely formed? *15.4, 15.5, 15.6*

Self-Quiz (Answers in Appendix V)

1. Development cannot proceed properly unless each stage is successfully completed before the next begins, starting with
_____.
   a. gamete formation   d. gastrulation
   b. fertilization   e. organ formation
   c. cleavage   f. growth, tissue specialization

2. During cleavage, the _____ is converted to a ball of cells, which in turn is transformed into the _____.
   a. zygote; blastocyst   c. ovum; embryonic disk
   b. trophoblast; embryonic disk   d. blastocyst; embryonic
                                        disk

3. In the week following implantation, cells of the _____ will give rise to the embryo.
   a. blastocyst   c. embryonic disk
   b. trophoblast   d. zygote

4. The developmental process called _____ produces the shape and structure of particular body regions.

5. _____ is the gene-guided process by which cells in different locations in the embryo become specialized.
   a. Implantation   c. Cell differentiation
   b. Neurulation   d. Morphogenesis

6. In a human zygote, the cell divisions of cleavage produce an embryonic stage known generally as a _____.
   a. zona pellucida   c. blastocyst
   b. gastrula   d. larva

7. Match each developmental stage with its description.
   _____ cleavage   _____ cell differentiation
   _____ gamete formation   _____ gastrulation
   _____ organ formation   _____ fertilization
   a. egg and sperm mature in parents
   b. sperm, egg nuclei fuse
   c. formation of germ layers
   d. zygote becomes a ball of cells called a morula
   e. cells come to have specific structures and functions
   f. starts when germ layers split into subpopulations of cells

8. Parts of the _____, an extraembryonic membrane, give rise to the embryo's digestive tube.
   a. yolk sac   c. amnion
   b. allantois   d. chorion

9. The _____, a fluid-filled sac, surrounds and protects the embryo from mechanical shocks and keeps it from drying out.
   a. yolk sac   c. amnion
   b. allantois   d. chorion

10. Blood vessels of the _____, an extraembryonic membrane, transport oxygen and nutrients to the embryo.
    a. yolk sac   c. amnion
    b. allantois   d. chorion

11. Substances are exchanged between the embryo and mother through the _____, which is composed of maternal and embryonic tissues.
    a. yolk sac   d. amnion
    b. allantois   e. chorion
    c. placenta

Critical Thinking: You Decide (Key in Appendix VI)

1. How accurate is the statement "A pregnant woman must do everything for two"? Give some specifics to support your answer.

Figure 15.23 The future mother of a healthy child—maybe.

2. The renowned developmental biologist Lewis Wolpert once observed that birth, death, and marriage are not the most important events in human life—rather, Wolpert said, gastrulation is. In what sense was he correct?

3. One of your best friends—the woman in Figure 15.23—tells you that she and her husband think she might be pregnant. She's planning to do some reading and check out some web sites to find out when she should begin taking vitamins and "eating for two." That way, she feels, she can wait until she's several months along before finding an obstetrician. *You* think she could use some medical advice sooner, and you tactfully suggest she discuss her plans, habits, and lifestyle with a physician as soon as possible. What are you concerned about?

Selected Key Terms

allantois *15.3*	germ layer *15.1*
amnion *15.3*	implantation *15.2*
blastocyst *15.2*	inner cell mass *15.2*
cell differentiation *15.1*	lactation *15.7*
chorion *15.3*	mesoderm *15.1*
cleavage *15.1*	morphogenesis *15.1*
ectoderm *15.1*	organogenesis *15.1*
embryonic disk *15.3*	ovum *15.2*
endoderm *15.1*	parturition *15.7*
extraembryonic membranes *15.3*	placenta *15.3*
fertilization *15.1*	trophoblast *15.2*
fetus *15.5*	umbilical cord *15.3*
gamete *15.1*	yolk sac *15.3*
gastrulation *15.1*	zygote *15.1*

Readings

Austad, S. 1997. *Why We Age—What Science Is Discovering about the Body's Journey through Life*. New York: Wiley.

Riddle, R. D. and C. Tabin. February 1999. "How Limbs Develop." *Scientific American*.

Smith, R. March 1999. "The Timing of Birth." *Scientific American*.

LIFE AT RISK: INFECTIOUS DISEASE

What Goes Around Comes Around

A few years ago *The Coming Plague* rocketed to the top of the nonfiction best-seller lists. Written by a respected science writer named Laurie Garrett, the book looked at changing patterns in the worldwide spread of microbes. It also posed the chilling question of whether humanity would be, or even *could* be, prepared for an onslaught of new, emerging, and renewed infectious diseases.

Twenty years after AIDS first shocked the world, the causative virus, HIV (Figure 16.1*a*), has infected more than 50 million people. At this writing, AIDS has claimed the lives of an estimated 16 million worldwide. As you will read later in this chapter, there is no sign that the virus will stop its deadly advance, and there is no cure. Current treatments only suppress the symptoms and are expensive—and therefore are available only to a relative few. In fact, AIDS has become a leading cause of death among Americans ages 25–44. Its incidence is steadily increasing among certain groups, including women and teenagers. In developing countries of Africa and elsewhere, this modern plague has begun to overwhelm health care systems and to erode needed economic progress. Hundreds of millions more are at risk in India, China, and some other parts of Southeast Asia.

It's not just HIV we have to worry about, though. To see for yourself, just plug into the World Wide Web and type the words "infectious disease" into your favorite search engine.

You might come across a report on *hepatitis C*. This disease, caused by any of *one hundred* known strains of the HCV virus (for *hepatitis C virus*), causes severe liver cirrhosis (Section 6.5) and cancer. HCV had infected more than 170 million people—four times the global toll from HIV—by the beginning of the new century. Public health officials fear that in the United States, in the next few years its death toll will surpass that of AIDS. Very little is known about HCV, except that it seems to be transmitted in blood and can be present in the body for years before disease symptoms appear. It *may* be spread by sexual contact; it definitely is spread by contaminated needles shared by IV drug abusers.

Your Web search might also turn up a variety of overlapping criteria, groupings, and categories scientists use as they try to make sense of the bewildering array of infectious diseases. They can be grouped by *reservoirs*—the environments where different pathogens flourish, such as the bodies of certain animals, water or soil, and the like. For example, humans are the only reservoirs for pathogens that cause the common cold and measles. For infectious diseases lumped as **zoonoses**, animals other than humans are the main reservoirs, but the pathogens can spread to humans. Strong evidence suggests that HIV is a zoonosis that "made the leap" from other primates to our own species. Rabies is another, as are viral influenza, some

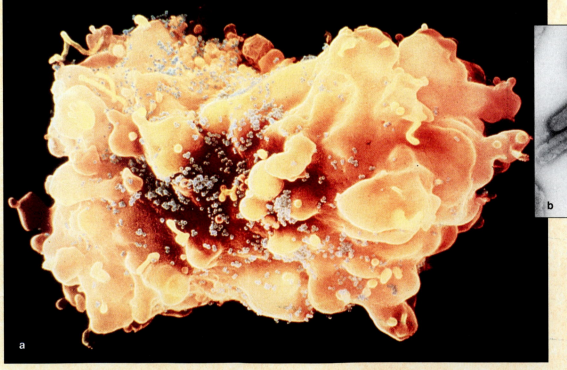

Figure 16.1 (**a**) Micrograph of a white blood cell being attacked by particles of the human immuno-deficiency virus, HIV. The devastation this virus wreaks on the human immune system leaves it unable to counterattack invading pathogens and some rare forms of cancer. (**b**, *above*) Electron micrograph of Ebola virus particles.

types of encephalitis, anthrax, tapeworm infections, and Lyme disease—and many other diseases as well.

Another way of grouping infectious diseases is to look at how they are transmitted. Extremely common are the **sexually transmitted diseases** such as genital herpes, chlamydial infections, and gonorrhea. *Salmonella* and the *E. coli* strain O157:H7, both of which can cause devastating *gastroenteritis*, typically enter the body in contaminated food or water.

If it seems to you that there are an awful lot of "new" diseases turning up, in a way you're right. Heading the long list of worries for many public health officials are two disease categories, **emerging diseases** and **reemerging** ones. Today, emerging diseases include *hantavirus pulmonary syndrome* (caused by an inhaled virus that lives normally in rodents), *Lyme disease*, and *hemorrhagic fevers* such as Ebola (Figure 16.1*b*), Lassa, and dengue fever ("hemorrhagic" because they cause massive bleeding). The reemerging diseases include tuberculosis and cholera (Section 3.7). At one time both were relatively well controlled in many parts of the world, but now are making a comeback.

Over the years, we have made impressive advances in combating infectious diseases such as smallpox and polio. So why is all this happening? Many interrelated factors are at work, but broadly speaking, a few stand out. For one, there are simply many more humans on the planet, interacting with their environment and with each other. Each person is a potential target for many pathogens. Also, technology and the growth of global economies have dramatically increased opportunities for travel. That is the main reason why, for example, hemorrhagic fevers that once were limited to tiny areas of Africa now are beginning to turn up in Europe and North America.

Another key factor is the misuse and overuse of antibiotics, drugs that once could vanquish bacterial infections in a few days. No more. Many of the worst of the modern "super bugs" are bacteria that have genetically outwitted efforts to wipe them out.

It can be awfully sobering to take a hard look at the worldwide problem of infectious disease, as we do in this chapter. However, as the saying goes, perhaps it's what we *don't* know that can hurt us the most. With that perspective, the following pages provide an introduction to the culprits—viruses, bacteria, and other infectious agents—and examples of how they can disrupt body functioning as they cause disease. We discuss sexually transmitted diseases at length, not only because they afflict hundreds of millions, but also because they are so easily preventable. You also will get glimpses of how biologists around the world are working to develop treatments and preventive vaccines and to make other advances that offer hope in the fight against infectious disease.

KEY CONCEPTS

1. Most infectious diseases are caused by viruses, bacteria, and single-celled microbes called protozoa ("first animals"). Others are caused by infectious proteins, fungi, and other organisms.

2. A virus is a noncellular infectious particle. Viruses multiply by pirating the metabolic machinery of a specific type of host cell. Bacteria are microscopic prokaryotic cells.

3. Antibiotics are effective against bacteria, but not against viruses. Many dangerous pathogens are increasingly becoming genetically resistant to antibiotics, creating a major global health crisis.

4. Protozoa and worms that cause serious disease in humans are often spread in contaminated water or food.

5. Infectious diseases are transmitted by direct or indirect contact with a pathogen. Disease outbreaks may be sporadic (occasional) or endemic (ongoing in a population). A disease is epidemic when it occurs at a level above the predicted rate, and pandemic when epidemics of the same disease break out in several regions of the world at once.

6. The virulence of a pathogen is a measure of its relative ability to cause serious disease.

7. Sexually transmitted diseases (**STD**s) are spread from infected to uninfected people during sexual intercourse.

8. Some sexually transmitted diseases are incurable. This group includes AIDS (caused by HIV) and genital herpes. STDs such as chlamydial infection, gonorrhea, and syphilis also can cause serious health problems.

9. A person's chances of contracting an STD can be reduced by behaviors that prevent or significantly reduce the likelihood of being exposed to STD pathogens.

CHAPTER AT A GLANCE

VIRUSES AND INFECTIOUS PROTEINS

Viruses aren't high on the "most popular microbe" list. The Nobel Prize–winning biologist Sir Peter Medawar once called them "bad news wrapped in protein." The term virus comes from a Latin word meaning slime or poison. Today we define a **virus** as a piece of genetic material packaged inside a protein coat. It is not a cell, but it can infect the cells of other organisms. This ability to get inside a host cell is vital, for a virus cannot reproduce itself—it depends on the host cell's metabolic machinery to do that.

We classify viruses by several criteria. One is the general type of host organism infected. Some viruses infect animals, others plants. Still others, the bacteriophages, infect bacteria. Bacteria and bacteriophages are used as research tools in genetic engineering (Chapter 21).

Viruses also are classified by structure. In general, a virus particle consists of a nucleic acid core inside a protein coat called a capsid. Sometimes the capsid also is enclosed, in a lipid envelope.

The genetic material of a virus is DNA *or* RNA (but never both). The coat around it consists of one or more types of protein subunits, often organized into a rodlike or many-sided (polyhedral) shape (Figure 16.2). The coat protects the viral genetic material during the journey from one host cell to another. It also contains proteins that can bind with specific receptors on host cells. Some viruses have coats that are enclosed in an envelope built mainly from the remnants of the plasma membrane of a previously infected, dead cell.

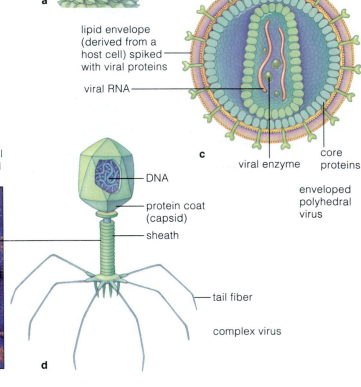

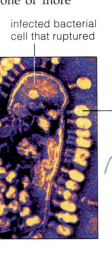

Figure 16.2 Examples of virus structure. (**a**) Sketch of a helical virus, which has a rod-shaped protein coat around its nucleic acid. (**b**) A polyhedral virus, represented here by an adenovirus, has a many-sided coat. Adenoviruses mainly infect the respiratory and digestive systems. (**c**) An enveloped virus, which has an envelope around a polyhedral or helical coat. HIV is the example shown here. Section 8.11 shows a photograph of an enveloped virus that causes influenza. (**d**) A complex virus, as in this example of a T-even bacteriophage, has other structures attached to the coat.

How Viruses Multiply

Different viruses multiply in different ways. However, in most cases there is a basic six-step cycle. The steps are: (1) *attachment* to the host cell; (2) *entry* of the virus or its genetic material into the cell's cytoplasm; (3) *replication* of the viral DNA or RNA and (4) *synthesis* of enzymes and other proteins, using the host cell's metabolic machinery; (5) *assembly* of the viral nucleic acids and viral proteins into new virus particles; and (6) *release* of the new virus particles from the cell. Figure 16.3 diagrams these steps for an enveloped DNA virus infecting an animal cell.

A cell is a potential host if a virus can chemically recognize and lock onto receptors at the cell surface. The differences in receptors on different cell types are why a given virus often can attack only one type of cell; an influenza virus can attack cells of respiratory epithelium but not liver cells, and a virus that causes hepatitis attacks liver cells but can't infect cells of the airways.

When the multiplication cycle of a virus comes to a close, the host cell usually dies and new virus particles are released. A viral enzyme synthesized in the cycle triggers apoptosis (the programmed cell death described in Section 8.6). Not all viral infections follow this sequence, however. Sometimes the virus does not kill the host cell outright. Instead, the virus enters a period of **latency**, in which its genes inside the host cell are not active. An example that occurs in nearly everyone's body is type 1 Herpes simplex, which causes cold sores. It remains latent inside a ganglion (a cluster of neuron cell bodies) in facial tissue. Stressors such as a sunburn, falling ill, or emotional upsets can reactivate the virus. Then, virus particles move down the neurons to their tips near the skin. From there they infect epithelial cells and cause painful eruptions on

the skin. Repeat outbreaks of genital herpes, caused by type 2 Herpes simplex, happen the same way.

Another herpes virus, the Epstein-Barr virus (EBV), is commonly transmitted by kissing. An EBV infection can lead to *infectious mononucleosis*. Symptoms are overwhelming fatigue, a sore throat, and a mild fever. It can take a month or more to recover.

A **retrovirus** is an RNA virus that infects animal cells. Retroviruses establish yet another relationship with the host cell. This group's name comes from its "reverse" mode of pirating the host's genetic material. After the viral RNA chromosome enters the host cell cytoplasm, a viral enzyme called *reverse transcriptase* uses the RNA as a template from which it synthesizes a DNA molecule. The new DNA molecule, called a *provirus*, then integrates into the host's DNA. This is what happens when a person becomes infected with HIV.

Prions—Infectious Proteins

Some infectious agents are even smaller and simpler than viruses. For example, a few rare, fatal degenerative diseases of the central nervous system are linked to small infectious proteins called **prions** (PREE-ons). One example is *Creutzfeldt-Jakob disease* (CJD). This prion disease can be hereditary, but some cases are thought to be transmitted by way of transplanted tissue or contaminated surgical instruments. What's going on? Section 2.12 described how proteins are folded into various shapes. Prions are versions of otherwise normal proteins that are present on brain neurons and some other types of cells. For some reason (possibly a gene mutation in inherited cases), they are *mis*folded. Once a prion forms, or enters the body from elsewhere, it can cause normal versions of the protein to misfold and become prions also.

Prions stick together in fibrous clumps that damage the cell's plasma membrane so badly that the cell dies. CJD generally has a long latent period. When enough fibrils accumulate there is nerve damage, resulting in the slow destruction of muscle coordination and brain function. Most people with CJD are middle-aged or older. A few cases of an unusual form that affects younger people have been linked to the outbreak of mad cow disease (*bovine spongiform encephalitis*) in Britain and France, raising the suspicion that CJD may be a human form of that prion disease.

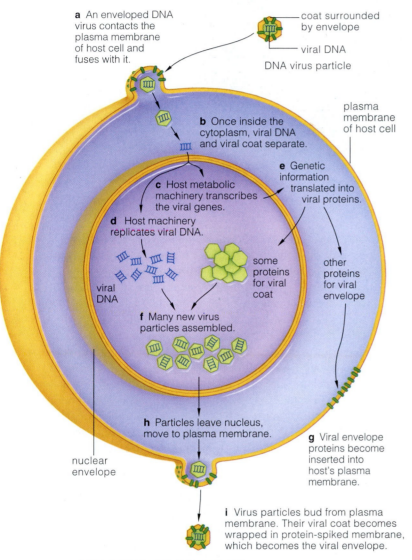

Figure 16.3 Multiplication cycle of an enveloped DNA virus infecting an animal cell. When the viral envelope fuses with the plasma membrane of the host cell, its genetic material enters the cytoplasm. The cell's metabolic machinery is subverted into churning out viral DNA and proteins. Some of the viral proteins become embedded in the internal membranes of the host cell; others are used to assemble new virus particles.

Before they bud from the cell, the new particles acquire an envelope, spiked with viral proteins, by becoming wrapped in a bit of the cell's nuclear envelope or plasma membrane.

a An enveloped DNA virus contacts the plasma membrane of host cell and fuses with it.

coat surrounded by envelope

viral DNA

DNA virus particle

b Once inside the cytoplasm, viral DNA and viral coat separate.

plasma membrane of host cell

e Genetic information translated into viral proteins.

c Host metabolic machinery transcribes the viral genes.

d Host machinery replicates viral DNA.

viral DNA

some proteins for viral coat

other proteins for viral envelope

f Many new virus particles assembled.

h Particles leave nucleus, move to plasma membrane.

nuclear envelope

g Viral envelope proteins become inserted into host's plasma membrane.

i Virus particles bud from plasma membrane. Their viral coat becomes wrapped in protein-spiked membrane, which becomes the viral envelope.

j The finished particle is equipped to infect a new potential host cell.

A virus is a noncellular infectious particle. It consists of nucleic acid enclosed in a protein coat and sometimes an outer envelope. Viruses can multiply only by pirating the metabolic machinery of a specific type of host cell. The infectious proteins called prions cause various degenerative diseases of the nervous system.

BACTERIA—THE UNSEEN MULTITUDES

Bacteria were the first living organisms on Earth. You may remember from Section 3.1 that they are prokaryotes. Unlike eukaryotic cells, such as your own, they have no nucleus or other membrane-bound organelles. (Their metabolic reactions take place at the plasma membrane.) Most bacteria have a *cell wall* outside their plasma membrane (Figure 16.4). In many types of bacteria the wall is a mesh of peptidoglycan. This is a molecule in which complex polysaccharide strands are cross-linked by peptides (small proteins). Peptidoglycan mesh makes the wall strong and fairly rigid, and helps maintain the shape of the bacterium. Although bacteria come in a variety of shapes, three basic ones are common. A ball shape is a coccus (plural, cocci; from a word that means berry). A rod shape is a bacillus (plural, bacilli, which means small staffs). A bacterial cell with one or more twists to it is a spirillum ("spiral"; plural, spirilla). Sometimes these "corkscrew"-shaped bacteria are called *spirochetes*.

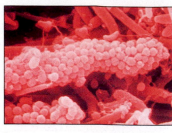

b

Figure 16.5 (**a**) *Helicobacter pylori* cell. Recall from Chapter 6 that this pathogen can colonize the stomach lining. If untreated, infection can lead to ulcers, gastritis, and possibly stomach cancer. (**b**) Surface view of bacilli and cocci on a human tooth.

coccus   bacillus   spirillum

a

The micrographs in Figure 16.5 show several kinds of bacteria common in the human body, including one that is a potentially dangerous pathogen. Figure 16.7 on the facing page shows one of the clinical signs of Lyme disease, which is caused by the spirochete *Borrelia burgdorferi*.

Some kinds of bacteria have threadlike structures anchored to the cell wall and plasma membrane. One type is a *bacterial flagellum*, a stiff protein filament that rotates like a propeller, moving the cell through its fluid surroundings. Some kinds of bacteria have filaments

called *pili* (singular: pilus) that help the cell stick to surfaces. For instance, the gonorrhea bacterium uses pili to attach to mucous membranes of the reproductive tract.

Bacteria reproduce by *prokaryotic fission* (Figure 16.6). In this kind of cell division, a cell's DNA is copied, then the cell divides into two genetically identical daughter cells. Bacterial cells have only one circular DNA molecule (a chromosome) to copy and parcel out to daughter cells. Prokaryotic fission often is rapid; under ideal conditions, some bacteria can divide roughly every 20 minutes.

When some species of bacteria reproduce, the daughter cells can inherit one or more small circles of extra DNA. These DNA circles, which include a few genes, are called **plasmids.** One type, a "fertility" plasmid, carries genes that enable the bacterium to briefly unite with another bacterial cell and transfer plasmid DNA to it. Bacteria with this ability to transfer genes include some pathogens, such as *Salmonella* (a source of food-borne illness), strains of *E. coli*, and species of *Streptococcus* that cause various respiratory infections and *strep throat*.

Human Interactions with Bacteria

We humans have many relationships with bacteria. We harness the metabolic machinery of some species to make cheeses, therapeutic drugs, and other useful items. As you will read in Chapter 21, bacteria (and plasmids) have been our partners in genetic engineering. For the most part, however, we tend to think of bacteria as the enemy because they cause so many diseases.

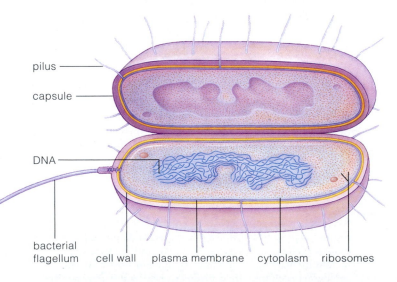

pilus

capsule

DNA

bacterial flagellum   cell wall   plasma membrane   cytoplasm   ribosomes

Figure 16.4 Generalized cell structure of a bacterium.

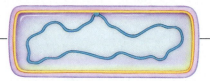

a Bacterium (cutaway view) before its DNA is copied.

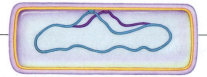

b Replication begins and proceeds in two directions, away from some point on the DNA molecule.

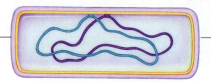

c The DNA copy is attached at a site close to the attachment site of the parent DNA molecule.

d Membrane growth occurs between the two attachment sites and moves the two DNA molecules apart.

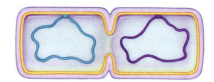

e New membrane and wall material start growing through the cell midsection.

f Membrane and wall material deposited at the cell midsection divide the cytoplasm in two.

Figure 16.6 Bacterial reproduction by prokaryotic fission, a cell division mechanism.

Once antibiotics were discovered in the mid-twentieth century, they were quickly harnessed to fight disease. An antibiotic is a substance that can destroy or inhibit the growth of bacteria and some other microorganisms. Bacteria and fungi produce most antibiotics. Those such as the penicillins, tetracyclines, and streptomycin kill microbes by interfering with different metabolic functions. For example, the penicillins break apart bonds necessary to hold molecules together in the cell walls of susceptible bacteria.

You may already know that antibiotics do not work against viruses, which are not cells and so do not have a metabolism to disrupt. The body's own defenses, such as the antiviral proteins called interferons (Section 8.10), may block a viral infection or replication of the virus inside cells. In addition, there are various *antiviral* drugs, which generally interfere with the ability of a virus to enter or replicate inside potential host cells.

Today there are scores of prescribed antibiotics. All are potent drugs. Different ones ones can have side effects such as triggering an allergic response or reducing the effectiveness of birth control pills. Even more serious, however, is the emergence of antibiotic-resistant microbes.

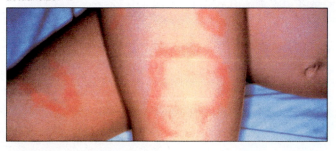

Figure 16.7 The bulls-eye rash that is a symptom of Lyme disease, now the most common tick-borne disease in the United States.

actual size: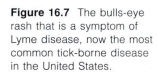

Biological Backlash to Antibiotics

Over the years, some doctors prescribed antibiotics for patients who had viral—not bacterial—illnesses. In some countries, antibiotics have not been prescription drugs, so some people take the drugs whenever they don't feel well. Some people stop taking an antibiotic when they start to feel better, even though the recommended course of treatment may be several days longer. With time, all these practices have contributed to the proliferation of bacteria that are genetically resistant to antibiotics that might otherwise kill them. Why? In basic terms, ill-advised and improper use of antibiotics wipes out susceptible bacteria, leaving behind resistant "super bugs." As that happens over and over, soon the only disease-causing bacteria left are those equipped to fend off one or more antibiotics.

The current list of drug-resistant microbes includes strains that cause some cases of tuberculosis, urinary tract infections, gonorrhea, dysentery, childhood ear infections, and many surgical-wound infections. Chance genetic mutations give bacteria resistance. And because bacteria can reproduce so rapidly, or can transfer genes via plasmids, just one or two resistant microbes can rapidly become a thriving multitude.

Staphylococcus aureus, or staph A, is a bacterium that can cause pneumonia and wound infections that rapidly destroy tissue. It is one of several microbes that may soon be resistant to *all* available antibiotics. Researchers are racing to learn more about how resistance develops, in the hope that their search will point the way to antibiotics that work against microbes in new, effective ways.

Bacteria are microscopic prokaryotic cells. They generally have one circular chromosome and often have extra DNA in the form of plasmids.

Antibiotics are effective against bacteria and certain other microorganisms, but not against viruses.

Like bacteria, protozoa ("first animals") are single cells. However, unlike bacteria they are eukaryotes—they have a nucleus that contains their DNA, and other organelles as well. It's a tremendously diverse group of "critters," and while some don't infect humans, others are serious disease threats. Some are **parasites**: they live on or in a host organism for at least part of its life cycle, feeding on specific tissues.

Figure 16.8 shows three parasitic protozoa of concern to humans. *Entamoeba histolytica*, an amoeba, forms a cyst at one stage of its life cycle. The cyst, which is a tough body covering, helps the organism wait out unfavorable environmental conditions, such as a lack of food. Where public sanitation is poor, food or water might contain the cysts. Once inside the human intestine, *E. histolytica* completes its life cycle. In the process it causes the infectious disease *amoebic dysentery*, a common type of "traveler's diarrhea." However, in regions where untreated human sewage regularly spews cysts of *E. histolytica* into the water supply, amoebic dysentery is a leading cause of death among infants and toddlers.

Giardia lamblia (Figure 16.8b) also forms cysts, which enter bodies of water or food supplies in contaminated feces. Anyone who has ever had a case of *giardiasis* is all too familiar with the awful "rotten egg" belches, explosive diarrhea, and other symptoms it brings. Now present in water supplies the world over, including "pristine" wilderness, *Giardia* regularly wreaks havoc in the guts of people from mountain climbers to children in urban day care centers.

Trypanosoma brucei causes *African trypanosomiasis*. Its common name is *sleeping sickness* (Figure 16.8c). This is a severe form of encephalitis transmitted from person to person by bites of the tsetse fly. After traveling in the bloodstream to various organs, the parasite may invade the central nervous system. Untreated, the disease is fatal. At any given time, at least a million Africans are infected.

In the early 1990s, when more than 400,000 people in Milwaukee, Wisconsin, developed the watery diarrhea and other miserable symptoms of *cryptosporidosis*, an obscure protozoan called *Cryptosporidium parvum* became big news. Now cryptosporidosis is on the growing list of emerging infectious diseases in the United States. *C. parvum* is a tough character. Its oocytes (immature eggs) can resist treatment with chlorine—they can even survive a 2-hour soaking with full-strength bleach! As with the cysts of *Giardia* and amoebas, the oocytes are spread in water or food contaminated with the feces of humans or some other animal hosts.

Poor sanitation isn't just a factor in diseases caused by protozoa; it also is how many worm infections spread. Another, less common route is eating contaminated meat or fish. For people in developed countries, such as the United States, pinworm infections are the most familiar. A pinworm is a small, white roundworm; it looks rather like a quarter-inch fleck of white thread. The infection route is oral-fecal; that is, first the person consumes worm eggs, then the eggs hatch and worms develop, then female pinworms lay *their* eggs just outside the anus, then those eggs find their way to a person's mouth, and so on.

Pinworms are only a nuisance. Other infectious worms, such as tapeworms, hookworms, whipworms, and the large *Ascaris* roundworms, can do serious damage to body tissues and organs. Serious worm infections are most common in the tropics, and can be treated with therapeutic drugs. Even so, billions of people in various parts of the world share their bodies with worms, often without knowing it.

Some species of protozoa are pathogens that cause serious diseases in humans. Often, they are spread by way of water or food contaminated with animal feces.

Infectious worms, such as the large roundworm *Ascaris*, can do serious damage to tissues and organs.

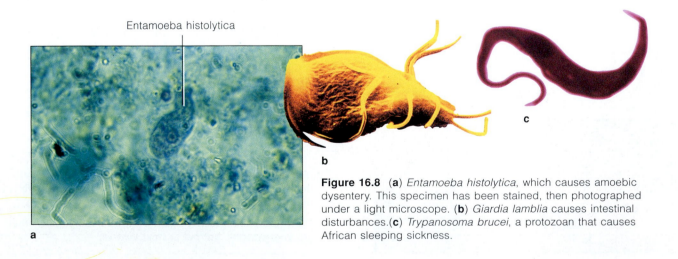

Entamoeba histolytica

a

b

c

Figure 16.8 (**a**) *Entamoeba histolytica*, which causes amoebic dysentery. This specimen has been stained, then photographed under a light microscope. (**b**) *Giardia lamblia* causes intestinal disturbances.(**c**) *Trypanosoma brucei*, a protozoan that causes African sleeping sickness.

16.4 MALARIA—HERE TODAY . . . AND TOMORROW AND TOMORROW

THE PROBLEM   By the grim tallies of the World Health Organization, every year *malaria* takes the lives of close to 3 million people, most of them children. The direct cause of malaria is a parasite, the protozoan *Plasmodium*. One of its life stages, a sporozoite, lives in the salivary glands of the female *Anopheles* mosquito. When the insect sucks blood from a person, *Plasmodium* travels from her salivary glands into the person's bloodstream. When it reaches the liver, severe illness and often a lifetime of intermittent misery begin.

Shaking, chills, a high fever, and drenching sweats are the classic symptoms of malaria. They begin after infected liver cells burst open and release a *Plasmodium* life stage called a merozoite. Some merozoites enter the person's red blood cells and develop into a trophozoite stage that uses hemoglobin in blood as a food source (*troph-* loosely means "feeder"). When this stage later reproduces, the offspring are merozoites. Eventually the infected red blood cells rupture and release the merozoites, which can infect other red blood cells.

Plasmodium has a complex life cycle (Figure 16.9). While some merozoites are invading red blood cells, others develop into a life stage called a gametocyte (a gamete-producing cell). If a female *Anopheles* mosquito sucks blood from an infected person, gametocytes enter her gut with the blood meal. Eventually the infective life stage moves into her salivary glands, and the cycle is set to continue.

After the initial bout of illness, malaria symptoms subside for weeks or months. Every so often, however, relapses occur as dormant parasites become active. In time, a malaria sufferer may develop anemia (from the loss of red blood cells) and a grossly enlarged liver. Some people, including those who are, or are descended from, West African Blacks, may carry the gene for *sickle-cell anemia*—and this genetic heritage can help protect them from malaria. In people who inherit only one copy of the gene, red blood cells respond to merozoite infection in a way that cuts short the parasite life cycle. Chapter 19 takes a closer look at this phenomenon.

THE DIFFICULT SEARCH FOR SOLUTIONS   Throughout human history, tropical Africa has been malaria's main stronghold. Today, however, its incidence is rising rapidly in parts of the United States, South America, Central Asia, and Eastern Europe. As bad luck would have it, some strains of *Plasmodium* are now resistant to once-useful antimalarial drugs, and newer drugs seem to be meeting the same fate. Another option, mosquito control and eradication efforts, are bumping up against insecticide-resistant *Anopheles*. At this writing several malaria vaccines are in clinical trials, although they are difficult to develop. *Plasmodium*'s complex life cycle is its "secret weapon," for no one has yet been able to create a vaccine that is effective against all stages of the cycle.

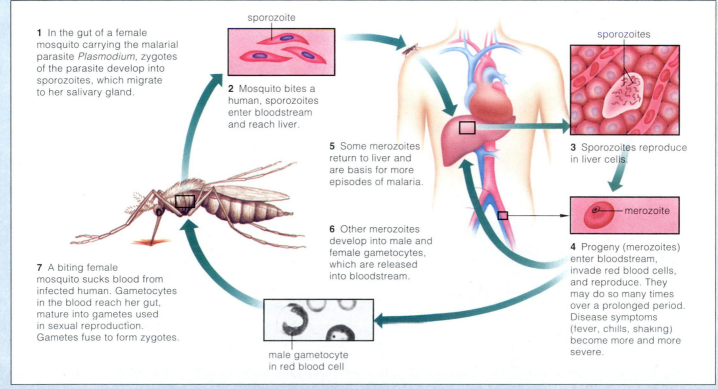

1 In the gut of a female mosquito carrying the malarial parasite *Plasmodium*, zygotes of the parasite develop into sporozoites, which migrate to her salivary gland.

sporozoite

2 Mosquito bites a human, sporozoites enter bloodstream and reach liver.

sporozoites

3 Sporozoites reproduce in liver cells.

5 Some merozoites return to liver and are basis for more episodes of malaria.

merozoite

4 Progeny (merozoites) enter bloodstream, invade red blood cells, and reproduce. They may do so many times over a prolonged period. Disease symptoms (fever, chills, shaking) become more and more severe.

6 Other merozoites develop into male and female gametocytes, which are released into bloodstream.

7 A biting female mosquito sucks blood from infected human. Gametocytes in the blood reach her gut, mature into gametes used in sexual reproduction. Gametes fuse to form zygotes.

male gametocyte in red blood cell

Figure 16.9 Life cycle of the parasite that causes malaria.

TRANSMISSION AND PATTERNS OF INFECTIOUS DISEASE

Modes of Transmission

The hallmark of an infectious disease is that it can be transmitted from person to person. Usually, such diseases are transmitted in one of four ways:

1. Direct contact with a pathogen, as by touching open sores or body fluids from an infected person. (This is where "contagious" comes from; the Latin *contagio* means touch or contact.) Infected people can transfer pathogens from their hands, mouth, or genitals and so transmit the agent through a handshake or a kiss.

2. Indirect contact, as by touching doorknobs, tissues (or handkerchiefs), diapers, or other objects previously in contact with an infected person. As already noted, food and water can be contaminated by pathogens, including those responsible for amoebic dysentery, bacterial dysentery (*Salmonella* and *Shigella*), typhoid fever, and hepatitis A.

3. Inhaling pathogens that have been spewed into the air by coughs and sneezes (Figure 16.10). This is the most common mode of transmission. Influenza is one of the more common infectious diseases transmitted this way (Section 8.11).

4. Encounters with vectors. Common examples include mosquitoes, flies, fleas, and ticks. A **disease vector** carries a pathogen from an infected person or contaminated

Figure 16.10 A full-blown sneeze. Inhaled pathogens spewed out during an unprotected sneeze can transmit colds, influenza, and other infectious diseases.

material to new hosts. In some cases, part of the pathogen's life cycle must take place inside the vector, which is an *intermediate host*. Mosquitoes serve this function for the *Plasmodium* parasites that cause malaria (Section 16.4).

Every year 5 to 10 percent of hospitalized people come down with a **nosocomial infection**—an infection that is acquired in a hospital, usually by direct contact. For example, despite the best precautions, bacteria may enter a person's urinary tract during a catherization procedure. Why are nosocomial infections so common? Anyone sick enough to be hospitalized may have a compromised (and therefore less effective) immune system, and invasive medical procedures give bacteria easy access to tissues. Also, the intensive use of antibiotics in hospitals increases the likelihood that antibiotic-resistant pathogens will be present there. Hospitals generally take extreme care to monitor patients who are likely to be most vulnerable to nosocomial infection.

Patterns in Which Diseases Occur

You may have heard infectious diseases described in terms of the patterns by which they occur. A **sporadic disease**, such as whooping cough, breaks out irregularly and affects relatively few people. By contrast, an **endemic disease** occurs more or less continuously. Many of the diseases noted in Table 16.1 are endemic in various parts of the world. So is the fungal disease called *ringworm*, and the common cold.

Table 16.1 Infectious Dieases: Global Health Threats*		
Disease	Type of Pathogen	Estimated Deaths per Year
Tuberculosis	Bacterium	3.1 million
Malaria	Protozoan	2.7 million
Hepatitis (includes A, B, C, D, E)	Virus	1–2 million
Various respiratory infections (pneumonia, viral influenza, diphtheria, strep infections)	Virus, bacteria	7 million
Diarrheas (includes amoebic dysentery, cryptosporidiosis)	Protozoa, virus, and bacteria	3+ million
Measles	Virus	220,000
Schistosomiasis	Worm	200,000
Whooping cough	Bacterium	100,000
Hookworm	Worm	50,000+

*Does not include AIDS-related deaths.
Source: National Institute of Allergy and Infectious Disease, 2000.

Further reading: Student Guide to InfoTrac on web site →

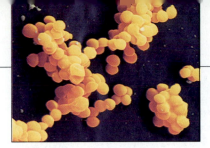

Figure 16.11 False-color micrograph of *Staphylococcus aureus.*

During an **epidemic**, a disease rate increases to a level that is above what we would predict, based on experience. When cholera broke out all through Peru in 1991, that was an epidemic. The bubonic plague epidemic in 14th-century Europe killed 25 million people. When epidemics break out in several countries around the world in a given time span, they collectively are called a **pandemic**. AIDS is pandemic; by 2000, approximately 36 million people were infected.

Virulence

Virulence is the name for the relative ability of a pathogen to cause serious disease. It depends on how fast the pathogen can invade tissues, how severe is the damage it causes, and which tissues it targets. For example, viruses that can cause pneumonia are more virulent than viruses causing the common cold. Rabies viruses are considered highly virulent because they target the brain.

Pathogens cause disease in various ways. Symptoms of a viral infection usually develop when infected cells begin to malfunction and die. A few viruses trigger cancer when they infiltrate the DNA in a host cell's chromosomes (Chapter 22). Paradoxically, our immune system's *responses* to viral infection also provoke symptoms we associate with disease. Two common examples are fever and fatigue. A controlled inflammatory response helps rid the body of pathogens (Section 8.3), but if the response goes awry it can lead to serious damage, especially if a major organ or organ system is involved. For instance, if the testicles of a male who contracts the sexually transmitted disease *gonorrhea* become inflamed, the inflammation can progress to an immune response that leaves the man sterile.

Most disease-causing bacteria secrete toxins that harm cells. If invading bacteria multiply faster than the immune system can respond to them, a large amount of toxin may enter tissues. Bacteria that release toxins into the bloodstream are especially dangerous because the toxin moves rapidly through the body, threatening the nervous system and vital organs.

Antibiotic resistance in certain bacteria, such as strains of *Staphylococcus aureus* (Figure 16.11), has made those microbes highly virulent. Infectious disease specialists have instituted a worldwide surveillance system to identify new resistant strains before they can become established.

Infectious diseases are transmitted by direct or indirect contact with pathogens and occur in different patterns. Some are becoming resistant to available antibiotic treatments.

TB, CONTINUED

A century ago *tuberculosis* (TB) was a devastating disease (then known as *consumption*) that caused debilitating pneumonialike symptoms, among other ills. Worldwide it probably killed up to 30 percent of adults under the age of sixty-five. Mummies thousands of years old show evidence of its ravages. Some decades ago, though, improved living conditions, better hygiene, and medical advances—including antibiotics—seemed to have put TB near the bottom of the heap of infectious diseases we needed to worry about. Then, in 1993, the hardworking people who gather health statistics got a shock: For January of that year, just under 800 new cases were reported, but the following December's number was a whopping 5,130 cases. The reemergence of TB as a scourge of humankind had begun.

Today just under a third of the world's people, about 1.6 billion souls, are infected with *Mycobacterium tuberculosis* (Figure 16.12), the bacillus that causes TB. The bacteria are transmitted in airborne droplets that are produced by coughing or sneezing. In an otherwise healthy person, the immune system kills invading bacilli. If small lesions (tubercles) do form in the lungs, their healing will leave a scar visible on a chest X ray.

Why is TB reemerging as major health threat? Several factors probably have a role. One is simply the increased movement of people from one country to another. People who have had to live for any length of time in crowded, unsanitary conditions (such as refugee camps and urban slums) are at much higher risk of contracting TB, and of taking it with them if they manage to improve their circumstances. Many recent cases have developed in people infected with HIV, whose immune systems are weakened. Intravenous drug abusers also account for a growing number of reported cases of active TB.

Antibiotics can cure TB. To be effective, however, the drugs must be taken regularly for at least a year—a regimen that is not easy for many people to follow. Not surprisingly, drug-resistant strains of the TB bacterium have turned up. Fortunately, they are not widespread—yet. TB screening programs try to ferret out early-stage cases for treatment. Meanwhile, 3 to 5 million new cases are diagnosed every year.

Figure 16.12 False-color micrograph of *Mycobacterium tuberculosis.*

AIDS is a constellation of diseases caused by infection with the **human immunodeficiency virus**, or **HIV**. By the end of the twentieth century, acquired immune deficiency syndrome (AIDS) had become the fourth leading cause of death in the world. In Africa, where HIV arose, it was the number one killer. If you look back at the introduction to this chapter, you will see a micrograph of HIV particles on the surface of a helper T cell.

HIV destroys T cells, crippling the immune system. It leaves the body vulnerable to infections and some rare forms of cancer. Physicians diagnose AIDS if a patient has a severely depressed immune system, tests positive for HIV, and has one or more of twenty-six "indicator diseases," including types of pneumonia, cancer, recurrent yeast infections, and drug-resistant tuberculosis.

HIV is transmitted when body fluids, especially blood and semen, of an infected person enter another person's tissues. The virus can enter the body through any kind of cut or abrasion, anywhere on or in the body. HIV-infected blood also can be present on toothbrushes and razors; on needles used to inject drugs intravenously, pierce ears, do acupuncture, or create tattoos; and on contaminated dental or surgical equipment.

Contaminated blood supplies once accounted for some HIV infections of surgery patients and people with the genetic blood disease *hemophilia* (which requires the patient to use donated blood products). HIV also can be transmitted from infected mothers to their infants during pregnancy, birth, and breast-feeding. In several countries, HIV has spread through reuse of unsterile needles by health care providers. However, there is *no* evidence that casual contact, insect bites, hugging, nonsexual touching, or sharing food can spread the virus.

The virus has been isolated from various body fluids. However, apparently only infected blood, semen, vaginal secretions, and breast milk contain HIV in concentrations that are high enough for successful transmission. Having another sexually transmitted disease (such as syphilis) that produces sores or other lesions increases risk of HIV infection, because any break in the skin affords the virus easy entry into the body. HIV usually cannot survive for more than about two hours outside an infected cell. Virus particles on needles and other objects are readily destroyed by disinfectants, including household bleach.

In developed countries, HIV spread at first among male homosexuals; among hemophiliacs and others who received tainted blood products; and among intravenous drug abusers who shared needles contaminated with infected blood. At this writing, almost half of the adults worldwide who are HIV-infected are women (Table 16.2). Some of those infections are due to intravenous drug abuse, but 75 percent are the result of sexual contact with infected men. According to the Centers for Disease Control and Prevention, in the United States about 15 percent of new infections in men come from heterosexual contact. Young people are also being hit hard; in recent years, more young adults in the United States have died from AIDS than from any other single cause.

Table 16.2 Statistical Snapshot of HIV/AIDS	
33.6 million	People infected as of January 2000; 32.4 million adults, 1.2 million children
14.9 million	Number of women worldwide living with HIV/AIDS
16.3 million	Cumulative AIDS deaths, 1981–1999
5.6 million	Number of new HIV infections in 1999, equal to 15,000 infections a day
650,000–900,000	Number of U.S. residents living with HIV infection in 1999
8 to 20 percent	Number of HIV-infected adults in various countries of Africa
80 percent	Conservative estimate of number of HIV infections in adults due to heterosexual intercourse
2.7 million	Number of new HIV infections in 1999 in people aged 15–24; 20,000 of these were in the United States.

Adapted from 2000 HIV/AIDS Statistics Fact Sheet, National Institutes of Health.

HIV Structure and Replication

HIV is a retrovirus—its genetic material is ribonucleic acid, RNA, instead of deoxyribonucleic acid, DNA. A protein core surrounds the RNA and several copies of the viral enzyme reverse transcriptase (Section 16.1). The core itself is wrapped in a lipid envelope pirated from the plasma membrane of a host helper T cell. Once inside a host cell, the reverse transcriptase uses the viral RNA as a template for making DNA. This DNA provirus (Section 16.1) then is integrated into a host chromosome, which contains the host's DNA.

Eventually the provirus's genetic instructions for making new viral particles are read out. A process called *transcription* rewrites the genetic message in DNA as RNA, and these RNA instructions then are "translated" into protein (Chapter 20). Figure 16.13 summarizes how this process generates new HIV particles in an infected cell. Some efforts to develop anti-HIV drugs focus on ways to chemically interfere with replication—and so prevent HIV from reproducing once it enters a host.

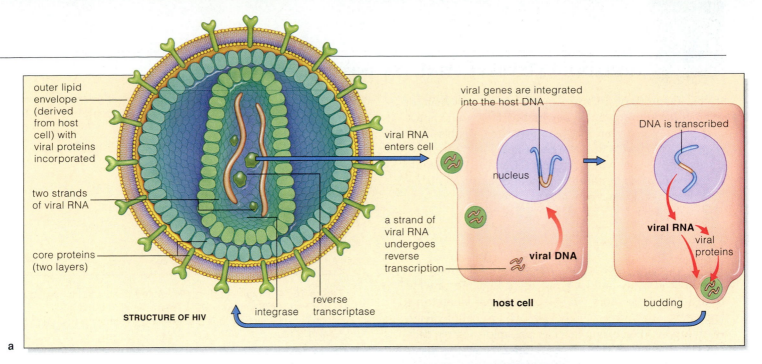

outer lipid envelope (derived from host cell) with viral proteins incorporated

two strands of viral RNA

core proteins (two layers)

STRUCTURE OF HIV

integrase

reverse transcriptase

viral RNA enters cell

a strand of viral RNA undergoes reverse transcription

viral genes are integrated into the host DNA

nucleus

viral DNA

host cell

DNA is transcribed

viral RNA viral proteins

budding

From Attack to Infection to AIDS

HIV infects antigen-presenting macrophages and helper T cells (Section 8.4). Some helper T cells are called *CD4 cells*, because they bear a receptor known as CD4. This receptor has a central role in what has been called the "secret handshake" by which HIV infects helper T cells. First, a viral glycoprotein docks with a CD4 receptor on a helper T cell. The docking then allows the protein to bind with a second receptor, called a *chemokine receptor*, on the helper T cell. Only when this second interaction takes place can HIV enter the helper T cell.

Soon after HIV successfully infects a person, it begins to replicate and circulate in the bloodstream. At this stage, many people have a bout of flulike symptoms. In response to HIV antigenic proteins, B cells make antibodies, which can be detected by diagnostic tests for HIV infection. Armies of helper T cells and cytotoxic T cells also form. During certain phases of infection, however, the virus infects an estimated *2 billion* helper T cells and produces *100 million to 1 billion new HIV particles* each day. They bud from the plasma membrane of the helper T cell or are released when the membrane ruptures (Figure 16.13*b*).

As the battle proceeds, huge reservoirs of HIV and masses of infected T cells accumulate in lymph nodes. The number of circulating virus particles rises and the body produces fewer and fewer helper T cells to replace the ones it has lost. As the population of healthy helper T cells is depleted, the person may lose weight and experience symptoms that range from fatigue and nausea to heavy night sweats, enlarged lymph nodes, and a series of minor infections. Eventually, one or more of the typical AIDS indicator diseases begin to appear. Men also may develop *Kaposi's sarcoma*. This deadly cancer, evidenced by bluish-brown spots on the arms and legs, only strikes males. It is extremely rare in males who are not HIV-infected.

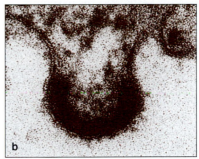

b

Figure 16.13 (**a**) Infection cycle of HIV. For clarity, the drawing shows only a single HIV budding from the host cell. In fact, many particles will be formed and exit from each infected cell. (**b**) An HIV particle budding from a host cell's plasma membrane.

About 5 percent of HIV-infected people are "non-progressors" in whom no symptoms of illness develop even 10 years or more after diagnosis. They may be infected with a mutated, less virulent strain of HIV. Also, it seems that a few HIV-infected people were born with a rare genetic mutation that results in the absence of a type of chemokine receptor on their macrophages. Lacking the receptor (needed for the secret handshake), the macrophages resist HIV infection. The few people having the mutation are *not* immune to HIV and AIDS. However, the mutation does slow the rate at which HIV infects cells, and thus the onset of AIDS. This delay allows time for a counterattack by anti-HIV treatment options described in the next section.

AIDS is a constellation of disorders caused by infection with the human immunodeficiency virus (HIV).

HIV is transmitted only when blood, semen, or certain other body fluids of an infected person enter another person's tissues.

HIV is a retrovirus. When it infects a cell, its genetic material is integrated into the host cell's DNA. Eventually the host cell begins producing new HIV particles.

If you live in Africa, Latin America, Southeast Asia, or another developing region hard-hit by HIV, your chance of dying within a year or two of infection is distressingly high. In the United States and other developed countries, the outlook is somewhat brighter (Figure 16.14), mainly because more and better treatment options are available. Still, there is no known cure for HIV infection.

Drugs and Vaccines

As the previous section described, both HIV infection and replication are multistep processes. Looking for ways to disrupt one or more vital steps, research laboratories have developed a reasonably effective arsenal of chemical weapons. For example, it is possible to chemically block the action of HIV protease, an enzyme that has a key role in processing of HIV proteins and the assembly of new virus particles. Unlike most retroviruses, HIV genes also include instructions for proteins that are synthesized early in replication and help regulate subsequent steps. The development of *protease inhibitors* that can prevent the normal processing of HIV proteins and so halt reproduction of the virus has helped advance the clinical treatments of HIV-positive people, as well as those who have full-blown AIDS. The search also is on for compounds that might disrupt the ability of HIV to bind to the chemokine receptors by which it gains entry to cells. Anti-HIV drugs have names like lamivudine or 3TC, zidovudine or AZT, zalcitabine or ddC. There are several others, as well. These drugs block the action of reverse transcriptase. They are not a cure, however, because they cannot eliminate the HIV genes that have become incorporated into the patient's DNA. Moreover, experience has shown that HIV rapidly develops resistance to these and other drugs. At present, the preferred treatment is a drug "cocktail" consisting of a protease inhibitor and two anti-HIV drugs. In some patients, such a triple-drug regimen can suppress the virus so much that standard tests can't detect it in the blood. Other tests reveal that HIV still is present in the person's body, however.

What about an HIV vaccine? As it turns out, that is a tall order. Recall from Chapter 8 that a vaccine confers protection by stimulating the immune system to respond to viral antigens—proteins produced by the invading virus. The big problem with HIV is that it mutates extremely rapidly. Even in the same person, it can exist in many different genetic forms—and, therefore, different *antigenic* forms. Any vaccine development program would have to keep pace with those antigenic variants.

Despite the challenges, the rapid spread of HIV has created pressure to develop anti-HIV vaccines. One approach is to base a vaccine on HIV particles from which key genetic information has been removed. In theory, the

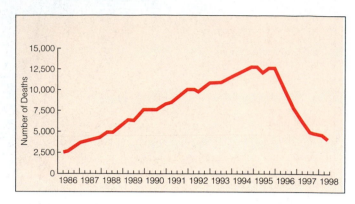

Figure 16.14 Graph charting HIV-related deaths in the United States from 1986 through 1998. The downward trend is continuing due largely to improved drug treatments. *Source: HIV/AIDS Statistics, NIAID Fact Sheet.*

altered virus would still be able to stimulate the immune system to produce a protective response but could not replicate or cause disease. Although this approach raises important safety questions, several dozen vaccines of various types have already been tested on small groups of HIV-infected people to determine if the vaccination can bolster their weakened immune systems. Larger-scale trials involving uninfected people are being planned or implemented in several countries. To help prevent discrimination, trial participants (who will thereafter test positive for HIV if the vaccine provokes B cells to make antibodies), carry an explanatory identification card.

AIDS Prevention

Every sexually active person should be aware of the danger of HIV infection. A person can minimize the risk of exposure in various simple and effective ways, as described in the *Focus on Health* essay on the facing page.

In addition to promoting safer sex, officials in hard-hit cities have implemented "clean needle" strategies, such as supplying IV-drug abusers with sterile needles. Though controversial in some places, such action is based on the premise that society has a huge stake in halting the spread of HIV. Free or low-cost confidential testing for AIDS is available through public health facilities. There are also over-the-counter blood tests for HIV infection, but they may not be as reliable as testing done by medical personnel.

Anti-HIV drugs, including protease and reverse transcriptase inhibitors, can be quite effective in reducing the amount of circulating virus and in forestalling disease. However, because there is no cure for an HIV infection, prevention is the best defense.

16.9 PROTECTING YOURSELF—AND OTHERS—FROM SEXUAL DISEASE

The old saying "An ounce of prevention is worth a pound of cure" has never been truer than with sexually transmitted diseases (Figure 16.15). The only people who are not at risk of STDs are those who are celibate (never have sex) or who are in a long-term, mutually monogamous relationship in which no disease is present. Otherwise, health care professionals recommend the following guidelines to help you minimize your risk of acquiring or spreading an STD. These guidelines aren't meant to be "preachy." They do reflect solid *scientific* advice in the realm of STD prevention.

GUIDELINES FOR SAFER SEX

1. Use a latex condom, or make sure that your partner uses one. Doing so during genital *or* oral sex will significantly reduce your risk of exposure to the pathogens that cause HIV, gonorrhea, herpes, and other diseases. With the condom, use a spermicide that contains nonoxynol-9, which may help kill virus particles. Condoms are now available for men and women.

2. Limit yourself to one partner who has sex only with you.

3. Get to know a prospective partner before you have sex. Having a friendly but frank discussion of your sexual histories, including any previous exposure to an STD, is very helpful. The safest policy is to assume you are at risk and to take appropriate precautions.

4. If you decide to become sexually intimate, be alert to the presence of sores, scabs, a discharge, or any other sign of possible trouble in the genital area. With so much at stake, there's nothing embarrassing about looking.

5. Keep yourself and your immune system healthy by getting sufficient rest, eating properly, and learning strategies for coping with stress.

6. If for no other reason than protection from STDs, avoid the abuse of alcohol and drugs. Studies show that alcohol and drug abuse both are correlated with unsafe sex practices.

COPING WITH INFECTION   Humans have probably been getting and giving sexually transmitted diseases for thousands of years. And for all those years people have probably felt worried, ashamed, guilty, embarrassed, or angry—or all of the above—about getting an STD. Admitting to a partner that you have an STD—and that *both* of you may require treatment—can be difficult. Yet most people recognize that they have an ethical obligation to inform partners when an STD enters the picture. In addition to simply being honest, dealing realistically with an infection involves the following:

1. Learn about and be alert for symptoms of STDs. If you have reason to think you have been exposed, abstain from sexual contact until a medical checkup rules out any problems. Self-treatment won't help. See a doctor or visit a clinic.

2. Take all prescribed medication. *Do not* share medication with a partner. Unless both of you take a full course of medication, chances of reinfection will be great. Your partner may need to be treated even if he or she does not show symptoms.

3. Avoid sexual activity until medical tests confirm that you are free from infection.

Figure 16.15 (*above*) Students at a workshop on STD prevention.

Chlamydial Infections and PID

In the United States, by far the most common bacterial STD is *chlamydial infection*, by *Chlamydia trachomatis* (Figure 16.16). According to the National Institutes of Health, each year an estimated 3 million Americans are infected, about two-thirds of them under age 25. Globally, *C. trachomatis* infects an estimated 90 million people annually. At least 30 percent of newborns who are treated for eye infections and pneumonia developed those disorders after being infected with *C. trachomatis* during birth.

The bacterium infects cells of the genital and urinary tract. Infected men may have a discharge from the penis and a burning sensation when they urinate. Women may have a vaginal discharge as well as burning and itching sensations. Often, however, *C. trachomatis* is a "stealth" STD; there frequently is no outward evidence of infection. Eighty percent of infected women and 40 percent of infected men don't have noticeable symptoms—yet they can still pass the bacterium to others.

Once a chlamydial infection is under way, the bacteria migrate to the person's lymph nodes, which become enlarged and tender. Impaired lymph drainage can lead to swelling of the surrounding tissues.

Chlamydial infections can be treated with antibiotics. However, because so many people are unaware they're infected, this STD does a lot of damage. Between 20 and 40 percent of women with genital chlamydial infections develop *pelvic inflammatory disease* (PID). PID strikes about 1 million women each year, most often sexually active women in their teens and twenties.

Although PID can arise when microorganisms that normally inhabit the vagina ascend into the pelvic region (typically as a result of excessive douching), it is also a serious complication of both chlamydial infection and gonorrhea. Usually, a woman's uterus, oviducts, and ovaries are affected. Pain may be so severe that infected women often think they are having an attack of acute appendicitis. If the oviducts become scarred, additional "complications," such as ectopic pregnancy (Section 15.2), chronic pelvic pain, and even sterility can result. In fact, PID is the leading cause of infertility among young women. An affected woman may also have chronic menstrual problems.

As soon as PID is diagnosed, a woman usually will be prescribed a course of antibiotics. However, advanced cases can require hospitalization and hysterectomy (removal of the uterus). A woman's partner should also be treated, even if the partner has no symptoms.

Gonorrhea

Like chlamydial infection, gonorrhea can be cured if it is diagnosed and treated promptly. Gonorrhea is caused by *Neisseria gonorrhoeae* (Figure 16.17a). This bacterium (also called gonococcus) can infect epithelial cells of the genital tract, the rectum, eye membranes, and the throat. Each year in the United States there are about 650,000 new cases reported; there may be up to 10 million unreported cases. Part of the problem is that the initial stages of the disease can be so uneventful that, as with the *Chlamydia* bacterium, a carrier may be unaware of being infected.

In early stages of infection, males are more likely than females to notice that something is wrong because the symptoms are easier to detect. Within a week, yellow pus begins to ooze from the penis. Urinating is more frequent and may be painful.

In women, the early stages of gonorrhea can be dangerously asymptomatic. For example, a female may or may not experience a burning sensation while urinating. She may or may not have a vaginal discharge; even if there is a slight discharge, she may not perceive it as abnormal. In the absence of worrisome symptoms, a woman's gonorrhea infection may well go untreated—and all the while, the bacteria may spread into her oviducts. Eventually, the woman may experience violent cramps, fever, and vomiting—and even become sterile due to scarring and blocking of her oviducts from pelvic inflammatory disease.

As noted earlier, a man can become sterile when untreated gonorrhea leads to inflammation of his testicles or scarring of the vas deferens.

Antibiotics can kill the gonococcus and prevent complications of gonorrhea. Penicillin was once the most commonly used drug treatment. Unfortunately, antibiotic-resistant strains of gonococcus have developed. As a result, many doctors now order testing to determine

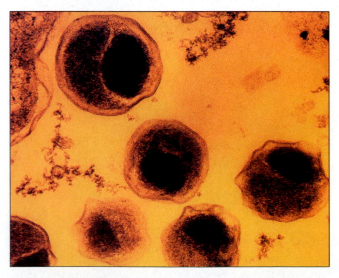

Figure 16.16 Color-enhanced micrograph of *Chlamydia trachomatis* bacteria.

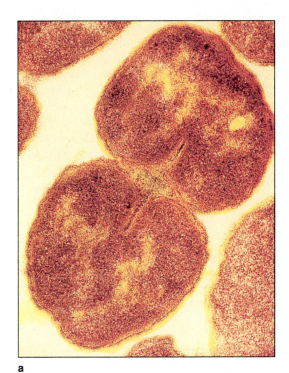

a

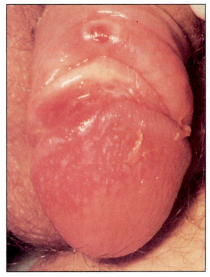

b

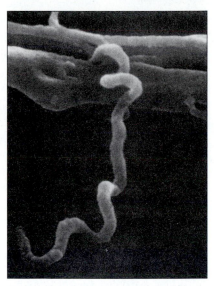

c

Figure 16.17 (**a**) *Neisseria gonorrhoeae*, or gonococcus, a bacterium that typically is seen as paired cells, as shown here. Threadlike pili not evident in this electron micrograph help the bacterium attach to its host, upon which it bestows gonorrhea. (**b**) Ulcer called a chancre ("shanker"), a sign of the first stage of syphilis. It appears anytime from about 9 days to 3 months after infection, on the genitals or near the anus or the mouth. It teems with infective bacteria, *Treponema pallidum* (**c**).

the strain responsible for a particular patient's illness and then treat the infection with an appropriate antibiotic.

Many people believe that once cured of gonorrhea, they can't be reinfected. Not true! People who have multiple sexual partners can use condoms to help avoid becoming infected. Using a condom can also prevent a person who has gonorrhea from infecting others.

Syphilis

Syphilis is caused by a motile, spirochete bacterium, *Treponema pallidum*. Each year in the United States about 70,000 new cases are reported; globally there are an estimated 12 million new cases every year.

The bacterium is transmitted by sexual contact. After it has penetrated exposed tissues, it produces a type of ulcer called a chancre (pronounced "shanker," Figure 16.17*b*) that teems with treponeme offspring. Usually the chancre is flat rather than bumpy, and it is not painful. It becomes visible between one and eight weeks following infection and is a symptom of the *primary stage* of syphilis. Using a technique called immunofluorescence, treponemes (Figure 16.17*c*) can be identified—and syphilis diagnosed—in a cell sample taken from a chancre. By then, however, bacterial cells have already moved into the lymph vascular system and bloodstream.

The *secondary stage* of syphilis begins about one to two months after the chancre appears. Lesions can develop in an affected person's mucous membranes, eyes, bones, and central nervous system. A blotchy rash breaks out over much of the body. After the rash subsides, the infection enters a latent stage, which can last for many years. In the meantime, the disease does not produce significant outward symptoms, and it can be detected only by laboratory tests.

Usually, the *tertiary stage* of syphilis begins from 5 to 20 years after infection. Lesions may develop in the skin and internal organs, including the liver, bones, and aorta. Scars form; the walls of the aorta can weaken. Treponemes also damage the brain and spinal cord in ways that lead to various forms of insanity and paralysis. Infected women who become pregnant typically have miscarriages, stillbirths, or sickly and syphilitic infants.

Penicillin effectively cures syphilis during the early stages. Even so, potential health damage is so serious that no one who is sexually active should take the disease lightly.

Chlamydial infection is the most common STD caused by a bacterium. It is curable with antibiotics, as are gonorrhea and syphilis. However, there are now some antibiotic-resistant strains of the gonorrhea bacterium.

Pelvic inflammatory disease is a dangerous complication of chlamydial infection and gonorrhea.

Genital Herpes

Infections with herpes simplex viruses, or HSV, are extremely contagious. HSV is transmitted when any part of a person's body comes into direct contact with active viruses or sores that contain them (Figure 16.18). Mucous membranes of the mouth or genital area are especially susceptible to invasion, as is broken or damaged skin. Transmission seems to require direct contact; the virus does not survive for long outside the body.

In 2000 the National Institutes of Health estimated that in the United States, one in five people over the age of twelve—roughly 45 million people—have the virus that causes **genital herpes**. The many strains of sexually transmitted HSV are classified as types 1 and 2. Type 1 strains infect mainly the lips, tongue, mouth, and eyes. Type 2 strains cause most (but not all) genital infections. That said, *type 1 strains also can cause genital lesions, and type 2 strains can cause oral lesions*. Also, an infected person can have both types. Symptoms most often develop within two weeks after infection, although sometimes they are mild or absent. Usually, small, painful blisters appear on the penis, vulva, cervix, urethra, or anal tissues. The sores can also occur on the buttocks, thighs, or back. During the initial flare-up a person may have a fever and flulike symptoms for several days. Within three weeks the sores crust over and heal.

Sporadic reactivation of the virus can produce new, painful sores at or near the original site of infection. Recurrences can be triggered by sexual intercourse, emotional stress, menstruation, a rise in body temperature, or other infections.

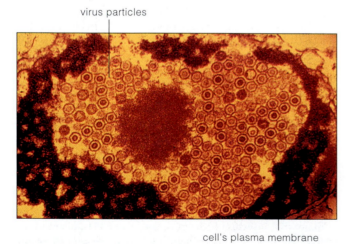

virus particles

cell's plasma membrane

Figure 16.18 Particles of herpesvirus in an infected cell. Herpes infections below the waist may involve either type 1 or type 2 strains, but most genital herpes infections are type 2.

If a pregnant woman has genital herpes, her infant can become infected during a vaginal delivery if the mother has active lesions. For the baby, the infection can be lethal or can lead to mental retardation; lesions in the infant's eyes can cause blindness. For women who have a history of genital herpes, physicians often recommend a cesarean section—surgically removing the baby through an incision in the mother's abdomen. This avoids any chance the newborn will come into contact with a contaminated cervix, vagina, vulva, or nearby surfaces.

There is no cure for HSV infection. Between flare-ups, the virus is latent in nervous tissue. However, several antiviral drugs inhibit the virus's ability to reproduce. They also reduce the shedding of virus particles from sores, and sores are often less painful and heal more rapidly. As an affected person gets older, recurrences often become both less frequent and less severe.

Human Papillomavirus

Genital warts are painless growths caused by infection of epithelium by the **human papillomavirus** (HPV). The warts can develop months or years after exposure to the virus and usually they occur in clusters on the penis, the cervix, or around the anus. After years of correlating the incidence of cervical cancer with females who have genital warts on the cervix, or who have had a partner who has genital warts, researchers now believe that certain forms of HPV are the source of more than 80 percent of cases of invasive cervical cancer—an uncommon but serious form of cervical cancer that kills around 5,000 women each year in the United States (Chapter 22). Any woman who has a history of genital warts should tell her physician, who may want to schedule an annual *Pap smear*. This is a painless test for abnormal cell growth on the cervix.

Genital warts can be difficult to diagnose because they are not always visible. When detected, the warts are usually removed surgically, by freezing (with liquid nitrogen) or burning, or by application of a drug that destroys the wart tissue.

Type B Viral Hepatitis

The **hepatitis B** virus (HBV) is a DNA virus. Like HIV, it is transmitted in blood or body fluids such as saliva, vaginal secretions, and semen. However, HBV is far more contagious than HIV. Sexually transmitted hepatitis B is increasingly common; in the United States, about 750,000 people are living with the disease, and about 80,000 new cases are reported each year. The virus attacks liver cells.

Further reading: Student Guide to InfoTrac on web site →

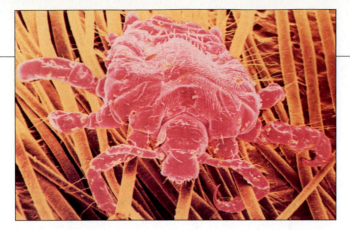

Figure 16.19 Magnified 120 times, this crab louse looks rather vicious. Crab lice are tiny but large enough to be visible on the skin, generally as mobile brownish dots. The mite uses claws at the end of its appendages to cling tenaciously to a hair shaft.

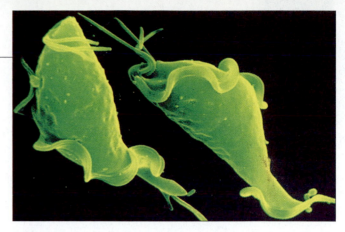

Figure 16.20 Flagella propel the protozoan parasite *Trichomonas vaginalis* through its fluid environment in the vagina. Structurally, flagella of animal cells such as this one differ from bacterial flagella, but all aid locomotion.

A key symptom is jaundice, yellowing of the skin and whites of the eyes as the liver loses its ability to process bilirubin pigments produced by the breakdown of hemoglobin from red blood cells. In about 10 percent of cases the HBV infection becomes chronic. Carriers are people who don't have symptoms but who can easily spread infection to their intimate contacts. Chronic hepatitis can lead to liver cirrhosis or cancer.

A blood test can reveal HBV or antibodies to it (which indicate that a person has been infected). The only treatment is rest. However, people at known risk for getting the disease (such as health care workers and anyone who requires repeated blood transfusions) can be vaccinated against the virus. An HBV vaccination is usually given to infants, along with other routine vaccinations. The antiviral agent interferon has been used to successsfully treat some cases of HBV.

Chancroid

The bacterium *Haemophilus ducreyi* causes **chancroid**. Once mainly a tropical disease, international travel is increasing the incidence of chancroid in the United States and Europe. Soft, painful ulcers on the external genitals are the main symptom. Untreated, the infection can spread to lymph nodes in the pelvic area and destroy tissues there. Chancroid can be treated by antibiotics.

Pubic Lice and Scabies

Two tiny ectoparasites (*ecto-* means external) can be transmitted by close body contact. Both are arthropods, "jointlegged" relatives of crabs and spiders. *Pubic lice*, also called crab lice or simply "crabs" (Figure 16.19), usually turn up in the pubic hair, although they can make their way to any hairy spot on the body. They cling tenaciously to individual hairs and attach their small, whitish eggs ("nits") to the base of the hair shaft. Itching and irritation can be intense when the parasites bite into the skin and suck blood.

A mite causes *scabies*. As this parasite burrows in the skin and lays eggs, it creates dark, undulating lines in the skin of the pubic area, armpits, around the nipples, or elsewhere. When the eggs hatch, they cause tremendous irritation and extreme itching. Antiparasitic drugs get rid of both pubic lice and scabies.

Vaginitis

The warm, moist environment of the vagina can provide a hospitable environment for a range of organisms, although the vagina's rather acidic pH usually keeps pathogens in check. When certain types of vaginal infections do occur, they can be transmitted to a sex partner during intercourse. Any event that alters the usual chemical balance of the vagina (such as taking an antibiotic) can trigger overgrowth of *Candida albicans*, a type of yeast (a fungus) that can be a common inhabitant of the vagina. Symptoms of a vaginal yeast infection (candidiasis) include a white "cottage cheesy" discharge and itching and irritation of the vulva. An affected male may notice itching, redness, and flaking skin on the penis. Yeast infections are easily treated by over-the-counter and prescription antifungal medications. Both partners may need to be treated to prevent reinfection.

Trichomonas vaginalis, a small protozoan parasite (Figure 16.20), can cause *trichomoniasis*, severe inflammation of the vaginal epithelium. Symptoms include a greenish, frothy, foul-smelling vaginal discharge and burning and itching of the vulva. A male who contracts *T. vaginalis* may experience painful urination and a discharge from the penis, the results of an inflamed urethra. Nearly always, both partners are treated with an antibiotic.

Common sexually transmitted diseases in the United States include genital warts (caused by human papillomavirus), genital herpes, type B hepatitis, chancroid, and infections by certain fungi and parasites.

SUMMARY

1. The global spread of infectious diseases has become a worldwide human health treat. The so-called emerging diseases, such as Lyme disease, are ones that have never before affected large numbers of people around the world. The reemerging diseases, such as tuberculosis, have been well controlled in the past but are again being diagnosed in large numbers of people.

2. Viruses are noncellular infectious particles. They consist of nucleic acid (DNA or RNA) enclosed in a protein coat; some have an additional outer envelope. Viruses multiply by taking over the metabolic machinery of a host cell. Viruses cannot be killed by antibiotics, but they may be susceptible to body defenses (interferons) and drugs that interfere with their ability to attach to host cells or reproduce inside them. Prion diseases are caused by infectious proteins smaller than viruses.

3. Bacteria are microscopic prokaryotic cells. They typically have a single, circular chromosome (Table 16.3). They also may have extra DNA in plasmids. Overuse and misuse of antibiotics have contributed to the emergence of bacteria resistant to many of those drugs.

4. Some protozoa (single-celled eukaryotes) are pathogens that cause serious disease in humans. Often they are spread by way of water or food contaminated with feces from infected humans or other animals. Some kinds of worm infections, such as by the roundworm *Ascaris*, also seriously harm tissues and organs.

5. An infectious disease may be transmitted by direct contact with a pathogen; by indirect contact (e.g., handling a contaminated object); by inhaling the pathogen; or by transmission through a vector that carries pathogens from infected people or contaminated material to new hosts.

6. An epidemic disease outbreak is one that affects more people than experience would have predicted. A pandemic occurs when epidemics break out more or less simultaneously in various places around the world. A pathogen's virulence is its relative ability to cause serious disease. HIV and rabies are examples of highly virulent pathogens.

7. AIDS is a constellation of diseases caused by infection with the human immunodeficiency virus (HIV). HIV destroys T cells, crippling the immune system. HIV is transmitted when blood, semen, or another contaminated body fluid of an infected person enters the body. Sexual contact and IV drug abuse are the most common modes of HIV transmission.

8. Viral STDs also include genital herpes, genital warts (HPV), and type B viral hepatitis. Bacteria cause chlamydial infections, gonorrhea, syphilis, and chancroid (Table 16.4).Untreated STDs can have serious complications. Also, sexually transmitted infections may be interconnected. For example, genital herpes and syphilis increase the risk of HIV infection during sexual encounters.

9. Only people who abstain from sexual contact or who are in an infection-free monogamous relationship can be sure of not being exposed to an STD. Ethical behavior with respect to STDs requires that an infected person inform his or her partner(s), obtain proper medical treatment, and avoid activity that might infect others.

Table 16.3 Characteristics of Bacterial Cells

1. All bacterial cells are prokaryotic; they do not have a membrane-bound nucleus.

2. Bacterial cells in general have a single chromosome (a circular DNA molecule); many species also have plasmids.

3. Most bacteria have a cell wall of peptidoglycan.

4. Most bacteria reproduce by prokaryotic fission.

Table 16.4 Some Pathogens That Cause Sexually Transmitted Diseases

Viruses	STD
DNA VIRUSES	
Herpesviruses	
Herpes simplex types 1 and 2	Cold sores on genitals, lips, eyes, buttocks
Hepadnaviruses	Hepatitis B
Papovaviruses	
Human papillomavirus	Genital warts
RNA VIRUSES	
Retroviruses	
HIV	AIDS
Bacteria	
Chlamydia trachomatis	Chlamydial infections
Neisseria gonorrhoeae	Gonorrhea
Treponema pallidum	Syphilis

Review Questions

1. What is a virus? What is a retrovirus? What is the difference between a virus and a prion? *16.1*

2. Define the difference between a virus and a bacterium, and briefly describe how bacteria reproduce. *16.2*

3. What is an antibiotic? List the factors that have contributed to the emergence of antibiotic-resistant microbes. *16.2*

4. How does a parasitic protozoan such as *Giardia lamblia* differ from a bacterium? *16.3*

5. How do disease pathogens spread? What do we mean by virulence? *16.5*

6. How does HIV cause disease? *16.7*

Self-Quiz *(Answers in Appendix V)*

1. A _____ is described as a noncellular infectious agent.
   a. virus
   b. bacterium
   c. prion
   d. protozoan

2. _____ is a eukaryotic cell that causes disease when it is ingested in contaminated food or water.
   a. An adenovirus
   b. *Chlamydia trachomatis*
   c. *Entamoeba histolytica*
   d. A prion

3. Most pathogenic bacteria _____.
   a. cause disease by secreting toxins that damage body cells
   b. can enter the body in contaminated water or other substances
   c. may be able to multiply so rapidly they overwhelm the immune system
   d. a and b only
   e. a, b, and c are all correct

4. Antibiotic resistance is a result of _____.
   a. use of antibiotics against viruses
   b. patients' failure to take a full course of the drug
   c. self-prescribing of the drugs
   d. b and c only
   e. a, b, and c

5. An infectious disease may *not* be transmitted by:
   a. a gene mutation
   b. simply shaking hands with an infected person
   c. being bitten by a flea
   d. surgical instruments in a hospital

6. Most HIV infections are transmitted through _____. In the vast majority of cases, the virus enters the body of a new host in infected _____, _____, _____, or _____.

7. Match the following STD concepts:
   ____ HPV
   ____ syphilis
   ____ HIV
   ____ genital herpes
   ____ PID
   ____ gonorrhea

   a. diminished immune response
   b. produces a sore called a chancre
   c. may be caused by either of two types of the responsible virus
   d. e. causes genital warts
   f. early stages may not be noticed
   g. a serious complication of chlamydial infection, other STDs

Critical Thinking: You Decide *(Key in Appendix VI)*

1. A janitor in the cafe where you work has been diagnosed as HIV-positive, and a server has come down with type B hepatitis. Some other employees start a petition demanding that both people be required to wear a mask over the mouth and nose, not handle soiled dishes or food, and not use the employee restroom. Both infected employees strenuously object to the plan. You are asked to lead a discussion aimed at resolving the issue, and you decide to prepare a handout giving the scientific basis for making a decision in each case. What does the handout say?

2. You've been dating someone for a while, and the relationship is getting serious. Everything seems fine, except that your potential partner refuses to consider using a condom if the two of you engage in sex. To allay your fears, the person shows you a lab report indicating a negative HIV blood test. You refuse to forgo condom use and decide to end the relationship. Why?

3. A woman with a history of STDs and chronic menstrual problems is having difficulty getting pregnant. Based on your reading in this chapter, which might the source of her apparent infertility be?

4. The micrograph in Figure 16.21 shows bacteria on the head of a pin. What general category of bacteria is shown? Thinking back to Chapter 8, what might occur in your body if you prick yourself with the pin and these bacteria are pathogenic?

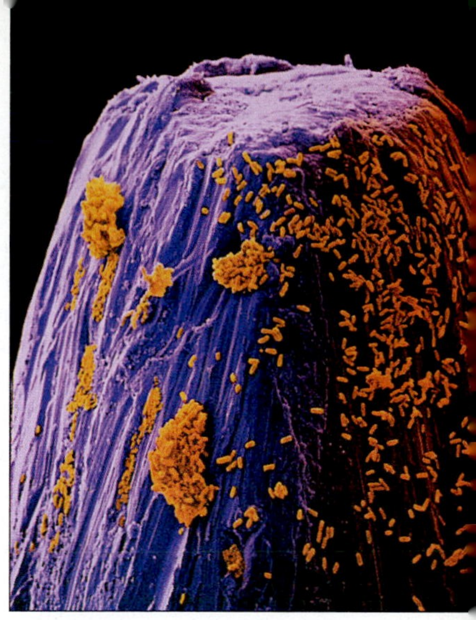

Figure 16.21 Bacteria on the tip of a pin.

5. The mild pneumonia sometimes called "walking pneumonia" is caused by one species of an unusual group of bacteria called mycoplasmas. *Mycoplasma pneumoniae* is usually transmitted by respiratory droplets, and causes pneumonia in about 10 percent of the people it infects. Antibiotics kill the pathogen. How would you classify (1) its mode of transmission, and (2) its virulence?

Selected Key Terms

antibiotic *16.2*
disease vector *16.5*
epidemic *16.5*
human immunodeficiency
   virus (HIV) *16.7*
latency *16.1*
nosocomial infection *16.5*

pandemic *16.5*
parasite *16.3*
prion *16.1*
sexually transmitted disease
   (STD) *I*
virulence *16.5*
virus *16.1*

Readings

Essex, M. October 1999. "The New AIDS Epidemic." *Harvard Magazine*. An eye-opening discussion of the spread of HIV through heterosexual sex.

Levy, S. March 1998. "The Challenge of Antibiotic Resistance." *Scientific American*.

2 July 1999. "The Scientific Challenge of Hepatitis C." A not-too-technical, informative story of how researchers are tackling an emerging infectious disease that affects some 170 million people and is causing the rate of liver disease to soar. From the editors of *Science* magazine.

Additional, current information on STDs, including HIV infections and AIDS, may be obtained online from the Centers for Disease Control and Prevention.

Sperm on the surface of an egg. Only one sperm can successfully fertilize an egg. Their union into a single cell brings together genetic instructions from mother and father, a DNA blueprint that will guide the development and functioning of the new individual.

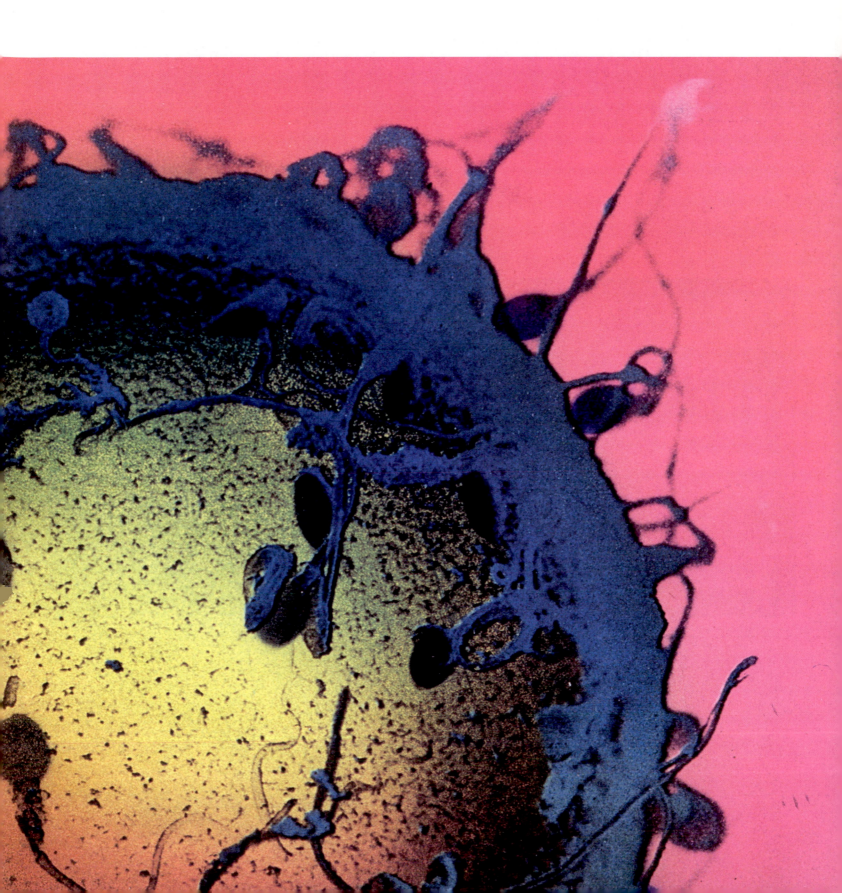

17

CELL REPRODUCTION

Trillions from One

Starting with the fertilized egg in your mother's body, a single cell divided in two, then the two into four. Cells continued dividing until billions were growing and developing in specialized ways, and dividing at different times to produce the genetically prescribed parts of your body (Figure 17.1). Today your body is composed of trillions of cells. Cells still divide inside it. Every five days, for example, the division of cells replaces the entire lining of your small intestine.

Depending on the end result of cell division, its early phase proceeds in one of two ways. One mechanism, called mitosis, occurs in all dividing cells except those that give rise to gametes—sperm and eggs. The other mechanism, meiosis, occurs only in the reproductive organs, in gamete-producing stem cells. Mitosis and meiosis clearly resemble each other. But as we'll see and as Table 17.1 suggests, they are fundamentally different in the way they parcel out the bundles

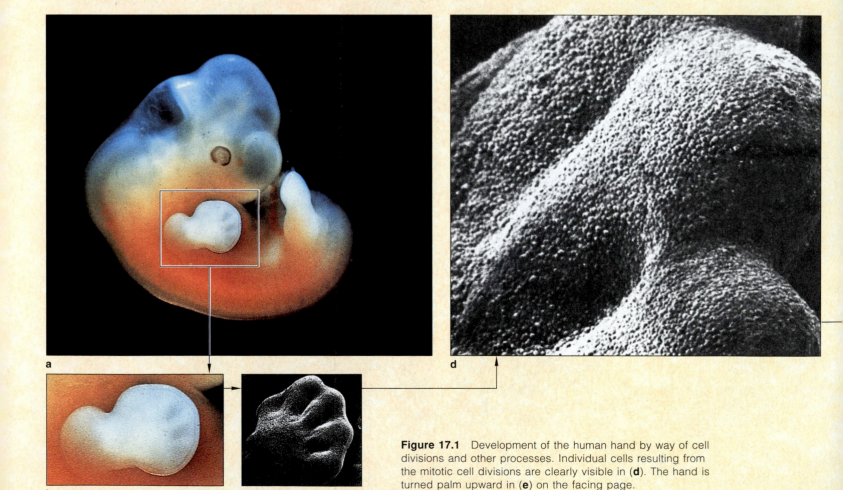

a

b

c

d

Figure 17.1 Development of the human hand by way of cell divisions and other processes. Individual cells resulting from the mitotic cell divisions are clearly visible in (**d**). The hand is turned palm upward in (**e**) on the facing page.

of DNA and protein we call chromosomes. Thanks to mitosis, the body grows larger, wounds heal, and many tissues are renewed. It is meiosis, followed by the union of sperm and egg at fertilization, that makes sexual reproduction possible.

Microscopic structures called chromosomes carry the instructions—genes—that determine which traits a person inherits. How are chromosomes and their genes distributed into daughter cells? In this chapter and the next, we consider the answers (and best guesses) to questions about cell reproduction and other aspects of human heredity.

Table 17.1 Cell Division Mechanisms	
Mechanisms	Functions
Mitosis, cytoplasmic division	Body growth and tissue repair
Meiosis, cytoplasmic division	Gamete formation and sexual reproduction

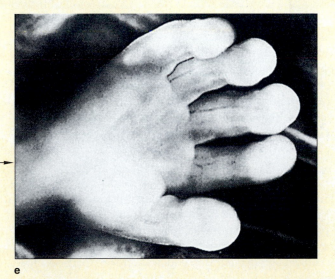

e

DIVIDING CELLS: THE BRIDGE BETWEEN GENERATIONS

Overview of Division Mechanisms

In biology, **reproduction** means producing a new generation of individuals *or* a new generation of cells. Reproduction is part of a **life cycle**, a recurring series of events in which individuals grow, develop, maintain themselves, and reproduce. The instructions that guide the human life cycle are encoded in our DNA, which each of us inherits from our parents. Reproduction typically begins with the division of single cells. It follows this basic rule: Each cell of a new generation must receive a copy of the parent cell's DNA and enough cytoplasmic machinery to start up its own operation.

DNA contains the genetic instructions for making proteins. Some proteins are structural materials. Many serve as enzymes during the synthesis of carbohydrates, lipids, and other building blocks of the cell or in crucial reactions such as glycolysis and the Krebs cycle. Unless new cells receive the necessary DNA instructions, they will not grow or function properly.

Of course, the cytoplasm of the parent cell already has operating machinery—enzymes, organelles, and so forth. Therefore, when a daughter cell inherits what might look like a blob of cytoplasm, it really is getting "start-up" machinery for its operation until it has time to use its inherited DNA for growing and developing on its own.

The cells of humans (and other eukaryotic organisms) divide their DNA by **mitosis** or **meiosis**. Both these mechanisms sort out and package DNA molecules into new nuclei for forthcoming daughter cells. In other words, mitosis and meiosis do this by dividing up the parent cell's chromosomes. A different mechanism divides the cytoplasm, splitting a parent cell into daughter cells.

In contrast to mitosis, meiosis occurs only in **germ cells** set aside for sexual reproduction—the oogonia in ovaries and spermatogonia in testes (Chapter 14). In these cells, recall, meiosis must take place before gametes (sperm and eggs) form. In order to understand the basic elements of mitosis and meiosis, you must first know more about chromosomes—the structures in the cell nucleus that carry hereditary information.

Some Key Points about Chromosomes

You may recall from Chapter 3 that the DNA in your cells comes packaged in chromosomes. A **chromosome** is one very long DNA molecule combined with a roughly equal amount (by weight) of protein. There are forty-six of these chromosomes in the nucleus of a human body cell, and their physical organization differs at different times in a cell's life. When a cell is dividing, each chromosome is tightly coiled, or *condensed*, into a rodlike shape. Between divisions, the chromosomes are not as tightly coiled and are dispersed throughout the nucleus. The

DNA and protein together in a chromosome are called *chromatin*. Each chromosome is still uncondensed and dispersed when it duplicates prior to cell division, and the two duplicate threads of the chromosome remain attached as **sister chromatids**:

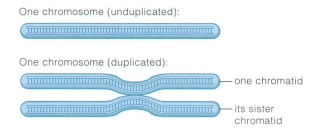

One chromosome (unduplicated):

One chromosome (duplicated):

one chromatid

its sister chromatid

Notice how the two chromatids of a duplicated chromosome are constricted at one small region. This region, the **centromere**, has sites where microtubules will attach and help move the chromosome when the nucleus divides:

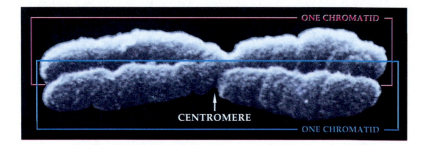

As Figure 17.2 suggests, the centromere's location may vary, depending on the chromosome. Also, the two parallel strands that make up a DNA molecule are twisted together repeatedly like a spiral staircase (a "double helix") and are then condensed further together with proteins. They are also much longer than shown here. You will read more about DNA structure in Chapter 20.

The Chromosome Number

Human cells other than sperm or eggs normally each have two full sets of the chromosomes characteristic of our species—one set from each parent. Your parents each contributed a full set of twenty-three chromosomes to the fertilized egg that became you, so there are forty-six chromosomes in each of your **somatic cells** (somatic means "of the body"). Any cell having two of each type of chromosome is said to be a **diploid** cell. Diploid cells, with two of each type of chromosome, are referred to as $2n$. The n stands for the number of chromosomes in one complete set. The term **chromosome number** refers to the number of each type of chromosome normally present in a cell.

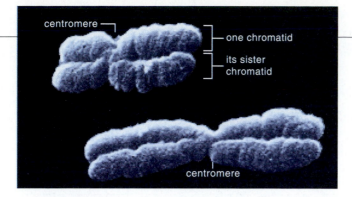

Figure 17.2 Photomicrograph of two chromosomes, each in the duplicated state. The organization of these chromosomes into tightly coiled, rodlike structures indicates that they are from a dividing cell.

Figure 17.3 Forty-six chromosomes from a human male. Each chromosome is in the duplicated state. There are pairs of chromosomes (two of each type), which tells you that they came from a diploid cell. One member of each pair contains genetic instructions inherited from the father. The other member contains instructions from the mother.

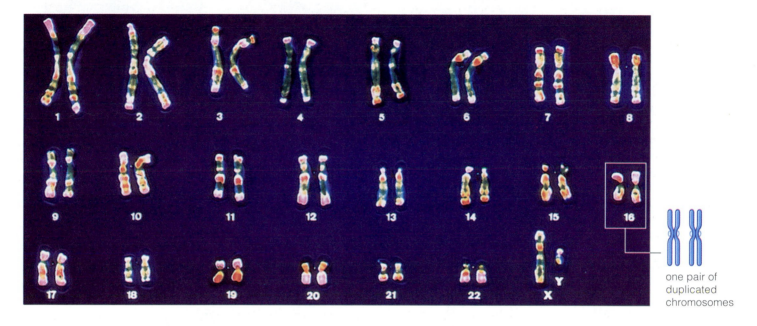

one pair of duplicated chromosomes

With mitosis, a diploid parent cell produces two diploid daughter cells. One member of each chromosome pair is maternal in origin, and the other is paternal in origin. Shortly before a diploid cell divides, the genetic material is duplicated, so that the diploid number of chromosomes doubles (yielding the sister chromatids just described). Mitosis allots half of this doubled genetic material to each new cell, so each daughter cell receives the full diploid number of chromosomes.

Figure 17.3 shows all the chromosomes in a human somatic cell. The chromosomes have been lined up as twenty-three pairs, an arrangement called a *karyotype* (the Greek root *karyo-* refers to a somatic cell nucleus). Except for the sex chromosomes, X and Y, the two members of each chromosome pair are the same length and shape; both carry hereditary instructions for the same traits. Such corresponding chromosomes, one from each parent, are called **homologous chromosomes**, or simply *homologues* (from a Greek word meaning "to agree"). In general, homologues are the same length, have the same shape,

and carry genes for the same traits. The X and Y sex chromosomes are exceptions; although they are considered homologues, they are of different size and form and for the most part they carry different genes. For example, genes that code for certain products, such as clotting factors in the blood, are located on the X chromosome). Only a few traits are influenced by Y-linked genes. Chapter 19 describes such traits more fully.

Mitosis and meiosis sort chromosomes, and therefore DNA, into new nuclei for daughter cells.

In mitosis, the chromosome number remains constant, division after division, from one cell generation to the next. So, if a parent cell is diploid, its two daughter cells will be diploid also. Mitosis takes place in somatic cells. The body grows and tissues are repaired by way of mitosis.

Meiosis takes place only in germ cells set aside for sexual reproduction. It is the mechanism that produces gametes.

THE CELL CYCLE

Mitosis is only one phase of the **cell cycle**. The cycle begins every time a new cell is produced, and it ends when the cell completes its own division. The cycle starts again for each new daughter cell, as sketched in Figure 17.4. Usually, the longest phase of the cell cycle is **interphase**. During this three-part phase, a cell increases its mass, roughly doubles the number of components in its cytoplasm, then duplicates its DNA. Section 17.7 will give a more complete picture of interphase. The parts of the cell cycle are often abbreviated this way:

G1 Of interphase, a "Gap" (interval) before the onset of DNA synthesis

S   Of interphase, the time when DNA is synthesized

G2 Of interphase, a second "Gap" between the completion of DNA replication and the onset of mitosis

M   *Mitosis; partitioning of chromosomes, commonly followed by cytoplasmic division*

The cell cycle lasts about the same length of time for cells of the same type. Its duration varies among cells of different types. For instance, the cycle lasts eighteen hours in bone marrow cells and twenty-five hours in epithelial cells in your stomach lining. New red blood cells form and replace your worn-out ones at an average rate of 2 to 3 million each second.

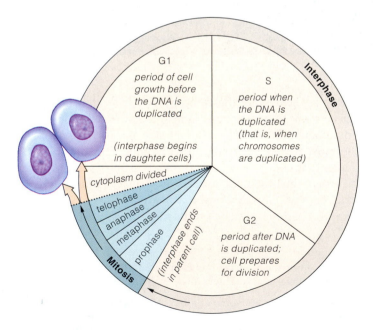

Figure 17.4 Generalized cell cycle. The duration of each phase differs among different cell types.

A CELL AT INTERPHASE:

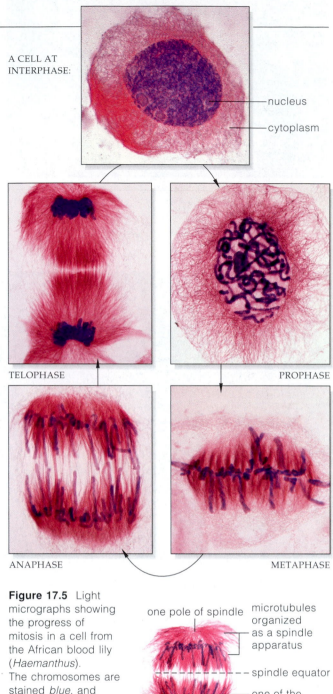

nucleus

cytoplasm

TELOPHASE

PROPHASE

ANAPHASE

METAPHASE

Figure 17.5 Light micrographs showing the progress of mitosis in a cell from the African blood lily (*Haemanthus*). The chromosomes are stained *blue*, and microtubules that are moving them about are stained *red*.

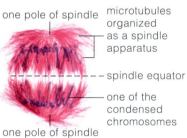

one pole of spindle

microtubules organized as a spindle apparatus

spindle equator

one of the condensed chromosomes

one pole of spindle

Figure 17.5 provides a glimpse of the changes in a cell as it leaves interphase and enters mitosis. The four stages of mitosis (prophase, metaphase, anaphase, and telophase) are the subject of Section 17.4.

A cell cycle begins at interphase, when a new cell (formed by mitosis and cytoplasmic division) increases its mass and the number of its cytoplasmic components, then duplicates its chromosomes. The cycle ends when the cell divides.

17.3 IMMORTAL CELLS—THE GIFT OF HENRIETTA LACKS

Each human starts out as a single fertilized egg. At birth a human body has about a trillion cells. Even in an adult, many cells are still dividing. Cells in your stomach lining divide every day. Liver cells usually don't divide—but if part of the liver becomes injured or diseased, some cells will divide repeatedly and produce new cells until the damaged part is replaced. Until recently, biologists believed that the cell cycle of each neuron stopped forever, once the neuron was fully differentiated for its specialized role in the body. We now know, however, that under the right circumstances neurons *can* divide again. Research aimed at finding better treatments for spinal cord injuries (Section 13.9) may benefit from this discovery.

In 1951, George and Margaret Gey of Johns Hopkins University were trying to develop a way to keep human cells dividing outside the body. Why? With such cells, researchers could study basic life processes, such as how processes in the cell nucleus translate the genetic message of DNA into proteins. They also could study cancer and other diseases, without having to experiment directly on humans. Local physicians had provided the Geys with both normal and diseased human cells from patients. But every time the Geys started a cell culture in their laboratory, it died out within a few weeks.

Mary Kubicek, one of their assistants, was about to give up after dozens of failed attempts. Still, one day in 1951, she dutifully prepared another sample of cancer cells for culture. Kubicek gave the sample the code name **HeLa cells**, for the first two letters of the patient's first and last names: Henrietta Lacks (Figure 17.6*a*).

The cells, which had been removed from a cancer tumor in the patient's uterus, were placed in a culture dish and provided with basic nutrients. They soon began to divide. And divide. And divide again. By the fourth day there were so many cells that they had to be subdivided into more culture dishes (Figure 17.6*b*).

As months passed, the HeLa culture continued to thrive. Unfortunately, the cancer cells inside the patient's body were just as vigorous. Six months after Henrietta Lacks was first diagnosed as having uterine cancer, tumor cells had spread through her body. Two months later, Henrietta Lacks, a young woman from Baltimore, was dead.

Although Henrietta was gone, some of her cells lived on in the Geys' laboratory as the first successful human cell culture. As word of the culture spread in the scientific community, HeLa cells were soon being shipped to other researchers, who passed cells on to others. With time, HeLa cells came to live in laboratories all over the world. They still do. Some even traveled into space aboard the *Discoverer XVII* satellite. Every year, research reported in hundreds of scientific papers is based on laboratory experiments with HeLa cells.

Today, sophisticated laboratory techniques, and improved understanding of cell growth factors and their effects, all contribute to the ability of scientists to probe the workings of cells using HeLa cultures.

Henrietta Lacks was only 31 years old when runaway cell divisions killed her. Now, many decades later, her legacy is still benefiting humans everywhere, in cells that are still alive and dividing, day after day after day.

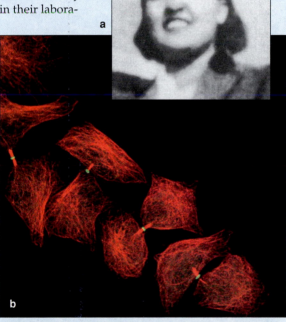

Figure 17.6 (**a**) Henrietta Lacks, a casualty of cancer whose contribution to science is still helping others. (**b**) Dividing HeLa cells.

A TOUR OF THE STAGES OF MITOSIS

When a cell makes the transition from interphase to mitosis, it has stopped making new cell parts and the DNA has been replicated. Inside that cell, profound changes now take place smoothly, one after the other, through four stages. The sequential stages of mitosis are **prophase**, **metaphase**, **anaphase**, and **telophase**.

Figure 17.7 shows mitosis in an animal cell. By comparing its series of photographs with those of Figure 17.4, you see the chromosomes are changing positions. They don't move independently, however. A **spindle apparatus** moves them. A spindle consists of microtubules arranged in two sets. Each set extends from one of the two poles (end points) of the spindle. The two sets overlap each other a little at the spindle equator, midway between the poles. Formation of this bipolar, microtubular spindle will establish the final destinations of chromosomes during mitosis, as you will see shortly.

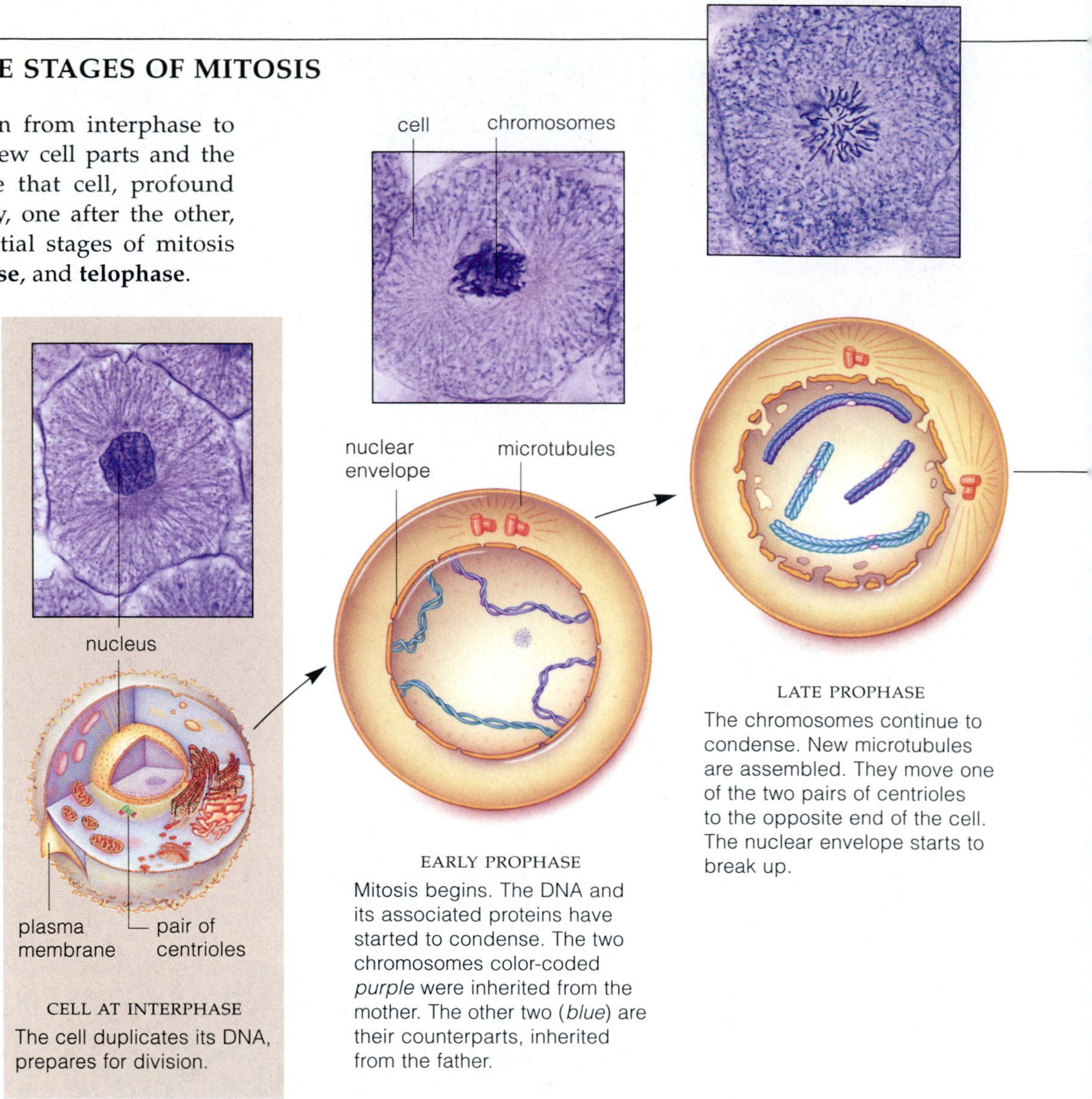

Prophase: Mitosis Begins

We know a cell is in prophase when chromosomes become visible in the light microscope as threadlike forms. ("Mitosis" comes from the Greek *mitos*, for thread.) Each chromosome was duplicated earlier, during interphase, so each is now two sister chromatids joined at the centromere. Early in prophase, the sister chromatids of every chromosome twist and fold into a more compact form. By late prophase, all of the chromosomes will be condensed into thicker, rod-shaped forms.

Meanwhile, in the cytoplasm, most microtubules of the cytoskeleton are breaking apart into their tubulin subunits (Section 3.3). Near the nucleus, the subunits reassemble as *new* microtubules of the spindle. Many of the new microtubules will extend from one spindle pole or the other to the centromere of a chromosome. The remainder will not interact at all with the chromosomes. They will extend from the poles and overlap each other.

While new microtubules are assembling, the nuclear envelope physically prevents them from interacting with the chromosomes inside the nucleus. However, as

CELL AT INTERPHASE
The cell duplicates its DNA, prepares for division.

nucleus

plasma membrane — pair of centrioles

cell   chromosomes

nuclear envelope   microtubules

EARLY PROPHASE
Mitosis begins. The DNA and its associated proteins have started to condense. The two chromosomes color-coded *purple* were inherited from the mother. The other two (*blue*) are their counterparts, inherited from the father.

LATE PROPHASE
The chromosomes continue to condense. New microtubules are assembled. They move one of the two pairs of centrioles to the opposite end of the cell. The nuclear envelope starts to break up.

Figure 17.7 Mitosis. This mechanism assures that daughter cells will have the same chromosome number as the parent cell. For clarity, the diagram shows only two pairs of chromosomes from a diploid (2n) animal cell. Most cells have more than two pairs of chromosomes, as you can see in the micrographs of mitosis in a whitefish cell.

prophase draws to a close, the nuclear envelope starts to break up.

Many cells have two barrel-shaped centrioles. Each centriole was duplicated during interphase, so there are two pairs of them when prophase is under way. Microtubules start moving one pair to the opposite pole of the newly forming spindle. Centrioles, recall, also give rise to flagella or cilia. If you observe them in the cells of an organism, you can bet that flagellated cells (such as sperm) or ciliated cells develop during some part of its life cycle.

Transition to Metaphase

So much happens between prophase and metaphase that researchers give this transitional period its own

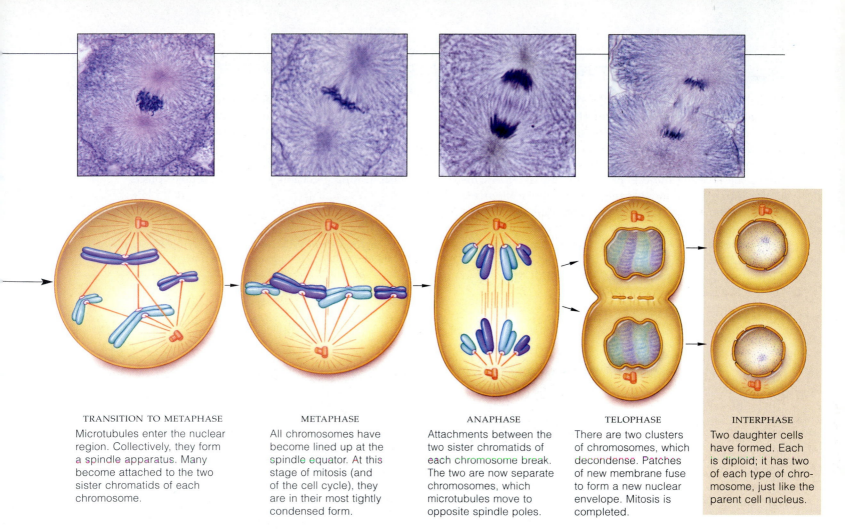

TRANSITION TO METAPHASE
Microtubules enter the nuclear region. Collectively, they form a spindle apparatus. Many become attached to the two sister chromatids of each chromosome.

METAPHASE
All chromosomes have become lined up at the spindle equator. At this stage of mitosis (and of the cell cycle), they are in their most tightly condensed form.

ANAPHASE
Attachments between the two sister chromatids of each chromosome break. The two are now separate chromosomes, which microtubules move to opposite spindle poles.

TELOPHASE
There are two clusters of chromosomes, which decondense. Patches of new membrane fuse to form a new nuclear envelope. Mitosis is completed.

INTERPHASE
Two daughter cells have formed. Each is diploid; it has two of each type of chromosome, just like the parent cell nucleus.

name, "prometaphase." The nuclear envelope breaks up completely into numerous tiny, flattened vesicles. Now the chromosomes are free to interact with microtubules that are extending toward them, from the poles of the forming spindle. Microtubules from both poles harness each chromosome and start pulling on it. The two-way pulling orients the chromosome's two sister chromatids toward opposite poles. Meanwhile, overlapping spindle microtubules ratchet past each other and push the poles of the spindle apart. The push–pull forces are balanced when the chromosomes reach the spindle's midpoint.

When all the duplicated chromosomes are aligned midway between the poles of a completed spindle, we call this metaphase (*meta-* means midway between). The alignment is crucial for the next stage of mitosis.

From Anaphase through Telophase

At anaphase, sister chromatids of each chromosome separate from each other and move to opposite poles by two mechanisms. First, the microtubules attached to the centromere regions shorten and *pull* the chromosomes to the poles. Second, the spindle elongates as overlapping microtubules continue to ratchet past each other and *push* the two spindle poles even farther apart. Once each

chromatid is separated from its sister, it has a new name. It is a separate chromosome in its own right.

Telophase gets under way as soon as each of two clusters of chromosomes arrive at a spindle pole. The chromosomes, no longer harnessed to microtubules, return to threadlike form. Vesicles derived from the old nuclear envelope fuse and form patches of membrane around the chromosomes. Patch joins with patch, and soon a new nuclear envelope separates each cluster of chromosomes from the cytoplasm. If the parent cell was diploid, each cluster contains two chromosomes of each type. With mitosis, remember, each new nucleus has the same chromosome number as the parent nucleus. Once two nuclei form, telophase is over—and so is mitosis.

Prior to mitosis, each chromosome in a cell's nucleus is duplicated, so that it consists of two sister chromatids.

Mitosis proceeds through four consecutive stages, called prophase, metaphase, anaphase, and telophase.

A microtubular spindle moves sister chromatids of every chromosome apart, to opposite spindle poles. Around each of two clusters of chromosomes, a new nuclear envelope forms. Both daughter nuclei formed this way have the same chromosome number as the parent cell's nucleus.

17.5 DIVISION OF THE CYTOPLASM

Division of the cytoplasm, called **cytokinesis**, usually coincides with the period from late anaphase through telophase. For most animal cells, material accumulates and forms a ringlike layer around microtubules at the dividing cell's midsection. A shallow depression appears above the layer, at the cell surface (Figure 17.8). At this depression, called a **cleavage furrow**, microfilaments made of the contractile protein actin pull the plasma membrane inward and cut the cell in two. Organelles in the divided cytoplasm are distributed to daughter cells.

This concludes our picture of mitotic cell division. Look now at your hands—and think of all the cells in your palms, thumbs, and fingers. Imagine all of the divisions of all the cells that preceded them when you were developing early on, inside your mother. Figure 17.1 at the beginning of this chapter gave an inkling of this unfolding process. It is difficult not to be in awe of the astonishing precision with which it takes place.

Mitosis is a small part of the cell cycle. As it draws to a close, the mechanism of cytokinesis cuts the cytoplasm into two daughter cells, each with a daughter nucleus.

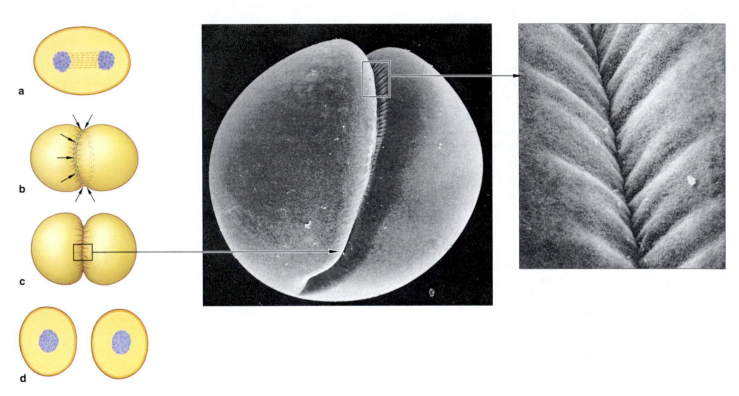

Figure 17.8 Cytokinesis in an animal cell. (**a**) Mitosis is complete and the spindle is disassembling. (**b**) Just beneath the plasma membrane, microfilament rings at the former spindle equator contract, like a purse string closing. (**c**,**d**) Continuing contractions divide the cell in two. The micrographs show how the plasma membrane sinks inward, defining the plane of cleavage.

Further reading: Student Guide to InfoTrac on web site

17.6

CONCERNS AND CONTROVERSIES OVER IRRADIATION

What do a routine dental X ray and an irradiated side of beef have in common? Both are examples of ways we humans have harnessed ionizing radiation for our own purposes, from health care to killing pathogens on food. Like some other technologies, however, this one can be a double-edged sword and even fuel serious controversy.

IRRADIATION EFFECTS ON THE BODY   Ionizing radiation includes various potentially harmful types of electromagnetic energy. Electromagnetic energy can take many forms—for instance, radio waves, visible light, microwaves, cosmic rays from outer space, and radioactive radon gas in rocks and soil. Forms that can harm living cells, including radon and X rays, are those that have enough energy to remove electrons from atoms and change them to positively charged ions (Section 2.4).

When ionizing radiation enters an organism, it may break apart chromosomes, alter genes, or both. The result can be disastrous. For example, if the chromosomes in an affected cell have been broken into fragments, the spindle apparatus will not be able to harness and move the fragments when the cell divides. The cell or its daughters may then die. When ionizing radiation damage occurs in germ cells, the resulting gametes can carry damaged DNA. Therefore, an infant who inherits the DNA may have a genetic defect. If "only" somatic cells are affected, only the person exposed to the radiation will suffer damage, but the health effects can still be severe.

When a person receives a sudden, large dose of ionizing radiation, it typically destroys cells of the immune system, epithelial cells of the skin and intestinal lining, and red blood cells, among other cell types. The results are raging infections, intestinal hemorrhages, anemia, and wounds that do not heal.

Small doses of ionizing radiation over a long period of time apparently cause less damage than the same total dosage given all at once. This may be due in part to the body's ability to repair damaged DNA. That said, ionizing radiation is associated with miscarriages, eye cataracts, and various cancers (Chapter 22).

Medical X rays and diagnostic technologies such as magnetic resonance imaging (MRI) and PET scanning (Section 2.2) are valuable uses of ionizing radiation in health care. So is radiation therapy used in treating some cancers. However, there is general agreement that the best policy is to avoid unnecessary exposure, using X rays only when they are essential for health or diagnostic purposes.

IRRADIATED FOOD   Foods ranging from grains and potatoes to fruits, spices, beef, pork, and other meats may be irradiated. Why? Just as living body cells can be damaged or killed by the radiation, so can potentially harmful bacteria, fungi, and other microorganisms. Some root crops, such as potatoes, may be irradiated to prevent them from sprouting, thereby prolonging the "shelf life" of a valuable food commodity. Irradiated food sold in the United States carries an identifying logo (Figure 17.9).

Irradiated food *is not radioactive*, and some people are quite comfortable eating it because there is no scientific evidence that it presents a health hazard. In addition, irradiation lowers food costs because it limits spoilage, and proponents argue that it may reduce the incidence of food-borne illnesses. On the other hand, opponents worry that irradiation might promote the development of radiation-resistant microbes. Some also are concerned that irradiation may chemically change food in ways that could harm consumers. For the time being, however, there is no scientific evidence to support that fear.

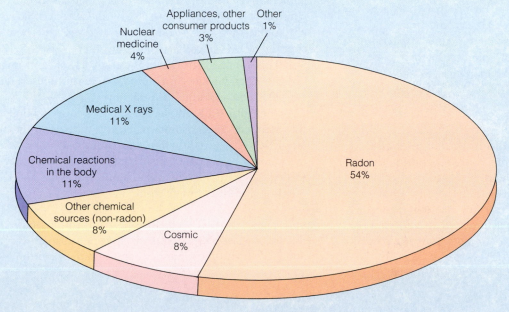

Appliances, other consumer products 3%

Other 1%

Nuclear medicine 4%

Medical X rays 11%

Chemical reactions in the body 11%

Other chemical sources (non-radon) 8%

Cosmic 8%

Radon 54%

Figure 17.9 Sources of radiation exposure for the population of the United States. (*Above*) Logo required to be placed on irradiated food products in the United States.

A CLOSER LOOK AT THE CELL CYCLE

The Wonder of Interphase

If you could coax the DNA molecules from the nucleus of just one of your somatic cells to stretch in a single line, one after another, that line would extend past the finger tips of your outstretched arms. The wonder is, enzymes and other proteins in the cell selectively scan all of the DNA, switch protein-building instructions on and off, and even make precise copies of each DNA molecule—all during interphase.

Interphase has three parts, G1, S, and G2 (see Figure 17.4 in Section 17.2). In each part of interphase, the cell makes specific biological molecules. Thus, during G1, most of the carbohydrates, lipids, and proteins for a cell's own use and for export are assembled. During the S period, the cell copies its DNA and manufactures the proteins that will be put together into structural scaffolding for the chromosomes. Finally, during G2, the cell produces proteins that will drive mitosis to its completion.

Once S begins and the copying of a cell's DNA gets under way, events normally continue all the way through mitosis. The pace at which this sequence takes place is about the same in all your dividing somatic cells. Gene-governed control mechanisms set the pace, and they determine whether or not a cell will divide. An analogy is the way a driver can use a car's hand brake to stop it from rolling downhill. If genetic commands "apply the brake" that can operate in G1, the cell cycle will stop there. If another genetic order releases the brake, the cycle usually continues on through mitosis and cytokinesis.

Our understanding of genetic controls over cell division is growing rapidly, in part due to research aimed at finding ways to stimulate differentiated cells—which under normal circumstances might divide rarely or not at all—to enter mitosis. For instance, as noted in previous chapters, efforts are under way to learn if growth factors can stimulate cell division in neurons of the brain and spinal cord. This kind of research might one day contribute to effective treatments for spinal cord injuries or ills such as Parkinson's and Alzheimer's disease.

Chromosomes and Spindles

Mitosis parcels out DNA for forthcoming daughter cells with impressive precision. That precision depends on the physical organization of DNA and proteins in a cell's chromosomes. It also depends on interactions among microtubules (Section 3.3) and certain proteins called *motor proteins*. Motor proteins are attached to and stick out from microtubules and microfilaments that take part in cell movements.

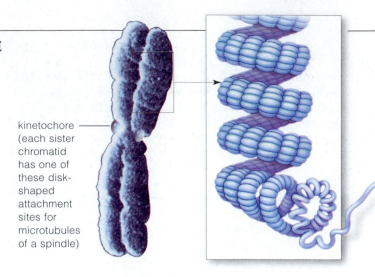

kinetochore (each sister chromatid has one of these disk-shaped attachment sites for microtubules of a spindle)

a A duplicated human chromosome at metaphase, when it is most condensed. Interactions among some chromosomal proteins may keep loops of DNA tightly packed in a "supercoiled" array.

Figure 17.10 One model of the levels of organization in a human chromosome at metaphase.

ORGANIZATION OF METAPHASE CHROMOSOMES DNA has many proteins bound tightly to it. Some of these proteins belong to a group called **histones**. Many histones are like spools for winding up small stretches of DNA. Each histone-DNA "spool" is a single structural unit called a **nucleosome**. Other histones stabilize the spools. During mitosis (and meiosis also), interactions between histones and DNA make the chromosome coil back on itself over and over again. The coiling tremendously increases the chromosome's diameter. Other proteins help form a structural scaffold when the DNA folds even more—for example, into a series of loops, as shown in Figure 17.10.

The characteristic size and shape of each chromosome come about late in prophase, when condensation of the chromosome is nearly complete. By then, each of its sister chromatids has at least one constricted region. The most prominent one, recall, is the centromere (Section 17.1). Small disk-shaped structures at the surface of centromeres are the docking sites for spindle microtubules (Figure 17.10*a*). They are called **kinetochores.**

When threadlike molecules of DNA are condensing into compact chromosome structures, why don't they get tangled up? Actually, they apparently do. However, laboratory experiments have revealed that an enzyme (called DNA topoisomerase) chemically recognizes the tangling and puts things right.

SPINDLES COMING AND GOING All through anaphase, the next-to-last phase of mitosis, the spindle microtubules attached to chromosomes stay in the same position—yet

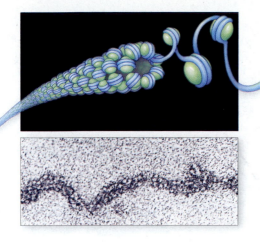

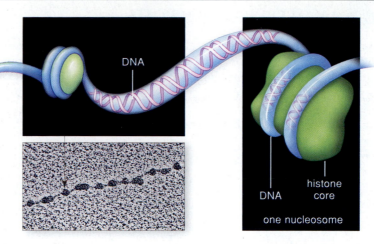

b At a deeper level of structural organization, the chromosomal proteins and the DNA are arranged as a cylindrical fiber.

c Beads-on-a-string organization of a chromosome. The "string" is one DNA molecule. Each "bead" is a nucleosome.

d A nucleosome consists of a double loop of DNA around a core of histones. Other histones stabilize the structural array.

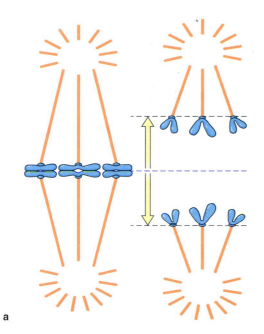

a

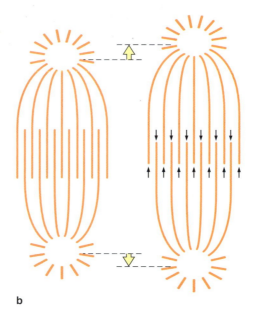

b

Figure 17.11 Models of two of the mechanisms that separate sister chromatids of a chromosome from each other at anaphase. In (**a**), the microtubules attached to the chromatids shorten. This decreases the distance between the kinetochores and spindle poles. In (**b**), overlapping microtubules ratchet past each other. In this way they move the spindle poles apart and increase the distance between sister chromatids of each chromosome.

the distance shrinks between those microtubules and the spindle poles. How? When a kinetochore slides over them, the microtubules shorten by disassembling (Figure 17.11). A kineochore is like a train chugging along a railroad track, except the track falls apart after the "train" has passed. Two motor proteins in it may drive the sliding motion.

Some spindle microtubules extending from both poles aren't attached to kinetochores. They, too, have motor proteins and actively slide past each other where they overlap in a spindle.

Any chemical that interferes with spindle microtubules can effectively derail mitosis. One powerful microtubule "poison" is colchicine, a chemical derived from certain plants. Cancer researchers and other cell biologists use it

in their work because of its power to stop mitosis within seconds or minutes after a cell is exposed to it.

Once the S stage of interphase begins, the cell cycle moves ahead at about the same rate in all cells of a given type, all the way through mitosis. Gene-based controls govern whether a cell enters S—and so control the rate of cell division.

The condensed form of metaphase chromosomes comes about by interactions among DNA and structural proteins, including histones, that associate with DNA throughout the cell cycle.

During mitosis, motor proteins act on microtubules of the spindle to produce the movements of chromosomes.

Meiosis takes place in dividing germ cells in ovaries or testes (Figure 17.12). Remember from Chapter 14 that human germ cells are *spermatogonia* in males and *oogonia* in females. Like other body cells, spermatogonia and oogonia are diploid. However, *unlike* other body cells, they can give rise to haploid gametes (sperm or eggs) by way of meiosis.

A mature human sperm or oocyte typically contains a single set of twenty-three chromosomes, rather than pairs of homologous chromosomes. This is because meiosis is a **reductional division**. It halves the diploid number of chromosomes (2*n*) to a **haploid** number (*n*). And not just any half: *Each haploid gamete ends up with one partner from each pair of homologous parent chromosomes.*

Two Divisions, Not One

Meiosis resembles mitosis in some respects, even though the outcome is different. Before interphase gives way to meiosis, a germ cell duplicates its DNA. At that point, each duplicated chromosome consists of two DNA molecules, one in each sister chromatid. The two DNA molecules are attached to one another, and as long as they stay attached they are sister chromatids:

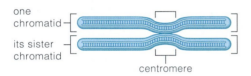

As in mitosis, microtubules of a spindle apparatus move the chromosomes in prescribed directions. In meiosis, however, there are two consecutive divisions of the

chromosomes, which end with the formation of four haploid nuclei. The two divisions are called meiosis I and meiosis II. During meiosis I, each duplicated chromosome lines up with its partner, homologue to homologue; then the partners separate from each other:

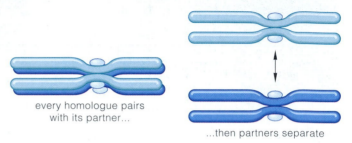

every homologue pairs with its partner...   ...then partners separate

The cytoplasm typically divides after meiosis I is completed. The two daughter cells are haploid, with only one of each type of chromosome. But remember, those chromosomes are still duplicated.

During meiosis II, *the two sister chromatids of each chromosome separate from each other:*

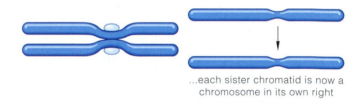

...each sister chromatid is now a chromosome in its own right

After the four nuclei form, the cytoplasm divides once more. The result is four haploid cells.

Meiosis in the Human Life Cycle

As with other multicellular animals, the human life cycle proceeds from meiosis to gamete formation, fertilization, then growth of the new individual by way of mitosis:

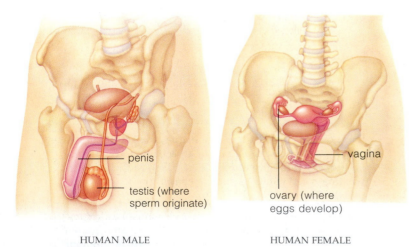

HUMAN MALE   HUMAN FEMALE

Figure 17.12 Structures that produce gametes in the human body. Germ cells in the ovaries and testes are the sources of eggs and sperm, respectively.

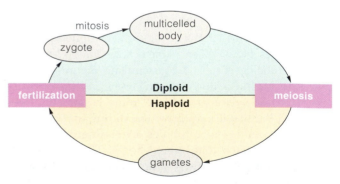

Spermatogenesis is the term for meiosis and gamete formation in males (Figure 17.13). First, a diploid germ cell increases in size. The resulting large, immature cell (a

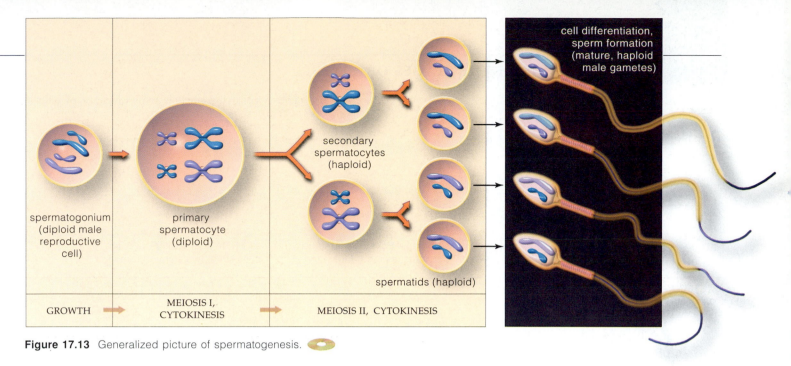

Figure 17.13 Generalized picture of spermatogenesis.

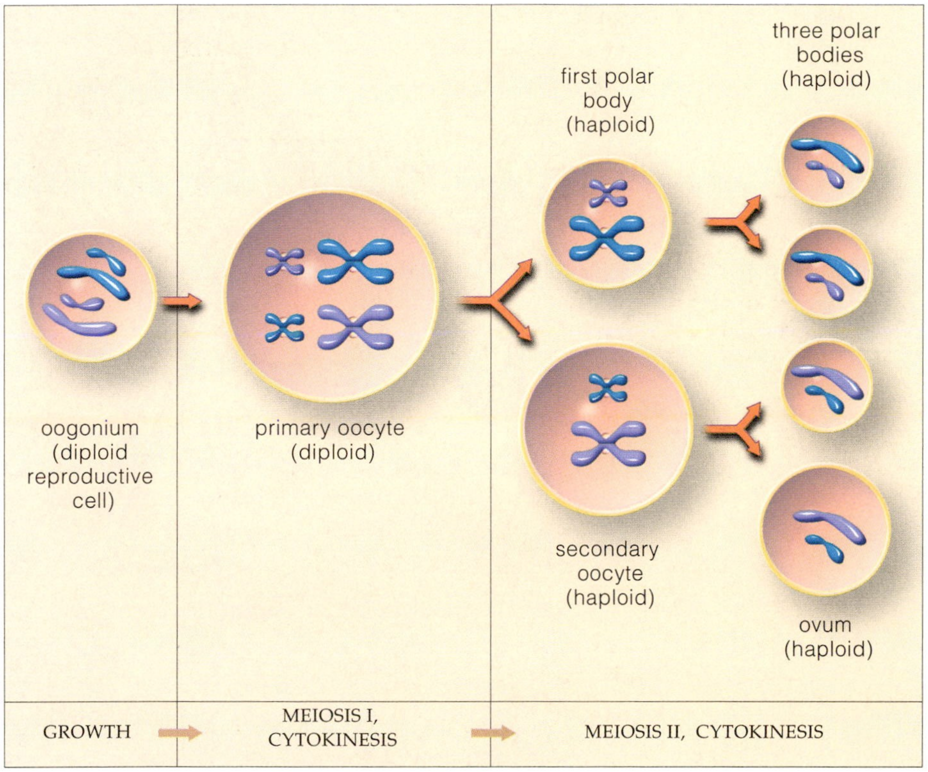

spermatogonium (diploid male reproductive cell)

primary spermatocyte (diploid)

secondary spermatocytes (haploid)

spermatids (haploid)

cell differentiation, sperm formation (mature, haploid male gametes)

GROWTH → MEIOSIS I, CYTOKINESIS → MEIOSIS II, CYTOKINESIS

Figure 17.14 Generalized picture of oogenesis. The diagram is not at the same scale as Figure 17.13. An egg becomes much larger than a sperm cell. Also, the polar bodies are extremely small compared to a mature egg.

oogonium (diploid reproductive cell)

primary oocyte (diploid)

first polar body (haploid)

three polar bodies (haploid)

secondary oocyte (haploid)

ovum (haploid)

GROWTH → MEIOSIS I, CYTOKINESIS → MEIOSIS II, CYTOKINESIS

The early stages of oogenesis unfold in a developing female embryo. Recall from Chapter 14, however, that until a girl reaches puberty, her primary oocytes are arrested in prophase I. Then, each month, meiosis resumes in (usually) one oocyte that is ovulated. This cell, the secondary oocyte, receives nearly all the cytoplasm; the other, much smaller cell is a polar body. Both cells enter meiosis II, but the process is arrested again at metaphase II. If the secondary oocyte is fertilized, meiosis II continues. It results in one large cell and (often) three extremely small polar bodies. The polar bodies are "dumping grounds" for three sets of chromosomes so that the future egg ends up with only the necessary haploid number. The large cell develops into the mature egg (ovum). Its cytoplasm contains components that will help guide development of an embryo.

Gametes produced in oogenesis and spermatogenesis are available for fertilization, the next stage in the human life cycle. As you may remember from Chapter 15, fertilization restores the diploid number of chromosomes in a forthcoming individual.

primary spermatocyte) undergoes meiosis. The resulting four haploid spermatids then change in form, develop tails, and become sperm—mature male gametes.

In females, meiosis and gamete formation are called **oogenesis** (Figure 17.14). As you might expect, oogenesis differs from spermatogenesis in some important ways. For example, compared to a primary spermatocyte, many more cytoplasmic components accumulate in a primary oocyte—the female germ cell that undergoes meiosis. Also, in females the cells formed after meiosis are of different sizes and have different functions.

Meiosis reduces the parental chromosome number by half—to the haploid number.

Meiosis is the first step leading to the formation of gametes, which are required for sexual reproduction. In males, meiosis and gamete formation are called spermatogenesis. In females they are called oogenesis.

A VISUAL TOUR OF THE STAGES OF MEIOSIS

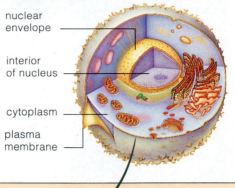

nuclear
envelope

interior
of nucleus

cytoplasm

plasma
membrane

Figure 17.15 Meiosis: the mechanism by which the parental number of chromosomes is reduced by half (to the haploid number) for forthcoming gametes. Only two pairs of homologous chromosomes are shown. Maternal chromosomes are shaded *purple*, and paternal ones *blue*.

A GERM CELL AT INTERPHASE

A germ cell with a diploid chromosome number (2*n*) is about to leave interphase and undergo the first division of meiosis. The cell's DNA is already duplicated, so each chromosome is in the duplicated state; it consists of two sister chromatids.

MEIOSIS I

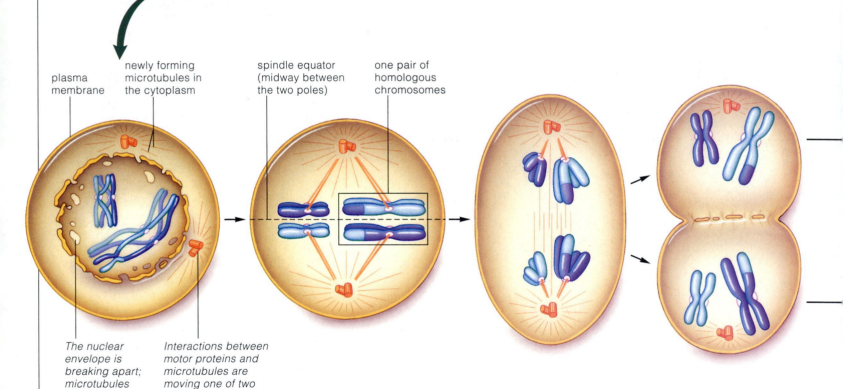

plasma
membrane

newly forming
microtubules in
the cytoplasm

spindle equator
(midway between
the two poles)

one pair of
homologous
chromosomes

The nuclear envelope is breaking apart; microtubules will be able to penetrate the nuclear region.

Interactions between motor proteins and microtubules are moving one of two pairs of centrioles toward the opposite spindle pole.

PROPHASE I

Each duplicated chromosome is in threadlike form, but now it starts to twist and fold into more condensed form. It pairs with its homologue, and the two typically swap segments. The exchange, called crossing over, is indicated by the break in color on the pair of larger chromosomes. Each chromosome becomes attached to some microtubules of a newly forming spindle.

METAPHASE I

Motor proteins have been driving the movement of microtubules that became attached to the kinetochores of chromosomes. As a result, the chromosomes have been pushed and pulled into position, midway between the spindle poles. Now the spindle is fully formed, owing to dynamic interactions of motor proteins, microtubules, and the chromosomes themselves.

ANAPHASE I

Microtubules extending from the poles and overlapping at the spindle equator *lengthen* and push the poles apart. At the same time, the microtubules extending from the poles to the kinetochores of chromosomes *shorten*, and each chromosome is thereby pulled away from its homologous partner. These motions move the homologous partners to opposite poles.

TELOPHASE I

At some point, the cytoplasm of the germ cell divides. Two cells, each with a haploid chromosome number (*n*), result. That is, the cells have one of each type of chromosome that was present in the parent (2*n*) cell. All chromosomes are still in the duplicated state.

Of the four haploid cells that form by way of meiosis and cytoplasmic division, one or all may develop into gametes and function in sexual reproduction.

MEIOSIS II

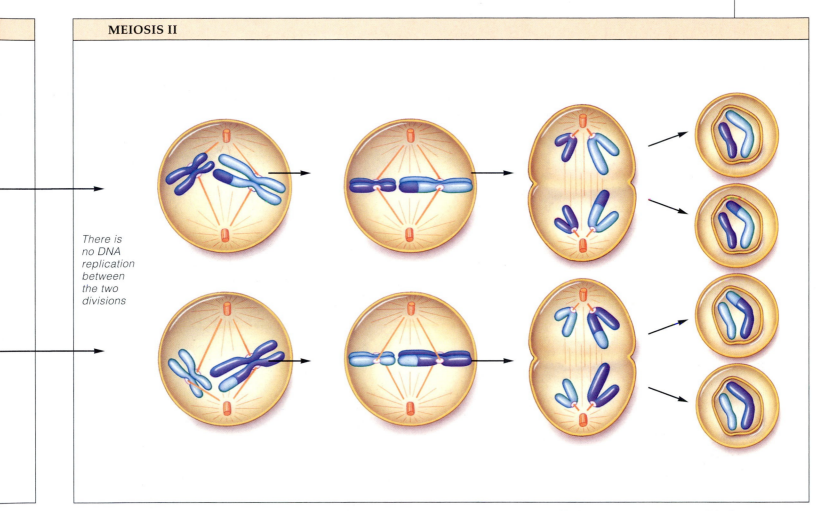

There is no DNA replication between the two divisions

PROPHASE II

In each of the two daughter cells, microtubules have already moved one member of the centriole pair to the opposite pole of the spindle during the transition to prophase II. Now, at prophase II, microtubules attach to the kinetochores of chromosomes, and motor proteins drive the movement of chromosomes toward the spindle's equator.

METAPHASE II

Now, in each daughter cell, interactions among motor proteins, spindle microtubules, and each duplicated chromosome have moved all the chromosomes so that they are positioned at the spindle equator, midway between the two poles.

ANAPHASE II

The attachment between the two chromatids of each chromosome breaks. The former "sister chromatids" are now chromosomes in their own right. Motor proteins interact with kinetochore microtubules to move the separated chromosomes to opposite poles of the spindle.

TELOPHASE II

By the time telophase II is completed, there will be four daughter nuclei. Also, at the time when the division of the cytoplasm is completed, each new daughter cell will have a haploid chromosome number (n). All of those chromosomes will be in the unduplicated state.

KEY EVENTS DURING MEIOSIS I

Sections 17.8 and 17.9 emphasized the main function of meiosis—to reduce the diploid number of chromosomes by half, so gametes will have a *haploid* number of chromosomes. However, as you'll now read, two other events that occur during prophase and metaphase of meiosis I accomplish another function—genetic variation.

What Goes On in Prophase I?

Genetic variation can be a major advantage for sexually reproducing organisms, like humans. This is because, as later chapters describe more fully, variations in traits governed by genes can lead to modifications in body structures and functions. Passed from one generation to the next, such changes may eventually enable a population of organisms to adapt to changes in the environment.

Some genetic variations can come about during prophase I of meiosis. Why? This is a time when genes on the chromosomes in germ cells are rearranged. Notice Figure 17.16a, which shows two chromosomes condensed to threadlike form. All the chromosomes in a germ cell condense this way. As they do, each is drawn close to its homologue. It is as if the homologues are stitched together point by point along their length, without much space between them. (The X and Y chromosomes are "stitched" at one end only.) This close, side-by-side alignment favors

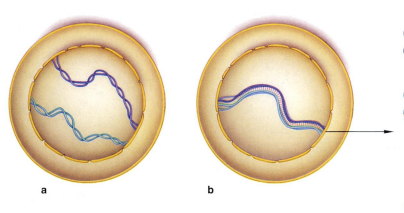

a   b

Figure 17.16 Key events during prophase I, the first stage of meiosis. For clarity, this diagram of a cell shows only one pair of homologous chromosomes and one crossover event. Here, the paternal chromosome is blue and its maternal homologue is purple.

a Both of the chromosomes were duplicated earlier, during interphase. Early in prophase I, the two sister chromatids of each duplicated chromosome are in thin, threadlike form. They are all positioned so close together they look like a single thread.

b Each chromosome becomes "zippered" to its homologue, so all four chromatids are intimately aligned. When the two sex chromosomes have different forms (X paired with Y) they still get zippered together, but only for a small part of their length.

c,d Here we show the pair of chromosomes as if they were already condensed and teased apart so that it is easier to visualize what goes on. Their double-stranded DNA molecules are still tightly aligned at this stage. The intimate contact allows one (or more, usually) crossovers to occur at intervals along the length of nonsister chromatids.

e Nonsister chromatids exchange segments. As prophase I ends, the chromosomes continue to condense into thicker, rodlike forms. Then they unzip from each other except at chiasmata, the places where they physically cross each other. A chiasma does not last long, but it is indirect evidence that a crossover occurred at some point in the chromosomes.

f What is the function of crossing over? It breaks up old combinations of alleles and puts new ones together in pairs of homologous chromosomes.

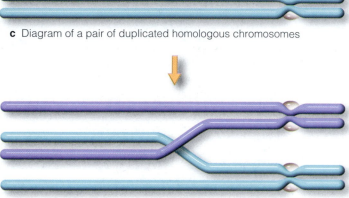

c Diagram of a pair of duplicated homologous chromosomes

d Crossover between nonsister chromatids of the two chromosomes.

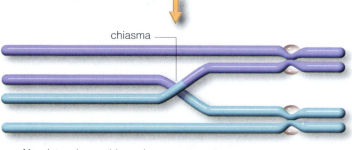

chiasma

e Nonsister chromatids exchange segments.

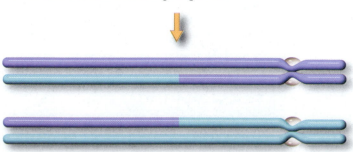

f Homologues have new combinations of alleles.

nonsister chromatids of a pair of homologues. In a crossover, nonsister chromatids break at the same places along their length. At these break points, they exchange corresponding segments—that is, genes. As diagrammed in Figure 17.16*e*, X-shaped configurations called chiasmata (singular: chiasma, "cross") are evidence of crossovers.

The exchange of chromosome pieces is called **genetic recombination**. It leads to variation in the traits offspring inherit. How? Genes can come in alternative forms called **alleles**. For example, the gene for earlobe shape has two alleles—one calls for *attached* earlobes, and the other calls for *detached* earlobes. Often, many of the alleles on one chromosome are not exactly identical to the corresponding alleles on its homologue. Every time a crossover takes place, new combinations of alleles may form.

Metaphase I Alignments

Whole chromosomes begin to be shuffled during the transition from prophase I to metaphase I, which is the second stage of meiosis. Suppose this is happening at this moment, in one of your germ cells. By now, crossovers have made genetic mosaics of the chromosomes, but put this aside to simplify tracking. As in Figure 17.17, think of the twenty-three chromosomes from your mother as the *maternal* chromosomes and their twenty-three homologues from your father as the *paternal* chromosomes.

Kinetochore microtubules have already oriented one chromosome of each pair toward one spindle pole and its homologue toward the other (compare Section 17.9). Now they are moving all the chromosomes, which soon will be positioned midway between the spindle poles.

Have all maternal chromosomes become attached to one spindle pole and all paternal chromosomes to the other? Probably not. Remember, the first contacts between kinetochore microtubules and the chromosomes are random. For this reason, the eventual position of a maternal or paternal chromosome at the spindle equator at metaphase I follows no particular pattern. Now we can carry this thought a step further. Either one of each pair of homologous chromosomes can end up at either pole of the spindle after they move apart at anaphase I.

Think about the possibilities when there are only three pairs of homologues. As Figure 17.17 shows, by metaphase I, three pairs of homologues may be arranged in any one of four possible positions. In this case, eight combinations (2^3) of maternal and paternal chromosomes are possible for the forthcoming gametes.

Of course, a human germ cell has twenty-three pairs of homologous chromosomes, not just three. So a grand total of 2^{23}, or *8,388,608 combinations* of maternal and paternal chromosomes is possible every time a germ cell gives rise to a gamete!

In each sperm or egg, the genetic instructions from

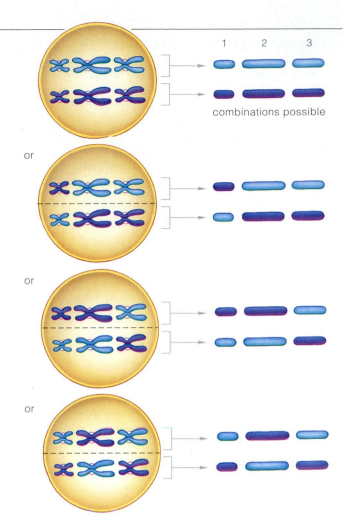

combinations possible

Figure 17.17 Possible outcomes of the random alignment of three pairs of homologous chromosomes at metaphase I of meiosis. Maternal chromosomes are *purple*; paternal ones *blue*.

the mother might differ slightly from those from the father. Chapter 18 explains more fully why this is so. But are you beginning to see why striking mixes of traits can show up even in the same family?

During anaphase I, each homologue is separated from its partner. This event is called **disjunction**. Each member of a set of sister chromatids moves to one pole of the spindle. So long as the separation takes place, it doesn't matter which sister chromatid moves to which pole. However, as you will learn in Chapter 19, the failure of homologues to separate during meiosis can lead to birth defects.

Crossing over is an interaction between a pair of homologous chromosomes. It breaks up old combinations of alleles and puts new ones together during prophase I of meiosis.

In metaphase I, the random attachment and later positioning of each pair of maternal and paternal chromosomes lead to varied combinations of parental traits in offspring.

MEIOSIS AND MITOSIS COMPARED

In this chapter, our focus has been on two different mechanisms that divide the DNA in cell nuclei. Mitosis occurs in somatic cells, and it is the mechanism that underlies body growth and tissue repair. Meiosis occurs only in germ cells; it is the underlying mechanism for the formation of gametes and sexual reproduction. Figure 17.18 summarizes the similarities and difference between mitosis and meiosis.

The end results of the two mechanisms differ in a crucial way. *Mitotic cell division only produces clones—genetically identical copies of a parent cell. Meiotic cell division, together with fertilization, promotes variation in traits among offspring.* First, crossing over at prophase I of meiosis puts new combinations of alleles in chromosomes. Second, the random assignment of either member of a pair of homologous chromosomes to either pole of a spindle at metaphase I affects gametes, which end up with different mixes of maternal and paternal alleles. And third, different combinations of alleles are brought together simply by chance at fertilization.

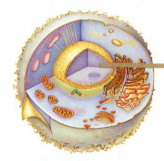

A *somatic cell* with a diploid chromosome number (2n) is at interphase. Before mitotic division begins, its DNA is replicated (all chromosomes are duplicated).

Figure 17.18 Comparison of mitosis and meiosis. This diagram is arranged to help you compare the similarities and differences between the two mechanisms. Maternal chromosomes are shaded *purple*, and paternal chromosomes are *blue*.

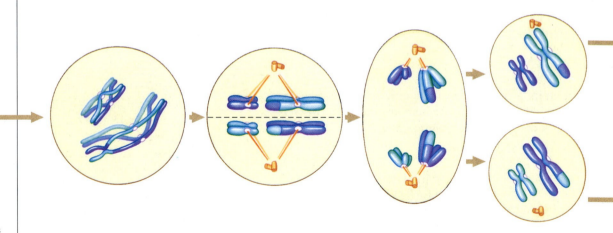

MEIOSIS I

A *germ cell* with a diploid chromosome number (2n) is at interphase. Before mitotic division begins, its DNA is replicated (all chromosomes are duplicated).

PROPHASE I

Each duplicated chromosome (consisting of two sister chromatids) condenses to threadlike form, then rodlike form. *Crossing over* occurs. Each chromosome unzips from its homologue. Each gets attached to the spindle in transition to metaphase.

METAPHASE I

All chromosomes are now positioned at the spindle's equator.

ANAPHASE I

Each chromosome is separated from its homologue. They are moved to opposite poles of the spindle.

TELOPHASE I

When the cytoplasm divides, there are two cells. Each has a haploid (n) number of chromosomes, but these are still in the duplicated state.

MITOSIS

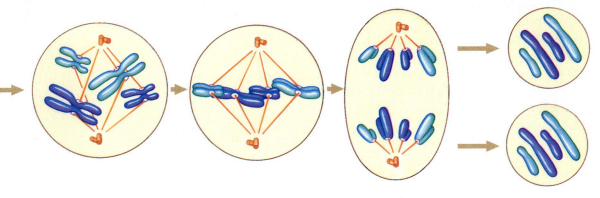

PROPHASE

Each duplicated chromosome (consisting of two sister chromatids) condenses from threadlike to rodlike form. Each gets attached to the spindle during the transition to metaphase.

METAPHASE

All chromosomes are now positioned at the spindle's equator.

ANAPHASE

Sister chromatids of each chromosome are separated from each other. These new, daughter chromosomes are moved to opposite poles of the spindle.

TELOPHASE

When the cytoplasm divides, there are two cells. Each is diploid (2n)—*it has the same chromosome number as the parent cell*.

MEIOSIS II

There is no DNA replication between the two divisions.

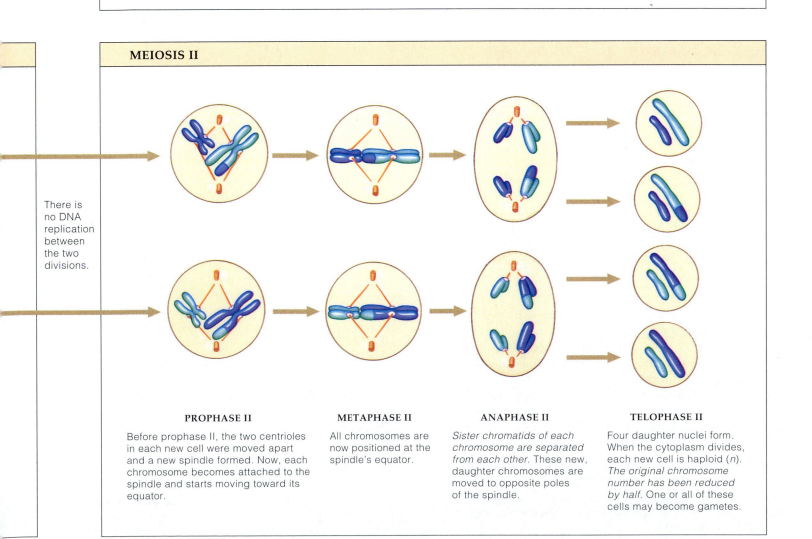

PROPHASE II

Before prophase II, the two centrioles in each new cell were moved apart and a new spindle formed. Now, each chromosome becomes attached to the spindle and starts moving toward its equator.

METAPHASE II

All chromosomes are now positioned at the spindle's equator.

ANAPHASE II

Sister chromatids of each chromosome are separated from each other. These new, daughter chromosomes are moved to opposite poles of the spindle.

TELOPHASE II

Four daughter nuclei form. When the cytoplasm divides, each new cell is haploid (n). *The original chromosome number has been reduced by half*. One or all of these cells may become gametes.

SUMMARY

1. Each cell of a new generation must receive a duplicate of all the parental DNA and enough cytoplasmic elements to start up its own operation.

2. Somatic (body) cells usually have a diploid number of chromosomes—that is, two of each type of chromosome characteristic of the species. Except for sex chromosomes, (X and Y) pairs of homologous chromosomes are alike in length, shape, and the genes they carry.

3. A somatic cell spends a major portion of the cell cycle in interphase. Interphase includes a primary growth stage (G1), during which the cell increases in mass and roughly doubles the number of cytoplasmic components; a stage (S), during which its chromosomes are duplicated; and a final, brief growth stage (G2). An unduplicated chromosome consists of one DNA molecule and associated proteins. A duplicated chromosome consists of two DNA molecules, temporarily attached to each other as sister chromatids.

4. Mitosis maintains the diploid number of chromosomes in the two daughter nuclei. Division of the cytoplasm, called cytokinesis, occurs toward the end of mitosis or at some point afterward.

5. Mitosis advances through four continuous stages:

a. Prophase. Duplicated, threadlike chromosomes condense into rodlike structures; new microtubules start to assemble in organized arrays near the nucleus; they will form a spindle apparatus. The nuclear envelope disappears.

b. Metaphase. Spindle microtubules orient the sister chromatids of each chromosome toward opposite spindle poles. The chromosomes align at the spindle equator.

c. Anaphase. Sister chromatids of each chromosome separate. Both are now independent chromosomes, and they move to opposite poles.

d. Telophase. Chromosomes decondense and a new nuclear envelope forms around the two clusters of chromosomes. Mitosis is completed. Cytokinesis divides the cytoplasm; the result is two diploid cells.

6. Meiosis is a reductional division mechanism that halves the parental diploid chromosome number to the haploid number (Figure 17.19). It consists of two rounds of chromosome partitioning that sort out the chromosomes in a germ cell. Microtubules of a spindle apparatus move the chromosomes. In meiosis I, each chromosome pairs with and then separates from its homologue. In meiosis II, the sister chromatids of each chromosome separate.

7. The following key events occur during meiosis I:

a. At prophase I, homologues pair up. Crossing over breaks up old combinations of alleles and puts together new ones in the chromosomes. This genetic recombination leads to variation in traits among offspring.

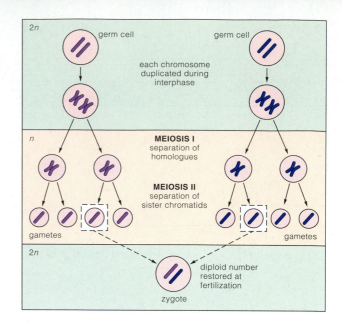

Figure 17.19 Changes in the chromosome number in meiosis.

Table 17.2 Basics of Mitosis and Meiosis

	Mitosis	Meiosis
Function	Growth, including repair and maintenance	Gamete production (sperm/eggs)
Occurs in	Somatic (body) cells	Germ cells in gonads (testes and ovaries)
Mechanism	One round of chromosome partitioning plus cytokinesis	Two rounds of chromosome partitioning and cytokinesis
Outcome	Maintains diploid chromosome number ($2n \rightarrow 2n$)	Reduces diploid chromosome number ($2n \rightarrow n$)
Effect	Two diploid daughter cells	Four haploid daughter cells

b. At metaphase I, all pairs of homologous chromosomes are aligned at the spindle equator.

c. At anaphase I, paired homologous chromosomes separate from one another and move to opposite spindle poles.

8. At metaphase II, all chromosomes are moved to the spindle equator. At anaphase II, the sister chromatids of each chromosome separate and move to opposite poles. Once separated, they are chromosomes in their own right. Following cytokinesis (cytoplasmic division), there are four haploid cells, which may function as gametes.

9. Mitotic cell division produces genetically identical copies of a parent cell (Table 17.2); meiotic cell division, together with fertilization, promotes variation in traits among offspring.

Review Questions

1. Define the two types of chromosome partitioning mechanisms that occur in the human body. What is cytokinesis? *17.1, 17.5*

2. Define somatic cell and germ cell. Which type of cell can undergo mitosis? *17.1*

3. What is a chromosome? What is the difference between a diploid cell and a haploid cell? *17.1*

4. What are homologous chromosomes? *17.1*

5. Describe the spindle apparatus and its general function in chromosome partitioning. *17.4*

6. Name the four main stages of mitosis, and describe the main features of each stage. *17.4*

7. In a paragraph, summarize the similarities and differences between mitosis and meiosis. *17.1, 17.11*

8. Explain the significance of fertilization. *17.8*

Self-Quiz *(Answers in Appendix V)*

1. DNA is distributed to daughter cells by _____ or _____, both of which are mechanisms for allocating chromosomes.

2. Each kind of organism contains a characteristic number of _____ in each cell; each of those structures is composed of a _____ molecule with its associated proteins.

3. A pair of chromosomes that are similar in length, shape, and the traits they govern are called _____.
    a. diploid chromosomes
    b. mitotic chromosomes
    c. homologous chromosomes
    d. germ chromosomes

4. Somatic cells usually have a _____ number of chromosomes.

5. Interphase is the stage when _____.
    a. a cell ceases to function
    b. a germ cell forms its spindle apparatus
    c. a cell grows and duplicates its DNA
    d. mitosis proceeds

6. After mitosis, each daughter cell contains genetic instructions that are _____ and _____ chromosome number of the parent cell.
    a. identical to the parent cell's; the same
    b. identical to the parent cell's; one-half the
    c. rearranged; the same
    d. rearranged; one-half the

7. All of the following are stages of mitosis *except* _____.
    a. prophase
    b. interphase
    c. metaphase
    d. anaphase

8. A duplicated chromosome has _____ chromatids.
    a. one   b. two   c. three   d. four

9. Crossing over in meiosis _____.
    a. alters the chromosome alignments at metaphase
    b. occurs between sperm DNA and egg DNA at fertilization
    c. leads to genetic recombination
    d. occurs only rarely

10. Because of the _____ alignment of homologous chromosomes at metaphase I, gametes can end up with _____ mixes of maternal and paternal chromosomes.
    a. unvarying; different   c. random; duplicate
    b. unvarying; duplicate   d. random; different

11. Match each stage of mitosis with the following key events.
    _____ metaphase
    _____ prophase
    _____ telophase
    _____ anaphase

    a. sister chromatids of each chromosome separate and move to opposite poles
    b. chromosomes condense and a microtubular spindle forms
    c. chromosomes decondense, daughter nuclei re-form
    d. all chromosomes align at spindle equator

Critical Thinking: You Decide *(Key in Appendix VI)*

1. Under normal circumstances you can't inherit both copies of a homologous chromosome from the same parent. Why? Assuming that no crossing over has occurred, what is the chance that one of your non-sex chromosomes is an exact copy of the same chromosome possessed by your maternal grandmother?

2. Suppose you have a means of measuring the amount of DNA in a single cell during the cell cycle. You first measure the amount during the G1 phase. At what points during the remainder of the cycle would you predict changes in the amount of DNA per cell?

3. Adam has a pair of alleles (alternate forms) of a gene that influences whether a person is right-handed or left-handed. One allele says "right" and its partner says "left." Visualize one of his germ cells, in which chromosomes are being duplicated prior to meiosis. Visualize what happens to the chromosomes during anaphase I and II. (It might help to use toothpicks as models of the sister chromatids of each chromosome.) What fraction of Adam's sperm will carry the allele for right-handedness? For left-handedness?

4. Fresh out of college, Maria has her first job teaching school. When she goes for a pre-employment chest X ray required by the school district, the technician places a lead-lined apron over her abdomen but not over any other part of Maria's body. The apron prevents electromagnetic radiation from penetrating into the protected body area. What cells is the lead shield designed to protect, and why?

Selected Key Terms

allele *17.10*
anaphase *17.4*
cell cycle *17.2*
centromere *17.1*
chromosome *17.1*
chromosome number *17.1*
cleavage furrow *17.5*
crossing over *17.10*
cytokinesis *17.5*
diploid *17.1*
genetic recombination *17.10*
germ cell *17.1*
haploid *17.8*
homologous chromosome *17.1*
interphase *17.2*
kinetochore *17.7*
meiosis *17.1*
metaphase *17.4*
mitosis *17.1*
oogenesis *17.8*
prophase *17.4*
reductional division *17.8*
sister chromatid *17.1*
somatic cell *17.1*
spermatogenesis *17.8*
spindle apparatus *17.4*
telophase *17.4*

Readings

Cummings, M. R. 2000. *Human Heredity: Principles and Issues*. Fifth edition. Pacific Grove, California: Brooks/Cole.

OBSERVABLE PATTERNS OF INHERITANCE

A Smorgasbord of Ears and Other Traits

Basketball ace Charles Barkley has them. So does actor Tom Cruise. Actress Joan Chen doesn't, and neither did a monk named Gregor Mendel. To see how *you* fit in with these folks, use a mirror to check your ears. Is the fleshy lobe at the base of each ear attached to the side of your head? If so, you and Barkley and Cruise have something in common. Or is the fleshy lobe unattached, so that you can flap it back and forth? If so, you are like Chen and Mendel (Figure 18.1).

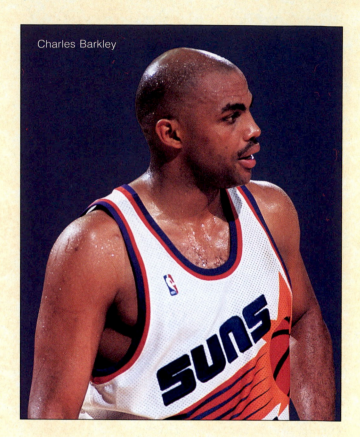
Charles Barkley

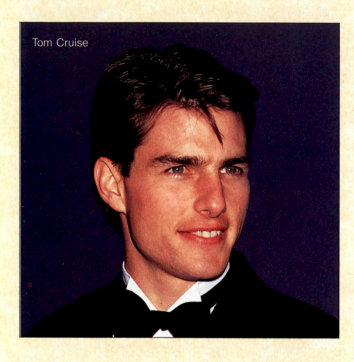
Tom Cruise

Whether a person is born with detached or attached earlobes depends on one particular gene. That gene comes in alternate forms, which differ slightly in their chemical makeup. Each alternate form of a gene is called an **allele**. In this example, one allele has information conferring detached earlobes. The information is put to use while a human body is developing inside the mother. It calls for a death signal, which is sent to all the cells positioned between the newly forming lobes and the head. Without the signal, the cells don't die, and the earlobes don't detach.

We all have genes for thousands of traits, including earlobes, cheeks, eyelashes, and eyeballs. Most of the traits vary in their details from person to person. Remember, humans inherit *pairs* of genes, on pairs of chromosomes. In some pairings, one allele has powerful effects and overwhelms the other's contribution to a trait. The overwhelmed allele is said to be recessive to the dominant one. If you have *detached* earlobes, *dimpled* cheeks, *long* lashes, or *large* eyeballs, you carry at least one and possibly two dominant alleles that influence the trait in a particular way.

When both alleles of a pair are recessive, nothing masks their effect on a trait. You get *attached* earlobes with one pair of recessive alleles (and *flat* feet with another, a *straight* nose with another, and so on).

How did we discover such remarkable things about our genes? It all started with Gregor Mendel, a monk who analyzed pea plants generation after generation. Mendel found indirect but *observable* evidence of how parents transmit units of hereditary information—genes—to offspring. This chapter focuses on Mendel's experimental methods and results. They are a classic example of how a scientific approach can reveal to us important secrets about the natural world. And to this day, they serve as the foundation for modern genetics.

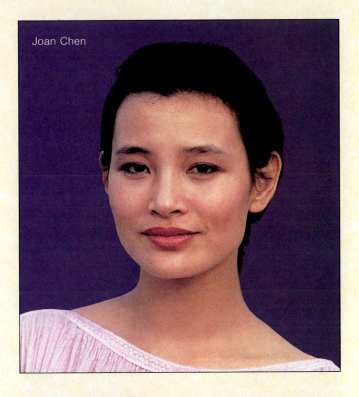

Joan Chen

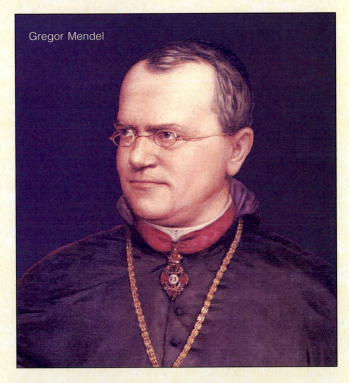

Gregor Mendel

Figure 18.1 Attached and detached earlobes of representative humans. This sampling provides observable evidence of a trait governed by a single gene, which exists in alternate forms (alleles) in the human population. Which version of the trait do you have? It depends on which form(s) of the gene you inherited from your mother and your father. As Gregor Mendel perceived, such easily observable traits can be used to identify patterns of inheritance that exist from one generation to the next.

KEY CONCEPTS

1. Genes are units of information about heritable traits. Alleles, which are different forms of the same gene, specify different versions of the same trait.

2. Each gene occurs in a particular location on a particular chromosome. This location is the gene's *locus*.

3. Humans inherit two copies of each gene. Both may be the same allele, or they may be different alleles of the gene. The two copies of a given gene are located at equivalent loci on one pair of homologous chromosomes.

4. When a pair of homologous chromosomes moves apart during meiosis, the two copies of each gene (one on each homologue) also segregate—that is, they also move apart, and they end up in different gametes. Gregor Mendel found indirect evidence of this phenomenon, called segregation, when he crossbred pea plants showing different versions of one trait, such as flower color.

5. When the two chromosomes that make up a homologous pair are separated in meiosis, each member of the pair (and the genes it carries) is sorted out for distribution into a gamete independently of how the other pairs of homologous chromosomes are sorted. Mendel found indirect evidence of this independent assortment of homologues when he tracked plants having observable differences in two traits, such as flower color and height.

6. Genetic instructions for the contrasting traits that Mendel studied were carried by different forms (alleles) of the gene in question. One allele was dominant, in that its effect on a trait masked the effect of a recessive allele paired with it.

7. Not all traits have such clearly dominant or recessive forms. One allele of a pair may be fully or partially dominant over its partner or codominant with it. Also, two or more gene pairs may influence the same trait, and some single genes influence many traits. Environmental conditions may result in further variation in traits.

CHAPTER AT A GLANCE

PATTERNS OF INHERITANCE

The Origins of Genetics

Having read about meiosis in Chapter 17, you already know a bit about the mechanisms of sexual reproduction. That is more than Gregor Mendel knew. Mendel was born into a farm family in what is now the Czech Republic. By the 1850s he had become a university-educated monk, having studied both theology and what was then called "natural history." Even so, Mendel did not know about chromosomes, so he did not know that the chromosome number is reduced by half in gametes, then restored at fertilization. Yet Mendel was interested in discovering how traits were inherited in the offspring of sexually reproducing organisms, and to pursue this question he experimented with the garden pea plant. This plant is self-fertilizing: Its flowers produce sperm (pollen) *and* eggs, and fertilization occurs in the same flower. Based on his research, Mendel hypothesized that fertilization united some sort of "factors" from each parent that were the units of heredity. Today we call such factors genes.

Some Terms Used in Genetics

The following list expresses some of Mendel's ideas in modern terms (see also Figure 18.2):

1. **Genes** are units of information about specific traits, and they are passed from parents to offspring. Each gene is a segment of DNA and it has a specific location, or locus, on a chromosome.

2. Diploid cells have two copies of each gene, on pairs of homologous chromosomes.

3. Although both copies of a gene deal with the same trait, they may vary in their information about it. This happens when there are slight differences in the DNA segments (sequences of nucleotides) that make up a gene. Each version of the gene is called an **allele**. Much of the variation we see in human traits is the result of contrasting alleles. For example, one allele of a gene specifies a straight hairline, and another allele of the same gene specifies a widow's peak.

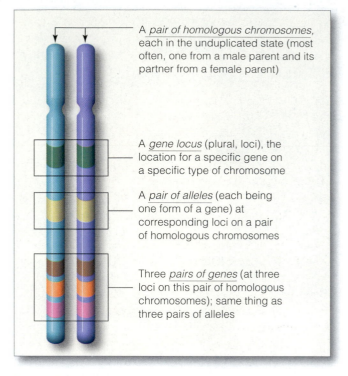

Figure 18.2 A few genetic terms illustrated. Diploid organisms such as humans have pairs of genes on pairs of homologous chromosomes. One chromosome of each pair is inherited from the mother and the other from the father.

Each gene we inherit may have slightly different forms; such alternate forms of a gene are called alleles. Different alleles carry instructions for slightly different versions of the same trait. An allele on a chromosome may or may not be identical to its counterpart on the homologous chromosome.

4. If the two copies of a gene are identical alleles, this is a **homozygous** condition (*homo:* same; *zygo:* joined together). If the two copies are different alleles, this is a **heterozygous** condition (*hetero:* different).

5. An allele is **dominant** when its effect on a trait masks that of any **recessive** allele paired with it. An uppercase letter represents a dominant allele, and a lowercase letter represents a recessive allele (for instance, *A* and *a* or *C* and *c*).

6. A *homozygous dominant* individual has a pair of dominant alleles (*AA*) for the trait being studied. A *homozygous recessive* individual has a pair of recessive alleles (*aa*). A *heterozygous* individual has a pair of nonidentical alleles (*Aa*).

7. Two terms help keep the distinction clear between genes and the traits they specify. **Genotype** refers to the alleles present in an individual. **Phenotype** refers to an individual's observable traits.

Table 18.1 summarizes the relationship between genotype and phenotype.

Table 18.1	Genotype and Phenotype Compared	
Genotype	Described as	Phenotype
CC	homozygous dominant	chin fissure
Cc	heterozygous (one of each allele; dominant form of trait observed)	chin fissure
cc	homozygous recessive	smooth chin

MENDEL'S THEORY OF SEGREGATION

Mendel's idea that there are units of inheritance (now called genes) was just the beginning of his search to understand heredity. Building on this concept, he proposed two hypotheses: (1) each diploid organism inherits two units for each trait, one from each parent; and (2) in parents, different units assort independently when a parent makes gametes. For the moment, we will consider the strategy Mendel pioneered to test the first hypothesis.

To probe the question of how a single trait is passed from parent to offspring, Mendel devised what is now called a **monohybrid cross** (*mono-* means one). In a monohybrid cross, the parents have different alleles for the gene in question. Although for ethical reasons we do not carry out controlled crosses between individual humans, monohybrid matings do occur naturally, and they show patterns of inheritance at work.

The following example illustrates a monohybrid cross for the configuration of a person's chin. This trait is governed by a gene that has two allele forms. One allele, which calls for an indentation called a chin fissure (Figure 18.3), is dominant when it is present; we can represent this allele as *C*. The recessive allele, which codes for a smooth chin, is *c*. In this example, one parent is homozygous for the *C* form of the gene and has a chin fissure, and the other is homozygous for the *c* form and has a smooth chin. The *CC* parent produces only *C* gametes, and the *cc* parent produces only *c* gametes:

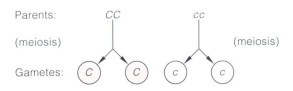

Each gamete receives one allele for the trait because each gamete contains only one copy of each chromosome. We know already that the genes on homologues match up into equivalent pairs. So, when homologues separate into different gametes after meiosis II (Section 17.8), the two genes of each pair separate as well (Figure 18.4). This is known as the principle of **segregation**. Because meiosis also reduces the diploid chromosome number to the haploid number, each gamete contains one member of each chromosome pair—and the genes it carries.

The two copies of each gene in a diploid organism segregate from each other during meiosis in germ cells. As a result, each gamete contains only one copy of each gene.

Figure 18.3 (**a**) The chin fissure, a heritable trait arising from a rather uncommon allele of a gene. Actor Michael Douglas received a gene that influences this trait from each of his parents. At least one of those genes was dominant. (**b**) What Mr. Douglas's chin might have looked like if he had inherited identical alleles for "no chin fissure" instead.

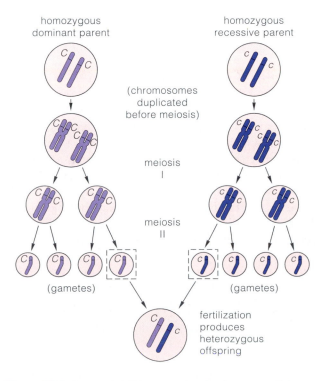

Figure 18.4 A monohybrid cross, showing how one allele of a pair segregates from the other. Two parents that are each homozygous for a different version of a trait give rise only to heterozygous offspring.

DOING GENETIC CROSSES AND FIGURING PROBABILITIES

The parental generation of a cross is designated by a **P**. Offspring of a monohybrid cross are called the **F₁** or *first filial* generation. In the present example, each child will inherit a pair of differing alleles for the trait, one from each homozygous parent. The children will thus each be heterozygous for the chin genotype, or *Cc*. Because *C* is dominant, each child will have a chin fissure.

Suppose now that one of the *Cc* children grows up and marries another *Cc* person. From this second cross we derive the **F₂** (second filial) generation. Because half of each parent's gametes (sperm or eggs) are *C* and half are *c* (due to segregation at meiosis), four outcomes are possible every time a sperm fertilizes an egg:

Possible event:	Probable outcome:	
sperm *C* unites with egg *C*	1/4 *CC* offspring	
sperm *C* unites with egg *c*	1/4 *Cc*	chin fissure
sperm *C* unites with egg *c*	1/4 *Cc* — or 1/2 *Cc*	
sperm *c* unites with egg *c*	1/4 *cc*	smooth chin

The diagrams in Figure 18.5 show how to construct a **Punnett square**, a convenient tool for determining the probable outcome of genetic crosses. In this case, there is a 75 percent chance that a child from a cross between two *Cc* parents will have at least one dominant *C* allele and a chin fissure. When a large number of offspring are involved, a ratio of 3:1 is likely (Figure 18.6).

Fertilization is a chance event, which is why rules of probability apply to crosses. **Probability** is a number between zero and one that expresses the likelihood of a particular event. For example, an event that has a probability of one will always occur; an event with a probability of zero will never occur, and an event with a probability of one-half (or 50 percent) is expected to occur in about half of all opportunities (Figure 18.7).

Having a chin fissure doesn't affect a person's health. However, a fair number of human genetic disorders, including cystic fibrosis and sickle-cell anemia, result from single-gene defects and so follow a Mendelian inheritance pattern. Genetic disorders are a major topic of Chapter 19. For the moment, it is important to realize two things:

1. The outcomes (fractions) predicted by probability do not necessarily turn up in a single family. For instance, it's common to see families in which the parents have produced several children, all of the same sex. Fractions predicted by probability turn up consistently only when

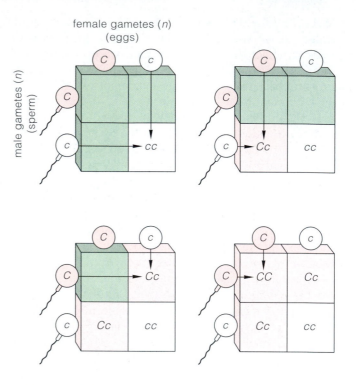

Figure 18.5 Punnett-square method of predicting the probable outcome of a genetic cross. In this example, the cross is between two heterozygous individuals. Circles represent gametes. Letters on gametes represent dominant or recessive alleles. The different squares depict the different genotypes possible among offspring.

a large number of events is analyzed. You can test this for yourself by flipping a coin. Probability predicts that heads and tails should each come up about half the time, but you may have to flip the coin a hundred times to consistently achieve a ratio close to 1:1. Likewise, probability predicts that parents with two or more children will have equal numbers of girls and boys, but this does not always happen.

2. In a given genetic situation, *probability is constant*. The likelihood that a certain genotype will occur—say, a baby with the genotype for cystic fibrosis—is the same for every child no matter how many children a couple has. Based on the parents' genotypes, if the probability that a child will inherit a certain genotype is one in four, then each child of those parents has a one-in-four (25 percent) chance of inheriting the genotype. If the parents have three children without the trait, the fourth child still has only a one-in-four chance of inheriting it.

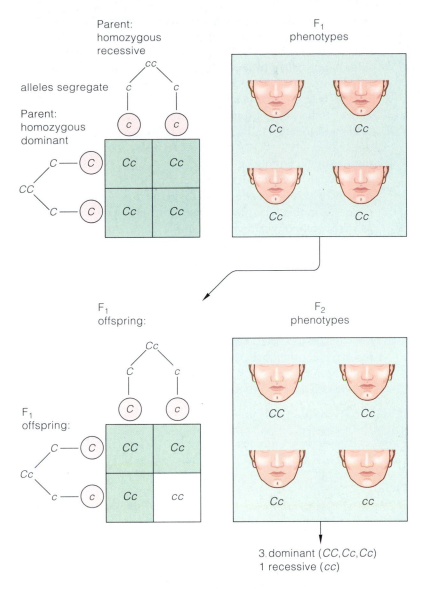

Parent: homozygous recessive

alleles segregate

Parent: homozygous dominant

CC

F₁ phenotypes

Cc Cc

Cc Cc

F₁ offspring:

Cc

F₁ offspring:

Cc

F₂ phenotypes

CC Cc

Cc cc

3 dominant (CC, Cc, Cc)
1 recessive (cc)

Figure 18.6 Results from a monohybrid cross, in which one parent is homozygous dominant for a trait and the other is homozygous recessive for the same trait. Notice that the dominant-to-recessive ratio is 3:1 for the second generation (F₂) offspring.

How to Calculate Probability

Step 1. Actual genotypes of parental gametes

In the cross $Cc \times Cc$, gametes have a 50–50 chance of receiving either allele (C or c) from each parent. Said another way, the probability that a particular sperm or egg will be C is 1/2, and the probability that it will be c is also 1/2:

probability of C: 1/2

probability of c: 1/2

Step 2. Probable genotypes of offspring

Offspring receive one allele from each parent. Three different combinations of alleles are possible in this cross. To figure the probability that a child will receive a particular allele combination, simply multiply the probabilities of the individual alleles:

probability of CC: $1/2 \times 1/2 = 1/4$

probability of Cc: $1/2 \times 1/2 = 1/4$ ⎫
                                                    ⎬ 1/2
probability of cC: $1/2 \times 1/2 = 1/4$ ⎭

probability of cc: $1/2 \times 1/2 = 1/4$

Step 3. Probable phenotypes

Chin fissure: $1/4 + 1/4 + 1/4 = 3/4$
(CC, Cc, cC)

Smooth chin: 1/4
(cc)

Figure 18.7 Calculating probabilities. Simple multiplication lets you figure the probability that a child will inherit alleles for a particular phenotype.

Rules of probability apply to the inheritance of single gene traits. Thus, if the genotypes of parents are known, it is possible to establish a potential child's chances for inheriting a particular genotype and thus for having a particular phenotype (trait).

18.4 THE TESTCROSS: A TOOL FOR DISCOVERING GENOTYPES

In working with nonhuman organisms, researchers often use a simple technique called a **testcross** to figure out an unknown genotype. One individual in the testcross is known to be homozygous for the recessive allele of the trait being studied. The other has a dominant phenotype that could be the result of a homozygous condition *or* a heterozygous one (Figure 18.8a). The phenotypes of the offspring reveal the genotype of the parent with the dominant phenotype.

We can find similar situations in human genetics that shed light on the genotype of a human parent. Suppose, for example, that a woman has smooth cheeks and her husband has dimples. "No dimples" is a recessive trait, so the woman must be *dd*. As Figure 18.8b diagrams, if a child is born with no dimples, then the father must be a heterozygote for this trait, with a genotype of *Dd*; that is the only way he could father a *dd* child. If the child has dimples, the father can be either *DD* or *Dd*. If he is *Dd*, a heterozygote, the probability that he will have a dimpled child is 1/2, a 50–50 chance, every time. If he were *DD*, every child would be dimpled.

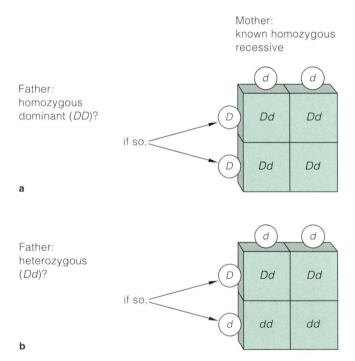

a

b

Figure 18.8 A testcross "works backward" from the phenotypes of offspring to help identify the genotype of a parent. In this example, the father's genotype can be determined if he mates with a woman known to be homozygous recessive for the trait in question *and* if they produce a child who shows the recessive phenotype. Then, the father must be a heterozygote. If the child shows the dominant phenotype, then all we can say for sure is that the father is either a heterozygote or homozygous dominant for the trait.

18.5 INDEPENDENT ASSORTMENT

Mendel also addressed the question of how genes for two *different* traits are passed to offspring. Based on the results of experiments, he established another basic principle of inheritance, called **independent assortment**. This general rule states that *alleles for different traits are sorted into gametes independently of one another* (Figure 18.9). Because genes on the same chromosome can be linked (they tend to remain together), the rule of independent assortment does not apply to genes that are very close to each other on the same chromosome.

Mendel found evidence for independent assortment by way of a method called the **dihybrid cross**. As its name suggests, in this type of cross two traits are studied. And as with a monohybrid cross, "simple dominance" exists: that is, there are two contrasting alleles of each gene, one dominant and one recessive. The parents have different combinations of alleles. Independent assortment becomes apparent when we follow two generations of a dihybrid cross.

Let's consider an example using dominant alleles *C* for chin fissure and *D* for dimples and *c* and *d* as their recessive counterparts. In this case, let's say that the P generation consists of a man who is *ccdd* and a woman who is *CCDD*. Each parent can produce just one type of gamete:

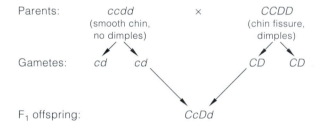

As Mendel would have predicted, all of this couple's children (the F$_1$ generation) are *CcDd* and have a chin fissure and dimples.

What happens in a second generation if two *CcDd* individuals mate? The man and woman can each produce four types of gametes in equal proportions:

1/4 *CD*   1/4 *Cd*   1/4 *cD*   1/4 *cd*

This outcome is possible because the copies of genes on different chromosomes assort independently of one another during meiosis.

Further reading: Student Guide to InfoTrac on web site →

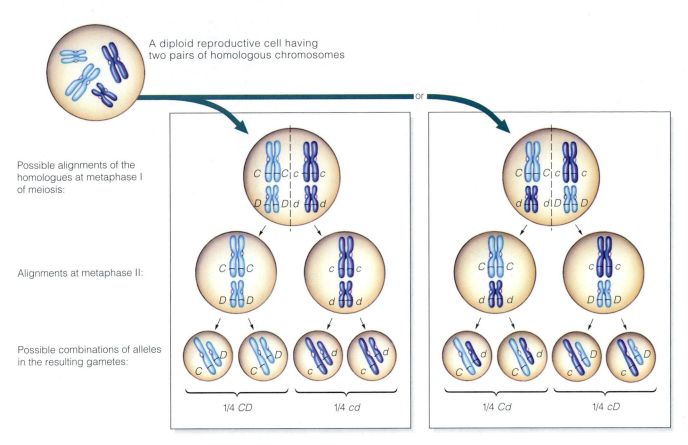

A diploid reproductive cell having two pairs of homologous chromosomes

or

Possible alignments of the homologues at metaphase I of meiosis:

Alignments at metaphase II:

Possible combinations of alleles in the resulting gametes:

1/4 CD   1/4 cd   1/4 Cd   1/4 cD

Figure 18.9 Example of independent assortment, showing just two pairs of homologous chromosomes. Because either chromosome of the pair can move to one spindle pole or the other during meiosis (Section 17.10), different gametes can end up with different mixes of alleles. Both options shown here are equally probable.

To work out the preceding result yourself, assume that the genes for chin fissure and dimples are located on two different pairs of homologous chromosomes. During metaphase I of meiosis, all the germ cell's chromosome pairs are positioned at random at the spindle equator. The chromosome with allele *C* may ultimately move to either spindle pole, and on into any one of four gametes. The same is true for its homologue with the *c* allele. And the same is true for the chromosomes with alleles *D* and *d*. These outcomes will be our starting point in the following section, which takes the dihybrid cross a step further. It explores (as Mendel did with successive generations of pea plants) what kinds of phenotypes are most likely in offspring when their parents are both heterozygous for *two* traits.

Because of events during meiosis, either copy of a gene may end up in a particular gamete.

A dihybrid cross can reveal evidence of independent assortment. This basic principle of inheritance states that a gene pair on one chromosome assorts into gametes independently of other gene pairs of other chromosomes.

A CLOSER LOOK AT INDEPENDENT ASSORTMENT

Simple multiplication (four kinds of sperm times four kinds of eggs) tells us that sixteen different gamete unions are possible when each parent in a dihybrid cross is heterozygous for the two genes in question. The Punnett square in Figure 18.10 below diagrams the possibilities, using the chin fissure and dimple traits. It assumes that each parent is heterozygous at both gene loci. That is, both parents have the genotype *CcDd*. Notice that when such

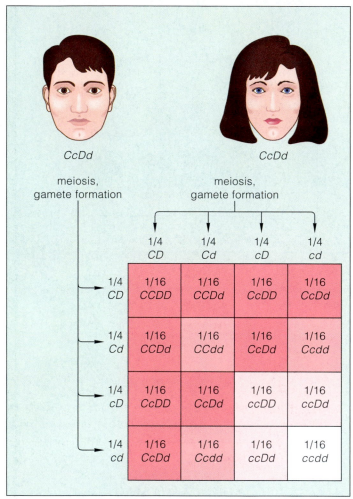

Adding up the combinations possible:

- 9/16 or 9 chin fissure, dimples
- 3/16 or 3 chin fissure, no dimples
- 3/16 or 3 smooth chin, dimples
- 1/16 or 1 smooth chin, no dimples

Figure 18.10 Results from a mating in which both parents are heterozygous at both loci. *C* and *c* represent dominant and recessive alleles for chin fissure. *D* and *d* represent dominant and recessive alleles for dimples. Rules of probability predict that certain combinations of phenotypes among offspring of this type of cross occur in a 9:3:3:1 ratio, on average.

Probability in a Mating Where Both Parents Are Heterozygous at Two Loci

A dihybrid cross considers two traits. If both parents are heterozygous for both traits, a dihybrid cross produces the following 9:3:3:1 phenotype ratio (Figure 18.10):

9/16 or 9 chin fissure, dimples

3/16 or 3 chin fissure, no dimples

3/16 or 3 smooth chin, dimples

1/16 or 1 smooth chin, no dimples

Individually, these phenotypes have the following probabilities:

probability of chin fissure (12 of 16)   = 3/4
probability of dimples (12 of 16)   = 3/4
probability of smooth chin (4 of 16)   = 1/4
probability of no dimples (4 of 16)   = 1/4

To figure the probability that a child will show a particular *combination* of phenotypes, multiply the probabilities of the individual phenotypes in each possible combination:

Trait combination		Probability
Chin fissure, dimples	3/4 × 3/4	9/16
Chin fissure, no dimples	3/4 × 1/4	3/16
Smooth chin, dimples	1/4 × 3/4	3/16
Smooth chin, no dimples	1/4 × 1/4	1/16

Figure 18.11 Probability applied to independent assortment.

---

individuals mate, there are nine possible ways for gametes to unite that produce a chin fissure and dimples, three for a chin fissure and no dimples, three for a smooth chin and dimples, and one for a smooth chin and no dimples. The probability of any one child having a chin fissure and dimples is 9/16; a chin fissure and no dimples, 3/16; a smooth chin and dimples, 3/16; and a smooth chin and no dimples, 1/16.

Figure 18.11 describes how to calculate the probability that a child will inherit genes for a particular set of two traits on different chromosomes.

If both parents are heterozygous for two particular genes, sixteen genotypes and four phenotypes are possible. Following the principles of probability, these phenotypes have probabilities of 9/16, 3/16, 3/16, and 1/16 associated with them.

MULTIPLE EFFECTS OF SINGLE GENES

Genes are chemical instructions for building proteins. We say that a gene is "expressed" when its instructions are implemented and the protein in question is synthesized. Expression of the alleles at just a single location on a chromosome may have positive—or negative—effects on one or more trait. This outcome of a single gene's activity is called **pleiotropy** (after the Greek *pleio-*, meaning more, and *-tropic*, meaning to change). For instance, a mutant allele in a single gene causes a disorder called *osteogenesis imperfecta*. Babies with this disorder typically are deaf and develop extremely fragile bones, weak tendons and ligaments, and other abnormalities.

Another example, *sickle-cell anemia*, arises when a person is homozygous for a recessive allele. The normal allele (*HbA*) provides the instructions for building functional hemoglobin, the oxygen-transporting protein in red blood cells. When a person is homozygous for the recessive mutant allele, *HbS*, he or she has sickle-cell anemia. Red blood cells, normally biconcave disks, become deformed into a sickle shape when the oxygen content of blood falls below a certain level. The sickled cells clump in blood capillaries and can rupture. Blood flow can be so disrupted that the person's oxygen-deprived tissues suffer severe damage. Over time, mounting effects of the defective hemoglobin can be extremely serious (Figure 18.12). Homozygotes (*HbS/HbS*) often die relatively young. Heterozygotes (*HbA/HbS*), on the other hand, have *sickle-cell trait*. Under most environmental conditions they have few symptoms because the one HbA allele provides enough normal hemoglobin to prevent red blood cells from sickling.

During a crisis, sickle-cell anemia patients may receive blood transfusions, oxygen, antibiotics, and painkilling drugs. There is evidence that the food additive butyrate can reactivate "dormant" genes responsible for fetal hemoglobin, an efficient oxygen carrier that normally is produced only before birth. For this reason, some states require hospitals to screen newborn infants for sickle-cell anemia so that appropriate action can be taken right away.

As described earlier, people who are heterozygous for a trait have two contrasting alleles for that trait. Sometimes, *both* alleles are expressed. We see a classic example of this **codominance** in people who are heterozygotes for alleles that confer A and B blood types. If you have type

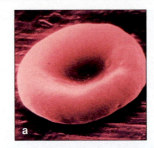
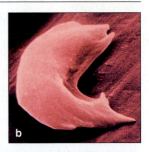

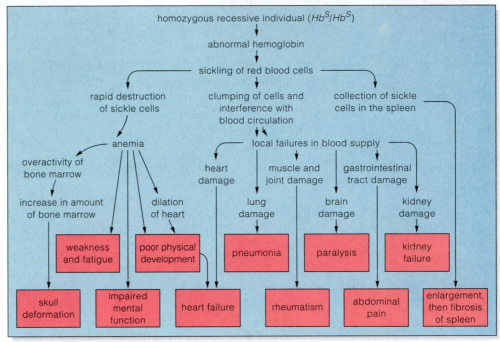

c

Figure 18.12 Scanning electron micrographs of normal (**a**) and sickled (**b**) red blood cells. (**c**) The range of symptoms characteristic of sickle-cell anemia.

AB blood, for instance, you have a pair of codominant alleles that are both expressed in the stem cells that give rise to your red blood cells. You may remember from Chapter 7 that a polysaccharide (a sugar) on the surface of red blood cells has different molecular forms that determine blood type. ABO blood typing (Section 7.4) can reveal which form(s) of the polysaccharide a person has.

The gene specifying an enzyme that synthesizes this polysaccharide has three alleles, which influence the sugar molecule's form in different ways. Two alleles, *IA* and *IB*, are codominant when paired with each other. A third allele, *i*, is recessive. When paired with either *IA* or *IB*, the effect of *i* is masked. A gene that has three or more alleles is called a **multiple allele system**.

The alleles at a single gene location may have positive or negative effects on two or more traits.

The effects may not be simultanous, but may have repercussions over time as one altered trait changes another trait, and so on, as in sickle-cell anemia.

HOW CAN WE EXPLAIN LESS PREDICTABLE VARIATIONS?

Penetrance and Expressivity

A gene can have an all-or-nothing effect on a trait—either you have dimples or you don't. In some other cases, gene interactions as well as environmental factors may alter the *degree* to which a gene is actually expressed.

Geneticists use the term **penetrance** to refer to the percentage of people in which the phenotype associated with a particular genotype is expressed. The recessive allele that produces cystic fibrosis is completely penetrant; 100 percent of children who are homozygous for it develop the disease. The dominant allele for having extra fingers or toes (called *polydactyly*) is incompletely penetrant. Some people who inherit the allele have the usual number of digits, and others have more (Figure 18.13). Some people with *campodactyly* have immobile, bent fingers on both hands. In others, the traits show up on one hand only. When expression of an allele can produce such a range of phenotypes, we say the allele shows "variable expressivity." The allele for campodactyly also is incompletely penetrant. In some people who inherit it, the traits don't show up at all.

From studies of some rare genetic disorders, we have learned that a trait may differ from person to person,

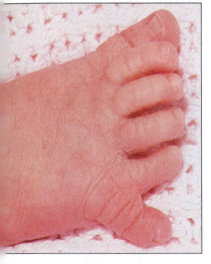

Figure 18.13 An example of polydactyly, a dominant trait.

Figure 18.14 Samples from the range of continuous variation in human eye color. Different alleles of more than one gene interact to produce and deposit melanin. Among other things, this pigment helps color the eye's iris. Different combinations of alleles result in small differences in eye color. So the frequency distribution for the eye color trait appears to be continuous over the range from black to light blue.

depending on which parent provided the allele(s) responsible for it. This phenomenon, which is called *genomic imprinting*, is the topic of this chapter's *Science Comes to Life* on the facing page.

Some of the variations we observe in gene expression probably result from gene interactions. With respect to this point, bear in mind that the path from most precursors to their products is actually a series of metabolic steps, each controlled by enzymes, which are the products of genes.

Figure 18.15 Students organized according to height—a vivid example of continuous variation.

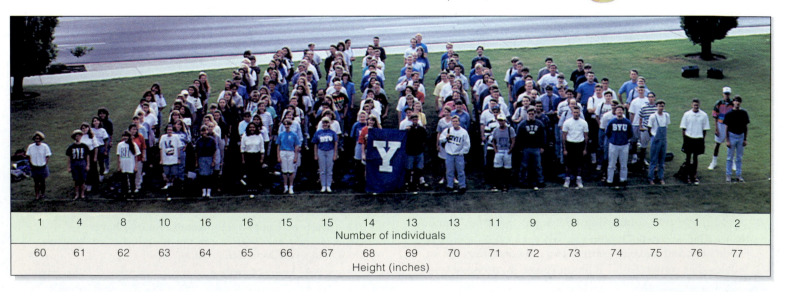

Number of individuals	1	4	8	10	16	16	15	15	14	13	13	11	9	8	8	5	1	2
Height (inches)	60	61	62	63	64	65	66	67	68	69	70	71	72	73	74	75	76	77

Genes or their products interact at many steps. Hence, if different people have varying combinations of alleles for those genes, their phenotypes may vary also.

Polygenic Traits

Polygenic traits result from the combined expression of several genes. For example, skin-color and eye-color phenotypes are the cumulative result of many genes involved in the stepwise production and distribution of melanin. Black eyes have abundant melanin deposits in the iris. Dark-brown eyes have less melanin, and light-brown or hazel eyes have still less (Figure 18.14). Green, gray, and blue eyes don't have green, gray, or blue pigments. Instead, they have so little melanin that we readily see blue wavelengths of light being reflected from the iris. Hair color probably results from the interactions of several genes. This explains our real-world observation that there is a tremendous variation in human hair color.

For many polygenic traits a population may show **continuous variation**. That is, its members show a range of continuous, rather than incremental, differences in some trait. Continuous variation is especially evident in traits that are easily measurable, such as height (Figure 18.15).

Do Genes "Program" Behavior?

Identical twins have identical genes and look alike. In addition, they have many behaviors in common. Are these parallels coincidence, clever hoaxes, or evidence that aspects of behavior are inherited characteristics, like hair and eye color? Although the question is intriguing, clear answers have proven very difficult to come by.

There is strong evidence that certain basic human behaviors, such as smiling to indicate pleasure and the crying of an infant when it is hungry, are indeed genetically programmed universals. Recently, scientists have also begun to look for connections between genes and alcoholism, some types of mental illness, violent behavior, and even sexual orientation. Such studies raise controversial social issues, and so far their greatest impact has been to point out how little we know about the biological basis of human behavior. In general, most human behavior is so complex that it is extraordinarily difficult to scientifically test hypotheses about genetic links. For the time being, we can only marvel at this remarkable complexity and ponder the intriguing possibilities of human inheritance.

Gene expression can sometimes vary, so that the resulting trait (phenotype) is unpredictable. Examples include alleles that are incompletely penetrant and polygenic traits that result from the combined expression of two or more genes.

Gregor Mendel probably would have been fascinated by the research of modern geneticists, who have the technical ability to explore our genetic heritage in ways Mendel could never have dreamed of. Consider the case of "mom and pop genes."

For decades, a basic tenet of Mendelian genetics was that the phenotype produced by a given form of a gene is always the same, regardless of the parent from whom it is inherited. Generally, this is still true. In the 1980s, however, researchers probing some rare disorders began to suspect a different scenario. They wondered if, in a few cases, whether a gene is expressed depends on its source—whether it is inherited from the mother or from the father.

As things turned out, this hypothesis was correct. The phenomenon is called *genomic imprinting*. A genome is the sum total of genes a person (or any other organism) inherits from her or his parents. And it now seems that certain genes are somehow marked, or imprinted, as "from mother" or "from father."

Tantalizing evidence for genomic imprinting has come from odd inheritance patterns for several disorders. For example, a rare inherited neck tumor (paraganglioma) is caused by a DNA segment on chromosome 11. However, genetic analysis shows that children develop the tumors only if they inherit the tumor-causing gene from their father. The same connection between a paternal imprint and disease may account for several types of childhood cancer, including osteosarcoma, a bone cancer.

Two extremely rare genetic diseases that cause different types of mental retardation also have shed light on genomic imprinting. In *Prader-Willi syndrome*, or PWS, an affected child has failed to inherit a particular section of chromosome 15 from the father. (The whole chromosome or the crucial region of it may be missing.) In *Angelmann syndrome* (AS) an affected child has failed to inherit the same region of chromosome 15 from its *mother*. These findings tell us that normal development demands that we have normal copies of chromosome 15 from both parents.

Imprinting only affects the *expression* of genes, not genes themselves. And it seems to be a factor with just a handful of chromosomes—possibly only six or seven. Its exact mechanism has not yet been deciphered, but it must be related to chemical changes in DNA that track the parental source of a chromosome from early in embryonic development.

SUMMARY

1. Mendel's studies with pea plants provided evidence that genes are specific units passed to offspring.

2. Monohybrid crosses (between parents with different versions of a single trait) demonstrate that a gene can have different forms, called alleles. Some alleles are dominant over other, recessive ones.

3. Individuals who are homozygous dominant for a trait have two dominant alleles (*AA*) for that trait. Individuals who are homozygous recessive have two recessive alleles (*aa*). Heterozygotes have two contrasting alleles (*Aa*).

4. Matings between two heterozygous individuals (*Cc* × *Cc*) produces these combinations of alleles in F₂ offspring:

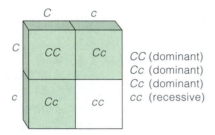

CC (dominant)
Cc (dominant)
Cc (dominant)
cc (recessive)

This results in a probability of 3/4 that any one child will have the dominant phenotype and 1/4 that the child will have the recessive phenotype.

5. Monohybrid crosses provide evidence for the principle of segregation: (1) diploid organisms have two copies of each gene, one on each of the two chromosomes of a homologous pair, and (2) the two copies of each gene segregate from each other during meiosis, so each gamete formed ends up with one gene or the other.

6. A dihybrid cross occurs between two homozygous parents showing different versions of two traits. The result in the F₁ generation is:

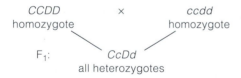

```
    CCDD          ×          ccdd
  homozygote               homozygote

     F₁:          CcDd
              all heterozygotes
```

A mating between two heterozygous parents results in the following probable phenotypes:

```
   CcDd    ×    CcDd

   9 dominant for both traits
   3 dominant C, recessive for d
   3 dominant D, recessive for c
   1 recessive for both traits
```

7. Dihybrid crosses provide evidence of independent assortment during meiosis. Each pair of homologous chromosomes sorts into gametes independently of how other chromosome pairs sort. Therefore, the genes on the chromosomes also assort independently.

8. There are degrees of dominance among some alleles of the same gene. Genes can also interact to produce an effect on a trait. In some cases a single gene can influence many seemingly unrelated traits.

Review Questions

1. Define the difference between: (a) gene and allele, (b) dominant allele and recessive allele, (c) homozygote and heterozygote, and (d) genotype and phenotype. *18.1*

2. State the theory of segregation. Does segregation occur during mitosis or meiosis? *18.2*

3. Distinguish between monohybrid and dihybrid crosses. What is a testcross, and why is it useful in genetic analysis? *18.3, 18.4*

4. What is independent assortment? Does independent assortment occur during mitosis or meiosis? *18.5*

Self-Quiz *(Answers in Appendix V)*

1. Alleles are _____.
   a. alternate forms of a gene
   b. different molecular forms of a chromosome
   c. always homozygous
   d. always heterozygous

2. A heterozygote has _____.
   a. only one of the various forms of a gene
   b. a pair of identical alleles
   c. a pair of contrasting alleles
   d. a haploid condition, in genetic terms

3. The observable traits of an organism are its _____.
   a. phenotype   c. genotype
   b. sociobiology   d. pedigree

4. Offspring of a monohybrid cross *AA* × *aa* are _____.
   a. all *AA*   d. 1/2 *AA* and 1/2 *aa*
   b. all *aa*   e. none of the above
   c. all *Aa*

5. Second-generation offspring from a cross between two homozygotes are the _____.
   a. F₁ generation   c. hybrid generation
   b. F₂ generation   d. none of the above

6. Assuming complete dominance, offspring of the cross *Aa* × *Aa* will show a phenotypic ratio of _____.
   a. 1:2:1   c. 9:1
   b. 1:1:1   d. 3:1

7. Which statement best fits the principle of segregation?
   a. Units of heredity are transmitted to offspring.
   b. Two genes of a pair separate from each other during meiosis.
   c. Members of a population become segregated.
   d. A segregating pair of genes is sorted out into gametes independently of how gene pairs located on other chromosomes are sorted out.

8. Dihybrid crosses of heterozygous individuals (*AaBb* × *AaBb*) lead to F₂ offspring with phenotypic ratios close to _____.
   a. 1:2:1   c. 3:1
   b. 1:1:1:1   d. 9:3:3:1

9. Match each genetic term appropriately.
   ____ dihybrid cross   a. *AA* × *aa*
   ____ monohybrid cross   b. *Aa*
   ____ homozygous condition   c. *AABB* × *aabb*
   ____ heterozygous condition   d. *aa*

Critical Thinking: Genetics Problems
(Answers in Appendix IV)

1. One gene has alleles *A* and *a*. Another has alleles *B* and *b*. For each genotype listed, what type(s) of gametes can be produced? (Assume independent assortment occurs.)
   a. *AABB*   c. *Aabb*
   b. *AaBB*   d. *AaBb*

2. Still referring to Problem 1, what will be the possible genotypes of offspring from the following matings? With what frequency will each genotype show up?
   a. *AABB* × *aaBB*   c. *AaBb* × *aabb*
   b. *AaBB* × *AABb*   d. *AaBb* × *AaBb*

3. A gene on one chromosome governs a trait involving tongue movement. If you have a dominant allele of that gene, you can curl the sides of your tongue upward (see photo). If you are homozygous for recessive alleles of the gene, you cannot roll your tongue. A gene on a different chromosome controls whether your earlobes are attached or detached. People with detached earlobes have at least one dominant allele of the gene. These two genes are on different chromosomes and so they assort independently. Suppose a tongue-rolling woman with detached earlobes marries a man who has attached earlobes and can't roll his tongue. Their first child has attached earlobes and can't roll its tongue.
   a. What are the genotypes of the mother, father, and child?
   b. What is the probability that a second child will have detached earlobes and won't be a tongue-roller?

4. Go back to Problem 1, and assume you now study a third gene having alleles *C* and *c*. For each genotype listed, what type(s) of gametes can be produced?
   a. *AABBCC*   c. *AaBBCc*
   b. *AaBBcc*   d. *AaBbCc*

5. When you decide to breed your purebred Labrador retriever Molly and sell the puppies, you discover that two of Molly's four siblings have developed a hip disorder that is traceable to the action of a single recessive allele. Molly herself shows no sign of the disorder. If you breed Molly to a male Labrador that does not carry the recessive allele, can you assure a purchaser that the puppies will also be free of the condition? Explain your answer.

6. The ABO blood system has been used to settle cases of disputed paternity. Suppose, as a geneticist, you must testify during a case in which the mother has type A blood, the child has type O blood, and the alleged father has type B blood. How would you respond to the following statements:
   a. *Man's attorney*: "The mother has type A blood, so the child's type O blood must have come from the father. Because my client has type B blood, he could not be the father."
   b. *Mother's attorney*: "Further tests prove this man is hetero-zygous, so he must be the father."

7. Soon after a couple marries, tests show that both the man and the woman are heterozygotes for the recessive allele that causes sickling of red blood cells; they are both Hb^A/Hb^S. What is the probability that any of their children will have sickle-cell trait? Sickle-cell anemia?

8. A man is homozygous dominant for 10 different genes that assort independently. How many genotypically different types of sperm could he produce? A woman is homozygous recessive for 8 of these genes and is heterozygous for the other 2. How many genotypically different types of eggs could she produce? What can you conclude about the relationship between the number of different gametes possible and the number of heterozygous and homozygous gene pairs that are present?

9. As is the case with the mutated hemoglobin gene that causes sickle-cell anemia, certain dominant alleles are crucial to normal functioning (or development). Some are so vital that when the mutant recessive alleles are homozygous, the combination is lethal and death results before birth or early in life. However, such recessive alleles can be passed on by heterozygotes (*Ll*). In many cases, these are not phenotypically different from homozygous normals (*LL*). If two heterozygotes mate (*Ll* × *Ll*), what is the probability that any surviving progeny will be heterozygous?

10. Bill and Marie each have flat feet, long eyelashes, and "achoo syndrome" (chronic sneezing). All are dominant traits. The genes for these traits each have two alleles, which we can designate as follows:

	Dominant	Recessive
Foot arch	*A*	*a*
Sneezing	*S*	*s*
Eyelash length	*E*	*e*

Bill is heterozygous for each trait. Marie is homozygous for all of them. What is Bill's genotype? What is Marie's genotype? If they have four children, what is the probability that each child will have the same phenotype as the parents? What is the probability that a child will have short lashes, high arches, and no achoo syndrome?

11. You decide to breed a pair of guinea pigs, one black and one white. In guinea pigs, black fur is caused by a dominant allele (*B*) and white is due to homozygosity for a recessive allele (*b*) at the same locus. Your guinea pigs have 7 offspring, 4 black and 3 white. What are the genotypes of the parents? Why is there a 1:1 ratio in this cross?

Selected Key Terms

Chromosome Chronicles

Some years ago scientists identified the genetic cause of the disease cystic fibrosis. The breakthrough made news because CF is devastating. In the United States roughly one in every 2,000 babies is born with the disease, and historically people who have CF are likely to die from it before they reach thirty.

Several different mutations in a gene labeled *CFTR* can lead to the CF phenotype. The gene is carried on human chromosome number 7. When a mutated form is expressed, the outcome is a defective channel protein in the plasma membrane of cells in exocrine glands. You may remember from Section 4.1 that our exocrine glands secrete a variety of substances, including some digestive enzymes (in the pancreas), sweat, and mucus in the lungs and digestive tract. In cystic fibrosis, the genetic flaw affects the flow of certain ions into and out of cells. As a result, an affected person's exocrine glands secrete an abnormally thick, gluey mucus. Ducts of the pancreas become so clogged with this mucus that enzymes required for proper digestion of proteins, lipids, and carbohydrates never make it to the small intestine. Until the disease is diagnosed, an affected child may suffer from severe malnutrition.

Many CF children begin the day with physical therapy that is designed to literally pound mucus from their lungs (Figure 19.1a). Why? The most dangerous, frightening effect of CF occurs in the lungs. Cells lining respiratory passages crank out huge quantities of the sticky mucus, overwhelming the normal cleansing action of cilia. Breathing then becomes extremely difficult, and the bronchi and lungs are susceptible to infections such as pneumonia. Most CF deaths result from infection and respiratory failure.

Thousands of disorders, including cancers, can arise from errors in our genes or chromosomes. The first abnormal chromosome to be associated with cancer was named the *Philadelphia chromosome* after the city in which it was discovered. It shows up in cells of people who have a chronic type of leukemia. As Chapter 22 will describe, leukemias develop when stem cells in bone marrow overproduce the white blood cells needed for the body's housekeeping and defense.

We can identify chromosomes using a method called karyotyping, in which chromosomes are viewed in a microscope when they are in metaphase—their most condensed form (Section 17.1). Section 19.1

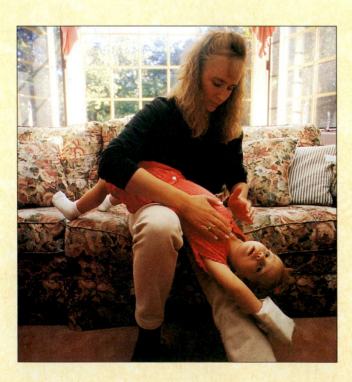

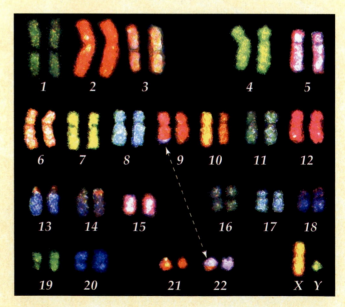

Figure 19.1 (**a**) A daily chest thumping for a child with cystic fibrosis. (**b**) A spectral karyotype of the forty-six chromosomes in a human diploid cell. The arrow indicates the bit of chromosome 22 fused to the Philadelphia chromosome—chromosome number 9 before it was altered.

explains how to construct a karyotype diagram. In a variation on this procedure, called *spectral karyotyping*, the chromosomes are artificially colored in a way that clearly reveals certain differences among them. That method was used to make a karyotype that showed a Philadelphia chromosome, as you can see in Figure 19.1*b*. As it turned out, the Philadelphia chromosome is longer than its normal counterpart, chromosome 9. The extra length is actually a piece of chromosome 22!

What happened? By chance, both chromosomes broke in a stem cell in bone marrow. Then, each broken piece reattached to the *wrong* chromosome—and a gene located at the end of chromosome 9 fused with a gene in chromosome 22. The altered gene's instructions lead to the synthesis of an abnormal protein. In some way that we do not yet understand, that protein stimulates the unrestrained division of white blood cells.

As part of this unit on heredity, this chapter delves more deeply into the topic of chromosomes and the genes they carry. Patterns of inheritance run the gamut from unforeseeable changes in genetic instructions—as with the Philadelphia chromosome—to the predictable appearance of a trait through many generations of a family. Fortunately, as you will learn in this chapter and the two following ones, our rapidly growing understanding of genetic disorders has tremendous promise for improving lives—but brings with it some sobering ethical concerns as well.

KEY CONCEPTS

1. Genes are segments of DNA. Genes follow one another along the length of a chromosome, and each gene has its own position in that sequence.

2. There are two types of chromosomes, autosomes and sex chromosomes, labeled X and Y. Diploid human cells contain twenty-two pairs of autosomes and one pair of sex chromosomes. In normal females the sex chromosome pair consists of two X chromosomes. In normal males the sex chromosome pair consists of an X and a Y.

3. The combination of alleles (alternate forms of a gene) in a chromosome may not stay the same during meiosis and the formation of gametes (sperm or eggs). Through the event called crossing over, some alleles in the sequence exchange places with their partners on the homologous chromosome. The alleles that are exchanged may or may not be identical.

4. The structure of a chromosome may change. For example, a chromosome segment may be deleted, duplicated, inverted, or moved to a new location.

5. The chromosome number may change as a result of improper separation of duplicated chromosomes during meiosis or mitosis.

6. Chromosome structure and the parental chromosome number rarely change. When a change does occur, it often leads to a genetic abnormality or disorder.

CHAPTER AT A GLANCE

THE CHROMOSOMAL BASIS OF INHERITANCE

Genes and Their Locations on Chromosomes

Earlier chapters gave you a general idea of the structure of chromosomes and what happens to them during meiosis. We can now put this information together with some concepts that help explain the major patterns of human inheritance:

1. Genes are units of information about heritable traits. The genes are distributed among a person's chromosomes, and each gene has its own location (called a locus) in one type of chromosome.

2. A cell with a diploid chromosome number (2n) has inherited pairs of **homologous chromosomes**. All but one pair are identical in their length, shape, and gene sequence. The exception is the sex chromosomes, which are designated X and Y.

3. In general, each chromosome has the same genes, in the same locations, as its partner. However, this rule does not apply to the X and Y chromosomes. *They* are homologous only in one small region.

4. A pair of homologous chromosomes can carry identical or nonidentical alleles at a particular locus. **Alleles** are slightly different forms of the same gene. They come about through mutation. Although many different forms of a gene may have arisen in a population, a diploid cell can have only two of them.

5. There is no set pattern to the way that maternal and paternal chromosomes are attached to the microtubular spindle just before metaphase I of meiosis, so different combinations are possible at anaphase I. As a result, gametes and then new individuals receive a mix of alleles of maternal and paternal chromosomes.

6. Genes that are close to one another on the same chromosome are linked. The farther apart two linked genes are, the more likely it is that they will undergo **crossing over**. Remember from Chapter 18 that this is an event by which homologous chromosomes exchange corresponding segments. Crossing over results in **genetic recombination**—combinations of linked alleles in gametes, then in offspring, that were not present in a parent.

7. On rare occasions, abnormal events during meiosis or mitosis change the structure of chromosomes or the chromosome number inherited from a parent.

8. For any trait, the particular combination of alleles inherited by the new individual may be beneficial, harmful, or neutral in effect.

As you will see, these points will help you make sense of inheritance patterns described in later sections.

Autosomes and Sex Chromosomes

It has been said that "men are from Mars, women from Venus. "While that is a topic that goes far beyond biology class, there *are* biological differences between males and females. In the late 1800s, microscopists discovered a chromosome difference between the sexes. In twenty-two of the twenty-three pairs of chromosomes in a human diploid cell, the maternal and paternal chromosomes are roughly the same size and shape. They also carry genes for the same traits. These are **autosomes**. The remaining pair of chromosomes are **sex chromosomes**, and they differ between males and females. A female has two sex chromosomes, both designated "X." Males have one X chromosome and a physically different chromosome that is designated "Y."

Today, special dyes can be used to delineate human autosomes and sex chromosomes at metaphase, when they are most condensed. At that point, each type has a certain length, banding pattern, and so forth. (The bands appear because some regions condense more than others and absorb dyes differently.) As the chapter introduction noted, these features are used to create a **karyotype**—a cut-up, rearranged photograph of the chromosomes of a cell at metaphase. To obtain chromosomes, a technician puts a sample of cells, usually from blood, into a glass container. The container holds a solution that stimulates cell growth and then mitosis. Then a chemical is added that arrests mitosis at metaphase, providing the highly condensed metaphase chromosome for analysis. The chromosomes are arranged in order, essentially from largest to smallest (Figure 19.2). The sex chromosomes are placed last.

Autosomes carry the vast majority of the estimated tens of thousands of human genes. The X chromosome carries more than 2,300 genes. Like other chromosomes, it carries some genes associated with sexual traits, such as the distribution of body fat. But most of the genes on the X chromosome are concerned with nonsexual traits, such as blood-clotting functions. These can be expressed in males as well as in females (males, remember, also carry one X chromosome).

A Y chromosome is quite a bit smaller than an X chromosome, and it carries far fewer genes. (As we write this, about two dozen have been identified.) Biologists know that these genes include instructions for male sexual development, and roles of other Y chromosome genes are being investigated. Despite their differences, the X and Y chromosomes can synapse (be "zippered" together briefly; Section 17.10) in a small region along their length. This allows them to function as homologues during meiosis.

Genes on sex chromosomes may be called "sex-linked genes." However, researchers now use the more precise designations **X-linked gene** or **Y-linked gene**. As you'll read, X-linkage is a central factor in some genetic disorders.

Gene Mutations on Chromosomes

Genes are segments of DNA on chromosomes. As you know, DNA consists of various types of nucleotides linked by chemical bonds (Section 2.13). A **mutation** is a change in one or more of the nucleotides that make up a particular gene. Mutations can arise in various ways, which we consider in Chapter 20. A mutation may have no apparent effect on an individual, it may code for a new, beneficial phenotype, or—as in many of the disorders we consider later in this chapter—it may call for a phenotype that seriously disrupts healthy functioning.

Chromosomes also can become rearranged in ways that are not normal, as described in Section 19.9. Such changes tend to be devastating to the affected individual, with outcomes ranging from mental retardation to death.

Autosomes are the pairs of chromosomes that are the same in males and females. One other pair—the sex chromosomes—govern a new individual's gender.

The two sex chromosomes of females are both X chromosomes. Males have one X chromosome and one Y chromosome.

Genes on the sex chromosomes are sometimes referred to as X-linked or Y-linked, respectively.

Genetic disorders can result from mutations of genes on chromosomes or from an abnormal rearrangement of a chromosome itself.

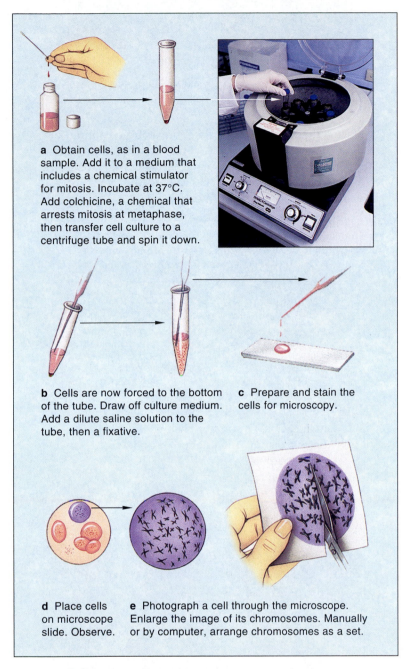

a Obtain cells, as in a blood sample. Add it to a medium that includes a chemical stimulator for mitosis. Incubate at 37°C. Add colchicine, a chemical that arrests mitosis at metaphase, then transfer cell culture to a centrifuge tube and spin it down.

b Cells are now forced to the bottom of the tube. Draw off culture medium. Add a dilute saline solution to the tube, then a fixative.

c Prepare and stain the cells for microscopy.

d Place cells on microscope slide. Observe.

e Photograph a cell through the microscope. Enlarge the image of its chromosomes. Manually or by computer, arrange chromosomes as a set.

Figure 19.2 How to prepare a karyotype (**a**–**e**). (**f**) A human karyotype. Human somatic cells have twenty-two pairs of autosomes and one pair of sex chromosomes (XX or XY). These are metaphase chromosomes; each is in the duplicated state.

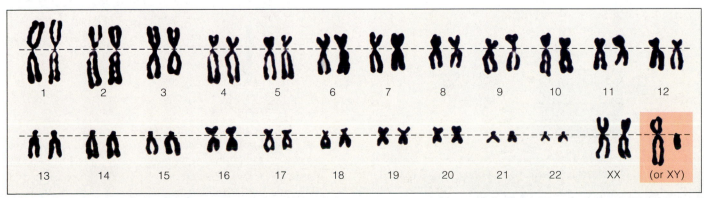

1 2 3 4 5 6 7 8 9 10 11 12

13 14 15 16 17 18 19 20 21 22 XX (or XY)

f

SEX DETERMINATION AND X INACTIVATION

All gametes produced by a female (XX) carry an X chromosome. Half the gametes produced by a male (XY) carry an X and half carry a Y chromosome. Hence, a new individual's sex is determined by the father's sperm because the mother's oocytes can contain only an X chromosome, while the father's sperm can carry either an X or a Y (Figure 19.3). At fertilization, when both the sperm and the egg carry an X chromosome, the new individual will develop into a female. On the other hand, when the sperm carries a Y chromosome, the new individual will be XY and develop into a male.

Recall from Chapter 14 that the few genes on the Y chromosome include a "sex-determining gene" (SRY). Its product causes an embryo to develop testes. In the absence of that gene product, ovaries will develop. Testes and ovaries both produce hormones that govern the development of particular sexual traits.

Since females have two X chromosomes and males have only one, do females have twice as many X-linked genes as males do and therefore have a double dose of their gene products? Generally no, because a compensating phenomenon called **X inactivation** occurs in females. Although the mechanism is not fully understood, it appears that most or all of the genes on one of a female's X chromosomes are "switched off" soon after the first cleavages of the zygote. In a cell, inactivation can occur in either one of the two X chromosomes. The inactivated X is condensed into a *Barr body* (Figure 19.4) that can be seen under the microscope. From then on, descendants of each cell will have the same inactivated X chromosome.

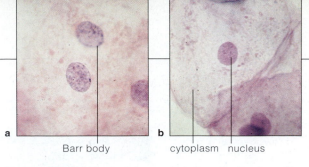

a   b
Barr body   cytoplasm   nucleus

Figure 19.4 (**a**) Nucleus from a female cell, showing a Barr body. (**b**) Nucleus from a male cell, which has no Barr body.

Once X inactivation takes place, the embryo continues to develop. Typically, a female's body has patches of tissue where the genes of the maternal X chromosome are expressed, and other patches where the genes of the paternal X chromosome are expressed.

Human sex is determined by the father's sperm, which can carry either an X chromosome or a Y chromosome. XY embryos develop as males, and XX embryos as females.

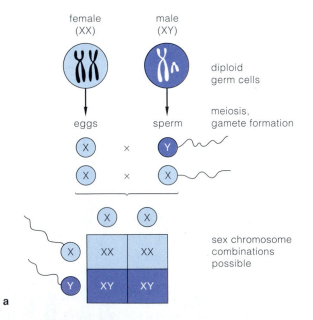

Figure 19.3 (**a**) Pattern of sex determination in humans. Males transmit their Y chromosome to sons but not daughters. Males get their X chromosome from their mother. (**b**) Duct system in the early embryo that develops into a male or female reproductive system.

female (XX)   male (XY)

diploid germ cells

eggs   sperm   meiosis, gamete formation

X × Y
X × X

sex chromosome combinations possible

	X	X
X	XX	XX
Y	XY	XY

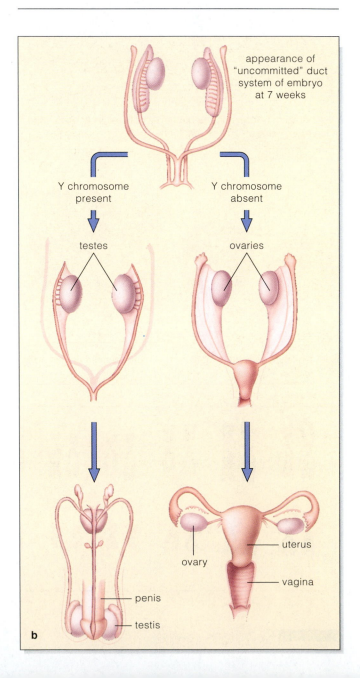

appearance of "uncommitted" duct system of embryo at 7 weeks

Y chromosome present   Y chromosome absent

testes   ovaries

uterus

ovary

vagina

penis

testis

b

Early geneticists, who worked with fruit flies, realized that some traits were inherited as a group from one parent or the other. Today, the term *linkage* is used to describe the tendency of genes located on the same chromosome to end up together in the same gamete. Linked genes are called a **linkage group**.

Linked genes are susceptible to crossing over (Figure 19.5). In fact, in most eukaryotic organisms, humans included, meiosis cannot be completed properly unless *every* pair of homologous chromosomes takes part in at least one crossover.

Imagine any two genes at two different locations on the same chromosome. The probability that a crossover will break up their linkage is proportional to the distance between them. Suppose that genes *A* and *B* are twice as far apart as genes *C* and *D:*

We would expect crossing over to disrupt the linkage between A and B much more often.

Two genes are very closely linked when the distance between them is small; they nearly always end up in the same gamete (Figure 19.6). Linked genes are more susceptible to crossing over when the distance between

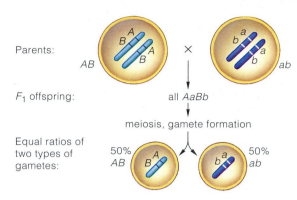

a Full linkage between two genes (no crossovers): half of the gametes have one parental genotype and half have the other.

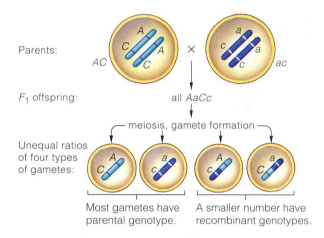

b Incomplete linkage; crossing over affected the outcome.

Figure 19.6 Examples of results from two hybrid crosses (**a**) when gene linkage is complete and (**b**) when crossing over affects the outcome.

them is greater. When two genes are very far apart, crossing over disrupts linkage so often that those genes assort independently of each other into gametes.

The patterns in which genes are distributed into gametes are so regular that they can be used to determine the positions of the various genes on a chromosome. This procedure is called "chromosome mapping." Is it possible to map the tens of thousands of genes on the twenty-three human chromosomes? That has been the goal of the Human Genome Project, and you will read about those exciting efforts in Chapter 21.

The farther apart two genes are on a chromosome, the greater will be the frequency of crossing over and recombination between them.

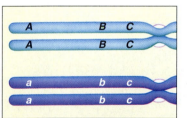

a One pair of homologous chromosomes in the duplicated state (each consists of two sister chromatids). One is shaded *blue*, the other *purple*. Three different genes are shown. At all three gene loci, the two alleles differ.

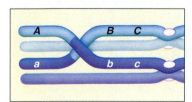

b In prophase I of meiosis, two nonsister chromatids exchange segments. The exchange represents one crossover event.

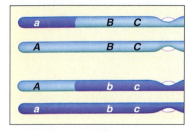

c This is the outcome of the crossover: genetic recombination between nonsister chromatids (which are shown after meiosis, as unduplicated, separate chromosomes).

Figure 19.5 Crossing over and genetic recombination.

HUMAN GENETIC ANALYSIS

Pedigrees: Genetic Connections

Organisms such as fruit flies lend themselves to genetic analysis. They grow and reproduce rapidly in small spaces, under controlled conditions. It doesn't take long to track a trait through many generations.

Humans are another story. We live under variable conditions in diverse environments. We select our own mates and reproduce if and when we want to. Human subjects live as long as the geneticists who study them, so tracking traits through generations can be rather tedious. Most human families are so small there aren't enough offspring for easy inferences about inheritance.

To get around some of the problems associated with analyzing human inheritance, geneticists put together **pedigrees**. A pedigree is a chart that is constructed, by standardized methods, to show the genetic connections among relatives. Figure 19.7 is an example, and it includes definitions of a few of the standardized symbols.

When analyzing pedigrees, geneticists rely on their knowledge of probability and Mendelian inheritance patterns, which may yield clues to a trait's genetic basis.

For instance, they might determine that the responsible allele is dominant or recessive or that it is located on an autosome or a sex chromosome. Gathering many family pedigrees increases the numerical base for analysis. When a trait shows a simple Mendelian inheritance pattern, a geneticist may be justified in predicting the probability with which it will occur among children of prospective parents. We will return to this topic later in the chapter.

A person who is heterozygous for a recessive trait can be designated as a *carrier*. A carrier shows the dominant phenotype (no disease symptoms) but can produce gametes with the recessive allele and potentially pass on that allele to a child. If both parents are carriers for a genetic disorder, a child has a 25 percent chance of being homozygous for the harmful recessive allele.

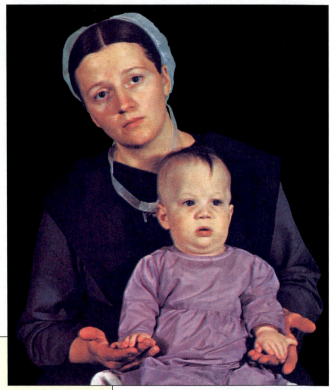

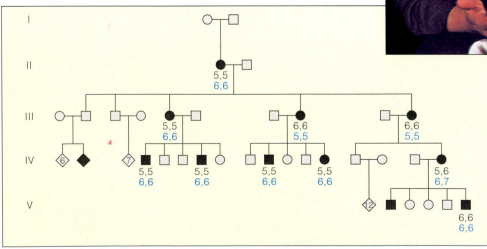

Figure 19.7 (**a**) Some symbols used in constructing pedigree diagrams. (**b**) This is an example of a pedigree for a family in which *polydactyly* occurs. A person with this condition has extra fingers, extra toes, or both. Expression of the gene governing polydactyly can vary from one person to the next. Here, *black* numerals designate the number of fingers on each hand. *Blue* ones designate the number of toes on each foot.

Keep in mind that "abnormality" and "disorder" are not necessarily the same thing. *Abnormal* simply means deviation from the average, such as having six toes on each foot instead of five. A *genetic disorder* is an inherited condition that causes mild to severe medical problems.

Every human being carries an average of three to eight harmful recessive alleles. Why don't alleles that cause severe disorders simply disappear from a population? *First,* rare mutations put new copies of the alleles in the population. *Second,* in heterozygotes, a recessive allele is paired with a normal dominant one that prevents phenotypic expression of the recessive allele, but allows it to be transmitted to offspring.

You may hear a genetic disorder called a disease (cystic fibrosis is a case in point), but the terms aren't always interchangeable. For instance, in some cases, a person's genes increase susceptibility or weaken the response to infection by a virus, bacterium, or some other disease agent. Strictly speaking, however, the resulting illness isn't a genetic disease.

Genetic Counseling

Sometimes prospective parents suspect they are very likely to produce a severely afflicted child. Their first child or a close relative may have a genetic disorder, and they wonder if future children also will be affected. When this is the case, clinical psychologists, geneticists, and other consultants may be asked to provide information and emotional support to parents at risk.

Counseling often begins with a diagnosis of the genotypes of a couple who are contemplating a pregnancy; this may reveal the potential for a child to have a specific disorder. Biochemical tests or DNA analysis can be used to detect many genetic disorders, even in an embryo (see the *Choices* essay). Detailed family pedigrees can aid the diagnosis. For disorders that follow a simple Mendelian inheritance pattern, it is possible to predict the chances a given child will be affected—but not all follow Mendelian patterns. And those that do can be influenced by other factors, some identifiable, others not. Even when the extent of risk has been determined with some confidence, prospective parents must know that the risk is the same for *each* pregnancy. If a pregnancy has one chance in four of producing a child with a genetic disorder, the same odds apply to every subsequent pregnancy.

For many genes, pedigree analysis may reveal Mendelian inheritance patterns that provide information about the probability the genes may be transmitted to children.

A genetic abnormality is an uncommon version of an inherited trait. A genetic disorder is an inherited condition that results in mild to severe medical problems.

19.5 PROMISE—AND PROBLEMS—OF GENETIC SCREENING

Some years ago in Great Britain, a couple who had a child with cystic fibrosis underwent embryo screening (Chapter 15) to make sure that their next baby would not have the disease. The woman's ovaries were hormonally stimulated to produce several eggs at once, and then the eggs were fertilized in the laboratory with the father's sperm. When the resulting embryos reached the eight-cell stage (photograph), genetic engineering methods were used to test each embryo for the presence of CF alleles. Embryos that tested positive were discarded, while two normal embryos were placed in the mother's uterus. One of them implanted and developed into a baby girl, named Chloe, who today is a healthy young woman.

In theory, more than 4,000 genetic disorders can be detected in the early embryo. Researchers already are screening embryos for sickle-cell anemia, Tay-Sachs disease, hemophilia, and both Duchenne and myotonic muscular dystrophy, among other disorders. The procedures are costly. Yet there are couples who are carriers for such disorders and wish desperately to have a child. The catch is, they only want to have a child if it can be guaranteed to be free of the genetic defect.

Procedures that can detect harmful genes in carriers or in offspring come under the heading of *genetic screening*. Common prenatal screening options include amniocentesis and chorionic villus sampling (Section 15.10). The screening of embryos has fueled controversy over how to handle our growing ability to detect defects in genes or chromosomes.

All of this raises the question: Is it wrong to abort a fetus or discard a developing embryo because it carries a genetic disorder—perhaps one with highly damaging effects? Some people answer with a resounding "Yes!" Others believe, on humanitarian grounds, that parents faced with the certainty of a seriously affected child should have the option of *not* bringing such a child into the world.

In the real world of *embryo screening*, there are many potential moral and ethical dilemmas. Consider the fact that physicians who conduct *in vitro* fertilization for otherwise-infertile couples routinely must discard some healthy embryos, simply because the method often produces more embryos than can successfully be implanted in the mother's uterus. Often, embryos are stored (frozen) before being implanted. What happens to them if the couple decides to divorce? Courts now are having to deal with contested "ownership" of frozen embryos. Do these real-life situations raise an ethical red flag for you? Or do you consider them to be unavoidable outcomes of a valuable reproductive technology?

PATTERNS OF AUTOSOMAL INHERITANCE

Autosomal Recessive Inheritance

For some traits, inheritance patterns reveal two clues that point to a recessive allele on an autosome. *First*, if both parents are heterozygous, any child of theirs will have a 50 percent chance of being heterozygous and a 25 percent chance of being homozygous recessive (Figure 19.8). *Second*, if both parents are homozygous recessive, any child of theirs will be too.

Phenylketonuria (PKU) results from abnormal buildup of the amino acid phenylalanine. An affected person is homozygous for a recessive mutant allele that fails to provide instructions for an enzyme that converts phenylalanine to tyrosine, so excess phenylalanine accumulates. If excess phenylalanine is diverted into other metabolic pathways, phenylpyruvic acid and other compounds may be produced. At high levels, phenylpyruvic acid can cause mental retardation. Fortunately, a diet that is low in phenylalanine will alleviate PKU symptoms. Many diet soft drinks and other products are artificially sweetened with aspartame, which contains phenylalanine. Such products must carry warning labels so people with phenylketonuria can avoid using them.

Cystic fibrosis (CF) is another autosomal recessive condition. It is one of the most common genetic disorders among Caucasians. As the chapter introduction noted, major symptoms are the buildup of huge amounts of sticky mucus in the lungs and ducts of the pancreas.

In recent years, various trials of a gene therapy for CF have been carried out. In some, patients inhale a nasal spray containing genetically engineered virus particles that carry functional genes for the affected protein. The

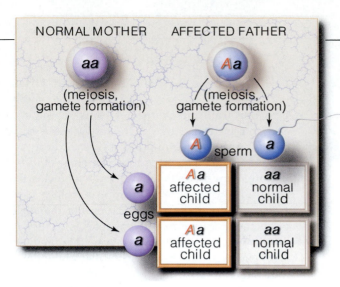

Figure 19.9 A pattern for autosomal dominant inheritance. In this example, the dominant allele (*red*) is fully expressed in the carriers.

idea is that the virus will infect cells in the patients' airways and thereby shuttle the functional gene into the cells, where it is integrated into the DNA and expressed. Thus far, results of the trials are mixed, and some patients have developed immunity to the genetically engineered viruses. We'll look again at gene therapy in Chapter 21.

In some autosomal recessive disorders, the defective gene product is an enzyme required in lipid metabolism. Infants born with *Tay-Sachs disease* lack hexosaminidase A, which is an enzyme required for the metabolism of sphingolipids, a type of lipid that is especially abundant in the plasma membrane of cells in nerves and the brain. Affected babies seem normal at birth, but they steadily lose motor functions and become deaf, blind, and mentally retarded. Most die in the first three years of life.

Tay-Sachs disease is most common among children of eastern European Jewish descent. Biochemical tests before conception can determine whether either member of a couple carries the recessive allele.

Autosomal Dominant Inheritance

Two clues of a different sort indicate that an autosomal dominant allele is responsible for a trait. First, the trait usually appears in each generation, for the allele is usually expressed even in heterozygotes. Second, if one parent is heterozygous and the other is homozygous for the normal, recessive allele, there is a 50 percent chance that any child of theirs will be heterozygous (Figure 19.9). A few dominant alleles that cause severe genetic disorders persist in populations. Some result from spontaneous mutations. In other cases, expression of a dominant allele may not prevent reproduction, or affected people have children before the disorder's symptoms become severe.

An example is *Huntington disorder*, in which the basal nuclei of the brain (Section 11.8) degenerate. In about

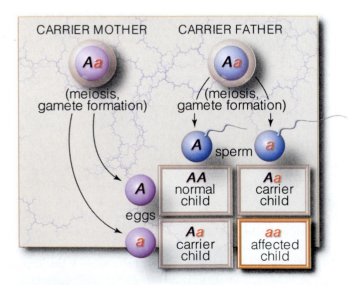

Figure 19.8 A pattern for autosomal recessive inheritance. In this case both parents are heterozygous carriers of the recessive allele (shaded *red* here).

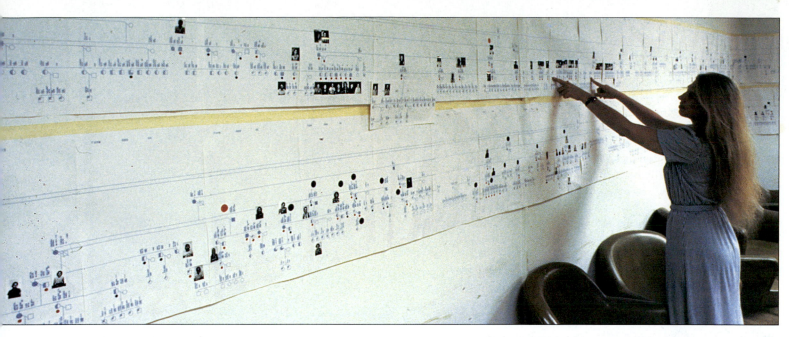

Figure 19.10 From the human genetic researcher Nancy Wexler, a pedigree for Huntington disorder, by which the nervous system progressively degenerates. Wexler's team constructed an extended family tree for nearly 10,000 people in Venezuela. Wexler has a special interest in the disorder because she herself has a 50 percent chance of developing it.

half the cases, symptoms emerge after age 30—when the person may already have had children (hence the disorder persists in the human population). Homozygotes for the Huntington allele die as embryos, so affected adults are always heterozygous. A person who has one parent with the disorder thus has a 50 percent chance of carrying the dominant allele (Figure 19.10). Testing can reveal if an individual has the disease-causing allele, which causes an abnormal sequence of nucleotides in a gene on chromosome 4. Because of the nature of the Huntington defect, there is no cure. Some at-risk people opt not to have the diagnostic test, and many elect to remain childless to avoid passing on the disorder. The harmful mechanism that leads to Huntington's disorder also is responsible for a condition called fragile X syndrome. Section 20.8 examines in detail how both disorders come about.

Another example, *achondroplasia*, affects about 1 in 10,000 people. Homozygous dominant infants usually are stillborn. Heterozygotes can reproduce. However, while they are young and their limb bones are forming, the cartilage elements of those bones cannot form properly. For that reason, at maturity affected people have abnormally short arms and legs (Figure 19.11). Adults who have achondroplasia are under 4 feet, 4 inches tall. Often, the dominant allele has no other phenotypic effects.

About 1 person in 500 is heterozygous for a dominant autosomal allele that causes *familial hypercholesterolemia*. People with this allele have dangerously elevated blood cholesterol because the allele fails to encode the normal number of cell receptors for low-density lipoproteins (LDLs). Recall from Chapter 7 that LDLs bind cholesterol in the blood and thereby help remove it from the cir-

Figure 19.11 A painting by Velazquez of Infanta Margarita Teresa of the Spanish court and her maids, including the achondroplasic woman at the far right.

culation. Rare individuals who are homozygous for the allele may develop severe cholesterol-related heart disease as children and usually die in early adulthood.

In autosomal recessive inheritance, if both parents are heterozygous carriers of the recessive allele, there is a 25 percent chance that a child will be homozygous for the trait and exhibit the recessive phenotype.

In autosomal dominant inheritance, the trait may appear in each generation, because the allele is expressed even in heterozygotes.

X-Linked Recessive Inheritance

Distinctive clues often show up when a recessive allele on an X chromosome causes a trait. *First*, males show the recessive phenotype far more often than females do. A recessive allele can be masked in females, who may inherit a dominant allele on their other X chromosome. It cannot be masked in males, who have only one X chromosome (Figure 19.12). *Second*, a son cannot inherit the recessive allele from his father. A daughter can. If a daughter is heterozygous, there is a 50 percent chance that each son of hers will inherit the allele.

Two forms of the bleeding disorder *hemophilia* are inherited as X-linked recessive traits. The most common one is hemophilia A, which is caused by a mutation in the gene that encodes a protein for blood clotting (see Section 7.12), called factor VIII. Males with a recessive allele on their X chromosome are always affected. They risk death from anything that causes bleeding, including untreated bruises or cuts. The blood clots normally in heterozygous females, because the nonmutated gene on their normal X chromosome produces enough factor VIII to cover the required function. Symptoms are similar in hemophilia B, but the clotting factor is factor IX.

Hemophilia A affects only about 1 in 7,000 males. However, the frequency of the recessive allele was atypically high among 19th-century European royal families. Queen Victoria of England was a carrier, as were two of her daughters. In a pedigree developed some years ago,

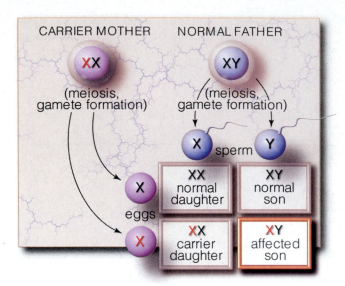

Figure 19.12 One pattern for X-linked inheritance. This example shows the outcomes possible when the mother carries a recessive allele on one of her X chromosomes (*red*).

more than 15 of her 69 descendants at that time were affected males or female carriers (Figure 19.13).

Diseases lumped under "muscular dystrophy" all involve progressive wasting of muscle tissue. One X-linked type, *Duchenne muscular dystrophy* (DMD), affects 1 in 3,500 males, usually in early childhood. As muscles degenerate, affected boys become weak and unable to walk. They

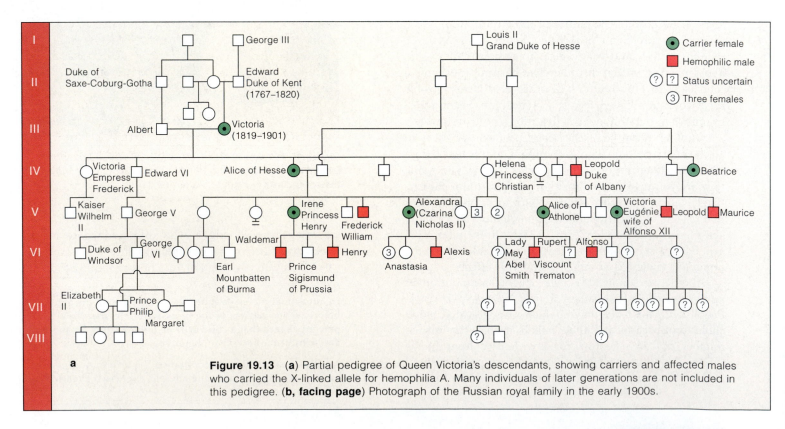

Figure 19.13 (a) Partial pedigree of Queen Victoria's descendants, showing carriers and affected males who carried the X-linked allele for hemophilia A. Many individuals of later generations are not included in this pedigree. (b, facing page) Photograph of the Russian royal family in the early 1900s.

usually die by age 20 from cardiac or respiratory failure. The normal gene that is mutated in DMD encodes the protein *dystrophin*, which gives structural support to the muscle cell plasma membrane. Lacking dystrophin, muscle fibers cannot withstand the physical stress of contraction and break down. Eventually the whole muscle tissue is destroyed.

Red/green color blindness is an X-linked recessive trait. About 8 percent of males in the United States have this condition, which arises from mutation of an allele that codes for a protein (opsin) that binds visual pigments in cone cells of the retina. Females also can have red/green color blindness, but this occurs rarely because a girl must inherit the recessive allele from both parents.

X-Linked Inheritance of Dominant Mutant Alleles

The *faulty enamel trait* is one of a few known examples of a trait caused by a dominant mutant allele that is X-linked. With this disorder, the hard, thick enamel coating that normally protects teeth fails to develop properly (Figure 19.14). The inheritance pattern resembles that for X-linked recessive alleles, except that the trait is expressed in heterozygous females. An affected father transmits the allele to all of his daughters and none of

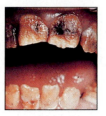

Figure 19.14 The discolored, abnormal tooth enamel of two people affected by the faulty enamel trait.

his sons. If a woman is heterozygous, she will transmit the allele to half her offspring, regardless of their sex.

Another Rare X-Linked Abnormality

Rarely, someone whose sex chromosomes are XY develops as a female. An example is *testicular feminizing syndrome*, or "androgen insensitivity." In an affected person, a mutation in a gene on the X chromosome leads to defective receptors for androgens (male sex hormones), including testosterone and DHT (dihydroxytestosterone). Normally, cells in the testes and other male reproductive organs bind one or more of the hormones and then proceed to develop. With defective receptors, however, they can't bind the hormones and the embryo develops externally as a female. The person is female externally, but has no uterus or ovaries.

A Few Qualifications

The preceding examples of autosomal and X-linked traits give you a general idea of the kinds of clues that hold meaning for trained geneticists. Genetic analysis is usually not an easy task, however. Before diagnosing a case, geneticists often find it necessary to pool together many pedigrees. Typically they make detailed analyses of clinical data and keep abreast of current research. Why? One reason is that more than one type of gene may be responsible for a given phenotype. Geneticists already know of dozens of conditions that can arise from a mutated gene on an autosome *or* a mutated gene on the X chromosome. They know of genes on autosomes that are dominant in males and recessive in females—so they may initially appear to be due to X-linked recessive inheritance, even though they are not.

Czarina Alexandra (a carrier; descendant of Queen Victoria)

Czar Nicholas II (free of allele for hemophilia A)

Alexis (hemophilic son)

b The Russian royal family members. All are believed to have been executed near the end of the Russian Revolution. They were recently exhumed from their hidden graves, but DNA fingerprinting indicated that the remains of Alexis and one daughter, Anastasia, were not among them. At this writing, the search is on for other graves, where the two missing children might have been buried.

A trait that shows up most often in males and that a son can inherit only from his mother most likely is passed on through X-linked recessive inheritance.

A few rare mutant traits are passed to offspring via X-linked dominant inheritance. A heterozygous mother will pass the allele to half her offspring. An affected father will pass the allele only to his daughters.

Some traits are *sex-influenced*—they appear more frequently in one sex than in the other, or the phenotype differs depending on whether the individual is male or female. The difference may reflect the varying influences of male and female sex hormones on gene expression. Genes for such traits are carried on autosomes, not on sex chromosomes. Pattern baldness, shown in Figure 19.15, is an intriguing example. We can designate the "no baldness" form of the responsible gene b^+ and the "baldness" form b. A man will develop pattern baldness if he is homozygous bb and *also* if he is heterozygous b^+b. Females develop pattern baldness only if they are bb, usually later in life and to a lesser degree than men.

Sex-influenced traits appear more frequently in one sex, *or* the phenotype differs among males and females.

Genotype	Male Phenotype	Female Phenotype
b^+b^+	Hair	Hair
b^+b	Bald	Hair
bb	Bald	Bald

Figure 19.15 Pattern baldness. The bald area on the top of this man's head is a typical instance of pattern baldness. Geneticists usually denote the alleles for a sex-influenced trait with lowercase letters. Here, b^+ is "dominant" and b is recessive. Because of the influence of sex hormones, female heterozygotes have a full head of hair. Male heterozygotes, and all homozygous recessive individuals, show pattern baldness, though it is not as extensive in women.

On rare occasions, a chromosome changes structurally through deletion, duplication, inversion, or translocation of one or more of its segments.

Deletion

A **deletion** is the loss of a chromosome region. It may occur spontaneously or it may be caused by a virus, by irradiation, chemical assaults, or some other environmental factor. One or more genes may be lost, as diagrammed in Figure 19.16:

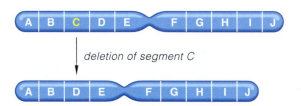

deletion of segment C

Figure 19.16 Simple diagram of a chromosome deletion.

A deletion can occur in any region of a chromosome. The loss may even occur at one end, which destabilizes the chromosome. Wherever a deletion happens, it permanently removes one or more of the chromosome's genes. Gene loss can mean serious problems. For example, one deletion from human chromosome 5 leads to mental retardation and an abnormally shaped larynx. When an affected infant cries, the sounds produced resemble meowing—hence the name of the disorder, *cri-du-chat* (cat cry). Figure 19.17 shows an affected child.

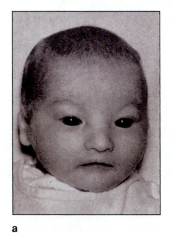
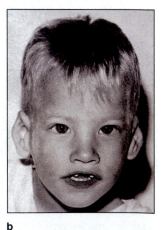

a   b

Figure 19.17 Cri-du-chat syndrome. (**a**) One affected boy just after birth. (**b**) The same boy 4 years later, showing how facial features change. By this age, affected individuals no longer make the mewing sounds typical of the cri-du-chat syndrome.

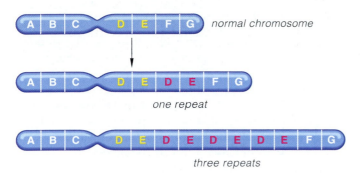

one repeat

three repeats

Figure 19.18 Simple diagrams of chromosome duplications.

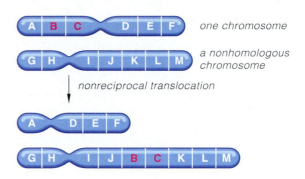

one chromosome

a nonhomologous chromosome

nonreciprocal translocation

Figure 19.20 Simple diagram of a chromosomal translocation.

You might be wondering whether, given that humans have diploid cells, genes on the affected chromosome's homologue compensate for the loss. In fact, this is often the case if a segment deleted from one chromosome is present—and normal—on the homologous chromosome. However, if the remaining, homologous segment is abnormal or carries a harmful recessive allele, nothing will mask *its* effects.

Duplications

Even normal chromosomes contain the changes called **duplications**, which are gene sequences that are repeated several to many times. Often the same gene sequence has been repeated thousands of times. Such a duplication can take several forms, as you can see in Figure 19.18.

No human genetic disorder has yet been linked to a chromosome duplication, although examples are known from other organisms.

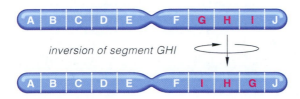

inversion of segment GHI

Figure 19.19 Simple diagram of a chromosome inversion.

Inversions and Translocations

An **inversion** is a chromosome segment that separated from the chromosome and then was inserted at the same place—but in reverse orientation. The reversal alters the position and order of the chromosome's genes (Figure 19.19). We know little about the effects of inversions in human genetics, possibly because this type of change in chromosome structure is rarely detected in humans.

Usually, a **translocation** is part of one chromosome that has exchanged places with a corresponding part of another chromosome that is *not* its homologous partner (Figure 19.20). In certain cancers, for example, a region of chromosome 8 has been translocated to chromosome 14. The disease develops because genes in that region are no longer properly regulated, a topic we return to in Chapter 22.

Genetic disorders can arise from changes in the structure of one or more of a person's chromosomes.

Factors such as viral infection and irradiation can delete a chromosome region. Sometimes the genes of a normal segment on the affected chromosome's homologue can compensate for the loss.

In duplication, a gene sequence is repeated in a chromosome.

In inversion, a chromosome segment separates from the chromosome and then is reinserted in the same place but in reverse orientation.

In translocation, a chromosome region is transferred to a nonhomologous chromosome.

Categories and Mechanisms of Change

Various abnormal cellular events can put too many or too few chromosomes into gametes. New individuals end up with the wrong chromosome number. The effects range from minor physical changes to lethal disruption of body function. More often, affected individuals are miscarried, or spontaneously aborted before birth.

With **aneuploidy**, new individuals have too many or too few chromosomes. This condition is a major cause of human reproductive failure, affecting possibly half of all fertilized eggs. Autopsies have shown that many human fetuses that were miscarried (spontaneously aborted before pregnancy reached full term) were aneuploids.

With **polyploidy**, new individuals have three or more of each type of chromosome. This condition is lethal for humans. All but 1 percent of human polyploids die before birth. The rare newborns die soon after birth.

Chromosome numbers can change during mitosis or meiosis or at fertilization. For instance, a cell cycle might advance through DNA duplication and mitosis, then be arrested before the cytoplasm divides. The cell then is tetraploid, with four of each type of chromosome. Alternatively, one or more pairs of chromosomes might fail to separate during mitosis or meiosis, a phenomenon called **nondisjunction**. Some or all of the resulting cells end up with too many or too few chromosomes (Figure 19.21).

If a gamete with an extra chromosome ($n + 1$) unites with a normal gamete at fertilization, the new individual will be **trisomic**, with three of one type of chromosome ($2n + 1$). If the gamete is missing a chromosome, then the individual will be **monosomic** ($2n - 1$).

Changes in the Number of Autosomes

Most changes in the number of autosomes arise through nondisjunction when gametes are forming. One of the most common of the resulting disorders is trisomy 21—having three copies of chromosome 21. A newborn with trisomy 21 will show the effects of Down syndrome. ("Syndrome" means a set of symptoms characterizing a disorder; typically the symptoms occur together.)

Symptoms of Down syndrome vary greatly, but most affected people are moderately to severely mentally retarded. About 40 percent develop heart defects. Because of abnormal skeletal development, older children have shortened body parts, loose joints, and poorly aligned hip, finger, and toe bones. Their muscles and muscle reflexes are weak, and their motor functions develop slowly. Yet with special training, people with Down syndrome often engage in normal activities (Figure 19.22) and take much enjoyment from life.

Down syndrome is one of the genetic disorders that can be detected by prenatal diagnosis (Section 15.10). About one child of every 1,100 live births has Down syndrome. The probability that a woman will conceive an embryo with Down syndrome rises steeply after age 35. Why? Recall from Chapter 14 that a female's primary oocytes are arrested in meiosis I until many years later, when they are ovulated. Over time, environmental or metabolic factors may damage some oocytes before they can complete meiosis. Another hypothesis is that, as a woman ages, controls that would normally prevent an abnormal embryo from implanting don't operate as well.

Figure 19.21 Example of nondisjunction. Here, chromosomes fail to separate during anaphase I of meiosis and so change the chromosome number in resulting gametes.

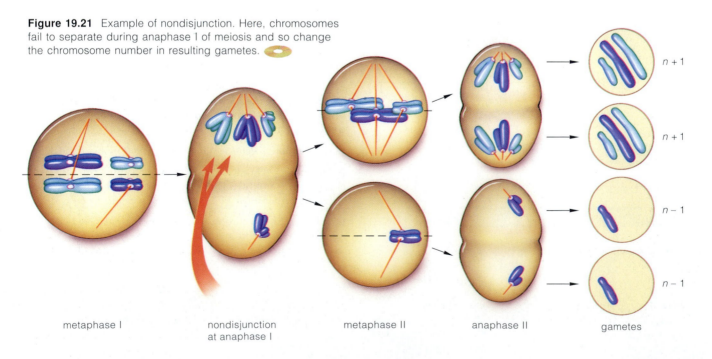

metaphase I   nondisjunction at anaphase I   metaphase II   anaphase II   gametes

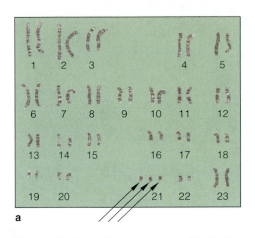

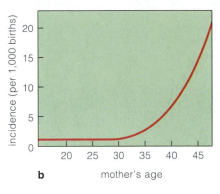

Figure 19.22 (**a**) Karyotype of a girl with Down syndrome; the three arrows identify the trisomy of chromosome 21. (**b**) Relationship between the frequency of Down syndrome and the mother's age. Results are from a study of 1,119 children with the disorder who were born in Victoria, Australia, between 1942 and 1957. (**c**) A child with Down syndrome.

Change in the Number of Sex Chromosomes

Most sex chromosome abnormalities come about as a result of nondisjunction during gamete formation. Let's look at a few phenotypic outcomes.

TURNER SYNDROME AND XXX FEMALES About 1 in every 5,000 newborns is destined to have *Turner syndrome*. Through a nondisjunction, affected individuals have a chromosome number of 45 instead of 46 (Figure 19.23).

They are missing a sex chromosome (and a Barr body), so this is a type of sex chromosome abnormality. It is symbolized as XO. Turner syndrome occurs less often than other sex chromosome abnormalities, probably because most XO embryos are miscarried early in pregnancy. Affected people are female and have a webbed neck and other phenotype abnormalities. Their ovaries do not function, they are sterile, and secondary sexual traits don't develop at puberty. Those with Turner syndrome often age prematurely and have shortened life expectancies.

Roughly 1 in 1,000 females has three X chromosomes. Two of these X chromosomes are condensed to Barr bodies, and most XXX females develop normally.

KLINEFELTER SYNDROME Nondisjunction can give rise to males who show Klinefelter syndrome. In most cases the genotype is XXY (Figure 19.23). This sex chromosome abnormality occurs at a frequency of about 1 in 500 liveborn males. XXY males have low fertility and usually some mental retardation. Their testes are much smaller than normal, body hair is sparse, and there may be some breast enlargement. Injections of testosterone can reverse some aspects of the phenotype, but not the infertility or mental retardation.

XYY CONDITION About 1 in every 1,000 males has one X and two Y chromosomes, an XYY condition. XYY males tend to be taller than average, but have a normal male phenotype.

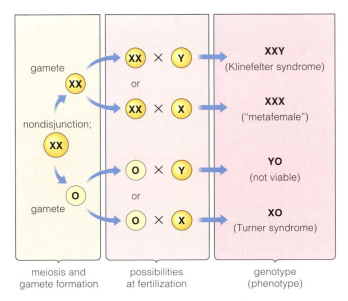

Figure 19.23 Genetic disorders that result from nondisjunction of X chromosomes followed by fertilization by normal sperm.

Most changes in chromosome number arise as a result of nondisjunction during meiosis and gamete formation.

SUMMARY

1. Genes, the units of instruction for heritable traits, are arranged linearly on a chromosome.

2. Human diploid cells have two chromosomes of each type, one from each parent. The two are homologues and pair at meiosis. Sex chromosomes (X and Y) differ from each other in size, shape, and the genes they carry. By contrast, autosomes are roughly the same in size and shape and carry genes for the same traits.

3. Genes on the same chromosome tend to be linked— they tend to stay together during meiosis and end up in the same gamete. However, the farther apart two genes are on a chromosome, the greater will be the frequency of crossing over and recombination between them.

4. A family pedigree is a chart of genetic relationships. It helps establish inheritance patterns and track genetic abnormalities through several generations. Table 19.1 lists some common genetic disorders.

5. In disorders involving autosomal recessive inheritance, a person who is homozygous for a recessive allele shows the recessive phenotype. Heterozygotes generally have no symptoms.

6. In autosomal dominant inheritance, a dominant allele usually is expressed to some extent.

7. Many genetic disorders are X-linked—the mutated gene occurs on the X chromosome. Males, who inherit only one X chromosome, typically are affected.

8. Chromosome structure can be altered by deletions, duplications, inversions, or translocations. Chromosome number can be altered by nondisjunction, in which chromosomes don't separate during meiosis or mitosis.

Review Questions

1. How do X and Y chromosomes differ? *19.1*

2. Why do harmful genes (alleles) persist in a population? *19.6*

3. What evidence indicates that a trait is coded by a dominant allele on an autosome? *19.6*

4. How do X-linked traits and sex-influenced traits differ? *19.7, 19.8*

Self-Quiz *(Answers in Appendix V)*

1. _____ segregate during _____ .
   a. Homologues; mitosis
   b. Genes on one chromosome; meiosis
   c. Homologues; meiosis
   d. Genes on one chromosome; mitosis

2. The alleles of a gene on homologous chromosomes end up in separate _____ .
   a. body cells
   b. gametes
   c. nonhomologous chromosomes
   d. offspring
   e. both b and d are possible

3. Genes on the same chromosome tend to remain together during _____ and end up in the same _____ .
   a. mitosis; body cell   d. meiosis; gamete
   b. mitosis; gamete   e. both a and d
   c. meiosis; body cell

4. The probability of a crossover occurring between two genes on the same chromosome is _____ .
   a. unrelated to the distance between them
   b. increased if they are closer together on the chromosome
   c. increased if they are farther apart on the chromosome
   d. impossible

5. Chromosome structure can be altered by _____ .
   a. deletions   d. translocations
   b. duplications   e. all of the above
   c. inversions

6. Nondisjunction can be caused by _____ .
   a. crossing over in meiosis
   b. segregation in meiosis
   c. failure of chromosomes to separate during meiosis
   d. multiple independent assortments

7. A gamete affected by nondisjunction could have _____ .
   a. a change from the normal chromosome number
   b. one extra or one missing chromosome
   c. the potential for a genetic disorder
   d. all of the above

8. Genetic disorders can be caused by _____ .
   a. gene mutations
   b. changes in chromosome structure
   c. changes in chromosome number
   d. all of the above

9. Match the following chromosome terms appropriately.
   _____ crossing over
   _____ deletion
   _____ nondisjunction
   _____ translocation
   _____ gene mutation

   a. a chemical change in DNA that may affect genotype and phenotype
   b. movement of a chromosome segment to a nonhomologous chromosome
   c. disrupts gene linkages during meiosis
   d. causes gametes to have abnormal chromosome numbers
   e. loss of a chromosome segment

Critical Thinking: Genetics Problems
(Answers in Appendix IV)

1. A female runner is disqualified from a race because testing shows that this individual is XY. Later, a medical exam reveals "androgen insensitivity," an abnormality in which an embryo's cells lack receptors for the male hormone testosterone. As a result, female sex characteristics develop, and indeed the person's phenotype is clearly female. Do you agree or disagree with the disqualification, and why?

2. If a couple has six boys, what is the probability that a seventh child will be a girl?

3. Human sex chromosomes are XX for females and XY for males.
   a. From which parent does a male inherit his X chromosome?
   b. With respect to an X-linked gene, how many different types of gametes can a male produce?
   c. If a female is homozygous for an X-linked allele, how many different types of gametes can she produce with respect to this allele?
   d. If a female is heterozygous for an X-linked allele, how many different types of gametes can she produce with respect to this allele?

Table 19.1 Examples of Human Genetic Disorders and Genetic Abnormalities

Disorder or Abnormality*	Main Consequences	Disorder or Abnormality*	Main Consequences
AUTOSOMAL RECESSIVE INHERITANCE		**X-LINKED RECESSIVE INHERITANCE**	
Albinism *19 CT*	Absence of pigmentation	Red-green color blindness *19.7; 12.9*	Inability to distinguish all or some colors of visible light
Cystic fibrosis *19 CI; 19.6*	Oversecretion of mucus leading to organ damage	Duchenne muscular dystrophy *19.7*	Progressive wasting of muscles
Phenylketonuria (PKU) *19.6*	Mental retardation	Hemophilia *19.7*	Impaired blood-clotting
Sickle-cell anemia *18.7*	Harmful pleiotropic effects on organs throughout body	Testicular feminizing syndrome *19.7, CT*	XY individual has some female traits and is sterile
		Fragile X syndrome *20.9*	Mental retardation
AUTOSOMAL DOMINANT INHERITANCE		**SEX-INFLUENCED INHERITANCE**	
Achondroplasia *19.6*	One form of dwarfism	Pattern baldness *19.8*	Loss of hair on the top and upper sides of the head
Achoo syndrome *18 CT*	Chronic sneezing		
Amyotrophic lateral sclerosis (ALS) *22.8*	Loss of all muscle function	**CHANGES IN CHROMOSOME NUMBER**	
Familial hypercholesterolemia *19.6*	High cholesterol levels in blood; eventually clogged arteries	Down syndrome *19.10*	Mental retardation, heart defects
Huntington disorder *19.6, 20.9*	Nervous system degenerates progressively, irreversibly	Klinefelter syndrome *19.10*	Sterility, retardation in many cases
Polydactyly *19.4*	Extra fingers, toes, or both	Turner syndrome *19.10*	Sterility; abnormal ovaries, abnormal sexual traits
Tay-Sachs disease *19.6*	Progressive deterioration of the nervous system	XYY condition *19.10*	Mild retardation or no symptoms
Marfan syndrome *19 CT*	Absence or abnormal formation of connective tissue	**CHANGES IN CHROMOSOME STRUCTURE**	
X-LINKED DOMINANT INHERITANCE		Cri-du-chat syndrome *19.9*	Mental retardation, abnormal larynx
Faulty enamel trait *19.7*	Abnormal tooth enamel	Chronic myelogenous leukemia *19 CI* (Philadelphia chromosome)	Overproduction of white blood cells followed by organ malfunction

*Italic numbers indicate sections in which a disorder is described. *CI* signifies Chapter Introduction. *CT* signifies an end-of-chapter Critical Thinking question.

4. Marfan syndrome is a genetic disorder with pleiotropic effects—expression of a single mutant allele has widespread effects throughout the body. Connective tissue that normally is present in many organs is absent or malformed. Affected persons are usually tall, thin, and have double-jointed fingers. Their arms, fingers, and lower limbs are disproportionately long. The spine has a pronounced curvature. Often they are nearsighted, and the lenses of their eyes have a tendency to dislocate.

The most serious effects are cardiovascular. Heart valves may open the wrong way when the heart muscle contracts, and the heartbeat is irregular. The aorta, the main artery carrying blood away from the heart, is fragile. Over time its diameter increases and its wall may tear—a deadly rupture.

Marfan's syndrome affects an estimated 40,000 people in the United States, including some prominent athletes. The allele arises by spontaneous mutation. Genetic analysis shows that it follows a pattern of autosomal dominant inheritance. Given this finding, what is the chance that a child will inherit the allele if one parent is heterozygous for it?

5. People with Down syndrome have an extra chromosome 21, for a total of 47 chromosomes in body cells. However, in a few cases of Down syndrome, 46 chromosomes are present. This total includes two normal-appearing chromosomes 21, one normal chromosome 14, and a longer-than-normal chromosome 14. Interpret this observation. How can these individuals have 46 chromosomes?

6. If a trait appears only in males, is this good evidence that the trait is due to a Y-linked allele? Explain why you answered as you did.

7. A woman unaffected by hemophilia A whose father had hemophilia A marries a man who also has hemophilia A. If their first child is a boy, what is the probability he will have the disorder?

8. Among people of European descent, about 4 percent have the allele for cystic fibrosis. Yet only about 1 in 2,500 people actually have the disorder. What is the most likely reason for this finding?

9. People with albinism have very pale skin, white hair, and pink eyes. This phenotype, typically caused by a recessive allele, is due to the absence of melanins, which impart color to the skin, hair, and eyes. Suppose a person with albinism marries a person with typical pigmentation and they have one child with albinism and three with typical pigmentation. What is the genotype of the parent with typical pigmentation? Why is the ratio of this couple's offspring 3:1?

Selected Key Terms

aneuploidy *19.10*	nondisjunction *19.10*
autosome *19.1*	pedigree *19.4*
deletion (chromosomal) *19.9*	polyploidy *19.10*
duplication *19.9*	sex chromosome *19.1*
genetic recombination *19.1*	translocation *19.9*
inversion *19.9*	X inactivation *19.2*
karyotype *19.1*	X-linked gene *19.1*
linkage group *19.3*	Y-linked gene *19.1*

Readings

Cummings, M. R. 2000. *Human Heredity: Principles and Issues*. Fifth edition. Pacific Grove, California: Brooks/Cole.

DNA STRUCTURE AND FUNCTION

Cardboard Atoms and Bent-Wire Bonds

A hundred years ago, biologists were grappling with one of the most fundamental questions in all science: What molecule serves as the book of genetic information in every living cell?

For decades researchers assumed that proteins, which could consist of virtually limitless combinations of amino acids, were the genetic material. That hypothesis made sense, given the awesome diversity of heritable traits. But like all good detective stories, this one would soon develop a twist.

By the late 1940s, there was something about another cellular substance—DNA—that tugged at more than a few good minds. Not that there weren't

obstacles to that line of thinking. Recall from Chapter 2 that DNA consists of only four kinds of subunits. For DNA to be the "master molecule of life," the subunits would have to be arranged in a way that could carry a vast amount of information. And there would have to be a mechanism to copy that structure in a way that would faithfully transfer hereditary information from one generation to the next. Research published in 1950 suggested strongly that DNA was indeed the

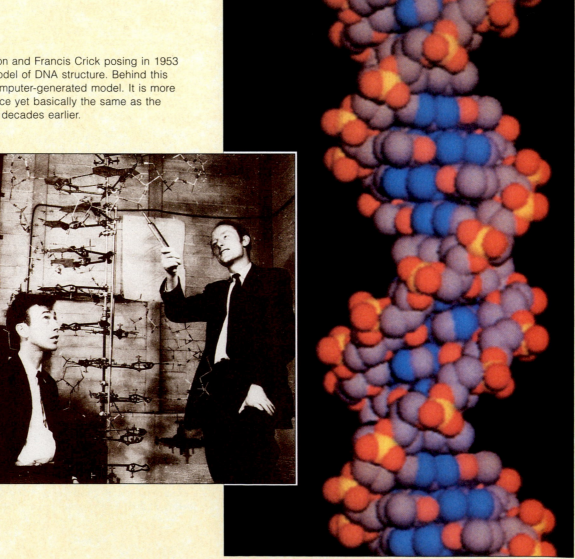

Figure 20.1 James Watson and Francis Crick posing in 1953 by their newly unveiled model of DNA structure. Behind this photograph is a recent computer-generated model. It is more sophisticated in appearance yet basically the same as the prototype built nearly four decades earlier.

genetic material, but still unknown were the crucial structural details that would explain how it functioned.

A small cadre of brilliant researchers dreamed in 1950 of winning the race to elucidate the structure of DNA. Biochemist Linus Pauling—who had already discovered that a single gene could cause sickle-cell anemia—came very, very close. In the end, however, the glory went mostly to James Watson, a 25-year-old postdoctoral student from Indiana University, and to Francis Crick, a Cambridge University researcher who was working on his Ph.D. Watson and Crick spent long hours arguing over everything they had read about the size, shape, and bonding requirements of DNA's subunits. They arranged and rearranged cardboard cutouts of the subunits and pestered chemists to identify potential bonds they might have overlooked.

In 1953, crucial evidence came from the laboratory of Maurice Wilkins, where a researcher named Rosalind Franklin had been analyzing DNA samples using X-ray diffraction—a method that can reveal vital details of a molecule's structure. Aided by clues in the X-ray images, Watson and Crick painstakingly fashioned models of bits of metal held together with wire "bonds" bent at chemically correct angles. In short order, they devised a model that fit what was known about DNA and the pertinent biochemical rules (Figure 20.1).

Watson and Crick had found the structure of DNA and ushered in the golden age of molecular genetics. Together with Maurice Wilkins, they won the 1962 Nobel Prize for medicine. Rosalind Franklin, who had died in 1958, was not considered for the prize.

The remarkable work of these scientists, and of others before and since, is the foundation for the concepts you will study in this chapter. As you read, keep in mind DNA's two main functions. First, it stores the vast number of hereditary instructions required for the building and functioning of each and every body cell, tissue, and organ. Second, by way of *self-replication*, it copies those hereditary instructions in a way that faithfully conveys them to a new generation. The key to both these remarkable feats lies in DNA's structure, and so that is the point at which we will begin.

KEY CONCEPTS

1. DNA is the cell's storehouse of information about heritable traits.

2. Hereditary information is encoded in the nucleotides that make up the DNA molecule. There are four kinds of nucleotides in DNA. They differ in one component, a nitrogen-containing base.

3. In a DNA molecule, two strands of nucleotides twist together like a spiral stairway, forming a double helix. Hydrogen bonds connect the bases of one strand to bases of the other.

4. Before a cell divides, its DNA is replicated. Replication takes place with the help of enzymes and other proteins. Each double-stranded DNA molecule starts unwinding, and then a new, complementary strand is assembled on the exposed bases of each parent strand.

5. There is only one DNA molecule in an unduplicated chromosome. A large number of proteins are attached to the DNA and function in its structural organization.

6. In each of the body's proteins, the sequence of amino acids corresponds to a gene—which is a sequence of nucleotide bases in DNA.

7. The path leading from genes to proteins has two steps, called transcription and translation. In transcription an RNA molecule is assembled on the exposed bases of one of the DNA strands. In translation, instructions in RNA direct the linking of one amino acid after another, in the sequence required to produce a specific polypeptide chain.

8. A permanent change in a gene's base sequence is called a mutation. Mutations are the source of genetic variation—they lead to alterations in the structure or the function of proteins, or both. In turn, the alterations in proteins may lead to differences in traits among individuals.

CHAPTER AT A GLANCE

DNA STRUCTURE

Every chromosome consists of a single DNA molecule and its associated proteins (Section 17.1). A molecule of DNA, in turn, is built from four kinds of **nucleotides**, the building blocks of nucleic acids (Figure 20.2).

A DNA nucleotide is built of a five-carbon sugar (deoxyribose), a phosphate group, and one of the following four nitrogen-containing bases:

adenine	guanine	thymine	cytosine
(A)	(G)	(T)	(C)

Many researchers were in the race to discover DNA's structure, but Watson and Crick were the first to realize that DNA consists of two strands of nucleotides twisted into a double helix. Nucleotides in a strand are linked together, like boxcars in a train, by strong covalent bonds. Weaker hydrogen bonds link the bases of one strand with bases of the other. The two strands run in opposite directions (an *antiparallel* arrangement), as sketched in Figure 20.3 on the facing page.

The shapes of the bases and their sites available for hydrogen bonding determine which bases can pair up. Adenine pairs with thymine, and guanine pairs with cytosine. Therefore, two kinds of **base pairs** occur in DNA: A—T and G—C. In a double-stranded DNA molecule, the amount of adenine equals the amount of thymine, and the amount of guanine equals the amount of cytosine.

At various times in a cell's life, regions of the double helix unwind so that a cell gains access to specific genes. A **gene** is a region of DNA that codes for a specific polypeptide chain. Polypeptide chains, remember, are the basic structural units of proteins. (Later sections describe *how* genes give rise to polypeptide chains.) This chapter's *Science Comes to Life* (Section 20.2) describes a laboratory technique that enables researchers to visualize segments of DNA.

The two long nucleotide strands of the double helix also unwind and separate from each other when DNA replicates. As described next, the exposed bases of each strand serve as a pattern, or *template*, upon which a new strand is built.

All chromosomes in a cell contain DNA. What does DNA contain? Four kinds of nucleotides, A, G, T, and C. Here are the structural formulas for those nucleotides:

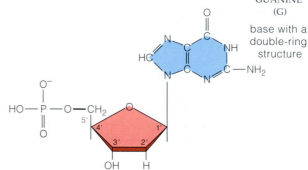

phosphate group

sugar (ribose)

ADENINE (A)
base with a double-ring structure

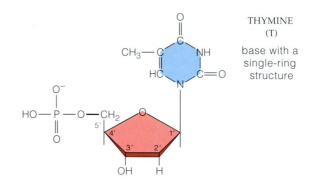

GUANINE (G)
base with a double-ring structure

THYMINE (T)
base with a single-ring structure

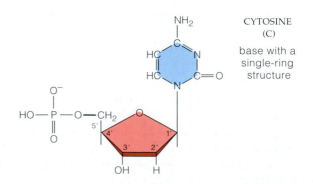

CYTOSINE (C)
base with a single-ring structure

Figure 20.2 The four kinds of nucleotides in DNA. A nucleotide has a five-carbon sugar (shaded *red*), which has a phosphate group attached to its ring structure. Also attached is one of four kinds of nitrogen-containing bases (shaded *blue*). Small numerals on the structural formulas identify the carbon atoms to which these and other parts of the molecule are attached.

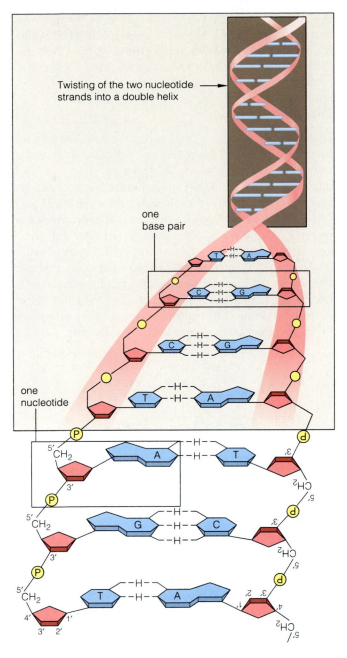

Twisting of the two nucleotide strands into a double helix

one base pair

one nucleotide

Figure 20.3 Arrangement of bases in the DNA double helix.

DNA holds the information for producing heritable traits.

A chromosome consists of a single DNA molecule and its associated proteins. A DNA molecule consists of two strands of nucleotides.

Nucleotides in DNA are each built of the sugar deoxyribose, a phosphate group, and one of the nitrogen-containing bases adenine (A), guanine (G), thymine (T), and cytosine (C).

Hydrogen bonds between pairs of nucleotide bases in DNA hold the DNA strands together in a double helix.

20.2

A WINDOW ON DNA

Each of our genes codes for the amino acid sequence of one polypeptide chain. Is this statement just another fact to memorize for a biology test? Not really. Experiments leading to this basic principle of biology were crucial to the discovery that genes are distinct pieces of DNA molecules.

It took some painstaking—and competitive—scientific sleuthing to provide the necessary clues. In the early 1900s research into the cause of a metabolic disorder suggested that each of the body's enzymes is specified by a different gene (then called a "unit of inheritance"). This was called the "one-gene, one-enzyme" hypothesis.

In the 1940s, as the race to figure out what genes are was heating up, biochemists started using a revolutionary new laboratory technique to obtain information about the chemical differences between different proteins. Called **gel electrophoresis**, the technique is still a staple tool of many laboratories, which use it to distinguish between proteins or between DNA fragments (Figure 20.4). It uses an electric field to force molecules or molecule fragments through a thick gel. The size of the molecule or fragment, and its net surface charge, determine how far it will move through the gel in a given time period.

Back in the 1940s, it was gel electrophoresis that allowed biochemists Linus Pauling (who, recall, was trying to win the DNA race) and Harvey Itano to discover the chemical difference between normal hemoglobin (Hb^A), the oxygen-transport protein in blood, and the defective hemoglobin (Hb^S) associated with sickle-cell anemia (see Section 18.7). Hb^S, they soon learned, has fewer negatively charged amino acids. Later, another researcher spelled out the exact difference—where normal hemoglobin has the amino acid glutamate, the sickle-cell hemoglobin molecule has the amino acid valine.

This discovery suggested that two different genes coded for the two forms of hemoglobin, one for each kind of polypeptide chain. From there it was a small intellectual step to the idea that *genes code for proteins in general*, not just enzymes. And if genes coded for proteins, then to Pauling, Watson, and Crick it could not have seemed overly far-fetched to wonder if genes might consist of some other substance—like DNA.

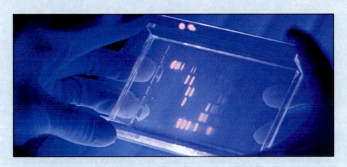

Figure 20.4 Photograph of DNA fragments visualized through gel electrophoresis.

How Is a DNA Molecule Duplicated?

The discovery of DNA structure was a turning point in studies of inheritance. Until then, no one could explain **DNA replication**. Recall from Chapter 17 that when DNA replicates, the result is a pair of duplicated chromosomes (sister chromatids) that are parceled out to daughter cells in mitosis and meiosis.

Hydrogen bonds hold together the two nucleotide strands making up a DNA double helix. Certain enzymes can break those bonds. Through enzyme action, one strand starts to unwind from the other at distinct sites, leaving the bases exposed. Free nucleotides pair with the exposed bases, following the base-pairing rule (A with T and G with C). They are linked by hydrogen bonds. Each parent strand remains intact while a new companion strand is assembled on it, nucleotide by nucleotide.

As replication advances, the newly formed double-stranded molecule twists back into a double helix. One strand is from the starting molecule, so *that* strand is said to be conserved. Only the second strand has been freshly synthesized—so each DNA molecule is really half new and half "old." The DNA replication mechanism is thus called **semiconservative replication.** Figures 20.5 and 20.6 sketch out the basic mechanism.

DNA Repair

During replication, enzymes and other proteins unwind the DNA molecule, keep the two strands separated at unwound regions, and assemble a new strand on each one. For example, **DNA polymerases** carry out the assembly of individual nucleotides on a parent strand. DNA polymerases, DNA ligases, and other enzymes also repair DNA. There is plenty of opportunity for errors and damage to occur. For example, DNA replication takes place at tremendous speed—between ten and twenty nucleotides per second are added at a replication site. Every time a cell replicates its store of DNA before dividing, at least 3 billion nucleotides must be assembled in the proper arrangements. It's hardly any wonder,

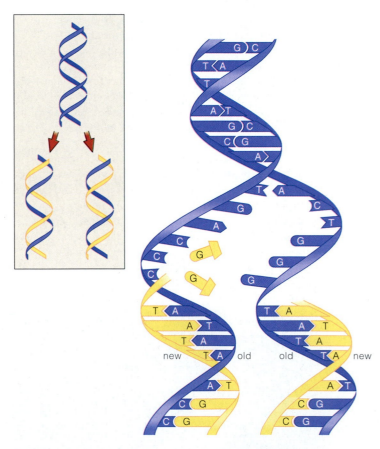

Figure 20.5 Semiconservative nature of DNA replication. The original two-stranded DNA molecule is shown in *blue*. A new strand (*yellow*) is assembled on each parent strand.

a Parent DNA molecule; two complementary strands of base-paired nucleotides.

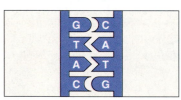

b Replication begins; the two strands unwind and separate from each other at specific sites along the length of the DNA molecule.

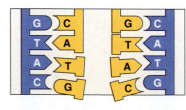

c Each "old" strand serves as a structural pattern (a template) for the addition of bases according to the base-pairing rule.

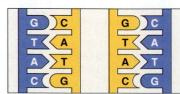

d Bases positioned on each old strand are joined together into a "new" strand. Each half-old, half-new DNA molecule is just like the parent molecule.

Figure 20.6 A closer look at DNA replication.

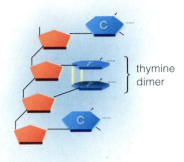

Figure 20.7 Sketch of a thymine dimer. Covalent bonds have formed between two thymines. As a result, the two nucleotides to which the thymines belong form an abnormal bulky structure that may well interfere with replication of the DNA molecule.

then, that "mistakes are made." It has been estimated that a human cell must repair breaks in a strand of its DNA up to 2 million times every hour! Fortunately, if an error occurs during replication, enzymes may detect and correct the problem, restoring the proper DNA sequence. When the error is not corrected, the result is a mutation (Section 20.8).

Our DNA also is vulnerable to damage from certain chemicals and ultraviolet light (as in sunlight or the rays of tanning lamps). The formation of *thymine dimers* is one type of damage. UV light causes two neighboring thymine bases to become linked (forming a dimer; Figure 20.7) and the new structure distorts the affected DNA molecule in a way that increases the chance that replication errors will take place.

When a repair mechanism doesn't work properly (for instance, if mutation changes one or more of the genes that govern it), the impact can be serious. Many researchers are investigating damage to a gene called p53. As described in Chapter 22, the p53 protein responds to cell injury in a way that halts the cell cycle until the problem can be fixed—even if, as sometimes happens, the "fix" is the cell's death. Unrepaired mutations in the p53 gene are thought to be responsible for roughly 50 percent of all cancers. As another example, people who have the rare genetic disorder *xeroderma pigmentosum* have inherited a mutated gene that encodes a DNA repair enzyme. They are highly susceptible to skin cancer because their skin cells cannot repair the damage caused by ultraviolet light.

DNA replication is semiconservative. After the double helix unwinds, each parent strand stays intact and enzymes assemble a new, complementary strand on it.

Enzymes involved in replication also may repair damage in a DNA molecule.

GENE VACCINES

Vaccines harness the weapons of our immune systems, protecting us against influenza, polio, measles, and some other scourges. However, the pathogens that cause such diseases as herpes, AIDS, and malaria still defy our best efforts to hold them at bay. At least, that is the case *now*. Within the next few years, though, medical research may give us the keys to a potent "genetic medicine chest" of DNA vaccines.

The basic idea is simple. You take a bacterial plasmid—a circle of double-stranded DNA (Section 16.1)—and use genetic engineering methods to craft it into a powerful source of antigen. An antigen, recall, is the nonself "red flag" that calls the body's immune defenses into battle (Section 8.4).

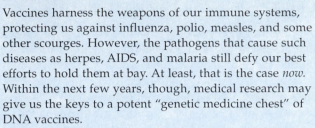

If you remember your reading in Chapter 8, different kinds of invaders elicit differing responses. Those that remain outside cells may trigger the production of antibodies, which circulate in the bloodstream, while those (such as viruses) that infect cells may draw the attention of killer (cytotoxic T) cells. The antigens that evoke an immune response usually are proteins that are specified by the invader's genetic material.

A vaccine is safest and most effective when it calls up both arms of the immune system—antibodies *and* killer cells—without exposing the patient to risk of contracting the disease in question or experiencing some other harmful reaction to foreign proteins. No vaccine available today can meet those demanding standards, and that is where DNA vaccines come in.

As we have learned to decipher the genetic code in DNA, and learned how enzymes snip out particular DNA regions, it has become possible to fashion plasmids that contain certain genes and no others. (Chapter 21 explains the basic laboratory techniques.) Those genes encode pathogen proteins that trigger immune responses—and, of course, the formation of memory cells that confer long-term immunity. In experiments, vaccines consisting of such made-to-order DNA sequences have provoked strong antibody-mediated and cell-mediated immune responses. The plasmids do not contain genes that the source pathogen needs in order to replicate, so they cannot cause disease.

A few experimental DNA vaccines, including ones against HIV, malaria, and hepatitis C, have made their way into early clinical trials on humans. Thus far they seem to be safe, if only modestly effective. Of the many challenges remaining, two of the most important will be learning which genes from various pathogens do the best job of generating immunity and how to make the engineered plasmids survive longer in the body.

DNA INTO RNA—PROTEIN SYNTHESIS BEGINS

A Brief Overview of Protein Synthesis

The path from genes to proteins involves two processes, called transcription and translation. In both, molecules of ribonucleic acid, or **RNA**, have crucial roles. Most often, RNA consists of a single strand. Structurally, it is much like a strand of DNA. Its nucleotides each consist of a sugar (ribose), a phosphate group, and a nitrogen-containing base. However, its bases are adenine, cytosine, guanine, and **uracil**. To sum up, the differences between DNA and RNA are:

	DNA	RNA
Sugar:	deoxyribose	ribose
Bases:	adenine, cytosine, guanine, thymine	adenine, cytosine, guanine, uracil

Like the thymine in DNA, the uracil in RNA base-pairs with adenine.

In **transcription**, molecules of RNA are produced on DNA templates in the nucleus. In **translation**, which you'll read about in Section 20.7, RNA molecules move from the nucleus into the cytoplasm, where they become templates for assembling polypeptide chains. When translation is complete, one or more polypeptide chains are folded into protein molecules. The proteins serve as enzymes in biosynthesis, as membrane receptors and channels, as transport proteins, and so on. They also include most of the enzymes and other proteins that take part in DNA replication, RNA synthesis, and protein synthesis.

In short, DNA guides the synthesis of RNA, then RNA guides the synthesis of proteins.

Genes are transcribed into three types of RNA:

ribosomal RNA (rRNA)   a nucleic acid chain that combines with certain proteins to form a **ribosome**, a structure on which a polypeptide chain is assembled

messenger RNA (mRNA)   a linear sequence of nucleotides that carries protein-building instructions; this "code" is delivered to the ribosome for translation into a polypeptide chain

transfer RNA (tRNA)   another nucleic acid chain that can pick up a specific amino acid *and* pair with an mRNA code word for that amino acid

An important point to remember is that only mRNA eventually is translated into a protein product. The other two types of RNA play key roles in translation.

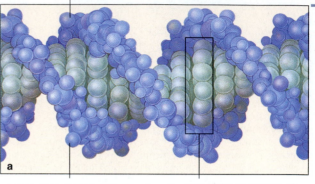

sugar-phosphate backbone of one strand of nucleotides in a DNA double helix

sugar-phosphate backbone of the other strand of nucleotides

part of the sequence of base pairs in DNA

Figure 20.8 The process of gene transcription, by which an RNA molecule is assembled on a DNA template. The sketch in (**a**) shows a gene in part of a DNA double helix. The base sequence of one of the two nucleotide strands is about to be transcribed into an RNA molecule, as shown in (**b**) through (**e**).

Transcription

In transcription, an RNA strand is assembled on a DNA template according to the base-pairing rules:

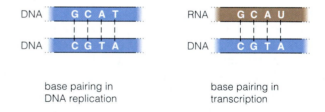

base pairing in DNA replication

base pairing in transcription

Transcription takes place in the cell nucleus, but it is *not* the same as DNA replication. One key difference is that only the gene serves as the template—not the whole DNA strand. A second difference is that enzymes called **RNA polymerases** are involved. Also, transcription produces a single-stranded molecule, not one with two strands.

Transcription starts at a **promoter**—a sequence of bases that signals the start of a gene. Proteins help position an RNA polymerase on the DNA so that it binds with the promoter. As transcription starts, a nucleotide "cap" is added to the beginning of the mRNA for protection. This capped end, which is designated 5', is also where the mRNA will bind to a ribosome when the time comes for translation. The trailing end of the forming mRNA molecule is designated 3'.

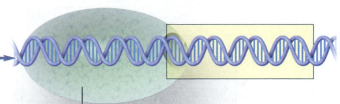

RNA Polymerase

b A molecule of RNA polymerase binds with a promoter region in the DNA. It will recognize the base sequence located downstream from that site as a template for linking together the nucleotides adenine, cytosine, guanine, and uracil into a strand of RNA.

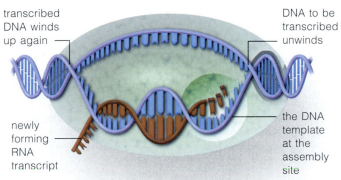

transcribed DNA winds up again

DNA to be transcribed unwinds

newly forming RNA transcript

the DNA template at the assembly site

c All through transcription, the DNA double helix is unwound in front of the RNA polymerase. Short lengths of the newly forming RNA strand briefly wind up with its DNA template strand. New stretches of RNA unwind from the template (and the two DNA strands wind up again).

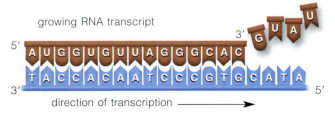

growing RNA transcript

direction of transcription ⟶

d What happened at the assembly site? RNA polymerase catalyzed the base-pairing of RNA nucleotides, one after another, with exposed bases on the DNA template strand.

e At the end of the gene region, the last stretch of the new mRNA transcript is unwound and released from the DNA.

As RNA polymerase moves along the DNA, it joins nucleotides together (Figure 20.8). When it reaches a termination sequence of bases, the RNA transcript is released. A protective "tail" is added to its 3' end.

The newly formed transcript is an unfinished molecule, called "pre-mRNA." It must be modified before its protein-

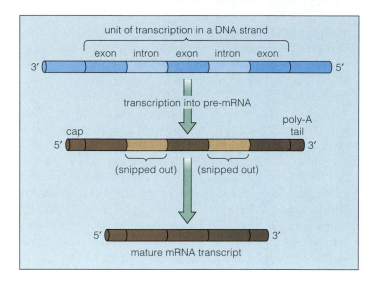

Figure 20.9 Transcription and modification of newly formed mRNA. The cap, a nucleotide, is a site for ribosome attachment. The tail, called poly-A, is a string of adenine nucleotides.

building instructions can be used. For example, when many human genes are transcribed, the pre-mRNA contains portions called introns—in some cases including as many as 100,000 nucleotides! At this writing, **introns** seem to be a sort of genetic gibberish; despite years of looking, researchers have not discovered introns that code for proteins.

All new mRNA transcripts also contain regions called exons. Unlike introns, **exons** are the nucleotide sequences that carry DNA's vital protein-building instructions. Before an mRNA leaves the nucleus, its introns are snipped out and its exons are spliced together (Figure 20.9). The result is a mature mRNA ready to enter the cell cytoplasm and be translated into a protein.

Protein synthesis has two steps, transcription and translation.

In transcription, a sequence of bases in one strand of a DNA molecule is the template for assembling a strand of RNA. The RNA strand is built according to base-pairing rules (adenine pairs with uracil, cytosine with guanine).

Before leaving the nucleus, new RNA transcripts are modified into their final form.

FROM mRNA TO PROTEINS

The Genetic Code

Like a DNA strand, mRNA is a linear sequence of nucleotides. What are the protein-building "words" encoded in its sequence? We now know that enzymes recognize nucleotide bases *three at a time*, as triplets. In mRNA, the base triplets are called **codons**. Figure 20.10 shows how the order of different codons in an mRNA molecule dictates the order in which particular amino acids are added to a growing polypeptide chain. If you count the codons in Figure 20.11, you will see that there are sixty-four kinds. That set of sixty-four different codons is what we call the **genetic code**. The code provides the basic instructions for protein synthesis.

Most of the twenty kinds of amino acids can be "ordered up" by more than one codon. (For example, glutamate corresponds to the code words GAA *or* GAG.) The codon AUG also establishes the reading frame for translation. That is, ribosomes start their "three-bases-at-a-time" selections at an AUG that serves as the "start" signal in an mRNA strand. Three codons (UAA, UAG, and UGA) are stop signals. They stop ribosomes from adding any more amino acids to a new polypeptide chain.

a Base sequence of a gene region in DNA:

| G | C | A | C | C | A | A | T | A | A | C | C | A | T | A |

b Part of an mRNA strand, transcribed from the DNA:

| C | G | U | G | G | U | U | A | U | | U | A | U |

c What the amino acid sequence will be when the mRNA is translated into a polypeptide chain:

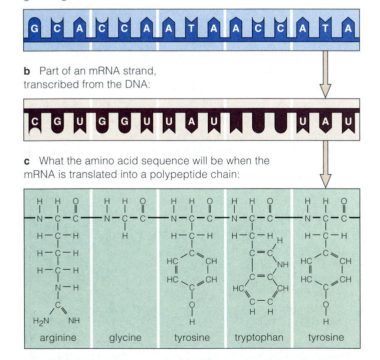

arginine | glycine | tyrosine | tryptophan | tyrosine

Figure 20.10 The steps from genes to proteins. (**a**) This region of a DNA double helix was unwound during transcription. (**b**) Exposed bases on one strand served as a template for assembling an mRNA strand. In the mRNA, every three nucleotide bases equaled one codon. Each codon called for one amino acid in this polypeptide chain. (**c**) Using Figure 20.11, can you fill in the blank codon for tryptophan in the chain?

FIRST BASE	Amino acids that correspond to base triplets: SECOND BASE OF A CODON				THIRD BASE
	U	C	A	G	
U	phenylalanine	serine	tyrosine	cysteine	U
	phenylalanine	serine	tyrosine	cysteine	C
	leucine	serine	STOP	STOP	A
	leucine	serine	STOP	tryptophan	G
C	leucine	proline	histidine	arginine	U
	leucine	proline	histidine	arginine	C
	leucine	proline	glutamine	arginine	A
	leucine	proline	glutamine	arginine	G
A	isoleucine	threonine	asparagine	serine	U
	isoleucine	threonine	asparagine	serine	C
	isoleucine	threonine	lysine	arginine	A
	methionine (or START)	threonine	lysine	arginine	G
G	valine	alanine	aspartate	glycine	U
	valine	alanine	aspartate	glycine	C
	valine	alanine	glutamate	glycine	A
	valine	alanine	glutamate	glycine	G

Figure 20.11 The genetic code. The codons in mRNA are nucleotide bases, "read" in blocks of three. Sixty-one of these base triplets correspond to specific amino acids. Three others serve as signals that stop translation. The left column of the diagram shows the first of the three nucleotides in each codon in mRNA. The middle columns show the second nucleotide. The right column shows the third. Reading from left to right, for instance, the triplet U G G corresponds to tryptophan. Both U U U and U U C correspond to phenylalanine.

Roles of tRNA and rRNA

There are unattached amino acids and tRNA molecules in the cell cytoplasm. Each tRNA has a molecular "hook," an attachment site for a specific amino acid. A tRNA also has an **anticodon**, a nucleotide triplet that can base-pair with codons (Figure 20.12). When a series of tRNAs bind to a series of codons, the matching up of codons and anticodons automatically lines up the amino acids attached to tRNAs in the order that is specified by mRNA.

A cell has more than sixty kinds of codons but fewer kinds of tRNAs. How do the needed match-ups take place? Remember that by the base-pairing rules, adenine pairs with uracil, and cytosine with guanine. However, for codon–anticodon interactions, the rules loosen up for

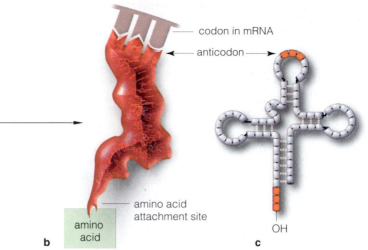

codon in mRNA

anticodon

amino acid attachment site

b amino acid

OH

c

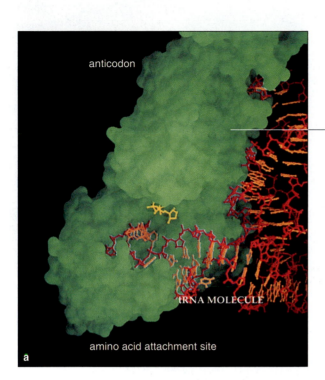

anticodon

tRNA MOLECULE

amino acid attachment site

a

Figure 20.12 (**a**) Computer-generated, three-dimensional model of one type of tRNA molecule. The tRNA (*reddish brown*) is shown attached to a bacterial enzyme (*green*), along with an ATP molecule (*gold*). This particular enzyme attaches amino acids to tRNAs. (**b**) Simplified model of tRNA that is used in later illustrations. The "hook" at one end is the site to which a specific amino acid can be attached. (**c**) Structural features common to all tRNAs.

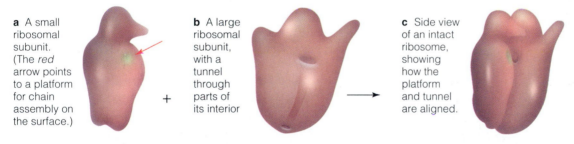

a A small ribosomal subunit. (The *red* arrow points to a platform for chain assembly on the surface.)

+

b A large ribosomal subunit, with a tunnel through parts of its interior

c Side view of an intact ribosome, showing how the platform and tunnel are aligned.

Figure 20.13 Model of eukaryotic ribosomes. Polypeptide chains are assembled on the small subunit's platform. Newly forming chains may move through the large subunit's tunnel.

the third base. For example, CCU, CCC, CCA, and CCG all specify proline but require only two tRNAs. Such freedom in codon–anticodon pairing at the third base is called the "wobble effect."

A cell's tRNAs interact with mRNA at binding sites on the surface of ribosomes. Each ribosome has two subunits (Figure 20.13). The subunits are built in the nucleus from rRNA and proteins; then they are shipped separately to the cytoplasm. There they will combine as functional units only during translation, as described in the next section.

Instructions for building proteins are encoded in the nucleotide sequence of DNA and mRNA. The genetic code is a set of sixty-four base triplets (nucleotide bases, read in blocks of three). A codon is a base triplet in mRNA.

Different combinations of codons specify the amino acid sequences of polypeptide chains.

mRNAs are the only molecules that carry protein-building instructions from a cell's DNA to its cytoplasm.

Specific tRNA anticodons bind specific codons in mRNA. This binding lines up amino acids in the order specified by mRNA. Thus tRNAs translate mRNA into a corresponding sequence of amino acids.

rRNAs are components of ribosomes, where amino acids are assembled into polypeptide chains.

HOW IS mRNA TRANSLATED?

The protein-building code that is built into mRNA transcripts is translated at ribosomes in the cytoplasm. Translation occurs in the cytoplasm. Translation has three stages, called initiation, elongation, and termination.

During the stage called *initiation*, a particular tRNA that can start transcription and an mRNA are both loaded

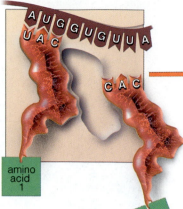

c As the final step of the initiation stage, a large ribosomal subunit joins with the small one. Once this initiation complex has formed, chain *elongation*—the second stage of translation—is about to get under way.

d This close-up of the small ribosomal subunit's platform shows the relative positions of binding sites for an mRNA transcript and for tRNAs that deliver amino acids to the intact ribosome.

e The initiator tRNA has become positioned in the first tRNA binding site (designated *P*) on the ribosome platform. Its anticodon matches up with the START codon (AUG) of the mRNA, which also has become positioned in *its* binding site. Another tRNA is about to move into the platform's second tRNA binding site (designated *A*). It is one that can bind with the codon following the START codon.

ELONGATION

b *Initiation*, the first stage of translating the mRNA transcript, is about to begin. An initiator tRNA (one that can start this stage) is loaded onto the platform of a small ribosomal subunit. The small subunit/tRNA complex attaches to the 5' end of the mRNA. It moves along the mRNA and "scans" it for an AUG START codon.

INITIATION

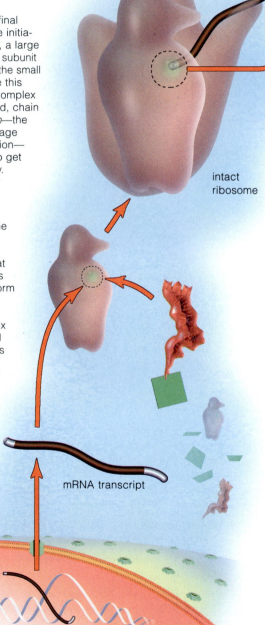

intact ribosome

mRNA transcript

a A mature mRNA transcript leaves the nucleus by passing through pores in the nuclear envelope. It enters the cytoplasm, which contains pools of many free amino acids, tRNAs, and ribosomal subunits.

Figure 20.14 Translation, the second step of protein synthesis.

onto a ribosome. First, the initiator tRNA binds with the small ribosomal subunit. AUG, the start codon for the transcript, matches up with this tRNA's anticodon. At the same time, the AUG binds with the small subunit. Next, a large ribosomal subunit binds with the small subunit. When joined together in this way, the three elements form an initiation complex (Figure 20.14*b*). Now the next stage can begin.

In the *elongation* stage of translation, a polypeptide chain forms as the mRNA strand passes between the ribosomal subunits, like a thread moving through the eye of a needle. Some proteins in the ribosome are enzymes. They join amino acids together in the sequence dictated by the codon sequence in the mRNA molecule. Figure 20.14*f–i* shows how a peptide bond forms between the most recently attached amino acid and the next amino being delivered to the ribosome. (Section 2.11 explains how a peptide bond forms.)

During the last stage of translation, *termination*, a STOP codon in the mRNA moves onto the ribosome platform, and no tRNA has a corresponding anticodon. Now proteins called release factors bind to the ribosome. They trigger enzyme action that detaches the mRNA *and* the chain from the ribosome (Figure 20.14*j–l*).

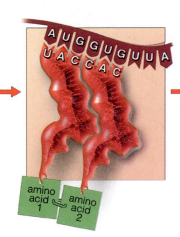

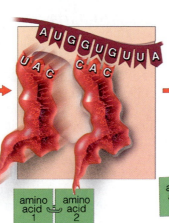

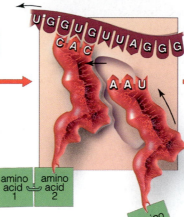

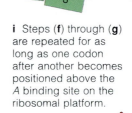

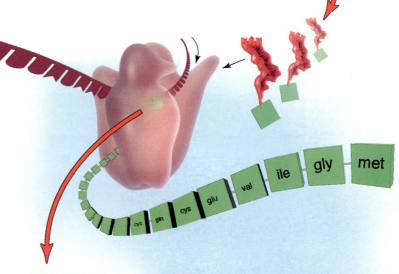

f Enzyme action breaks the bond between the initiator rRNA and the amino acid hooked to it. At the same time, enzyme action catalyzes the formation of a peptide bond between that amino acid and the one hooked to the second tRNA. Then the initiator tRNA is released from the ribosome.

g Now the first amino acid is attached only to the second one—which is still hooked to the second tRNA. This tRNA is about to move into the ribosomal platform's *P* site and slide the mRNA along with it by one codon. This will align the third codon in the *A* site.

h A third tRNA is about to move into the vacated *A* site. Its anticodon is able to base-pair with the third codon of the mRNA transcript. Next, through enzyme action, a peptide bond will form between amino acids 2 and 3.

i Steps (**f**) through (**g**) are repeated for as long as one codon after another becomes positioned above the *A* binding site on the ribosomal platform.

What Happens to the New Polypeptides?

Some cells, such as unfertilized oocytes, may have to rapidly synthesize many copies of different proteins. Such cells stockpile mRNA transcripts in their cytoplasm. In cells that are rapidly using or secreting proteins (such as hormones), *polysomes* are often present. Each polysome is a cluster of many ribosomes that are translating the same mRNA transcript at the same time. The transcript threads through all of them, one after another.

After the new polypeptide chains are complete, many of them become part of the pool of free proteins in a cell's cytoplasm. Many others enter the rough ER of the cytomembrane system (Section 3.4). There they are modified into their final form before they are shipped to their ultimate destinations inside or outside the cell.

Translation is initiated when a small ribosomal unit and an initiator tRNA arrive at an mRNA's start codon and a large ribosomal subunit binds to them.

tRNAs deliver amino acids to the ribosome in the order dictated by the sequence of mRNA codons, to which the tRNA anticodons base pair. A polypeptide chain grows as peptide bonds form between the amino acids.

Translation ends when a stop codon triggers events that cause the chain and the mRNA to detach from the ribosome.

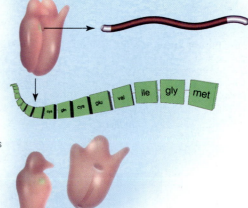

TERMINATION

j A STOP codon moves onto the ribosomal assembly platform. It is the signal to release the mRNA transcript from the ribosome.

k The newly formed polypeptide chain also is released from the ribosome. It is free to join the pool of proteins in the cytoplasm or to enter rough ER of the cytomembrane system.

l The two ribosomal subunits separate.

Whenever a cell puts its genetic code into action, it is making precisely the proteins it needs for its structure and functions. If something changes a gene's code words, the resulting protein may also change. If the protein has a key role in a cell's structure or metabolism, we can expect the outcome to be an abnormal cell.

Every so often, genes *do* change. Sometimes one base gets substituted for another in the nucleotide sequence. At other times, an extra base is inserted into the sequence or a base is lost from it. Such small-scale changes in the nucleotide sequence of genes in a DNA molecule are **point mutations**, often simply called "gene mutations." There is some leeway here, since more than one codon may specify the same amino acid. For example, if a mutation were to change UCU to UCC, it probably would not have dire effects; both codons specify the amino acid serine. More often, however, mutations give rise to proteins with altered or lost functions. Another point to remember is that while mutations can occur in any cell, they are inherited only when they occur in the germline, the cells in gonads that give rise to gametes.

Common Mutations

Figure 20.15 shows a common gene mutation. In this example, one base (adenine) is wrongly paired with another (cytosine) during DNA replication. Proofreading enzymes can recognize an error in a newly replicated strand and fix it. If they don't, a mutation will become established in one DNA molecule (a chromosome) in the next round of replication. As a result of this mutation, a **base-pair substitution**, one amino acid might replace another during protein synthesis.

That is what has happened in people who have sickle-cell anemia; they carry a mutated gene, which specifies the amino acid sequence of Hb^S hemoglobin. The mutation affects the DNA strand coding for the beta chain of hemoglobin (Section 2.12). In a chain 150 amino acids long, only one amino acid is substituted for another—yet, as we now know, the consequences of the substitution can be severe. The harmful effects on organs throughout the body are described in Section 18.7.

Some diseases, including some cases of heart disease, may arise from mutations in the DNA in mitochondria. As you might guess, the damage that results is due to defects in the production of ATP or the cell's ability to use energy.

Figure 20.16 shows a different kind of mutation. Here an *extra* base was inserted into a gene. Recall that RNA polymerases read a nucleotide sequence in blocks of three. In this example, the insertion shifts this three-at-a-time reading frame by one base; hence the name "frameshift mutation." Because the gene is not read correctly, an abnormal protein is synthesized. Frameshift mutations fall into broader categories of gene mutation called **insertions** and **deletions**. In those cases, one to several base pairs are inserted into a DNA molecule or deleted from it.

Figure 20.15 Example of a base-pair substitution.

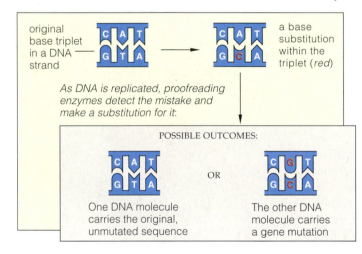

original base triplet in a DNA strand

a base substitution within the triplet (*red*)

As DNA is replicated, proofreading enzymes detect the mistake and make a substitution for it.

POSSIBLE OUTCOMES:

OR

One DNA molecule carries the original, unmutated sequence

The other DNA molecule carries a gene mutation

Figure 20.16 Example of an insertion, a mutation in which an extra base gets inserted into a gene. This insertion has caused a frameshift; it has changed the reading frame for base triplets in the DNA and in the mRNA transcript of the gene. As a result, the wrong amino acids will be put in place when the mRNA transcript is translated into protein.

a

mRNA transcribed from the DNA

PART OF PARENTAL DNA TEMPLATE

ARGININE   GLYCINE   TYROSINE   TRYPTOPHAN   ASPARAGINE   resulting amino acid sequence

b

altered message in mRNA

A BASE INSERTION (RED) IN DNA

ARGININE   GLYCINE   LEUCINE   LEUCINE   GLUTAMATE   the altered amino acid sequence

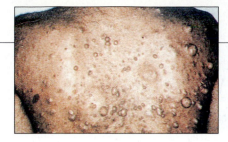

Figure 20.17 Soft skin tumors of a person with neurofibromatosis.

Causes of Gene Mutations

Some mutations result from exposure to mutagens. These are agents that increase the risk of inheritable alterations in the structure of DNA. Common mutagens include ultraviolet light, ionizing radiation, and some chemicals in tobacco smoke. Spontaneous mutations can result when segments of DNA called transposable elements move around. These bits of DNA can move from one location to another in the same DNA molecule or in a different one. Often, if a transposable element is inserted into a gene, the gene is inactivated. The DNA of a human diploid cell typically contains hundreds of thousands of copies of one transposable element (called Alu). When it is inserted in a particular location, a genetic disorder called *neurofibromatosis* results (Figure 20.17).

Figure 20.18 shows an example of yet another category of gene mutation, called **expansion mutations**. Within a gene, a particular nucleotide sequence is repeated over and over again—sometimes hundreds of times. At least nine human genetic disorders are caused by expansion mutations. They include *fragile X syndrome*, in which the brain does not develop properly, and *Huntington disorder* (see Section 19.6). Usually, this degenerative condition results from an expansion of a region of CAG repeats (cytosine, adenine, and guanine).

Whether a mutation ultimately is harmful, neutral, or beneficial depends on a variety of factors, including how the resulting protein affects body functions. We will revisit these ideas in Chapter 23.

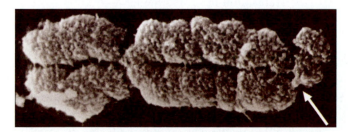

Figure 20.18 A human chromosome showing the constriction characteristic of fragile X syndrome (arrow).

A gene mutation is an alteration in one or more bases in the nucleotide sequence of DNA.

A mutation may have harmful, neutral, or beneficial effects on body structures and functions.

Regulatory Proteins

Most of the cells of your body carry the same genes. Many of those genes carry instructions for synthesizing proteins that are essential to any cell's structure and functioning. Yet each type of cell also uses a small subset of genes in specialized ways. For example, every cell carries the genes for hemoglobin, but only the precursors of red blood cells activate those genes. Only the white blood cells known as lymphocytes activate genes for antibodies. *Each cell determines which genes are active and which gene products appear, when, and in what amounts.* Some genes might be switched on and off throughout a person's life. Other genes might be turned on only in certain cells and only at certain times.

Genes are regulated by way of proteins, hormones, and other molecules that interact with DNA, RNA, or gene products. For example, **regulatory proteins** enhance or suppress the rate of transcription.

A gene may be associated with one or more DNA sequences that do not code for a protein. Instead, the sequences *interact* with regulatory proteins. When a regulatory protein binds with one of these noncoding DNA sequences, it may trigger the transcription of an adjacent gene into mRNA (or shut it down altogether). For instance, think back to the discussion of steroid hormones in Section 13.2. Once inside a target cell, molecules of steroid hormones such as estrogen and testosterone enter the nucleus. There they bind to receptor proteins. Each hormone-receptor complex then can bind to a DNA sequence that does not code for a protein—instead, it acts as a regulator. Specifically, the sequence activates the promoter of one or more genes. As noted in Section 20.5, it is the promoter that triggers the transcription of a gene.

Hormones have widespread effects on gene activity in many cell types. As described in Section 13.3, for instance, the pituitary gland secretes growth hormone. Most cells have receptors for growth hormone, which helps control the synthesis of proteins required for cell division and, ultimately, body growth. On the other hand, as with estrogen and testosterone, some hormones affect only certain cells at certain times. Another example is prolactin, which activates genes in mammary gland cells whose sole task is milk production. Liver cells and heart cells have the same genes, but they do not have the regulatory sequences that would allow them to respond to signals from prolactin.

Although most cells have the same genes, some of those genes are activated or suppressed in different ways to produce differences in cell structure or function.

SUMMARY

1. A gene is a sequence of nucleotide bases in DNA. The bases are adenine, thymine, guanine, and cytosine (A, T, G, and C). Most genes contain protein-building instructions; their nucleotide sequence corresponds to the sequence of amino acids in a polypeptide chain. The path from genes to proteins has two steps, called transcription and translation (Figure 20.19):

2. In transcription, a DNA template is used to synthesize RNA. A double-stranded DNA is unwound at a gene region. Enzymes use the exposed bases as a template to assemble a strand of ribonucleic acid (RNA) from nucleotides present in the cell. Base-pairing rules determine which bases pair up. In RNA, guanine pairs with cytosine and uracil (not thymine) pairs with adenine:

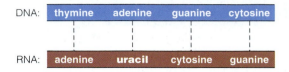

3. Different gene regions are templates for assembling different RNA molecules:

   a. Messenger RNA (mRNA), the only class of RNA that carries protein-building instructions.

   b. Ribosomal RNA (rRNA); together with proteins, (rRNA) becomes a component of ribosomes, the physical units on which amino acids are assembled into polypeptide chains.

   c. Transfer RNA (tRNA), the "go-between" molecules of translation. Different kinds of tRNA pick up different amino acids and deliver them to ribosomes in the sequences specified by mRNAs.

4. RNA transcripts are processed into final form before they are shipped into the cytoplasm.

5. In translation, the three classes of RNAs interact in the synthesis of polypeptide chains. Amino acids are linked one after another, in the sequence required to produce a specific polypeptide chain.

   a. Translation follows the genetic code, a set of sixty-four base triplets. Triplets are a series of nucleotide bases. Ribosomal proteins "read" the series of bases three at a time.

   b. Each base triplet in an mRNA molecule is called a codon. An anticodon is a complementary triplet in a tRNA molecule. A given combination of codons specifies the amino acid sequence of a polypeptide chain, start to finish.

6. Translation has three stages:

   a. Initiation. A small ribosomal subunit and an initiator tRNA bind with an mRNA transcript and move along it until they come to an AUG start codon. The small subunit binds with a large ribosomal subunit.

   b. Chain elongation. tRNAs deliver amino acids to the ribosome. Their anticodons base-pair with mRNA codons. The amino acids are linked (by peptide bonds) to form a new polypeptide chain.

   c. Chain termination. After an mRNA stop codon moves onto the ribosomal platform, the polypeptide chain and mRNA detach from the ribosome.

7. Gene mutations are heritable, small-scale alterations in the nucleotide sequence of DNA. They may be harmful, neutral, or beneficial, depending on how the proteins they specify interact with other genes and with the environment.

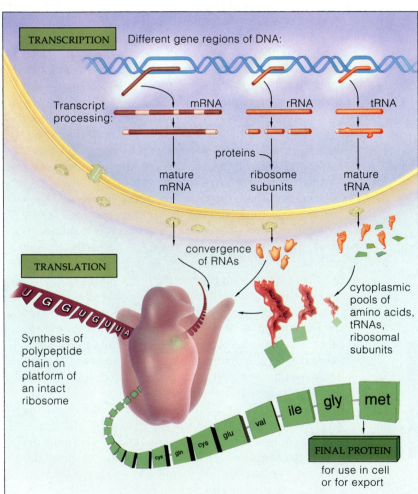

Figure 20.19 Summary of transcription and translation, the two steps leading to protein synthesis. DNA is transcribed into RNA in the nucleus. RNA is translated into proteins in the cytoplasm.

Review Questions

1. Explain why DNA replication is called "semiconservative." *20.1*

2. Are the polypeptide chains of proteins specified by DNA assembled *on* DNA? If not, tell where they are assembled and on which molecules. *20.5*

3. Name the three classes of RNA; define their functions. *20.5*

4. In what key respect does an RNA molecule differ from a DNA molecule? *20.5*

5. Before an mRNA transcript leaves the nucleus, are its introns or exons snipped out? *20.5*

6. Distinguish between a codon and an anticodon. *20.6*

7. Describe the three steps of translation. *20.7*

8. Describe one kind of mutation that leads to an altered protein. What determines whether the altered protein will have beneficial, neutral, or harmful effects on the individual? *20.8*

Self-Quiz *(Answers in Appendix V)*

1. Nucleotide bases, read _____ at a time, serve as the "code words" of genes.

2. DNA contains different genes that are transcribed into _____.
   a. proteins   d. tRNAs
   b. mRNAs   e. b, c, and d
   c. rRNAs

3. RNA molecules are _____.
   a. a double helix   c. always double-stranded
   b. usually single-stranded d. usually double-stranded

4. mRNA is produced by _____.
   a. replication   c. transcription
   b. duplication   d. translation

5. _____ carries coded instructions for an amino acid sequence to the ribosome.
   a. DNA   c. mRNA
   b. rRNA   d. tRNA

6. tRNA _____.
   a. delivers amino acids to ribosomes
   b. picks up genetic messages from rRNA
   c. synthesizes mRNA
   d. all of the above

7. Except for stop codons, a codon calls for a specific _____.
   a. protein   c. amino acid
   b. polypeptide   d. carbohydrate

8. Assuming that the reading frame begins after the start signal AUG, how many amino acids are coded for in this mRNA sequence: AUGUUACACCGUCAC?
   a. three   d. seven
   b. four   e. more than seven
   c. six

9. An anticodon pairs with the nitrogen-containing bases of _____.
   a. mRNA codon   c. tRNA anticodon
   b. DNA codons   d. amino acids

10. The loading of mRNA onto the small ribosomal subunit occurs during _____.
   a. initiation of transcription   c. translation
   b. transcript processing   d. chain elongation

11. Use the genetic code (Figure 20.11) to translate the mRNA sequence AUGCGCACCUCAGGAUGAGAU. (All human reading frames start with AUG.) Which amino acid sequence is being specified?
   a. meth-arg-thr-ser-gly-stop-asp . . .
   b. meth-arg-thr-ser-gly . . .
   c. meth-arg-tyr-ser-gly-stop-asp . . .
   d. none of the above

12. Match the terms related to protein building.

   _____ alters genetic   a. initiation, elongation,
           instructions   termination
   _____ codon   b. conversion of genetic
   _____ transcription   messages in RNA into
   _____ translation   polypeptide chains
   _____ stages of   c. base triplet for an amino acid
           transcription and   d. RNA synthesis
           translation   e. mutation

Critical Thinking: You Decide *(Key in Appendix VI)*

1. Which mutation would be more harmful: a mutation in DNA or one in mRNA? Explain your answer.

2. Some genes in humans are pleiotropic—they have multiple phenotypic effects, as described in Section 18.7. Does this phenomenon contradict the hypothesis that each gene codes for one protein (Section 20.2)? Explain your answer.

3. A DNA polymerase made an error during the replication of an important gene region of DNA. None of the DNA repair enzymes detected or repaired the damage. A portion of the DNA strand with the error is shown here:

After the DNA molecule is replicated and two daughter cells have formed, one cell carries a mutation and the other cell is normal. Develop a hypothesis to explain this observation.

Selected Key Terms

anticodon *20.6*   promoter *20.5*
codon *20.6*   regulatory protein *20.9*
DNA replication *20.3*   ribosomal RNA (rRNA) *20.5*
exon *20.5*   ribosome *20.5*
gel electrophoresis *20.2*   RNA *20.5*
genetic code *20.6*   RNA polymerase *20.5*
intron *20.5*   transcription *20.5*
messenger RNA (mRNA) *20.5*   transfer RNA (tRNA) *20.5*
mutagen *20.8*   translation *20.7*
nucleotide *20.1*   uracil *20.5*
point mutation *20.8*

Readings

Collins, F. and K. Jegalian. December 1999. "Deciphering the Code of Life." *Scientific American.*

RECOMBINANT DNA AND GENETIC ENGINEERING

Mr. Jefferson's Genes

Thomas Jefferson (Figure 21.1*a*) was the third president of the United States, a post he held from 1801 to 1809. Just a few years earlier, he had been the main author of the Declaration of Independence. Renowned in his lifetime as a patriot, statesman, and inventor, two centuries later Jefferson was still making news—but this time in a riveting public discussion of his personal life. At the heart of the debate was the discovery of strong genetic evidence that Mr. Jefferson had fathered at least one child—and possibly as many as seven—with one of his slaves, Sally Hemings.

Some of the evidence for Jefferson's relationship with Hemings was circumstantial. A widower from the age of 39 onward, Jefferson employed Sally Hemings in his household for years. And while Jefferson traveled widely, historians had been able to place him and Sally in the same locations nine months before each of

her children were born. There were also consistent oral traditions in a few African-American families—histories that, generation by generation, enumerated more than 200 years of Jefferson–Hemings descendants.

Until 1998 little other information was available to shed light on the purported, and highly controversial, liaison. Then, British researchers dropped a bombshell. They reported that, using a genetic engineering method, they had established with close to 100 percent certainty that a modern male descendant of Sally Hemings has a Y chromosome that had been passed down to him through the generations from Thomas Jefferson.

The British DNA detectives had used a method called DNA fingerprinting, or *RFLP analysis*, which you will learn more about later in this chapter. They worked

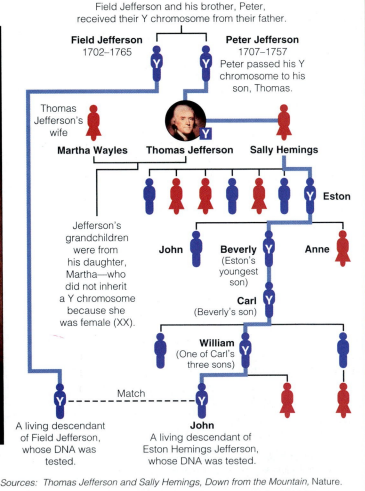

Figure 21.1 (**a**) Thomas Jefferson. (**b**) Pedigree showing the line of descent from Jefferson's postulated union with Sally Hemings. The line was traced by identifying genetic markers, called RFLPs, on the Y chromosome.

with DNA in blood samples gathered from known male descendants of Jefferson's line and from known male descendants of Hemings. Males were chosen for the study because only they have a Y chromosome—and, conveniently, the Y has relatively few genes. Except for rare, random mutations, the Y is passed from father to son virtually unchanged. Like other human chromosomes, it has characteristic stretches of noncoding DNA between genes (the introns discussed in Chapter 20). The nucleotide sequences of these noncoding regions serve as *genetic markers* that can be compared when Y chromosomes from different males are examined. There will be few or no differences in the markers passed down from father to son, even through many generations.

Jefferson received his Y chromosome from his father, Peter. He didn't have any sons, but his father's brother, Field Jefferson, did. For the British study, the DNA tested came from a male descendant of Field Jefferson, who would have had the same Y chromosome as Peter—and as Thomas. They also had Y chromosome DNA from a descendant of Eston Hemings, one of Sally's sons. Nineteen markers were compared. As a control, the analysis included the same markers from Y chromosomes of 1,200 other males. As you can see in Figure 21.1*b*, the researchers were able to track markers on the Jefferson Y chromosome all the way to modern descendants of Eston Hemings and Field Jefferson.

While biotechnology can solve genetic mysteries, its most staggering potential lies in medicine, industry, and agriculture. As you read these words, some biologists are using such methods to insert human genes into mice as part of the search for treatments for cancer, AIDS, and other diseases. Others are exploring the frontiers of gene therapy for inherited disorders. And more than likely, your local drugstore carries at least some bioengineered products, including the hormone insulin required by people with type 1 diabetes—insulin that is produced by vats of genetically altered bacteria.

Around the world, at a mind-bending pace, researchers are learning how to mix and match genes of various species. At the same time, other people are sounding notes of caution, even fear, about what this "brave new world" could bring. With so many actual and potential applications to human health and well-being, we are well advised to understand the basics of biotechnology. In this chapter we'll also consider some of the ecological, social, and ethical questions such methods raise.

KEY CONCEPTS

1. Genetic experiments have been occurring in nature for billions of years through gene mutations, crossing over and recombination, and other events.

2. Humans are now purposefully bringing about genetic changes by way of recombinant DNA technology. Such enterprises are called genetic engineering.

3. With this technology, researchers isolate, cut, and splice together genes from different species, then greatly amplify the number of copies of the genes that interest them. The genes and, in some cases, the proteins they specify are produced in quantities large enough for research and for practical applications.

4. Three activities are at the heart of recombinant DNA technology. First, procedures based on specific enzymes are used to cut DNA molecules into fragments. Second, the fragments are inserted into cloning tools, such as plasmids. Third, fragments containing the genes of interest are identified, then copied rapidly, over and over.

5. In genetic engineering, genes from an organism are isolated, modified, and inserted back into the same organism or into a different one. The goal is to beneficially modify traits that the genes influence. Human gene therapy, which focuses on controlling or curing genetic disorders, is an example.

6. The new technology raises social, legal, ecological, and ethical questions about its benefits and risks.

CHAPTER AT A GLANCE

A TOOLKIT FOR MAKING RECOMBINANT DNA

For at least 3 billion years, nature has been conducting countless genetic experiments, through mutation, crossing over, and other events that introduce changes in genetic messages. This is the source of life's diversity.

For many thousands of years, we humans have been changing genetically based traits of species. Through artificial selection, we produced modern crop plants and new breeds of cattle, birds, dogs, and cats from wild ancestral stocks. We developed meatier turkeys and sweeter oranges, larger corn, and other useful plants. We produced hybrids such as the tangelo (tangerine × grapefruit) and mule (donkey × horse).

Researchers now use **recombinant DNA technology** to analyze genetic changes. With this technology, they cut and splice DNA from different species, then insert the modified molecules into bacteria or other types of cells that can rapidly replicate genetic material and divide. The cells copy the foreign DNA along with their own. In short order, huge populations produce useful quantities of recombinant DNA molecules. The new technology also is the basis of **genetic engineering**, in which genes are isolated, modified, and inserted back into the same organism or into a different one.

Restriction Enzymes

Recombinant DNA technology is made possible by the genetic workings of bacteria. As you may remember from Chapter 16, bacteria have only one chromosome, a circular DNA molecule with all the genes needed for growth and reproduction. Many bacteria also have plasmids (Figure 21.2), which are small, circular molecules of "extra" DNA

that contain a few genes (Section 16.2). The bacterium's replication enzymes can copy and reproduce plasmid DNA, just as they do with chromosomal DNA.

Over time, to combat invasion by viruses, bacteria evolved defenses that included restriction enzymes. A **restriction enzyme** can recognize and cut apart specific sequences of four to eight bases in DNA. Today, plasmids and restriction enzymes are essential parts of a toolkit for doing genetic recombination in the laboratory.

Many restriction enzymes make staggered cuts. The cuts leave single-stranded "tails" on the end of a DNA fragment (shown below). Depending on the particular molecule being cut, the resulting fragments may be quite short; others will be tens of thousands of bases long. That is long enough to be useful for studying the organization of a **genome**, which is all the DNA in a species' haploid number of chromosomes. The human genome is about 3.2 billion base pairs long.

Modification Enzymes

DNA fragments with staggered cuts have what are called "sticky" ends. "Sticky" means that a restriction fragment's single-stranded tail can base-pair with a complementary tail of any other DNA fragment or molecule cut by the same restriction enzyme. If you mix together some DNA fragments cut by the same restriction enzyme, the sticky ends of any two fragments that have complementary base sequences will base-pair. They form a recombinant DNA molecule:

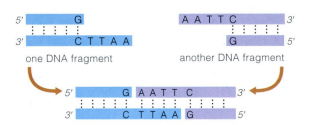

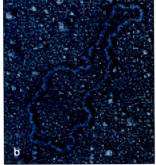

In the sketch below, notice the nicks where such fragments base-pair. A modification enzyme called **DNA ligase** can seal these nicks:

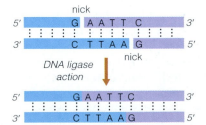

Figure 21.2 (**a**) Plasmids (*blue* arrows) released from a ruptured *Escherichia coli* cell. (**b**) A plasmid at high magnification.

Further reading: Student Guide to InfoTrac on web site →

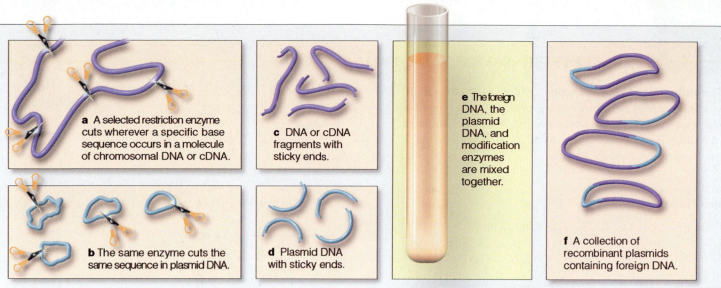

a A selected restriction enzyme cuts wherever a specific base sequence occurs in a molecule of chromosomal DNA or cDNA.

b The same enzyme cuts the same sequence in plasmid DNA.

c DNA or cDNA fragments with sticky ends.

d Plasmid DNA with sticky ends.

e The foreign DNA, the plasmid DNA, and modification enzymes are mixed together.

f A collection of recombinant plasmids containing foreign DNA.

g Host cells able to divide rapidly take up recombinant plasmids.

Figure 21.3 (**a–f**) Formation of recombinant DNA—in this case, a collection of chromosomal DNA fragments or cDNA that are sealed into bacterial plasmids. (**g**) Recombinant plasmids are inserted into host cells that can rapidly amplify the foreign DNA of interest.

Cloning Vectors for Amplifying DNA

In the laboratory, we can use restriction and modification enzymes to insert foreign DNA into bacterial plasmids. The result is called a DNA clone, because when a bacterium replicates a plasmid, it makes many identical, "cloned" copies of it. A modified plasmid that can accept foreign DNA is called a **cloning vector**. It serves as a taxi for delivering foreign DNA into a host cell that can divide rapidly (such as a bacterium or a yeast cell). This can be the start of a cloning factory—a population of rapidly dividing descendants, all with identical copies of the foreign DNA (Figure 21.3). As they divide they *amplify*—make much more of—the foreign DNA.

Reverse Transcriptase to Make cDNA

Researchers analyze genes to unlock secrets about gene products (proteins) and how they are put to use. However, even if a host cell takes up a foreign gene, it might not be able to synthesize the protein encoded by the gene. For example, recall from Chapter 20 that most eukaryotic genes have noncoding sequences called introns. New mRNA transcripts of those genes cannot be translated until introns are snipped out and coding regions (exons) are spliced together. Bacteria do not have the chemical means to remove introns, so they often can't translate human genes into proteins.

Researchers solved this problem with **cDNA**, a DNA strand copied from a mature mRNA transcript. Reverse transcriptase, an enzyme found in RNA viruses (Section 16.1), catalyzes transcription in reverse. That is, it assembles a complementary DNA strand on an mRNA transcript (Figure 21.4). Other enzymes remove the RNA from the

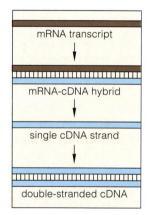

mRNA transcript

mRNA-cDNA hybrid

single cDNA strand

double-stranded cDNA

a A mature mRNA transcript of a gene of interest is used as a template to assemble a single strand of DNA. The enzyme reverse transcriptase catalyzes the assembly. An nRNA–cDNA hybrid molecule is the result.

b By enzyme action, the mRNA is removed and a second strand of DNA is assembled on the DNA strand remaining. The outcome is a double-stranded cDNA, as "copied" from an mRNA template.

Figure 21.4 Formation of cDNA from an RNA transcript.

hybrid mRNA-cDNA molecule, make a complementary DNA strand, and so make double-stranded cDNA.

Double-stranded cDNA can also be modified. For instance, chemical signals for transcription and translation can be added to it before it is inserted into a plasmid for amplification. Then the recombinant plasmids can be inserted into bacteria. If all goes well, the bacteria will follow the cDNA instructions and synthesize the desired protein.

Restriction enzymes and modification enzymes can cut apart chromosomal DNA or cDNA and splice it into plasmids and other cloning vectors. The recombinant plasmids can be inserted into bacteria or other rapidly dividing cells that can make multiple, identical copies of the foreign DNA.

PCR—A FASTER WAY TO AMPLIFY DNA

The **polymerase chain reaction**, widely known as **PCR**, is another way to amplify fragments of chromosomal DNA or cDNA. These copy-making reactions occur in test tubes, not in bacterial cloning factories. Primers get them going.

What Are Primers?

Primers are man-made, short nucleotide sequences (about ten to thirty nucleotides long) that base-pair with any complementary sequences in DNA. The workhorses of DNA replication—the DNA polymerases—chemically recognize primers as START tags. Following a computer program, machines synthesize a primer one step at a time. How do researchers decide on the order of those nucleotides? First, they must identify short nucleotide sequences located just before and just after a DNA region from a cell that interests them. Then they build primers that have the complementary sequences.

What Are the Reaction Steps?

For PCR, the enzyme of choice is a DNA polymerase extracted from *Thermus aquaticus*, a bacterium that lives in hot springs (even in water heaters). The enzyme is not destroyed at the elevated temperatures required to unwind a DNA double helix. (Such high temperatures will denature and destroy the activity of most DNA polymerases.)

Researchers mix together primers, the polymerase, DNA from an organism, and nucleotides. Next, they expose the mixture to precise temperature cycles. During each temperature cycle, the two strands of all the DNA molecules in the mixture unwind from each other.

Primers become positioned on exposed nucleotides at the targeted site according to base-pairing rules (Figure 21.5). Each round of reactions doubles the number of DNA molecules amplified from the target site. For example, if there are 10 such molecules in the test tube, there soon will be 20, then 40, 80, 160, 320, and so on. A target region from a single DNA molecule can be swiftly amplified to billions of molecules.

In short, PCR amplifies samples that contain even tiny amounts of DNA. At this writing, the procedure is used in thousands of laboratories all over the world. As you will see in this chapter's *Science Comes to Life*, such samples can be obtained from even a single hair follicle or drop of blood left at the scene of a crime.

PCR is a method of amplifying DNA inside test tubes. Compared to cloning methods, PCR is amazingly rapid.

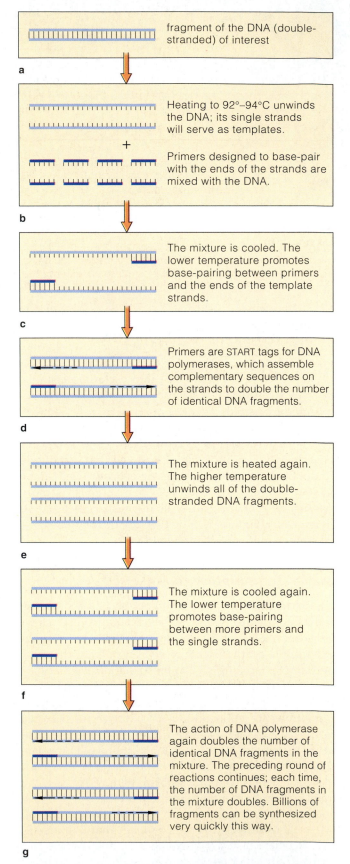

a   fragment of the DNA (double-stranded) of interest

b   Heating to 92°–94°C unwinds the DNA; its single strands will serve as templates.

+

Primers designed to base-pair with the ends of the strands are mixed with the DNA.

c   The mixture is cooled. The lower temperature promotes base-pairing between primers and the ends of the template strands.

d   Primers are START tags for DNA polymerases, which assemble complementary sequences on the strands to double the number of identical DNA fragments.

e   The mixture is heated again. The higher temperature unwinds all of the double-stranded DNA fragments.

f   The mixture is cooled again. The lower temperature promotes base-pairing between more primers and the single strands.

g   The action of DNA polymerase again doubles the number of identical DNA fragments in the mixture. The preceding round of reactions continues; each time, the number of DNA fragments in the mixture doubles. Billions of fragments can be synthesized very quickly this way.

Figure 21.5 The polymerase chain reaction (PCR).

21.3 DNA FINGERPRINTS

Other than identical twins, no two people have exactly the same sequence of bases in their DNA. By detecting the differences in DNA sequences, a scientist can distinguish one person from another. As you know, each human has a unique set of fingerprints, a marker of his or her identity. Each person also has a **DNA fingerprint**—a unique array of DNA fragments, inherited from parents in a Mendelian pattern. Forensic scientists can use DNA fingerprinting to identify criminals and crime victims from the DNA in blood, semen, or small amounts of tissue left at a crime scene. Because DNA fragments that make up a DNA fingerprint are inherited, scientists can use them to resolve questions of parentage. In fact, DNA fingerprints are so accurate that even full siblings are easily distinguished from one another.

More than 99 percent of the DNA is exactly the same in all humans. But DNA fingerprinting focuses only on the part that tends to differ from one person to the next. Throughout the human genome are tandem repeats—short regions of repeated DNA—that are very different from person to person. (Tandem means "two together"; the repeated DNA region is double-stranded.) For example, the five bases TTTTC are repeated anywhere from four to fifteen times in tandem in different people, and three bases (CGG) are repeated five to fifty times in tandem.

During DNA replication, the number of repeats may increase or decrease. Mutation rates in tandem repeats are much higher than the rates at most other sites in our DNA. Because such mutations occurred over many generations in countless family lineages, each person is usually heterozygous for a repeat number at any given tandem-repeat locus. That is, the number of times bases are repeated in each DNA strand is usually different. By examining many tandem-repeat sites, researchers learned that each person carries a unique combination of repeat numbers.

We detect differences in tandem-repeat sites with gel electrophoresis, the method decribed in Section 20.2. Recall that this laboratory procedure uses an electric field to move molecules through a viscous gel. In this case, it separates the DNA fragments according to their length. Size alone dictates how far each fragment moves through the gel, so the tandem repeats of different sizes migrate at different rates through the gel.

A gel is immersed in a buffered solution, and then DNA fragments from different people are added to the gel. When an electric current is applied to the solution, one end of the gel takes on a negative charge and the other end a positive charge. DNA molecules carry a negative charge (because of the negatively charged phosphate groups), so they migrate through the gel toward the positively charged pole. They do so at different rates,

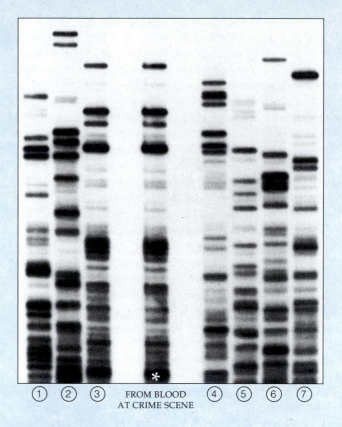

Figure 21.6 DNA fingerprint from a crime-scene bloodstain, with DNA fingerprints from blood of seven suspects (circled numbers). Which one of the seven was a match?

and so they separate into bands according to length. The smaller the fragment, the farther it will migrate through the gel. After a set time, researchers identify fragments of different lengths by staining the gel or by specifically highlighting fragments that contain tandem repeats.

Figure 21.6 shows a series of tandem-repeat DNA fragments that were separated by gel electrophoresis. They are DNA fingerprints from blood collected at a crime scene. Notice how much the DNA fingerprints differ—and how only one exactly matches the pattern from the crime scene.

DNA fingerprinting is now firmly established as an accurate way to identify individuals. The variation in tandem repeats also can be detected as *restriction fragment length polymorphisms*, or RFLPs ("riff-lips"). These are DNA fragments of different sizes that have been cut out of DNA by restriction enzymes. In the case of tandem repeats, a restriction enzyme cuts the DNA on either side of the repeat. Differences in the size of the fragments, which reveal genetic differences, can be detected with electrophoresis.

HOW IS DNA SEQUENCED?

In 1995, researchers accomplished a feat that was little more than a dream a few decades ago. They determined the entire DNA sequence of an organism—*Haemophilus influenzae*, a bacterium that causes respiratory infections. Not long afterward, the genomes of several other species were fully sequenced. Recently, the hotly pursued race to determine the sequence of all 3.2 billion nucleotides of the human genome ended when scientists in a private laboratory announced they had reached that milestone.

How did they do it? The key was **automated DNA sequencing**. This laboratory technique can reveal the sequence of either cloned DNA or PCR-amplified DNA in a few hours. The technique has all but replaced the time-consuming, expensive methods that preceded it.

For automated DNA sequencing, researchers use the four standard nucleotides (T, C, A, and G). They also use four modified versions, which we can represent as **T***, **C***, **A***, and **G***. Each modified version has been labeled. That is, attached to it is a molecule that fluoresces (lights up) a particular color when it passes through a laser beam. Each time a modified nucleotide is incorporated in a growing DNA strand, it blocks DNA synthesis.

Before the reactions, researchers mix together the eight kinds of nucleotides. To that mixture they add the millions of copies of the DNA to be sequenced, a primer, and DNA polymerase. Then they separate the DNA into single strands, and the reactions begin.

The primer binds with its complementary sequence on one of the strands. DNA polymerase synthesizes a new DNA strand behind the primer. One by one, it adds nucleotides in the order dictated by the exposed sequence. Each time, one of the standard nucleotides, or one of the modified versions, may be attached.

If DNA polymerase encounters, say, a T in a template strand, it will catalyze the base-pairing of either A or **A*** to it. If A is added to the growing strand, replication will continue. But if **A*** is added, replication will stop; the modified nucleotide will *block* the addition of more nucleotides to that DNA strand. The same thing happens at each nucleotide in a template strand. If a standard nucleotide is added, replication will continue; if a modified version is added, replication stops.

Remember, the starting mixture contained millions of identical copies of a DNA sequence. Because either a standard or a modified nucleotide could be added at each exposed base, the new strands ended at different places in the sequence. The mixture now contains millions of copies of tagged fragments of different lengths. These can be separated by length into *sets* of fragments, with each fragment set corresponding to just one of the nucleotides in the entire base sequence.

The automated DNA sequencer is a machine that separates the sets of fragments by gel electrophoresis. The

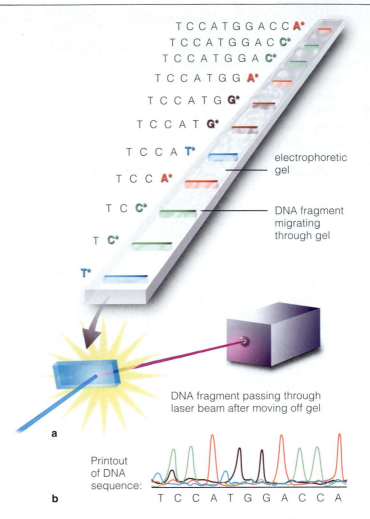

electrophoretic gel

DNA fragment migrating through gel

DNA fragment passing through laser beam after moving off gel

a

Printout of DNA sequence:

T C C A T G G A C C A

b

Figure 21.7 Automated DNA sequencing. (**a**) DNA fragments from the genome of an organism are labeled at their ends with a modified nucleotide that fluoresces a particular color. (**b**) Printout of the DNA sequence in this example. Each peak indicates a particular labeled nucleotide.

set having the shortest fragments migrates fastest through the gel and peels away from it first. The last set to peel away has the longest fragments. Because its fragments have a modified nucleotide at the 3′ end, the set fluoresces a particular color as it passes through a laser beam. The automated sequencer detects the color and thus determines which nucleotide is on the end of the fragments in each set. It assembles the information from all the nucleotides in the sample and reveals the entire DNA sequence.

Figure 21.7 shows the printout from an automated DNA sequencer. Each peak in the tracing represents the detection of a particular color as the sets of fragments reached the end of the gel.

With automated DNA sequencing, the order of nucleotides in a cloned or amplified DNA fragment can be determined.

FROM HAYSTACKS TO NEEDLES—ISOLATING SPECIFIC GENES

Any genome consists of thousands of genes. There may be more than 100,000 in the human genome, for example. What if you wanted to learn about or modify the structure of any one of those genes? First you would have to isolate that gene from all others in the genome.

There are several ways to do this. If part of the gene sequence is already known, you can design primers to amplify the whole gene or part of it with PCR. Often, though, researchers must isolate and clone a gene. First they make a **gene library**. This is a mixed collection of bacteria that house different cloned DNA fragments, one of which is the gene of interest. A *genomic* library contains cloned DNA fragments from an entire genome. A *cDNA* library contains DNA derived from mRNA. In general, a cDNA library is the most useful, because there are no introns (noncoding sequences) in it. Even then, however, the gene of interest is like a needle in a haystack. How can you isolate it from all the others in the library? One way is to use a nucleic acid probe.

What Are Probes?

A probe is a very short stretch of DNA, labeled with a radioisotope so that it can be distinguished from other DNA molecules in a given sample. Part of a probe must be able to base-pair with some portion of the gene of interest. Any base-pairing that takes place between sequences of DNA (or RNA) from different sources is called **nucleic acid hybridization**.

Where does a suitable probe come from? Sometimes part of the gene or a closely related one has already been cloned, in which case it can serve as a probe. If the gene's structure is a mystery, it still may be possible to work backward from the amino acid sequence of its protein product, if that is available. By using the genetic code as a guide (Section 20.5), you might be able to build a DNA probe that is more or less similar to the gene of interest.

Screening for Genes

Once you have a DNA library and a suitable probe, you are ready to isolate the gene. Figure 21.8 shows the steps of one isolation method. The first step is to take bacterial cells of the library and spread them on a petri plate. When that is done properly, individual bacteria will divide. Each cell starts a colony of genetically identical cells. The bacterial colonies appear as hundreds of tiny white spots on the surface of a culture medium.

After colonies appear, you lay a nylon filter on top of the colonies. Some cells stick to the filter at locations that mirror the sites of the original colonies. After chemical solutions rupture the cells, the released DNA sticks to the filter. Now, another chemical treatment makes the DNA

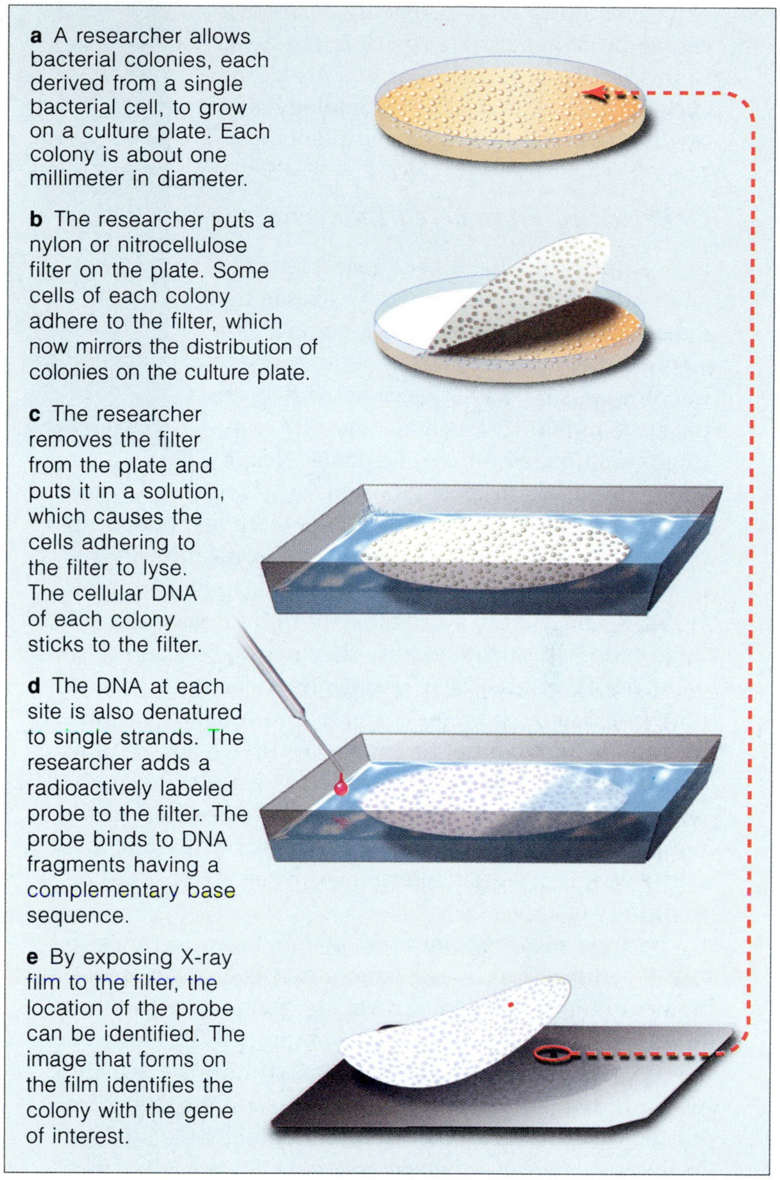

a A researcher allows bacterial colonies, each derived from a single bacterial cell, to grow on a culture plate. Each colony is about one millimeter in diameter.

b The researcher puts a nylon or nitrocellulose filter on the plate. Some cells of each colony adhere to the filter, which now mirrors the distribution of colonies on the culture plate.

c The researcher removes the filter from the plate and puts it in a solution, which causes the cells adhering to the filter to lyse. The cellular DNA of each colony sticks to the filter.

d The DNA at each site is also denatured to single strands. The researcher adds a radioactively labeled probe to the filter. The probe binds to DNA fragments having a complementary base sequence.

e By exposing X-ray film to the filter, the location of the probe can be identified. The image that forms on the film identifies the colony with the gene of interest.

Figure 21.8 Use of a probe to identify colonies of bacteria that have taken up a gene library.

unwind to single strands; then the probes are added. The probes hybridize only with DNA from the colony that took up the gene of interest. If the probe-hybridized DNA is exposed to X-ray film, the radioactivity forms a pattern that will identify that colony. With this information you can now culture cells from that one colony, which will replicate only the cloned gene.

Probes may be used to identify a particular gene among many in gene libraries. Bacterial colonies that have taken up the library can be cultured to isolate the genes.

ENGINEERING BACTERIA AND PLANTS

Any genetically engineered organism that carries one or more foreign genes is *transgenic*. In this section we take a look at some of the current applications of biotechnology that involve transgenic bacteria and plants.

Genetically Engineered Bacteria

For centuries we have been using bacteria and yeasts as biochemical workhorses in the manufacture of cheeses, bread, yogurt, and alcoholic beverages. In recent decades, these microorganisms have been harnessed to produce antibiotics (such as penicillin from fungi) and many other medicinal drugs. Today, plasmids in bioengineered bacteria carry a range of human genes, and those genes are expressed to produce large quantities of clinically useful human proteins. Many of these proteins, such as human growth hormone, once were available only in tiny amounts and were extremely costly because they had to be chemically extracted from glandular tissues. In addition to growth hormone, human proteins currently produced by bacteria include insulin and interferons (Figure 21.9). Insulin, once available only from the pancreas glands of slaughtered pigs and cattle, helps regulate glucose metabolism (Section 13.6), and it is essential to the survival of people with type 1 diabetes. Interferons, recall, have roles in immunity (Chapter 8).

Bacteria may also be used as miniature factories to manufacture vaccines, substances that trigger the development of antibodies to a particular antigen (such as an influenza virus) and so provide immunity to that antigen (Section 8.10). Vaccines have traditionally been made using weakened or killed microbes, which carry on their surface proteins that trigger antibody production by the infected individual. However, as described in Section 20.4, it is

Figure 21.10 (**a**) Three aspen seedlings genetically engineered for a percent higher ratio of cellulose, compared to the control plant at left. Wood harvested from such trees might make it easier to manufacture paper and some clean-burning fuels, such as ethanol. The altered trees also grew 25–30 percent faster than unaltered ones.

(**b**) *Left*: Unmodified cotton plant. *Right*: Genetically engineered cotton plant with a gene for herbicide resistance. Both plants had been sprayed with a weedkiller that is widely applied in cotton fields.

possible to make a vaccine by inserting the gene for the antigen protein into a plasmid, then growing the transgenic bacteria to produce large quantities of the protein. The protein is used as a vaccine, which can then mobilize a person's immune system against that particular antigen.

Genetically engineered bacteria may also become major weapons in efforts to clean up pollution, a process called *bioremediation*. *Focus on Our Environment* (Section 21.7) describes some examples of this "environmental biotechnology."

Designer Plants

Many years ago, plant scientists who were culturing cells from carrot plants managed to get some of them to develop into small embryos. Some of the embryos even grew into whole plants. The feat was big news, in those days. Today, researchers routinely regenerate crop plants and many other plant species from cultured cells. They use various methods to pinpoint a gene in a culture that contains, say, millions of cells. What happens if the researcher includes a toxin from a plant pathogen in the culture medium? If any cells carry a gene that confers resistance to the toxin, they will be the only ones left.

Once whole plants are regenerated from preselected cultured cells, researchers can hybridize them with other varieties. The idea is to transfer desired genes to a specific type of plant to improve traits (such as resistance to pests or herbicides) and crop yields.

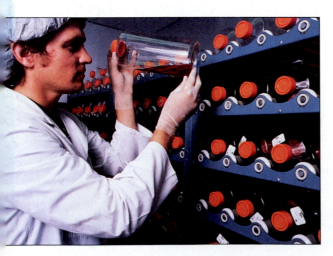

Figure 21.9
Racks of commercially produced interferon.

Given that possibility, botanists comb the world for seeds of the wild ancestors of potatoes, corn, and other valued plants. They send the prizes—seeds that carry the plants' genetic heritage—to seed banks. The safe storage facilities preserve the genetic diversity of the food supply for most of the human population. That supply is quite vulnerable, for farmers focus mainly on genetically similar varieties of high-yield crop plants. Genetic uniformity makes our food crops dangerously vulnerable to many kinds of pathogenic fungi, viruses, and bacteria.

How Are Genes Transferred into Plants?

A simple example gives insight into how genes can be transferred into plants. Genetic engineers often insert new or modified genes into a Ti plasmid, which is extracted from *Agrobacterium tumefaciens*. This bacterium infects many species of flowering plants. Some plasmid genes invade a plant's DNA, then induce a tumor to form. When the Ti plasmid is used to ferry genes, however, the tumor-inducing genes are first removed and desired ones substituted. Plant cells then are placed in a culture that contains the modified bacteria. When some cells take up the gene, whole plants may be regenerated from the cells' descendants. In some cases, researchers use electric shocks or chemicals to deliver modified genes into plant cells, or blast microscopic particles coated with DNA into the cells. The expression of foreign genes in plants (or in other organisms) is not always a sure thing, however.

Similar methods have already produced practical successes, including trees that produce improved wood for paper manufacturing and cotton plants genetically engineered to resist a common herbicide (Figure 21.10).

Also on the horizon are engineered plants that can serve as factories for pharmaceuticals. A few years ago, researchers began inserting human genes for certain proteins into plants. One of the proteins, serum albumin, is widely used in preparations given to burn patients and other ill people to replace lost body fluids. Researchers have been able to extract useful amounts of the protein from genetically altered potato and tobacco plants.

Despite some very real potential benefits, genetic modification of plants, especially those used for human food, is not without potential hazards. Concerns about so-called GM foods has triggered serious controversy—a subject we will consider in Section 21.9.

Genetic engineering of microorganisms and plants is yielding a range of useful products, including medically valuable proteins and crop plants with new beneficial traits.

BACTERIA THAT CLEAN UP POLLUTANTS

As a group, bacteria have a rather astonishing quality. They are so diverse metabolically that they can gain energy for life processes from breaking down substances that are highly toxic, even carcinogenic. In the process, many noxious substances—including crude oil, toxic heavy metals such as mercury, and pesticides—are taken up by the bacterium or chemically altered to a safer form.

"Oil-eating" bacteria, for example, oxidize the hydrocarbons in petroleum to carbon dioxide (Figure 21.11). Mercury-resistant bacteria metabolically process molecules of the metal (which causes severe neurological damage in humans) into a nontoxic compound.

Often, the genes that code for these metabolic feats are carried on plasmids—and so are ready candidates for bioengineering. A number of "de-tox" genes have been isolated, cloned, and transferred to other bacteria. Recently, for example, plasmids with genes that code for enzymes that break down hydrocarbons were successfully transferred to marine bacteria. The bacteria have since been used to help clean up oil spills off Texas and elsewhere.

Recombinant plasmids also have been put together using genes obtained from bacteria that normally break down various contaminants in toxic waste dumps. The goal is to engineer microbes that can be used to treat industrial wastes and possibly help clean up polluted soils. The U.S. Environmental Protection Agency has identified more than 2,000 seriously contaminated toxic-dump sites in the United States, many of them endangering groundwater supplies. Biotechnology may be an invaluable tool for restoring such places to environmental health.

Figure 21.11 Bioremediation, the removal of oil by natural biodegradation, is accelerated by the application of fertilizers that provide oil-eating bacteria with nitrogen and phosphorus nutrients to augment the oil's natural carbon content. Here a cleanup worker sprays a shoreline with a liquid fertilizer (Inipol™).

Super Mice and Biotech Barnyards

Animal cells do not accept plasmids, but they can be "micro-injected" with foreign DNA. Some time ago, researchers introduced the gene for growth hormone from rats into fertilized mouse eggs. The mice grew much larger than their normal littermates and had dramatically higher blood concentrations of the hormone. Why? The rat gene had become integrated into the mouse DNA and was being expressed. Later, when the gene for human growth hormone was successfully introduced and expressed in mice, the result was the "super rodent" shown in Figure 21.12a.

Similar experiments with large domesticated animals have begun to show some success. Transgenic sheep, pigs, and cows may become sources of important therapeutic drugs. Experimental goats have been nurtured to adulthood from fertilized eggs that were injected with recombinant DNA consisting of goat gene sequences spliced to human genes for tissue plasminogen activator, or tPA. This protein dissolves blood clots in heart attack and stroke patients, and the goats' milk contains it. Although tPA and other human proteins are currently made by recombinant bacteria, so-called barnyard biotechnology is exciting because in some cases transgenic livestock have the potential to produce large quantities of drugs faster and more cheaply than bacteria cultured in huge industrial vats. Recombinant DNA technology also has been used to transfer human and cow growth hormones into pigs—which grow much faster as a result.

People who have hemophilia A now can obtain the needed blood-clotting factor VIII from a drug produced by Chinese hamster ovary cells in which genes for human factor VIII have been inserted. Factor VIII produced in this way eliminates the need to obtain it from human blood—and so eliminates the risk of transmitting blood-borne pathogens such as HIV.

In 1996 researchers made a genetic duplicate—a clone—of an adult ewe. They inserted the nucleus from one of her mammary gland cells into an egg (from another ewe) from which the nucleus had been removed. Eventually, a cluster of embryonic cells developed and was implanted into a surrogate mother. The resulting lamb, named Dolly, today is a thriving adult ewe herself (Figure 21.12b). As experiments with transgenic animals continue around the globe, they may well be laying the groundwork for commercial development of animals with custom-made genetic characteristics—including, for instance, tissues and organs for use in treating various human diseases.

The success of efforts to clone mammals such as sheep and pigs has raised the jarring prospect that someone may attempt to clone human beings. In the United States, the government has imposed a ban on federal funding of research into human cloning.

Mapping and Using the Human Genome

By now, you probably have heard of the Human Genome Project. This massive international effort has tapped the

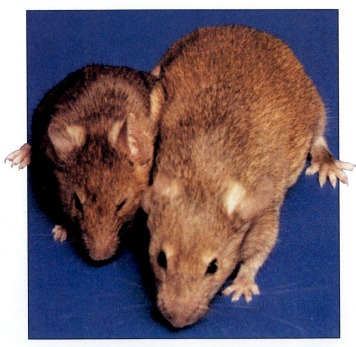

a

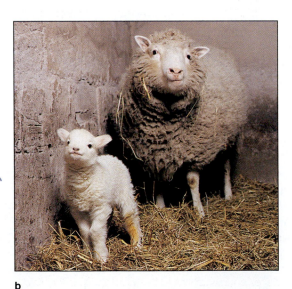

b

Figure 21.12 (a) Ten-week-old mouse littermates. The larger mouse grew from a fertilized egg into which the gene for human somatotropin (growth hormone) had been inserted. (b) The cloned sheep Dolly with her first lamb. The lamb was conceived the old-fashioned way.

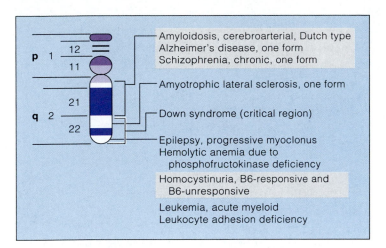

Figure 21.13 Map of human chromosome 21 showing genes correlated with specific diseases. In all, researchers have mapped 225 genes to this chromosome. The upper arm of the chromosome is marked p and the lower arm, q. In this case, each arm consists of a single region, labeled 1 and 2, respectively. In other chromosomes, each arm may have two or more regions. Stains used in chromosome analysis produce a series of bands within each region. The small numbers just to the left of each arm indicate specific bands. A combination of letters and numbers indicates the chromosome region where a given gene is found; for instance, the gene for one form of amyotrophic lateral sclerosis (ALS) is 21q2.22.

creativity—and spurred the highly competitive spirit—of laboratories all over the world. In 2000, a private research company announced it had won the race to determine the sequence of nucleotides in the entire complement of human DNA.

Another aim of the Human Genome Project is to create a physical map of all human chromosomes—that is, to identify the locations of specific genes on each chromosome. Recently, maps of all the genes on chromosomes 21 and 22 were completed. (Those chromosomes were good candidates for the initial mapping effort because they are the two smallest human chromosomes.) At the same time, tens of thousands of markers have been mapped to these and other chromosomes. For instance, we now know that chromosome 21 carries genes for early-onset Alzheimer's disease, some forms of epilepsy, and one type of ALS (*amyotrophic lateral sclerosis*), a disease that gradually destroys motor neurons. Figure 21.13 shows where those genes are located.

Today, anyone with access to the World Wide Web can obtain information about human gene sequences from several huge databases. One of the most useful databases, called EST (for *expressed sequence tags*), includes sequences of human cDNAs from various tissues.

Information that comes from the mapping of our chromosomes will undoubtedly increase understanding of the genetic basis of human traits and diseases. However, it also has raised thorny issues. For example, some people are concerned that detailed chromosome maps will lead to relatively easy testing of individuals for genetic predispositions to many disorders and diseases—testing that could result in discrimination by insurance companies or employers. Other critics complain about laboratories that attempt to patent the information they uncover—presumably for future commercial gain. These debates are likely to intensify along with the rising tide of our understanding of the human genome.

Altering Gene Expression

Experimental approaches to altering gene activity also are emerging from biotechnological research. Several aim to develop drugs that act on DNA or RNA in specific ways. For example, genetic engineering methods are being used to create drugs that can block the synthesis of undesirable proteins by shutting down transcription of DNA to messenger RNA (Section 20.5). One target is the virus that causes herpes infection: Scientists are working to develop a drug that can enter an infected cell and shut down transcription of viral DNA without harming expression of the cell's own genes. Research on such *transcription factors* could also lead to drugs that can turn off oncogenes responsible for various cancers (Section 22.2) or that shut down mutated alleles for various diseases.

Another approach is to block protein synthesis by way of molecules that can prevent mRNA from being translated in the cell cytoplasm. Engineered "antisense" RNA molecules are one option. These are short stretches of nucleic acid that can bind to specific mRNAs and prevent translation. Alternatively, **ribozymes** are bits of RNA that act like enzymes and, when inserted into a cell, cut up specific mRNA sequences before they can be translated. Ribozymes will probably first be used in antiviral drugs.

Applications of genetic engineering include efforts to develop transgenic animals or animal cells capable of producing medically useful substances.

Researchers have determined the sequence of nucleotides in the entire human genome.

The Human Genome Project is mapping the locations of specific genes on human chromosomes. The project holds great promise for scientific and medical research.

METHODS AND PROSPECTS FOR GENE THERAPY

Generally speaking, **gene therapy** is the insertion of one or more normal genes into a person's body cells in order to correct a genetic defect. Numerous experiments in human gene therapy are now under way, involving desperately ill victims of skin cancer, cystic fibrosis, and other disorders.

As you know from previous chapters, diseases such as cystic fibrosis and hemophilia result when a single gene is mutated, producing a nonfunctional protein. Chapter 22 will explain how many cancers develop when mutations upset normal gene regulation, and several genes in a cell turn on or off at the wrong time. Gene therapy aims to replace mutated genes with normal ones, which will encode functional proteins, or to insert genes that restore normal controls over gene activity. An inserted gene might also code for a protein that blocks a cell surface receptor associated with a disorder. For example, it might be possible to treat arthritis by inserting genes coding for proteins that block the binding of substances that cause inflammation. Scientists around the world are actively exploring these and other possibilities.

Strategies for Transferring Genes

The size of a gene (the number of its base pairs) helps determine what sort of mechanism can be used to insert it into a host cell. Animal cells do not take up plasmids, so plasmids cannot be directly used as vectors in gene therapy. Smaller genes can be carried into animal cells by viruses; larger ones must enter a host cell some other way.

In *transfection*, cells cultured in the laboratory are directly exposed to DNA segments that contain a gene of interest. Typically, some of the foreign DNA becomes integrated into the host cell's genome. Exposing the host cells to a weak electric current, called *electroporation*, seems to help. Even so, this strategy is inefficient: Sometimes only one cell in 10 million incorporates a new gene.

A small gene can be inserted into a virus, which can then "infect" a host cell and carry the desirable gene with it. Many gene therapy trials use retroviruses in this way because such RNA viruses easily infect animal cells. Once inside the host cell, the viral RNA is transcribed into DNA (by way of reverse transcriptase), and usually the foreign DNA is integrated into the host cell's DNA. The first step in gene therapy is to remove certain segments of genetic instructions from the virus, so that it cannot replicate and cause disease. In their place, a researcher substitutes the gene to be transferred. Next the virus is allowed to infect target cells (Figure 21.14). Retroviruses are efficient at integrating into the host DNA, increasing the odds that the new gene will become active. Some gene therapy experiments use other viruses that typically infect the kinds of host cells being manipulated.

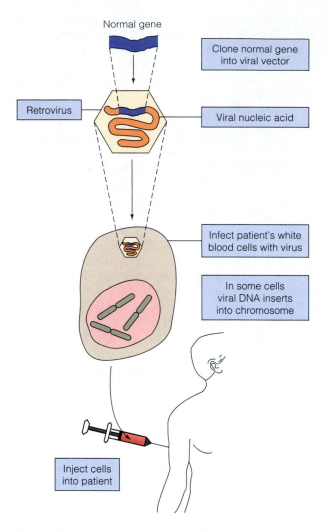

Figure 21.14 One method of gene therapy. In this method, a sample of white blood cells is removed from a patient with a genetic disorder. Next, a retrovirus is used as a vector to insert a normal gene into the DNA. After the normal gene begins to function and produce its protein product, the cells are placed back into the patient. If enough of the normal protein is present in the patient, symptoms of the disorder will diminish. Because the genetically altered blood cells live only for a few months, however, the procedure must be repeated regularly.

There are pitfalls involved with using viruses to introduce genes. Suppose that a virus is engineered to infect cells with a "good" copy of a mutated gene. When such a virus integrates into a host cell's DNA, there is no guarantee that the new gene will be in its normal position—and if it is not, the gene may not be transcribed into protein. Sometimes changes to an altered virus render any genes they carry ineffective.

At present, only a few types of human cells can successfully be removed, induced to take up engineered genes, grown in the laboratory, and then reinserted into the body. Thus, this method can be used only with genes that are expressed in those tissues. There are no reliable procedures for making large quantities of certain vectors that are commonly used to carry new DNA into host cells. In addition, many introduced genes turn off within a few days or weeks. Finally, a few patients in gene therapy trials have died from complications of the procedure, raising serious questions about safety. In spite of these obstacles, research on various gene therapy methods is continuing at a rapid pace.

Early Results of Gene Therapy

Chapter 19 noted gene therapy trials for patients with the inherited disease cystic fibrosis. Gene therapy researchers also are targeting numerous other disorders. We mention just a few of them here to give you an idea of the kinds of efforts under way.

ADA About a decade ago, a little girl named Ashanthi De Silva became the subject of the first federally approved gene therapy test for humans. Born with a defective gene that codes for the enzyme adenosine deaminase (ADA), Ashanthi suffers from severe combined immune deficiency (SCID). This set of disorders develops from a drastic reduction in or the complete absence of T and B lymphocytes (Section 8.1).

Using recombinant DNA methods, researchers introduced copies of the ADA gene into some of the child's T cells, which were then stimulated to divide. Next, about a billion copies of the genetically engineered cells were delivered through a plastic tube into Ashanthi's bloodstream. After a series of monthly infusions, Ashanthi began receiving one treatment a year. Tests show that about half of her circulating T cells carry the normal gene. However, in some other patients, at most about 1 percent of circulating T cells are normal.

Another research team extracted lymphocyte stem cells (called *lymphoblasts*) from umbilical cord blood of three newborns with ADA deficiency. They then used a recombinant retrovirus as a vector to introduce functional ADA genes into the stem cells. The altered cells were cultured with substances that stimulated growth and division, then reinserted into the newborn patients. However, only about 10 percent of the children's T cells carried the normal gene. Since no ADA patient other than Ashanthi De Silva has "grown" an immune system with gene therapy, all patients (including Ashanthi) receive regular doses of a synthetic form of ADA.

CYSTIC FIBROSIS Cystic fibrosis was an early candidate for gene therapy. Researchers had high hopes of introducing normal copies of the defective gene into cells of the respiratory system, using a viral vector in a nasal spray. To date, in roughly a dozen gene therapy trials, only about 5 percent of affected cells have taken up the normal gene. In this and other gene therapy efforts, one major problem has seemed to be that the retrovirus vectors used are not effective enough at delivering the new genes to cells that need them. In addition, as already noted, investigators still cannot ensure that an inserted gene will be incorporated into the patient's DNA in a way that will allow the gene to be expressed. Despite such obstacles, work is continuing, focusing on the development of new, more effective vectors.

CANCERS At this writing, gene therapy has been most successful in treating cancer. Early trials have targeted the deadly skin cancer malignant melanoma, leukemia, a fast-growing form of lung cancer, and cancers of the brain, ovaries, and other organs. In some approaches, tumor cells are first removed from a patient and grown in the laboratory. Genes for an interleukin (which helps activate white blood cells of the immune system) are then introduced into the cells, and the cells are returned to the body. In theory, interleukins produced by the tumor cells may act as "suicide tags" that stimulate T cells of the immune system to recognize cancerous cells and attack them. An exciting variation on this theme involves structures called "lipoplexes," laboratory-made packets in which a plasmid is encased in a lipid coat that facilitates its entry into a cell. The plasmid carries a gene encoding a protein marker that can trigger an immune system attack on cancer cells. Thus far, several dozen melanoma patients have been treated with lipoplex therapy, with encouraging results.

Other trials are adding the gene for the powerful antitumor agent called tumor necrosis factor (TNF) to tumor-infiltrating lymphocytes (TILs). As their name suggests, TILs home in specifically on tumors. One goal of such experiments is to determine whether TILs carrying the TNF gene can deliver enough TNF to kill a targeted tumor without harming surrounding tissues.

At present, gene therapy is highly experimental, extremely costly, and available to only a few. Authorities agree that the key to its widespread use is the development of vectors, such as specially engineered viruses and lipoplexes, that can deliver genes to target cells safely and reliably. Without a doubt, as methods of gene therapy are refined, its use will become more widespread, and gene therapy will play a major role in future efforts to treat and cure disease.

In gene therapy, one or more normal genes are inserted into body cells to correct a genetic defect or enhance the activity of specific genes.

ISSUES FOR A BIOTECHNOLOGICAL SOCIETY

Some people say that no matter what the species of organism, DNA should never be altered. But as we noted earlier, nature alters DNA all the time. The real issue is how to bring about beneficial changes without harming ourselves or the environment.

In recent years, a host of issues have been brought to the table for discussion. For example, some hotly debated concerns include the following:

• Transgenic bacteria or viruses could mutate, possibly becoming new pathogens in the process. However, as described in Section 21.6, most recombinant microbes are altered in ways that will prevent them from reproducing outside the laboratory.

• Bioengineered plants could escape from test plots and pose an ecological risk by becoming "superweeds" resistant to herbicides and other control measures.

• Crop plants with added insect resistance could set in motion an evolutionary seesaw in which new, even more formidable insect pests arise.

• Transgenic fish that feed voraciously and grow rapidly could displace natural species, wrecking the delicate ecological balance in streams and lakes.

• Major ethical problems could arise if people undertake "eugenic engineering"—using biotechnology to insert genes into normal people (or sperm or eggs) simply for the purpose of creating people that are "more desirable."

• Genetic screening could provide a means of discrimination by insurance companies and others against people who carry a gene predisposing them to a disorder, even if such a person shows no sign of ill health.

Experts point out that some of these possibilities—mutation of transgenic bacteria or viruses into pathogenic strains or escape of bioengineered plants as "superweeds"—are not very likely. However, others, such as discrimination on the basis of genetic screening, are issues we already face.

Genetically engineered food is a particularly hot topic these days. In the United States, it is probably impossible to avoid such foods—a whopping 38 percent of soybeans and 25 percent of corn crops are engineered to withstand weedkillers or make their own pesticides. For years, the corn and soybeans have found their way into breakfast cereals, soy sauce, vegetable oils, beer, soft drinks, and other food products. They also are fed to farm animals.

In Europe, especially Britain, public resistance to genetically engineered food is high. Many people speak out

Figure 21.15 British protesters uprooting genetically modified crop plants.

against what the tabloids call "Frankenfoods." Protesters routinely vandalize crops (Figure 21.15). Worries abound that such foods may be more toxic, have lower nutritional value, and promote antibiotic resistance.

Biotechnologists envision a new Green Revolution. They argue that designer plants can hold down food production costs, reduce dependence on pesticides and herbicides, enhance crop yields, and offer improved flavor and nutritional value, even salt tolerance and drought tolerance.

Yet the chorus of critics in some quarters is growing louder. There are serious threats of boycotts of genetically modified food exports from the United States—a 50 billion dollar enterprise annually. Even in the U.S., some upscale restaurants and "gourmet" food producers have banned such foods from their kitchens. Restrictions on genetic engineering will profoundly impact United States agriculture, and inevitably the impact will trickle down to what you eat and how much you pay for it.

We invite you to read up on scientific research on this issue and form your own opinions. The alternatives are to be swayed by catchy, scary phrases (such as Frankenfood) or by biased reports from groups (such as chemical manufacturers) with their own agendas. With advances coming every day, it may be time for all of us to think about the benefits *and* risks of living in a biotechnological society.

SUMMARY

1. Genetic "experiments" have been occurring in nature for billions of years. Mutation, crossing over and recombination at meiosis, and other natural events have all contributed to the current diversity that we see among organisms.

2. Humans have been manipulating the genetic character of different species for thousands of years. The emergence of recombinant DNA technology in the past few decades has enormously expanded our capacity to control genetic change. Recombinant DNA technology uses procedures by which DNA molecules can be cut into fragments, inserted into plasmids or some other cloning tool, then multiplied in a population of rapidly dividing cells.

3. A DNA clone is a foreign DNA sequence that has been introduced and amplified in dividing cells. DNA sequences also can be amplified in test tubes by the polymerase chain reaction.

4. Recombinant DNA technology and genetic engineering have enormous potential for research and applications in medicine, agriculture, and industry. As with any new technology, potential benefits must be weighed against potential risks, including ecological and social disruptions.

5. Although the new technology has not developed to the extent that human genes can readily be modified, the social, legal, ecological, and ethical questions it raises should be explored in detail before such an application is possible.

Review Questions

1. What is a plasmid? What is a restriction enzyme? Do such enzymes occur naturally in organisms? *21.1*

2. Recombinant DNA technology involves producing DNA restriction fragments, amplifying the DNA, and identifying modified host cells. Briefly describe one of the methods used in each of these categories. *21.1, 21.2*

Self-Quiz *(Answers in Appendix V)*

1. Gene mutations, crossing over and recombination during meiosis, and other natural events are the basis of the _____ observed in present-day organisms.

2. Causing genetic change by deliberately manipulating DNA is known as _____.

3. _____ are small circles of bacterial DNA that are separate from the bacterial chromosome.

4. Genetic researchers use plasmids as

5. Rejoined cut DNA fragments from different organisms are best known as _____.
   a. cloning genes
   b. mapping genes
   c. recombinant DNA
   d. conjugating DNA

6. Repeated DNA replications and cell divisions of the host cells yield _____.

7. Using the metabolic machinery of a bacterial cell to produce multiple copies of genes carried on recombinant plasmids is _____.
   a. DNA fingerprinting
   b. bacterial conjugation
   c. mapping a genome
   d. DNA amplification

8. A collection of DNA fragments produced by restriction enzymes, incorporated into plasmids, and introduced into bacterial host cells is a _____.
   a. DNA clone   c. hybridized sequence
   b. gene library   d. gene map

9. The polymerase chain reaction _____.
   a. is a natural reaction in bacterial DNA
   b. cuts DNA into fragments
   c. amplifies DNA sequences in test tubes
   d. inserts foreign DNA into bacterial DNA

10. Recombination is made possible by _____ that can make cuts in DNA molecules.

Critical Thinking: You Decide *(Key in Appendix VI)*

1. Having read about examples of genetic engineering in this chapter, can you think of some additional potential benefits of this technology? Can you envision other potential problems?

2. Jimmie's Produce put out a bin of tomatoes having beautiful red color and looking "just right" in terms of ripeness. A sign above the bin identified them as genetically engineered produce. Most shoppers selected unmodified tomatoes in the neighboring bin, even though those tomatoes were pale pink and hard as a rock. Which ones would you pick? Why?

Selected Key Terms

cDNA *21.1*
cloning vector *21.1*
DNA fingerprint *21.3*
DNA ligase *21.1*
gene library *21.5*
gene therapy *21.9*
genetic engineering *21.1*

genome *21.1*
nucleic acid hybridization *21.5*
plasmid *21.1*
polymerase chain reaction (PCR) *21.2*
primer *21.2*
recombinant DNA technology *21.1*
restriction enzyme *21.1*

Readings

Aldridge, S. 1997. *The Thread of Life: The Story of Genes and Genetic Engineering.* New York: Cambridge University Press.

Mirsky, S., and J. Rennie. June 1997. "What Cloning Means for Gene Therapy." *Scientific American.*

Wingerson, L. 1998. *Unnatural Selection: The Promise and Power of Human Gene Research.* New York: Bantam.

CANCER: A CASE STUDY OF GENES AND DISEASE

The Body Betrayed

The tiny mass of tissue, only a few hundred thousand cells strong, may go unnoticed for ten, twenty, even thirty years. In fact, not until the mass becomes a clump of perhaps 10 billion cells will it be felt, fretted over, and finally—perhaps reluctantly—discussed with a doctor.

Cancer is a grim fact of human life. It strikes one in three people in the United States and kills one in four—according to the American Cancer Society, about 1,500 cancer deaths every day, over half a million in a year. On average, cancer strikes more males than

Figure 22.1 One out of three people in the United States will be diagnosed with cancer during their lifetime. These young women are taking steps to limit their risk of lung cancer, the chief cancer killer of both men and women.

females, but the pattern varies depending on the type of cancer involved. Thus the number of new lung cancer cases is increasing faster among women due to a corresponding increase in the number of women who began smoking cigarettes in the last several decades (Figure 22.1). Adult leukemias and bladder cancers occur more often in men; on the other hand, males and females are about equally susceptible to colon cancer. Cancer strikes more often in older age groups, although it is the leading cause of death among children under the age of 15. Fortunately, intensive research is rapidly increasing our understanding of and ability to treat many kinds of cancer, including those of the breast, ovary, colon, and skin.

Cancer typically begins with cells in which the controls over cell division are lost. In this sense cancer is a betrayal, a bit of the body that turns *against* the body. More than one hundred specific cancer types are known. Their names, such as retinoblastoma and fibrosarcoma, typically identify the type of tissue in which the first cancer cells develop.

Some cancers arise from genetic changes caused by viral infection; some come from an inherited mutation; others are the end result of gene mutations triggered by environmental factors, including certain chemicals and radiation. Establishing causes is not a simple matter, however. Multiple triggering events almost certainly are the root of many cancers.

Statistics hint at the impact of cancer on human beings, but they don't tell us much about the biological causes and effects of cancers, and they don't reveal recent advances in our understanding of the genetic events that underlie this group of diseases. These are the topics we consider in this chapter, along with cancer treatments and lifestyle choices that can affect personal cancer risk. Your main objective in reading this chapter should be to come away with knowledge that will help you recognize early symptoms. Most cancers are readily treatable, and many are curable—but only if the disease is discovered early.

KEY CONCEPTS

1. Cancer arises when the genetic controls over cell division are lost.

2. Cancer develops through a multistep process. It requires changes in a series of genes, including tumor suppressor genes, that normally operate in cells.

3. Environmental factors may trigger cancer. These factors include infection by certain viruses, exposure to mutagens such as ultraviolet radiation, and chemical carcinogens that switch on oncogenes (cancer-causing genes), turn off tumor suppressor genes, or do both.

4. Lifestyle choices can limit a person's risk of developing many types of cancer. The most important choices appear to involve diet, exposure to sunlight, and tobacco use.

CHAPTER AT A GLANCE

Characteristics of Tumors

As genes switch on and off, they determine what a cell's specialized function will be, when and how rapidly the cell will grow and divide, even when it will *stop* dividing and when it will die. If cells overgrow, the result is a defined mass of tissue called a **tumor**. The name for this mass is *neoplasm*, which literally means new growth.

A tumor may not be "cancer." As Figure 22.2*a* shows, the cells of a *benign* tumor are often enclosed by a capsule of connective tissue, and inside the capsule they are organized in an orderly array. They also tend to grow slowly and to be well differentiated (structurally specialized), much like normal cells of the same tissue (Section 15.1). Benign tumors usually stay put in the body, push aside but don't invade surrounding tissue, and generally can be easily removed by surgery. Benign tumors *can* threaten health, as when they occur in the brain. Nearly everyone has at least several of the benign tumors we call moles. Most of us also have or have had some other type of benign neoplasm, which is usually destroyed by the immune system. Table 22.1 compares the main features of malignant and benign tumors.

Dysplasia ("bad form") is an abnormal change in the sizes, shapes, and organization of cells in a tissue. It is often a precursor to cancer. Under the microscope, the edges of a cancerous tumor often look ragged (Figure 22.2*b*), and its cells form a disorganized clump. Cancer cells usually also have characteristics that enable them to behave very differently from normal body cells.

Characteristics of Cancer Cells

STRUCTURAL ABNORMALITIES A cancer cell generally has an abnormally large nucleus and less cytoplasm than usual. It also is *poorly differentiated*. That is, cancer cells often do not have the clear structural specializations of cells in mature body tissues. The extent of differentiation of cancer cells can be medically important. In general, the less differentiated cancer cells are, the more readily they break away from the primary tumor and spread the disease.

When a normal cell becomes transformed into a cancerous cell, additional changes occur. The cytoskeleton shrinks, becomes disorganized, or does both. Proteins of the plasma membrane are lost or altered, and new, different ones appear. These changes are passed on to the cell's descendants: When a transformed cell divides, its daughter cells are cancerous cells too.

Table 22.1	Comparison of Benign and Malignant Tumors	
	Malignant Tumor	Benign Tumor
Rate of growth	Rapid	Slow
Nature of growth	Invades surrounding tissue	Expands within tissue
Spread	Metastasis by way of bloodstream and lymphatic system	Stays localized
Cell differentiation	Usually poor	Nearly normal

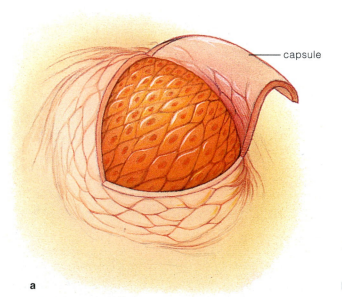

capsule

a

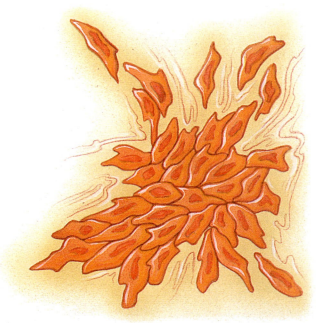

b

Figure 22.2 (**a**) Sketch of a benign tumor. Cells appear nearly normal, and the tumor mass is encapsulated within connective tissue. (**b**) A cancerous neoplasm. Due to the abnormal growth of cancer cells, the tumor is a disorganized heap of cells, some of which break off and invade surrounding tissues (metastasis).

Further reading: Student Guide to InfoTrac on web site ➡

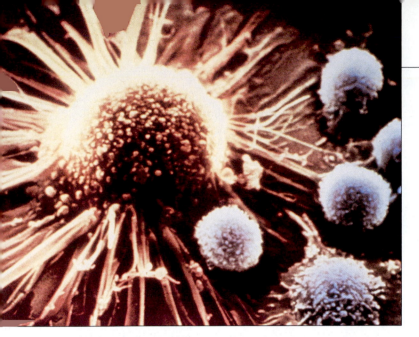

Figure 22.3 Scanning electron micrograph of a cancer cell surrounded by some of the body's white blood cells that may or may not be able to destroy it.

UNCONTROLLED GROWTH Cancer cells lack normal controls over cell division. Contrary to popular belief, cancer cells do not necessarily divide more rapidly than normal cells do, but they do increase in number faster. Why? Normally, the death of cells closely balances the production of new ones through mitosis, so that the cells are arranged in an orderly tissue. In a cancerous tumor, however, at any given moment more cells are dividing than are dying. As this rampant cell division continues, the cancer cells do not respond to crowding—that is, the normal *contact inhibition* does not occur. A normal cell stops dividing once it comes into contact with another cell, but a cancerous cell keeps on dividing. Therefore, cancer cells accumulate in a disorganized heap—which is why cancerous tumors are often lumpy.

Cancer cells also lack strong cell-to-cell adhesion with neighboring cells. As Figure 22.3 shows, they may form extensions (pseudopodia, "false feet") that enable them to move about. This property permits **metastasis**—cancer cells break away from the parent tumor and invade other tissues, including the lymphatic system and circulatory system (Figure 22.4). The ability to metastasize is what makes cancers *malignant*.

Some kinds of cancer cells produce the hormone HCG, human chorionic gonadotropin. (Recall from Chapter 15 that HCG maintains the uterine lining when a pregnancy begins. It also helps prevent the mother's immune system from attacking the fetus as "foreign.") The presence of HCG in the blood can serve as a red flag that a cancer exists somewhere in a person's body (Section 22.4).

Some cancer cells produce interleukin-2, a cytokine that stimulates cell division (Chapter 8), and also display receptors for it. Cancer cells also secrete a growth factor, *angiogenin*, that encourages new blood vessels to grow

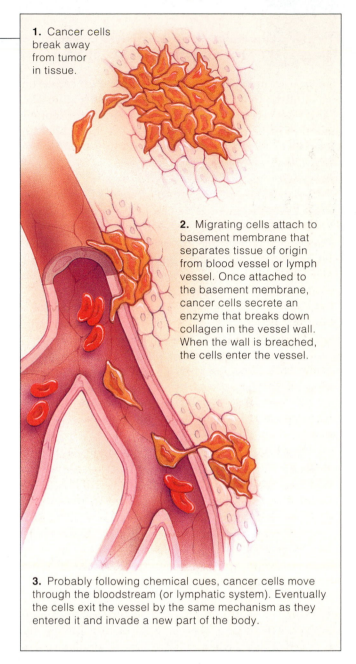

1. Cancer cells break away from tumor in tissue.

2. Migrating cells attach to basement membrane that separates tissue of origin from blood vessel or lymph vessel. Once attached to the basement membrane, cancer cells secrete an enzyme that breaks down collagen in the vessel wall. When the wall is breached, the cells enter the vessel.

3. Probably following chemical cues, cancer cells move through the bloodstream (or lymphatic system). Eventually the cells exit the vessel by the same mechanism as they entered it and invade a new part of the body.

Figure 22.4 Stages in metastasis.

around the tumor. Tumor growth requires a large supply of nutrients and oxygen supplied via the bloodstream. Some of the most exciting current cancer research is focused on developing drug therapies that "starve" tumors to death by blocking angiogenin's effects.

Cancer cells lack normal controls over cell division and organization into tissues. They are also structurally abnormal.

Cancer cells are often poorly differentiated. They can leave a primary tumor, invade surrounding tissue, and metastasize to other parts of the body.

The transformation of a normal cell into a cancerous one is a multistep process called **carcinogenesis** (Figure 22.5). Cancer develops in most vertebrates, some invertebrates, and even some plants. Fossils show that some dinosaurs developed cancer. Wherever it turns up, *cancer almost always develops through two or more steps in which genetic changes alter normal controls over cell division.*

Certain genes are called **proto-oncogenes**. (*Proto* means before.) An **oncogene** is a gene that can induce cancer. The distinction is important. Proto-oncogenes are normal genes that regulate cell growth and development. They encode proteins that include growth factors (signals sent by one cell to trigger growth in other cells), proteins that regulate cell adhesion, and the protein signals for cell division. If some event alters the structure or expression of a proto-oncogene, it may be changed into an oncogene that does not respond to controls over cell division.

An oncogene acting alone does not cause malignant cancer. That usually requires mutations in several genes, including at least one **tumor suppressor gene**. Such genes encode proteins that are produced when cells are stressed. They typically halt cell growth and division, preventing cancers from occurring.

Several lines of evidence have shed light on the roles of specific tumor suppressor genes. Studies of the childhood eye cancer *retinoblastoma* have revealed that the disease is likely to develop when a child inherits only one functional copy of a tumor suppressor gene called *Rb*. If a mutation alters a functional allele to a nonfunctional form, then retinoblastoma is likely. Colorectal cancer and perhaps several other common cancers have also been linked to damage in a tumor suppressor gene.

A tumor suppressor gene called p53 is particularly interesting because it appears to prevent cancerous changes. This gene codes for a regulatory protein that stops cell division when cells are stressed or damaged, as often happens in cancer cells. However, when the p53 protein cannot be produced because there is a mutation in the p53 gene, the controls don't operate. Then, cell division may continue unchecked. In addition, when the p53 gene mutates, the resulting faulty protein may actually *promote* the development of cancer, possibly by activating an oncogene. More than half of cancers involve a mutated or absent p53 gene. This discovery has spurred the development of tests that screen for p53 mutations. There also are promising experimental cancer treatments in which normal copies of the gene are inserted into cancerous tumors—and seem to be able to slow the tumor's growth.

Oncogene Activation

An oncogene may be present in a cell and never cause trouble because some condition in the cell, such as the presence of a suppressor gene, prevents the oncogene from being expressed. This state of affairs can change, though. The oncogene or a related suppressor gene may mutate in a way that triggers expression. Change in a chromosome's structure may move an oncogene away from a regulatory nucleotide sequence that would otherwise prevent it from being expressed. Or new genetic material may enter a cell (as by viral infection) and disrupt controls.

Other Routes to Carcinogenesis

INHERITED SUSCEPTIBILITY TO CANCER   Heredity plays a major role in about 5 percent of cancers. If a mutation exists in a germ cell (sperm or egg), and if it alters a proto-oncogene or tumor suppressor gene, the defect can be passed on to offspring. The first step toward cancer has now occurred, so that an affected person may be more susceptible to cancer if mutations occur in other proto-oncogenes and tumor suppressor genes. Patterns of familial breast, colon, and lung cancer suggest that several genes are involved, including genes that control aspects of cell metabolism and responses to hormones.

VIRUSES   Viruses cause some cancers. Sometimes a viral infection can alter a proto-oncogene when the viral DNA is inserted at a certain position in the host cell DNA. For example, the viral gene could take the place of a regulatory sequence that normally prevents a proto-oncogene from switching on (or off) at the wrong time. Other viruses simply carry oncogenes as part of their genetic material and insert them into the host's DNA. Most viruses linked to human cancer are DNA viruses, but RNA retroviruses are associated with some types of leukemia.

CHEMICAL CARCINOGENS   There are thousands of known chemical **carcinogens**, cancer-causing substances that can cause DNA damage and a subsequent mutation in DNA. A carcinogen that directly causes DNA damage is sometimes called an "initiator." The list includes many compounds that are by-products of the industrialization of human societies, such as asbestos, coal tar, vinyl chloride, and benzene. The list also includes hydrocarbons in cigarette smoke (as well as other types of smoke) and on the charred surfaces of barbecued meats and carcinogenic substances in dyes and pesticides (*Focus on Our Environment*, Section 22.3). One of the first carcinogens to be recognized was chimney soot (more accurately, substances that soot contains), which frequently caused cancer of the scrotum in chimney sweeps. Plants and fungi can produce carcinogens; aflatoxin, a metabolic by-product of a fungus that attacks stored grain and other seeds, causes liver cancer. For this reason, some authorities advise against eating "raw" peanut butter. (Commercial peanut butters are safe because processing kills the aflatoxin fungus.)

Some chemicals may be "pre-carcinogens" that cause gene changes only *after* they have been altered by metabolic activity in the body or cell. Other substances are cancer "promoters"—alone they are not carcinogenic, but they can make a carcinogen more potent if the carcinogen *and* the promoter act on a cell at the same time.

RADIATION As you know from previous chapters, radiation can damage DNA, causing mutations that lead to cancer. Common sources include ultraviolet radiation from sunlight and tanning lamps, medical and dental X rays, and some radioactive materials used to diagnose diseases. Other sources are background radiation from cosmic rays and radon gas in soil and water, and the gamma rays emitted from nuclear reactors and radioactive wastes. Sun exposure is probably the greatest radiation risk factor for most people. (Skin cancer is the most frequently diagnosed cancer in the Northern Hemisphere.) It is also wise to avoid unnecessary X rays.

BREAKDOWNS IN IMMUNITY Typically, when a normal cell becomes cancerous, certain proteins at the cell surface become altered and function like foreign antigens—the "nonself" tags that mark a cell for destruction by cytotoxic T cells and natural killer cells (see Section 8.4). Research has uncovered clear evidence that a healthy immune system regularly detects and destroys some types of cancer cells. This protective function can break down as a person ages. Some researchers have hypothesized that if the immune system is suppressed by any of a range of factors, including some therapeutic drugs and mental states such as anxiety and severe depression, the risk of cancer rises. As yet, however, there is no hard evidence to support this view.

The presence of a growing cancer may also suppress the immune system. In addition, cancerous cells that bear tumor antigens may lack other molecules that must bind with receptors on lymphocytes in order to trigger an immune response. In some cases tumor antigens may be chemically disguised or masked. For whatever reason, the transformed cells are free to divide uncontrollably.

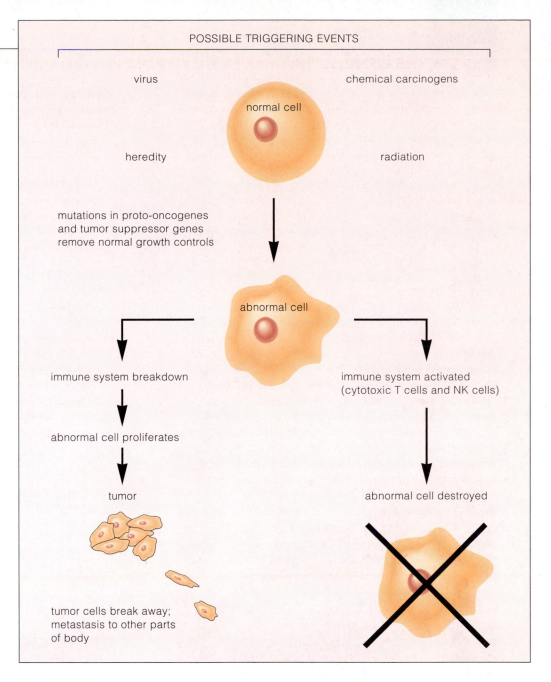

Figure 22.5 Overview of steps in carcinogenesis.

Cancer develops through a multistage process in which gene changes remove normal controls over cell division.

Oncogenes have a major role in inducing cancer in a normal cell. Proto-oncogenes may become oncogenes if some event changes their structure or the way they are expressed.

The development of cancer typically also requires the absence or mutation of at least one tumor suppressor gene.

22.3

ASSESSING THE CANCER RISK FROM ENVIRONMENTAL CHEMICALS

According to the American Cancer Society, factors in our environment lead to about half of all cancers. This statistic includes exposure to UV light and radiation, and it also includes agricultural and industrial chemicals. How are people exposed to these chemicals? And how dangerous are they? Let's begin with the first question.

Government statistics indicate that about 40 percent of the food in American supermarkets contains detectable residues of one or more of the active ingredients in commonly used pesticides. The residues are especially likely to be found in tomatoes, grapes, apples, lettuce, oranges, potatoes, beef, and dairy products. Imported crops, such as fruits, vegetables, coffee beans, and cocoa, can also carry significant pesticide residues—sometimes including pesticides, such as DDT, that are banned in the United States. The pesticide category includes chemicals used as fungicides, insecticides, and herbicides. There are roughly 600 different chemicals in this category, which are used alone or in combination.

Avoiding exposure to pesticides is difficult. Although residues of some pesticides can (and should) be removed from the surfaces of fruits and vegetables by washing before eating, it can be difficult to avoid coming into contact with pesticides used in community spraying programs to control mosquitoes and other pests, or used to eradicate animal and plant pests in golf courses and along roadsides. We have more control over the chemicals we use in gardens and on lawns (Figure 22.6).

Agricultural chemicals are not the only potential threats to human health. A variety of industrial chemicals also have been linked to cancer. In one way or another, the industrial chemicals in Table 22.2 all can cause carcinogenic mutations in DNA.

Some years ago biochemist Bruce Ames developed a test that could be used to assess the ability of chemicals to cause mutations. This Ames test uses *Salmonella* bacteria as the "guinea pigs," because chemicals that cause mutations in bacterial DNA may also have the same effect on human DNA. After extensive experimentation, Ames arrived at some interesting conclusions. First, he found that more than 80 percent of known cancer-causing chemicals do produce mutations. However, Ames testing at many different laboratories has *not* revealed a "cancer epidemic" caused by synthetic chemicals.

Ames's findings do not mean we should carelessly expose our crops, our gardens, or ourselves to environmental chemicals. For instance, the National Academy of Sciences has warned that the active ingredients in 90 percent of all fungicides, 60 percent of all herbicides, and 30 percent of all insecticides in use in the United States *have the potential* to cause cancer in humans. At the same time responsible scientists recognize that it is virtually impossible to determine that a certain level of a specific chemical caused a particular cancer or some other harmful effect. Given these facts, it is probably prudent to err on the side of caution and, as much as possible, limit our exposure to the potential carcinogens in our increasingly chemical world.

Figure 22.6 A common source of pesticide exposure. Home garden chemicals are just one avenue by which mutagenic or carcinogenic substances can come into contact with the human body.

Table 22.2 Selected Industrial Chemicals Linked to Cancer	
Chemical/Substance	Type of Cancer
Benzene	Leukemias
Vinyl chloride	Liver, various connective tissues
Various solvents	Bladder, nasal epithelium
Ether	Lung
Asbestos	Lung, epithelial linings of body cavities
Arsenic	Lung, skin
Radioisotopes	Leukemias
Nickel	Lung, nasal epithelium
Chromium	Lung
Hydrocarbons in soot, tar smoke	Skin, lung

Further reading: Student Guide to InfoTrac on web site →

DIAGNOSING CANCER

Table 22.3 lists seven common cancer warning signs. These can help you spot cancer in its early stages, when treatment is most effective. Routine cancer screening becomes important as a person ages; some recommended cancer screening tests are listed in Table 22.4.

Early and accurate diagnosis of cancer is extremely important to maximize the chances that a cancer can be cured. To confirm—or rule out—cancer, various types of tests can refine the diagnosis. Blood tests can detect **tumor markers**, substances produced by specific types of cancer cells or by normal cells in response to the cancer. For example, as we noted earlier, the hormone HCG is a highly specific marker for certain cancers. Prostate-specific antigen, or PSA, is a highly useful marker for detecting prostate cancer (see Section 22.7). Radioactively labeled monoclonal antibodies, which home in on tumor antigens, are useful for pinpointing the location and sizes of tumors of the colon, brain, bone, and some other tissues. *Medical imaging* of tumors includes methods such as magnetic resonance imaging (MRI; shown in Figure 22.7), X rays, ultrasound, and computerized tomography (CT). The definitive cancer detection tool is **biopsy**. A small piece of suspect tissue is removed from the body through a hollow needle or exploratory surgery. A pathologist then microscopically examines cells of the tissue sample for the characteristic features of cancer cells.

A *DNA probe* is a segment of radioactively labeled DNA. Such probes can be used to locate gene mutations or alleles associated with some types of inherited cancers.

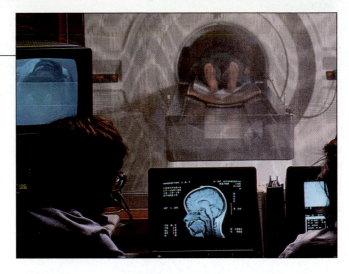

Figure 22.7 MRI scanning. MRI does not use X rays. The patient is placed inside a chamber that is surrounded by a magnet. The machine generates a magnetic field in which nuclei of hydrogen and some other atoms in the body align and absorb energy. A computer analyzes the information and uses it to generate an image of soft tissues.

The procedure is expensive, and most people don't have health insurance that will cover it. Some researchers point out, however, that using DNA probes for widespread screening could allow people with increased genetic susceptibility to make medical and lifestyle choices that could significantly reduce their cancer risk.

Biopsy is a definitive cancer detection tool. Other diagnostic procedures include blood testing for substances produced by cancer cells and medical imaging.

Table 22.3 The Seven Warning Signs of Cancer*
Change in bowel or bladder habits and function
A sore that does not heal
Unusual bleeding or bloody discharge
Thickening or lump
Indigestion or difficulty swallowing
Obvious change in a wart or mole
Nagging cough or hoarseness

* Notice that the first letters of the signs spell the advice CAUTION. *Source:* American Cancer Society.

Table 22.4 Recommended Cancer Screening Tests				
Test or Procedure	Cancer	Sex	Age	Frequency
Breast self-examination	Breast	Female	20+	Monthly
Mammogram	Breast	Female	40–49 50+	Every 1–2 years Yearly
Testicle self-examination	Testicle	Male	18+	Monthly
Sigmoidoscopy	Colon	Male, Female	50+	Every 3–5 years
Fecal occult blood test	Colon	Male, Female	50+	Yearly
Digital rectal examination	Prostate, colorectal	Male, Female	40+	Yearly
Pap test	Uterus, cervix	Female	18+ and all sexually active women	Every other year until age 35; yearly thereafter
Pelvic examination	Uterus, ovaries, cervix	Female	18–39 40+	Every 1–3 years w/Pap, yearly
Endometrial tissue sample	Endometrium	Female	At menopause for women at high risk	Once
General checkup		Male, Female	20–39 40+	Every 3 years Yearly

TREATING AND PREVENTING CANCER

Surgery, drugs, and irradiation of tumors have long been the major weapons against cancer. Surgery may be a cure when a tumor is fully accessible and has not spread.

Chemotherapy uses drugs to kill cancer cells. Most anticancer drugs are designed to kill dividing cells. They disrupt DNA replication during the S phase of the cell cycle or prevent mitosis by interfering with formation of the mitotic spindle. The drugs fluorouracil and vincristine are two compounds that work this way. Unfortunately, such drugs are also toxic to rapidly dividing healthy cells such as hair cells, stem cells in bone marrow, lymphocytes of the immune system, and epithelial cells of the gut lining. That is why chemotherapy patients can suffer side effects such as hair loss, nausea and vomiting, anemia, and reduced immune responses. Radiation likewise kills cancer cells *and* healthy cells in the irradiated area. A treatment option called *adjuvant therapy* (adjuvant means helping) combines surgery and a less toxic dose of chemotherapy. A patient might receive enough chemotherapy to shrink a tumor, for instance, then have surgery to remove what remains.

Researchers are also using **monoclonal antibodies** as "magic bullets" to deliver lethal doses of anticancer drugs to tumor cells while sparing healthy cells. As described in Section 8.10, the antibodies are engineered to target abnormal cell surface markers (antigens) on various types of tumors. The idea is to link tumor-specific monoclonal antibodies with lethal doses of cytotoxic drugs. Recent experiments that combine monoclonal antibodies with anticancer drugs have shown promising results in a few patients with *chronic myelogenous leukemia*. Similar studies are targeting a type of breast cancer and certain *gliomas* (cancers that arise in glia, the support cells of the central nervous system described in Chapter 11).

Another promising prospect for cancer treatment is **immunotherapy**—giving patients substances that will trigger a strong immune response against cancer cells. Recall from Chapter 8 that many cells can produce and release interferons, which in turn can activate cytotoxic T cells and natural killer cells. These can recognize and kill various types of cancer cells. So far, therapies that involve administering interferon have been useful only against some rare forms of cancer. Another approach is to develop cancer "vaccines" that stimulate cytotoxic T cells to recognize and destroy body cells bearing an abnormal surface protein. In some patients with malignant melanoma, doses of an experimental vaccine have caused tumors to regress, at least temporarily. Yet another approach is to develop drugs that trigger apoptosis (cell suicide) in cancer cells but not in healthy body cells.

Interleukins—signaling molecules produced by the immune system's lymphocytes—are also potential anticancer weapons. Cytotoxic T cells grown in culture with

Figure 22.8 A low-fat diet that includes vegetables in the cabbage family, such as broccoli, may help limit personal cancer risk.

IL-2 become activated. In clinical trials, several hundred patients with advanced kidney cancer have received doses of such activated cells, combined with large amounts of more IL-2. Although the treatment can have serious side effects, enough patients have responded favorably that regulators have approved it for limited use.

None of us can control factors in our heredity or biology that might lead one day to cancer, but each of us can make lifestyle decisions that promote health. The American Cancer Society recommends the following strategies for limiting your cancer risk:

1. Avoid tobacco in any form, including secondary smoke from others.

2. Maintain a desirable weight. Being more than 40 percent overweight increases the risk of several cancers.

3. Eat a low-fat diet that includes plenty of vegetables and fruits (Figure 22.8). As noted in Chapter 6, antioxidants such as vitamin E may help prevent some kinds of cancer.

4. Drink alcohol in moderation. Heavy alcohol use, especially in combination with smoking, increases risk for cancers of the mouth, larynx, esophagus, and liver.

5. If you are a woman entering menopause, consult with your doctor about using estrogen, which *may* increase the risk of breast and endometrial cancer.

6. Learn whether your job or residence exposes you to such industrial agents as nickel, chromate, vinyl chloride, benzene, asbestos, and agricultural pesticides, which are associated with various cancers.

7. Protect your skin from excessive sunlight.

SOME MAJOR TYPES OF CANCER

In general, a cancer is named according to the type of tissue in which it first forms. For instance, cancers of connective tissues such as muscle and bone are *sarcomas*. Various types of *carcinomas* arise from epithelium, including cells of the skin and epithelial linings of internal organs. When a cancer begins in a gland or its ducts, it is called an *adenocarcinoma*. *Lymphomas* are cancers of lymphoid tissues in organs such as lymph nodes, and cancers arising in blood-forming regions—mainly stem cells in bone marrow—are *leukemias*.

The remainder of this chapter surveys some common cancers in the United States. Figure 22.9 summarizes the most recent data for males and females.

A cancer is categorized according to the tissue in which it arises. Examples include sarcomas (connective tissue), carcinomas (epithelium), adenocarcinomas (glandular tissue), lymphomas (lymphoid tissues), and leukemias (blood-forming regions such as bone marrow).

Cancer Incidence by Site and Sex*		Cancer Deaths by Site and Sex	
MALE	**FEMALE**	**MALE**	**FEMALE**
prostate 180,400	breast 182,800	lung and bronchus 89,300	lung and bronchus 67,600
lung and bronchus 89,500	lung and bronchus 74,600	prostate 31,900	breast 40,800
colon and rectum 63,600	colon and rectum 66,600	colon and rectum 27,800	colon and rectum 28,500
urinary bladder 38,300	uterine corpus 36,100	pancreas 13,700	pancreas 14,500
non-Hodgkin lymphoma 31,700	ovary 23,100	non-Hodgkin lymphoma 13,700	ovary 14,000
melanoma of the skin 27,300	non-Hodgkin lymphoma 23,200	leukemia 12,100	non-Hodgkin lymphoma 12,400
oral cavity 20,200	melanoma of the skin 20,400	esophagus 9,200	leukemia 9,600
kidney 18,800	urinary bladder 14,900	liver 8,500	uterine corpus 6,500
leukemia 16,900	pancreas 14,600	urinary bladder 8,100	brain 5,900
pancreas 13,700	thyroid 13,700	stomach 7,600	stomach 5,400
all sites 619,700	all sites 600,400	all sites 284,100	all sites 268,100

* Excludes basal and squamous cell skin cancer and carcinomas in situ except urinary bladder.

Figure 22.9 Summary of annual incidence of and deaths from common cancers, by site and sex. Data are estimates for the United States, 2000. Courtesy American Cancer Society.

CANCERS OF THE BREAST AND REPRODUCTIVE SYSTEM

Breast Cancer

In the United States, about one woman in nine develops breast cancer. Of all cancers in women, breast cancer currently ranks second only to lung cancer as a cause of death. Breast cancer also occurs in males, but much more rarely. In women, obesity, late childbearing, early puberty, late menopause, and excessive levels of estrogen (and perhaps certain other hormones) seem to play roles in breast cancer development. A family history of breast cancer increases the risk; it may indicate the presence of one of several mutated genes thought to be responsible for many cases of familial breast cancer. Although 80 percent of breast lumps are *not* cancer, a woman should seek medical advice about any breast lump, thickening, dimpling, breast pain, or discharge.

Chances for cure are excellent if breast cancer is detected early and treated promptly. Hence a woman should examine her breasts every month, about a week after her menstrual period. Figure 22.10 shows the steps of a self-exam, as recommended by the American Cancer Society. Low-dose mammography (breast X ray) is the most effective method for detecting small breast cancers. It is 80 percent reliable. The American Cancer Society recommends an annual mammogram for women over 50 and for younger women at high risk.

Treatment depends mainly on the extent of the disease. In *modified radical mastectomy,* the affected breast tissue, overlying skin, and nearby lymph nodes are removed, but muscles of the chest wall are left intact. If a breast tumor is small, the preferred treatment may be *lumpectomy,* which is less disfiguring than mastectomy because it leaves some breast tissue in place. In both cases, removed lymph nodes are examined to determine the need for further treatment and to predict the prospects of a cure.

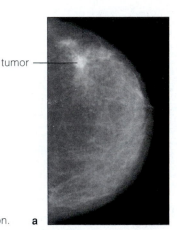

tumor

Figure 22.10 (**a**, *right*) Mammogram showing a tumor, which a biopsy revealed to be cancer. The white patches at the front of the breast are milk ducts and fibrous tissue. (**b**, *below*) How to perform a breast self-examination.

a

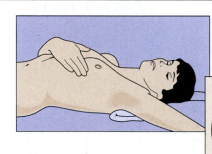

1. Lie down and put a folded towel under your left shoulder, then put your left hand behind your head. With the right hand (fingers flat), begin the examination of your left breast by following the outer circle of arrows shown. Gently press the fingers in small, circular motions to check for any lump, hard knot, or thickening. Next, follow the inner circle of arrows. Continue doing this for at least three more circles, one of which should include the nipple. Then repeat the procedure for the right breast. For a complete examination, repeat the procedure while standing in a shower. Hands glide more easily over wet skin.

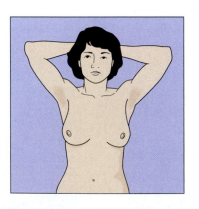
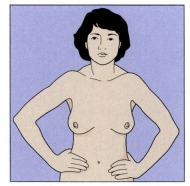
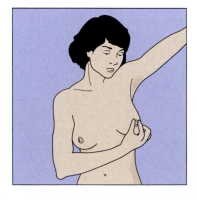

2. Stand before a mirror, lift your arms over your head, and look for any unusual changes in the contour of your breasts, such as a swelling, dimpling, or retraction (inward sinking) of the nipple. Also check for any unusual discharge from the nipple.

If you discover a lump or any other change during a breast self-examination, it's important to see a physician at once. Most changes are not cancerous, but let the doctor make the diagnosis.

b

Further reading: Student Guide to InfoTrac on web site →

Increasingly, drugs are being used in the fight against breast cancer. In some cases *tamoxifen* can shrink tumors; researchers also are exploring its use in preventing disease in women at high risk for breast cancer (although that treatment can involve serious side effects). At this writing, two other promising drugs, *herceptin* and *rituxan*, are in clinical trials. They are coupled to monoclonal antibodies, which carry them directly to the targeted cancer cells.

Cancers of the Reproductive System

UTERINE AND OVARIAN CANCER Cancers of the uterus most often affect the endometrium (uterine lining) and the cervix. Various types are treated by surgery, radiation, or both. The incidence of uterine cancers is falling, in part because precancerous phases of cervical cancer can be easily detected by the *Pap smear* that is part of a routine gynecological examination. The risk factors for cervical cancer include having many sex partners, early age of first intercourse, cigarette smoking, and genital warts (Section 16.11). Endometrial cancer is more common during and after menopause; women who take estrogen supplements during menopause may be at higher risk.

Ovarian cancer is often lethal because symptoms, mainly an enlarged abdomen, do not appear until the cancer is advanced and has already metastasized. In a few women the first sign is abnormal vaginal bleeding or vague abdominal discomfort. Risk factors include family history of the disease, childlessness, and a history of breast cancer. New hope for victims of ovarian (and breast) cancers has come from the discovery of taxol, a compound originally derived from the Pacific yew tree. In 20–30 percent of advanced cases, taxol shrinks tumors significantly. How? It makes a tumor cell's microtubules highly stable, so they can't disassemble and reassemble—key operations, recall, in cell division. Hence, the cancer cells die just as they are about to divide.

CANCER OF THE TESTIS AND PROSTATE About 7,500 cases of cancer of the testis are diagnosed annually in the United States. In its early stages, testicular cancer is painless. However, it can spread to lymph nodes in the abdomen, chest, neck, and, eventually, the lungs. Once a month from high school onward, men should examine each testis separately after a warm bath or shower (when the scrotum is relaxed). The testis should be rolled gently between the thumb and forefinger to check for any unusual lump, enlargement, or hardening (Figure 22.11). Because the epididymis may be confused with a lump, the important thing is to *compare the two testes*. A lump may or may not cause discomfort, but only a physician can rule out the possibility of disease. Surgery is the usual treatment, and the success rate is high when the cancer is caught before it can spread.

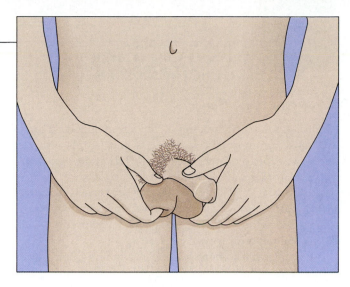

Figure 22.11 Method for testicular self-examination recommended by the American Cancer Society.

Do the exam when the scrotum is relaxed, as it is after a warm bath or shower. Simply roll each testicle between the thumb and forefinger, feeling for any lumps or thickening. By performing the exam regularly, you can detect changes early on and discuss them with your physician. As with breast lumps, most such changes are *not* cancer, but only a doctor can make the diagnosis.

Prostate cancer is the second leading cause of cancer deaths in men (lung cancer is first). There are no confirmed risk factors. Symptoms include various urinary problems, although these can also signal simply a noncancerous enlarged prostate. For men over 40, an annual digital rectal examination, which enables a physician to feel the prostate, is the first step in detecting unusual lumps. The PSA blood test can screen for suspiciously large amounts of that tumor marker. If a man's physician suspects cancer after these two tests have been performed, the next step is a tissue biopsy. The cure rate for prostate tumors detected early is over 90 percent.

Many prostate cancers grow slowly and seem to cause few problems. In such cases a physician may recommend simply monitoring the tumor, taking no other action until there is a clear threat to health. Like all medical decisions, however, this one can be made only after carefully weighing the risks and benefits.

Breast cancer affects about one woman in nine. Rarely, it also occurs in males.

The most common reproductive cancers in women develop in the ovaries and uterus. In males, cancers of the testis and prostate gland are the most common reproductive cancers.

Chances for a cure are best when cancer is detected early. Monthly self-examination is a crucial tool for early detection of breast cancer and testicular cancer.

A SURVEY OF OTHER COMMON CANCERS

Oral and Lung Cancers

Cancers of the mouth, tongue, salivary glands, and throat are most common among smokers and people who use smokeless tobacco (snuff), especially if they are also heavy alcohol drinkers. In 1999, the American Cancer Society reported nearly 30,000 new cases of oral cancer and over 8,000 deaths, mostly among men. Of people who are diagnosed with oral cancer, only about half survive for 5 years, even with treatment.

Lung cancer kills more people than any other cancer. Long-term tobacco smoking is the overwhelming risk factor. (Recent evidence suggests that one chemical in tobacco smoke, benzypyrene, specifically mutates the p53 tumor supressor gene described in Section 22.2.) Other risk factors are exposure to asbestos, to industrial chemicals such as arsenic, and to radiation. For a smoker, a combination of these factors greatly boosts the odds of developing cancer. For nonsmokers, especially spouses and children of smokers, inhaling tobacco smoke also poses a significant cancer risk. The Environmental Protection Agency estimates that lung cancer resulting from regularly breathing secondhand smoke kills 3,000 people in the United States each year.

In recent years, the incidence of lung cancer has decreased in men. However, a rise in the relative number of female smokers in the last several decades is now reflected in the fact that lung cancer has surpassed breast cancer as the leading cancer killer of women. Warning signs include a nagging cough, shortness of breath, chest pain, blood in coughed-up phlegm, unexplained weight loss, and frequent respiratory infections or pneumonia.

Four types of lung cancer account for 90 percent of cases. About one-third of lung cancers are *squamous cell carcinomas*, affecting squamous epithelium in the bronchi. Another 48 percent are either *adenocarcinomas* or *large-cell carcinomas* (Figure 22.12). A fourth type, called *small-cell carcinoma*, spreads rapidly and kills most of its victims within 5 years of diagnosis.

Cancers of the Digestive System and Related Organs

Cancers of the stomach and pancreas are usually adeno-carcinomas of duct cells. Often they are not detected until they have spread to other organs. Cigarette smoking is a risk factor for pancreatic cancer, while stomach cancer may be associated with heavy alcohol consumption and a diet rich in smoked, pickled, and salted foods. Liver cancer is uncommon in the United States, but growing evidence suggests that hepatitis B infection can trigger it.

Most colon cancers are adenocarcinomas. Warning signs include a change in bowel habits, rectal bleeding,

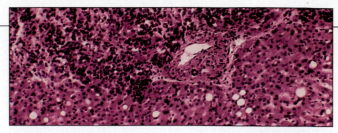

a

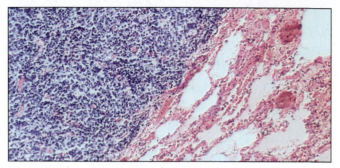

b

Figure 22.12 Photomicrographs of (**a**) adenocarcinoma and (**b**) large-cell carcinoma of the lung.

and blood in the feces. Family history of colon cancer (indicating a genetic component) or inflammatory bowel disease is a major risk factor.

Urinary System Cancers

Carcinomas of the bladder and kidney account for about 70,000 new cancer cases each year. The incidence is higher in males, and smoking and exposure to certain industrial chemicals are major risk factors for both. Kidney cancer easily metastasizes via the bloodstream to the lungs, bone, and liver. An inherited type, *Wilms tumor*, is one of the most common of all childhood cancers.

Cancers of the Blood and Lymphatic System

Lymphomas develop in lymphoid tissues in organs such as the lymph nodes, spleen, and thymus. They include the diseases known as *Hodgkin disease*, *non-Hodgkin lymphoma*, and *Burkitt lymphoma*. Risk seems to increase along with infections—such as HIV—that impair immune system functioning. Burkitt lymphoma is most common in parts of Africa, where it seems to develop especially in children who become infected with both the Epstein-Barr virus (which also causes mononucleosis) and malaria. Lymphoma symptoms include enlarged lymph nodes, rashes, weight loss, and fever. Intense itching and night sweats also are typical of Hodgkin disease. Chemotherapy and radiation are standard treatments, and new treatments using targeted monoclonal antibodies are being developed.

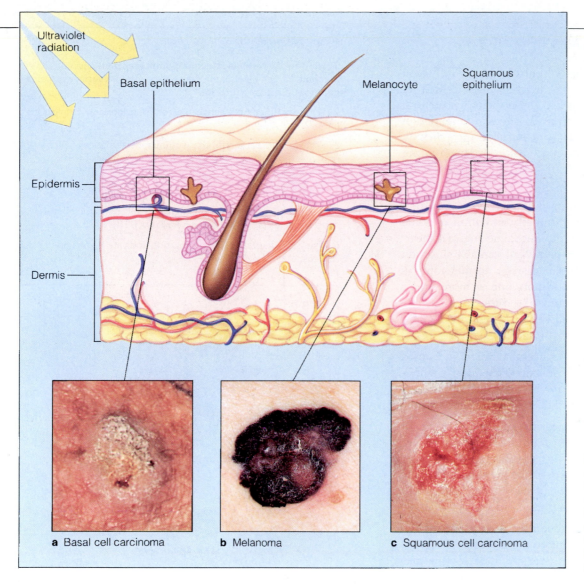

Figure 22.13 (**a**) Basal cell carcinoma, (**b**) malignant melanoma, and (**c**) squamous cell carcinoma. Each type of skin cancer has a distinctive appearance. Malignant melanoma can be deadly because it metastasizes aggressively.

Leukemias are cancers in which stem cells in bone marrow overproduce white blood cells. Some types are the most common childhood cancers, but other types are diagnosed in more than 70,000 adults each year. Risk factors include Down syndrome, exposure to chemicals such as benzene, and radiation exposure. Many leukemias can be effectively treated with chemotherapy using two compounds—vincristine and vinblastine—derived from species of periwinkle plants that grow only on the African island of Madagascar. Environmentalists often cite these drugs and other plant-derived compounds such as taxol as reasons why humans should attempt to preserve natural areas where as-yet-undiscovered medicinal plants may be present.

Skin Cancer

Skin cancers are the most common of all cancers. For all types, fair skin and exposure to ultraviolet (UV) radiation in sunlight (or tanning salons) are risk factors. UV damage to the DNA in melanocytes, the melanin-producing cells in the skin's epidermis, is the most dangerous kind, called malignant melanoma (see Section 4.9). Malignant melanoma is a particular threat because in its later stages it metastasizes aggressively.

Squamous cell (epidermal) carcinomas start out as scaly, reddened bumps (Figure 22.13*c*). They grow rapidly and can spread to adjacent lymph nodes unless they are surgically removed. Basal cell carcinomas begin as small, shiny bumps and slowly grow into ulcers with beaded margins. Basal cell carcinoma and squamous cell carcinoma are much more common than malignant melanoma, and usually they are easily treated by minor surgery in a doctor's office.

Common cancers include oral and lung cancers and those of the digestive system (especially colon cancer), the urinary system, the blood and lymphatic systems (leukemias and lymphoma), and the skin. Each year, these collectively account for more than 750,000 cases of cancer in the United States alone.

SUMMARY

1. Cancer arises when genetic controls over cell division are lost. Cancer cells have specific characteristics that set them apart from normal cells. These include structural abnormalities, including (typically) poor differentiation and altered surface proteins; uncontrolled growth with absence of contact inhibition; invasion of surrounding tissues; and the ability to metastasize to other body regions.

2. Cancer develops through a multistep process involving several genetic changes. Initially, mutation or some other event may alter a proto-oncogene into a cancer-causing oncogene. Infection by a virus can also insert an oncogene into a cell's DNA or disrupt normal controls over a proto-oncogene. In addition, one or more tumor suppressor genes probably must be missing or become mutated before a normal cell can be transformed into a cancerous one.

3. A predisposition to some cancers is inherited. Causes of carcinogenesis are viral infection, chemical carcinogens, radiation, faulty immune system functioning, and possibly breakdown in normal DNA repair mechanisms.

4. Common methods for cancer diagnosis include blood testing for the presence of tumor markers, which are substances produced either by specific types of cancer cells or by normal cells in response to the cancer. Medical imaging (as by magnetic resonance imaging) also can aid diagnosis. The definitive diagnostic procedure is biopsy.

5. In general, a cancer is named according to the type of tissue in which it arises. Common types include sarcomas (connective tissues such as muscle and bone), carcinomas (epithelium), adenocarcinoma (glands or their ducts), lymphomas (lymphoid tissues), and leukemias (blood-forming regions).

6. Standard cancer treatments include surgery, chemotherapy, and tumor irradiation. Target-specific monoclonal antibodies and immune therapy using interferons and interleukins are other treatment options currently under development.

7. Lifestyle choices such as the decision not to use tobacco, to maintain a low-fat diet, and to avoid overexposure to direct sunlight and chemical carcinogens can help limit personal cancer risk.

8. The incidence of various cancers varies among the sexes. Among adult females, the most prevalent sites are the breasts, colon and rectum, lungs, and uterus. Among adult males the most prevalent sites are the prostate gland, lung, colon and rectum, and bladder.

Review Questions

1. How are cancer cells structurally different from normal cells of the same tissue? What is the relevance of altered surface proteins to uncontrolled growth? *22.1*

2. What are differences between a benign tumor and a cancerous one? *22.1*

3. Write a short paragraph that summarizes the roles of proto-oncogenes, oncogenes, and tumor suppressor genes in carcinogenesis. *22.2*

4. List the four main categories of cancer tumors. *22.6*

5. What are the seven warning signs of cancer? (Remember the American Cancer Society's clue word, CAUTION.) *22.4*

6. Using the diagram below as a guide, indicate the major steps in cancer metastasis. *22.1*

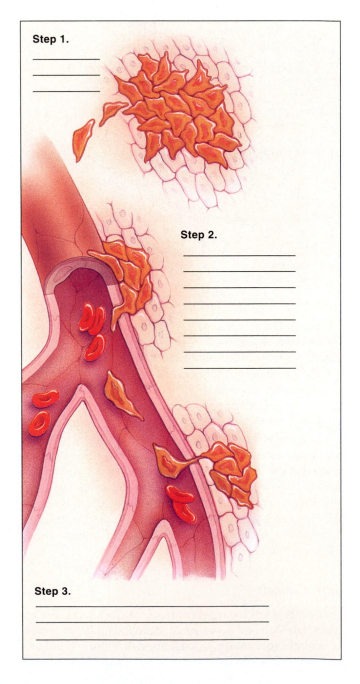

Step 1.

Step 2.

Step 3.

Self-Quiz (Answers in Appendix V)

1. A tumor is _____.
   a. malignant by definition
   b. always enclosed by connective tissue
   c. a mass of tissue that may be benign *or* malignant
   d. always slow-growing

2. Cancer cells _____.
   a. lack normal controls over cell division
   b. secrete the growth factor angiogenin
   c. display altered surface proteins
   d. do not respond to contact inhibition
   e. all of the above

3. The onset of cancer seems to require the activity of an oncogene plus the absence or mutation of at least one _____.

4. Chemical carcinogens _____.
   a. include viral oncogenes
   b. can damage DNA and cause a mutation
   c. must be ingested in food
   d. are not found in foods

5. So far as we know, carcinogenesis is *not* triggered by _____.
   a. breakdowns in DNA repair
   b. a breakdown in immunity
   c. radiation
   d. protein deficiency
   e. inherited gene defects

6. Tumor suppressor genes _____.
   a. occur normally in cells
   b. promote metastasis
   c. enter cells by way of viral infection
   d. probably affect the development of cancer only in rare cases

7. _____ is the definitive method for detecting cancer.
   a. Blood testing
   b. Physician examination
   c. Biopsy
   d. Medical imaging

8. The most common therapeutic approaches to treating cancer include all of the following except:
   a. chemotherapy
   b. irradiation of tumors
   c. surgery to remove cancerous tissue
   d. administering doses of vitamins

9. The goal of immune therapy is to _____.
   a. cause defective T cells in the thymus to disintegrate
   b. activate cytotoxic T cells
   c. dramatically increase the numbers of circulating macrophages
   d. promote the secretion of monoclonal antibodies

10. Currently, _____ cancer is the leading cause of death among adult females; _____ cancer is the leading cause of cancer death among adult males.
    a. lung; prostate
    b. breast; colon
    c. lung; lung
    d. breast; lung

Critical Thinking: You Decide (Key in Appendix VI)

1. Propose an explanation of the observation that higher rates of cancer are associated with increasing age.

2. A textbook on cancer contains the following statement: "Fundamentally, cancer is a failure of the immune system." Why do you think the author wrote this?

3. Ultimately, cancer kills because it spreads and disturbs homeostasis. Consider, for example, a kidney cancer that metastasizes to the lungs and liver. What are some specific homeostatic mechanisms that the spreading disease could disrupt?

4. Over the last few months, your best friend Mark has noticed a small, black-brown, raised growth developing on his arm. When you suggest that he have it examined by his doctor right away, he says he's going to wait and see if it gets any larger. You know that's not a very smart answer. Give at least three arguments that you can use to try to convince Mark to seek medical advice now.

Selected Key Terms

biopsy 22.4	metastasis 22.1
carcinogen 22.2	oncogene 22.2
carcinogenesis 22.2	proto-oncogene 22.2
chemotherapy 22.5	tumor 22.1
dysplasia 22.1	tumor marker 22.4
immunotherapy 22.5	tumor suppressor gene 22.2

Readings

Cancer Facts and Figures, 2000. American Cancer Society. Free pamphlet published annually; an excellent summary of the latest information on cancer, risk factors, and treatments.

Ott, W., and J. W. Roberts. February 1998. "Everyday Exposure to Toxic Pollutants." *Scientific American.*

Teeley, P., and P. Bashe. 2000. *The Complete Cancer Survival Guide.* New York: Doubleday. In addition to providing background on major types of cancer, this book explains how cancer patients can take advantage of clinical trials for potential new treatments.

Waldholz, M. 1997. *Curing Cancer: Solving One of the Greatest Medical Mysteries of Our Time.* New York: Simon & Schuster. This popular book's strength lies in its engrossing behind-the-scenes descriptions of how modern cancer research is carried out.

Weinberg, R. A. 1998. *One Renegade Cell: How Cancer Begins.* New York: Basic Books.

A smoke cloud over South America's Amazon River basin photographed by astronauts in the space shuttle Discovery *in 1988. The cloud was caused by the clearing and burning of tropical forests, pasture, and croplands, activities that continue today at a rapid pace. If the cloud were placed over North America, it would cover more than one-third of the area of the 48 contiguous United States. Human impacts on Earth's ecosystems are a central topic in Unit V.*

PRINCIPLES OF EVOLUTION

Unity and Diversity

Many years ago, four boys out for a romp stumbled into a cave near Lascaux, a town in the Perigord region of France. What they found inside the cave's tunnels stunned the world. Magnificent sketches, engravings, and paintings swept across the cave walls. The red, yellow, purple, and brown pigments were as vivid as if they had just been applied. Yet radioisotope measurements revealed that they were between 17,000 and 20,000 years old. The prehistoric artists had worked deep inside the cave, where sunlight could not fade the images, and winds and water could not wear them away. By the light of crude oil lamps, they had captured the graceful, dynamic lines of bison, stags, horses, ibexes, lions, a rhinoceros, and a large heifer, now known as the Great Black Cow (Figure 23.1a).

Caves throughout southern France, northern Spain, and Africa hold treasures from even earlier times. About 25,000 years ago, for example, prehistoric peoples scribed more than 150 imprints and outlines of their hands to walls in the cave of Gargas, in the Pyrenees.

Who were the people who did this? From their fossilized remains, we know that they were anatomically like us. According to one current hypothesis, they may have arrived in Europe from Africa roughly 40,000 years ago (Figure 23.1b). But the story of the human species, *Homo sapiens*, began far earlier—more than 60 million years ago—with the origin of primates. In its turn, the primate story began more than 250 million years ago, with the origin of mammals. *That* story extends back to the origin of animals at some time before 750 million years ago—and so on back in time, to the origin of the first living cells.

This great evolutionary drama is our subject in this chapter. In Latin, the word *evolutio* means "the act of unrolling," as a long scroll might unroll across a table. That image is an apt way to begin thinking of the countless millennia over which the story of life has been "unrolling" on Earth. Based on its Latin root, in biology the term **evolution** refers to genetic change within a line of descent through successive generations.

You have probably heard the history of life compared to a large tree with many branches. Each branch of the tree represents a line of descent. The branch points represent a divergence—in general, a gradual split—leading to new species. Evidence suggests that life emerged on Earth roughly 3.8 billion years ago; over that vast span of time the "tree of life" has come to represent thousands upon thousands of species. For example, the vertebrate

a

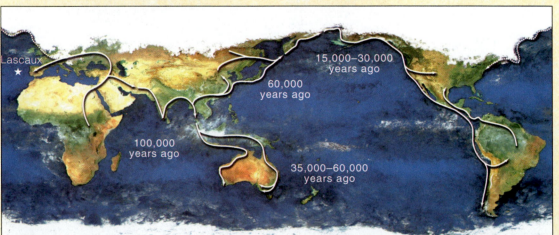

b

Figure 23.1 (a) The Big Black Cow, a prehistoric cave painting from Lascaux. (b) Estimated times when early populations of humans were colonizing different parts of the world, based on radiometric dating of fossils. The dispersal routes shown seem to support the hypothesis that our species, *Homo sapiens*, emerged in Africa.

branch of the tree includes nearly 50,000 existing species of fishes, amphibians, reptiles, birds, and mammals!

How can we organize the hundreds of thousands of known species in a meaningful way? For centuries naturalists (and, later, scientists) grouped organisms based on similarities of their body structures. Modern evolutionary biologists study such similarities, along with other information. Considered together, all the information can be an important part of the evidence for building a branching classification scheme that bases groupings on presumed evolutionary relatedness.

Classification requires the identification of each new species, which is assigned its own, unique name. In a binomial system devised long ago by Carolus Linnaeus, the name has two parts—as in *Homo sapiens*, the binomial used for humans. The first part is the genus name. A **genus** (plural: genera) encompasses all the species that are similar to one another and distinct from others in certain traits. (There are no other species in the modern genus *Homo*, although that was not always the case.) The second part of the name designates the particular species within the genus.

Classification systems organize species into a series of ever more inclusive groupings, called taxa (singular, *taxon*; hence this work is called *taxonomy*). Beginning with the species, these groupings typically include the taxa shown in Table 23.1 for humans. Each taxon above the genus level includes a larger, more diverse group of organisms that share more general features.

With this brief introduction, let's now consider the basic principles of evolution and the kinds of evidence evolutionary biologists use in their work. With those concepts as background, we'll also survey key trends in human evolution, and close with a look at what is currently known about the chemical origins of life on our planet—which, after all, is where the evolutionary story began.

Table 23.1	Classification of Humans
Kingdom	Animalia
Phylum	Chordata
Class	Mammalia
Order	Primates
Family	Hominidae
Genus	*Homo*
Species	*sapiens* (only living species of this genus)

KEY CONCEPTS

1. Biological evolution is genetic change in a line of descent (lineages) through successive generations.

2. Members of a population generally have the same number and kinds of genes, which give rise to the same assortment of traits.

3. In a population, each gene may exist in two or more slightly different forms (alleles). Different individuals don't necessarily inherit the same alleles, so they may differ in the details of their traits.

4. Natural selection is simply the result of a difference in survival and reproduction among individuals who differ in one or more traits.

5. An allele may become more or less common in a population, or it may even disappear. Such changes in a population's allele frequencies over time are called microevolution.

6. A species consists of one or more populations of individuals that can interbreed under natural conditions and produce fertile offspring, and that are reproductively isolated from other such populations.

7. New species evolve from variant individuals of existing species. Hence all species that have ever lived on Earth are related. The term macroevolution refers to patterns, trends, and rates of change in lineages over geologic time.

8. Evidence of macroevolution comes from fossils, the geologic record, and radioactive dating of rocks. Comparing the anatomy and biochemistry of different lineages helps us identify patterns of change through time.

9. Human evolution has been marked by trends that involved changes in our bones, muscles, teeth, sensory systems, and the brain.

10. Life apparently originated on Earth more than 3.8 billion years ago. Its origin and subsequent evolution have been linked to the physical and chemical evolution of the universe, the stars, and our solar system.

CHAPTER AT A GLANCE

A LITTLE EVOLUTIONARY HISTORY

a

In 1831, the origin of the Earth's amazing diversity of life forms was a matter of hot dispute amd more than a little confusion. A widely accepted view held that all species had come into existence at the same time in the distant past, at the same center of creation, and had not changed since. Yet, by the mid-19th century, several hundred years of exploration and advances in the sciences of geology and comparative anatomy had raised questions that were difficult to ignore. For example, why were some species found in particular isolated regions, and nowhere else? Why, on the other hand, were similar (but not identical) species found in widely separated parts of the world? Why did species as different as humans, whales, and bats have certain strikingly similar anatomical structures? What was the significance of geologists' discoveries of similar fossil organisms in similar layers of the Earth's sedimentary rocks—no matter where in the world the layers occurred?

Enter Charles Darwin, who in 1831 was 22 years old, had a freshly minted theology degree from Cambridge, and was wondering what to do with his life. He was *not* interested in becoming a clergyman. All he had ever wanted to do was hunt, fish, collect shells, or simply watch insects and birds. John Henslow, a botanist who befriended Darwin during his university years, saw the young man's real interests. He arranged for Darwin to work as a naturalist aboard the HMS *Beagle*, which was about to embark on a 5-year voyage around the world (Figure 23.2). The *Beagle* sailed first to South America, to complete work on mapping the coastline. During the Atlantic crossing, Darwin studied geology and collected and examined marine life. During stops along the coast and at various islands, he observed other species of organisms in environments ranging from sandy shores to high mountains. After returning to England in 1836, Darwin began talking with other naturalists about a topic that was on many scholars' minds—the growing evidence that life forms evolve, changing over time.

How could organisms evolve? One clue came from an essay by Thomas Malthus, a clergyman and economist, who proposed that any population tends to outgrow its resources, and so in time its members must compete for what is available. Darwin's observations also suggested that any population can produce more individuals than the environment can support, yet populations tend to remain stable over time. For instance, a starfish can release 2,500,000 eggs a year, but the oceans are not filled with starfish. What determines who lives and who dies as predators, starvation, and environmental insults take their toll on a population? Chance could be a factor, but a second important clue to the evolution puzzle came from observations by Darwin and others that even members of the same species vary slightly in their traits.

b

Figure 23.2 (**a**) A replica of the HMS *Beagle*, shown sailing off the coast of South America. (**b**) Charles Darwin and a blue-footed booby, one of the species he encountered during his 5-year voyage around the world.

As you will read in the following sections, Darwin's thoughtful melding of his observations of the natural world with the ideas of other thinkers led him to propose that evolution could occur by way of a process called *natural selection*. Widespread acceptance of this theory would not come until nearly 70 years later, when a new field, genetics, provided insights on the source of variation in traits. Now let's consider some current views of the mechanisms of evolution.

Combining observations of the natural world with ideas about interactions of populations with their environment, Darwin forged his theory of evolution by natural selection.

SOME BASIC PRINCIPLES OF EVOLUTION

The history of life on Earth spans nearly 4 billion years. It is a story of how species originated, survived or went extinct, and stayed put or radiated into new environments. The overall "plot" of the story is evolution, genetic change in lines of descent over time. **Microevolution** is the name for cumulative genetic changes that may give rise to new species. **Macroevolution** is the name for the large-scale patterns, trends, and rates of change among *groups* of species. Later sections provide a fuller picture of these two patterns of change. For the moment, we will examine the basic principles of evolutionary change.

Variation in Populations

As the chapter introduction noted, an individual does not evolve. Evolution occurs when there is change in the genetic makeup of *populations of organisms*. Biologists define a population very specifically. A **population** is a group of individuals of the same species occupying a given area. As you know from your own experience with other human beings, there is a great deal of genetic variation within and among populations of the same species.

Overall, the members of a population have similar traits—that is, phenotypes. They have the same general form and appearance (*morphological* traits), their body structures function in the same way (*physiological* traits), and they respond the same way to certain basic stimuli (*behavioral* traits). However, the details of traits vary greatly from one individual to another. For instance, individual humans vary in the color of their body hair, as well as in its texture, amount, and distribution over the body. As we know from previous chapters, this example only hints at the immense genetic variation in human populations (Figure 23.3). Populations of most other species of organisms show the same kind of variation.

Where Does Variation Come From?

The members of a population generally have inherited the same number and kinds of genes. As you know, though, a given gene may have slightly different forms, called alleles. Variations in traits in a population—hair color, say—come about as individuals inherit different combinations of alleles. Whether your hair is black, brown, red, or blond depends on *which* alleles of certain genes you inherited from your mother and father. Earlier chapters described five events that contribute to that mix of alleles, and we can review them here:

1. Gene mutation (produces new alleles)

2. Crossing over at meiosis (leads to new combinations of alleles in chromosomes)

Figure 23.3 A small sample of the outward variation in *Homo sapiens*. Variation in traits arises from different combinations of alleles carried by different members of populations.

3. Independent assortment at meiosis (leads to mixes of maternal and paternal chromosomes in gametes)

4. Fertilization (puts together combinations of alleles from two parents)

5. Changes in chromosome structure or number (leads to the loss, duplication, or alteration of alleles)

If you pour out a bag of chocolate candies that have different-colored sugar coatings, you will find that some colors turn up more or less frequently than others do. The same holds true for the alleles in a population. Different alleles are more or less abundant. The abundance of a given kind of allele in a population is called that allele's *frequency*. Small-scale changes in allele frequencies may be brought about by mutation, natural selection, and other processes. When such changes take place, the result is microevolution, the topic we turn to next.

A population is a group of individuals of the same species occupying a given area.

Although individuals in a population share many general phenotypes, in nearly all populations there is a great deal of underlying genetic variation.

Evolution is genetic change in lines of descent over time. It results from changes in the genetic makeup of populations of organisms.

PROCESSES OF MICROEVOLUTION

Mutation: Sole Source of New Alleles

Throughout our lives, our DNA is subject to mutations, each one a heritable change in DNA (Section 20.8). Most likely, you carry several mutations. Whether a particular mutation ends up being harmful, neutral, or beneficial depends on how its product (the protein it encodes) interacts with that of other genes *and* with the environment. In evolutionary terms, mutation is a significant event. Why? *Mutations are the source of alleles (alternative forms of genes) that have been accumulating in different lineages since the beginning of life on Earth.*

Because they are accidental changes in the DNA sequence, most mutations are probably harmful, altering traits such that an affected individual cannot survive or reproduce as well as other individuals. For example, humans live in environments in which small cuts or other minor injuries happen often. Before effective medical treatments existed, hemophiliacs, whose blood does not clot properly, could die at a young age from such minor injuries. As a result, the frequencies of the mutated alleles for the various hemophilias historically have been low. By contrast, a trait that is beneficial *improves* some aspect of an individual's functioning in the environment and so improves the chances of surviving and reproducing. A neutral trait, such as attached earlobes in humans, neither helps nor hinders survival under prevailing conditions.

Natural Selection

Natural selection probably accounts for more changes in allele frequencies than does any other microevolutionary process. As you have read, Darwin discovered this major process by correlating his understanding of inheritance with certain features of populations and the environment. In 1859 he published his ideas in a classic book, *On the Origin of Species.* Here we express the main points of Darwin's correlation in light of modern genetics:

1. The individuals of a population vary in their form, functioning, and behavior.

2. Much of the variation in a population is heritable. This means that more than one kind of allele exists for genes that give rise to the traits. These alleles can be transmitted from parents to offspring.

3. Some forms of a trait are more "fit" than others, improving chances of surviving and reproducing.

4. **Natural selection** is the *difference in survival and reproduction* that has occurred among individuals that differ in one or more traits. (This differential in survival and reproduction is sometimes informally referred to as "survival of the fittest.")

5. A population is *evolving* when some forms of a trait are becoming more or less common relative to the other forms. The shifts are evidence of changes in the relative abundances of alleles for that trait.

6. Life's diversity is the outcome of changes in allele frequencies in different lines of descent over time. Such genetic changes can result in the splitting of a single species into two related ones.

As natural selection occurs over time, organisms come to have characteristics that suit them to the conditions in a particular environment. We call this trend **adaptation**.

You have probably heard references to the "theory" of evolution. Yet decades of careful scientific observations have yielded so much evidence of evolution and natural selection that both ideas are more properly considered fundamental principles of the living world. Later we will consider a few examples of this evidence.

Genetic Drift and Gene Flow

Besides the nonrandom changes in allele frequency caused by natural selection, allele frequencies in a population also can change randomly because of chance events alone. This is called **genetic drift**. Often the change is most rapid in small populations. In one type of genetic drift, called the *founder effect*, a few individuals leave a population and establish a new one. Purely by chance, the frequencies of certain alleles in the new population are likely to differ from those in the old group. The founder effect probably explains why there are virtually no Native Americans with type B blood. Genes for that blood group must not have been present in the few people who first crossed the Bering Strait from Asia into what is now Alaska at least 10,000 years ago.

Allele frequencies also can change as individuals leave a population (emigration) or new individuals enter it (immigration). This physical movement of alleles, or **gene flow**, helps keep neighboring populations genetically similar. Over time it tends to counter the differences between populations that are brought about through mutation, genetic drift, and natural selection. In this age of international travel, the pace of gene flow among human populations has increased dramatically.

Reproductive Isolation and Speciation

For sexually reproducing organisms, a **species** is a genetic unit consisting of one or more populations of organisms that usually closely resemble each other physically and physiologically. Members of a species can interbreed and produce fertile offspring under natural conditions. No matter how diverse those individuals become, they remain members of the same species so long as they can continue to interbreed successfully and share a common pool of alleles. A woman lawyer in India may never meet up with an Icelandic fisherman, but there is no biological

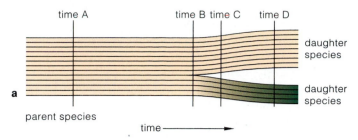

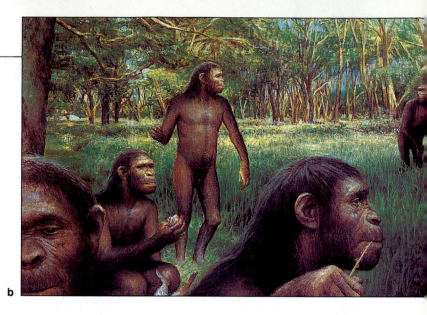

Figure 23.4 (**a**) Divergence, the first step on the path to speciation. Each horizontal line represents a different population. Because evolution is gradual, we cannot say at any one point in time that there are now two species rather than one. At time A there is only one species. At D there are two. At B and C the split has begun but is far from complete. (**b**) Artist's view of *Homo habilis* males in an East African woodland.

reason why the two could not mate and produce children. Neither person, however, could mate successfully with a chimpanzee, even though chimps and humans are closely related species and have more than 90 percent of their genes in common. In evolutionary terms, we would say that humans and chimpanzees are *reproductively isolated*. There are enough genetic differences between them that they cannot interbreed successfully.

Reproductive isolation develops when gene flow between two populations stops. This often occurs when two populations become separated geographically. For example, the fossil record indicates that early members of the genus *Homo* diverged genetically, possibly when one or more populations emigrated out of Africa toward Asia and Europe. Such newly separated populations may have to function in slightly different environments. In those different environments, mutation, natural selection, and genetic drift begin to operate independently in each population. These processes can alter the allele frequencies of each in different ways, and eventually the differences among separated populations may lead to **reproductive isolating mechanisms**. These mechanisms are aspects of structure, function, or behavior that reduce the likelihood of successful interbreeding. The two populations may breed in different seasons, for example, or they may have different mating rituals. There may be changes in body structure that physically interfere with mating. Other types of isolating mechanisms prevent zygotes or hybrid offspring from developing properly.

The build-up of differences in allele frequencies among isolated populations is called *divergence* (Figure 23.4*a*). Divergence may become the first step on the road to **speciation**, the evolutionary process by which species originate. When genetic differences between isolated populations are great enough, their members will not be able to interbreed successfully in nature. At this point, the populations are separate species. Figure 23.4*b* is an artist's rendition of an early human species, *Homo habilis*, now long extinct. As you can see simply by looking in the mirror, processes of microevolution have produced

many differences between *H. habilis* and our own species, *Homo sapiens*.

Rates of Evolutionary Change

We know that factors such as gene mutation and natural selection ultimately trigger speciation; even so, biologists disagree about the pace and timing of microevolution.

According to the traditional model, called *gradualism*, new species emerge through many small changes in form over long spans of time. In other words, microevolution is constantly going on in imperceptibly tiny increments; new species result. By contrast, according to the model called *punctuated equilibrium*, most evolutionary change occurs in bursts. That is, each species undergoes a spurt of changes in form when it first branches from the parental lineage, then changes little for the rest of its time on Earth.

There is evidence that the driving force behind rapid changes in species may be dramatic changes in climate or some other aspect of the physical environment. This type of change suddenly (in evolutionary terms) alters the physical conditions to which populations of organisms have become adapted. For example, the onset of an ice age changes the living conditions for many land-dwelling species in affected areas. A major shift in ocean currents (which help determine water temperature) might have the same effect on marine life forms. Punctuated equilibrium could help explain why the fossil record does not provide much evidence of a continuum of microevolution—one often cannot find forms between closely related species in the fossil record. Both models probably have a place in explaining the history of life.

Mutations are the source of new alleles, hence of heritable variation in traits.

A species is a unit of one or more populations of individuals that can interbreed under natural conditions and produce fertile offspring.

LOOKING AT FOSSILS AND BIOGEOGRAPHY

When Darwin developed his ideas on evolution, he was trying to come up with an idea that could explain certain observations of the natural world. Key information came from **fossils**, which are recognizable, physical evidence of ancient life (Figure 23.5). The more Darwin and others learned about the similarities and differences among fossils and living organisms, the stronger the evidence became that such diversity is the result of evolution—and specifically of evolution by natural selection as populations adapted to their surroundings.

Fossils: Evidence of Ancient Life

The most common fossils are bones, teeth, shells, seeds, and other hard parts of organisms. (When an organism dies, soft parts usually decompose first.) **Fossilization** begins with burial in sediments or volcanic ash. With time, water seeps into the organic remains, infusing them with dissolved metal ions and other inorganic compounds. As more and more sediments accumulate above the burial site, the remains are subjected to increasing pressure. Over great spans of time, the chemical changes and pressure transform them to stony hardness.

Organisms are more likely to be preserved when they are buried rapidly in the absence of oxygen. Entombment by volcanic ash or anaerobic mud satisfies this condition very well. Preservation also is favored when a burial site is not disturbed. Usually, though, fossils are broken, crushed, deformed, or scattered by erosion, rock slides, and other geologic events.

Fossil-containing layers of sedimentary rock formed long ago, when silt, volcanic ash, and other materials were gradually deposited, one above the other. This layering of sedimentary deposits is called stratification. Most sedimentary layers probably form horizontally, although geologic disturbances can tilt or rupture them.

Interpreting the Fossil Record

We currently have fossils of about 250,000 species, far more than in Darwin's day. However, judging from present diversity, there must have been millions of ancient, now-extinct species, and we will not be able to recover fossils for most of them. Why? There are several reasons. To begin with, most of the important, large-scale movements in the Earth's crust have wiped out evidence from crucial periods in the history of life. In addition, most members of ancient communities simply have not been preserved. For example, plenty of hard-shelled mollusks and bony fishes are represented in the fossil record. On the other hand, jellyfishes and soft-bodied worms are not, even though they may have been extremely common. Population density and body size further skew the record.

a

Figure 23.5 Examples of fossils. (**a**) Fossilized leaves from *Archeopteri*, which probably was on the evolutionary road that led to gymnosperms (such as pine trees) and angiosperms (flowering plants). (**b**) The complete skeleton of a bat that lived 50 million years ago. Fossils of this quality are rare. (**c**) Skulls of a prehuman lineage of African apes, called australopithecines. The skulls are part fossil, part reconstruction. Males of *A. afarensis*, *A. robustus*, and *A. boisei* had a crested skull, like a gorilla's. *A. afarensis* (far left) had a large face and projecting jaws, as apes do, but its canine and molar teeth were humanlike in some ways.

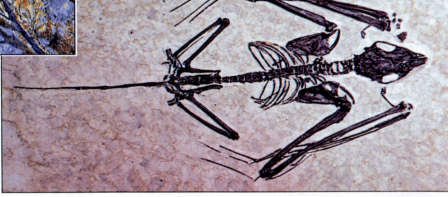

b

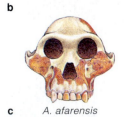   

c   *A. afarensis*   *A. africanus*   *A. robustus*   *A. boisei*

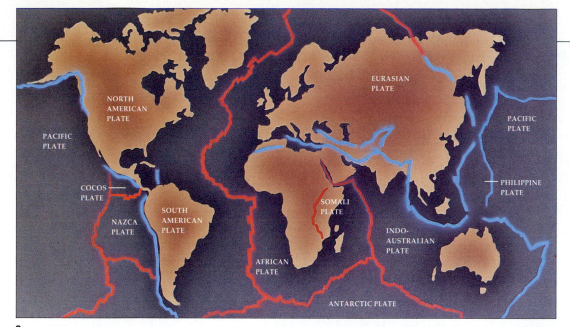

a

Figure 23.6 (a) Plate tectonics, the arrangement of the Earth's crust in rigid plates that split apart, move about, and collide with one another. (b) About 240 million years ago (mya) Earth's land masses were joined in a massive supercontinent, Pangaea. By about 40 million years ago plate movements had split Pangea into isolated land masses, including Africa, South America, Australia, and Eurasia.

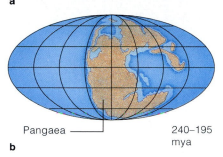
Pangaea —————————— 240–195 mya

b

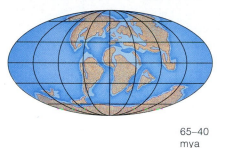
65–40 mya

10 mya

A population of ancient plants may have produced millions of spores in one growing season, while the earliest members of the human lineage lived in small groups and produced few offspring. Hence, the chance of finding a fossilized skeleton of an early human is small, compared to the chance of finding spores of plant species that lived at the same time.

The fossil record is also heavily biased toward certain environments. Most species for which we have fossils lived on land or in shallow seas that, through geologic uplifting, became part of continents. We have few fossils from sediments beneath the ocean, which covers three-fourths of the Earth's surface. Also, most fossils have been found in the Northern Hemisphere, because that's where most geologists have lived and worked.

Biogeography

Darwin also believed that the concept of evolution by natural selection could help shed light on the subject we know today as **biogeography**—the study of the world distribution of plants and animals. Biogeography asks the basic question of why certain species (and higher groupings) occur where they do. For example, why do Australia, Tasmania, and New Guinea have species of monotremes (egg-laying mammals such as the duckbilled platypus), while such animals are absent from other regions of the world where the living conditions are

similar? And why do we find the greatest diversity of life forms in tropical regions? The simplest explanation for such biogeographical patterns is that species occur where they do either because they evolved there from ancestral species or because they dispersed there from somewhere else.

Darwin probably would have been thrilled to learn of modern *plate tectonics*, the movement of plates of the Earth's crust. From studying evidence of such movements, we know that earlier in our planet's history all present-day continents, including Africa and South America, were parts of a massive "supercontinent" called Pangaea (Figure 23.6). By determining the locations of plates at different times in Earth's history, researchers can shed light on possible dispersal routes for some groups of organisms and on when (in geological history) such movements occurred.

Fossils are present in layers of sedimentary rocks. The deeper the layers, the older the fossils. The completeness of the fossil record varies, depending on the kinds of organisms represented, where they lived, and how stable their burial sites have been.

Biogeographical patterns can provide clues to where a species arose. Correlated with evidence from plate tectonics, the patterns also can shed light on possible dispersal routes for some groups of organisms.

COMPARING THE FORM AND DEVELOPMENT OF BODY PARTS

In addition to considering fossils and the distribution of life forms, early evolutionary thinkers also took note of the patterns we can observe in the form of body parts. More recently, biologists have shed light on evolutionary pathways by comparing stages of development in major groups of organisms.

Comparing Body Forms

Through **comparative morphology**, researchers can reconstruct evolutionary history using information that is contained in patterns of body form. When populations of a species branch out in different evolutionary directions, they diverge in appearance, functions, or both. Yet the related species also remain alike in many ways, for their evolution proceeds by modification of a shared body plan. In such species we see **homologous structures**. These are the same body parts modified in different ways in different lines of descent from a common ancestor (recall that *homo-* means "same").

For example, the same ancestral organism probably gave rise to most land-dwelling vertebrates, which have homologous structures. Apparently their common ancestor had four five-toed limbs. The limbs diverged in form and became wings in pterosaurs, birds, and bats (Figure 23.7). All these wings are homologous—they have the same parts. The five-toed limb also evolved into the flippers of porpoises and the anatomy of your own forearms and fingers.

Sometimes body parts in organisms that do *not* have a recent common ancestor come to resemble one another in form and function. These **analogous structures** (from *analogos*, meaning similar to one another) arise when different lineages evolve in the same or similar environments. Different body parts, which were put to similar uses, became modified through natural selection and ended up resembling one another. This pattern of change is called **morphological convergence**. For example, a dolphin, a fast-swimming marine mammal, has a sleek, torpedo-shaped torso—and so does a tuna, a fast-swimming fish.

Comparing Patterns of Development

Vertebrates include fishes, amphibians, reptiles, birds, and mammals. Despite the diversity of these groups, however, comparisons of the ways in which their embryos develop provide compelling evidence of their evolutionary connection to one another.

Early in development, the embryos of all the different vertebrate lineages go through strikingly similar stages (Figure 23.8). During vertebrate evolution, mutations that disrupted an early stage of development would have had devastating effects on the organized interactions required for later stages. Evidently, embryos of different groups

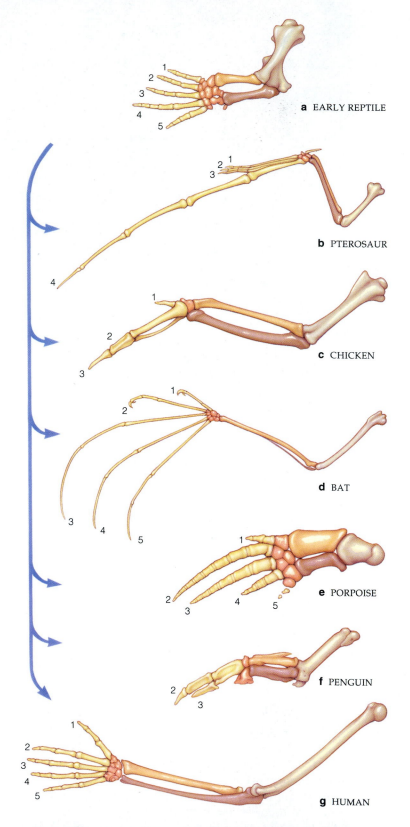

Figure 23.7 Morphological divergence in the vertebrate forelimb, starting with a generalized form of ancestral early reptiles. Diverse forms evolved even as similarities in the number and position of bones were preserved. The drawings are not to the same scale.

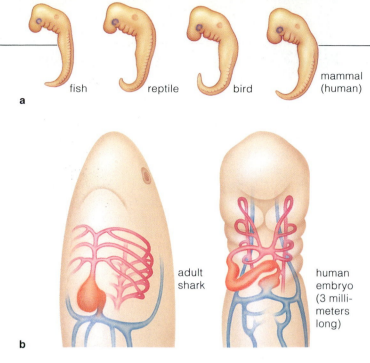

fish   reptile   bird   mammal (human)

a

adult shark

human embryo (3 millimeters long)

b

Figure 23.8 From comparative embryology, some evidence of evolutionary relationship among vertebrates. (**a**) Adult vertebrates show great diversity, yet the very early embryos retain striking similarities. This is evidence of change in a common program of development. (**b**) Fishlike structures still form in early embryos of reptiles, birds, and mammals. For example, a two-chambered heart (*orange*), certain veins (*blue*), and portions of arteries called aortic arches (*red*) develop in fish embryos and persist in adult fishes. The same structures form in an early human embryo.

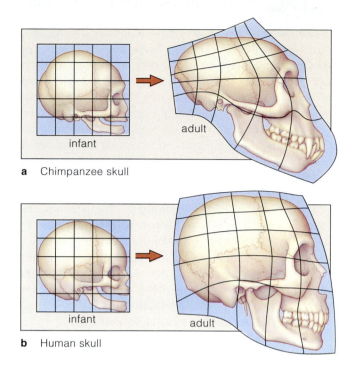

a Chimpanzee skull

infant   adult

b Human skull

infant   adult

Figure 23.9 Proportional changes in a chimpanzee skull (**a**) and a human skull (**b**). The skulls are quite similar in infants. Imagine that those represented here are paintings on a blue rubber sheet divided into a grid. Stretching the sheet deforms the grid's squares. For the adult skulls, differences in size and shape in corresponding grid sections reflect differences in growth patterns.

remained similar because mutations that altered early steps in development were selected against.

So how did the *adults* of different vertebrate groups come to be so different? At least some differences resulted from mutations that altered the onset, rate, or time of completion of certain developmental steps. Such mutations would bring about changes in shape through increases or decreases in the size of body parts. They also could lead to adult body forms that still have some juvenile features. For instance, Figure 23.9 illustrates how change in the growth rate at a key point in development may have led to differences in the proportions of chimpanzee and human skull bones, which are alike at birth. Later on, they change dramatically for chimps but only slightly for humans. Chimps and humans arose from the same ancestral stock. Their genes are nearly identical. However, at some point on the separate evolutionary road leading to humans, some regulatory genes probably mutated in ways that proved adaptive. Since then, instead of promoting the rapid growth required for dramatic changes in skull bones, the mutated genes have blocked it.

As Figure 23.10 indicates, the bodies of humans, pythons, and other organisms can have what seem like useless *vestigial* structures. For example, consider your own ear-wiggling muscles—which four-footed mammals (such as dogs) use to orient their ears. In humans, such body parts are left over from a time when more functional versions were important for an ancestor.

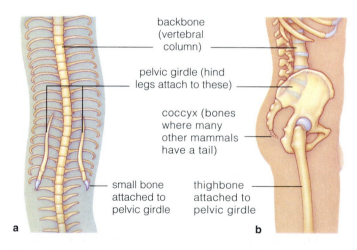

backbone (vertebral column)

pelvic girdle (hind legs attach to these)

coccyx (bones where many other mammals have a tail)

small bone attached to pelvic girdle

thighbone attached to pelvic girdle

a   **b**

Figure 23.10 Python bones (**a**) corresponding to the pelvic girdle of other vertebrates, including humans. (**b**) A snake has small vestigial hind limbs on its underside, remnants from a limbed ancestor. The human coccyx is a similar vestige from an ancestral species that had a bony tail, although muscles still attach to it.

Comparative morphology gleans evolutionary information from observed patterns of body form. Similarities in patterns of development may be clues to evolutionary relationships among organisms.

All species have a mix of traits—some that were present in ancestral species, some that were not. The kinds and numbers of traits species do or do not share are clues to how closely they are related. The same holds true for biochemical traits. Remember, the DNA of each species contains instructions for making the RNAs and proteins necessary to produce new individuals. This means that comparisons of DNA, RNA, and proteins from different species are additional ways of evaluating the evolutionary relationships among those species.

For example, from morphological studies alone you might suppose that monkeys, humans, chimpanzees, and other primates are related. You can test your idea by looking for differences in the amino acid sequence of a protein, such as hemoglobin, that occurs in all primates. You also could test it by determining whether the nucleotide sequences in their DNA match closely or not much at all. Logically, the species that are most similar in their biochemistry are the most closely related.

Suppose two species have the same conserved gene. The amino acid sequences of the gene's product are the same or nearly so. The absence of mutation implies that the species are closely related. What if the sequences differ quite a bit? Then many neutral mutations must have accumulated in them. A very long time must have passed since the species shared a common ancestor.

For instance, a wide range of organisms, from certain bacteria to corn plants to humans, produce cytochrome *c*, a protein of electron transport chains. Studies show that the gene coding for the protein has changed very little over vast spans of time. In humans cytochrome *c* has a primary structure of 104 amino acids. Chimps have the identical amino acid sequence. It differs by 1 amino acid in rhesus monkeys, 18 in chickens, 19 in turtles, and 56 in yeasts. On the basis of this biochemical information, would you assume humans are more closely related to a chimpanzee or to a rhesus monkey? A chicken or a turtle?

Methods that permit researchers to compare the sequences of nucleotides from various species also can be used to estimate the "when" of a divergence—the approximate time in history when it occurred. Neutral mutations are the key to this approach. Because they have little or no effect on survival or reproduction, neutral mutations can accumulate in DNA. By some calculations, neutral mutations in highly conserved genes (such as that for cytochrome *c*) accumulate at a regular rate. They are like a series of predictable ticks of a **molecular clock**. The total number of ticks will "unwind" down through the past, stopping at the approximate time when lineages diverged onto separate evolutionary paths.

Biochemical similarities are greatest among the most closely related species and weakest among the most distantly related.

The fossil record, development patterns, comparative morphology, and comparative biochemistry all provide insight into "who came from whom." Very often, information about this continuity of relationship among species is shown in the form of an "evolutionary tree." In such treelike diagrams, branches represent separate lines of descent from a common ancestor. As Figure 23.11 indicates, each branch point in a tree represents a time of divergence and speciation.

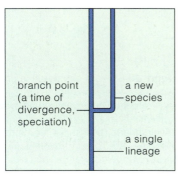
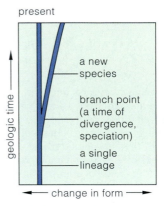

a Softly angled branching means speciation occurred through gradual changes in traits over geologic time.

b Horizontal branching means traits changed rapidly around the time of speciation. Vertical continuation of a branch means traits of the new species did not change much thereafter.

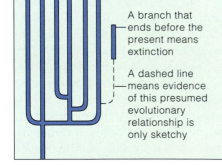

c Many branchings of the same lineage at or near the same point in geologic time mean that an adaptive radiation occurred.

A branch that ends before the present means extinction

A dashed line means evidence of this presumed evolutionary relationship is only sketchy

Figure 23.11 How to read evolutionary tree diagrams. The branch points represent times of divergence and speciation as brought about by microevolutionary processes.

Do changes from the original species form proceed slowly or rapidly? Many small changes may accumulate gradually within a lineage over long spans of time. (This is the premise of the *gradual* model of rates of change in an evolving lineage.) Conversely, species may originate through rapid spurts of change over hundreds or thousands of years, then change little for the next 2–6 million years. (This is the premise of the *punctuation* model, which has most changes occurring about the time of speciation.) ⬤

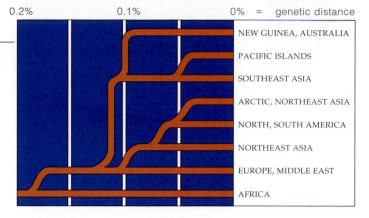

NEW GUINEA, AUSTRALIA

PACIFIC ISLANDS

SOUTHEAST ASIA

ARCTIC, NORTHEAST ASIA

NORTH, SOUTH AMERICA

NORTHEAST ASIA

EUROPE, MIDDLE EAST

AFRICA

Figure 23.13 One proposed family tree for populations of modern humans (*Homo sapiens*) that are native to different regions of the world. The diagram's branch points (suggested by biochemical studies) show presumed, small genetic divergences between populations.

Extinctions

Inevitably, a lineage loses a number of species as local conditions change. The rather steady rate at which species disappear over time is "background extinction." Regional extinctions occur when the background rate is exceeded in a local area. By contrast, a **mass extinction** is an abrupt, widespread rise in extinction rates above the background level. It is a catastrophic, global event in which major groups are wiped out simultaneously. Dinosaurs and many marine groups died out during a mass extinction that occurred about 65 million years ago—possibly the result of environmental changes that occurred after one or more large meteorites struck the Earth during a short period of time.

In the last few hundred years human activities have caused the extinction of thousands of species. The extinction rate is accelerating as we cut down forests, fill in wetlands, and otherwise destroy the habitats of other animals and plants with which we share the Earth. The *Focus on Our Environment* essay in Section 23.9 considers some current efforts to rescue endangered species, and we will return to this topic in Chapters 24 and 25.

Adaptive Radiations

In an **adaptive radiation**, new species of a lineage move into a wide range of habitats during bursts of micro-evolutionary events. Many adaptive radiations have occurred during the first few million years after a mass extinction. Fossil evidence suggests this happened after dinosaurs went extinct. Many new species of mammals arose and radiated into habitats the dinosaurs vacated.

Adaptive radiations also have occurred in the human lineage. The ancestors of modern humans, including a tool-using species called *Homo habilis* ("handy man") apparently remained in Africa until about 2 million years ago. Around that time, genetic divergence led to new species, including *Homo erectus*, a human species that the fossil record places on the evolutionary road to modern humans. *H. erectus* coexisted for a time with *H. habilis*. But while members of its sister species also were upright walkers, *H. erectus* did the name justice; its populations walked out of Africa, and turned left into Europe or right into Asia. Judging from fossils in the Middle East, *Homo sapiens* had evolved by 100,000 years ago.

What kinds of selection pressures triggered the adaptive radiation? We don't know, although this *was* a time of physical changes. One group of early humans, the Neandertals, had large brains and were massively built. Some Neandertal populations were the first humans to adapt to the coldest regions (Figure 23.12). Subsequent genetic modifications gave rise to anatomically modern humans. Figure 23.13 shows a possible family tree for those groups.

With this brief history in mind, we turn next to a survey of evolutionary trends that shaped pre-human ancestors and set the course for our own species' evolutionary journey.

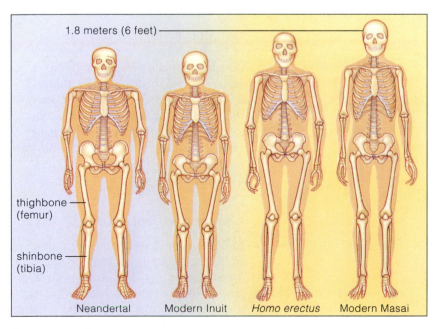

1.8 meters (6 feet)

thighbone (femur)

shinbone (tibia)

Neandertal   Modern Inuit   *Homo erectus*   Modern Masai

Figure 23.12 Climate and body build. Humans adapted to cold climates have a heat-conserving body—stockier, with shorter legs, compared to humans adapted to hot climates.

The branches of an evolutionary tree represent separate lines of descent from a common ancestor.

As lineages evolve, they may undergo adaptive radiation, in which new species rapidly (on a geological time scale) fill a range of habitats. Lineages also lose species to extinction.

TRENDS IN HUMAN EVOLUTION

A basic tenet of evolution is that *new* traits come about only through the modification of *existing* traits. For our human lineage, the primate heritage was the starting point for five trends that came to define our species. We briefly summarize those trends in this section, beginning with the key trait of walking upright on two legs.

Upright Walking

Of all primates, only humans can stride freely on two legs for long periods of time. This habitual two-legged gait, called **bipedalism**, emerged as elements of the ancestral primate skeleton were reorganized. Compared with apes and monkeys, humans have a shorter, S-shaped, and somewhat flexible backbone. The position and shape of the human backbone, knee and ankle joints, and pelvic girdle are the basis of bipedalism (Figure 23.14). By current thinking, the evolution of bipedalism was the pivotal modification in the origin of hominids.

Figure 23.14 Comparison of the skeletal organization of modern monkeys, apes (the gorilla is shown here), and humans. Modifications of the basic mammalian plan have allowed three distinct modes of locomotion. The quadrupedal monkeys climb and leap, and apes climb and swing by their forelimbs. Both modes of locomotion are well suited for life in the trees. On the ground, apes "knuckle walk." Humans are two-legged walkers.

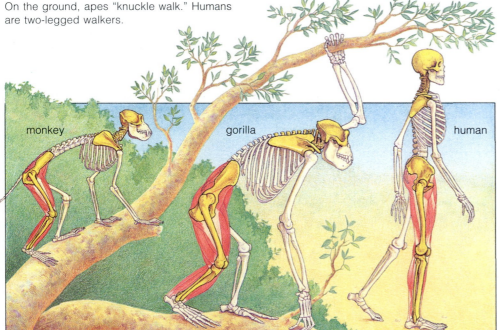

monkey
gorilla
human

Precision Grip and Power Grip

The first mammals spread their toes apart to help support the body as they walked or ran on four legs. Primates still spread their toes or fingers. Many also make cupping motions, as when a monkey lifts food to its mouth. Two other hand movements also developed in ancient tree-dwelling primates. Through alterations in handbones, fingers could be wrapped around objects (that is, *prehensile* hand movements were possible), and the thumb and tip of each finger could touch (opposable movements).

In time, hands also began to be freed from load-bearing functions—they were not needed to support the body. Later, when hominids evolved, refinements in hand movements led to the precision grip and power grip:

These hand positions gave early humans the capacity to make and use tools. They were a foundation for the development of early technologies and cultural development.

Enhanced Daytime Vision

Early primates had an eye on each side of the head. Later ones had forward-directed eyes, an arrangement that is better for sampling shapes and movements in three dimensions. Further modifications allowed the eyes to respond to variations in color and light intensity (dim to bright). Responsiveness to such complex visual stimuli is an advantage for life in the trees.

Changes in Dentition

Millions of years ago, modifications in the teeth and jaws of primates accompanied a shift from eating insects, to fruits and leaves, and on to a mixed diet. Later, rectangular jaws and long canine teeth came to be further defining features of monkeys and apes. Along the road leading to humans, a bow-shaped jaw and teeth that were smaller and all about the same length evolved.

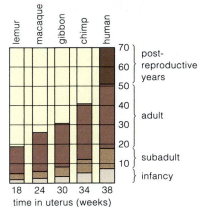

Figure 23.15 Trend toward longer life spans and longer dependency among primates.

(Chart labels: lemur, macaque, gibbon, chimp, human; vertical axis 10–70; categories: post-reproductive years, adult, subadult, infancy; horizontal axis "time in uterus (weeks)": 18 24 30 34 38)

Changes in the Brain and Behavior

Living on tree branches favored shifts in reproductive and social behavior. Imagine the advantages of single births over litters, for example, or of clinging longer to the mother. In many primate lineages, parents started to invest more effort in fewer offspring. They formed strong bonds with their young, maternal care became intense, and the learning period grew longer (Figure 23.15).

The brain case enlarged and brain regions involved in information processing began to expand. New behavior promoted development of those regions—which in turn stimulated more new behavior. In other words, brain modifications and behavioral complexity became closely linked. We see the links clearly in the parallel evolution of the human brain and culture. **Culture** is the sum total of behavior patterns of a social group, passed between generations by learning and symbolic behavior—especially language. The capacity for language arose among ancestral humans through changes in the skull bones and expansion of parts of the brain (Figure 23.16).

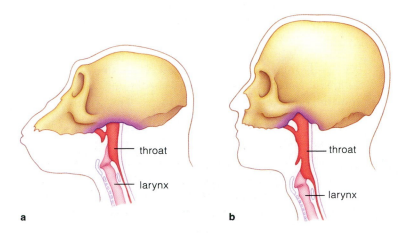

Figure 23.16 Structural basis of speech. (**a**) Like modern chimpanzees, early humans had a skull with a flattened base. Their larynx (the airway leading to the lungs) was not far below the skull, so the throat volume was small. (**b**) The base of a modern human's skull angles down sharply. This moves the larynx down also. The result is an increase in the volume of the throat, where sounds are produced.

Primate Origins

Primates evolved from ancestral mammals more than 60 million years ago. The first ones resembled small rodents. They may have foraged for insects, seeds, buds, and eggs near the forest floor. They had a long snout and a good sense of smell, useful for detecting predators or food. Between 54 and 38 million years ago, some primates were staying in the trees. Fossils give us evidence of larger brains, a shorter snout, enhanced daytime vision, and refined grasping movements. How did these traits evolve?

The features of life in the trees hint at an answer. Trees offered food and safety from ground-dwelling predators. They also were a habitat where natural selection would rigorously favor some traits over others. Imagine dappled sunlight, boughs swaying in the wind, colorful fruit hidden among the leaves, perhaps predatory birds. A long, odor-sensitive snout would not have been of much use in the trees, where air currents disperse odors. But a brain that assessed depth, shape, movement, and color would have been a definite plus. So would a brain that worked fast when its owner was running and leaping from branch to branch. It would be vital to be able to accurately estimate one's distance from the destination, the effects of obstacles or wind—and to be able to quickly make adjustments for miscalculations. Possibly shaped by such environmental factors, by 36 million years ago tree-dwelling *anthropoids* had evolved in tropical forests. One or more types were on or very close to the evolutionary road that would lead to monkeys, apes, and humans.

Some 25 to 5 million years ago, Earth's land masses were drifting into their current positions. The global climate was becoming cooler and drier, and as these changes took place an adaptive radiation of apelike forms—the first *hominoids*—occurred. Fossils tell us that by 13 million years ago, hominoids had spread widely. Most forms went extinct. Among the survivors, genetic studies point to three divergences that occurred between 10 million and 5 million years ago. Two of the branchings in groups of ancestral apes led to gorillas and chimpanzees. The third branching gave rise to *hominids*, including the ancestors of our own species.

It was the hominid lineage that ultimately gave rise to *Homo sapiens* 100,000 years ago. From the time that hominids such as the Neandertals disappeared and modern humans emerged, about 40,000 years ago, human evolution has been almost entirely cultural rather than biological—and so we leave the story there.

Five trends emerged along the evolutionary road leading to humans: upright walking, refined hand movements, generalized dentition, refined vision, and the interconnected development of a more complex brain and cultural behavior.

23.9 GENES OF ENDANGERED SPECIES

Barbara Durrant and many other biologists at zoos, private conservation centers, and wildlife refuges are scrambling to buy time for the world's most *endangered species*. Populations of such species are so small that they are at the brink of extinction. The biologists face daunting problems, including an absence of genetic diversity.

Consider the cheetah (Figure 23.17). Only 20,000 of these sleek, swift cats have survived to the present. It seems that cheetahs went through a severe bottleneck about 10,000 years ago. Parents mated with their own offspring when no other mates were available. As a result of inbreeding among the survivors and their descendants, all cheetahs now carry strikingly similar alleles. They are so genetically uniform that a patch of skin from one can be successfully grafted to another. This seldom works with other mammals.

Among their shared alleles are previously rare harmful ones. Some of the harmful, mutated alleles affect fertility. Typically, a male cheetah's sperm count is low, and 70 percent of the sperm that do form are abnormal. Other alleles are responsible for weakened resistance to disease. Infections that are seldom life-threatening to other cats can reach epidemic proportions among cheetahs. During one outbreak of *feline infectious peritonitis* in a wild animal park, the infectious agent (a virus) had little effect on captive lions, but it killed every one of the cheetahs.

Intense hunting and other human activities have endangered another big cat, the Florida panther. With fewer than 50 individuals, the last remaining population is highly inbred. Researchers at the National Zoo retrieve DNA from unfertilized eggs of "road-killed" female panthers. With genetic engineering, the DNA may find use in captive breeding programs.

Figure 23.17 A few of the world's remaining cheetahs, all of which carry some harmful alleles that made it through a severe bottleneck at some time in the species' history.

23.10 EVOLUTION AND EARTH HISTORY

When all of the diverse kinds of evidence for evolution are carefully pieced together, an important fact emerges: *The evolution of life has been linked, from its origin to the present, to the physical and chemical evolution of the Earth.*

Origin of the Earth

Cloudlike remnants of stars are mostly hydrogen gas. But they also contain water, iron, silicates, hydrogen cyanide, methane, ammonia, formaldehyde, and many other simple inorganic and organic substances. Most likely, the contracting cloud from which our solar system evolved was similar in composition. Between 4.6 and 4.5 billion years ago, the outer regions of the cloud cooled. Swarms of mineral grains and ice orbited around the sun. By electrostatic attraction, by gravitational pull, they started clumping together. In time, the larger, faster clumps started colliding and shattering. Some of the larger ones became more massive by sweeping up asteroids, meteorites, and other rocky remnants of collisions—and gradually they evolved into planets.

While the early Earth was forming, much of its inner rocky material melted. Most likely, stupendous asteroid impacts generated the required heat, although heat released during the radioactive decay of minerals contributed to it. As rocks melted, nickel, iron, and other heavy materials moved toward the interior, and lighter minerals floated toward the surface. This process of differentiation resulted in the formation of a crust of basalt, granite, and other types of low-density rock, a rocky region of intermediate density (the mantle), and a high-density, partially molten core of nickel and iron.

Four billion years ago, the Earth was a thin-crusted inferno (Figure 23.18). Yet within 200 million years, life had originated on its surface! We have no record of the event. As far as we know, movements in the mantle and crust, volcanic activity, and erosion obliterated all traces of it. Still, we can put together a plausible explanation of how life originated by considering three questions:

1. What were the prevailing physical and chemical conditions on Earth at the time of life's origin?

2. Based on physical, chemical, and evolutionary principles, could large organic molecules have formed spontaneously and then evolved into molecular systems displaying the characteristics of life?

3. Can we devise experiments to test whether living systems could have emerged by chemical evolution?

We begin with a brief look at Earth's early atmosphere, which must have laid the physical and chemical foundation for life's beginnings. Section 23.11 then considers plausible scenarios for the emergence of living cells.

Further reading: Student Guide to InfoTrac on web site →

Figure 23.18 Representation of the primordial Earth, about 4 billion years ago. Within another 500 million years, diverse types of living cells would be present on the surface.

Conditions on the Early Earth

When patches of crust were forming, heat and gases blanketed the Earth. This first atmosphere probably consisted of gaseous hydrogen (H_2), nitrogen (N_2), carbon monoxide (CO), and carbon dioxide (CO_2). Were gaseous oxygen (O_2) and water also present? Probably not. Rocks release very little oxygen during volcanic eruptions. Even if oxygen were released, those small amounts would have reacted at once with other elements, and any water would have evaporated because of the intense heat.

When the crust finally cooled and solidified, water condensed into clouds and the rains began. For millions of years, runoff from rains stripped mineral salts and other compounds from the Earth's parched rocks. Salt-laden waters collected in depressions in the crust and formed the early seas.

What we are describing here are events that were crucial to the beginning of life. Without an oxygen-free atmosphere, the organic compounds that started the story of life never would have formed on their own. Why? Oxygen would have attacked their structure and disrupted their functioning (as oxygen free radicals do in cells). Without liquid water, cell membranes would not have formed, because cell membranes take on their bilayer organization only in water.

As you know, cells are the basic units of life. Each has a capacity for independent existence. Cells could never simply have appeared one day on the early Earth. Their emergence required the existence of biological molecules built from organic compounds. It also required metabolic pathways, which could be organized and controlled within the confines of a cell membrane. These essential events are our next topic.

From its origin to the present, the evolution of life on Earth has been linked to the planet's physical and chemical evolution.

The oxygen-free atmosphere of the early Earth provided conditions necessary for the formation of organic compounds that would become the raw material for the first biological molecules. The accumulation of liquid water provided conditions in which the first cells could have formed.

THE ORIGIN OF LIFE

Figure 23.19 shows fossil cells discovered in the Warrawoona rock formation in Western Australia—a fossil that has been dated at 3.5 billion years old! The first living cells probably emerged between 4 billion and 3.8 billion years ago. Before something as complex as a cell was possible, however, biological molecules must have come about through *chemical evolution*. In fact, researchers have been able to put together a reasonable scenario for the emergence of life.

Synthesis of Biological Molecules

Rocks collected from Mars, meteorites, and the Earth's moon—which all formed at the same time as the Earth, from the same cosmic cloud—contain precursors of biological molecules. Possibly, sunlight, lightning, or heat escaping from the Earth's crust supplied enough energy to drive condensation of precursors into more complex organic molecules. In the early 1950s a chemistry graduate student named Stanley Miller took the likely constituents of the early atmosphere—hydrogen, methane, ammonia, and water—and mixed them in a reaction chamber (Figure 23.20). He recirculated the mixture and for several days bombarded it with a spark discharge to simulate lightning. Within a week, many amino acids and other organic compounds had formed.

In other experiments that simulated conditions on the early Earth, glucose, ribose, deoxyribose, and other sugars were produced from formaldehyde. Adenine was produced from hydrogen cyanide. Adenine plus ribose occur in ATP, NAD, and other nucleotides vital to cells.

If *complex* organic compounds formed in the seas, they would not have lasted long. In water, the spontaneous direction of the required reactions would have been toward hydrolysis, not condensation.

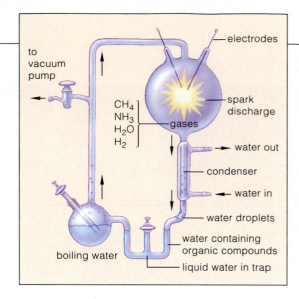

Figure 23.20 Sketch of Stanley Miller's apparatus used to study the synthesis of organic compounds under conditions believed to have been present on the early Earth. (The condenser cools circulating steam and causes water to condense into droplets.)

Perhaps more lasting bonds formed at the margins of seas. By one scenario, clay in the rhythmically drained muck of tidal flats and estuaries served as templates (structural patterns) for the spontaneous assembly of proteins and other complex organic compounds. Clay consists of thin, stacked layers of aluminosilicates with metal ions at its surface. Clay and metal ions attract amino acids. From experiments we know that when clay is first warmed by sunlight, then alternately dried out and moistened, it actually promotes condensation reactions that produce complex organic compounds.

In another hypothesis, complex organic compounds formed spontaneously near deep-sea hydrothermal vents. Today, species of primitive bacteria thrive in and around the vents. Tests have shown that when amino acids are heated and immersed in water, they order themselves into small protein-like molecules called "proteinoids."

However the first proteins formed, some may have had the shape and chemical behavior needed to function as weak enzymes in hastening bonds between amino acids. They would have had selective advantage over other templates in the chemical competition for available amino acids. Perhaps selection was at work before the origin of cells, favoring the chemical evolution of enzymes.

The First Metabolic Pathways

A defining characteristic of life is metabolism. The term refers to the reactions by which cells harness energy and use it to drive their activities, including biosynthesis. During the first 600 million years of Earth's history, enzymes, ATP, and other molecules could have assem-

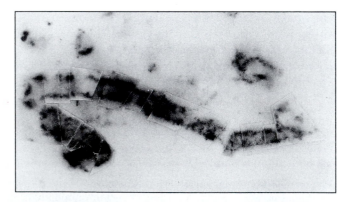

Figure 23.19 One of the oldest known fossils, dated at 3.5 billion years old. It is a string of walled cells unearthed in the Warrawoona rocks of Western Australia.

bled spontaneously in places where they were in close physical proximity. If so, their close association would have promoted chemical interactions—and the beginning of metabolic pathways.

Origin of Self-Replicating Systems

Another defining characteristic of life is a capacity for reproduction, which now starts with protein-building instructions in DNA. The DNA molecule is fairly stable, and it is easily replicated before each cell division. As you know from Chapter 20, arrays of enzymes and RNA molecules operate together to carry out DNA's encoded instructions.

Most existing enzymes are assisted by the small organic molecules or metal ions called coenzymes. Intriguingly, some types of coenzymes have a structure that is exactly like that of RNA nucleotides. And if you heat nucleotide precursors and short chains of phosphate groups together, they will self-assemble into strands of RNA. On the early Earth, energy from the sun or geothermal vents could have driven the spontaneous formation of RNA from such starting materials.

Simple, self-replicating systems of RNA, enzymes, and coenzymes have been created in some laboratories. Did RNA later become the information-storing templates for protein synthesis? Perhaps it initially did so, although existing RNA molecules are just too chemically fragile for such a role. Yet the use of RNA templates may have set up an **RNA world** that preceded DNA's dominance as the main informational molecule. Whatever the case, DNA eventually assumed this function, probably because it can form long nucleotide chains in more stable fashion.

We still don't know how DNA entered the picture. Until we identify the likely chemical ancestors of RNA and DNA, the story of life's origin will be incomplete.

Origin of the First Plasma Membranes

Before cells could exist, there must have been membrane-bound sacs that sheltered information-storing templates and various metabolic agents, such as amino acids, from the environment. Experiments show that such sacs can form spontaneously. In one classic study, Sidney Fox heated amino acids until they formed protein chains, which he placed in hot water. After the chains cooled, they self-assembled into small, stable spheres. Like membranes of cells, these proteinoid spheres were selectively permeable to various substances. In other experiments, the spheres picked up lipids from their surroundings and a lipid–protein film formed at their surface. In still other experiments, lipids self-assembled into small, water-filled sacs that had many properties of cell membranes.

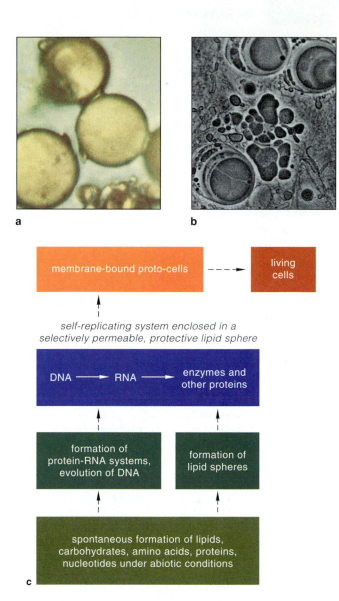

Figure 23.21 Microspheres of (**a**) proteins and (**b**) lipids that self-assembled under abiotic (nonliving) conditions. (**c**) One possible sequence of events that led to the first self-replicating systems, then to the first living cells.

Despite major gaps in our knowledge of life's origins, there is also good experimental evidence that chemical evolution could have given rise to the molecules and structures characteristic of life. Figure 23.21 summarizes some key events in that chemical evolution, which may have led to the first cells.

Many experiments provide indirect evidence that the complex organic molecules characteristic of life could have formed under conditions that existed on the early Earth.

Experiments suggest that chemical and molecular evolution could have given rise to proto-cells.

SUMMARY

1. Evolution proceeds by modifications of already existing species. Hence, there is a continuity of relationship among all species, past and present.

2. In a population, genetic variability gives rise to variations in traits. The frequencies of different genes (alleles) change as a result of four processes of microevolution: mutation, genetic drift, gene flow, and natural selection. The large-scale patterns, trends, and rates of change among groups of species (higher taxa) over time are called macroevolution.

3. Natural selection is a difference in survival and reproduction among members of a population that vary in one or more traits. That is, under prevailing conditions, one form of a trait may be more adaptive. Its bearers tend to survive and therefore to reproduce more often, so it becomes more common than other forms of the trait.

4. Speciation results when the differences between isolated populations become so great that their members are not able to interbreed successfully under natural conditions.

5. Evidence of evolutionary relationships comes from the fossil record and Earth history, from radioactive dating methods, from comparisons of body forms and development patterns, and also from comparative biochemistry.

6. Comparative morphology often reveals similarities in embryonic development that indicate an evolutionary relationship. They may reveal homologous structures, shared as a result of descent from a common ancestor. Comparative biochemistry relies on gene mutations that have accumulated in different species. Methods such as DNA-DNA hybridization reveal the degree of similarity among genes and gene products of different species.

7. Family trees may be constructed to show evolutionary relationships. The branches of such trees are lines of descent (lineages). The branch points are speciation events brought about by mutation and by natural selection and other microevolutionary processes (Table 23.2).

Table 23.2	Major Processes of Microevolution
Mutation	A heritable change in DNA
Genetic drift	Random fluctuation in allele frequencies over time, due to chance occurrences alone
Gene flow	Movement of alleles among populations through migration of individuals
Natural selection	Change or stabilization of allele frequencies due to differences in survival and reproduction among variant members of a population

8. A mass extinction is a catastrophic event in which major lineages perish abruptly. An adaptive radiation is a burst of evolutionary activity; a lineage rapidly produces many new species. Both events have changed the course of biological evolution many times.

9. Humans (*Homo sapiens*) are classified in the hominid family of the primate order, and are members of the only existing species of the genus *Homo*.

10. Important evolutionary trends in the primate lineage leading to *H. sapiens* included (1) a transition from a four-legged gait to bipedalism, with related changes in the skeleton; (2) increased manipulative skills related to structural modification of the hands, which were gradually freed from functions in locomotion and supporting body weight; (3) increased reliance on enhanced daytime vision, including color vision and depth perception; (4) transition away from specialized eating habits, with corresponding modification of dentition; and (5) the enlargement and reorganization of the brain. Larger, more complex brains correlate with increasingly sophisticated technology (tool use), and with the development of complex behaviors and cultural attributes such as language.

11. Every element of the solar system, the planet Earth, and life itself are products of the physical and chemical evolution of the universe and its stars.

12. Four billion years ago, the Earth had a high-density core, a mantle of intermediate density, and a thin, extremely unstable crust of low-density rocks. Probably gaseous hydrogen, nitrogen, carbon monoxide, and carbon dioxide made up the first atmosphere.

13. Many studies and experiments provide indirect evidence that life originated under conditions that presumably existed on the early Earth.

　a. Comparative analyses of the composition of cosmic clouds and of rocks from other planets and the Earth's moon suggest that precursors of the complex molecules associated with life were available.

　b. In laboratory tests that simulated primordial conditions, including the absence of free oxygen, the precursors spontaneously assembled into sugars, amino acids, and other organic compounds.

　c. Metabolic pathways could have evolved as a result of chemical competition for the limited supplies of organic molecules that had accumulated in the seas.

　d. Self-replicating systems of RNA, enzymes, and coenzymes have been synthesized in the laboratory. How DNA entered the picture is not yet understood.

　e. Lipid and lipid–protein membranes that show properties of cell membranes form spontaneously under the prescribed conditions. Life originated about 3.8 billion years ago.

Review Questions

1. Distinguish between microevolution and macroevolution. *23.2*

2. Explain how natural selection differs from adaptation. *23.3*

3. Explain the difference between:
   a. divergence and gene flow *23.3*
   b. homologous and analogous structures *23.5*

4. Explain the difference between a primate and a hominid. *23.8*

5. Describe the chemical and physical environment in which the first living cells may have evolved. *23.11*

Self-Quiz *(Answers in Appendix V)*

1. A_____is a genetic unit consisting of one or more populations of organisms that usually closely resemble each other physically and physiologically.

2. The frequencies of different genes (alleles) change as a result of four processes of microevolution:_____,_____,_____, and_____.

3. A difference in survival and reproduction among members of a population that vary in one or more traits is called_____.

4. The fossil record of evolution correlates with evidence from
   _____.
   a. the geologic record
   b. radioactive dating
   c. comparing development patterns and morphology
   d. comparative biochemistry
   e. all of the above

5. Comparative biochemistry_____.
   a. is based mainly on the fossil record
   b. often reveals similarities in embryonic development stages that indicate evolutionary relationships
   c. is based on mutations that have accumulated in the DNA of different species
   d. compares the proteins and the DNA from different species to reveal relationships
   e. both c and d are correct

6. Comparative morphology_____.
   a. is based mainly on the fossil record
   b. shows evidence of divergences and convergences in body parts among certain major groups
   c. compares the proteins and the DNA from different species to reveal relationships
   d. both b and c are correct

7. In_____, new species of a lineage move into a wide range of habitats by way of bursts of microevolutionary events.
   a. an adaptive radiation
   b. natural selection
   c. genetic drift
   d. punctuated equilibrium

8. The pivotal modification in hominid evolution was_____:
   a. the transition to bipedalism.
   b. hand modification that increased manipulative skills.
   c. a shift from omnivorous to specialized eating habits.
   d. less reliance on smell, more on vision.
   e. expansion and reorganization of the brain.

Selected Key Terms

adaptation *23.3*	macroevolution *23.2*
adaptive radiation *23.7*	mass extinction *23.7*
analogous structures *23.5*	microevolution *23.2*
biogeography *23.4*	morphological convergence *23.5*
bipedalism *23.8*	natural selection *23.3*
comparative morphology *23.5*	population *23.2*
evolution *CI*	reproductive isolating
fossil *23.4*	mechanism *23.3*
gene flow *23.3*	speciation *23.3*
genetic drift *23.3*	species *23.3*
homologous structure *23.5*	

Critical Thinking: You Decide *(Key in Appendix VI)*

1. Humans can inherit various alleles for the liver enzyme ADH (alcohol dehydrogenase), which breaks down ingested alcohol. People of Italian and Jewish descent commonly have a form of ADH that detoxifies alcohol very rapidly. People of northern European descent have forms of ADH that are moderately effective in alcohol breakdown, while people of Asian descent typically have ADH that is less efficient at processing alcohol. Explain why researchers have been able to use this information to help trace the origin of human use of alcoholic beverages.

2. In 1992 the frozen body of a Stone Age man was discovered in the Austrian Alps. Although the "Iceman" died about 5,300 years ago, his body is amazingly intact and researchers have analyzed DNA extracted from bits of his tissue. Can these studies tell us something about early human evolution? Explain your reasoning.

3. When it comes to human origins, different researchers "read" the fossil record in different parts of the world in different ways. Is this an indication that the primate fossils they are investigating might have nothing whatsoever to do with human origins? Why or why not?

4. The Finnish people have become a very interesting study population for evolutionary biologists and geneticists. Finnish is not related to any other major European language, and most Finns have distinctive light blond hair and blue eyes. They rarely get certain genetic diseases, such as cystic fibrosis, that are common in other European populations, but they *are* subject to more than two dozen other genetic disorders that are rare or nonexistent elsewhere. Evidence suggests that Finland was settled about 4,000 years ago, after which its people kept mostly to themselves for centuries. Researchers also suspect that the initial band of settlers was small. What kind of microevolutionary event(s) might explain these findings?

Readings

Dott, R., Jr., and R. Batten. 1998. *Evolution of the Earth.* 4th ed. New York: McGraw-Hill.

Tattersall, I. January 2000. "Once We Were Not Alone." *Scientific American.* An intriguing exploration of possible reasons why ours was the only hominid species to survive.

24 ECOSYSTEMS

Crêpes for Breakfast, Pancake Ice for Dessert

Think of Antarctica, and you think of ice. Mile-thick slabs of the stuff hide all but a small fraction of a continent whipped by fierce winds and kept frozen by murderously low temperatures, on the order of −100°F. And yet, on patches of exposed rocky soil and on nearby islands, mosses and lichens grow. There, during the breeding season, penguins as well as seals form great noisy congregations, then reproduce and raise offspring. They cruise offshore or venture out in the open ocean in pursuit of food—krill, fishes, and squids.

The first explorers of Antarctica called it "The Last Place on Earth." For all its harshness, though, the eerie, unspoiled beauty fired the imaginations of those first travelers and others who followed them.

Several decades ago, thirty-eight nations signed a treaty to set aside Antarctica as a reserve for scientific research. This was the start of scientific outposts—and of the trashing of Antarctica. At first, researchers only discarded a few oil drums and old tires. Then they erected prefabricated villages. Onto the ice or into the water went garbage and used equipment. Offshore, the marine life began to be exposed to human sewage and chemical wastes.

Antarctica also became a destination of cruise ships. Each summer thousands of tourists, fortified by three sumptuous meals a day plus snacks, began to arrive, ferried from ship to land across channels glistening with pancake ice (Figure 24.1). In the early years, they trampled vegetation and bobbed around the penguins, cameras clicking. Today, tour guides encourage more respect for the landscape, but it is impossible to prevent some damage from occurring.

We do not know what tourism's long-term impact will be, but it pales by comparison to other assaults. Nations looking for new sources of food started licking their chops over Antarctica's krill. Word also got out about potentially rich deposits of uranium, oil, and gold—and soon treaty nations were poised to authorize digs and drilling ventures. But oil spills or commercial harvesting of krill could easily destroy the fragile

Figure 24.1 (**a**) A boatload of tourists crossing a channel of "pancake ice" off the coast of Antarctica. (**b**) On the shore of an island near Antarctica, tourists get close to nature—maybe too close.

ecosystem. For example, the tiny, shrimplike krill are all that Adélie penguins eat. Krill are a key food for baleen whales and other marine animals that are, in turn, food for still other species.

In 1991, the treaty nations imposed a 50-year ban on mineral exploration. Research stations started burying or incinerating wastes, treating sewage, and taking other measures to curb pollution. Krill are not yet being harvested on a massive scale—but we can rest assured that the pressure is mounting.

Because Antarctica seems remote from the rest of the world, it is easier to see how species interconnect there. We can ask whether harm to one species or one habitat will lead to collapse of the whole, and be fairly sure of the answer. What about places not as sharply defined? Are they as vulnerable to disturbance or more resilient? We won't know for sure until researchers gain deeper insights into the evolutionary and ecological histories of species in those places.

In the meantime, biodiversity is rapidly declining in ecosystems on land and in the seas. Competition from exotic (non-native) species is one threat; the human propensity to overexploit species and destroy habitats is another. To counter the threats, scientists around the world are engaged in *conservation biology*. To appreciate the magnitude of the task before them, it helps to understand a few priniciples of how ecosystems work—the topic of this chapter.

b

KEY CONCEPTS

1. An ecosystem is an association of organisms and their physical environment, linked by a flow of energy and a cycling of materials through it.

2. Every ecosystem is an open system, with inputs and outputs of both energy and nutrients.

3. Over time, energy flows in one direction through an ecosystem. The flow usually begins when photosynthesizing organisms (mainly green plants) harness sunlight energy and convert it to forms that they and other organisms of the ecosystem can use. Photo-synthesizers are "primary producer" organisms for the ecosystem.

4. Energy-rich organic compounds that the primary producers manufacture are incorporated into their body tissues. The compounds are stored forms of energy, and they serve as the foundation for the ecosystem's food webs. Food webs consist of a number of interconnected food chains.

5. Each chain in a food web extends in a straight-line sequence from producers through arrays of consumers, decomposers, and detritivores ("detritus eaters"). Each chain also connects with other chains at different feeding levels.

6. Water and major nutrients move from the physical environment, through organisms, then back to the environment. Their movements, called biogeochemical cycles, take place on a global scale.

7. Human activities are disrupting the natural cycles of nature and so are endangering ecosystems.

CHAPTER AT A GLANCE

24.1 INTRODUCTION TO PRINCIPLES OF ECOLOGY

The living world encompasses the **biosphere**—those regions of the Earth's crust, waters, and atmosphere in which organisms live. The Earth's surface is remarkably diverse. In climate, soils, vegetation, and animal life, its deserts differ from hardwood forests, which differ from tropical rain forests, prairies, and arctic tundra. Oceans, lakes, and rivers differ in their physical properties and arrays of organisms (Figure 24.2).

Ecology is the study of the interactions of organisms with one another and with the physical environment. The general type of place in which a species normally lives is its **habitat**. For example, muskrats live in a stream habitat, damselfish in a coral reef habitat. The habitat of any organism characteristically has certain physical and chemical features. Every species also interacts with others that occupy the same habitat. Humans live in "disturbed" habitats, which we have deliberately altered for purposes such as agriculture and urban development. In any given habitat the populations of all species directly or indirectly associate with one another as a **community**.

Have you heard someone speak of an "ecological niche"? The **niche** ("nitch") of a species consists of all the physical, chemical, and biological conditions the species needs to live and reproduce in an ecosystem. Examples of those conditions include the amount of water, oxygen, and other nutrients a species needs, the temperature ranges it can tolerate, the places it finds food, and the type of food it consumes. *Specialist* species have narrow niches. They may be able to use only one or a few types of food or live only in one type of habitat. For example, the red-cockaded woodpecker builds its nest mainly in longleaf pines that must be at least 75 years old. Humans and houseflies are examples of *generalist* species with broad niches. Both can live in a range of habitats and eat many types of food.

An **ecosystem** consists of one or more communities of organisms interacting with one another *and* with the physical environment through a flow of energy and a cycling of materials. Figure 24.3 shows some of the interacting organisms in one type of ecosystem, arctic tundra.

Communities of organisms make up the *biotic*, or living, portions of an ecosystem. New communities may arise in habitats initially devoid of life or in disturbed but previously inhabited areas such as abandoned pastures. Immature or disturbed areas are ecologically unstable. Through a process called **succession**, the first species that thrive in a habitat are then replaced by others, which are replaced by others in orderly progression until the composition of species becomes steady as long as other conditions remain the same. This more or less stable array of species is a "climax community."

a

b

Figure 24.2 Major types of ecosystems on Earth are called "biomes." They include land biomes such as the warm desert near Tucson, Arizona, shown in (**a**), and aquatic realms such as the mountain lake in (**b**), which lies in the Canadian Rockies.

In *primary succession*, changes begin when pioneer species colonize a newly available habitat, such as a recently deglaciated region (Figure 24.4). In *secondary succession*, a community develops toward the climax state after parts of the habitat have been disturbed. For example, this pattern occurs in abandoned fields, where wild grasses and other plants quickly take hold when cultivation stops. Natural disasters (such as forest fires caused by lightning), changing climate, and other factors often interfere with succession toward climax conditions. Hence, truly stable climax conditions are rarely achieved.

The general type of place where a species normally lives is its habitat. A community encompasses the populations of all species in a given habitat.

An ecosystem consists of one or more communities of organisms interacting with one another and with the physical environment through a flow of energy and a cycling of materials.

Further reading: Student Guide to InfoTrac on web site →

Figure 24.3 Some of the producers, consumers, and decomposers of the arctic tundra: sedges, mosses, and other plants, along with the lemming (eater of plant parts); the snowy owl (eater of lemmings); and a fungal decomposer. As in all ecosystems, these organisms interact with one another and with the physical environment through a one-way flow of energy and a cycling of materials.

a

c

Figure 24.4 Primary succession in the Glacier Bay region of Alaska (**a**), where glaciers are receding and changes in newly deglaciated regions have been well documented. (**b**) The first plants here were small flowering plants that can grow and spread over glacial till. Within 20 years, young alders begin to flourish (**c**). After 80 years, a climax community of spruces crowds out the mature alders (**d**).

b

d

THE NATURE OF ECOSYSTEMS

Although there are many different types of ecosystems, they are all alike in many aspects of their structure and function. Nearly every ecosystem runs on energy from the sun. Plants and other photosynthetic organisms are **autotrophs** (self-feeders). They capture sunlight energy and convert it to forms they can use to build organic compounds from simple inorganic substances. By securing energy from the physical environment, autotrophs serve as the **producers** for the entire system (Figure 24.5).

All other organisms in the system are **heterotrophs**, "other-feeders." They depend directly or indirectly on energy stored in the tissues of producers. Some heterotrophs are **consumers**, which feed on the tissues of other organisms. Within this category, *herbivores* such as grazing animals and insects eat plants, *carnivores* such as lions and snakes eat animals, and *parasites* reside in or on living hosts and extract energy from them. Most humans are *omnivores*—they feed on a variety of food types. Other heterotrophs, the **detritivores**, get energy from partly decomposed particles of organic matter. Crabs, nematodes, and earthworms are examples. Still other heterotrophs, the **decomposers**, include fungi and bacteria that extract energy from the remains or products of organisms.

Autotrophs secure nutrients *and* energy for the entire system. As they grow, they take up water and carbon dioxide (which provide oxygen, carbon, and hydrogen) along with dissolved minerals including nitrogen and phosphorus. These materials are the building blocks for biological molecules. When detritivores and decomposers get their turn at this organic matter, they can break it down completely to inorganic bits. If those bits are not washed away or otherwise removed from the system, autotrophs can reuse them as nutrients.

It is important to remember that ecosystems are *open* systems. They are not self-sustaining. Ecosystems require a continual *energy input* (as from the sun) and often they require *nutrient inputs* (as when erosion carries minerals into a lake). Ecosystems also have energy and nutrient outputs. Energy can't be recycled; in time, most of the energy originally fixed by autotrophs is lost to the environment in the form of metabolic heat. Nutrients typically are cycled, but some are lost, as through soil leaching.

Feeding Relationships in Ecosystems

Each species in an ecosystem has its own position in a hierarchy of feeding relationships called **trophic levels** (*troph-* means nourishment). A key aspect of ecosystem functioning is the transfer of energy from one trophic level to another.

Primary producers, which gain energy directly from sunlight, make up the first trophic level. Photosynthetic autotrophs in a lake (including some bacteria and aquatic

Photosynthesis by primary producers captures and then transforms sunlight energy to chemical energy of organic compounds, which are used at once or stored.

AUTOTROPHS
(plants, other primary producers)

NUTRIENT CYCLING

HETEROTROPHS
(consumers, decomposers, detritivores)

Chemical energy is transferred from one organism to another. In time, all of the energy that entered the ecosystem leaves the ecosystem, mainly in the form of heat.

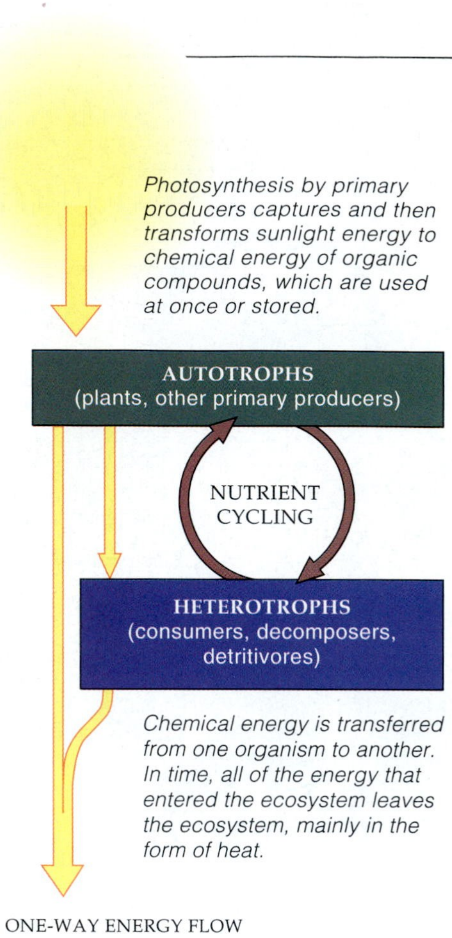

Figure 24.5 Simple model of the one-way flow of energy and the cycling of materials through ecosystems.

ONE-WAY ENERGY FLOW

plants) are examples. Snails and other herbivores that feed directly on the producers are at the next trophic level. Birds and other primary carnivores that prey directly on the herbivores form a third level. A hawk that eats a snake is a secondary carnivore. Decomposers, humans, and many other organisms can obtain energy from more than one source. They cannot be assigned to a single trophic level and are more like "trophic groups."

Food Webs

A linear sequence of who eats whom in an ecosystem is sometimes called a **food chain**. However, you will have a hard time finding such a simple, isolated sequence in nature. The same species usually belongs to more than one food chain, especially when it is at a low trophic level. It is more accurate to view food chains as *cross-connecting* with one another in **food webs**. Figure 24.6 shows an Antarctic food web.

A food web is a network of crossing, interlinked food chains encompassing primary producers and an array of consumers and decomposers.

By way of food webs, different species in an ecosystem are interconnected.

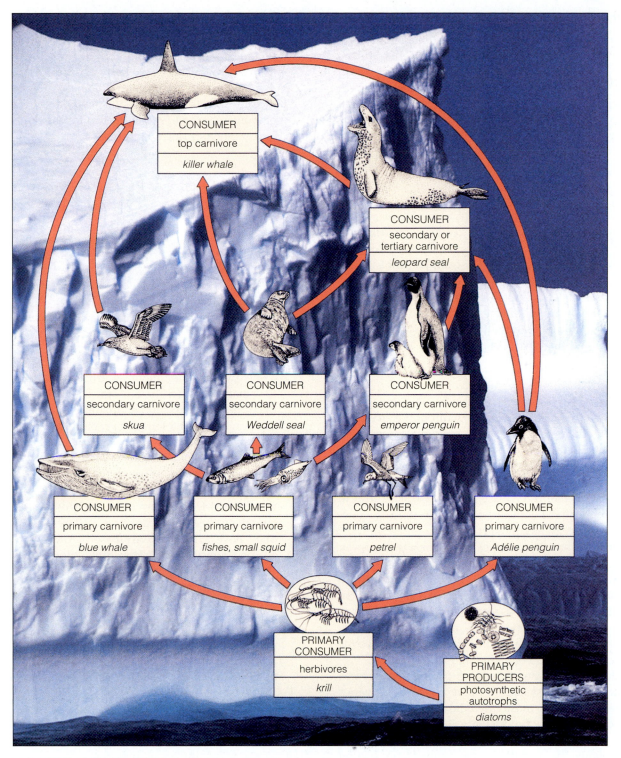

Figure 24.6 Simplified picture of a food web in the Antarctic. There are actually many more participants, including an array of decomposer organisms.

Primary Productivity

To get an idea of how energy flow is studied, consider a land ecosystem for which multicellular plants are the primary producers. The plants trap sunlight energy and, through photosynthesis, convert it into the chemical energy of organic compounds in their tissues. The rate at which primary producers capture and store energy is an ecosystem's *gross* **primary productivity**. How much energy is stored depends on how many plants are present and on the balance between photosynthesis and energy used in aerobic respiration by the plants. This energy "bottom line" is *net* primary productivity.

Other factors affect the amount of net primary productivity, its seasonal patterns, and its distribution through the habitat. This is true of ecosystems on land and in the seas (Figure 24.7a). For example, the amount of new plant growth can vary depending on the availability of minerals, the temperature range, and the amount of sunlight and rainfall during the growing season. The harsher the environment, the fewer plant shoots will be produced—and the lower the productivity.

Major Pathways of Energy Flow

What is the direction of energy flow through ecosystems on land? Only a small part of the energy from sunlight becomes fixed in plants. The plants themselves use up to half of what they fix, and they lose metabolically generated heat. Other organisms tap into the energy stored in plant tissues, remains, or wastes. They, too, lose heat to the environment. All of these heat losses represent a one-way flow of energy out of the ecosystem.

Figure 24.7 (parts *b* and *c*) diagrams this one-way flow of energy through two kinds of food webs. In **grazing food webs**, the energy flows from plants to herbivores, then through an array of carnivores. In **detrital food webs**, it flows mainly from plants through detritivores and decomposers. Usually the two kinds cross-connect, as when a crab of a detrital food web is eaten by a herring gull of a grazing food web.

The amount of energy moving through food webs differs from one ecosystem to the next and often varies with the seasons. In most cases, however, the greatest portion of net primary productivity passes through detrital food webs. For instance, when cattle graze heavily on pasture plants, about half the net primary production enters a grazing food web. But cattle don't use all the stored energy they consume. Quantities of undigested plant parts and other material in cattle feces become available for decomposers and detritivores. In marshes, most of the stored energy is not used until plant parts die and become available for detrital food webs.

Ecological Pyramids

Often the trophic structure of an ecosystem is represented as an **ecological pyramid**, in which producers form a base for successive tiers of consumers above them. Some pyramids are based on biomass (the weight of all the individuals at each trophic level). Here is a pyramid of biomass (grams/square meter) for the aquatic ecosystem at Silver Springs, Florida:

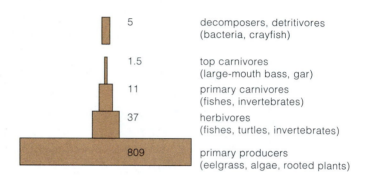

5	decomposers, detritivores (bacteria, crayfish)
1.5	top carnivores (large-mouth bass, gar)
11	primary carnivores (fishes, invertebrates)
37	herbivores (fishes, turtles, invertebrates)
809	primary producers (eelgrass, algae, rooted plants)

Some pyramids of biomass are "upside down," with the smallest tier on the bottom. Think of phytoplankton (including diatoms) and zooplankton (including rotifers) in a pond. A small biomass of rapidly growing and reproducing phytoplankton may support a greater biomass of larger zooplankton that grow more slowly and consume less energy per unit of weight.

An energy pyramid is another useful way to depict an ecosystem's trophic structure. Such pyramids show the energy losses at each transfer to a different trophic level. They have a large energy base at the bottom and are always "right-side up." They give a more accurate picture of ever diminishing amounts of energy flowing through successive trophic levels of the ecosystem.

Energy flows into food webs of ecosystems from an outside source—usually, the sun. Energy leaves ecosystems mainly by loss of metabolic heat, which each organism generates.

Gross primary productivity is the total energy stored by an ecosystem's photosynthesizers in a specified interval. Net primary productivity is the rate of energy storage in plant tissues minus energy used in the plants' aerobic respiration.

Living tissues of photosynthesizers are the basis of grazing food webs. The nonliving remains of photosynthesizers and consumers are the basis of detrital food webs.

The amount of useful energy flowing through consumer trophic levels declines at each energy transfer as metabolic heat is lost and as some food energy is shunted into organic wastes.

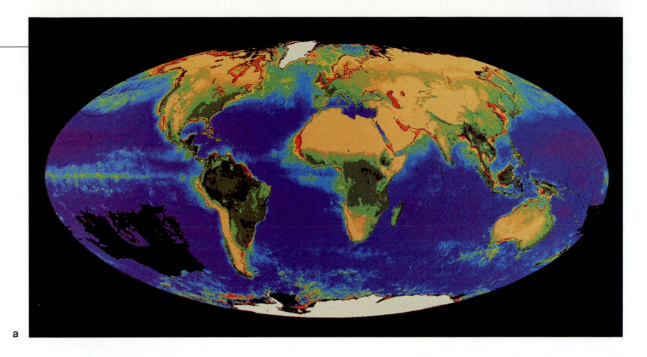

a

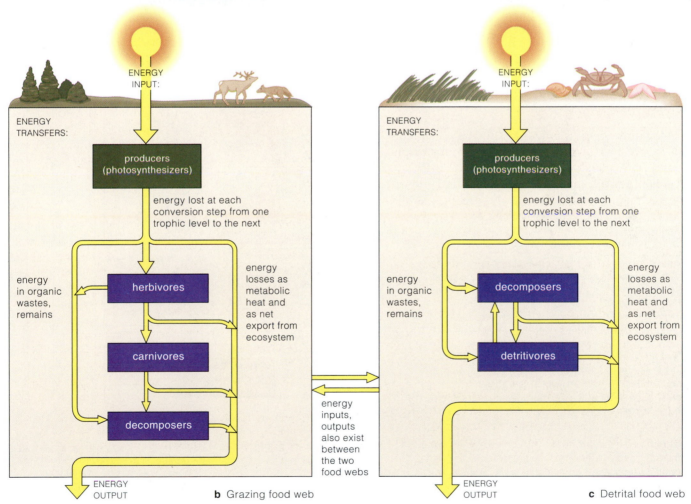

b Grazing food web

c Detrital food web

Figure 24.7 Energy bases and energy flow in ecosystems. (**a**) Summary of 3 years of satellite data on the Earth's primary productivity. *Dark green* denotes rain forests and other highly productive regions. *Yellow* denotes deserts (low productivity). Productivity in oceans, ranging from high to low, is coded *red* down through *orange, yellow, green,* and *blue.* (**b,c**) The one-way flow of energy through two kinds of cross-connected food webs in ecosystems.

24.4

ENERGY FLOW AT SILVER SPRINGS

To gather data for an energy pyramid, researchers measure the energy that each type of individual in the ecosystem takes in, burns during metabolism, and stores in body tissues. They also measure the energy remaining in its waste products. They then multiply the energy per individual by population size. Energy inputs and outputs are calculated so that energy flow can be expressed per unit of land (or water) per unit of time.

The pyramid in Figure 24.8a is based on a long-term study of a grazing food web in an aquatic ecosystem in Florida. Figure 24.8b shows some of the calculations used to construct it. Such studies tell us that, given the metabolic demands of organisms and the amount of energy shunted into organic wastes, only about 6 to 16 percent of the energy entering one trophic level becomes available to organisms at the next level. Because the efficiency of these energy transfers is so low, ecosystems generally have no more than four consumer trophic levels.

Figure 24.8 Measurements of energy flow through an aquatic ecosystem—Silver Springs, Florida.

(**a**) The pyramid of energy flow, as calculated for 1 year. The numerals represent the number of individuals in each category.

(**b**) Annual energy flow, in kilocalories/square meter/year. Producers in this small spring are mostly aquatic plants. Carnivores are insects and small fishes; top carnivores are larger fishes. The energy source (sunlight) is available all year. Detritivores and decomposers cycle organic compounds from other trophic levels.

The producers trap 1.2 percent of incoming solar energy, and only a little more than a third of this amount is fixed in new plant biomass (4,245 + 3,368). The producers use more than 63 percent of the fixed energy for their own metabolism. About 16 percent of it is transferred to herbivores, and most of this is used for metabolism or transferred to detritivores and decomposers. Only 11.4 percent of the energy transferred to herbivores reaches the next trophic level (carnivores), and the carnivores use all but about 5.5 percent, which is transferred to top carnivores. In time, all of the 5,060 kilocalories transferred through the system appear as metabolically generated heat.

This diagram is oversimplified, for no community is isolated from others. Organisms and materials constantly drop into the springs. Organisms and materials are slowly lost by way of a stream that leaves the springs.

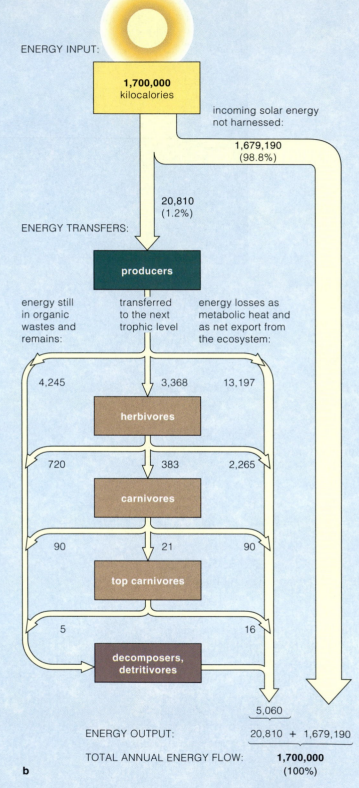

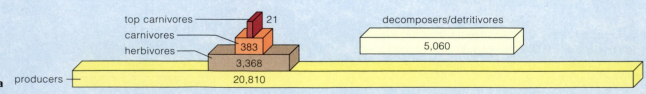

Further reading: Student Guide to InfoTrac on web site →

BIOGEOCHEMICAL CYCLES—AN OVERVIEW

The availability of nutrients as well as energy profoundly influences the structure of ecosystems. Photosynthetic producers require carbon, oxygen, and hydrogen, which they obtain from water and air. They require nitrogen, phosphorus, and other mineral ions. Because mineral deficiencies lower primary productivity, they adversely affect the ecosystem at large.

In a **biogeochemical cycle,** ions or molecules of a nutrient are transferred from the environment to organisms, then back to the environment—part of which serves as a reservoir for them. They generally move slowly through the reservoir, compared to their rapid exchange between organisms and the environment.

Figure 24.9 is a model of the relationship between most ecosystems and the biogeochemical part of the cycles. This model is based on four factors.

First, mineral elements that producer organisms use as nutrients usually are available in the form of mineral ions, such as ammonium (NH_4^+). *Second*, inputs from the physical environment, together with nutrient cycling activities of all of the decomposers and detritivores, maintain the ecosystem's nutrient reserves. *Third*, the actual amount of a nutrient that is being cycled on through most major ecosystems is greater than the amount entering or leaving in a given year. *Fourth*, rainfall, snowfall, and the slow weathering of rocks are common sources of environmental inputs into an ecosystem's nutrient reserves. So are the combined effects of metabolic activities, such as nitrogen fixation.

Ecosystems on land also have typical outputs from the nutrient reserves. The loss of mineral ions by way of runoff from irrigated cropland is an example.

There are three categories of biogeochemical cycles, based on the part of the environment that contains the greatest portion of the ion or molecule being considered. As you will see shortly, in the *hydrologic* (water) cycle, oxygen and hydrogen move in the form of water molecules. This movement is also known as the global

water cycle. In *atmospheric* cycles, a large percentage of the nutrient is in the form of an atmospheric gas. This is true of carbon and nitrogen, for example. *Sedimentary* cycles involve phosphorus and other solid nutrients that do not have gaseous forms. Solid nutrients move from land to the seafloor and return to dry land only through

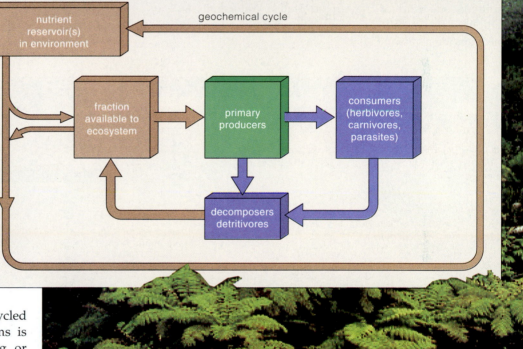

Figure 24.9 Generalized model of nutrient flow through a land ecosystem. The overall movement of nutrients from the physical environment, through organisms, and back to the environment constitutes a biogeochemical cycle.

geological uplifting, which may take millions of years. The largest storehouse for sedimentary cycles is the Earth's crust.

In a biogeochemical cycle, ions or molecules of a nutrient move from the environment to organisms, then back to the environmental reservoir for them.

Nutrients generally move slowly through the environment but rapidly between organisms and the environment.

THE HYDROLOGIC CYCLE

Driven by ongoing inputs of solar energy, the Earth's waters move slowly, on a vast scale, from the ocean into the atmosphere, to land, and back to the ocean—the main reservoir. Water evaporating into the lower atmosphere initially stays aloft in the form of vapor, clouds, and ice crystals (Figure 24.10). It returns to the Earth as precipitation—mostly rain and snow. Ocean currents and prevailing wind patterns influence the global **hydrologic cycle**, as shown in Figure 24.11.

Water is vital for all organisms. It also is a transport medium; it moves nutrients into and out of ecosystems. Its role in moving nutrients became clear in long-term studies of watersheds. A **watershed** is any region in which the precipitation becomes funneled into a single stream or river. Watersheds may be any size. For example, the Mississippi River watershed extends across roughly one-third of the United States. Watersheds at Hubbard Brook Valley in New Hampshire average 14.6 hectares (36 acres).

Figure 24.10 Cycling of water during a thunderstorm in the American Midwest.

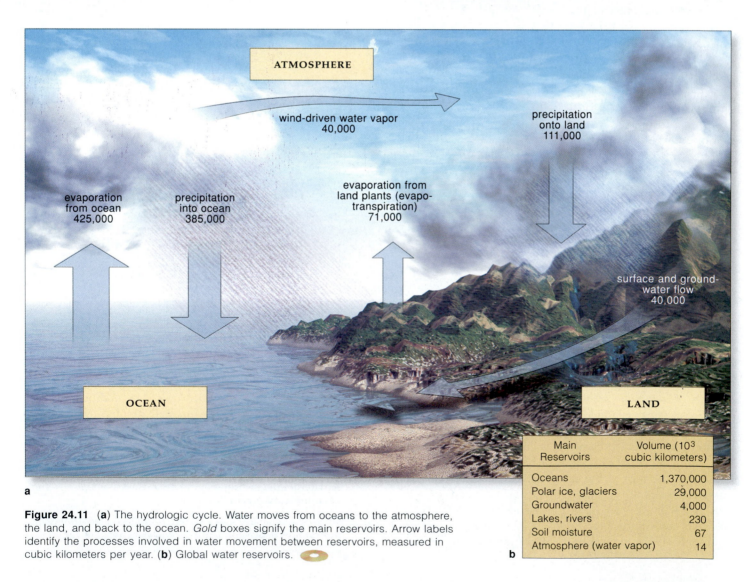

a

Figure 24.11 (a) The hydrologic cycle. Water moves from oceans to the atmosphere, the land, and back to the ocean. *Gold* boxes signify the main reservoirs. Arrow labels identify the processes involved in water movement between reservoirs, measured in cubic kilometers per year. (b) Global water reservoirs.

Main Reservoirs	Volume (10^3 cubic kilometers)
Oceans	1,370,000
Polar ice, glaciers	29,000
Groundwater	4,000
Lakes, rivers	230
Soil moisture	67
Atmosphere (water vapor)	14

b

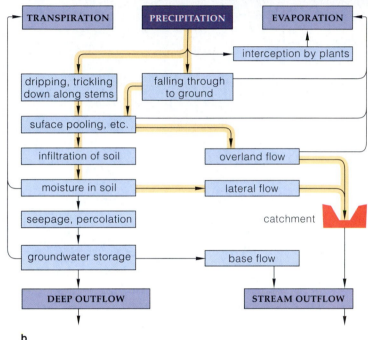

b

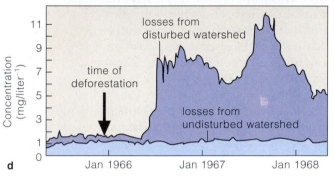

d

Figure 24.12 Results from experiments involving deliberate disturbances to forests of experimental watersheds in Hubbard Brook Valley (**a**). (**b**) In this model of the movement of water through watersheds, the *yellow* lines indicate the water flow studied in these experiments.

(*Dark blue* indicates inputs to the watershed; *medium blue*, distribution within the watershed; and *light blue*, outputs from it. Transpiration, noted at the upper left, is the evaporation of water from plants. The plants take up water in the soil.)

Researchers stripped the vegetation cover in forested areas but did not disturb the soil. They applied herbicides to the soil for three years to prevent regrowth. All water that drained from each watershed was measured as it flowed over the V-shaped concrete catchments, as in (**c**). The arrow in (**d**) marks the time of deforestation. Concentrations of calcium and other minerals were compared against those in water passing over a control catchment in an undisturbed area. Calcium losses were six times greater from the deforested area.

Most of the water entering a watershed seeps into soil or becomes surface runoff that moves into streams (Figure 24.12). Plants take up water and dissolved minerals from soil, then lose water by transpiration.

Measurements of water inputs and outputs at watersheds have practical application. Cities that depend on surface water supplies in watersheds can adjust water use based on seasonal variations. Watershed studies have also shown how much plants influence the movement of nutrients through the ecosystem phase of biogeochemical cycles. For example, you might think that water draining a watershed would rapidly leach minerals, such as calcium. But in young, undisturbed forests in Hubbard Brook watersheds, each hectare lost only about 8 kilograms of calcium. Rainfall and weathering of rocks brought in some new calcium. And tree roots also "mined" the soil, so calcium was being stored in a growing biomass of tree tissues.

In experimental watersheds in the Hubbard Brook Valley, deforestation caused a shift in nutrient outputs. The results, shown in Figure 24.12c, are sobering. Calcium and other nutrients cycle so slowly that deforestation may disrupt nutrient availability for entire ecosystems. This is especially the case for forests that cannot rapidly regenerate themselves. The conifer forests of the Pacific Northwest, which require decades to grow to maturity, are like this.

In the hydrologic cycle, water slowly moves from the oceans (the main reservoir), through the atmosphere, onto land, then back to the oceans.

On land, plants stabilize the soil and absorb dissolved minerals, minimizing the loss of soil nutrients in runoff.

THE CARBON CYCLE

In a vital atmospheric cycle, carbon moves through the atmosphere and food webs on its way to and from the oceans, sediments, and rocks (Figure 24.13). Its global movement is called the **carbon cycle.** Sediments and rocks hold most of the carbon, followed by the ocean, soil, atmosphere, and land biomass. It enters the atmosphere as cells engage in aerobic respiration, as fossil fuels burn, and when volcanoes erupt and release it from rocks in the Earth's crust. Most atmospheric carbon occurs as carbon dioxide (CO_2). Most carbon dissolved in the ocean is in the forms of bicarbonate and carbonate.

When you watch bubbles escaping from a glass of carbonated soda left in the sun, you might wonder: Why doesn't all the CO_2 dissolved in warm ocean surface waters escape to the atmosphere? Driven by winds and regional diffferences in water density, water makes a gigantic loop from the surface of the Pacific and Atlantic Oceans to the Atlantic and Antarctic seafloors. There its CO_2 moves into deep storage reservoirs before water loops up (Figure 24.14). It is a factor in carbon's distribution and global budget, as listed below:

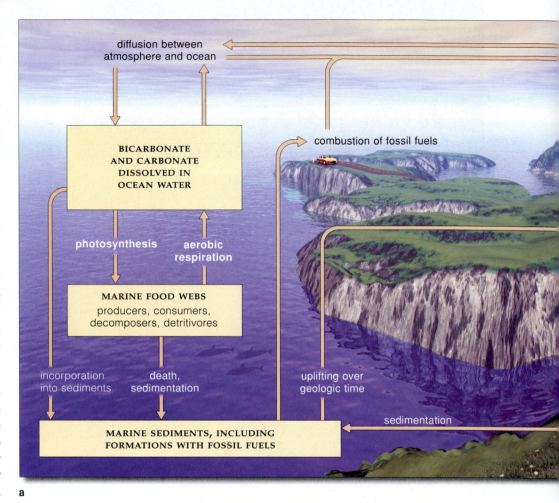

a

MAIN CARBON RESERVOIRS AND HOLDING STATIONS:

Sediments and rocks	77,000,000 (10^{15} grams)
Ocean (dissolved forms)	39,700
Soil	1,500
Atmosphere	750
Biomass on land	715

ANNUAL FLUXES IN GLOBAL DISTRIBUTION OF CARBON:

From atmosphere to plants (carbon fixation)	120
From atmosphere to ocean	107
To atmosphere from ocean	105
To atmosphere from plants	60
To atmosphere from soil	60
To atmosphere from fossil fuel burning	5
To atmosphere from net destruction of plants	2
To ocean from runoff	0.4
Burial in ocean sediments	0.1

Photosynthesizers capture billions of metric tons of carbon atoms in organic compounds every year. However, the average length of time that a carbon atom is held in any given ecosystem varies quite a bit. For example, organic wastes and remains decompose rapidly

b

Figure 24.13 (**a**) Global carbon cycle. The part of the diagram on this page shows carbon's movement through typical marine ecosystems. The facing page shows carbon's movement through ecosystems on land. *Gold* boxes indicate key reservoirs.

(**b**) A Los Angeles freeway system under its self-generated blanket of smog at twilight. Each day, fossil fuel burning (by vehicles, industries, and homes) puts carbon and other substances into the air.

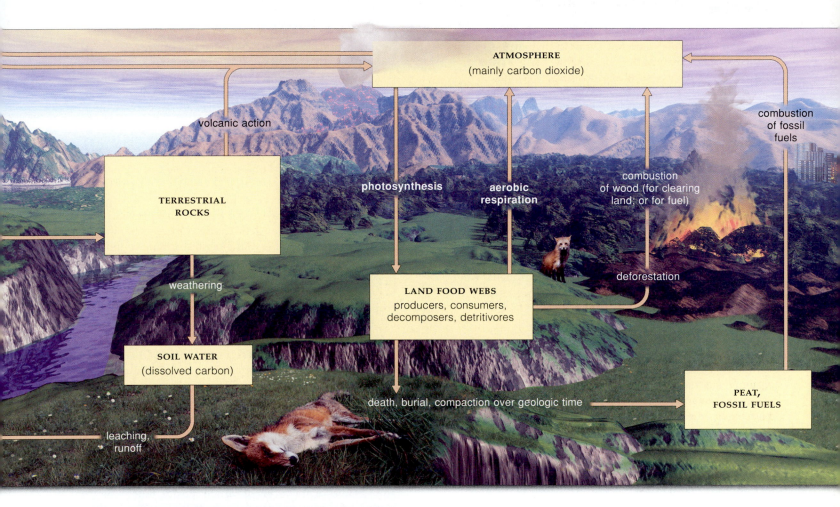

ATMOSPHERE
(mainly carbon dioxide)

volcanic action

combustion
of fossil
fuels

TERRESTRIAL
ROCKS

photosynthesis

aerobic
respiration

combustion
of wood (for clearing
land; or for fuel)

weathering

deforestation

LAND FOOD WEBS
producers, consumers,
decomposers, detritivores

SOIL WATER
(dissolved carbon)

death, burial, compaction over geologic time

PEAT,
FOSSIL FUELS

leaching,
runoff

Warm, less salty, shallow current

Cold, salty, deep current

Figure 24.14 Loop of ocean water that delivers carbon dioxide to its deep ocean reservoir. It sinks in the cold, salty North Atlantic and rises in the warmer Pacific.

in tropical rain forests, so not much carbon accumulates at the surface of soils. In marshes, bogs, and other anaerobic areas, decomposers cannot break down organic compounds completely, so carbon gradually builds up in peat and other forms of compressed organic matter.

Also, in food webs of ancient aquatic ecosystems, carbon was incorporated in shells and other hard parts. The shelled organisms died and sank through water, then became buried in sediments. The same things are happening today. Carbon remains buried for many millions of years in deep sediments until part of the seafloor is uplifted above the ocean surface through geologic forces. Other buried carbon is slowly converted to long-standing reserves of gas, petroleum, and coal, which we tap for use as fossil fuels.

Human activities, including the worldwide burning of fossil fuels, are putting more carbon into the atmosphere than can be cycled to the ocean reservoir. This is amplifying the greenhouse effect. For that reason, it may be contributing to global warming. *Science Comes to Life* in Section 24.8 describes this effect and some possible outcomes of increases—or decreases—in it.

The ocean and atmosphere interact in the global cycling of carbon. Fossil fuel burning and other human activities may be contributing to imbalances in the global carbon budget.

24.8 FROM GREENHOUSE GASES TO A WARMER PLANET

GREENHOUSE EFFECT   Atmospheric concentrations of gaseous molecules play a profound role in shaping the average temperature near the surface of the Earth. That temperature has enormous effects on the global climate.

Countless molecules of carbon dioxide, water, ozone, methane, nitrous oxide, and chlorofluorocarbons are key players in interactions that dictate the global temperature. Collectively, the gases act somewhat like a pane of glass in a greenhouse—hence their name, "greenhouse gases." Wavelengths of visible light can pass around them and reach the Earth's surface. But greenhouse gases impede the escape of longer, infrared wavelengths—heat—from the Earth into space. How? Gaseous molecules can absorb the wavelengths and reradiate much of the absorbed energy back toward the Earth, as shown in Figure 24.15.

The constant reradiation of heat energy from the greenhouse gases, together with the constant bombardment and absorption of wavelengths from the sun, causes heat to build up in the lower atmosphere. The name for this warming action is the **greenhouse effect**.

GLOBAL WARMING DEFINED   Without the action of greenhouse gases, the Earth would be cold and lifeless. However, there can be too much of a good thing. Largely as a result of human activities, greenhouse gases are building to atmospheric levels that are higher than they were in the past (Figure 24.16). As many researchers suspect, the increase may be contributing to long-term higher temperatures at the Earth's surface, an effect called **global warming**.

What is so alarming about a warmer planet? Suppose the temperature of the lower atmosphere were to rise by only 4°C (7°F). Sea levels could rise by about 2 feet, or 0.6 meter. Why? The warming would increase ocean surface temperatures—and water expands when heated. Global warming also could make glaciers and polar ice sheets melt faster, so low coastal regions could flood.

Think of a long-term rise in sea level, combined with high tides and storm waves. Waterfronts of Vancouver, Boston, San Diego, Galveston, and other coastal cities would be submerged. So would agricultural lowlands and deltas in India, China, and Bangladesh, where much of the world's rice is grown. Huge tracts of Florida and Louisiana would face saltwater intrusions. Besides this, global warming could disturb regional patterns of precipitation and temperature. Crop yields would decline in currently productive regions, including parts of Canada and the United States, and increase in others.

Some predicted effects of climate change are emerging. For example, in 1996, we found that the water several hundred feet below the Arctic icecap is 1°C warmer than it was just five years earlier. If the rapid warming continues, the icecap may disappear before the end of this century. In addition, during the 1990s, record-breaking hurricanes, flooding, and prolonged droughts have adversely affected crop production around the world. Sandy beaches have been diminishing by almost 2 feet a year.

EVIDENCE OF AN INTENSIFIED GREENHOUSE EFFECT
In the late 1950s, researchers on Hawaii's highest volcano started measuring concentrations of different greenhouse gases, and their monitoring activities are still going on. They chose that remote site because it is almost free of local airborne contamination and is representative of conditions for the Northern Hemisphere. Monitoring there continues.

Consider what they found out about carbon dioxide alone. Atmospheric levels of carbon dioxide follow the annual cycle of plant growth in the Northern Hemisphere. They decline in summer, when photosynthesis rates are highest. They rise in winter, when aerobic respiration continues and photosynthesis slows.

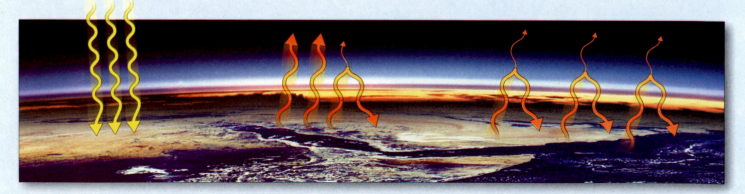

a Rays of sunlight penetrate the lower atmosphere and warm the Earth's surface.

b The surface radiates heat (infrared wavelengths) to the lower atmosphere. Some heat escapes into space. But greenhouse gases and water vapor absorb some infrared energy and radiate a portion of it back toward Earth.

c Increased concentrations of greenhouse gases trap more heat near Earth's surface. Sea surface temperature rises, more water evaporates into the atmosphere, and Earth's surface temperature rises.

Figure 24.15 The greenhouse effect—the result mainly of carbon dioxide, water, ozone, methane, nitrous oxide, and chlorofluorocarbons in the atmosphere.

Figure 24.16 Changes in atmospheric carbon dioxide levels (*white* graph line) correlated with past ice ages and interglacials. The 1985 mean temperature was used as a baseline for tracking the global temperature changes. (*red* graph line). *Red* arrow: start of most recent ice age. *Yellow* arrow: start of last interglacial. *White* arrow: start of industrial revolution.

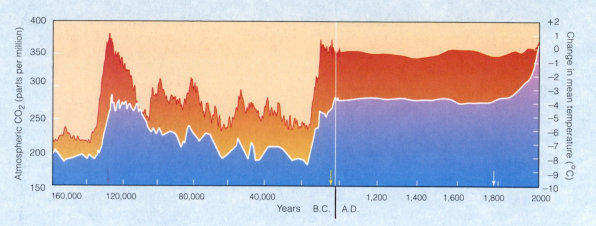

The troughs and peaks around the graph line in Figure 24.17*a* represent annual lows and highs. As you can see, this midline steadily increased from 1960 to 1995. For the first time, scientists saw the integrated effects of the carbon balances for the land and water ecosystems of an entire hemisphere. Many take this as evidence of a buildup of carbon dioxide levels. They predict that it will intensify the greenhouse effect over the coming decades.

The global burning of fossil fuels is probably contributing most to increasing carbon dioxide levels. Deforestation adds to it; carbon is released when wood burns. Today, vast tracts of forests throughout the world are being cleared and burned at a rapid rate. More important, the plant biomass is plummeting—and this affects global absorption of carbon dioxide in photosynthesis.

Will atmospheric levels of greenhouse gases continue to increase until the middle of this century? Will global temperature rise by several degrees? If the warming trend is already in motion, we will not be able to reverse it now by stopping fossil fuel burning and deforestation. So there is widespread agreement that we should begin preparing for the consequences. For example, we might step up genetic engineering studies to develop drought-resistant and salt-resistant plants. Such plants may prove crucial in regions of saltwater intrusions and climatic change.

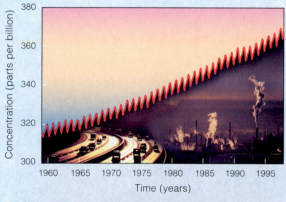

a CARBON DIOXIDE (CO_2). Of all human activities, the burning of fossil fuels and deforestation contribute the most to increasing atmospheric levels.

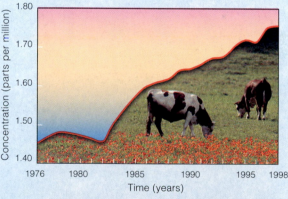

c METHANE (CH_4). Termite activity and anaerobic bacteria living in swamps, landfills, and the stomachs of cattle and other ruminants produce great amounts of methane as a by-product of metabolism.

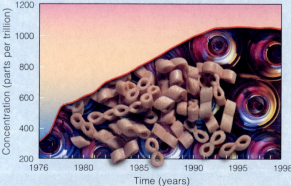

b CFCS. Until restrictions were in place, CFCs were widely used in plastic foams, refrigerators, air conditioners, and industrial solvents.

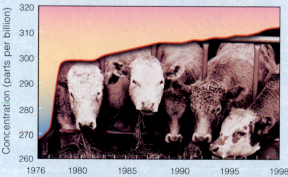

d NITROUS OXIDE (N_2O). Denitrifying bacteria produce N_2O in metabolism. Also, fertilizers and animal wastes release enormous amounts; this is especially the case for vast livestock feedlots.

Figure 24.17 Recently documented increases in atmospheric concentrations of four greenhouse gases: carbon dioxide, CFCs, methane, and nitrous oxide.

THE NITROGEN CYCLE

Since the time of life's origin, the atmosphere and oceans have contained nitrogen. This component of all proteins and nucleic acids moves in an atmospheric cycle called the **nitrogen cycle**. Gaseous nitrogen (N_2) makes up about 80 percent of the atmosphere, the largest nitrogen reservoir. Triple covalent bonds hold its two atoms together ($N{\equiv}N$), and few organisms can break them. Only certain bacteria, volcanic action, and lightning convert N_2 into forms that enter food webs.

Of all nutrients required for plant growth, nitrogen often is the scarcest. Today, nearly all nitrogen in soils has been put there by nitrogen-fixing organisms. Ecosystems lose it through the activities of bacteria that "unfix" the fixed nitrogen. Land ecosystems lose more through leaching of soils, although this provides nitrogen inputs to aquatic ecosystems such as streams, lakes, and the oceans (Figure 24.18).

Cycling Processes

Let's follow nitrogen atoms through the ecosystem part of the nitrogen cycle. They move through organisms by way of processes of nitrogen fixation, assimilation and biosynthesis, decomposition, ammonification, and nitrification.

In **nitrogen fixation**, a few kinds of bacteria convert N_2 to ammonia (NH_3), which dissolves quickly in the cytoplasm to form ammonium (NH_4^+). Certain microorganisms (cyanobacteria) are nitrogen fixers of aquatic ecosystems. Others fix nitrogen in many land ecosystems. Collectively these organisms fix about 200 million metric tons of nitrogen each year! Plants assimilate and use this nitrogen in the biosynthesis of amino acids, proteins, and nucleic acids. Plant tissues serve as the only nitrogen source for animals, including humans, which feed directly or indirectly on plants.

Through **decomposition** and **ammonification**, bacteria and fungi break down nitrogen-containing wastes and remains of organisms. The decomposers use part of the released proteins and amino acids for their own metabolism. But most of the nitrogen is still in the decay products, in the form of ammonia or ammonium, which plants take up. Nitrifying bacteria also act on ammonia or ammonium. In **nitrification**, they strip these compounds of electrons, and nitrite (NO_2^-) is the result. Other bacteria use the nitrite in metabolism and produce nitrate (NO_3^-), which plants take up.

Certain plants are better than others at securing nitrogen. Peas, beans, clover, and other legumes have mutually beneficial associations with nitrogen-fixing

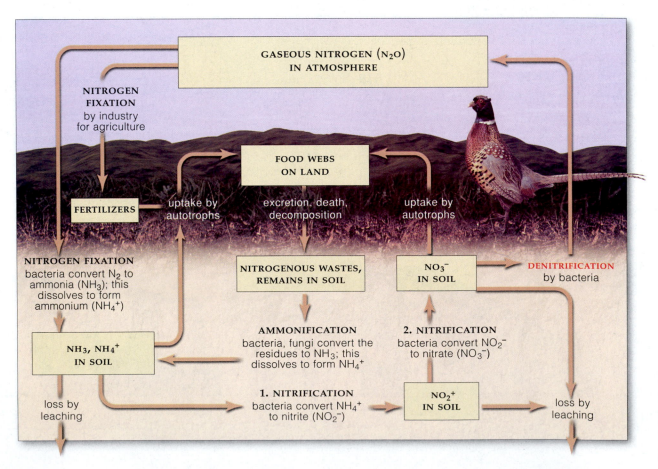

Figure 24.18 The nitrogen cycle in an ecosystem on land.

Further reading: Student Guide to InfoTrac on web site →

bacteria. In addition, most land plants have similar associations with fungi, forming specialized roots known as mycorrhizae ("fungus roots") that incorporate fungal tissues. The mycorrhizae enhance nitrogen uptake by the plant.

Nitrogen Scarcity

Given the cycling processes, you would think that land plants could get enough nitrogen. In fact, however, the ammonium, nitrite, and nitrate that form during the cycle are highly vulnerable to leaching and runoff. With leaching, recall, soil water moves out of an area, which thereby loses the nutrients dissolved in the water.

Also, some nitrogen is lost to the air by **denitrification**. Here, bacteria convert nitrate or nitrite to N_2 and a bit of nitrous oxide (N_2O). Ordinarily, most denitrifying bacteria rely on aerobic respiration. When soil is waterlogged and poorly aerated, they switch to anaerobic pathways and use nitrate, nitrite, or nitrous oxide as the final electron acceptor instead of oxygen. In these reactions, fixed nitrogen is converted to N_2, much of which escapes into the atmosphere.

Besides this, nitrogen fixation comes at high metabolic cost to plants that are associated with nitrogen fixers. In exchange for nitrogen, they give up sugars and other substances. Such plants do have the competitive edge in nitrogen-poor soil. In nitrogen-rich soil, however, species that do not have to pay the metabolic price often displace them.

Human Impact on the Nitrogen Cycle

Humans are altering the cycling of nitrogen in natural ecosystems. Consider how air pollutants contribute to changes in soil. Vehicles, power plants that burn fossil fuels, and nitrogenous fertilizers are sources of pollutants, including oxides of nitrogen. These contribute to soil acidity—which reduces the amounts of magnesium, calcium, and potassium that plants can take up. Figure 24.19 merely hints at the ion interactions in soil water.

And what about those nitrogen fertilizers? To be sure, nitrogen losses from soil are enormous in agricultural regions. With each harvest, nitrogen departs from the fields (in tissues of harvested plants). Soil erosion and leaching remove more. In Europe and North America, farmers traditionally have rotated crops, as when they alternate wheat with legumes. Along with other conservation practices, crop rotation has helped keep soils stable and productive, sometimes for thousands of years. Today's intensive agriculture is based on the use of nitrogen-rich fertilizers. New strains of crop plants are bred for their ability to take up fertilizers, and crop

Figure 24.19 Dead and dying spruce trees in a forest in Germany. They are cited as casualties of extremely high concentrations of nitrogen oxides and other forms of air pollution.

yields per hectare have doubled and even quadrupled over the past 40 years. Whether pest control and soil management technologies can sustain high yields indefinitely remains uncertain, for reasons that will become apparent in Chapter 25.

We can't get something for nothing. Fertilizer production requires huge amounts of energy from fossil fuels—not from free, unending sunlight. Few once believed that fossil fuel supplies might run out, so there was little concern about fertilizer costs. It still is common to pour more energy into soil (as in fertilizers) than we get out of it (in the form of food). As long as the human population continues to grow exponentially, farmers will be engaged in a race to grow as much food as they can for as many people as possible. Soil enrichment with nitrogen-containing fertilizers is part of the race, as it is now being run.

Nitrogen, a component of all proteins and of nucleic acids, moves in an atmospheric cycle.

Nitrogen atoms move through ecosystems by nitrogen fixation, assimilation and biosynthesis, decomposition, ammonification, and nitrification.

The cycling of nitrogen in natural ecosystems depends on the activity of nitrogen-fixing bacteria and on mycorrhizae. Human disruptions to that cycling may harm the ecosystems.

THE PHOSPHORUS CYCLE

We conclude our look at biogeochemical cycling with an example of a sedimentary cycle. In the **phosphorus cycle**, phosphorus moves from land to sediments in the seas and then back to the land (Figure 24.20). The Earth's crust is the main storehouse for this mineral and for others, including calcium and potassium.

Phosphorus is typically present in rock formations on land, in the form of phosphates. Through the natural processes of weathering and erosion, phosphates enter rivers and streams, which eventually transport them to the ocean. There, mainly on the continental shelves, phosphorus accumulates with other minerals as insoluble deposits. Millions of years pass. Where crustal plates collide, part of the seafloor may be uplifted and drained. The seafloor, with its mineral deposits, thereby becomes exposed as new land surfaces. Over geologic time, weathering releases phosphates from the rocks—and the geochemical phase of the phosphorus cycle begins again.

The ecosystem phase of the cycle is far more rapid than the long-term geochemical phase. Living organisms require small amounts of phosphorus. It is a key component of ATP, NADPH, phospholipids, nucleic acids, and other organic compounds. Plants have the metabolic means to take up dissolved, ionized forms of phosphorus. Actually, they do this so rapidly and efficiently that they often reduce soil concentrations of phosphorus to extremely low levels. Herbivores obtain phosphorus only by eating the plants; carnivores obtain it by eating herbivores. Herbivores and carnivores excrete phosphorus as a waste product in urine and feces. Phosphorus is also released to the soil by the decomposition of organic matter. The plants then take up phosphorus and so recycle it rapidly within the ecosystem.

The Earth's crust is the main storehouse for phosphorus and other minerals that move through ecosystems as part of sedimentary cycles. The geochemical phase of these cycles proceeds extremely slowly.

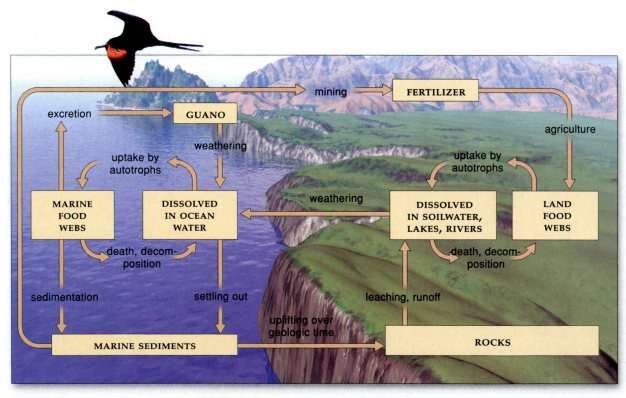

Figure 24.20 The phosphorus cycle. This is an example of a sedimentary cycle.

TRANSFER OF HARMFUL COMPOUNDS THROUGH ECOSYSTEMS

DDT, the first of the synthetic organic pesticides, was first used during World War II. In mosquito-infested regions of the tropical Pacific, people were vulnerable to a dangerous disease, malaria. DDT helped control the mosquitoes that were transmitting the sporozoan disease agents (*Plasmodium japonicum*). In war-ravaged cities of Europe, people were suffering from the crushing headaches, fevers, and rashes associated with typhus. DDT helped control the body lice that were transmitting *Rickettsia rickettsii*, the bacterial agent of this terrible disease. After the war, it seemed like a good idea to use DDT against insects that were agricultural or forest pests, transmitters of pathogens, or merely nuisances in homes and gardens.

DDT is a relatively stable hydrocarbon compound. It is nearly insoluble in water, so you might think that it would act only where applied. But winds can carry DDT in vapor form; water can transport fine particles of it. DDT also is highly soluble in fats, so it accumulates in the tissues of organisms. Thus, as we now know, DDT can show **biological magnification**. The term refers to an increase in concentration of a nondegradable (or slowly degradable) substance in organisms as it is passed upward through food chains (Figure 24.21). Most of the DDT from all the organisms that a consumer eats during its lifetime will become concentrated in its tissues. Besides this, many organisms have the means to partially metabolize DDT to DDE and other modified compounds with different but still disruptive effects. Both DDT and the modified compounds are toxic or physiologically disruptive to *many* aquatic and terrestrial animals.

After the war, DDT began to move through the global environment, infiltrate food webs, and affect organisms in ways that no one had predicted. In cities where DDT was sprayed to control Dutch elm disease, songbirds started dying. In streams flowing through forests where DDT was sprayed to control spruce budworms, salmon started dying. In croplands sprayed to control one kind of pest, new kinds of pests moved in. DDT was indiscriminately killing off the natural predators that had been keeping pest populations in check! It took no great leap of the imagination to make the connection. All of those organisms were dying at the same time and the same places as the DDT applications.

Then side effects of biological magnification started showing up in places far removed from the areas of DDT application—and much later in time. Most devastated were species at the end of food chains, including bald eagles, peregrine falcons, ospreys, and brown pelicans. One product of DDT breakdown interferes with physiological processes. As one consequence, birds produced eggs with brittle shells—and many of the chick embryos didn't make it to hatching time. Some species were at the brink of extinction.

For decades now, DDT has been banned in the United States, except for restricted applications where public health is endangered. Many hard-hit species have partially recovered their numbers. Even today, however, some birds lay thin-shelled eggs. They pick up DDT at their winter ranges in Latin America. As recently as 1990, the California State Department of Health recommended that a fishery off the coast of Los Angeles be closed. DDT from industrial waste discharges that ended 20 years before is still contaminating that ecosystem.

Figure 24.21 An example of biological magnification. In 1987, scientists discovered a variety of toxic industrial chemicals in the tissues of a coelacanth (**top**), a "living fossil" that inhabits extremely deep waters off the west coast of Africa. The huge fish must have ingested other organisms that had been exposed to the chemical barrage.

1. An ecosystem is an entire complex of producers, consumers, detritivores, and decomposers and their physical environment, all interacting through a flow of energy and a cycling of materials.

2. Ecosystems are open systems, with inputs and outputs of energy and nutrients.

 a. With few exceptions, sunlight is the source of energy, and photosynthesizing autotrophs are the primary producers. They convert the energy of sunlight to ATP and other forms that can be used to synthesize large organic compounds from simple inorganic substances.

 b. Primary producers also assimilate many of the nutrients that are eventually transferred to other members of the system.

3. Directly or indirectly, primary producers nourish an array of heterotrophs in an ecosystem.

 a. The heterotrophs include consumers: herbivores that feed on algae and plants, carnivores that feed on animals, and omnivores that have eclectic diets.

 b. The heterotrophs also include decomposers (mainly certain fungi and bacteria that digest organic substances) and detritivores (such as crabs and earthworms, which feed on particles of dead or decomposing material).

4. Feeding relationships within ecosystems are structured as trophic levels: a hierarchy of energy transfers, sometimes referred to as "who eats whom."

 a. Primary producers make up the first trophic level, herbivores make up the next, carnivores the next, and so on.

 b. Decomposers, humans, and many other organisms get energy from more than one source and cannot be assigned to a single trophic level.

5. An isolated food chain (a straight-line sequence of who eats whom in an ecosystem) is rare in nature. Instead, food chains cross-connect with one another, forming food webs.

6. Ecosystems generally are most open for inputs and outputs of energy, water, and carbon.

 a. The rate at which primary producers capture and store a given amount of energy in a given time interval is called the primary productivity. (The total rate is called the *gross* primary productivity. The rate of energy storage in plants in excess of the rate of aerobic metabolism by the plants is the *net* primary productivity.)

 b. Most mineral nutrients are cycled within a natural ecosystem.

7. Energy fixed by photosynthesizers passes through grazing food webs and detrital food webs. Both types of

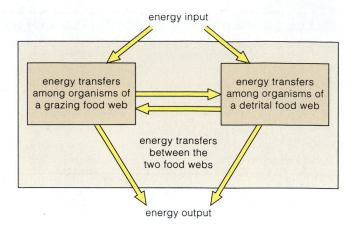

Figure 24.22 Summary of the one-way flow of energy through two kinds of cross-connected food webs in ecosystems.

food webs typically are interconnected in the same ecosystem (Figure 24.22).

 a. Both types of food webs lose energy (as heat) through metabolism of all organisms in the ecosystem.

 b. In either type of food web, the amount of useful energy flowing through consumer levels declines at each energy transfer. It declines through the loss of metabolically generated heat and as food energy is shunted into organic wastes.

8. In biogeochemical cycles, substances move from the physical environment, to organisms, then back to the environment.

 a. Water moves through the hydrologic cycle. Substances that exist primarily in gaseous phases move through atmospheric cycles. Phosphorus and other minerals move through sedimentary cycles.

 b. Nutrients generally move slowly through the geochemical phase of these cycles but rapidly between organisms and the environment.

9. Ecosystems on land have predictable rates of nutrient losses that generally increase when the land is cleared or otherwise disturbed.

10. Fossil fuel burning and conversion of natural ecosystems to cropland or grazing land are contributing to increased atmospheric concentrations of carbon dioxide. The increase may be contributing to a global warming trend.

11. Nitrogen availability is often a limiting factor for the total net primary productivity of land ecosystems. Gaseous nitrogen is abundant in the atmosphere, but it must be converted to ammonia and to nitrates that primary producers can use. Some bacteria as well as volcanic action and lightning can cause the conversion.

Review Questions

1. Define ecosystem, and name the central roles that autotrophs play in all ecosystems. *24.1, 24.2*

2. Define and give examples of trophic levels in ecosystems. *24.2*

3. Distinguish between food chain and food web. *24.2*

4. If you were growing a vegetable garden, what variables might affect its net primary production? *24.3*

5. Characterize grazing and detrital food webs. Indicate how energy leaves each one and how the two are interconnected. *24.3*

6. Describe the greenhouse effect. Make a list of 20 agricultural products and manufactured goods that you depend on. Are any implicated in the amplification of the greenhouse effect? *24.8*

7. Describe the reservoirs and organisms involved in one of the biogeochemical cycles. *24.5, 24.7, 24.9, 24.10*

8. Define nitrogen fixation, nitrification, ammonification, and denitrification. *24.9*

Self-Quiz *(Answers in Appendix V)*

1. _____ can be thought of as an ecosystem.
   a. A freshwater spring   c. A city
   b. Antarctica   d. All of the above

2. Ecosystems have _____.
   a. energy inputs and outputs   c. one trophic level
   b. nutrient cycling but not outputs   d. a and b

3. Trophic levels can be described as _____.
   a. structured feeding relationships
   b. who eats whom in an ecosystem
   c. a hierarchy of energy transfers
   d. all of the above

4. A feeding relationship that proceeds from algae to a fish, then to a fisherman, and then to a shark is _____.
   a. a food chain   c. a and b
   b. a food web

5. Primary productivity is affected by _____.
   a. photosynthesis and respiration by plants
   b. how many plants are neither eaten nor decomposed
   c. rainfall
   d. temperatures
   e. all of the above

6. Match the ecosystem terms with the suitable description.
   ____ primary producers   a. herbivores, carnivores, omnivores, parasites
   ____ consumers   b. feed on partly decomposed organic particles
   ____ decomposers
   ____ detritivores   c. break down remains or products of other organisms
      d. autotrophs

7. Match the ecosystem terms with the suitable description.
   ____ nitrogen availability   a. movement of water or nutrients from the environment, to organisms, and back
   ____ ecosystem components
   ____ phosphorus   b. form of nitrogen plants can take up
   ____ ammonium   c. key limiting factor for net primary production in land ecosystems
   ____ biogeochemical cycle   d. producers, consumers, detritivores, decomposers
      e. moves through a sedimentary cycle

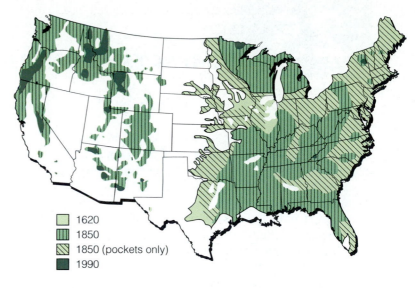

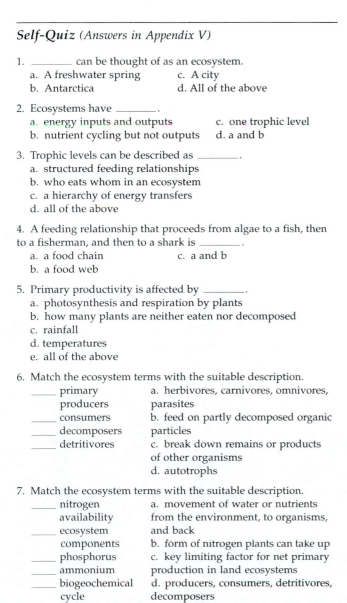

▢	1620
▨	1850
▨	1850 (pockets only)
■	1990

Figure 24.23 Map showing the extent of deforestation in the United States from 1620 through 1990.

Critical Thinking: You Decide *(Key in Appendix VI)*

1. Imagine and describe an extreme situation whereby you would be a participant (not the apex predator) in a food chain.

2. In 1995 biologists Reed Noss, J. Michael Scott, and Edward LaRoe issued a report on endangered ecosystems in the United States, including the once-vast forests (Figure 24.23). The report has prompted researchers and policymakers to ask whether we as a nation should protect entire ecosystems, rather than just endangered species. One objection is that doing so might endanger property values. Would you favor setting aside large tracts of land for ecological restoration? Or would you consider such efforts too intrusive on individual rights of property ownership?

Selected Key Terms

ammonification *24.9*	global warming *24.8*
autotrophs *24.2*	grazing food web *24.3*
biogeochemical cycle *24.5*	greenhouse effect *24.8*
biological magnification *24.11*	habitat *24.1*
biosphere *24.1*	heterotroph *24.2*
carbon cycle *24.7*	hydrologic cycle *24.6*
community *24.1*	niche *24.1*
consumer *24.2*	nitrification *24.9*
decomposer *24.2*	nitrogen cycle *24.9*
denitrification *24.9*	nitrogen fixation *24.9*
detrital food web *24.3*	phosphorus cycle *24.10*
detritivore *24.2*	primary productivity *24.3*
ecological pyramid *24.3*	producer *24.2*
ecology *24.1*	succession *24.1*
ecosystem *24.1*	trophic level *24.2*
food chain *24.2*	watershed *24.6*
food web *24.2*	

Readings

Krebs, C. 1994. *Ecology*. 4th ed. New York: HarperCollins.

25

IMPACTS OF THE HUMAN POPULATION

Tales of Nightmare Numbers

When the 21st century began, there were more than 6 billion people on Earth, about 2 billion of them living in grinding poverty. Both numbers grow with every passing day. Each year, 40 million more humans join the ranks of the starving.

Next to China, India is the most populous country, with roughly 1 billion inhabitants (Figure 25.1). By 2010, India's population may reach 1.82 billion. Forty percent of those people live in rat-infested shantytowns, without enough food or water. They are forced to wash their clothes and dishes in open sewers. Land available to raise their food shrinks by 365 acres a day, on average. Why? Irrigated soil becomes too salty when it drains poorly, and India does not get enough rain to flush away the salts.

Recognizing that India's huge and rapidly growing population creates social problems and severely strains national resources, for decades the Indian government has sponsored family-planning programs. With what results? To begin with, administering such programs has been difficult, because many millions of people live in remote villages. Because of widespread illiteracy, information often must be conveyed by word of mouth. And because disease kills many children, villagers often resist limiting family size. Without a large family, they ask, who would survive and help tend fields? Who will go to the cities and earn money to send back home? Who will care for aging parents when they are no longer able to work?

Can wealthier, less densely populated nations help? After all, they use most of the world's resources. Maybe they should learn to get by more efficiently, on less. For example, people might limit their meals to cereal grains and water; give up their private cars, living quarters,

a

Figure 25.1 (**a**) This crowded scene in India shows a sampling of the more than 6 billion humans on Earth. In this chapter we turn to the principles governing the growth and sustainability of populations, including our own. (**b**, *right*) Billboard in China promoting family planning.

air conditioners, televisions, computers, and dishwashers; stop taking vacations and stop laundering so often; close all the malls, restaurants, and theaters at night; and so on.

Perhaps wealthier nations also should donate more surplus food than they already donate to less fortunate ones. Then again, would such donations help, or would they encourage dependency and spur more increases in population size? And what if surpluses run out?

It is a monumental dilemma. At one extreme, the redistribution of resources on a global scale would allow the greatest number of people to survive, but at the lowest comfort level—and would demand sacrifices that many people, and many nations, may be unwilling to make. (How many of the "luxuries" listed above are you willing to give up?) At the other extreme, limiting foreign aid to nations that restrict population growth would allow fewer human beings to be born and improve the quality of life. However, you probably already can envision the kinds of political, social, and religious controversies such an approach might provoke.

The issues we have been describing are real, and will have to be dealt with—probably sooner than many would like to think. And regardless of the positions that nations ultimately take with regard to the growth and ecological impacts of Earth's human populations, we can be sure of one fact. For humans and all other organisms, *ecological principles govern the growth and sustainability of populations over time*. This chapter describes these principles, then shows how they apply to the past, present, and future of the human species.

b

KEY CONCEPTS

1. Certain ecological principles govern the growth and sustainability of populations over time, including Earth's human population.

2. All populations face limits to growth, because no environment can indefinitely sustain a continuously increasing number of individuals.

3. Population growth generally follows certain patterns. When a population grows exponentially, it increases in size by ever larger amounts per unit of time.

4. Now and in the foreseeable future, we have attained the population size, the technology, and the cultural inclination to use energy and modify our collective environment at astonishing rates.

5. The accumulation of pollutants generated by human activity is disrupting complex interactions among the atmosphere, oceans, and land in ways that may have the most serious kinds of consequences in the near future.

CHAPTER AT A GLANCE

CHARACTERISTICS OF POPULATIONS

You may remember from Chapter 23 that the biological definition of a population is a group of individuals of the same species occupying a given area. Ecologists gather many kinds of information in their study of populations, including genetic data. Our focus here is on the insights provided by **demographics**—that is, the vital statistics of a population, such as its size, density, distribution, and age structure.

Population size is the number of individuals in the population's gene pool. **Population density** is the total number of individuals in some specified volume or area of habitat, such as the number of minnows in each liter of water in a small stream or the number of people who live within a hectare of land. **Population distribution** is the general pattern in which the population's members are distributed through its habitat. We humans are social animals. Our populations tend to be clumped in villages, towns, and cities, where we interact with one another and have access to jobs and other resources.

A population's **age structure** is the relative proportion of individuals of each age. These are often divided into prereproductive, reproductive, and postreproductive age categories. Individuals in the first category have the capacity to produce offspring when they reach sexual maturity. Together with the actually and potentially reproducing individuals in the second category, they help make up the population's **reproductive base**. Figure 25.2 graphs the age structure diagrams for human populations growing at different rates. In these graphs, ages 15 to 44 are the average range of childbearing years (for both sexes). The population of the United States has a narrow base and is an example of slow growth. Figure 25.3 tracks its 78 million *baby boomers*, a group that formed in 1946 after soldiers returned home from World War II and many couples settled down to start families. By contrast, age structure diagrams for a rapidly growing population have a broad base.

A country's prereproductive base—the proportion of individuals younger than 15—is the most telling statistic,

Figure 25.2 (a) General age structure diagrams for countries with rapid, slow, zero, and negative population growth rates. Prereproductive years are *green* bars, reproductive years *purple*, and post-reproductive years *light blue*. The vertical axes divide each graph into males (*left*) and females (*right*). Bar widths correspond to the proportion of individuals in each age group. (b) 1997 age-structure diagrams for representative countries. Population sizes are measured in millions.

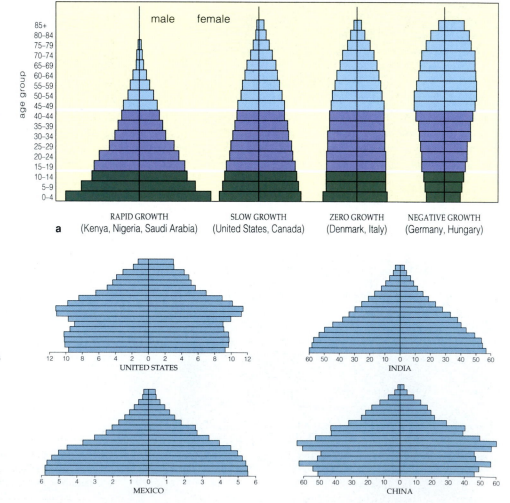

because it reflects both recent population growth *and* potential future growth when those individuals mature. Mexico's prereproductive base is broad, for example; not only is that nation's population growing rapidly now, but we can also expect such growth to continue.

Population Size and Patterns of Growth

Over a given time period, a population's size depends on two factors: how many people enter it by birth or immigration and how many leave it by death or emigration:

$$\begin{pmatrix} \text{amount of} \\ \text{population} \\ \text{growth} \end{pmatrix} = \begin{pmatrix} \text{births} \\ + \\ \text{immigration} \end{pmatrix} - \begin{pmatrix} \text{deaths} \\ + \\ \text{emigration} \end{pmatrix}$$

If immigration and emigration are balanced, then population size is stable when the birth rate is balanced over the long term by the death rate. When this balance exists, there is **zero population growth**.

In reality, immigration may contribute heavily to a nation's population growth. In the United States, for example, immigration (legal and otherwise) accounts for more than 30 percent of annual population growth.

Populations increase in size when the number of births exceeds the number of deaths. But how fast and how much can they increase in a given period? Imagine a population that in a given year has 1,000 members ($N = 1,000$). In each succeeding year, for each 1,000 members, 10 members die but 50 are born. The population's rate of increase (*r*) would be:

$$r = \frac{\text{births} - \text{deaths}}{N} = \frac{50 - 10}{1,0000} = 0.04 \text{ or } 4\% \text{ (here, 40) per year}$$

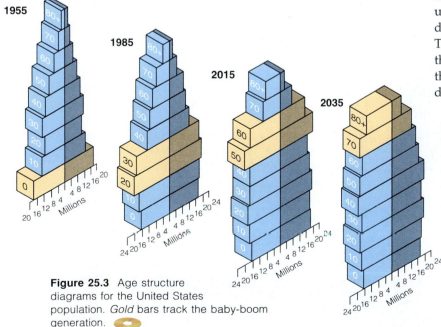

1955

1985

2015

2035

Figure 25.3 Age structure diagrams for the United States population. *Gold* bars track the baby-boom generation.

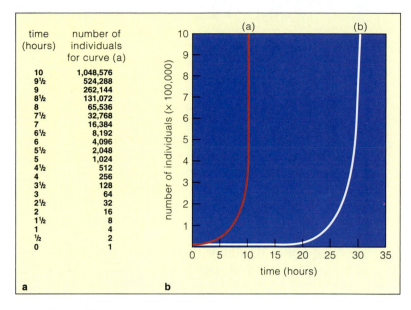

time (hours)	number of individuals for curve (a)
10	1,048,576
9½	524,288
9	262,144
8½	131,072
8	65,536
7½	32,768
7	16,384
6½	8,192
6	4,096
5½	2,048
5	1,024
4½	512
4	256
3½	128
3	64
2½	32
2	16
1½	8
1	4
½	2
0	1

Figure 25.4 (**a**) Exponential growth for a bacterial population that is dividing by fission every half hour. (**b**) Exponential growth of the population when division occurs every half hour but when 25 percent die between divisions. Deaths slow the rate of increase, but cannot stop exponential growth.

As long as *r* holds constant, any population will show **exponential growth**. The number of its individuals increases in *doubling* increments—from 2 to 4, then 8, 16, 32, 64, and so on. For a population with a 4 percent rate of increase, this doubling time is 17.5 years.

We can observe a limited period of exponential growth in the laboratory by putting a single bacterium into a culture flask with a supply of nutrients. After about 30 minutes the bacterium divides into two, and 30 minutes later the two divide into four. If no cells die between divisions, the population will double every 30 minutes. The larger the population becomes, the more bacteria there are to divide. After only 9½ hours (19 doublings), the population will exceed 500,000. After 10 hours (20 doublings) it will soar past 1 million!

In a population undergoing unrestricted, exponential growth, a J-shaped curve results when population size is plotted against time (Figure 25.4). Populations of this type increase in size by ever larger amounts per unit of time. When nothing stops its growth, such a population will grow exponentially, even when the birth rate only slightly exceeds the death rate. Only the time scale changes. Doubling still occurs—it simply takes longer.

> As long as the birth rate remains even slightly above the death rate, a population will grow. If the rates remain constant, it will grow exponentially.

A CLOSER LOOK AT GROWTH PATTERNS

Biotic Potential

The **biotic potential** of a population is its *maximum* rate of growth under ideal conditions. It is the rate that might be achieved when food and living space are abundant, when other organisms aren't interfering with access to those resources, and when no predators or disease agents are present.

Biotic potential is not the same for every species. For many bacteria, it is 100 percent every half hour; for humans and other large mammals, it is between 2 and 5 percent per year. The differences arise through variations in (1) how soon individuals start reproducing, (2) how often reproduction occurs, and (3) how many offspring are born each time.

A population may grow exponentially even when it is not expressing its full biotic potential. It is biologically possible for human females to bear 20 children or more, although few have done so. Yet the human population has been growing exponentially since the mid-18th century. At any given time, the *actual* rate of increase is influenced by environmental circumstances affecting human society.

Limiting Factors and Carrying Capacity

Most often, environmental circumstances prevent a population from reaching its full biotic potential. For instance, when an essential resource such as food or water is in short supply, it becomes a **limiting factor** on population growth. Predation (as by disease organisms), competition for living space, and buildup of toxins are other examples of limiting factors. The number of such factors can be enormous, and their effects can vary.

The concept of limiting factors is important because it defines the **carrying capacity**—the number of individuals of a given species that can be sustained indefinitely by the resources in a given area. Some experts believe that Earth has the resources to support from 7 to 12 billion humans, with a reasonable standard of living for many. Others believe that the current human population of 5.7 billion is already exceeding its carrying capacity through soil erosion, deforestation, pollution of groundwater supplies, and changes in global climate. All these viewpoints share the premise that *overpopulation is the root of many, if not most, of the environmental problems the world now faces.*

A low-density population starts growing slowly, then goes through a rapid growth phase, and then growth levels off once the carrying capacity is reached. This pattern is called **logistic growth**. A plot of logistic growth gives us an S-shaped curve (Figure 25.5). This curve is only a simple approximation of what goes on in nature, however. Because environmental conditions vary, carrying capacity also can vary over time.

Checks on Population Growth

When a growing population's density increases, high density and overcrowding result in competition for resources. They also put individuals at increased risk of being killed by infectious diseases and parasites, which are more easily spread in crowded living conditions. These are **density-dependent controls** on population growth. Once such factors take their toll on a population and its density decreases, the pressures ease and the population may grow once more.

A classic example is the *bubonic plague* that killed 25 million Europeans—about one third of the population—during the 14th century. *Yersinia pestis*, the bacterium responsible, normally lives in wild rodents; fleas transmit it to new hosts. It spread like wildfire through the cities of medieval Europe because human dwellings were crowded together, sanitary conditions were poor, and rats were abundant. In 1994, bubonic plague and a related disease, pneumonic plague, raced through rat-infested cities in India where garbage and animal carcasses had piled up for months in the streets. Some fleeing residents carried the diseases as far away as London, England. Only concerted efforts by public health officials averted a pandemic.

Density-independent controls can also operate. These are events such as floods, earthquakes, or other natural disasters that cause deaths or births regardless of whether the members of a population are crowded or not.

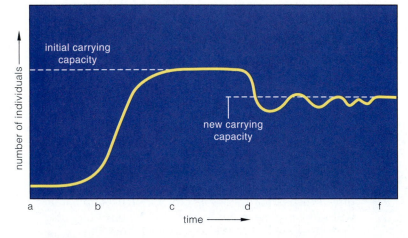

Figure 25.5 Idealized S-shaped curve characteristic of logistic growth. Following a rapid growth phase, growth slows and the curve flattens out as the carrying capacity is reached. Variations can occur in S-shaped growth curves, as when changed environmental conditions bring about a decrease in the carrying capacity. This happened to the human population in Ireland in the late 19th century, when a fungus wiped out the potatoes that were the mainstay of the diet.

Table 25.1 Life Table for the U.S. Human Population, 1989*

Age Interval (category for individuals between the two ages listed)	Survivorship (number alive at start of age interval, per 100,000 individuals)	Mortality (number dying during the age interval)	Life Expectancy (average lifetime remaining at start of age interval)	Number of Reported Live Births for Total U.S. Population
0–1	100,000	986	75.3	
1–5	99,104	192	75.0	
5–10	98,912	117	71.1	
10–15	98,975	132	66.2	11,486
15–20	98,663	429	61.3	506,503
20–25	98,234	551	56.6	1,077,598
25–30	97,683	606	51.9	1,263,098
30–35	97,077	737	47.2	842,395
35–40	96,340	936	42.5	253,878
40–45	95,404	1,220	37.9	44,401
45–50	94,184	1,766	33.4	1,599
50–55	92,148	2,727	28.9	
55–60	89,691	4,234	24.7	
60–65	85,357	6,211	20.8	
65–70	79,146	8,477	17.2	
70–75	70,669	11,470	13.9	
75–80	59,199	14,598	10.9	
80–85	44,601	17,448	8.3	
85 +	27,153	27,153	6.2	
				Total: 4,040,958

*Compiled by Marion Hansen, based on data from U.S. Bureau of the Census, *Statistical Abstract of the United States, 1992* (edition 112).

Life History Patterns

Like all species, humans have a characteristic life span, although few people reach the maximum age possible. Death is more probable at some ages and less so at others. Also, individuals are more likely to reproduce at some ages than at others, and these ages vary from one species to the next. The study of such age-specific patterns is the subject of *demography*.

A **life table** (Table 25.1) is a summary of the age-specific patterns of birth and death for a given population in a given area. Originally developed by insurance companies, such tables are used to help set the price of life or health insurance for people of different ages.

The biotic potential of a population is its maximum rate of increase under ideal conditions. For any population, the actual rate of increase is influenced by environmental circumstances.

Carrying capacity is the number of individuals of a species that can be sustained indefinitely by resources in a given area.

Factors that limit human population growth include availability of various types of resources, predation by disease organisms, and effects of pollution.

Because these factors vary in their effects over time, both the carrying capacity and the population size can fluctuate.

HUMAN POPULATION GROWTH

In 2000, the human population totaled 6 billion (Figure 25.6). In the previous year, roughly 100 million babies were born—an average of 1.9 million per week, 273,000 per day. This growth is the consequence of advances in agriculture, industrialization, sanitation, and health care. It took *2 million years* for the human population to reach the first billion. It took only 130 years to reach the second billion, 30 years to reach the third, 15 years to reach the fourth, *and only 12 years to reach the fifth.*

The **demographic transition model** (Figure 25.7) links changes in population growth with changes that unfold during four stages of economic development. In the first, *preindustrial stage,* there is little population growth because living conditions are harsh, and birth and death rates are both high. In the *transitional stage* industrialization begins, food production rises, and sanitation and health care improve. Death rates drop, but birth rates remain high, so the population grows rapidly over a long period. Then growth starts to level off as living conditions improve.

In the *industrial stage*—when industrialization is in full swing—population growth slows, mostly because urban couples regulate the size of their families. Many decide that raising children is expensive and that having too many puts them at an economic disadvantage. In the *postindustrial stage,* zero population growth is reached. Then the death rate exceeds the birth rate. When that happens, the population size slowly decreases.

Today, the United States, Canada, Australia, Japan, New Zealand, and most countries of Europe are in the industrial stage, and their growth is slowly decreasing. Eighteen countries, including Sweden, the United Kingdom, Germany, and Hungary, are close to, at, or slightly below zero population growth.

Developing countries, such as Mexico, are in the transitional stage. Many may not stay there, however. In many countries in this stage, population growth outpaces economic growth, and it is difficult if not impossible for such nations to afford the costs of fossil fuels and other resources that drive industrialization. Some transitional countries and their populations may return to the harsh conditions of the preceding stage.

Figure 25.8*a* shows the average annual growth rate for different parts of the world in 1997. The average rate of 1.47 percent is expected to decline. Even so, the world population is still projected to reach more than 9 billion by 2050 (Figure 25.8*b*). It may be impossible to achieve corresponding increases in food production, drinkable

Figure 25.6 Growth curve (*red*) of the human population. The diagram's vertical axis represents world population, in billions. (The slight dip between the years 1347 and 1351 is the time when 25 million people died in Europe as a result of bubonic plague.) The growth pattern over the past two centuries has been exponential, sustained by agricultural revolutions, industrialization, and improvements in health care. The list in the *blue* box tells how long it took for the human population to double in size at different times in history.

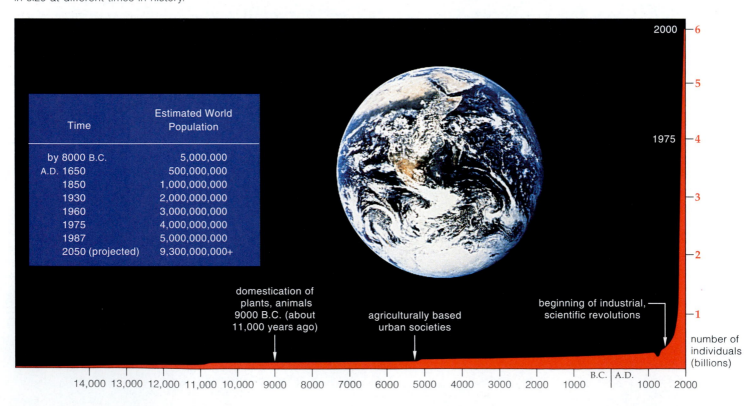

Time	Estimated World Population
by 8000 B.C.	5,000,000
A.D. 1650	500,000,000
1850	1,000,000,000
1930	2,000,000,000
1960	3,000,000,000
1975	4,000,000,000
1987	5,000,000,000
2050 (projected)	9,300,000,000+

domestication of plants, animals 9000 B.C. (about 11,000 years ago)

agriculturally based urban societies

beginning of industrial, scientific revolutions

number of individuals (billions)

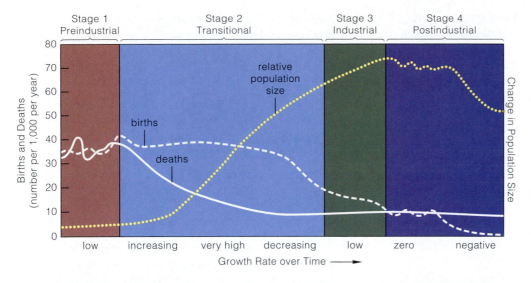

Figure 25.7 The demographic transition model of changes in the growth characteristics and size of populations, correlated with changing economic development.

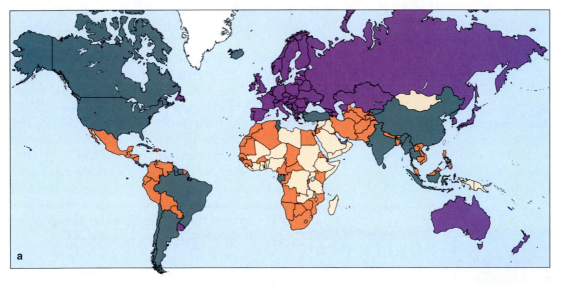

a

Figure 25.8 (a) The average annual population growth rate in different regions of the world. (b) Population sizes in 1996 (*orange* bars) and projected for 2025 (*blue* bars).

water, energy reserves, and all the wood, steel, and other materials we use to meet everyone's basic needs—something we are not doing even now. As we see in the next section, there is evidence that harmful by-products of our activities—pollutants—are changing the land, the seas, and the atmosphere in ominous ways. From what we know of the principles governing population growth, unless there are spectacular technological breakthroughs, it is realistic to expect an increase in human death rates. Although at this writing our stupendously accelerated growth continues, it cannot be sustained indefinitely.

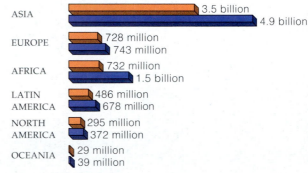

b

	1996	2025
ASIA	3.5 billion	4.9 billion
EUROPE	728 million	743 million
AFRICA	732 million	1.5 billion
LATIN AMERICA	486 million	678 million
NORTH AMERICA	295 million	372 million
OCEANIA	29 million	39 million

Differences in population growth among countries correlate with economic development. By 2025, the human population will reach a level that will severely strain resources.

AIR POLLUTION

Let's start this survey of the impact of human activities by defining pollution, which is central to our problems. **Pollutants** are substances that adversely affect the health, activities, or survival of a population. Air pollutants are prime examples. As you can see in Table 25.2, they include carbon dioxide, oxides of nitrogen and sulfur, and chlorofluorocarbons (CFCs). Among them also are photochemical oxidants formed as the sun's rays interact with certain chemicals. The United States alone releases more than 700,000 metric tons of air pollutants every day. Whether these substances remain concentrated at the source or dispersed during a given interval depends on the local climate and topography, as you will now see.

Smog

During a **thermal inversion**, weather conditions trap a layer of cool, dense air under a layer of warm air. If the trapped air contains pollutants, winds cannot disperse them, and they may accumulate to dangerous levels. Thermal inversions have been key factors in some of the worst air pollution disasters, because they intensify an atmospheric condition called smog (Figure 25.9).

Two types of smog form in major cities. Where winters are cold and wet, **industrial smog** forms as a gray haze over industrialized cities that burn coal and other fossil fuels for manufacturing, heating, and generating electric power. The burning releases airborne dust, smoke, ashes, soot, asbestos, oil, bits of lead and other heavy

Table 25.2   Major Classes of Air Pollutants

Carbon oxides	Carbon monoxide (CO), carbon dioxide (CO_2)
Sulfur oxides	Sulfur dioxide (SO_2), sulfur trioxide (SO_3)
Nitrogen oxides	Nitric oxide (NO), nitrogen dioxide (NO_2), nitrous oxide (N_2O)
Volatile organic compounds	Methane (CH_4), benzene (C_6H_6), chlorofluorocarbons (CFCs)
Photochemical oxidants	Ozone (O_3), peroxyacyl nitrates (PANs), hydrogen peroxide (H_2O_2)
Suspended particles	Solid particles (dust, soot, asbestos, lead, etc.), liquid droplets (sulfuric acid, oils, dioxins, pesticides)

metals, and sulfur oxides. Industrial smog caused 4,000 deaths in 1952 during an air pollution disaster in London. New York, Pittsburgh, and Chicago once were gray-air cities until they restricted coal burning. Now, most industrial smog forms in cities of developing countries, including China and India, as well as in countries of eastern Europe.

In warm climates, **photochemical smog** forms as a brown, smelly haze over large cities. It becomes concentrated when the surrounding land forms a natural basin, as in Los Angeles and Mexico City. The key culprit is nitric oxide. After it is released from vehicles, nitric oxide reacts with oxygen in the air to form nitrogen dioxide. When exposed to sunlight, nitrogen dioxide can react with hydrocarbons (such as spilled or partly burned gasoline) to form photochemical oxidants. The main oxidants in smog are ozone and PANs (peroxyacyl nitrates). PANs resemble tear gas; even traces can sting the eyes, irritate lungs, and damage crops.

Acid Deposition

Oxides of sulfur and nitrogen are among the worst air pollutants. Coal-burning power plants, factories, and metal smelters emit most sulfur dioxides. Motor vehicles, power plants that burn gas and oil, and nitrogen-rich fertilizers produce nitrogen oxides. In dry weather, fine particles of oxides may briefly stay airborne, then fall to Earth as **dry acid deposition**. When dissolved in atmospheric water, they form weak solutions of sulfuric and nitric acids. Winds may disperse them over great distances. If they fall to Earth in rain and snow, we call this wet acid deposition, or **acid rain.** Acid rain can be much more acidic than normal rainwater, sometimes becoming as acidic as lemon juice (pH 2.3). The deposited acids eat away at marble, metals, mortar, plastic, even nylon stockings. They also seriously disrupt the physiology of organisms and the chemistry of ecosystems.

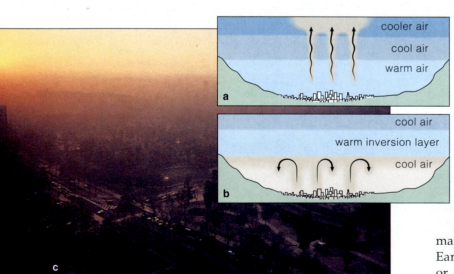

Figure 25.9 (**a**) Normal pattern of air circulation in smog-forming regions. (**b**) Air pollution trapped under a thermal inversion layer. (**c**) Mexico City on a sunny morning. Topography, huge numbers of people and vehicles, and industry combine to make its air among the world's smoggiest.

Further reading: Student Guide to InfoTrac on web site ➛

Figure 25.10 (**a**) Ice clouds above Antarctica that have a role in the ozone thinning each spring. Seasonal ozone thinning above Antarctica in (**b**) 1979, (**c**) 1991, and (**d**) 1996. Lowest ozone values are coded *magenta* and *purple*.

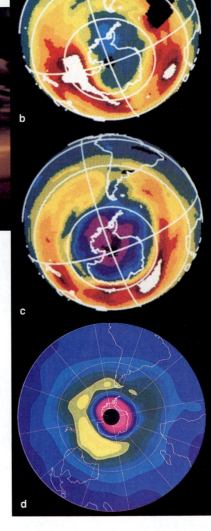

The precipitation in much of eastern North America is 30 to 40 times more acidic than it was several decades ago, and croplands and forests are suffering. Fish have vanished from hundreds of lakes in New York state. Acidic pollutants from the industrial regions of England and Germany are key factors in the destruction of large tracts of forests in northern Europe. They also are a serious problem in heavily industrialized parts of Asia, Latin America, and Africa.

Damage to the Ozone Layer

Ozone is a molecule of three oxygen atoms (O_3). It occurs in two regions of Earth's atmosphere. In the troposphere, the region closest to the Earth's surface, ozone is part of smog and can damage the respiratory system as well as other organisms. However, ozone in the *next* atmospheric layer—the stratosphere (17–48 kilometers, or 11–30 miles above Earth)—intercepts harmful ultraviolet radiation that can cause skin cancer and eye cataracts. As you may know, this protective screen has been thinning. Each year, from early September through mid-October, an ozone "hole" appears over the Antarctic, extending over an area about the size of the continental United States. Satellites and high-altitude planes have been monitoring the ozone hole (Figure 25.10). Since 1987 the ozone layer over the Antarctic has been thinning by about half every year; a new hole, over the Arctic, appeared in 2000. For several decades ozone has steadily decreased over some populated regions. By 2050 the ozone layer over Asia, North America, and Europe may thin by 10–25 percent.

As the ozone layer thins, more harmful ultraviolet radiation reaches Earth's surface. Already, the incidence of skin cancer is increasing dramatically. In addition, ultraviolet radiation can weaken the immune system, making people more vulnerable to infections. Reducing the ozone layer also may damage phytoplankton, microscopic photosynthetic organisms that are the basis of food webs in freshwater and marine ecosystems. Phytoplankton are also a factor in maintaining the composition of the atmosphere because they act to remove carbon dioxide from surface waters and release oxygen.

Chlorofluorocarbons (CFCs) are the chief culprits in ozone depletion. These compounds of chlorine, fluorine, and carbon have been used as propellants in aerosol sprays, as coolants in refrigerators and air conditioners, and for other industrial and commercial uses. They are also released during volcanic activity.

When a CFC in the stratosphere absorbs ultraviolet light, it gives up a chlorine atom. The chlorine can react with ozone (O_3), yielding oxygen (O_2) and chlorine monoxide. Reactions between oxygen and chlorine monoxide release more chlorine atoms—each one able to destroy up to 10,000 molecules of ozone!

Another major ozone eater is methyl bromide, a fungicide. It will account for about 15 percent of the thinning in future years unless production stops.

Under international agreement, CFC production has been phased out in the developed countries. Methyl bromide production is supposed to stop by 2010. Assuming that international goals are met, it will still be at least 50 years before the ozone layer is restored to 1985 levels and another 100 to 200 years to total recovery—that is, to pre-1950 levels. Meanwhile, you and all the children and grandchildren of future generations will be living with the destructive effects of air pollutants. And if levels of greenhouse gases continue to rise, the stratosphere would cool enough to increase the size and duration of seasonal ozone depletion over the poles.

A pollutant is a substance that adversely affects the health, activities, or survival of a population. Air pollution can have global repercussions, as when CFCs and other compounds contribute to a thinning of the ozone layer that shields life from the sun's ultraviolet radiation.

A GLOBAL WATER CRISIS

Three of every four humans do not have enough clean water to meet basic needs. Most of Earth's water is salty (in oceans). Of every million liters of water on our planet, only 6 liters are readily usable for human activities. As our population grows exponentially, so do demands and impacts on this limited water supply.

Impact of Large-Scale Irrigation

About a third of the world's food grows on irrigated land (Figure 25.11). Water is piped into agricultural fields from groundwater or diverted from lakes or rivers. Irrigation

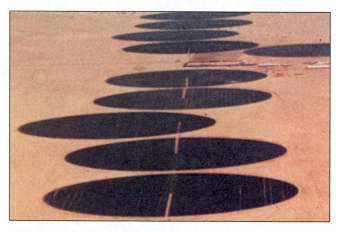

Figure 25.11 Irrigation-dependent crops growing in the Sahara Desert of Algeria.

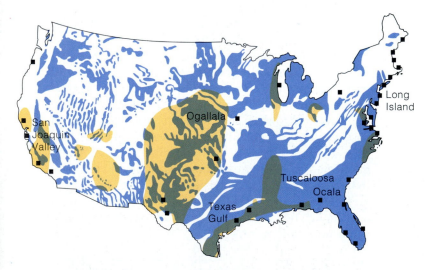

Figure 25.12 Major underground aquifers (shown in *blue*) containing 95 percent of all freshwater in the United States. These aquifers are being depleted in many areas (*gold*) and contaminated elsewhere through pollution and saltwater intrusion. The black boxes indicate areas of saltwater intrusion.

water often contains large amounts of mineral salts, and where soil drainage is poor, evaporation may cause salt buildup, or **salinization**. Soil salinization can decrease yields and eventually kill crops. Worldwide, salinization is estimated to have reduced yields on 25 percent of all irrigated cropland. Improperly drained irrigated lands also can become waterlogged, so that soil around plant roots becomes saturated with toxic saline water.

Large-scale irrigation is also a major factor in groundwater depletion. For instance, farmers from South Dakota to Texas withdraw so much water from the Ogallala aquifer (Figure 25.12) that the annual overdraft (the amount not replenished) nearly equals the yearly flow of the Colorado River! The already-low water tables in much of the region are dropping rapidly, and stream and underground spring flows are dwindling. *Subsidence—* the sinking of land when groundwater is withdrawn—is increasing. In coastal areas, overuse of groundwater can cause saltwater intrusion into human water supplies.

Problems with Water Quality

In many regions, agricultural runoff pollutes public water sources with sediments, pesticides, and fertilizers. Power plants and factories pollute water with chemicals (including known carcinogens), radioactive materials, and excess heat (thermal pollution). Such pollutants may accumulate in lakes, rivers, and bays before reaching their ultimate destination, the oceans. Contaminants from human activities have begun to turn up even in supposedly "pure" water in underground aquifers.

Many people view the oceans as convenient refuse dumps. Cities throughout the world dump untreated sewage, garbage, and other noxious debris into coastal waters. Cities along rivers and harbors maintain shipping channels by dredging the polluted muck and barging it out to sea. We do not yet know the full impact of such practices on fisheries from which humans obtain food.

The United States has facilities for treating the liquid wastes from about 70 percent of the population and 87,000 industries. The remaining wastes, mostly from suburban and rural populations, are treated in lagoons or septic tanks or discharged—untreated—directly into waterways. Studies show that water pollution intensifies as rivers flow toward the oceans and that water treatment often does not remove toxic wastes from industries upstream from a city. Do you know where your own water supply has been?

Worldwide, human water supplies are threatened by overuse and by pollution with agricultural and industrial wastes. Parts of the world's oceans also are becoming contaminated by wastes and toxins produced by human activities.

Further reading: Student Guide to InfoTrac on web site →

WHERE TO PUT SOLID WASTES? WHERE TO GROW FOOD?

Solid Wastes

Billions of metric tons of solid wastes are dumped, burned, and buried annually in the United States alone. This includes 50 billion nonreturnable cans and bottles. Paper products make up one-half of the total volume of solid wastes.

Paper is relatively easily recycled. In the Netherlands, more than half of the nation's wastepaper is recycled, and in Japan the figure is about the same. In the United States recycling efforts by individuals and businesses are improving. Simply recycling the nation's Sunday newspapers would save 500,000 trees per week. Using recycled paper products also reduces air pollution that results from paper manufacturing by 95 percent, lowers water pollution by 35 percent, and takes 30–50 percent less energy than making new paper.

In natural ecosystems, solid wastes are recycled, but we bury them in landfills or burn them in incinerators. Incinerators can add heavy metals and other toxic pollutants to the air and leave a highly toxic ash that must be disposed of safely. Land available and acceptable for landfills is scarce and becoming scarcer. All landfills eventually "leak," posing a threat to groundwater supplies. That is one reason why communities increasingly take the "not in my backyard" approach to landfills. More and more people now participate in recycling programs, but it seems that few want to take responsibility for the nonrecyclable garbage they generate.

A transition from a throwaway mentality to one based on recycling and reuse is affordable, technically feasible, and environmentally essential. Consumers can play a role by refusing to buy goods that are lavishly wrapped and boxed, packaged in indestructible containers, or designed for one-time use. Individuals can ask the local post office to turn off their daily flow of junk mail. They can also engage in curbside recycling (or, in rural areas, using local recycling centers), by which they presort recyclable wastes. Curbside recycling is more efficient than using huge "resource recovery centers." These require so much trash to turn a profit that owners may end up encouraging the throwaway mentality.

Converting Marginal Lands for Agriculture

Today, human population growth is forcing expansion of agriculture onto marginally productive land. Almost 21 percent of the Earth's land is now being used for agriculture. Another 28 percent may be suitable for cropland or grazing land, but its potential productivity is so low that conversion may not be worth the cost.

Scientists have made valiant efforts to improve crop production on existing land. Under the banner of the **green revolution**, research has been directed toward improving the varieties of crop plants for higher yields and exporting modern agricultural practices and equipment to developing countries. Unfortunately, the green revolution involves intensive, mechanized agriculture. It is based on massive inputs of fertilizers and pesticides and ample irrigation to sustain high-yield crops. It is based also on fossil fuel energy to drive farm machines. Crop yields *are* four times as high as from traditional methods. But the modern practices use up 100 times more energy and minerals. Also there are signs that limiting factors are coming into play to slow down further increases in crop yields.

Overgrazing of livestock on marginal lands is a prime cause of **desertification**—the conversion of grasslands, rain-fed cropland, or irrigated cropland to a less productive, desertlike state (Figure 25.13). It occurs as vegetation is removed, leaving few or no plant roots to prevent wind and water from eroding the soil away. The subsoil then bakes rock-hard, and plants can no longer grow there. Worldwide, about 9 million square kilometers have become desertified over the past 50 years, and the trend is continuing.

Figure 25.13 Desertification in the Sahel, a region of West Africa that forms a belt between the dry Sahara Desert and tropical forests. This savanna country is rapidly undergoing desertification as a result of overgrazing and overfarming.

DEFORESTATION—AN ASSAULT ON FINITE RESOURCES

The world's great forests profoundly influence the biosphere. Like giant sponges, forested watersheds absorb, hold, and gradually release water. By intervening in the downstream flow of water, forests help control soil erosion, flooding, and sediment buildup in rivers, lakes, and reservoirs.

Deforestation is the name for removal of all trees from large tracts of land for logging, agricultural, or grazing operations. The loss of vegetation exposes the soil, and this promotes leaching of nutrients and erosion, especially on steep slopes. Figures 25.14 and 25.15 provide close-up and panoramic views of deforestation in South America's Amazon basin.

In the tropics, clearing forests for agriculture leads to a long-term loss in productivity. The irony is that tropical forests are one of the worst places to grow crops and raise pasture animals. The high temperatures and heavy, frequent rainfall favor decomposition. In intact forests, organic remains and wastes decompose too fast for litter

to build up. The forest trees and other plants absorb and assimilate nutrients as they become available.

Shifting cultivation (once called slash-and-burn agriculture) disrupts the forest ecosystem. People cut and burn trees, then till the ashes into the soil. The nutrient-rich ashes sustain crops for one to several seasons. Then, as a result of leaching, the cleared plots become infertile and are abandoned. Shifting cultivation on small, widely scattered plots may not damage forest ecosystems much, but fertility plummets when large areas are cleared and when plots are cleared again at shorter intervals.

Shifting rates of evaporation, transpiration, and runoff may even disrupt regional patterns of rainfall. Between 50 and 80 percent of the water vapor above tropical forests alone is released from the trees. Without trees, annual precipitation declines. Rain rapidly runs off the bare soil. As the region gets hotter and drier, soil fertility and moisture decline even more. In time, sparse

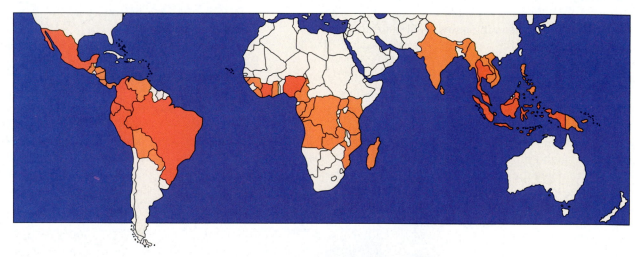

Figure 25.14 Countries permitting the greatest destruction of tropical forests. *Red* shading denotes where 2,000 to 14,800 square kilometers are deforested annually. *Orange* denotes "moderate" deforestation (100 to 1,900 square kilometers). The photograph shows the "slash-and-burn" clearing of a parcel of tropical forest.

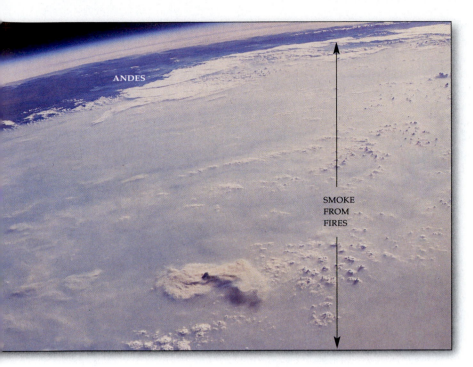

Figure 25.15 The vast Amazon River basin of South America in September 1988. Smoke from fires that had been deliberately set to clear tropical forests, pasturelands, and croplands completely obscured its features. The smoke extended to the Andes Mountains on the western horizon, about 650 miles (1,046 kilometers) away. That smoke cover was the largest that astronauts had ever observed, extending almost 175 million square kilometers (1,044,000 square miles).

The smoke plume near the center of the photograph covered an area comparable to that of the huge forest fire in Yellowstone National Park in that same year.

Massive deforestation is not confined to equatorial regions. For example, in the past century, 2 million acres of redwood forests along California's coast have been logged over. Most of the destruction of such temperate forests has been occurring since 1950, owing to the widespread use of chainsaws and tractors and the practice of exporting the logs to lumbermills overseas, where wages are low.

grassland or even desertlike conditions prevail instead of a rich tropical forest.

Widespread tropical forest destruction may have global repercussions. These forests absorb much of the sunlight reaching equatorial regions of the Earth's surface. When the forests are cleared, the land becomes shinier, so to speak, and reflects more incoming energy back into space. Also, by their photosynthetic activity, the great numbers of trees in these vast forest biomes help sustain the global cycling of carbon and oxygen. When trees are harvested or burned, carbon stored in their biomass is released to the atmosphere in the form of carbon dioxide—and this may be amplifying the greenhouse effect.

Figure 25.16 Wangari Maathai, a Kenyan who organized the internationally acclaimed Green Belt Movement in 1977. The 50,000 members of this women's group are committed to establishing nurseries, raising seedlings, and planting a tree for each of the 27 million Kenyans. Together with half a million schoolchildren, they had planted more than 10 million trees by 1990. Their success inspired similar programs in more than a dozen countries in Africa. Dr. Maathai's efforts are not appreciated by her own government. Kenyan police have jailed her twice, and in 1992 they severely beat her because of her efforts on behalf of the environment.

About half the world's tropical forests have been destroyed for cropland, grazing land, timber, and fuelwood. Deforestation is greatest in Brazil, Indonesia, Colombia, and Mexico. If clearing and destruction continue at present rates, only Brazil and Zaire will have large tropical forests in 2010. By 2035, most of their forests will be gone.

Conservation biologists are attempting to reverse the trend. In Brazil, a coalition of 500 groups is dedicated to preserving the country's remaining tropical forests. In India, women have already built and installed 300,000 inexpensive, smokeless wood stoves. Over the past decade, the stoves saved more than 182,000 metric tons' worth of trees by reducing demand for fuelwood. In Kenya, women have planted millions of trees to hold soil in place and to provide fuelwood (Figure 25.16).

Because forests profoundly influence the global cycling of carbon and oxygen, soil productivity, and the functioning of watersheds, large-scale deforestation can have serious repercussions on the biosphere.

Both tropical forests and temperate forests are rapidly being destroyed as trees are cut for timber and fuel and to clear land for agriculture and livestock grazing.

CONCERNS ABOUT ENERGY USE

Paralleling the J-shaped curve of human population growth in Section 25.3 is a steep rise in total and per capita energy consumption. The increase is due to increased numbers of energy users and to extravagant consumption and waste. For example, in one of the most pleasant of all climates, a major university constructed seven- and eight-story buildings with narrow, sealed windows. The windows can't be opened to catch prevailing ocean breezes. The buildings and windows were not designed or aligned to use sunlight for passive solar heating and breezes for passive cooling. Massive energy-demanding cooling and heating systems were installed.

When you hear talk of abundant energy supplies, keep in mind that there is a huge difference between the *total* and the *net* amounts available. Net energy is that left over after subtracting the energy used to locate, extract, transport, store, and deliver energy to consumers. Some sources, such as direct solar energy, are renewable. Others, such as coal and petroleum, are not (Figure 25.17).

Fossil Fuels

Fossil fuels are the legacy of forests that disappeared many hundreds of millions of years ago, so they are nonrenewable resources. Over time, the carbon-containing remains of the plants were buried in sediments, compacted, and chemically transformed into coal, petroleum (oil), and natural gas.

Even with strict conservation, known petroleum and natural gas reserves may be used up in this century. As known reserves run out in accessible areas, we explore wilderness areas in Alaska and other fragile environments, such as continental shelves. Net energy declines when the cost of extraction and transportation to and from remote areas increases. The environmental costs of extraction and transportation escalate. The widespread damage and clean-up costs following the 11-million-gallon spill from the supertanker *Valdez*, off Alaska's coast, are a classic example.

What about coal? In theory, world reserves can meet the energy needs of the human population for at least several centuries. But coal burning has been the main source of air pollution. Most coal reserves contain low-quality, high-sulfur material. Unless sulfur is removed before or after burning, sulfur dioxides are released into the air. They add to the global problem of acid deposition. Fossil fuel burning also releases carbon dioxide and adds to the greenhouse effect.

Extensive strip mining of coal reserves close to the Earth's surface carries its own problems. It removes land from agriculture, grazing, and wildlife. Restoration is difficult and expensive in arid and semiarid lands, where much of the strip mining is proceeding.

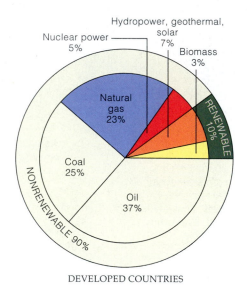

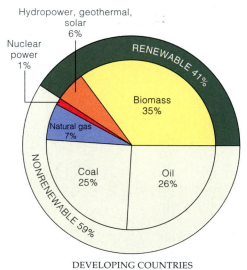

Figure 25.17 Energy consumption in the developed and developing countries, which differ greatly in sources of energy and average per capita energy use. The values indicated do not take into account energy from the sun used in agriculture.

Nuclear Energy

Today, nuclear power plants dot the landscape. At one time, proponents of nuclear power predicted that by 2000 it would meet the energy needs of 70 percent of the world's people. Although industrialized nations that are poor in energy resources now depend heavily on nuclear power, in most countries plans to develop more nuclear energy have been delayed or canceled. What happened? We started to question the operating cost, efficiency, safety record, and environmental impact of nuclear power.

By 1990 nuclear power was generating electricity at a cost only slightly above that of coal-burning plants. Putting aside other factors, its electricity-generating costs are now lower. However, soon solar energy with natural gas backup also should cost less.

What about safety? Compared to coal-burning plants of the same capacity, a nuclear plant emits less radioactivity and carbon dioxide, and no sulfur dioxide. However, there is greater danger associated with the potential for a **meltdown** if the reactor core becomes overheated. As happened at Chernobyl in Ukraine in 1986, the reactor core could melt through its thick concrete containment slab and into the Earth, contaminating groundwater and releasing lethal amounts of radiation. Thirty-one people died immediately, and several hundred others died later of radiation sickness. Inhabitants of entire villages were relocated; their former homes were bulldozed under. The fallout of radioactive material put

Figure 25.18 Harnessing solar energy. (**a**) Electricity-producing photovoltaic cells in panels that collect sunlight energy. (**b**) Wind turbines, which exploit the air circulation patterns that arise from latitudinal variations in the intensity of incoming sunlight.

an estimated 300 to 400 million people throughout Europe at increased risk of leukemia and other radiation-induced disorders. Throughout Europe, hundreds of millions of dollars were lost when the fallout made crops and livestock unfit for consumption. By 1998, the rate of thyroid abnormalities in children living immediately downwind of Chernobyl was nearly seven times as high as for those living upwind; their thyroid glands had collected iodine radioisotopes from the fallout.

Unlike coal, nuclear fuel cannot be burned to harmless ashes. The fuel elements are spent after about three years, but they still contain uranium fuel and hundreds of new radioisotopes formed during the reactions. The wastes are extremely radioactive and dangerous. As they undergo radioactive decay, they become extremely hot. They are plunged at once into cooling pools, which also keep radioactive material from escaping. After several months, the remaining isotopes are still lethal. Some must be isolated for at least 10,000 years. If one kind of plutonium isotope (239^{Pu}) is not removed, the wastes must be kept isolated for a quarter of a million years!

After nearly 50 years of research, scientists still do not agree on the best way to store high-level radioactive wastes. Even if they could do so, there is no politically acceptable solution. No one wants radioactive wastes anywhere near where they live.

Alternative Energy Sources

Various alternative energy sources are on the drawing boards. For instance, we might develop low-cost solar cells that can harness sunlight energy, then use it to generate the electricity required to produce hydrogen gas (Figure 25.18a). **Solar-hydrogen energy** would assure the human population of an affordable, renewable energy source, and the cleansing of the environment could begin in earnest. Such energy is stored efficiently for as long as required. It costs less to distribute hydrogen (H_2) than electricity, and water is the only by-product of using it. Space satellites run on it. Here on Earth, fossil fuels are still technically cheaper—although the calculation does not take their environmental impacts into account.

As you know, solar energy also is converted into the mechanical energy of winds. Where prevailing winds travel faster than 7.5 meters per second, we find **wind farms** (Figure 25.18b). These arrays of cost-effective turbines exploit wind patterns that arise from variations (by latitude) in the intensity of incoming sunlight. One percent of California's electricity comes from wind farms. Possibly the winds of North and South Dakota alone can meet all but 20 percent of the current energy needs of the United States. Wind energy also has potential for islands and other remote areas far from utility grids. One drawback: Winds don't blow on a regular schedule, so they can't be an exclusive or major energy source.

Fusion power is another alternative. The idea is to fuse hydrogen atoms to form helium atoms, a reaction that releases considerable energy. Researchers confine a certain fuel (a heated gas of two isotopes of hydrogen) in magnetic fields, then bombard it with lasers. The fuel implodes, it is greatly compressed to extremely high densities—and energy is released. The more "high-powered" the lasers, the greater the compression, and the more the fuel will burn.

The bad news is, although the amount of energy released by fusion reactors has been steadily increasing, it will be at least fifty years before fusion reactors might be operating, and the costs will probably be high. The good news is, that's about the time fossil fuels will start running out. For now and the near future, however, we must all reduce our energy consumption by using more energy-efficient vehicles, architecture, heating and cooling systems, and appliances.

Increasing demand for safe, renewable, cost-efficient energy supplies is focusing attention on alternatives to fossil fuels and nuclear power.

25.9 A PLANETARY EMERGENCY: LOSS OF BIODIVERSITY

On March 24, 1900, in Ohio, a young boy shot the last known wild passenger pigeon. More recently, fisheries biologist John Musick of the Virginia Institute of Marine Science reported his latest findings on the decline in populations of sharks (Figure 25.19). Over the past 30 years, Musick has tracked the numbers of shark species that spend much of their lives off the mid-Atlantic coast of the United States and have become popular as human food. Musick's numbers show that, like codfish in parts of the North Atlantic, today several species of sharks have all but disappeared. Based on similar evidence from many scientists, the National Marine Fisheries Service has imposed catch restrictions on commercial shark fishing operations, over the strong protests of fleet operators. Since then, additional fish species have been recognized as threatened, including numerous kinds of salmon, the bluefin tuna, and others.

Sooner or later all species become extinct, but humans have become a primary factor in the premature extinction of more and more of them. Already our actions are leading to the premature extinction of at least six species per hour.

Many biologists consider this epidemic of extinction an even more serious problem than depletion of stratospheric ozone or global warming because it is happening faster and is irreversible. Such rapid extinction cannot be balanced by speciation because it takes many thousands of generations for new species to evolve.

Globally, tropical deforestation is the greatest killer of species, followed by destruction of coral reefs. However, plant extinctions can be more important ecologically than animal extinctions since most animal species depend directly or indirectly on plants for food, and often for shelter as well. Humans also have traditionally depended on plants as sources of medicines. Indeed, 40 percent of prescription drug sales in the United States involve natural

Figure 25.20 The rosy periwinkle found in the threatened tropical forests of Madagascar. The plant is a source of two anticancer drugs.

plant products. One example is the rosy periwinkle of Madagascar (Figure 25.20). This plant is the source of the anticancer drugs vincristine and vinblastine. As it happens, humans have destroyed 90 percent of the vegetation on Madagascar, so we may never know if it was home to plant sources of other life-saving drugs as well. Meanwhile, in the United States we are going about the business of habitat destruction at a dizzying pace. Among other things, we have cut down 92 percent of old-growth forests and drained half the wetlands, which filter human water supplies and provide homes for waterfowl and juvenile fishes. At least 500 species native to these areas have been driven to extinction, and dozens more are endangered.

The underlying causes of wildlife extinction are human population growth and economic policies that fail to value the environment. Instead, they promote unsustainable exploitation. As our population grows we clear, occupy, and damage more land to supply food, fuel, timber, and other resources. In wealthy countries, affluence leads to greater than average resource use per person. In less affluent nations, the combination of rapid population growth and poverty push the poor to cut forests, grow crops on marginal land, overgraze grasslands, and poach endangered animals.

How can we help stem the tide of extinctions? Among other strategies, individuals can support efforts to reduce deforestation, projected global warming, ozone depletion, and poverty—the greatest threats to Earth's wildlife *and* to the human species.

Figure 25.19 A shark in its natural habitat. Worldwide, sharks are increasingly at risk of overexploitation by human enterprises.

THE IDEA OF SUSTAINABLE LIVING

You may have read or heard the expressions "sustainable agriculture" or "sustainable development." A **sustainable society** is one that manages its population growth and economic activities in ways that prevent serious damage to the environment. Every life form must obtain energy and other resources from its environment in order to grow and reproduce. The key to sustainable living is simple: We must not take or use resources in a manner or amount that irreparably harms ecosystems—or causes them to collapse altogether.

Elsewhere in this chapter you have read of human activities that in the long run are unsustainable. Our heavy dependence on fossil fuels is an obvious one, because we know that the Earth has a limited supply of fossil fuels and eventually the supply will run out. Most likely, we also cannot long continue widespread deforestation, overfishing of the oceans, and pollution of the atmosphere with greenhouse gases and ozone-destroying chemicals (among other activities). Environmental experts disagree about the precise level of degradation the Earth's land and water ecosystems and its atmosphere can sustain. Most agree, however, that without significant changes in human behavior and technologies, our planet will become unfit for human habitation.

The large and growing numbers of humans and our disregard for the limits of ecosystems have put our species at grave risk. The final straw could be a prolonged global food shortage, irrevocable poisoning of water supplies, passing a critical threshold for ozone depletion, or some other nightmarish disaster. However, one reason we include a chapter such as this in a human biology text is that we firmly believe that *it is not too late*. Humans have the capacity to anticipate events *before* they happen. We are not locked into responding only after irreversible change has begun. We all have the capacity for adapting to a future that we can partly shape. We can, for example, learn to live with less. Far from being a return to primitive simplicity, this could be one of the most intelligent behaviors our species ever engaged in.

It is probably not too extreme to say that our species' survival depends on coming up with some alternatives for the future (Figure 25.21). It depends on designing and constructing human ecosystems that are in harmony not only with what we define as basic human values, but also with biological reality. Our values and expectations must be adapted to the altered world in which we now live, because the principles of energy flow and resource use that govern the survival of all systems of life do not change.

In the final analysis, we must come to terms with these principles and ask ourselves what will be our long-term contribution to the world of life.

Figure 25.21 A biogas digester in a village in India. An example of an *appropriate technology*, this device produces methane gas. Its raw material is animal dung mixed with anaerobic bacteria. The microorganisms derive nutrients from the dung and produce methane gas as a by-product. The methane can be used as a household fuel. In addition to more than 750,000 of the devices operating in India, another 6 million provide usable fuel in rural areas of China. Farmers use solid residues of the process as fertilizer on food crops or trees.

SUMMARY

1. The growth rate of a population depends on the birth rate, the death rate, and the rates of immigration and emigration. If the birth rate exceeds the death rate by a constant amount, the population will grow exponentially (assuming immigration and emigration remain zero).

2. Carrying capacity is the number of individuals of a species that can be sustained indefinitely by the available resources in a given area.

3. Population size is determined by carrying capacity, competition, and other factors that limit population growth. Limiting factors vary in their relative effects and over time, so population size also can change over time.

4. Some limiting factors, such as competition for resources, disease, and predation, are density-dependent.

5. Currently, human population growth varies from zero in some of the more developed countries to more than 4 percent per year in some of the less developed countries.

6. Rapid growth of the human population during the past two centuries was possible largely because of our capacity to expand into new environments and because of agricultural and technological developments that increased the carrying capacity.

7. Accompanying the exponential growth of the human population are increased energy demands and environmental pollution.

8. Many pollutants are an outcome of human activities, and they adversely affect the health, activities, or survival of human populations.

9. Industrial smog and photochemical smog are examples of local air pollution. Acid rain, thinning of the ozone layer, and possible enhancement of the greenhouse effect are examples of global air pollution.

10. Global freshwater supplies are limited, yet they are being polluted by agricultural runoff (which includes sediments as well as pesticides and fertilizers), industrial wastes, and human sewage.

11. Human populations are damaging land surfaces by accumulation of vast amounts of solid wastes and by conversion of marginal lands for agriculture. Widespread desertification and the destruction of tropical and temperate forests may be altering regional soils and patterns of rainfall.

12. Energy supplies in the form of fossil fuels are non-renewable, dwindling, and environmentally costly to extract and use. Nuclear energy in itself is less polluting, but the costs and risks associated with fuel containment and with storing radioactive wastes are enormous. The challenge is to develop affordable alternatives based on renewable resources.

Review Questions

1. Why do populations that are not restricted in some way tend to grow exponentially? *25.1*

2. If the birth rate equals the death rate, what happens to the growth rate of a population? If the birth rate remains slightly higher than the death rate, what happens? *25.1*

3. At current growth rates, how many years will elapse before another billion people are added to the human population? *25.3*

4. Write a short essay about a hypothetical population that shows one of the following age structures. What might happen to younger and older age groups when members move into new categories? *25.1*

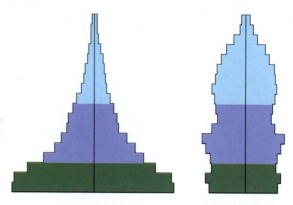

5. If a third of the world population is now below age 15, what effect will this age distribution have on the growth rate of the human population? What sorts of humane recommendations would you make that would encourage this age group to limit the number of children they plan to have? *25.1*

Self-Quiz *(Answers in Appendix V)*

1. _____ is the study of how organisms interact with one another as well as with their physical and chemical environment.

2. The rate at which a population grows or declines depends on the _____.
   a. birth rate
   b. death rate
   c. immigration rate
   d. emigration rate
   e. all of the above

3. Populations grow exponentially when _____.
   a. birth rate remains above death rate and neither changes
   b. death rate remains above birth rate
   c. immigration and emigration rates are equal (a zero value)
   d. emigration rates exceed immigration rates
   e. both a and c combined are correct

4. The number of individuals of a species that can be sustained indefinitely by the resources in a given region is the _____.
   a. biotic potential
   b. carrying capacity
   c. environmental resistance
   d. density control

5. Which of the following factors does *not* affect sustainable population size?
   a. predation
   b. competition

Figure 25.22 City as ecosystem.

c. available resources
d. pollution
e. each of the above can affect population size

6. Population growth controls such as resource competition and disease are said to be _____.
   a. density independent
   b. population sustaining
   c. population dynamics
   d. density dependent

7. During the past two centuries, rapid growth of the human population has occurred largely because of _____.
   a. increased birth rate worldwide
   b. increased death rate worldwide
   c. carrying capacity reduction
   d. carrying capacity expansion

8. Match the following population ecology terms.
   _____ carrying capacity
   _____ exponential growth
   _____ population growth rate
   _____ density-dependent
         controls

   a. examples are diseases and predation
   b. dependent on birth, death, immigration, and emigration rates

   c. number of individuals of a given species that can be sustained indefinitely by the resources in a given area
   d. increases in population size by ever larger amounts per unit of time

---

Critical Thinking: You Decide *(Key in Appendix VI)*

1. How have humans increased the carrying capacity of their environments? Have they avoided some of the limiting factors on population growth, or is the avoidance an illusion?

2. The poet T. S. Eliot once wrote that the world would end "not with a bang but a whimper." Applying this grim thought to the topics of this chapter, design a 10-point plan for proving Eliot wrong.

3. Human populations (like those of other species) use resources and produce wastes. Use Figure 25.22 as a starting point for a brief essay on the accumulation and uses of energy and materials, including wastes, in a major city.

Selected Key Terms

acid rain *25.4*
age structure *25.1*
biotic potential *25.2*
carrying capacity *25.2*
deforestation *25.7*
demographics *25.1*
density-dependent control *25.2*
density-independent control *25.2*
desertification *25.6*
dry acid deposition *25.4*
exponential growth *25.1*
fossil fuel *25.8*
green revolution *25.6*
industrial smog *25.4*

life table *25.2*
limiting factor *25.2*
logistic growth *25.2*
meltdown *25.8*
photochemical smog *25.4*
pollutant *25.4*
population density *25.1*
reproductive base *25.1*
salinization *25.5*
shifting cultivation *25.7*
sustainable society *25.10*
thermal inversion *25.4*
zero population growth *25.1*

Readings

Cohen, J. November 1992. "How Many People Can the Earth Hold?" *Discover*.

Dixon, T. F., J. Boutwell, and G. Rathjens. February 1993. "Environmental Change and Violent Conflict." *Scientific American*.

Miller, G. T. 1999. *Environmental Science*. 7th ed. Belmont, California: Wadsworth.

Stolzenburg, W. July/August 1996. "10 Things (at Least) You Can Do to Save Life's Diversity." *Nature Conservancy*.

"Ten Years After." January 1997. Survey article by the editors of *Discover* on the lasting damage done by the nuclear meltdown at Chernobyl.

APPENDIX I.   CONCEPTS IN CELL METABOLISM

The Nature and Uses of Energy

Any time an object is not moving, it has a store of **potential energy**—a capacity to do work, simply owing to its position in space and the arrangment of its parts. If a stationary runner springs into action, some of the runner's potential energy is transformed into **kinetic energy**, the energy of motion.

Energy on the move does work when it imparts motion to other things—for example, when you throw a ball. In skeletal muscle cells in your arm, the energy currency ATP (adenosine triphosphate, Section 3.13) gave up some of its potential energy to molecules of contractile units and set them in motion. The combined motions in many muscle cells resulted in the movement of whole muscles. The transfer of energy from ATP also resulted in the release of another form of kinetic energy called **heat**, or *thermal energy*.

The potential energy of molecules is called **chemical energy** and is measured as kilocalories. A **kilocalorie** is the amount of energy it takes to heat 1,000 grams of water from 14.5°C to 15.5°C at standard pressure.

As noted in Chapter 3, cells use energy for chemical work, to stockpile, build, rearrange, and break apart substances. They channel it into *mechanical work*—to move cell structures and the whole body or parts of it. They also channel it into *electrochemical* work—to move charged substances into or out of the cytoplasm or an organelle compartment.

Laws of Thermodynamics

We cannot create energy from scratch; we must first get it from someplace else. Why? According to the **first law of thermodynamics**, the total amount of energy in the universe remains constant. More energy cannot be created; existing energy cannot vanish or be destroyed. It can only be converted from one form to some other form. For instance, when you eat, your cells extract energy from food and convert it to other forms, such as kinetic energy for moving about.

With each metabolic conversion, some of the energy escapes to your surroundings, as heat. Even when you "do nothing," your body gives off about as much heat as a 100-watt lightbulb because of conversions in your cells. The energy being released is transferred to atoms and molecules that make up the air, and in this way it heats up the surroundings, as shown in Figure A-1. In general, the body cannot recapture energy lost as heat, but the energy still exists in the environment outside the body. Overall, there is a one-way flow of energy in the universe.

The human body obtains its energy mainly from the covalent bonds in organic compounds, such as glucose and glycogen. When the compounds enter metabolic reactions, specific bonds break or are rearranged. For example, your cells release usable energy from glucose by breaking all of its covalent bonds. After many steps, six molecules of carbon dioxide and six of water remain. Compared with glucose, these leftovers have more stable arrangements of atoms, but chemical energy in their bonds is much less. Why? Some energy was lost at each breakdown step leading to their formation. This is why glucose is a much better source of usable energy than, for example, water is.

As the molecular events just described take place, some heat is lost to the surroundings and cannot be recaptured. Said another way, no energy conversion can ever be 100 percent efficient. Therefore, the total amount of energy in the universe is spontaneously flowing from forms rich in energy (such as glucose) to forms having less and less of it. This is the main point of the **second law of thermodynamics**.

Examples of Energy Changes

When cells convert one form of energy to another, there is a change in the amount of potential energy that is available to them. Cells of photosynthetic organisms, notably green plants, convert energy in sunlight into chemical energy, which is stored in the bonds of organic compounds. The outcome is a net increase in energy in the product molecule (such as glucose), as diagrammed

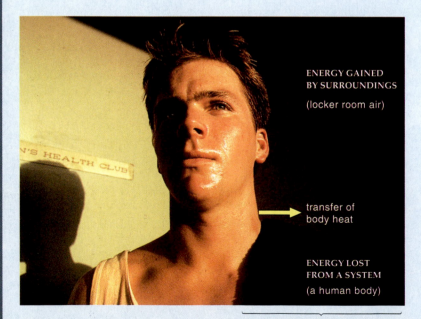

ENERGY GAINED
BY SURROUNDINGS

(locker room air)

transfer of
body heat

ENERGY LOST
FROM A SYSTEM

(a human body)

Figure A-1   NET ENERGY CHANGE = 0

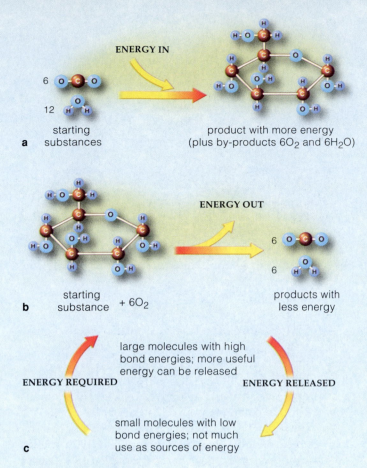

a starting substances → product with more energy (plus by-products $6O_2$ and $6H_2O$)

ENERGY IN

6 $O=C=O$
12 H_2O

b starting substance $+ 6O_2$ → products with less energy

ENERGY OUT

6 $O=C=O$
6 H_2O

c

ENERGY REQUIRED — large molecules with high bond energies; more useful energy can be released — ENERGY RELEASED

small molecules with low bond energies; not much use as sources of energy

Figure A-2

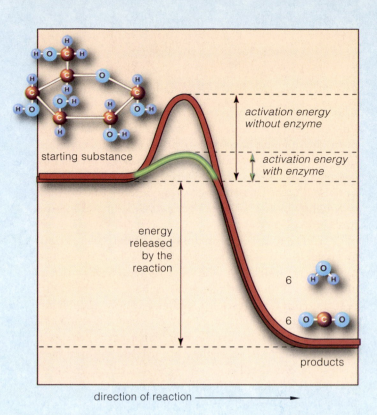

starting substance

activation energy without enzyme

activation energy with enzyme

energy released by the reaction

6 H_2O

6 $O=C=O$

products

direction of reaction ⟶

Figure A-3

in Figure A-2*a*. A reaction in which there is a net increase in energy in the product compound is an **endergonic reaction** (meaning energy in). By contrast, reactions in cells that break down glucose (or another energy-rich compound) release energy. They are called **exergonic reactions** (meaning energy out).

The Role of Enzymes in Metabolic Reactions

The catalytic molecules called **enzymes** are crucial actors in metabolism. To better understand why, it helps to begin with the idea that in cells, molecules or ions of substances are always moving at random. As a result of this random motion, they are constantly colliding. Metabolic reactions may take place when participating molecules collide—but only *if* the energy associated with the collisions is great enough. This minimum amount of energy required for a chemical reaction is called **activation energy**. Activation energy is a barrier that must be surmounted one way or another before a reaction can proceed.

Nearly all metabolic reactions are reversible. That is, they can run "forward," from starting substances to products, or in "reverse," from a product back to starting substances. Which way such a reaction runs depends partly on the ratio of reactant to product. When there is a high concentration of reactant molecules, the reaction is likely to run strongly in the forward direction. On the other hand, when the product concentration is high enough, more molecules or ions of the product are available to revert spontaneously to reactants. Any reversible reaction tends to run spontaneously toward **chemical equilibrium**—the point at which it will be running at about the same pace in both directions.

As just described, before reactants enter a metabolic reaction they must be activated by an energy input; only then will the steps leading to products proceed. And while random collisions *might* provide the energy for reactions, our survival depends on thousands of reactions taking place with amazing speed and precision. This is the key function of enzymes, for *enzymes lower the activation energy barrier* (Figure A-3). As Section 3.13 described, substrates and enzymes interact at the enzyme's active site (see Figure 3.24). According to the **induced fit model**, a surface region of each substrate has chemical groups that are almost but not quite complementary to chemical groups in an active site. However, as substrates settle into the site, the contact strains some of their bonds, making them easier to break. There also are interactions among charged or polar groups that prime substrates for conversion to an activated state. With these changes, substrates fit precisely in the enzyme's active site. They now are in an activated state, in which they will react spontaneously.

Glycolysis: The First Stage of the Energy-Releasing Pathway

Energy that is converted into the chemical bond energy of adenosine triphosphate—ATP—fuels cell activities. Cells make ATP by breaking down carbohydrates (mainly glucose), fats, and proteins. During the breakdown reactions, electrons are stripped from intermediates, then energy associated with the liberated electrons drives the formation of ATP.

Recall that cells rely mainly on **aerobic respiration**, an oxygen-dependent pathway of ATP formation. The main energy-releasing pathways of aerobic respiration all start with the same reactions in the cytoplasm. During this initial stage of reactions, called **glycolysis**, enzymes break apart and rearrange a glucose molecule into two molecules of pyruvate, which has a backbone of three carbon atoms. Following up on the discussion in Section 3.14, here you can track in a bit more detail on what happens to a glucose molecule in the first stage of aerobic respiration.

Glucose is one of the simple sugars. Each molecule of glucose contains six carbon, twelve hydrogen, and six oxygen atoms, all joined by covalent bonds (Figure A-4). The carbons make up the backbone. With glycolysis, glucose or some other carbohydrate in the cytoplasm is partially broken down, the result being two molecules of the three-carbon compound pyruvate:

glucose → glucose-6-phosphate → 2 pyruvate

The first steps of glycolysis require energy. As diagrammed in Figure A-5 on the facing page, they advance only when two ATP molecules each transfer a phosphate group to glucose and so donate energy to it. Such transfers, recall, are phosphorylations. In this case, they raise the energy content of glucose to a level that is high enough to allow the *energy-releasing* steps of glycolysis to begin.

The first energy-releasing step breaks the activated glucose into two molecules. Each of these molecules is called PGAL (phosphoglyceraldehyde). Next, each PGAL is converted to an unstable intermediate that allows ATP to form by giving up a phosphate group to ADP. The next intermediate in the sequence does the same thing.

Thus, a total of four ATP form by **substrate-level phosphorylation**. This metabolic event is the direct transfer of a phosphate group from a substrate of a reaction to some other molecule—in this case, ADP. Remember, though, two ATP were invested to jump-start the reactions. So the *net* energy yield is only two ATP.

Meanwhile, the coenzyme NAD⁺ picks up electrons and hydrogen atoms liberated from each PGAL, thus becoming NADH. When the NADH gives up its cargo at

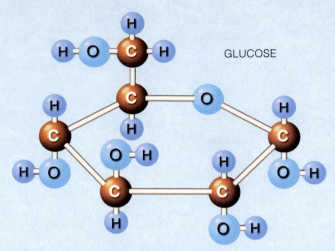

GLUCOSE

Figure A-4

a different reaction site, it reverts to NAD⁺. Said another way, like other coenzymes NAD⁺ is reusable.

In sum, glycolysis converts energy stored in glucose to a transportable form of energy, in ATP. NAD⁺ picks up electrons and hydrogen that are removed from each glucose molecule. The electrons and hydrogen have roles in the next stage of reactions. So do the end products of glycolysis—the two molecules of pyruvate.

Figure A-5 Glycolysis, first stage of the main energy-releasing pathways. The reaction steps proceed inside the cytoplasm of every living prokaryotic and eukaryotic cell. In this example, glucose is the starting material. By the time the reactions end, two pyruvate, two NADH, and four ATP have been produced. Cells invest two ATP to start glycolysis, however, so the *net* energy yield of glycolysis is two ATP.

Depending on the type of cell and on environmental conditions, the pyruvate may enter the second set of reactions of the aerobic pathway, which includes the Krebs cycle. Or it may be used in other reactions, such as a fermentation pathway (Section 3.15).

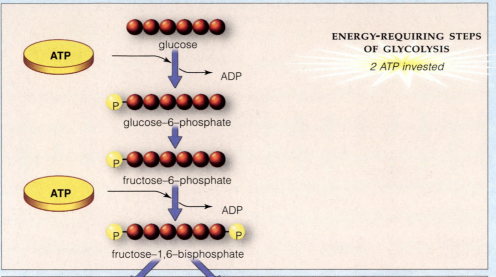

**ENERGY-REQUIRING STEPS
OF GLYCOLYSIS**

2 ATP invested

glucose

glucose–6–phosphate

fructose–6–phosphate

fructose–1,6–bisphosphate

a This diagram tracks the fate of the six carbon atoms (*red* spheres) of a glucose molecule, shown above. Glycolysis starts with an energy investment of two ATP.

b A phosphate-group transfer from one of the ATP molecules to glucose causes atoms in glucose to undergo rearrangements.

c A phosphate-group transfer from another ATP causes rearrangements that form the intermediate fructose–1,6–bisphosphate.

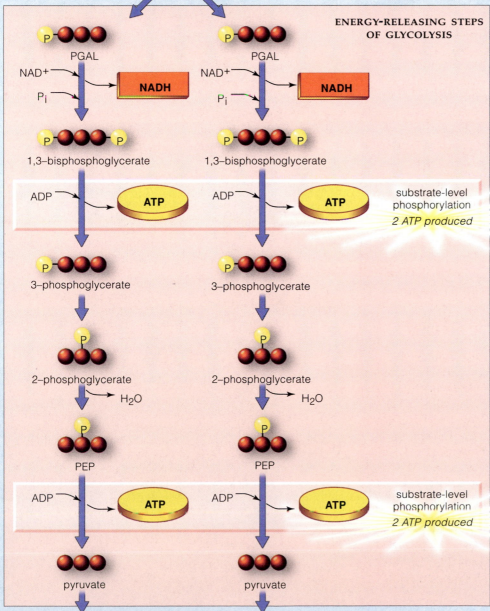

**ENERGY-RELEASING STEPS
OF GLYCOLYSIS**

PGAL

NAD^+

P_i

NADH

1,3–bisphosphoglycerate

ADP · **ATP**

substrate-level phosphorylation

2 ATP produced

3-phosphoglycerate

2-phosphoglycerate

H_2O

PEP

ADP · **ATP**

substrate-level phosphorylation

2 ATP produced

pyruvate

to second set of reactions

d The intermediate splits at once into two molecules, each with a three-carbon backbone. We can call these two PGAL.

e Each PGAL gives up two electrons and a hydrogen atom to NAD^+, thus forming two NADH.

f Each PGAL also combines with inorganic phosphate (P_i) in the cytoplasm and then makes a phosphate-group transfer to ADP.

g *Thus two ATP have formed by the direct transfer of phosphate from two intermediate molecules that served as substrates for the reactions.* Formation of two ATP means the original energy investment of two ATP has been paid off.

h During the next two enzyme-mediated reactions, the two intermediates each release a hydrogen atom and an —OH group, which combine to form water.

i Two 3–phosphoenolpyruvate (PEP) molecules result. Each PEP makes a phosphate-group transfer to ADP.

j *Once again, two ATP have formed by sub-strate-level phosphorylation.*

k In sum, the net energy yield from glycolysis is two ATP for each glucose molecule entering the reactions. Two molecules of pyruvate, the end product, may enter the next set of reactions in an energy-releasing pathway.

Second Stage of the Aerobic Pathway

When pyruvate molecules formed by glycolysis leave the cytoplasm and enter a mitochondrion, the scene is set for both the second and the third stages of the aerobic pathway. Figure A-6 diagrams these steps in detail.

Figure A-6 Second stage of aerobic respiration: the Krebs cycle and reaction steps that precede it. For each three-carbon pyruvate molecule entering the cycle, three CO_2, one ATP, four NADH, and one $FADH_2$ molecules form. The steps shown proceed *twice*, because each glucose molecule was broken down earlier to *two* pyruvate molecules.

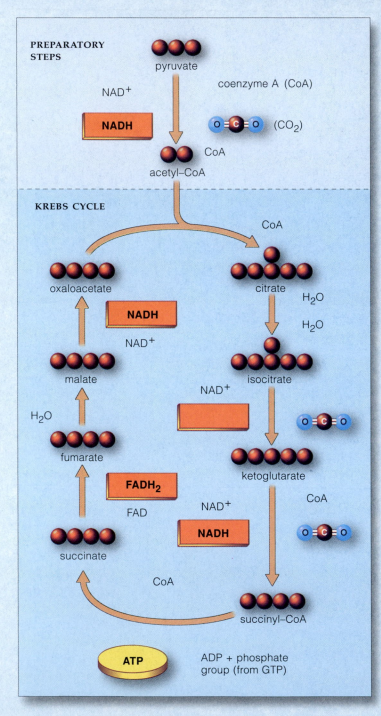

a Pyruvate from glucose enters a mitochondrion. It undergoes initial conversions before entering cyclic reactions (the Krebs cycle).

b Pyruvate is stripped of a carboxyl group (COO^-), which departs as CO_2. It also gives up hydrogen and electrons to NAD^+, forming NADH. Then a coenzyme joins with the remaining two-carbon fragment, forming acetyl–CoA.

c The acetyl–CoA transfers its two-carbon group to oxaloacetate, a four-carbon compound that is the entry point into the Krebs cycle. The result is citrate, with a six-carbon backbone. Addition and then removal of H_2O changes citrate to isocitrate.

d Isocitrate enters conversion reactions, and a COO^- group departs (as CO_2). Hydrogen and electrons are transferred to NAD^+, forming NADH.

e Another COO^- group departs (as CO_2) and another NADH forms. The resulting intermediate attaches to a coenzyme A molecule, forming succinyl–CoA. *At this point, three carbon atoms have been released, balancing out the three that entered the mitochondrion (in pyruvate).*

f Now the attached coenzyme is replaced by a phosphate group (donated by the substrate GTP). That phosphate group becomes attached to ADP. Thus, for each turn of the cycle, one ATP forms by substrate-level phosphorylation.

Third Stage of the Aerobic Pathway

Most ATP is produced in the third stage of the aerobic pathway. Electron transport systems and neighboring proteins called ATP synthases serve as the production machinery. They are embedded in the inner membrane that divides the mitochondrion into two compartments (Figure A-7). They interact with electrons and H^+ ions, which coenzymes deliver from reaction sites of the first two stages of the aerobic pathway.

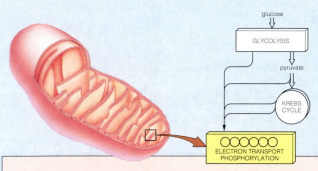

Figure A-7 Electron transport phosphorylation, the third and final stage of aerobic respiration. The reactions proceed at electron transport systems and at ATP synthases, a type of transport protein, in the inner mitochondrial membrane. Each electron transport system consists of specific enzymes, cytochromes, and other proteins that act in sequence.

The inner membrane functionally divides the mitochondrion into two compartments. The third-stage reactions start in the inner compartment, when NADH and $FADH_2$ give up electrons and hydrogen to transport systems. Electrons are transferred *through* the system, but electron transport proteins drive unbound hydrogen (H^+) *to* the outer compartment:

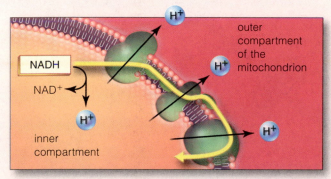

In short order, there is a higher concentration of H^+ in the outer compartment compared to the inner one. Concentration and electric gradients now exist across the membrane. The ions follow the gradients and flow across the membrane, through the interior of the ATP synthases. Energy associated with the flow drives the formation of ATP from ADP and unbound phosphate (P_i). Hence the name, electron transport *phosphorylation*:

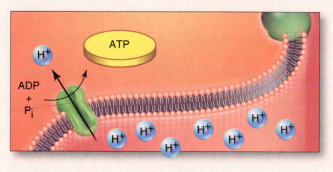

Summary of Glycolysis and Aerobic Respiration

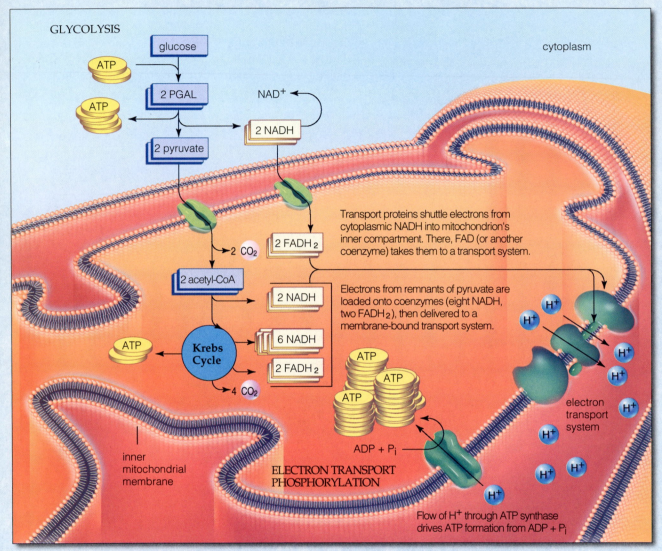

a Two ATP formed in first stage in cytoplasm (during glycolysis, by *substrate-level* phosphorylations).

b NADH that formed in cytoplasm during first stage delivers electrons and hydrogen that help drive the formation of four ATP during third stage at the inner mitochondrial membrane (by *electron transport* phosphorylations).

c Two ATP form at second stage in mitochondrion (by *substrate-level* phosphorylations of Krebs cycle).

d Coenzymes from Krebs cycle and its preparatory steps deliver electrons and hydrogen that drive formation of twenty-eight ATP at third stage (by *electron transport* phosphorylations at the inner mitochondrial membrane).

---

36 ATP

TYPICAL NET
ENERGY YIELD

Figure A-8 Summary of the harvest from the energy-releasing pathway of aerobic respiration. Commonly, thirty-six ATP form for each glucose molecule that enters the pathway. But the net yield varies according to shifting concentrations of reactants, intermediates, and end products of the reactions. It also varies among different types of cells.

Cells differ in how they use the NADH from glycolysis, which cannot enter mitochondria. At the outer mitochondrial membrane, these NADH give up electrons and hydrogen to transport proteins, which shuttle the electrons and hydrogen across the membrane. NAD^+ or FAD already inside the mitochondrion accept them, thus forming NADH or $FADH_2$.

Any NADH inside the mitochondrion delivers electrons to the highest possible entry point into a transport system. When it does, enough H^+ is pumped across the inner membrane to make *three* ATP. By contrast, any $FADH_2$ delivers them to a lower entry point. Fewer hydrogen ions can be pumped, so only *two* ATP can form.

In liver, heart, and kidney cells, for example, electrons and hydrogen from glycolysis enter the highest entry point of transport systems, so the energy harvest is thirty-eight ATP. More commonly, as in skeletal muscle and brain cells, they are transferred to FAD—so the harvest is thirty-six ATP.

APPENDIX II. PERIODIC TABLE OF THE ELEMENTS

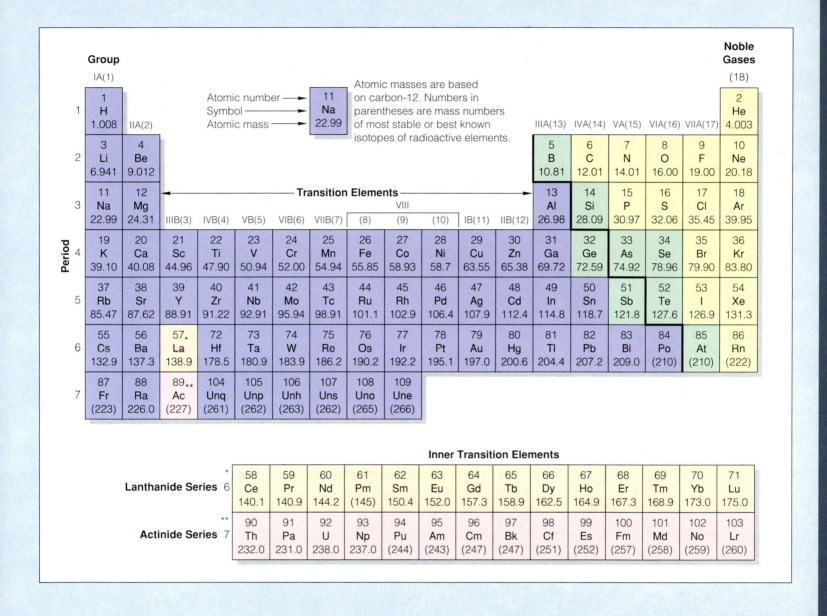

Group

Atomic number → 11
Symbol → Na
Atomic mass → 22.99

Atomic masses are based on carbon-12. Numbers in parentheses are mass numbers of most stable or best known isotopes of radioactive elements.

Noble Gases

Transition Elements

Inner Transition Elements

	Lanthanide Series 6													
58 Ce 140.1	59 Pr 140.9	60 Nd 144.2	61 Pm (145)	62 Sm 150.4	63 Eu 152.0	64 Gd 157.3	65 Tb 158.9	66 Dy 162.5	67 Ho 164.9	68 Er 167.3	69 Tm 168.9	70 Yb 173.0	71 Lu 175.0

	Actinide Series 7													
90 Th 232.0	91 Pa 231.0	92 U 238.0	93 Np 237.0	94 Pu (244)	95 Am (243)	96 Cm (247)	97 Bk (247)	98 Cf (251)	99 Es (252)	100 Fm (257)	101 Md (258)	102 No (259)	103 Lr (260)

APPENDIX III. UNITS OF MEASURE

Metric-English Conversions

Length

English		Metric
inch	=	2.54 centimeters
foot	=	0.30 meter
yard	=	0.91 meter
mile (5,280 feet)	=	1.61 kilometer

To convert	multiply by	to obtain
inches	2.54	centimeters
foot	30.00	centimeters
centimeters	0.39	inches
millimeters	0.039	inches

Weight

English		Metric
grain	=	64.80 milligrams
ounce	=	28.35 grams
pound	=	453.60 grams
ton (short) (2,000 pounds)	=	0.91 metric ton

To convert	multiply by	to obtain
ounces	28.3	grams
pounds	453.6	grams
pounds	0.45	kilograms
grams	0.035	ounces
kilograms	2.2	pounds

Volume

English		Metric
cubic inch	=	16.39 cubic centimeters
cubic foot	=	0.03 cubic meter
cubic yard	=	0.765 cubic meters
ounce	=	0.03 liter
pint	=	0.47 liter
quart	=	0.95 liter
gallon	=	3.79 liters

To convert	multiply by	to obtain
fluid ounces	30.00	milliliters
quart	0.95	liters
milliliters	0.03	fluid ounces
liters	1.06	quarts

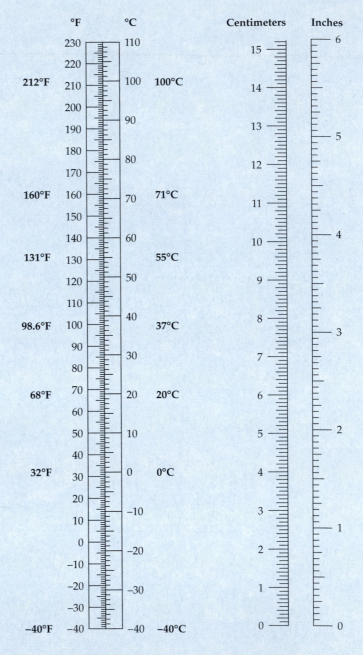

To convert temperature scales:

Fahrenheit to Celsius: $°C = 5/9(°F - 32)$

Celsius to Fahrenheit: $°F = 9/5(°C) + 32$

APPENDIX IV.   ANSWERS TO GENETICS PROBLEMS

CHAPTER 18

1. a: *AB*

   b: *AB* and *aB*

   c: *Ab* and *ab*

   d: *AB, aB, Ab,* and *ab*

2. a: *AaBB* will occur in all the offspring.

   b: 25% *AABB*; 25% *AaBB*; 25% *AABb*; 25% *AaBb*

   c: 25% *AaBb*; 25% *Aabb*; 25% *aaBb*; 25% *aabb*

   d: 1/16 *AABB* (6.25%)
      1/8 *AaBB* (12.5%)
      1/16 *aaBB* (6.25%)
      1/8 *AABb* (12.5%)
      1/4 *AaBb* (25%)
      1/8 *aaBb* (12.5%)
      1/16 *AAbb* (6.25%)
      1/8 *Aabb* (12.5%)
      1/16 *aabb* (6.25%)

3. a: Mother must be heterozygous for both genes; father is homozygous recessive for both genes. The first child is also homozygous recessive for both genes.

   b: The probability that the second child will not be able to roll the tongue and will have detached earlobes is 1/4 (25%).

4. a: ABC

   b: aBc, ABc

   c: ABC, ABc, aBc, aBC

   d: ABC, abc, aBC, AbC, Abc, ABc, abC, aBc

5. Because Molly does not exhibit the recessive hip disorder, she must be either homozygous dominant (HH) for this trait, or heterozygous (Hh). If the father is homozygous dominant (HH), then he and Molly cannot produce gametes that are homozygous recessive (hh), and so none of their offspring will have the undesirable phenotype. However, if Molly is a heterozygote for the trait, notice that the probability is 1/2 (50%) that a puppy will be heterozygous (Hh) and so carry the trait.

6. a: The mother must be heterozygous (I^Ai). The man having type B blood could have fathered the child if he were also heterozygous. (I^Bi).

   b: If the man is heterozygous, then he *could be* the father. However, because any other type B heterozygous male could also be the father, one cannot say that this particular man absolutely must be. Actually, any male who could contribute an O allele (*i*) could have fathered the child. This would include males with type O blood (*ii*) or type A blood who are heterozygous.

7. The probability is 1/2 (50%) that a child of this couple will be a heterozygote and have sickle cell trait. The probability is 1/4 (25%) that a child will be homozygous for the sickling allele and so will have sickle cell anemia.

8. For these ten traits, all the man's sperm will carry identical genes. He cannot produce genotypically different sperm. The woman can produce eggs with four genotypes. This example underscores the fact that the more heterozygous gene pairs that are present, the more genetically different gametes are possible.

9. The mating is *Ll* x *Ll*.

   Progeny genotypes: 1/4 *LL* + 1/2 *Ll* + 1/4 *ll*.

   Phenotypes: 1/4 homozyous survivors (*LL*)

   1/2 heterozygous survivors (*Ll*)

   1/4 lethal (*ll*) nonsurvivors

10. Bill's genotype: *Aa, Ss, Ee*
    Marie's genotype: *AA, SS, EE*

No matter how many children Bill and Marie have, the probability is 100% that each child will have the parents' phenotype. Because Marie can produce only dominant alleles, there is no way that a child could inherit a *pair* of recessive alleles for any of these three traits—and that is what would be required in order for the child to show the recessive phenotype. Thus the probability is zero that a child will have short lashes, high arches, and no achoo syndrome.

11. The white-furred parent's genotype is *bb*; the black-furred parent must be *Bb*; because if it were *BB* all offspring would be heterozygotes (*Bb*) and would have black fur. A monohybrid cross between a heterozygote and a homozygote typically yields a 1:1 phenotype ratio. 4 black and 3 white guinea pigs is close to a 1:1 ratio.

CHAPTER 19

1. Key: Among other complicated issues, here you must consider whether (in your view) a person's gender is determined by "genetic sex" or by reproductive structures. Also, review the discussion of sex determination at the beginning of Chapter 14. In addition, note that the secondary sex characteristics of normal males, which are due to the sensitivity of cells to testosterone, include greater muscle mass—which is why males typically are physically stronger than females.

2. The probability is the same for each child: 1/2, or 50%.

3. a: From his mother.

   b: A male can produce two types of gametes with respect to an X-linked gene. One type will possess only a Y chromosome and so lack this gene; the other type will have an X chromosome and will have the X-linked gene.

   c: A female homozygous for an X-linked gene will produce just one type of gamete containing an X chromosome with the gene.

   d: A female heterozygous for an X-linked gene will produce two types of gametes. One will contain an X chromosome with the dominant allele, and the other type will contain an X chromosome with the recessive allele.

4. Because this is a case of autosomal dominant inheritance and because one parent bears the allele, the probability of any child inheriting the allele is 50%.

5. Most of chromosome 21 has been translocated to chromosome 14. While this individual has 46 chromosomes, there are in fact three copies of chromosome 21. The third copy of chromosome 21 is attached to chromosome 14.

6. No. Many traits are sex-limited and influenced by genes on autosomes.

7. 50%. His mother is heterozygous for the allele, so there is a 50% chance that any male offspring will inherit the allele. Since males do not inherit an X chromosome from their fathers, the genotype of the father is irrelevant to this question.

8. The allele for cystic fibrosis is recessive. Most of the carriers are heterozygous for the allele and do not have the cystic fibrosis phenotype.

9. The parent with typical pigmentation is heterozygous for the albinism allele. The probability that any one child will have the albinism phenotype is 50%. However, with a small sample of only four offspring, there is a high probability of a deviation from a 1:1 ratio due to the random mixes of alleles that occur during meiosis and fertilization (discussed in Chapter 18).

APPENDIX V. ANSWERS TO SELF-QUIZZES

CHAPTER 1: 1. DNA, 2. Metabolism, 3. Homeostasis, receptors, 4. adaptations, 5. mutations, 6. c, 7. c, 8. b

CHAPTER 2: 1. carbon, 2. a, 3. c, 4. c, 5. d, 6. b, 7. b, 8. c, 9. c, e, b, d, a, 10. a (plus R group interactions)

CHAPTER 3: 1. e, 2. cytoskeleton, 3. c, 4. a, 5. c, 6. b, g, f, d, c, a, e , 7. d, 8. d, 9. c, e, d, a, b, 10. b, 11. b, a, c, 12. The electron transport systems and carrier proteins required for ATP formation are embedded in the membrane between the inner and outer compartment of the mitochondrion.

CHAPTER 4: 1. d, 2. a, 3. b, 4. b, 5. b, 6. a, 7. c, 8. d, 9. c, 10. a, 11. Receptors, integrator, effectors, 12. d, e, c, b, a

CHAPTER 5: 1. Skeletal and muscular, 2. skeletal, smooth, cardiac, 3. d, 4. b, 5. b, 6. d, 7. g, e, f, a, c, b, h, d

CHAPTER 6: 1. digestive, circulatory, respiratory, urinary, 2. digesting, absorbing, eliminating, 3. caloric, energy, 4. carbohydrates, 5. essential amino acids, essential fatty acids, 6. b, 7. c, 8. d, 9. c, 10. e, d, a, b, c

CHAPTER 7: 1. c, 2. d, 3. d, 4. c, 5. a, 6. d, 7. b, 8. d, 9. c, 10. d, b, a, c , 10. c, d, a, b, 11. c, d, a, b

CHAPTER 8: 1. d, 2. c, 3. f. , 4. d, 5. a, 6. d, 7. d, 8. b, 9. a, 10. c, b, a, e, d, f

CHAPTER 9: 1. d, 2. b, 3. d, 4. b, 5. c, 6. e, 7. c, 8. d, 9. d, 10. d, 11. f, a, e, d, b, c

CHAPTER 10: 1. d, 2. f, 3. a, 4. b, 5. c, 6. d, 7. a, 8. d, 9. b, d, e, c, a

CHAPTER 11: 1. stimuli, neurons, 2. action potentials, 3. neurotransmitter, 4. integration, 5. d, 6. c, 7. b, 8. c, 9. e, d, b, c, a

CHAPTER 12: 1. stimulus, 2. sensation, 3. Perception, 4. d, 5. b, 6. d, 7. c, 8. c, 9. d, 10. b, 11. c, 12. e, c, d, a, b

CHAPTER 13: 1. f, 2. d, 3. b, 4. d , 5. e, 6. b, 7. b, 8. b, 9. b, e, g, d, f, a, c, 10. d, a, b, e, c

CHAPTER 14: 1. hypothalamus, 2. b, 3. d, 4. c, 5. b, 6. d

CHAPTER 15: 1. a, 2. a, 3. c, 4. morphogenesis, 5. c, 6. c, 7. d, a, f, e, c, b, 8. a, 9. c, 10. b, 11. c

CHAPTER 16: 1. a, 2. c, 3. e, 4. e, 5. a, 6. unsafe sexual contact; blood, semen, vaginal secretions, breast milk, 7. e, b, a, c, g, f

CHAPTER 17: 1. mitosis; meiosis, 2. chromosomes; DNA, 3. c, 4. diploid, 5. c, 6. a, 7. b, 8. b, 9. c, 10. d, 11. d, b, c, a

CHAPTER 18: 1. a, 2. c, 3. a, 4. c, 5. b, 6. d, 7. b, 8. d, 9. c, a, d, b

CHAPTER 19: 1. c, 2. e, 3. e, 4. c, 5. e, 6. c, 7. d, 8. d, 9. c, e, d, b, a

CHAPTER 20: 1. three, 2. e, 3. b, 4. c, 5. c, 6. a, 7. c, 8. b, 9. a, 10. c, 11. a, 12. e, c, d, a, b

CHAPTER 21: 1. variation, 2. genetic engineering, 3. Plasmids, 4. vectors, 5. c, 6. cloned DNA, 7. d, 8. b, 9. c, 10. restriction enzymes

CHAPTER 22: 1. c, 2. e, 3. tumor suppressor gene, 4. b, 5. d, 6. a, 7. c, 8. d, 9. b, 10. c

CHAPTER 23: 1. species, 2. mutation, genetic drift, gene flow, and natural selection, 3. natural selection, 4. e, 5. e, 6. b, 7. a, 8. a

CHAPTER 24: 1. a, 2. d, 3. d, 4. a, 5. e, 6. d, a, c, b, 7. c, d, e, b, a

CHAPTER 25: 1. Ecology, 2. e, 3. e, 4. b, 5. e, 6. d, 7. a, 8. c, d, b, a

APPENDIX VI.   KEY TO CRITICAL THINKING QUESTIONS

CHAPTER 1

1. Key: The different approaches to "truth" inherent in scientific inquiry and in subjective spiritual, moral, and ethical approaches to issues. Another problem you might consider: Based on discussions in this chapter, can anyone—including scientists—ever claim to know the "whole" truth about a particular question?

2 and 3. Key: Consider experimental design utilizing the scientific method of investigation.

4. Review the steps scientists use in studying nature (Section 1.4).

CHAPTER 2

1. Key: The mathematical relationship between values on the pH scale.

2. Key: Cream is mainly animal fat, syrup is mainly water and a sugar such as fructose.

3. Review Section 2.8.

4. In your Web search, try key words such as acid rain and acid deposition.

CHAPTER 3

1. Review Figure 3.2.

2. Key: The biochemical limits on mechanisms by which cells, especially muscle cells, can obtain energy to do work. In explaining the second observation, think of a "nonaerobic" exercise as one that, unlike aerobic exercise, does not require the body to make adjustments (as in breathing rate) in order to supply cells with the oxygen needed to keep aerobic cellular respiration going.

3. Key: What are the roles of nucleic acids such as DNA and RNA?

4. Review Sections 3.5 and 3.10.

CHAPTER 4

1. Key: Glands are derived from which basic type of tissue?

2. Review Section 4.2

3. Key: Knowing the roles collagen plays in the body.

4. Key: Consider whether drinking water will reverse *or* briefly intensify the change (dehydration) in the hikers' bodies.

5. This is easy. Simply compare porphyria's symptoms with the purported behaviors of a vampire.

6. You might start with the National Institutes of Health or one of the popular medical websites.

CHAPTER 5

1. Key: Review the various types of muscle fibers, and relate their characteristics to the athletic requirements of sprinting versus the prolonged activity required of a marathoner's leg muscles.

2. Key: Production of GH and the maximum length of an individual's long bones are both genetically determined.

3. Key: Knowing the components of *living* bone.

4. Key: Consider that a developing fetus needs calcium.

CHAPTER 6

1. Review Sections 6.3 through 6.6.

2 and 3. Key: Review Sections 6.8 through 6.11. Consider the different caloric demands of an older, less active person compared to those of a normally active child. Also, keep in mind that developing youngsters require ample protein.

4. Key: Consider how diarrhea affects the body (Section 6.6).

CHAPTER 7

1. Key: The relationship between blood pressure and blood volume, cross-sectional area of systemic vessels, and cardiac output.

2. Key: Review the function(s) of the formed elements in blood.

3. Key: Review Section 7.11.

CHAPTER 8

1. Key: Recall the functions of the body's normal bacterial flora (Section 8.1).

2. Key: Review the mechanism by which memory cells recognize invaders.

3. Key: Consider illnesses and abnormal responses that result from faulty self-recognition.

4. Key: Consider the role of the immune system's memory cells.

5. Key: What is the basic function of the immune response?

CHAPTER 9

1. Key: Hemoglobin binds much more readily with carbon monoxide than it does with oxygen. Also: consider the partial pressure of oxygen in air versus that in a tank of pure oxygen.

2. Key: What blood gases do neural controls monitor?

3. Key: What is the physiological trigger for inspiration?

4. Key: Review Section 9.4.

CHAPTER 10

1. Key: See *Focus on Your Health*, Section 10.6.

2. Key: Review Section 10.4 on filtration and reabsorption.

3. Key: Review Section 10.10.

4. Key: Consider the anatomy of the urinary system.

CHAPTER 11

1. Key: Review the various ways neurotransmitters can affect post-synaptic neurons.

2. Key: Consider the blood-brain barrier, and the functions of cerebrospinal fluid.

3. Again, what is the function of the blood-brain barrier?

4. Key: What kind of neurons control skeletal muscle? To which major branch of the nervous system do they belong?

CHAPTER 12

1. Key: The role of olfaction in taste perception.

2. Key: Review the discussion of the sense of balance.

3. Key: Middle-ear anatomy and its role in sound transmission.

4. Key: Review the ear's anatomy, Section 12.5.

5. Key: Review the pathways of visual perception, Section 12.9.

CHAPTER 13

1. Key: The biological functions of melatonin.

2. Key: Review Section 13.6.

3. Key: Review the roles of anterior pituitary lobe secretions, including the effects on target cells in other endocrine glands.

4. Key: Consider insulin's effect on blood sugar levels.

CHAPTER 14

1. Key: What chemical conditions in the vagina are most conducive to fertilization?

2. Key: Review the role of estrogen at ovulation.

3. Key: Figure 14.8.

4. Key: Figures 14.8 and 14.9.

5. Key: Where does the DNA in the first cell of a new individual come from?

CHAPTER 15

1. Key: Take into account respiration, circulation, nutrient intake, waste excretion, immune functions, and functions of other basic body systems.

2. Key: What is the immediate outcome of gastrulation? Consider also the long-term developmental consequences if gastrulation goes awry.

3. Review Section 15.9.

CHAPTER 16

1. Key: Compare modes of transmission and relative "contagiousness" of HIV, hepatitis B.

2. Key: Review discussions of HIV transmission and prevention of infection.

3. See Section 16.10.

4. Key: Start by recalling that bacteria can be classified by shape.

5. Review Section 16.5.

CHAPTER 17

1. Key: Review stages of meiosis and discussions of sources of genetic variation in offspring (such as crossing over and random alignment of chromosomes at metaphase; also, keep in mind that in each new generation, offspring receive half their chromosomes from each parent.

2. Key: Review what happens to a cell's DNA at different points in the cell cycle.

3. Review Figure 17.15 (Section 17.9).

4. Key: Which cells give rise to gametes? How can radiation affect them?

CHAPTER 18

See Answers to Genetics Problems.

CHAPTER 19

See Answers to Genetics Problems.

CHAPTER 20

1. Key: Consider the links between DNA, protein synthesis, mutation, and inheritance.

2. Key: Review Sections 20.2 and 18.7.

3. Key: Review events of mitosis/meiosis (Chapter 17).

CHAPTER 21

1. Key: Genetic engineering and recombinant DNA technology are most applicable to plant and animal breeding, commercial and industrial microbiology (including use of bacteria and yeast in food processing), and to medical fields such as gene therapy and cancer treatments. Consider also potential uses of genetic engineering in manipulating human gametes and embryos.

2. Key: Keep in mind the scientific approach to analysis, which relies on answers to testable questions rather than on biases or emotions. Review the *Choices* essay (Section 21.10).

CHAPTER 22

1. Key: Among other factors, consider potential effects of lifestyle on the development of cancer.

2. Key: Consider current hypotheses about the roles of immune system cells in recognizing and eliminating cancerous cells.

3. Key: Review kidney functions that help maintain homeostasis (Chapter 10), as well as lung functions (Chapter 9) and liver functions (Chapter 6).

4. Review Sections 22.4 and 22.8.

CHAPTER 23

1. Key: Consider that (1) proteins, including enzymes, are encoded by genes; (2) different populations of a species may show differing gene (allele) frequencies, and (3) various lines of evidence shed light on the geographical movements of early human populations.

2. Key: Review the time frame in which human evolution has occurred (Section 23.8).

3. Review Table 23.1 in the chapter introduction and review Sections 23.4 and 23.8.

4. Review Section 23.3.

CHAPTER 24

1. Key: What animal could eat you?

2. Key: Your personal views will shape your answer, but you may find it useful to review Section 24.1 and the principles of ecosystem structure function described throughout the chapter.

CHAPTER 25

1. Key: Consider changes in health care, agriculture, and public sanitation in the past few decades and centuries; for example, before penicillin was discovered in the 1940s, people often died of infections we would today consider minor. Also: review the discussion of limits on population growth.

2. Key: To survive, our species must grapple with environmental problems with energy and imagination. Although different people will focus their efforts on different aspects of our current predicament, it is generally agreed that a key element of sustaining a livable Earth is controlling human population growth.

3. Key: Figure 25.22.

GLOSSARY OF BIOLOGICAL TERMS

ABO blood typing Method of characterizing an individual's blood according to whether one or both of two protein markers, A and B, are present at the surface of red blood cells. The O signifies that neither marker is present.

abortion Spontaneous or induced expulsion of the embryo or fetus from the uterus.

absorption The movement of nutrients, fluid, and ions across the gastrointestinal tract lining and into the internal environment.

accommodation In the eye, adjustments of the lens position that move the focal point forward or back so that incoming light rays are properly focused on the retina.

acetyl-CoA (uh-SEED-ul) Coenzyme A with a two-carbon fragment from pyruvate attached. In the second stage of aerobic respiration, it transfers the fragment to oxaloacetate for the Krebs cycle.

acid A substance that releases hydrogen ions in water.

acidity Of a solution, an excess of hydrogen ions relative to hydroxyl ions.

acid rain Wet acid deposition; falling of rain (or snow) rich in sulfur and nitrogen oxides.

acrosome An enzyme-containing cap that covers most of the head of a sperm and helps the sperm penetrate an egg at fertilization.

actin (AK-tin) A globular contractile protein. In muscle cells, actin interacts with another protein, myosin, to bring about contraction.

action potential An abrupt, brief reversal in the steady voltage difference (resting membrane potential) across the plasma membrane of a neuron.

active site A crevice on the surface of an enzyme molecule where a specific reaction is catalyzed.

active transport The pumping of one or more specific solutes through a transport protein that spans the lipid bilayer of a cell membrane. Most often, the solute is transported against its concentration gradient. The protein is activated by an energy boost, as from ATP.

adaptation [L. *adaptare*, to fit] In evolutionary biology, the process of becoming adapted (or more adapted) to a given set of environmental conditions. Of sensory neurons, a decrease in the frequency of action potentials (or their cessation) even when a stimulus is maintained at constant strength.

adaptive behavior A behavior that promotes the propagation of an individual's genes and that tends to increase in frequency in a population over time.

adaptive radiation A burst of speciation events, with lineages branching away from one another as they partition the existing environment or invade new ones.

adaptive trait Any aspect of form, function, or behavior that helps an organism survive and reproduce under a given set of environmental conditions.

adenine (AH-de-neen) A purine; a nitrogen-containing base in certain nucleotides.

ADH Antidiuretic hormone. Produced by the hypothalamus and released by the posterior pituitary, it stimulates reabsorption in the kidneys and so reduces urine volume.

adipose tissue A type of connective tissue having an abundance of fat-storing cells and blood vessels for transporting fats.

ADP/ATP cycle In cells, a mechanism of ATP renewal. When ATP donates a phosphate group to other molecules (and so energizes them), it reverts to ADP, then forms again by phosphorylation of ADP.

adrenal cortex (ah-DREE-nul) Outer portion of the adrenal gland; its hormones have roles in metabolism, inflammation, maintaining extracellular fluid volume, and other functions.

adrenal medulla Inner region of the adrenal gland; its hormones help control blood circulation and carbohydrate metabolism.

aerobic respiration (air-OH-bik) [Gk. *aer*, air, and *bios*, life] The main energy-releasing metabolic pathway of ATP formation, in which oxygen is the final acceptor of electrons stripped from glucose or some other organic compound. The pathway proceeds from glycolysis through the Krebs cycle and electron transport phosphorylation. A typical net yield is 36 ATP for each glucose molecule.

age structure Of a population, the number of individuals in each of several or many age categories.

agglutination (ah-glue-tin-AY-shun) The clumping together of foreign cells that have invaded the body (as pathogens or in tissue grafts or transplants). Clumping is induced by cross-linking between antibody molecules that have already bound antigen at the surface of the foreign cells.

aging A range of processes, including the breakdown of cell structure and function, by which the body gradually deteriorates. All organisms showing extensive cell differentiation undergo aging.

agranulocyte Class of white blood cells that lack granular material in the cytoplasm; includes the precursors of macrophages (monocytes) and lymphocytes.

AIDS Acquired immunodeficiency syndrome. A set of chronic disorders following infection by the human immunodeficiency virus (HIV), which destroys key cells of the immune system.

alcohol Organic compound that includes one or more hydroxyl groups (—OH); it dissolves readily in water. Sugars are examples.

alcoholic fermentation Anaerobic pathway of ATP formation in which pyruvate from glycolysis is broken down to acetaldehyde, which accepts electrons from NADH to become ethanol, and NAD is regenerated. Its net yield is two ATP.

aldosterone (al-DOSS-tuh-roan) Hormone secreted by the adrenal cortex that helps regulate sodium reabsorption.

allantois (ah-LAN-twahz) [Gk. *allas*, sausage] One of four extraembryonic membranes that form during embryonic development. In humans, it functions in early blood formation and development of the urinary bladder.

allele (uh-LEEL) For a given location on a chromosome, one of two or more slightly different molecular forms of a gene that code for different versions of the same trait.

allele frequency Of a given gene locus, the relative abundances of each kind of allele carried by the individuals of a population.

allergen Any normally harmless substance that provokes inflammation, excessive mucus secretion, and often immune responses.

allergy An immune response made against a normally harmless substance.

all-or-none principle Principle that states that individual cells in a muscle's motor units always contract fully in response to proper stimulation. If the stimulus is below a certain threshold, the cells do not respond at all.

alveolar sac (al-VEE-uh-lar) Any of the pouchlike clusters of alveoli in the lungs; the major sites of gas exchange.

alveolus (ahl-VEE-uh-lus), plural alveoli [L. *alveus*, small cavity] Any of the many cup-shaped, thin-walled outpouchings of respiratory bronchioles. A site where oxygen diffuses from air in the lungs to the blood, and carbon dioxide diffuses from blood to the lungs.

amine hormone A hormone derived from the amino acid tyrosine.

amino acid (uh-MEE-no) A small organic molecule having a hydrogen atom, an amino group, an acid group, and an R group covalently bonded to a central carbon atom. The subunit of polypeptide chains, which represent the primary structure of proteins.

ammonification (uh-MOAN-ih-fih-KAY-shun) Together with decomposition, a process by which certain bacteria and fungi break down nitrogen-containing wastes and remains of other organisms.

amnion (am-NEE-on) Of land vertebrates, one of four extraembryonic membranes. It becomes a fluid-filled sac in which the embryo (and fetus) can grow, move freely, and be protected from sudden temperature shifts and impacts.

anaerobic pathway (AN-uh-row-bik) [Gk. *an*, without, and *aer*, air] Metabolic pathway in which a substance other than oxygen serves as the final acceptor of

electrons that have been stripped from substrates.

analogous structures Body parts, once different in separate lineages, that were put to comparable uses in similar environments and that came to resemble one another in form and function. They are evidence of morphological convergence.

anaphase (AN-uh-faze) The stage at which microtubules of a spindle apparatus separate sister chromatids of each chromosome and move them to opposite spindle poles. During anaphase I of meiosis, the two members of each pair of homologous chromosomes separate. During anaphase II, sister chromatids of each chromosome separate.

aneuploidy (AN-yoo-ploy-dee) A change in the chromosome number following inheritance of one extra or one fewer chromosome.

antibiotic [Gk. *anti*, against] A normal metabolic product of certain microorganisms that kills or inhibits the growth of other microorganisms.

antibody Any of a variety of Y-shaped receptor molecules with binding sites for specific antigens. Only B cells produce antibodies, then position them at their surface or secrete them.

anticodon In a tRNA molecule, a sequence of three nucleotide bases that can pair with an mRNA codon.

antigen (AN-tih-jen) [Gk. *anti*, against, and *genos*, race, kind] Any molecular configuration that is recognized as foreign to the body and that triggers an immune response. Most antibodies are protein molecules at the surface of infectious agents or tumor cells.

antigen-MHC complex Unit consisting of fragments of an antigen molecule, bound to MHC proteins. MHC complexes displayed at the surface of an antigen-presenting cell such as a macrophage promote an immune response by lymphocytes.

antigen-presenting cell A macrophage or other cell that displays antigen-MHC complexes at its surface and so promotes an immune response by lymphocytes.

anus Terminal opening of the gastrointestinal tract.

aorta (ay-OR-tah) [Gk. *airein*, to lift, heave] Main artery of systemic circulation; carries oxygenated blood away from the heart to all body regions except the lungs.

aortic body Any of several receptors in artery walls near the heart that respond to changes in levels of carbon dioxide (and to some degree, oxygen) in arterial blood.

apoptosis (APP-oh-TOE-sis) Programmed cell death. Molecular signals lead to self-destruction in body cells that have finished their prescribed functions or have become altered, as by infection or transformation into a cancerous cell.

appendicular skeleton (ap-en-DIK-yoo-lahr) Bones of the limbs, hips, and shoulders.

appendix A slender projection from the cup-shaped pouch (cecum) at the start of the colon.

arrhythmia Irregular or abnormal heart rhythm, sometimes caused by stress, drug effects, or coronary artery disease.

arteriole (ar-TEER-ee-ole) Any of the blood vessels between arteries and capillaries. They are control points where the volume of blood delivered to different body regions can be adjusted.

artery Any of the large-diameter blood vessels that conduct deoxygenated blood to the lungs and oxygenated blood to all body tissues. The thick, muscular artery wall allows arteries to smooth out pulsations in blood pressure caused by heart contractions.

asexual reproduction Mode of reproduction by which offspring arise from a single parent and inherit the genes of that parent only.

atherosclerosis Condition in which an artery's wall thickens and loses its elasticity, and the vessel becomes clogged with lipid deposits. In the artery wall, abnormal smooth muscle cells accumulate, and there is an increase in connective tissue. Plaques consisting of lipids, calcium salts, and fibrous material extend into the artery lumen, disrupting blood flow.

atmosphere A region of gases, airborne particles, and water vapor enveloping the Earth; 80 percent of its mass is distributed within 17 miles of the Earth's surface.

atmospheric cycle A biogeochemical cycle in which the atmosphere is the largest reservoir of an element. The carbon and nitrogen cycles are examples.

atom The smallest unit of matter that is unique to a particular element.

atomic number The number of protons in the nucleus of each atom of an element; it differs for each element.

ATP Adenosine triphosphate (ah-DEN-uh-seen try-FOSS-fate) A nucleotide composed of adenine, ribose, and three phosphate groups. As the main energy carrier in cells, it directly or indirectly delivers energy to or picks up energy from nearly all metabolic pathways.

atrioventricular node (AV node) In the septum dividing the heart atria, a site that contains bundles of conducting fibers. Stimuli arriving at the AV node from the cardiac pacemaker (sinoatrial node) pass along the bundles and continue on via Purkinje fibers to contractile muscle cells in the ventricles.

atrioventricular valve One-way flow valve between the atrium and ventricle in each half of the heart.

atrium (AY-tree-um) Upper chamber in each half of the heart; the right atrium receives deoxygenated blood (from tissues) entering the pulmonary circuit of blood flow, and the left atrium receives oxygenated blood from pulmonary veins.

australopith (oss-TRAH-low-pith) [L. *australis*, southern, and Gk. *pithekos*, ape] Any of the earliest known species of hominids; that is, the first species of the evolutionary branch leading to humans.

autoimmune response Misdirected immune response in which lymphocytes mount an attack against normal body cells.

autonomic nervous system (ah-toe-NOM-ik) Those nerves leading from the central nervous system to the smooth muscle, cardiac muscle, and glands of internal organs and structures—that is, to the visceral portion of the body.

autosomal dominant inheritance Condition arising from the presence of a dominant allele on an autosome (not a sex chromosome). The allele is always expressed to some extent, even in heterozygotes.

autosomal recessive inheritance Condition arising from a recessive allele on an autosome (not a sex chromosome). Only recessive homozygotes show the resulting phenotype.

autosome Any of the chromosomes that are of the same number and kind in both males and females of the species.

autotroph (AH-toe-trofe) [Gk. *autos*, self, and *trophos*, feeder] An organism able to build its own large organic molecules by using carbon dioxide and energy from the physical environment. Compare heterotroph.

axial skeleton (AX-ee-uhl) In vertebrates, the skull, backbone, ribs, and breastbone (sternum).

axon Of a neuron, a long, cylindrical extension from the cell body, with finely branched endings. Action potentials move rapidly, without alteration, along an axon; their arrival at axon endings may trigger the release of neurotransmitter molecules that influence an adjacent cell.

B lymphocyte, or B cell The only white blood cell that produces antibodies, then positions them at the cell surface or secretes them as weapons in immune responses.

bacterial conjugation The transfer of plasmid DNA from one bacterial cell to another.

bacterial flagellum Of many bacterial cells, a whiplike motile structure that does not contain a core of microtubules.

bacteriophage (bak-TEER-ee-oh-fahj) [Gk. *baktērion*, small staff, rod, and *phagein*, to eat] Category of viruses that infect bacterial cells.

Barr body In the cells of female mammals, a condensed X chromosome that was inactivated during early embryonic development.

basal body A centriole that, after having given rise to the microtubules of a

flagellum or cilium, remains attached to its base in the cytoplasm.

basal metabolic rate Amount of energy required to sustain body functions when a person is resting, awake, and has not eaten for 12–18 hours.

base A substance that releases OH⁻ in water.

base pair A pair of hydrogen-bonded nucleotide bases in two strands of nucleic acids. In a DNA double helix, adenine pairs with thymine, and guanine with cytosine. When an mRNA strand forms on a DNA strand during transcription, uracil (U) pairs with the DNA's adenine.

base sequence The particular order in which one nucleotide base follows the next in a strand of DNA or RNA.

basement membrane Noncellular layer of mostly proteins and polysaccharides that is sandwiched in between an epithelium and underlying connective tissue.

basophil Fast-acting white blood cells that secrete histamine and other substances during inflammation.

biogeochemical cycle The movement of an element such as carbon or nitrogen from the environment to organisms, then back to the environment.

biogeography [Gk. *bios*, life, and *geographein*, to describe the surface of the Earth] The study of major land regions, each having distinguishing types of plants and animals and generally retaining its identity because of climate and geographic barriers to gene flow.

biological clock Internal time-measuring mechanism that has a role in adjusting an organism's daily activities, seasonal activities, or both in response to environmental cues.

biological magnification The increasing concentration of a nondegradable or slowly degradable substance in body tissues as it is passed along food chains.

biomass The combined weight of all the organisms at a particular trophic (feeding) level in an ecosystem.

biome A broad, vegetational subdivision of a biogeographic realm shaped by climate, topography, and composition of regional soils.

biopsy Diagnostic procedure in which a small piece of tissue is removed from the body through a hollow needle or exploratory surgery, and then examined for signs of a particular disease (often cancer).

biosphere [Gk. *bios*, life, and *sphaira*, globe] All regions of the Earth's waters, crust, and atmosphere in which organisms live.

biosynthetic pathway A metabolic pathway in which small molecules are assembled into lipids, proteins, and other large organic molecules.

biotic potential Of a population, the maximum rate of increase per individual under ideal conditions.

bipedalism A habitual standing and walking on two feet, as by ostriches and humans.

blastocyst (BLASS-tuh-sist) [Gk. *blastos*, sprout, and *kystis*, pouch] In mammalian development, a blastula stage consisting of a hollow ball of surface cells and an inner cell mass.

blastomere One of the small, nucleated cells that form during cleavage of a zygote.

blastula (BLASS-chew-lah) An embryonic stage consisting of a ball of cells produced by cleavage.

blood A fluid connective tissue composed of water, solutes, and formed elements (blood cells and platelets); it carries substances to and from cells and helps maintain an internal environment that is favorable for cell activities.

blood pressure Fluid pressure, generated by heart contractions, that keeps blood circulating.

blood-brain barrier Set of mechanisms that helps control which blood-borne substances reach neurons in the brain.

bone The mineral-hardened connective tissue of bones. *Bones* are organs that function in movement and locomotion, protection of other organs, mineral storage, and (in some bones) blood cell production.

Bowman's capsule Cup-shaped portion of a nephron that receives water and solutes being filtered from blood.

brain Organ that receives, integrates, stores, and retrieves information, and coordinates appropriate responses by stimulating and inhibiting the activities of different body parts.

brain case The eight bones that together surround and protect the brain.

brainstem The vertebrate midbrain, pons, and medulla oblongata, the core of which contains the reticular formation that helps govern activity of the nervous system as a whole.

bronchiole A component of the finely branched bronchial tree inside each lung.

bronchus, plural **bronchi** (BRONG-cuss, BRONG-kee) [Gk. *bronchos*, windpipe] Tubelike branchings of the trachea that lead into the lungs.

buffer system A weak acid and the base that forms when it dissolves in water. The two work as a pair to counter slight shifts in pH.

bulk A volume of fiber and other undigested material that absorption processes in the colon cannot decrease.

cancer A type of malignant tumor, the cells of which show profound abnormalities in the plasma membrane and cytoplasm, abnormal growth and division, and weakened capacity for adhesion within the parent tissue (leading to metastasis). Unless eradicated, cancer is lethal.

capillary [L. *capillus*, hair] A thin-walled blood vessel that functions in the exchange of gases and other substances between blood and interstitial fluid.

capillary bed Dense capillary networks containing true capillaries where exchanges occur between blood and tissues, and also thoroughfare channels that link arterioles and venules.

carbaminohemoglobin A hemoglobin molecule that has carbon dioxide bound to it; $HbCO_2$.

carbohydrate [L. *carbo*, charcoal, and *hydro*, water] A simple sugar or large molecule composed of sugar units. All cells use carbohydrates as structural

materials, energy stores, and transportable forms of energy. The three classes of carbohydrates include monosaccharides, oligosaccharides, and polysaccharides.

carbon cycle A biogeochemical cycle in which carbon moves from its largest reservoir in the atmosphere, through oceans and organisms, then back to the atmosphere.

carbonic anhydrase Enzyme in red blood cells that catalyzes the conversion of unbound carbon dioxide to carbonic acid and its dissociation products, thereby helping maintain the gradient that keeps carbon dioxide diffusing from interstitial fluid into the blood.

carcinogen (kar-SIN-uh-jen) An environmental agent or substance, such as ultraviolet radiation, that can trigger cancer.

carcinogenesis The transformation of a normal cell into a cancerous one.

cardiac conduction system Set of noncontractile cells in heart muscle that spontaneously produce and conduct the electrical events that stimulate heart muscle contractions.

cardiac cycle (KAR-dee-ak) [Gk. *kardia*, heart, and *kyklos*, circle] The sequence of muscle contraction and relaxation constituting one heartbeat.

cardiac pacemaker Sinoatrial (SA) node; the basis of the normal rate of heartbeat. The self-excitatory cardiac muscle cells that spontaneously generate rhythmic waves of excitation over the heart chambers.

cardiovascular system Organ system that is composed of blood, the heart, and blood vessels and that functions in the rapid transport of substances to and from cells.

carnivore [L. *caro, carnis*, flesh, + *vovare*, to devour] Animal that eats other animals.

carotid body Any of several sensory receptors that monitor oxygen (and to some extent, carbon dioxide) levels in blood; located at the point where carotid arteries branch to the brain.

carrier protein Type of transport protein that binds specific substances and changes shape in ways that shunt the substances across a plasma membrane. Some carrier proteins function passively, others require an energy input.

carrying capacity The maximum number of individuals in a population (or species) that can be sustained indefinitely by a given environment.

cartilage A type of connective tissue with solid yet pliable intercellular material that resists compression.

cartilaginous joint Type of joint in which cartilage fills the space between adjoining bones; only slight movement is possible.

cDNA Any DNA molecule copied from a mature mRNA transcript by way of reverse transcription.

cell [L. *cella*, small room] The smallest living unit; an organized unit that can survive and reproduce on its own, given DNA instructions and suitable environmental conditions, including appropriate sources of energy and raw materials.

cell count The number of cells of a given type in a microliter of blood.

cell cycle Events during which a cell increases in mass, roughly doubles its number of cytoplasmic components, duplicates its DNA, then undergoes nuclear and cytoplasmic division. It extends from the time a new cell is produced until it completes its own division.

cell differentiation The gene-guided process by which cells in different locations in the embryo become specialized.

cell-to-cell junction Of multicelled organisms, a point of contact that physically links two cells or that provides functional links between their cytoplasm.

cell theory A theory in biology, the key points of which are that (1) all organisms are composed of one or more cells, (2) the cell is the smallest unit that still retains a capacity for independent life, and (3) all cells arise from preexisting cells.

cell wall A rigid or semirigid wall outside the plasma membrane that supports a cell and imparts shape to it; a cellular feature of plants, fungi, protista, and most bacteria.

central nervous system The brain and spinal cord.

centriole (SEN-tree-ohl) A cylinder of triplet microtubules that gives rise to the microtubules of cilia and flagella.

centromere (SEN-troh-meer) [Gk. *kentron*, center, and *meros*, a part] A small, constricted region of a chromosome having attachment sites for microtubules that help move the chromosome during nuclear division.

cerebellum (ser-ah-BELL-um) [L. diminutive of *cerebrum*, brain] Hindbrain region with reflex centers for maintaining posture and refining limb movements.

cerebral cortex Thin surface layer of the cerebral hemispheres. Some regions of the cortex receive sensory input, others integrate information and coordinate appropriate motor responses.

cerebrospinal fluid Clear extracellular fluid that surrounds and cushions the brain and spinal cord.

cerebrum (suh-REE-bruhm) Part of the forebrain; the most complex integrating center.

chancroid Clinical term for the sexually transmitted disease caused by infection with the bacterium *Haemophilis ducreyi*. Infection results in soft, painful ulcers on the external genitals and can spread to pelvic tissues.

channel protein Type of transport protein that serves as a pore through which ions or other water-soluble substances move across the plasma membrane. Some channels remain open, while others are gated and open and close in controlled ways.

chemical bond A union between the electron structures of two or more atoms or ions.

chemical synapse (SIN-aps) [Gk. *synapsis*, union] A small gap, the synaptic cleft, that separates two neurons (or a neuron and a muscle cell or gland cell) and that is bridged by neurotransmitter molecules released from the presynaptic neuron.

chemiosmotic theory (keem-ee-oz-MOT-ik) Theory that an electrochemical gradient across a cell membrane drives

ATP formation. Metabolic reactions cause hydrogen ions (H^+) to accumulate in a compartment formed by the membrane. The combined force of the resulting concentration and electric gradients propels hydrogen ions down the gradient, through channel proteins. Through enzyme action at these proteins, ADP and inorganic phosphate combine to form ATP.

chemoreceptor (KEE-moe-ree-sep-tur) Sensory receptor that detects chemical energy (ions or molecules) dissolved in the surrounding fluid.

chemotherapy The use of therapeutic drugs to kill cancer cells.

chlamydia Sexually transmitted disease caused by infection by the bacteria *Chlamydia trachomatis*. The bacterium infects cells of the genital organs and urinary tract.

chlorofluorocarbon (klore-oh-FLOOR-oh-car-bun), or CFC. One of a variety of odorless, invisible compounds of chlorine, fluorine, and carbon, widely used in commercial products, that are contributing to the destruction of the ozone layer above the Earth's surface.

chorion (CORE-ee-on) Of placental mammals, one of four extraembryonic membranes; it becomes a major component of the placenta. Absorptive structures (villi) that develop at its surface are crucial for the transfer of substances between the embryo and mother.

chromatid Of a duplicated eukaryotic chromosome, one of two DNA molecules and its associated proteins. One chromatid remains attached to its "sister" chromatid at the centromere until they are separated from each other during a nuclear division; then each is a separate chromosome.

chromatin A cells' collection of DNA and all of the proteins associated with it.

chromosome (CROW-moe-soam) [Gk. *chroma*, color, and *soma*, body] Of eukaryotes, a DNA molecule with many associated proteins. A bacterial chromosome does not have a comparable profusion of proteins associated with the DNA.

chromosome number Of eukaryotic species, the number of each type of chromosome in all cells except dividing germ cells or gametes.

cilium (SILL-ee-um), plural **cilia** [L. *cilium*, eyelid] Of eukaryotic cells, a short, hairlike projection that contains a regular array of microtubules. Cilia serve as motile structures, help create currents of fluids, or are part of sensory structures. They typically are more profuse than flagella.

circadian rhythm (ser-KAYD-ee-un) [L. *circa*, about, and *dies*, day] Of many organisms, a cycle of physiological events that is completed every 24 hours or so, even when environmental conditions remain constant.

circulatory system An organ system consisting of the heart, blood vessels, and blood; the system transports materials to and from cells and helps stabilize body temperature and pH.

cladogram Branching diagram that represents patterns of relative relationships among organisms based on discrete morphological, physiological, and behavioral traits that vary among taxa being studied.

clavicle Long, slender "collarbone" that connects the pectoral girdle with the sternum (breastbone).

cleavage Stage of development when mitotic cell divisions convert a zygote to a ball of cells, the blastula.

cleavage furrow Of a cell undergoing cytoplasmic division, a shallow, ringlike depression that forms at the cell surface as contractile microfilaments pull the plasma membrane inward. It defines where the cytoplasm will be cut in two.

cleavage reaction Enzyme action that splits a molecule into two or more parts; hydrolysis is an example.

climate Prevailing weather conditions for an ecosystem, including temperature, humidity, wind speed, cloud cover, and rainfall.

climax community Following primary or secondary succession, the array of species that remains more or less steady under prevailing conditions.

clonal selection hypothesis Hypothesis that suggests that lymphocytes activated by a specific antigen will rapidly multiply and differentiate into huge subpopulations of cells, all having the parent cell's specificity against that antigen.

cloned DNA Multiple, identical copies of DNA fragments that have been inserted into plasmids or some other cloning vector.

cloning vector Plasmid that has been modified in the laboratory to accept foreign DNA.

cochlea Coiled, fluid-filled chamber of the inner ear. Sound waves striking the ear drum become converted to pressure waves in the cochlear fluid, and the pressure waves ultimately cause a membrane to vibrate and bend sensory hair cells. Signals from bent hair cells travel to the brain, where they may be interpreted as sound.

codominance Condition in which a pair of nonidentical alleles are both expressed, even though they specify two different phenotypes.

codon One of a series of base triplets in an mRNA molecule, most of which code for a sequence of amino acids of a specific polypeptide chain. (Of 64 codons, 61 specify different amino acids and three of these also serve as start signals for translation; one other serves only as a stop signal for translation.)

coenzyme A type of nucleotide that transfers hydrogen atoms and electrons from one reaction site to another. NAD^+ is an example.

coevolution The joint evolution of two or more closely interacting species; when one species evolves, the change affects selection pressures operating between the two species, so the other also evolves.

cofactor A metal ion or coenzyme; it helps catalyze a reaction or serves briefly as an agent that transfers electrons, atoms, or functional groups from one substrate to another.

colon (CO-lun) The large intestine.

community The populations of all species occupying a habitat; also applied to groups of organisms with similar lifestyles in a habitat.

compact bone Type of dense bone tissue that makes up the shafts of long bones and outer regions of all bones. Narrow channels in compact bone contain blood vessels and nerves.

comparative morphology [Gk. *morph*, form] Anatomical comparisons of major lineages.

complement system A set of about 20 proteins circulating in blood plasma with roles in nonspecific defenses and in immune responses. Some induce lysis of pathogens, others promote inflammation, and others stimulate phagocytes to engulf pathogens.

compound A substance in which the relative proportions of two or more elements never vary. Organic compounds have a backbone of carbon atoms arranged as a chain or ring structure. The simpler, inorganic compounds do not have comparable backbones.

concentration gradient A difference in the number of molecules (or ions) of a substance between two adjacent regions, as in a volume of fluid.

condensation reaction Two molecules become covalently bonded into a larger molecule, and water often forms as a by-product.

cone cell In the retina, a type of photoreceptor that responds to intense light and contributes to sharp daytime vision and color perception.

connective tissue A category of animal tissues, all having mostly the same components but in different proportions. These tissues contain fibroblasts and other cells, the secretions of which form fibers (of collagen and elastin) and a ground substance (of modified polysaccharides).

consumer [L. *consumere*, to take completely] Of ecosystems, a heterotrophic organism that obtains energy and raw materials by feeding on the tissues of other organisms. Herbivores, carnivores, omnivores, and parasites are examples.

continuous variation A more or less continuous range of small differences in a given trait among all the individuals of a population.

control group In a scientific experiment, a group used to evaluate possible side effects of a test involving an experimental group. Ideally, the control group should differ from the experimental group only with respect to the variable being studied.

core temperature The body's internal temperature, as opposed to temperatures of the tissues near its surface. Normal human core temperature is about 37°C (98.6°F).

cornea Transparent tissue in the outer layer of the eye, which causes incoming light rays to bend.

coronary artery Either of two arteries leading to capillaries that service cardiac muscle.

corpus callosum (CORE-pus ka-LOW-sum) A band of 200 million axons that functionally link the two cerebral hemispheres.

corpus luteum (CORE-pus LOO-tee-um) A glandular structure; it develops from cells of a ruptured ovarian follicle and secretes progesterone and some estrogen, both of which maintain the lining of the uterus (endometrium).

cortex [L. *cortex*, bark] In general, a rindlike layer; the kidney cortex is an example.

covalent bond (koe-VAY-lunt) [L. *con*, together, and *valere*, to be strong] A sharing of one or more electrons between atoms or groups of atoms. When electrons are shared equally, the bond is nonpolar. When electrons are shared unequally, the bond is polar—slightly positive at one end and slightly negative at the other.

cross-bridge The interaction between actin and myosin filaments that is the basis of muscle cell contraction.

crossing over During prophase I of meiosis, an interaction between a pair of homologous chromosomes. Their nonsister chromatids break at the same place along their length and exchange corresponding segments at the break points. Crossing over breaks up old combinations of alleles and puts new ones together in chromosomes.

culture The sum total of behavior patterns of a social group, passed between generations by learning and by symbolic behavior, especially language.

cyclic AMP Cyclic adenosine monophosphate. A nucleotide that has roles in intercellular communication, as when it serves as a second messenger (a cytoplasmic mediator of a cell's response to signaling molecules).

cytokinesis (sigh-toe-kih-NEE-sis) [Gk. *kinesis*, motion] Cytoplasmic division; the splitting of a parental cell into daughter cells.

cytomembrane system [Gk. *kytos*, hollow vessel] Organelles, functioning as a system to modify, package, and distribute newly formed proteins and lipids. Endoplasmic reticulum, Golgi bodies, lysosomes, and a variety of vesicles are its components.

cytoplasm (SIGH-toe-plaz-um) [Gk. *plassein*, to mold] All cellular parts, particles, and semifluid substances enclosed by the plasma membrane except for the nucleus.

cytosine (SIGH-toe-seen) A pyrimidine; one of the nitrogen-containing bases in nucleotides.

cytoskeleton A cell's internal "skeleton." Its microtubules and other components structurally support the cell and organize and move its internal components.

cytotoxic T cell Type of T lymphocyte that directly kills infected body cells and tumor cells by lysis.

decomposer [L. *de-*, down, away, and *companere*, to put together] A heterotroph that obtains energy by chemically breaking down the remains, products, or wastes of other organisms. Decomposers help cycle nutrients to producers in ecosystems. Certain fungi and bacteria are examples.

deductive logic Pattern of thinking by which a person makes inferences about specific consequences or specific predictions that must follow from a hypothesis.

deforestation The removal of all trees from a large tract of land, such as the Amazon Basin or the Pacific Northwest.

degradative pathway A metabolic pathway by which molecules are broken down in stepwise reactions that lead to products of lower energy.

deletion At the cellular level, loss of a segment from a chromosome. At the molecular level (in a DNA molecule), loss of one to several base pairs from a DNA molecule.

demographic transition model Model of human population growth in which changes in the growth pattern correspond to different stages of economic development. These are a preindustrial stage, when birth and death rates are both high; a transitional stage; an industrial stage; and a postindustrial stage, when the death rate exceeds the birth rate.

denaturation (deh-nay-chur-AY-shun) Of any molecule, the loss of three-dimensional shape following disruption of hydrogen bonds and other weak bonds.

dendrite (DEN-drite) [Gk. *dendron*, tree] A short, slender extension from the cell body of a neuron.

denitrification (dee-nite-rih-fih-KAY-shun) The conversion of nitrate or nitrite by certain bacteria to gaseous nitrogen (N_2) and a small amount of nitrous oxide (N_2O).

density-dependent controls Factors, such as predation, parasitism, disease, and competition for resources, that limit population growth by reducing the birth rate, increasing the rates of death and dispersal, or all of these.

density-independent controls Factors such as storms or floods that increase a population's death rate more or less independently of its density.

dentition The type, size, and number of an animal's teeth.

dermis The layer of skin underlying the epidermis, consisting mostly of dense connective tissue.

desertification (dez-urt-ih-fih-KAY-shun) The conversion of grasslands, rain-fed cropland, or irrigated cropland to desertlike conditions, with a drop in agricultural productivity of 10 percent or more.

detrital food web Of most ecosystems, the flow of energy mainly from plants through detritivores and decomposers.

detritivores (dih-TRY-tih-vorz) [L. *detritus*; after *deterere*, to wear down] Heterotrophs that consume dead or decomposing particles of organic matter. Earthworms, crabs, and nematodes are examples.

development The genetically guided emergence of specialized, morphologically distinct body parts.

diaphragm (DIE-uh-fram) [Gk. *diaphragma*, to partition] Muscular partition between the thoracic and abdominal cavities, the contraction and relaxation of which contributes to breathing. Also, a contraceptive device used temporarily to prevent sperm from entering the uterus during sexual intercourse.

diastole Relaxation phase of the cardiac cycle.

diffusion Net movement of like molecules (or ions) down their concentration gradient. In the absence of other forces, molecular motion and random collisions cause their net outward movement from one region into a neighboring region where they are less concentrated (because collisions are more frequent where the molecules are most crowded together).

digestion The breakdown of food particles, then into nutient molecules small enough to be absorbed.

digestive system An internal tube from which ingested food is absorbed into the internal environment; divided into regions specialized for food transport, processing, and storage.

dihybrid cross An experimental cross in which offspring inherit two gene pairs, each consisting of two nonidentical alleles.

diploid number (DIP-loyd) For many sexually reproducing species, the chromosome number of somatic cells and of germ cells prior to meiosis. Such cells have two chromosomes of each type (that is, pairs of homologous chromosomes). Compare haploid number.

directional selection Of a population, a shift in allele frequencies in a steady, consistent direction in response to a new environment or to a directional change in the old one. The outcome is that forms of traits at one end of the range of phenotypic variation become more common than the intermediate forms.

disaccharide (die-SAK-uh-ride) [Gk. *di*, two, and *sakcharon*, sugar] A type of simple carbohydrate, of the class called oligosaccharides; two monosaccharides covalently bonded.

disruptive selection Of a population, a shift in allele frequencies to forms of traits at both ends of a range of phenotypic variation and away from intermediate forms.

distal tubule The tubular section of a nephron most distant from the glomerulus; a major site of water and sodium reabsorption.

divergence Accumulation of differences in allele frequencies between populations that have become reproductively isolated from one another.

DNA Deoxyribonucleic acid (dee-OX-ee-rye-bow-new-CLAY-ik) For all cells (and many viruses), the molecule of inheritance. A category of nucleic acids, each usually consisting of two nucleotide strands twisted together helically and held together by hydrogen bonds. The nucleotide sequence encodes the instructions for assembling proteins, and, ultimately, a new individual.

DNA clone Many identical copies of foreign DNA that was inserted into plasmids and later replicated repeatedly after being taken up by a population of host cells (typically, bacteria).

DNA fingerprint Of each individual, a unique array of RFLPs, resulting from the DNA sequences inherited (in a Mendelian pattern) from each parent.

DNA ligase (LYE-gase) Enzyme that seals together the new base-pairings during DNA replication; also used by recombinant DNA technologists to seal base-pairings between DNA fragments and cut plasmid DNA.

DNA polymerase (poe-LIM-uh-rase) Enzyme that assembles a new strand on a parent DNA strand during replication; also takes part in DNA repair.

DNA repair Following an alteration in the base sequence of a DNA strand, a process that restores the original sequence, as carried out by DNA polymerases, DNA ligases, and other enzymes.

DNA replication Of cells, the process by which the hereditary material is duplicated for distribution to daughter nuclei. An example is the duplication of eukaryotic chromosomes during interphase, prior to mitosis.

dominant allele In a diploid cell, an allele that masks the expression of its partner on the homologous chromosome.

drug addiction Chemical dependence on a drug, which assumes an "essential" biochemical role in the body following habituation and tolerance.

dry acid deposition The falling to Earth of sulfur and nitrogen oxides; called *wet acid deposition*, or acid rain, when it occurs in rain or snow.

duplication A change in a chromosome's structure resulting in the repeated appearance of the same gene sequence.

dysplasia An abnormal change in the sizes, shapes, and organization of cells in a tissue.

early *Homo* A type of early hominid that may have been the maker of stone tools that date from about 2.5 million years ago.

ecology [Gk. *oikos*, home, and *logos*, reason] Study of the interactions of organisms with one another and with their physical and chemical environment.

ecosystem [Gk. *oikos*, home] An array of organisms and their physical environment, all of which interact through a flow of energy and a cycling of materials.

ecosystem analysis Method of predicting unforeseen effects of disturbances to an ecosystem, based on computer programs and models.

ectoderm [Gk. *ecto*, outside, and *derma*, skin] The outermost primary tissue layer (germ layer) of an embryo, which gives rise to the outer layer of the integument and to tissues of the nervous system.

effector A muscle (or gland) that responds to signals from an integrator (such as the brain) by producing movement (or chemical change) that helps adjust the body to changing conditions.

effector cell Of the differentiated sub-populations of lymphocytes that form during an immune response, the type of cell that engages and destroys the antigen-bearing agent that triggered the response.

egg A mature female gamete; also called an ovum.

El Niño Massive eastward movement of warm surface waters of the western equatorial Pacific that displaces cooler water off the coast of South America. A recurring event that disrupts the climate and ecosystems throughout the world.

electrolyte Any chemical substance, such as a salt, that ionizes and dissociates in water and is capable of conducting an electrical current.

electron Negatively charged unit of matter, with both particulate and wave-like properties, that occupies one of the orbitals around the atomic nucleus. Atoms can gain, lose, or share electrons with other atoms.

electron transfer A molecule donates one or more electrons to another molecule.

electron transport system An organized array of enzymes and cofactors, bound in a cell membrane, that accept and donate electrons in sequence. When such systems operate, hydrogen ions (H^+) flow across the membrane, and the flow drives ATP formation and other reactions.

element Any substance that cannot be decomposed into substances with different properties.

embryo (EM-bree-oh) [Gk. *en*, in, and probably *bryein*, to swell] Of animals generally, the stage formed by way of cleavage, gastrulation, and other early developmental events.

embryonic disk In early development, the oval, flattened cell mass that gives rise to the embryo shortly after implantation.

emerging pathogen A newly mutated strain of an existing pathogen or one that is now exploiting an increased availability of human hosts.

emulsification In chyme, a suspension of fat droplets coated with bile salts.

end product Substance present at the end of a metabolic pathway.

endergonic reaction Chemical reaction having a net gain in energy.

endocrine gland A ductless gland that secretes hormones, which usually enter interstitial fluid and then the blood-stream.

endocrine system System of cells, tissues, and organs that is functionally linked to the nervous system and that exerts control by way of its hormones and other chemical secretions.

endocytosis (en-doe-sigh-TOE-sis) Movement of a substance into cells; the substance becomes enclosed by a patch of plasma membrane that sinks into the cytoplasm, then forms a vesicle around it. Phagocytic cells also engulf pathogens or prey in this manner.

endoderm [Gk. *endon*, within, and *derma*, skin] The inner primary tissue layer, or germ layer, of an embryo, which gives rise to the inner lining of the gut and organs derived from it.

endometrium (en-doh-MEET-ree-um) [Gk. *metrios*, of the womb] Inner lining of the uterus consisting of connective tissues, glands, and blood vessels.

endoplasmic reticulum or ER (en-doe-PLAZ-mik reh-TIK-yoo-lum) An organelle that begins at the nucleus and curves through the cytoplasm. In rough ER (which has many ribosomes on its cytoplasmic side), many new polypeptide chains acquire specialized side chains. In many cells, smooth ER (with no attached ribosomes) is the main site of lipid synthesis.

endotherm Animal such as a human that maintains body temperature from within, generally by metabolic activity and controls over heat conservation and dissipation.

energy The capacity to do work.

energy carrier A molecule that delivers energy from one metabolic reaction site to another. ATP is the most widely traveled of these; it readily donates energy to nearly all metabolic reactions.

energy pyramid A pyramid-shaped representation of an ecosystem's trophic structure, illustrating the energy losses at each transfer to a different trophic level.

enzyme (EN-zime) One of a class of proteins that greatly speed up (catalyze) reactions between specific substances. The substances that each type of enzyme acts upon are called its substrates.

eosinophil Fast-acting, phagocytic white blood cell that takes part in inflammation but not in immune responses.

epidermis The outermost tissue layer of a multicelled animal.

epiglottis A flaplike structure at the start of the larynx, the position of which directs the movement of air into the trachea or food into the esophagus.

epinephrine (ep-ih-NEF-rin) Adrenal hormone that raises blood levels of sugar and fatty acids; increases the heart's rate and force of contraction.

epiphyseal plate Region of cartilage that covers either end of a growing long bone, permitting the bone to lengthen. The epiphyseal plate is replaced by bone when growth stops in late adolescence.

epithelium (ep-ih-THEE-lee-um) A tissue consisting of one or more layers of adhering cells that covers the body's external surfaces and lines its internal cavities and tubes. Epithelium has one free surface; the opposite surface rests on a basement membrane between it and an underlying connective tissue. Epidermis or skin is an example.

erosion Movement of land under the force of wind, running water, and ice.

erythrocyte (eh-RITH-row-site) [Gk. *erythros*, red, and *kytos*, vessel] Red blood cell.

esophagus (ee-SOF-uh-gus) Tubular portion of the digestive system that receives swallowed food and leads to the stomach.

essential amino acid Any of eight amino acids that the body cannot synthesize and must be obtained from food.

essential fatty acid Any of the fatty acids that the body cannot synthesize and must be obtained from food.

estrogen (ESS-tro-jen) A sex hormone that helps oocytes mature, induces changes in the uterine lining during the menstrual cycle and pregnancy, and maintains secondary sexual traits; also influences body growth and development.

eukaryotic cell (yoo-carry-AH-tic) [Gk. *eu*, good, and *karyon*, kernel] A type of cell that has a "true nucleus" and other distinguishing membrane-bound organelles. Compare prokaryotic cell.

evolution, biological [L. *evolutio*, act of unrolling] Change within a line of descent over time. A population is evolving when some forms of a trait are becoming more or less common relative to the other kinds of traits. The shifts are evidence of changes in the relative abundances of alleles for that trait, as brought about by mutation, natural selection, genetic drift, and gene flow.

evolutionary tree A treelike diagram in which branches represent separate lines of descent from a common ancestor.

excitatory postsynaptic potential or **EPSP** One of two competing signals at an input zone of a neuron; a graded potential that brings the neuron's plasma membrane closer to threshold.

excretion Any of several processes by which excess water, excess or harmful solutes, or waste materials leave the body by way of the urinary system or certain glands.

exergonic reaction Chemical reaction that shows a net loss in energy.

exocrine gland (EK-suh-krin) [Gk. *es*, out of, and *krinein*, to separate] Glandular structure that secretes products, usually through ducts or tubes, to a free epithelial surface.

exocytosis (ek-so-sigh-TOE-sis) Movement of a substance out of a cell by means of a transport vesicle, the membrane of which fuses with the plasma membrane, so that the vesicle's contents are released outside.

exon Of eukaryotic cells, any of the nucleotide sequences of a pre-mRNA molecule that are spliced together to form the mature mRNA transcript and are ultimately translated into protein.

experiment A test in which some phenomenon in the natural world is manipulated in controlled ways to gain insight into its function, structure, operation, or behavior.

expiration Expelling air from the lungs; exhaling.

exponential growth (ex-po-NEN-shul) Pattern of population growth in which greater and greater numbers of individuals are produced during the successive doubling times; the pattern that emerges when the per capita birth rate remains even slightly above the per capita death rate, putting aside the effects of immigration and emigration.

external respiration Movement of oxygen from alveoli into the blood, and of carbon dioxide from the blood into alveoli.

extinction Irrevocable loss of a species.

extracellular fluid In animals generally, all the fluid not inside cells; includes plasma (the liquid portion of blood) and interstitial fluid (which occupies the spaces between cells and tissues).

extracellular matrix A material, largely secreted, that helps hold many animal tissues together in certain shapes; it consists of fibrous proteins and other components in a ground substance.

extraembryonic membranes Membranes that form along with a developing embryo, including the yolk sac, amnion, allantois, and chorion.

F_1 (first filial generation) The offspring of an initial genetic cross.

F_2 (second filial generation) The offspring of parents who are the first filial generation from a genetic cross.

FAD Flavin adenine dinucleotide, a nucleotide coenzyme. When delivering electrons and unbound protons (H^+) from one reaction to another, it is abbreviated $FADH_2$.

family pedigree A chart of genetic relationships of the individuals in a family through successive generations.

fat A lipid with a glycerol head and one, two, or three fatty acid tails. The tails of saturated fats have only single bonds between carbon atoms and hydrogen atoms attached to all other bonding sites. Tails of unsaturated fats additionally have one or more double bonds between certain carbon atoms.

fatty acid A long, flexible hydrocarbon chain with a —COOH group at one end.

feedback inhibition Of cells, a control mechanism by which the production (or secretion) of a substance triggers a change in some activity that in turn shuts down further production of the substance.

femur Thigh bone; longest and strongest bone of the body.

fermentation [L. *fermentum*, yeast] A type of anaerobic pathway of ATP formation; it starts with glycolysis, ends when electrons are transferred back to one of the breakdown products or intermediates, and regenerates the NAD^+ required for the reaction. Its net yield is two ATP per glucose molecule broken down.

fertilization [L. *fertilis*, to carry, to bear] Fusion of a sperm nucleus with the nucleus of an egg, which thereupon becomes a zygote.

fetus Term applied to an embryo after it reaches the age of eight weeks.

fever Body temperature that has climbed above the normal set point, usually in response to infection. Mild fever promotes an increase in body defense activities.

fibrous joint Type of joint in which fibrous connective tissue unites the adjoining bones and no cavity is present.

filtration In urine formation, the process by which blood pressure forces water and solutes out of glomerular capillaries and into the cupped portion of a nephron wall (glomerular capsule).

flagellum (fluh-JELL-um), plural flagella [L., *whip*] Tail-like motile structure of many free-living eukaryotic cells; it has a distinctive 9 + 2 array of microtubules.

fluid mosaic model Model of membrane structure in which proteins are embedded in a lipid bilayer or attached to one of its surfaces. The lipid molecules give the membrane its basic structure, impermeability to water-soluble molecules, and (through packing variations and movements) fluidity. Proteins carry out most membrane functions, such as transport, enzyme action, and reception of signals or substances.

follicle (FOLL-ih-kul) In an ovary, a primary oocyte (immature egg) together with the surrounding layer of cells.

food chain A straight-line sequence of who eats whom in an ecosystem.

food web A network of cross-connecting, interlinked food chains encompassing primary producers and an array of consumers, detritivores, and decomposers.

forebrain Brain region that includes the cerebrum and cerebral cortex, the olfactory lobes, and the hypothalamus.

fossil Recognizable evidence of an organism that lived in the distant past. Most fossils are skeletons, shells, leaves, seeds, and tracks that were buried in rock layers before they decomposed.

fossilization How fossils form. An organism or traces of it become buried in sediments or volcanic ash. Water and dissolved inorganic compounds infiltrate the remains. Accumulating sediments exert pressure above the burial site. Over time, the pressure and chemical changes transform the remains to stony hardness.

free radical Any highly reactive molecule or molecule fragment having an unpaired electron.

FSH Follicle-stimulating hormone. The name comes from its function in females, in whom FSH helps stimulate follicle development in ovaries. In males, it acts in the testes as part of a sequence of events that trigger sperm production.

functional group An atom or group of atoms that is covalently bonded to the carbon backbone of an organic compound and that influences its behavior.

gallbladder Organ of the digestive system that stores bile secreted from the liver.

gamete (GAM-eet) [Gk. *gametēs*, husband, and *gametē*, wife] A haploid cell that functions in sexual reproduction. Sperm and eggs are examples.

ganglion (GANG-lee-un), plural ganglia [Gk. *ganglion*, a swelling] A distinct clustering of cell bodies of neurons in regions other than the brain or spinal cord.

gastrointestinal (GI) tract The digestive tube, extending from the mouth to the anus and including the stomach, small and large intestines, and other specialized regions with roles in food transport and digestion.

gastrulation (gas-tru-LAY-shun) The stage of embryonic development in which cells become arranged into two or three primary tissue layers (germ layers); in humans, the layers are an inner endoderm, an intermediate mesoderm, and a surface ectoderm.

gene A unit of information about a heritable trait that is passed on from parents to offspring. Each gene has a specific location on a chromosome.

gene flow A microevolutionary process; a physical movement of alleles out of a population as individuals leave (emigrate) or enter (immigrate); allele frequencies change as a result.

gene frequency More precisely, allele frequency: the relative abundances of all the different alleles for a trait that are carried by the individual members of a population.

gene library A mixed collection of bacteria that contain many different cloned DNA fragments.

gene locus A given gene's particular location on a chromosome.

gene pair In diploid cells, the two alleles at a given locus on a pair of homologous chromosomes.

gene pool Sum total of all genotypes in a population. More accurately, allele pool.

gene therapy Generally, the transfer of one or more normal genes into the body cells of an organism in order to correct a genetic defect.

genetic code [After L. *genesis*, to be born] The correspondence between nucleotide triplets in DNA (then in mRNA) and specific sequences of amino acids in the resulting polypeptide chains; the basic language of protein synthesis.

genetic disorder An inherited condition that results in mild to severe medical problems.

genetic divergence Gradual accumulation of differences in gene pools of populations or subpopulations of a species after a new geographic barrier separates them; thereafter, mutation, natural selection, and genetic drift are operating independently of one another.

genetic drift A microevolutionary process; a change in allele frequencies over the generations due to chance events alone.

genetic engineering Altering the information content of DNA through use of recombinant DNA technology.

genetic recombination Presence of a new combination of alleles in a DNA molecule compared to the parental genotype; the result of processes such as crossing over at meiosis, chromosome rearrangements, gene mutation, and recombinant DNA technology.

genital herpes Infection of tissues in the genital area by a herpes simplex virus; an extremely contagious sexually transmitted disease.

genome All the DNA in a haploid number of chromosomes of a given species.

genotype (JEEN-oh-type) Genetic constitution of an individual. Can mean a single gene pair or the sum total of the individual's genes. Compare phenotype.

genus, plural genera (JEEN-us, JEN-er-ah) [L. *genus*, race, origin] A taxon into which all species exhibiting certain phenotypic similarities and evolutionary relationship are grouped.

germ cell One of a cell lineage set aside for sexual reproduction; germ cells give rise to gametes. Compare somatic cell.

germ layer One of two or three primary tissue layers that forms during gastrulation and that gives rise to certain tissues of the adult body. Compare ectoderm; endoderm; mesoderm.

gland A secretory cell or multicelled structure derived from epithelium and often connected to it.

glomerular capillaries The set of blood capillaries inside the glomerular capsule of the nephron.

glomerulus (glow-MARE-you-luss) [L. *glomus,* ball] The first portion of the nephron, where water and solutes are filtered from blood.

glucagon (GLUE-kuh-gone) Hormone that stimulates conversion of glycogen and amino acids to glucose; secreted by alpha cells of the pancreas when the flow of glucose decreases.

glucocorticoid Hormone secreted by the adrenal cortex that influences metabolic reactions that help maintain the blood glucose level.

glyceride (GLISS-er-eyed) One of the molecules, commonly called fats and oils, that has one, two, or three fatty acid tails attached to a glycerol backbone. They are the body's most abundant lipids and its richest source of energy.

glycerol (GLISS-er-all) [Gk. *glykys,* sweet, and L. *oleum,* oil] A three-carbon molecule with three hydroxyl groups attached; together with fatty acids, a component of fats and oils.

glycogen (GLY-kuh-jen) In animals, a storage polysaccharide that is a main food reserve; can be readily broken down into glucose subunits.

glycolysis (gly-CALL-ih-sis) [Gk. *glykys,* sweet, and *lysis,* loosening or breaking apart] Initial reactions of both aerobic and anaerobic pathways by which glucose (or some other organic compound) is partially broken down to pyruvate with a net yield of two ATP. Glycolysis proceeds in the cytoplasm of all cells, and oxygen has no role in it.

glycoprotein A protein having linear or branched oligosaccharides covalently bonded to it. Most human cell surface proteins and many proteins circulating in blood are glycoproteins.

Golgi body (GOHL-gee) Organelle in which newly synthesized polypeptide chains as well as lipids are modified and packaged in vesicles for export or for transport to specific locations within the cytoplasm.

gonad (GO-nad) Primary reproductive organ in which gametes are produced.

gonorrhea Clinical term for the sexually transmitted disease caused by the bacterium *Neisseria gonorrhoeae.* This bacterium can infect epithelial cells of the genital tract, rectum, eye membranes, and the throat.

graded potential Of neurons, a local signal that slightly changes the voltage difference across a small patch of the plasma membrane. Such signals vary in magnitude, depending on the stimulus. With prolonged or intense stimulation, they may spread to a trigger zone of the membrane and initiate an action potential.

granulocyte Class of white blood cells that have a lobed nucleus and various types of granules in the cytoplasm; includes neutrophils, eosinophils, and basophils.

gray matter The dendrites, neuron cell bodies, and neuroglial cells of the spinal cord and cerebral cortex.

grazing food web Of most ecosystems, the flow of energy from plants to herbivores, then through an array of carnivores.

green revolution In developing countries, the use of improved crop varieties, modern agricultural practices (including massive inputs of fertilizers and pesticides), and equipment to increase crop yields.

greenhouse effect Warming of the lower atmosphere due to the presence of greenhouse gases: carbon dioxide, methane, nitrous oxide, ozone, water vapor, and chlorofluorocarbons.

ground substance The intercellular material made up of cell secretions and other noncellular components.

growth factor A type of signaling molecule that can influence growth by regulating the rate at which target cells divide.

guanine A nitrogen-containing base; present in one of the four nucleotide building blocks of DNA and RNA.

habitat [L. *habitare,* to live in] The type of place where an organism normally lives, characterized by physical features, chemical features, and the presence of certain other species.

hair cell Type of mechanoreceptor that may give rise to action potentials when bent or tilted.

haploid number (HAP-loyd) The chromosome number of a gamete that, as an outcome of meiosis, is only half that of the parent germ cell (it has only one of each pair of homologous chromosomes). Compare diploid number.

HCG Human chorionic gonadotropin. A hormone that helps maintain the lining of the uterus during the menstrual cycle and during the first trimester of pregnancy.

heart Muscular pump that keeps blood circulating through the animal body.

helper T cell Type of T lymphocyte that produces and secretes chemicals that promote formation of large effector and memory cell populations.

hemoglobin (HEEM-oh-glow-bin) [Gk. *haima,* blood, and L. *globus,* ball] Iron-containing, oxygen-transporting protein that gives red blood cells their color.

hemostasis (hee-mow-STAY-sis) [Gk. *haima,* blood, and *stasis,* standing] Stopping of blood loss from a damaged blood vessel through coagulation, blood vessel spasm, platelet plug formation, and other mechanisms.

hepatic portal vein Vessel that receives nutrient-laden blood from villi of the small intestine and transports it to the liver, where excess glucose is removed. The blood then returns to the general circulation via a hepatic vein.

hepatitis B An extremely contagious, sexually transmitted disease caused by infection by the hepatitis B virus. Chronic hepatitis can lead to liver cirrhosis or cancer.

herbivore [L. *herba,* grass, and *vovare,* to devour] Plant-eating animal.

heterotroph (HET-er-oh-trofe) [Gk. *heteros,* other, and *trophos,* feeder] Organism that cannot synthesize its own organic compounds and must obtain nourishment by feeding on autotrophs, each other, or organic wastes. Animals, fungi, many protista, and most bacteria are heterotrophs. Compare autotroph.

heterozygous condition (het-er-oh-ZYE-guss) [Gk. *zygoun,* join together] For a given trait, having nonidentical alleles at a particular locus on a pair of homologous chromosomes.

hindbrain One of the three divisions of the brain; the medulla oblongata, cerebellum, and pons; includes reflex centers for respiration, blood circulation, and other basic functions; also coordinates motor responses and many complex reflexes.

histone Any of a class of proteins that are intimately associated with DNA and that are largely responsible for its structural (and possibly functional) organization in eukaryotic chromosomes.

homeostasis (hoe-me-oh-STAY-sis) [Gk. *homo*, same, and *stasis*, standing] A physiological state in which the physical and chemical conditions of the internal environment are being maintained within tolerable ranges.

homeostatic feedback loop An interaction in which an organ (or structure) stimulates or inhibits the output of another organ, then shuts down or increases this activity when it detects that the output has exceeded or fallen below a set point.

hominid [L. *homo*, man] All species on the evolutionary branch leading to modern humans. *Homo sapiens* is the only living representative.

hominoid Apes, humans, and their recent ancestors.

Homo erectus A hominid lineage that emerged between 1.5 million and 300,000 years ago and that may include the direct ancestors of modern humans.

Homo sapiens The hominid lineage of modern humans that emerged between 300,000 and 200,000 years ago.

homologous chromosome (huh-MOLL-uh-gus) [Gk. *homologia*, correspondence] One of a pair of chromosomes that resemble each other in size, shape, and the genes they carry, and that line up with each other at meiosis I. The X and Y chromosomes differ in these respects but still function as homologues.

homologous structure The same body part, modified in different ways, in different lines of descent from a common ancestor.

homozygous condition (hoe-moe-ZYE-guss) Having two identical alleles at a given locus (on a pair of homologous chromosomes).

homozygous dominant condition Having two dominant alleles at a given locus (on a pair of homologous chromosomes).

homozygous recessive condition Having two recessive alleles at a given gene locus (on a pair of homologous chromosomes).

hormone [Gk. *hormon*, to stir up, set in motion] Any of the signaling molecules secreted from endocrine glands, endocrine cells, and some neurons that the bloodstream distributes to nonadjacent target cells (any cell having receptors for that hormone).

human genome project A basic research project in which researchers throughout the world work together to sequence the estimated 3 billion nucleotides present in the DNA of human chromosomes.

human immunodeficiency virus (HIV) A retrovirus; the pathogen that causes AIDS (acquired immune deficiency syndrome).

human papillomavirus (HPV) Virus that causes genital warts; HPV infection is suspected of having a role in the development of some cases of cervical cancer.

humerus The long bone of the upper arm.

hybrid offspring Of a genetic cross, offspring with a pair of nonidentical alleles for a trait.

hydrocarbon A molecule having only hydrogen atoms attached to a carbon backbone.

hydrogen bond Type of chemical bond in which an atom of a molecule interacts weakly with a neighboring atom that is already taking part in a polar covalent bond.

hydrogen ion A free (unbound) proton; a hydrogen atom that has lost its electron and so bears a positive charge (H^+).

hydrologic cycle A biogeochemical cycle, driven by solar energy, in which water moves slowly through the atmosphere, on or through surface layers of land masses, to the ocean and back again.

hydrolysis (high-DRAWL-ih-sis) [L. *hydro*, water, and Gk. *lysis*, loosening or breaking apart] Enzyme-mediated reaction in which covalent bonds break, splitting a molecule into two or more parts, and H^+ and OH^- (derived from a water molecule) become attached to the exposed bonding sites.

hydrophilic substance [Gk. *philos*, loving] A polar substance that is attracted to the polar water molecule and so dissolves easily in water. Sugars are examples.

hydrophobic substance [Gk. *phobos*, dreading] A nonpolar substance that is repelled by the polar water molecule and so does not readily dissolve in water. Oil is an example.

hydrosphere All liquid or frozen water on or near the Earth's surface.

hypertonic solution A fluid having a greater concentration of solutes relative to another fluid.

hypodermis A subcutaneous layer having stored fat that helps insulate the body; although not part of skin, it anchors skin while allowing it some freedom of movement.

hypothalamus [Gk. *hypo*, under, and *thalamos*, inner chamber, or possibly *tholos*, rotunda] A brain center that monitors visceral activities (such as salt–water balance, temperature control, and reproduction) and that influences related forms of behavior (as in hunger, thirst, and sex).

hypothesis A possible explanation of a specific phenomenon.

hypotonic solution A fluid that has a lower concentration of solutes relative to another fluid.

immune response A series of events by which B and T lymphocytes recognize a specific antigen, undergo repeated cell divisions that form huge lymphocyte populations, and differentiate into subpopulations of effector and memory cells. Effector cells engage and destroy antigen-bearing agents. Memory cells enter a resting phase and are activated during subsequent encounters with the same antigen.

immunization Various processes, including vaccination, that promote increased immunity against specific diseases.

immunoglobulin Any of the five classes of antibodies. Different immunoglobulins participate in specific ways in defense and immune responses. Examples are IgM antibodies (first to be secreted during immune responses) and IgG antibodies (which activate complement proteins and neutralize many toxins).

immunotherapy Procedures that enhance a person's immunological defenses against tumors or certain pathogens.

implantation Series of events in which a trophoblast (pre-embryo) invades the endometrium (lining of the uterus) and becomes embedded there.

in vitro fertilization Conception outside the body ("in glass" petri dishes or test tubes).

incomplete dominance Of heterozygotes, the appearance of a version of a trait that is somewhere between the homozygous dominant and recessive conditions.

independent assortment Mendelian principle that each gene pair tends to assort into gametes independently of other gene pairs located on nonhomologous chromosomes.

induced-fit model Model of enzyme action whereby a bound substrate induces changes in the shape of the enzyme's active site, resulting in a more precise molecular fit between the enzyme and its substrate.

inductive logic Pattern of thinking by which a person derives a general statement from specific observations.

industrial smog A type of gray-air smog that develops in industrialized regions when winters are cold and wet.

infection Invasion and multiplication of a pathogen in a host. Disease follows if defenses are not mobilized fast enough; the pathogen's activities interfere with normal body functions.

inflammation, acute In response to tissue damage or irritation, phagocytes and plasma proteins, including complement proteins, leave the bloodstream, then defend and help repair the tissue. Proceeds during both nonspecific and specific (immune) defense responses.

inheritance The transmission, from parents to offspring, of structural and functional patterns that have a genetic basis and are characteristic of each species.

inhibiting hormone A signaling molecule produced and secreted by the hypothalamus that controls secretions by the anterior lobe of the pituitary gland.

inhibitor A substance that can bind with an enzyme and interfere with its functioning.

inhibitory postsynaptic potential, or IPSP. Of neurons, one of two competing types of graded potentials at an input zone; tends to drive the resting membrane potential away from threshold.

inner cell mass In early development, a clump of cells in the blastocyst that will give rise to the embryonic disk.

insertion The end of a muscle attached to the bone that moves most when the muscle contracts.

inspiration The drawing of air into the lungs; inhaling.

insulin Pancreatic hormone that lowers the level of glucose in blood by causing cells to take up glucose; also promotes the synthesis of fat and protein and inhibits the conversion of protein to glucose.

integration, neural [L. *integrare,* to coordinate] Moment-by-moment summation of all excitatory and inhibitory synapses acting on a neuron; occurs at each level of synapsing in a nervous system.

integrator Of homeostatic systems, a control point where different bits of information are pulled together in the selection of a response. The brain is an example.

integument A protective body covering such as skin.

interferon Protein produced by T cells and that interferes with viral replication. Some interferons also stimulate the tumor-killing activity of macrophages.

interleukin One of a variety of communication signals, secreted by macrophages and by helper T cells, that drive immune responses.

intermediate Substance that forms between the start and end of a metabolic pathway.

intermediate filament A cytoskeletal component that consists of different proteins in different types of animal cells.

internal respiration Movement of oxygen into tissues from the blood, and of carbon dioxide from tissues into the blood.

interneuron Any of the neurons in the brain and spinal cord that integrate information arriving from sensory neurons and that influence other neurons in turn.

interphase Of cell cycles, the time interval between nuclear divisions in which a cell increases its mass, roughly doubles the number of its cytoplasmic components, and finally duplicates its chromosomes (replicates its DNA).

interstitial fluid (in-ter-STISH-ul) [L. *interstitus,* to stand in the middle of something] That portion of the extracellular fluid occupying spaces between cells and tissues.

intervertebral disk One of a number of disk-shaped structures containing cartilage that serve as shock absorbers and flex points between bony segments of the vertebral column.

intron A noncoding portion of a newly formed mRNA molecule.

inversion A change in a chromosome's structure after a segment separated from it was then inserted at the same place, but in reverse. The reversal alters the position and order of the chromosome's genes.

ion (EYE-on) An atom or a compound that has gained or lost one or more electrons and hence has acquired an overall negative or positive charge.

ionic bond An association between ions of opposite charge.

iris Of the eye, a circular pigmented region behind the cornea with a "hole" in its center (the pupil) through which incoming light enters.

isotonic solution A fluid having the same solute concentration as a fluid against which it is being compared.

isotope (EYE-so-tope) For a given element, an atom with the same number of protons as the other atoms but with a different number of neutrons.

J-shaped curve A curve, obtained when population size is plotted against time, that is characteristic of unrestricted, exponential growth.

joint An area of contact or near-contact between bones.

juxtaglomerular apparatus In kidney nephrons, a region of contact between the arterioles of the glomerulus and the distal tubule. Cells in this region secrete renin, which triggers hormonal events that stimulate increased reabsorption of sodium.

karyotype (CARRY-oh-type) Of eukaryotic individuals (or species), the number of metaphase chromosomes in somatic cells and their defining characteristics.

keratin A tough, water-insoluble protein manufactured by most epidermal cells.

keratinization (care-at-in-iz-AY-shun) Process by which keratin-producing epidermal cells of skin die and collect at the skin surface as keratinized "bags" that form a barrier against dehydration, bacteria, and many toxic substances.

kidney One of a pair of organs that filter mineral ions, organic wastes, and other substances from the blood and help regulate the volume and solute concentrations of extracellular fluid.

kilocalorie 1,000 calories of heat energy, or the amount of energy needed to raise the temperature of 1 kilogram of water by 1°C; the unit of measure for the caloric value of foods.

kinetochore A specialized group of proteins and DNA at the centromere of a chromosome that serves as an attachment point for several spindle microtubules during mitosis or meiosis. Each chromatid of a duplicated chromosome has its own kinetochore.

Krebs cycle Together with a few conversion steps that precede it, the stage of aerobic respiration in which pyruvate is completely broken down to carbon dioxide and water. Coenzymes accept the unbound protons (H^+) and electrons stripped from intermediates during the reactions and deliver them to the next stage.

lactate fermentation Anaerobic pathway of ATP formation in which pyruvate from glycolysis is converted to the three-carbon compound lactate, and NAD^+ (a coenzyme used in the reactions) is regenerated. Its net yield is two ATP.

lactation The production of milk by hormone-primed mammary glands.

lacteal Small lymph vessel in villi of the small intestine that receives absorbed triglycerides. Triglycerides move from the lymphatic system to the general circulation.

large intestine The colon; a region of the GI tract that receives unabsorbed food residues from the small intestine and concentrates and stores feces until they are expelled from the body.

larynx (LARE-inks) A tubular airway that leads to the lungs. It contains vocal cords, where sound waves used in speech are produced.

latency Of viruses, a period of time during which viral genes remain inactive inside the host cell.

lens Of the eye, a saucer-shaped region behind the iris containing multiple layers of transparent proteins. Ligaments can move the lens, which functions to focus incoming light onto photoreceptors in the retina.

Leydig cell In testes, cells in connective tissue around the seminiferous tubules that secrete testosterone and other signaling molecules.

LH Luteinizing hormone, secreted by the anterior lobe of the pituitary gland. In males it acts on interstitial cells of the testes and prompts them to secrete testosterone. In females, LH stimulates follicle development in the ovaries.

life cycle Recurring pattern of genetically programmed events from the time

individuals are produced until they themselves reproduce.

life table Tabulation of age-specific patterns of birth and death for a population.

ligament A strap of dense, regular connective tissue that connects two bones at a joint.

limbic system Brain regions that, along with the cerebral cortex, collectively govern emotions.

limiting factor Any essential resource that is in short supply and so limits population growth.

lineage (LIN-ee-age) A line of descent.

linkage The tendency of genes located on the same chromosome to end up in the same gamete. For any two of those genes, the probability that crossing over will disrupt the linkage is proportional to the distance separating them.

lipid A greasy or oily compound of mostly carbon and hydrogen that shows little tendency to dissolve in water, but that dissolves in nonpolar solvents (such as ether). Cells use lipids as energy stores and structural materials, especially in membranes.

lipid bilayer The structural basis of cell membranes, consisting of two layers of mostly phospholipid molecules. Hydrophilic heads force all fatty acid tails of the lipids to become sandwiched between the hydrophilic heads.

lipoprotein Molecule that forms when proteins circulating in blood combine with cholesterol, triglycerides, and phospholipids absorbed from the small intestine.

liver Glandular organ with roles in storing and interconverting carbohydrates, lipids, and proteins absorbed from the gut; maintaining blood; disposing of nitrogen-containing wastes; and other tasks.

local signaling molecule Cellular secretion that alters chemical conditions in the immediate vicinity where it is secreted, then is swiftly degraded.

locus (LOW-cuss) The specific location of a particular gene on a chromosome.

logic Thought patterns by which a person draws a conclusion that does not contradict evidence used to support that conclusion.

logistic growth (low-JIS-tik) Pattern of population growth in which a low-density population slowly increases in size, goes through a rapid growth phase, then levels off once the carrying capacity is reached.

loop of Henle The hairpin-shaped, tubular region of a nephron that functions in reabsorption of water and solutes.

lung Saclike organ that serves as an internal respiratory surface.

lymph (limf) [L. *lympha*, water] Tissue fluid that has moved into the vessels of the lymphatic system.

lymph capillary A small-diameter vessel of the lymph vascular system that has no obvious entrance; tissue fluid moves inward by passing between overlapping endothelial cells at the vessel's tip.

lymph node A lymphoid organ that serves as a battleground of the immune system; each lymph node is packed with organized arrays of macrophages and lymphocytes that cleanse lymph of pathogens before it reaches the blood.

lymph vascular system [L. *lympha*, water, and *vasculum*, a small vessel] The vessels of the lymphatic system, which take up and transport excess tissue fluid and reclaimable solutes as well as fats absorbed from the digestive tract.

lymphatic system An organ system that supplements the circulatory system. Its vessels take up fluid and solutes from interstitial fluid and deliver them to the bloodstream; its lymphoid organs have roles in immunity.

lymphocyte A T cell or B cell.

lymphoid organ The lymph nodes, spleen, thymus, tonsils, adenoids, and other organs with roles in immunity.

lysis [Gk. *lysis*, a loosening] Gross structural disruption of a plasma membrane that leads to cell death.

lysosome (LYE-so-sohm) The main organelle of digestion, with enzymes that can break down polysaccharides, proteins, nucleic acids, and some lipids.

lysozyme Infection-fighting enzyme that attacks and destroys various types of bacteria by digesting the bacterial cell wall. Present in mucous membranes that line the body's surfaces.

macroevolution The large-scale patterns, trends, and rates of change among groups of species.

macrophage One of the phagocytic white blood cells. It engulfs anything detected as foreign. Some also become antigen-presenting cells that serve as the trigger for immune responses by T and B lymphocytes. Compare antigen-presenting cell.

malnutrition A state in which body functions or development suffers due to inadequate or unbalanced food intake.

mammal A type of vertebrate; the only animal having offspring that are nourished by milk produced by mammary glands of females.

mass extinction An abrupt rise in extinction rates above the background level; a catastrophic, global event in which major taxa are wiped out simultaneously.

mass number The total number of protons and neutrons in an atom's nucleus. The relative masses of atoms are also called atomic weights.

mechanoreceptor Sensory cell or cell part that detects mechanical energy associated with changes in pressure, position, or acceleration.

medulla oblongata Part of the brainstem with reflex centers for respiration, blood circulation, and other vital functions.

meiosis (my-OH-sis) [Gk. *meioun*, to diminish] Two-stage nuclear division process in which the chromosome number of a germ cell is reduced by half, to the haploid number. (Each daughter nucleus ends up with one of each type of chromosome.) Meiosis is the basis of gamete formation.

memory The storage and retrieval of information about previous experiences; underlies the capacity for learning.

memory cell Any of the various B or T lymphocytes of the immune system that are formed in response to invasion

by a foreign agent and that circulate for some period, available to mount a rapid attack if the same type of invader reappears.

meninges Membranes of connective tissue that are layered between the skull bones and the brain and cover and protect the neurons and blood vessels that service the brain tissue.

menopause (MEN-uh-pozz) [L. *mensis*, month, and *pausa*, stop] End of the period of a human female's reproductive potential.

menstrual cycle The cyclic release of oocytes and priming of the endometrium (lining of the uterus) to receive a fertilized egg; the complete cycle averages about 28 days in female humans.

menstruation Periodic sloughing of the blood-enriched lining of the uterus when pregnancy does not occur.

mesoderm (MEH-so-derm) [Gk. *mesos*, middle, and *derm*, skin] In an embryo, a primary tissue layer (germ layer) between ectoderm and endoderm. Gives rise to muscle; organs of circulation, reproduction, and excretion; most of the internal skeleton (when present); and connective tissue layers of the gastro-intestinal tract and integument.

messenger RNA A linear sequence of ribonucleotides transcribed from DNA and translated into a polypeptide chain; the only type of RNA that carries protein-building instructions.

metabolic pathway An orderly sequence of enzyme-mediated reactions by which cells maintain, increase, or decrease the concentrations of particular substances.

metabolism (meh-TAB-oh-lizm) [Gk. *meta*, change] All controlled, enzyme-mediated chemical reactions by which cells acquire and use energy. Through these reactions, cells synthesize, store, break apart, and eliminate substances in ways that contribute to growth, survival, and reproduction.

metaphase Of mitosis or meiosis II, the stage when each duplicated chromosome has become positioned at the midpoint of the microtubular spindle, with its two sister chromatids attached to microtubules from opposite spindle poles. Of meiosis I, the stage when all pairs of homologous chromosomes are positioned at the spindle's midpoint, with the two members of each pair attached to opposite spindle poles.

metastasis The process in which cancer cells break away from a primary tumor and migrate (via blood or lymphatic tissues) to other locations, where they establish new cancer sites.

MHC marker Any of a variety of proteins that are self-markers. Some occur on all body cells of an individual; others are unique to the macrophages and lymphocytes.

micelle (my-CELL) Of fat digestion, tiny droplet of bile salts, fatty acids, and monoglycerides; plays a role in fat absorption from the small intestine.

microevolution Changes in allele frequencies brought about by mutation, genetic drift, gene flow, and natural selection.

microfilament [Gk. *mikros*, small, and L. *filum*, thread] One of a variety of cytoskeletal components. Actin and myosin filaments are examples.

micrograph Photograph of an image brought into view with the aid of a microscope.

microorganism Organism, usually single celled, too small to be observed without a microscope.

microtubular spindle A bipolar structure composed of organized arrays of microtubules that forms during nuclear division and that moves the chromosomes.

microtubule Hollow cylinder of mainly tubulin subunits; a cytoskeletal element with roles in cell shape, motion, and growth and in the structure of cilia and flagella.

microtubule organizing center MTOC; mass of substances in the cell cytoplasm that dictate the orientation and organization of the cell's microtubules.

microvillus (my-crow-VILL-us) [L. *villus*, shaggy hair] A slender, cylindrical extension of the animal cell surface that functions in absorption or secretion.

midbrain A brain region that evolved as a coordination center for reflex responses to visual and auditory input; together with the pons and medulla oblongata, part of the brainstem, which includes the reticular formation.

mineral An inorganic substance required for the normal functioning of body cells.

mineralocorticoid Type of hormone, secreted by the adrenal cortex, that mainly regulates the concentrations of mineral salts in extracellular fluid.

mitochondrion (my-toe-KON-dree-on), plural mitochondria. Organelle that specializes in ATP formation; it is the site of the second and third stages of aerobic respiration, an oxygen-requiring pathway.

mixture Atoms of two or more elements intermingled in proportions that can and usually do vary.

mitosis (my-TOE-sis) [Gk. *mitos*, thread] Type of nuclear division that maintains the parental chromosome number for daughter cells. It is the basis of bodily growth and, in many eukaryotic species, asexual reproduction.

molecular clock Model used to calculate the time of origin of one lineage or species relative to others. The underlying assumption is that neutral mutations accumulate in a lineage at predictable rates that can be measured as a series of ticks back through time.

molecule A unit of matter in which chemical bonding holds together two or more atoms of the same or different elements.

monoclonal antibody Antibody produced in the laboratory by a population of genetically identical cells that are clones of a single "parent" antibody-producing cell.

monohybrid cross [Gk. *monos*, alone] An experimental cross in which offspring inherit a pair of nonidentical alleles for a single trait being studied, so that they are heterozygous.

monomer A small molecule that is commonly a subunit of polymers, such as the sugar monomers of starch.

monosaccharide (mon-oh-SAK-ah-ride) [Gk. *monos*, alone, single, and *sakharon*, sugar] The simplest carbohydrate, with only one sugar unit. Glucose is an example.

monosomy Abnormal condition in which one chromosome of diploid cells has no homologue.

morphogenesis (more-foe-JEN-ih-sis) [Gk. *morphe*, form, and *genesis*, origin] Processes by which differentiated cells in an embryo become organized into tissues and organs, under genetic controls and environmental influences.

morphological convergence Lineages only remotely related; evolved in response tosimilar environmental pressures, they become similar in appearance, functions, or both. Analogous structures are evidence of this evolutionary pattern.

motor neuron A type of neuron; it delivers signals from the brain and spinal cord that can stimulate or inhibit the body's effectors (muscles, glands, or both).

motor unit A motor neuron and the muscle cells under its control.

multiple allele system Three or more different molecular forms of the same gene (alleles) that exist in a population.

muscle fatigue A decline in tension of a muscle that has been kept in a state of tetanic contraction as a result of continuous, high-frequency stimulation.

muscle tension A mechanical force, exerted by a contracting muscle, that resists opposing forces such as gravity and the weight of objects being lifted.

muscle tissue Tissue having cells able to contract in response to stimulation, then passively lengthen and so return to their resting stage.

muscle twitch Muscle response in which the muscle contracts briefly, then relaxes, when a brief stimulus activates a motor unit.

mutagen (MEW-tuh-jen) An environmental agent that can permanently modify the structure of a DNA molecule. Certain viruses and ultraviolet radiation are examples.

mutation [L. *mutatus,* a change] A heritable change in DNA due to the deletion, addition, or substitution of one to several bases in the nucleotide sequence.

myelin sheath Of many sensory and motor neurons, an axonal sheath that affects how fast action potentials travel; formed from the plasma membranes of Schwann cells that are wrapped repeatedly around the axon and are separated from each other by a small node.

myocardium The cardiac muscle tissue.

myofibril (MY-oh-fy-brill) One of many threadlike structures inside a muscle cell; each is functionally divided into sarcomeres, the basic units of contraction.

myosin (MY-uh-sin) A type of protein with a head and long tail. In muscle cells, it interacts with actin, another protein, to bring about contraction.

NAD⁺ Nicotinamide adenine dinucleotide; a nucleotide coenzyme. When carrying electrons and unbound protons (H^+) between reaction sites, it is abbreviated NADH.

NADP Nicotinamide adenine dinucleotide phosphate; a phosphorylated nucleotide coenzyme. When carrying electrons and unbound protons (H^+) between reaction sites, it is abbreviated $NADPH_2$.

nasal cavity The region of the respiratory system where air is warmed, moistened, and filtered of airborne particles and dust.

natural killer cell Cell of the immune system, possibly a type of lymphocyte, that kills tumor cells (by lysis) or infected cells identified as abnormal.

natural selection A microevolutionary process; a difference in survival and reproduction among members of a population that vary in one or more traits.

negative feedback mechanism A homeostatic feedback mechanism in which an activity changes some condition in the internal environment and so triggers a response that reverses the changed condition.

nephron (NEFF-ron) [Gk. *nephros,* kidney] Of the vertebrate kidney, a slender tubule in which water and solutes filtered from blood are selectively reabsorbed and in which urine forms.

nerve Cordlike communication line of the peripheral nervous system, composed of axons of sensory neurons, motor neurons, or both packed within connective tissue. In the brain and spinal cord, similar cordlike bundles are called nerve pathways or tracts.

nerve impulse *See* action potential.

nerve tract A bundle of myelinated axons of interneurons inside the spinal cord and brain.

nervous system System of neurons oriented relative to one another in precise message-conducting and information-processing pathways.

nervous tissue Tissue composed of neurons and (in the central nervous system) neuroglia.

net energy Of energy resources available to the human population, the amount of energy that is left over after subtracting the energy used to locate, extract, transport, store, and deliver energy to consumers.

neuroendocrine control center The portions of the hypothalamus and pituitary gland that interact to control many body functions.

neuroglial cell (nur-OH-glee-uhl) One of the cells that provide structural and metabolic support for neurons and that collectively represent about half the volume of the nervous system.

neuromodulator Type of signaling molecule that influences the effects of transmitter substances by enhancing or reducing membrane responses in target neurons.

neuromuscular junction Chemical synapses between axon terminals of a motor neuron and a muscle cell.

neuron A nerve cell; the basic unit of communication in nervous systems. Neurons collectively sense environmental change, integrate sensory inputs, then activate muscles or glands that initiate or carry out responses.

neurotransmitter Any of the class of signaling molecules that are secreted from neurons, act on immediately adjacent cells, and are then rapidly degraded or recycled.

neutral mutation A gene mutation that neither harms nor helps the individual's ability to survive and reproduce.

neutron Unit of matter, one or more of which occupies the atomic nucleus, that has mass but no electric charge.

neutrophil Phagocytic white blood cell that takes part in inflammatory responses against bacteria.

niche (nitch) [L. *nidas,* nest] Of a species, the full range of physical and biological conditions under which its members can live and reproduce.

nitrification (nye-trih-fih-KAY-shun) A chemosynthetic process in which certain bacteria strip electrons from ammonia or ammonium present in soil. The end product, nitrite (NO_2^-), is broken down to nitrate (NO_3^-) by different bacteria.

nitrogen cycle Biogeochemical cycle in which the atmosphere is the largest reservoir of nitrogen.

nitrogen fixation Process by which a few kinds of bacteria convert gaseous nitrogen (N_2) to ammonia. This dissolves rapidly in their cytoplasm to form ammonium, which can be used in biosynthetic pathways.

nociceptor A receptor, such as a free nerve ending, that detects any stimulus causing tissue damage.

nondisjunction Failure of one or more chromosomes to separate properly during mitosis or meiosis.

nongonococcal urethritis (NGU) An inflammation of the urethra; often caused by infection by the bacterium that causes chlamydia and considered a sexually transmitted disease.

nonsteroid hormone A type of water-soluble hormone, such as a protein hormone, that cannot cross the lipid bilayer of a target cell. These hormones enter the cell by receptor-mediated endocytosis, or they bind to receptors that activate membrane proteins or second messengers within the cell.

nuclear envelope A double membrane (two lipid bilayers and associated proteins) that is the outermost portion of a cell nucleus.

nucleic acid (new-CLAY-ik) A long, single- or double-stranded chain of four different kinds of nucleotides joined one after the other at their phosphate groups. They differ in which nucleotide base follows the next in sequence. DNA and RNA are examples.

nucleic acid hybridization The base-pairing of nucleotide sequences from different sources, as used in genetics, genetic engineering, and studies of evolutionary relationship based on similarities and differences in the DNA or RNA of different species.

nucleoid Of bacteria, a region in which DNA is physically organized apart from other cytoplasmic components.

nucleolus (new-KLEE-oh-lus) [L. *nucleolus*, a little kernel] Within the nucleus of a nondividing cell, a site where the protein and RNA subunits of ribosomes are assembled.

nucleosome (new-KLEE-oh-sohm) Of chromosomes, one of many organizational units, each consisting of a small stretch of DNA looped twice around a "spool" of histone molecules, which another histone molecule stabilizes.

nucleotide (new-KLEE-oh-tide) A small organic compound having a five-carbon sugar (deoxyribose), nitrogen-containing base, and phosphate group. Nucleotides are the structural units of adenosine phosphates, nucleotide co-enzymes, and nucleic acids.

nucleotide coenzyme A protein that transports hydrogen atoms (free protons) and electrons from one reaction site to another in cells.

nucleus (NEW-klee-us) [L. *nucleus*, a kernel] Of atoms, the central core of one or more positively charged protons and (in all but hydrogen) electrically neutral neutrons. In cells, a membranous organelle that physically isolates and organizes the DNA, out of the way of cytoplasmic machinery.

nutrient Element with a direct or indirect role in metabolism that no other element fulfills.

nutrition All those processes by which food is selectively ingested, digested, absorbed, and later converted to the body's own organic compounds.

obesity An excess of fat in the body's adipose tissues, caused by imbalances between caloric intake and energy output.

oligosaccharide A carbohydrate consisting of a short chain of two or more covalently bonded sugar units. One subclass, disaccharides, has two sugar units. Compare monosaccharide; polysaccharide.

omnivore [L. *omnis*, all, and *vovare*, to devour] An organism able to obtain energy from more than one source rather than being limited to one trophic level.

oncogene (ON-coe-jeen) Any gene having the potential to induce cancerous transformations in a cell.

oocyte An immature egg.

oogenesis (oo-oh-JEN-uh-sis) Formation of a female gamete, from a germ cell to a mature haploid ovum (egg).

orbital Volume of space around the nucleus of an atom in which electrons are likely to be at any instant.

organ A structure of definite form and function that is composed of more than one tissue.

organ of Corti Membrane region of the inner ear that contains the sensory hair cells involved in hearing.

organ system Two or more organs that interact chemically, physically, or both in performing a common task.

organelle Of cells, an internal, membrane-bounded sac or compartment that has a specific, specialized metabolic function.

organic compound In biology, a compound assembled in cells and having a carbon backbone, often with carbon atoms arranged as a chain or ring structure.

organogenesis Stage of development in which primary tissue layers (germ layers) split into subpopulations of cells, and different lines of cells become unique in structure and function; foundation for growth and tissue specializa-tion, when organs acquire specialized chemical and physical properties.

origin The end of a muscle that is attached to the bone that remains relatively stationary when the muscle contracts.

osmosis (oss-MOE-sis) [Gk. *osmos*, act of pushing] Of cells, the tendency of water to move through channel proteins that span a membrane in response to a concentration gradient, fluid pressure, or both. Hydrogen bonds among water molecules prevent water itself from becoming more or less concentrated; but a gradient may exist when the water on either side of the membrane has more substances dissolved in it.

osteocyte A living bone cell.

osteon A set of thin, concentric layers of compact bone tissue surrounding a narrow canal carrying blood vessels and nerves; arrays of osteons make up compact bone.

ovary (OH-vuh-ree) In females, the primary reproductive organ in which eggs form.

oviduct (OH-vih-dukt) Duct through which eggs travel from the ovary to the uterus. Formerly called Fallopian tube.

ovulation (ahv-you-LAY-shun) During each turn of the menstrual cycle, the release of a secondary oocyte (immature egg) from an ovary.

ovum (OH-vum) A mature female gamete (egg).

oxaloacetate A four-carbon compound with roles in metabolism(e.g., the point of entry into the Krebs cycle).

oxidation-reduction reaction An electron transfer from one atom or molecule to another. Often hydrogen is transferred along with the electron or electrons.

oxidative phosphorylation (foss-for-ih-LAY-shun) Final stage of aerobic respiration, in which ATP forms after hydrogen ions and electrons (from the Krebs cycle) are sent through a transport system that gives up the electrons to oxygen.

oxyhemoglobin A hemoglobin molecule that has oxygen bound to it; HbO_2.

ozone thinning Pronounced seasonal thinning of the Earth's ozone layer, as in the lower stratosphere above Antarctica.

palate Structure that separates the nasal cavity from the oral cavity. The bone-reinforced hard palate serves as a hard surface against which the tongue can press food as it mixes it with saliva.

pancreas (PAN-cree-us) Gland that secretes enzymes and bicarbonate into the small intestine during digestion, and that also secretes the hormones insulin and glucagon.

pancreatic islets Any of the two million clusters of endocrine cells in the pancreas, including alpha cells, beta cells, and delta cells.

parasite [Gk. *para*, alongside, and *sitos*, food] An organism that obtains nutrients directly from the tissues of a living host, which it lives on or in and may or may not kill.

parasympathetic nerve Of the autonomic nervous system, any of the nerves carrying signals that tend to slow the body down overall and divert energy to basic tasks; also work continually in opposition with sympathetic nerves to bring about minor adjustments in internal organs.

parathyroid glands (pare-uh-THY-royd) Endocrine glands embedded in the thyroid gland that secrete parathyroid hormone, which helps restore blood calcium levels.

parturition Birth.

passive immunity Temporary immunity conferred by deliberately introducing antibodies into the body.

passive transport Diffusion of a solute through a channel or carrier protein that spans the lipid bilayer of a cell membrane. Its passage does not require an energy input; the protein passively allows the solute to follow its concentration gradient.

pathogen (PATH-oh-jen) [Gk. *pathos*, suffering, and *-genēs*, origin] An infectious, disease-causing agent, such as a virus or bacterium.

pattern formation Mechanisms responsible for specialization and positioning of tissues during embryonic development.

PCR *See* polymerase chain reaction.

pectoral girdle Set of bones, including the scapula (shoulder blade) and clavicle (collarbone), to which the long bone of each arm attaches. The pectoral girdles form the upper part of the appendicular skeleton and are only loosely attached to the rest of the body by muscles.

pedigree A chart of genetic connections among individuals, as constructed according to standardized methods.

pelvic girdle Set of bones including coxal bones that form an open basin, the pelvis; the lower part of the appendicular skeleton. The upper portions of the two coxal bones are the "hipbones"; the thighbones (femurs) join the coxal bones at hip joints. The pelvic girdle bears the body's weight when a person stands and is much more massive than the pectoral girdle.

pelvic inflammatory disease (PID) Generally, a bacterially caused inflammation of the uterus, oviducts, and ovaries. Often a complication of gonorrhea, chlamydia, or some other sexually transmitted disease.

penetrance In a given population, the percentage of individuals in which a particular genotype is expressed (that is, the percentage of individuals who have the genotype and also exhibit the corresponding phenotype).

penis Male organ that deposits sperm into the female reproductive tract.

pepsin Any of several digestive enzymes that are part of gastric fluid in the stomach.

peptide hormone A hormone that consists of a chain of specific amino acids.

perception The conscious interpretation of some aspect of the external world created by the brain from nerve impulses generated by sensory receptors.

perforin A type of protein secreted by a natural killer cell of the immune system, and which creates holes (pores) in the plasma membrane of a target cell.

peripheral nervous system (per-IF-ur-uhl) [Gk. *peripherein*, to carry around] The nerves leading into and out from the spinal cord and brain and the ganglia along those communication lines.

peristalsis (pare-ih-STAL-sis) A rhythmic contraction of muscles that moves food forward through the gastrointestinal tract.

peritoneum A lining of the coelom that also covers and helps maintain the position of internal organs.

peritubular capillaries The set of blood capillaries that threads around the tubular parts of a nephron; they function in reabsorption of water and solutes back into the body and in secretion of hydrogen ions and some other substances in the forming urine.

peroxisome Enzyme-filled vesicle in which fatty acids and amino acids are digested first into hydrogen peroxide (which is toxic), then to harmless products.

PGA Phosphoglycerate (foss-foe-GLISS-er-ate). A key intermediate in glycolysis.

PGAL Phosphoglyceraldehyde. A key intermediate in glycolysis.

pH scale A scale used to measure the concentration of free hydrogen ions in blood, water, and other solutions; pH 0 is the most acidic, 14 the most basic, and 7, neutral.

phagocyte (FAYG-uh-sight) [Gk. *phagein*, to eat, and *-kytos*, hollow vessel] A macrophage or certain other white blood cells that engulf and destroy foreign agents.

phagocytosis (fayg-uh-sigh-TOE-sis) [Gk. *phagein*, to eat, and *-kytos*, hollow vessel] Engulfment of foreign cells or substances by amoebas and some white blood cells by means of endocytosis.

pharynx (FARE-inks) A muscular tube by which food enters the gastrointestinal tract; the dual entrance for the tubular part of the digestive tract and windpipe (trachea).

phenotype (FEE-no-type) [Gk. *phainein*, to show, and *-typos*, image] Observable trait or traits of an individual; arises from interactions between genes, and between genes and the environment.

pheromone (FARE-oh-moan) [Gk. *phero*, to carry, and *-mone*, as in hormone] A type of signaling molecule secreted by exocrine glands that serves as a communication signal between individuals of the same species.

phospholipid A type of lipid that is the main structural component of cell membranes. Each has a hydrophobic tail (of two fatty acids) and a hydro-philic head that incorporates glycerol and a phosphate group.

phosphorus cycle Movement of phosphorus from rock or soil through organisms, then back to soil.

phosphorylation (foss-for-ih-LAY-shun) The attachment of unbound (inorganic) phosphate to a molecule; also the transfer of a phosphate group from one molecule to another, as when ATP phosphorylates glucose.

photochemical smog A brown-air smog that develops over large cities when the surrounding land forms a natural basin.

photoreceptor A light-sensitive sensory cell.

phylogeny Evolutionary relationships among species, starting with most ancestral forms and including the branches leading to their descendants.

pigment A light-absorbing molecule.

pilomotor response Contraction of smooth muscle controlling the erection of body hair when outside temperature drops. This creates a layer of still air that reduces heat losses from the body. (It is most effective in mammals that have more body hair than humans do.)

pineal gland (py-NEEL) A light-sensitive endocrine gland that secretes melatonin, a hormone that influences reproductive cycles and the development of reproductive organs.

pioneer species Typically small plants with short life cycles that are adapted to growing in exposed, often windy areas with intense sunlight, wide swings in air temperature, and soils deficient in nitrogen and other nutrients. By improving conditions in areas they colonize, pioneers invite their own replacement by other species.

pituitary gland An endocrine gland that interacts with the hypothalamus to coordinate and control many physio-

logical functions, including the activity of many other endocrine glands. Its posterior lobe stores and secretes hypothalamic hormones; the anterior lobe produces and secretes its own hormones.

placenta (pluh-SEN-tuh) Of the uterus, an organ composed of maternal tissues and extraembryonic membranes (the chorion especially); it delivers nutrients to the fetus and accepts wastes from it, yet allows the fetal circulatory system to develop separately from the mother's.

plasma (PLAZ-muh) Liquid component of blood; consists of water, various proteins, ions, sugars, dissolved gases, and other substances.

plasma membrane Of cells, the outermost membrane. Its lipid bilayer structure and proteins carry out most functions, including transport across the membrane and reception of extracellular signals.

plasmid Of many bacteria, a small, circular molecule of extra DNA that carries only a few genes and replicates independently of the bacterial chromosome.

plate tectonics Arrangement of the Earth's outer layer (lithosphere) in slablike plates, all in motion and floating on a hot, plastic layer of the underlying mantle.

platelet (PLAYT-let) Any of the cell fragments in blood that release substances necessary for clot formation.

pleiotropy (pleye-ah-troe-pee) [Gk. *pleon*, more, and *trope*, direction] A type of gene interaction in which a single gene exerts multiple effects on seemingly unrelated aspects of an individual's phenotype.

pleura Thin, double membrane surrounding each lung.

polar body Any of three cells that form during the meiotic cell division of an oocyte; the division also forms the mature egg, or ovum.

pollutant Any substance with which an ecosystem has had no prior evolutionary experience in terms of kinds or amounts, and that can accumulate to disruptive or harmful levels. Can be naturally occurring or synthetic.

polymer (PAH-lih-mur) [Gk. *polus*, many, and *meris*, part] A molecule composed of three to millions of small subunits that may or may not be identical.

polymerase chain reaction (PCR). DNA amplification method; DNA containing a gene of interest is split into single strands, which enzymes (polymerases) copy; the enzymes also act on the accumulating copies, multiplying the gene sequence by the millions.

polymorphism (poly-MORE-fizz-um) [Gk. *polus*, many, and *morphe*, form] Of a population, the persistence through the generations of two or more forms of a trait, at a frequency greater than can be maintained by new mutations alone.

polypeptide chain Three or more amino acids joined by peptide bonds.

polyploidy (PAHL-ee-ployd-ee) A change in the chromosome number following inheritance of three or more of each type of chromosome.

polysaccharide [Gk. *polus*, many, and *sakharon*, sugar] A straight or branched chain of hundreds of thousands of covalently linked sugar units of the same or different kinds. The most common polysaccharides are cellulose, starch, and glycogen.

polysome Of protein synthesis, several ribosomes all translating the same messenger RNA molecule, one after the other.

pons Hindbrain region; traffic center for signals between centers of the cerebellum and forebrain.

population A group of individuals of the same species occupying a given area.

population density The number of individuals of a population that are living in a specified area or volume.

population distribution The general pattern of dispersion of individuals of a population throughout their habitat.

population size The number of individuals that make up the gene pool of a population.

positive feedback mechanism Homeostatic mechanism by which a chain of events is set in motion that intensifies a change from an original condition; after a limited time, the intensification reverses the change.

prediction A claim about what you can expect to observe in nature if a theory or hypothesis is correct.

primary immune response Actions by white blood cells and their products elicited by a first-time encounter with an antigen; includes both antibody-mediated and cell-mediated responses.

primary productivity Of ecosystems, *gross* primary productivity is the rate at which the producer organisms capture and store a given amount of energy during a specified interval. *Net* primary productivity is the rate of energy storage in the tissues of producers in excess of their rate of aerobic respiration.

primate The mammalian lineage that includes prosimians, tarsioids, and anthropoids (monkeys, apes, and humans).

primer Short nucleotide sequence designed to base-pair with any complementary DNA sequence; later, DNA polymerases recognize it as a START tag for replication.

prion Small infectious protein that causes rare, fatal degenerative diseases of the nervous system.

probability With respect to any chance event, the most likely number of times it will turn out a certain way, divided by the total number of all possible outcomes.

probe Very short stretch of DNA designed to base-pair with part of a gene being studied and labeled with an isotope to distinguish it from DNA in the sample being investigated.

producer, primary. Of ecosystems, any of the organisms that secure energy from the physical environment, as by photosynthesis or chemosynthesis.

progesterone (pro-JESS-tuh-rown) Female sex hormone secreted by the ovaries.

prokaryotic cell (pro-carry-OH-tic) [L. *pro*, before, and Gk. *karyon*, kernel] A bacterium; a single-celled organism that has no nucleus or any of the other membrane-bound organelles characteristic of eukaryotic cells.

promoter Of transcription, a base sequence that signals the start of a gene; the site where RNA polymerase initially binds.

prophase Of mitosis, the stage when each duplicated chromosome starts to condense, microtubules form a spindle apparatus, and the nuclear envelope starts to break up.

prophase I Of meiosis, the stage at which the microtubular spindle starts to form, the nuclear envelope starts to break up, and each duplicated chromosome also condenses and pairs with its homologous partner. At this time, their sister chromatids typically undergo crossing over and geneticize recombination.

prophase II Of meiosis, a brief stage after interkinesis during which each chromosome still consists of two chromatids.

prostaglandin Any of various lipids present in tissues throughout the body and that can act as local signaling molecules. Prostaglandins typically cause smooth muscle to contract or relax, as in blood vessels, the uterus, and respiratory airways.

prostate gland Gland in males that wraps around the urethra and ejaculatory ducts; its secretions become part of semen.

protein Large organic compound composed of one or more chains of amino acids held together by peptide bonds. Proteins have unique sequences of different kinds of amino acids in their polypeptide chains; such sequences are the basis of a protein's three-dimensional structure and chemical behavior.

proto-oncogene A gene sequence similar to an oncogene but that codes for a protein required in normal cell function; may trigger cancer, generally when specific mutations alter its structure or function.

proton Positively charged particle, one or more of which is present in the atomic nucleus.

proximal tubule Of a nephron, the tubular region that receives water and solutes filtered from the blood.

pulmonary circuit Blood circulation route leading to and from the lungs.

pulse Rhythmic pressure surge of blood flowing in an artery, created during each cardiac cycle when a ventricle contracts.

Punnett-square method A method to predict the possible outcome of a mating or an experimental cross in simple diagrammatic form.

purine Nucleotide base having a double ring structure. Adenine and guanine are examples.

pyrimidine (pie-RIM-ih-deen) Nucleotide base having a single ring structure. Cytosine and thymine are examples.

pyruvate (pie-ROO-vate) A compound with a backbone of three carbon atoms. Two pyruvate molecules are the end products of glycolysis.

r Designates net population growth rate; the birth and death rates are assumed to remain constant and so are combined into this one variable for population growth equations.

radioisotope An unstable atom that I has dissimilar numbers of protons and neutrons and that spontaneously decays (emits electrons and energy) to a new, stable atom that is not radioactive.

radius One of two long bones of the forearm that extend from the humerus (at the elbow joint) to the wrist. The radius runs along the "thumb side" of the forearm, parallel to the ulna.

reabsorption In the kidney, the diffusion or active transport of water and usable solutes out of a nephron and into capillaries leading back to the general circulation; regulated by ADH and aldosterone.

receptor, molecular A molecule at the surface of the plasma membrane or in the cytoplasm that binds molecules present in the extracellular environment. The binding triggers changes in cellular activities.

receptor, sensory A sensory cell or cell part that may be activated by a specific stimulus.

recessive (allele or trait) [L. *recedere*, to recede] In heterozygotes, an allele whose expression is fully or partially masked by expression of its partner; fully expressed only in the homozygous recessive condition.

recognition protein Protein at cell surface recognized by cells of like type; helps guide the ordering of cells into tissues during development and functions in cell-to-cell interactions.

recombinant DNA Any molecule of DNA that incorporates one or more non-parental nucleotide sequences. It is the outcome of microbial gene transfer in nature or resulting from recombinant DNA technology.

recombinant DNA technology Procedures by which DNA (genes) from different species may be isolated, cut, spliced together, and the new recombinant molecules multiplied in quantity in a population of rapidly dividing cells such as bacteria.

recombination Any enzyme-mediated reaction that inserts one DNA sequence into another.

rectum Final region of the gastrointestinal tract, which receives and temporarily stores undigested food residues (feces).

red blood cell Erythrocyte; an oxygen-transporting cell in blood.

red marrow A substance in the spongy tissue of many bones that serves as a major site of blood cell formation.

reductional division Mode of cell division represented by meiosis, in which daughter cells end up with one-half the normal diploid number of chromosomes.

reflex [L. *reflectere*, to bend back] A simple, stereotyped movement elicited directly by sensory stimulation.

reflex arc [L. *reflectere*, to bend back] Type of neural pathway in which signals from sensory neurons directly stimulate or inhibit motor neurons without intervention by interneurons.

refractory period Of neurons, the period following an action potential at a given patch of membrane when sodium gates are shut and potassium gates are open, so that the patch is insensitive to stimulation.

regulatory protein A protein that enhances or suppresses the rate at which a gene is transcribed.

releasing hormone A hypothalamic signaling molecule that stimulates or slows down secretion by target cells in the anterior lobe of the pituitary gland.

repressor protein Regulatory protein that provides negative control of gene activity by preventing RNA polymerase from binding to DNA.

reproduction In biology, processes by which a new generation of cells or multicelled individuals is produced. Sexual reproduction requires meiosis, formation of gametes, and fertilization. Asexual reproduction refers to the production of new individuals by any mode that does not involve gametes.

reproductive isolating mechanism Any aspect of structure, functioning, or behavior that restricts gene flow between two populations.

reproductive isolation An absence of gene flow between populations.

reproductive success Production of viable offspring by an individual.

respiration [L. *respirare*, to breathe] The overall exchange of oxygen from the environment for carbon dioxide wastes from cells by way of circulating blood. Compare aerobic respiration.

respiratory bronchiole Smallest airway in the respiratory system; opens onto alveoli.

respiratory surface In alveoli of the lungs, the thin, moist membrane across which gases diffuse.

respiratory system An organ system that functions in respiration.

resting membrane potential Of neurons and other excitable cells that are not being stimulated, the steady voltage difference across the plasma membrane.

restriction enzymes Class of bacterial enzymes that cut apart foreign DNA injected into them, as by viruses; also used in recombinant DNA technology.

reticular activating system A branch of the brain's reticular formation that controls the changing levels of consciousness by sending signals to the spinal cord, cerebellum, and cerebrum, as well as back to itself.

reticular formation Of the brainstem, a major network of interneurons that helps govern activity of the whole nervous system.

reverse transcription Assembly of DNA on a single-stranded mRNA molecule by viral enzymes.

RFLP Restriction fragment length polymorphism. Of DNA samples from different individuals, slight but unique differences in the banding pattern of fragments of the DNA that have been cut with restriction enzymes.

Rh blood typing A method of characterizing red blood cells on the basis of a protein that serves as a self-marker at their surface; Rh^+ signifies its presence and Rh^-, its absence.

rhodopsin Substance in rod cells of the eye consisting of the protein opsin and a side group, cis-retinal. When the side group absorbs incoming light energy, a series of chemical events follows that result in action potentials in associated neurons.

ribosomal RNA (rRNA) Type of RNA molecule that combines with proteins to form ribosomes, on which the polypeptide chains of proteins are assembled.

ribosome In all cells, the structure at which amino acids are strung together in specified sequence to form the polypeptide chains of proteins. An intact ribosome consists of two subunits, each composed of ribosomal RNA and protein molecules.

RNA Ribonucleic acid. A category of single-stranded nucleic acids that function in processes by which genetic instructions are used to build proteins.

rod cell Of the retina, a photoreceptor sensitive to very dim light and that contributes to coarse perception of movement.

S-shaped curve A curve, obtained when population size is plotted against time, that is characteristic of logistic growth.

salinization A salt buildup in soil as a result of evaporation, poor drainage, and often the importation of mineral salts in irrigation water.

salivary amylase Starch-degrading enzyme in saliva.

salivary gland Any of the glands that secrete saliva, a fluid that initially mixes with food in the mouth and starts the breakdown of starch.

salt Compound that releases ions other than H^+ and OH^- in solution.

saltatory conduction In myelinated neurons, rapid, node-to-node hopping of action potentials.

sampling error Error that develops when a an experimenter uses a sample (or subset) of a population, an event, or some other aspect of nature for an experimental group that is not large enough to be representative of whole.

sarcomere (SAR-koe-meer) The basic unit of muscle contraction; a region of myosin and actin filaments organized in parallel between two Z lines of a myofibril inside a muscle cell.

sarcoplasmic reticulum (sar-koe-PLAZ-mik reh-TIK-you-lum) In muscle cells, a membrane system that takes up, stores, and releases the calcium ions required for cross-bridge formation in sarcomeres, hence for contraction.

scapula Flat, triangular bone on either side of the pectoral girdle; the scapulae form the shoulder blades.

Schwann cell A specialized neuroglial cell that grows around a neuron axon, forming a myelin sheath.

second messenger A molecule inside a cell that mediates and generally triggers amplified response to a hormone.

secondary immune response Rapid, prolonged response by white blood cells, memory cells especially, to a previously encountered antigen.

secondary oocyte An oocyte (unfertilized egg cell) that has completed meiosis I; it is this haploid cell that is released at ovulation.

secondary sexual trait Trait associated with maleness or femaleness, but not directly involved with reproduction. Beard growth in males and breast development in females are examples.

secretion A cell acting on its own or as part of glandular tissue releases a substance across its plasma membrane, into the surroundings.

sedimentary cycle Biogeochemical cycle. An element having no gaseous phase moves from land, through food webs, to the seafloor, then returns to land through long-term uplifting.

segmentation In tubular organs, an oscillating movement produced by rings of muscle in the tube wall.

segregation, Mendelian principle of [L. *se-*, apart, and *grex,* herd] The principle that diploid organisms inherit a pair of genes for each trait (on a pair of homologous chromosomes) and that the two genes segregate during meiosis and end up in separate gametes.

selective gene expression Activation or suppression of a fraction of the genes in unique ways in different cells, leading to pronounced differences in structure and function among different cell lineages.

selective permeability Of a cell membrane, a capacity to let some substances but not others cross it at certain times. The capacity arises as an outcome of the membrane's lipid bilayer structure and its transport proteins.

semen [L. *serere,* to sow] Sperm-bearing fluid expelled from the penis during male orgasm.

semicircular canals Fluid-filled canals positioned at different angles within the vestibular apparatus of the inner ear and that contain sensory receptors that detect head movements, deceleration, and acceleration.

semiconservative replication [Gk. *hēmi,* half, and L. *conservare,* to keep] Reproduction of a DNA molecule when a complementary strand forms on each of the unzipping strands of an existing DNA double helix, the outcome being two "half-old, half-new" molecules.

semilunar valve In each half of the heart; during each heartbeat, it opens and closes in ways that keep blood flowing in one direction from the ventricle to the arteries leading away from it.

seminiferous tubules Coiled tubes inside the testes where sperm develop.

senescence (sen-ESS-cents) [L. *senescere,* to grow old] Sum total of processes leading to the natural death of an organism or some of its parts.

sensation The conscious awareness of a stimulus.

sensory neuron Any of the nerve cells that act as sensory receptors, detecting specific stimuli (such as light energy) and relaying signals to the brain and spinal cord.

sensory receptor Of the nervous system, a sensory cell or specialized cell adjacent to it that can detect a particular stimulus.

sensory system Element of the nervous system consisting of sensory receptors (such as photoreceptors), nerve pathways from the receptors to the brain, and brain regions that process sensory information.

septum Of the heart, a thick wall that divides the heart into right and left halves.

Sertoli cell Any of the cells in seminiferous tubules that nourish and otherwise aid the development of sperm.

sex chromosome A chromosome whose presence determines a new individual's gender. Compare autosomes.

shell model Model of electron distribution in which all orbitals available to electrons of atoms occupy a nested series of shells.

shifting cultivation The cutting and burning of trees, followed by tilling of ashes into the soil; once called "slash-and-burn agriculture."

sinoatrial node Region of conducting cells in the upper wall of the right atrium that generate periodic waves of excitation that stimulate the atria to contract.

sinus In the skull, an air-filled space lined with mucous membrane and that functions to lighten the skull.

sister chromatid Of a duplicated chromosome, one of two DNA molecules (and associated proteins) that remain attached at their centromere only during nuclear division. Each ends up in a separate daughter nucleus.

skeletal muscle Type of muscle that interacts with the skeleton to bring about body movements. A skeletal muscle typically consists of bundles of many long cylindrical cells encapsulated by connective tissue.

skull Bony structure that includes more than two dozen bones, including bones of the brain case and facial bones.

sliding filament model Model of muscle contraction in which myosin filaments physically slide along and pull two sets of actin filaments toward the center of the sarcomere, which shortens. The sliding requires ATP energy and cross-bridge formation between the actin and myosin.

small intestine The portion of the digestive system where digestion is completed and most nutrients absorbed.

smog, industrial Gray-colored air pollution that predominates in industrialized cities with cold, wet winters.

smog, photochemical Form of brown, smelly air pollution occurring in large cities with warm climates.

sodium-potassium pump A transport protein spanning the lipid bilayer of the plasma membrane. When activated by ATP, its shape changes and it selectively transports sodium ions out of the cell and potassium ions in.

solute (SOL-yoot) [L. *solvere*, to loosen] Any substance dissolved in a solution. In water, this means spheres of hydration surround the charged parts of individual ions or molecules and keep them dispersed.

solvent Fluid in which one or more substances is dissolved.

somatic cell (so-MAT-ik) [Gk. *somā*, body] Any body cell that is not a germ cell (which gives rise to gametes).

somatic nervous system Those nerves leading from the central nervous system to skeletal muscles.

somatic sensation Awareness of touch, pressure, heat, cold, pain, and limb movement.

speciation (spee-cee-AY-shun) The evolutionary process by which species originate. One speciation route starts with divergence of two reproductively isolated populations of a species. They become separate species when accumulated differences in allele frequencies prevent them from interbreeding successfully under natural conditions.

species (SPEE-ceez) [L. *species*, a kind] Of sexually reproducing organisms, a unit consisting of one or more populations of individuals that can interbreed under natural conditions to produce fertile offspring that are reproductively isolated from other such units.

sperm [Gk. *sperma*, seed] A type of mature male gamete.

spermatogenesis (sperm-at-oh-JEN-ih-sis) Formation of a mature sperm from a germ cell.

sphere of hydration Through positive or negative interactions, a clustering of water molecules around the individual molecules of a substance placed in water. Compare solute.

sphincter (SFINK-tur) Ring of muscle between regions of a tubelike system (as between the stomach and small intestine).

spinal cord Of the central nervous system, the portion threading through a canal inside the vertebral column and providing direct reflex connections between sensory and motor neurons as well as communication lines to and from the brain.

spindle apparatus A type of bipolar structure that forms during mitosis or meiosis and that moves the chromosomes. It consists of two sets of microtubules that extend from the opposite poles and that overlap at the spindle's equator.

spleen One of the lymphoid organs; it is a filtering station for blood, a reservoir of red blood cells, and a reservoir of macrophages.

spongy bone Type of bone tissue in which hard, needlelike struts separate large spaces filled with marrow. Spongy bone occurs at the ends of long bones and within the breastbone, pelvis, and bones of the skull.

stabilizing selection Of a population, a persistence over time of the alleles responsible for the most common phenotypes.

start codon Of protein synthesis, a base triplet in a strand of mRNA that serves as the start signal for mRNA translation.

stem cell Unspecialized cell that can give rise to descendants that differentiate into specialized cells.

sternum Elongated flat bone (also called the breastbone) to which the upper ribs attach and so form the rib cage.

steroid (STAIR-oid) A lipid with a backbone of four carbon rings and with no fatty acid tails. Steroids differ in their functional groups. Different types have roles in metabolism, intercellular communication, and cell membranes.

steroid hormone A type of lipid-soluble hormone synthesized from cholesterol. Many steroid hormones move into the nucleus and bind to a receptor for it there; others bind to a receptor in the cytoplasm, and the entire complex moves into the nucleus.

sterol A type of lipid with a rigid backbone of four fused carbon rings. Sterols occur in cell membranes; cholesterol is the main type in human tissues.

stimulus [L. *stimulus*, goad] A specific change in the environment, such as a variation in light, heat, or mechanical pressure, that the body can detect through sensory receptors.

stomach A muscular, stretchable sac that receives ingested food; the organ between the esophagus and intestine in which considerable protein digestion occurs.

stop codon Of protein synthesis, a base triplet in a strand of mRNA that serves as the stop signal for translation, so that no more amino acids are added to the polypeptide chain.

stromatolite Of shallow seas, layered structures formed from sediments and large mats of the slowly accumulated remains of photosynthetic populations.

substrate A reactant or precursor molecule for a metabolic reaction; a specific molecule or molecules that an enzyme can chemically recognize, bind briefly to itself, and modify in a specific way.

substrate-level phosphorylation The direct, enzyme-mediated transfer of a phosphate group from the substrate of a reaction to another molecule. An example is the transfer of phosphate from an intermediate of glycolysis to ADP, forming ATP.

succession, primary (suk-SESH-un) [L. *succedere,* to follow after] Orderly changes from the time pioneer species colonize a barren habitat through replacements by various species until the climax community, when the composition of species remains steady under prevailing conditions.

succession, secondary Orderly changes in a community or patch of habitat toward the climax state after having been disturbed, as by fire.

surface-to-volume ratio A mathematical relationship in which volume increases with the cube of the diameter, but surface area increases only with the square. Of growing cells, the volume of cytoplasm increases more rapidly than the surface area of the plasma membrane that must service the cytoplasm. Because of this constraint, cells generally remain small or elongated, or have elaborate membrane foldings.

survivorship curve A plot of the age-specific survival of a group of individuals in a given environment, from the time of their birth until the last one dies.

sympathetic nerve Of the autonomic nervous system, any of the nerves generally concerned with increasing overall body activities during times of heightened awareness, excitement, or danger; also work continually in opposition with parasympathetic nerves to bring about minor adjustments in internal organs.

synaptic integration (sin-AP-tik) The moment-by-moment combining of excitatory and inhibitory signals arriving at a trigger zone of a neuron.

syndrome A set of symptoms that may not individually be a telling clue but collectively characterize a disorder or disease.

synovial joint Freely movable joint in which adjoining bones are separated by a fluid-filled cavity and stabilized by straplike ligaments. An example is the ball-and-socket joint at the hip.

syphilis Clinical term for the sexually transmitted disease caused by infection by the spirochete bacterium *Treponema pallidum.* Untreated syphilis can lead to lesions in mucous membranes, the eyes, bones, skin, liver, and central nervous system.

systemic circuit (sis-TEM-ik) Circulation route in which oxygenated blood flows from the lungs to the left half of the heart, through the rest of the body (where it gives up oxygen and takes on carbon dioxide), then back to the right side of the heart.

systole Contraction phase of the cardiac cycle.

T lymphocyte A white blood cell with roles in immune responses.

target cell Any cell that has receptors for a specific signaling molecule and that may alter its behavior in response to the molecule.

tectorial membrane Inner ear structure against which sensory hair cells are bent, producing action potentials that travel to the brain via the auditory nerve.

telophase (TEE-low-faze) Of mitosis, the final stage when chromosomes decondense into threadlike structures and two daughter nuclei form. Of meiosis I, the stage when one of each pair of homologous chromosomes has arrived at one or the other end of the spindle pole. At telophase II, chromosomes decondense and four daughter nuclei form.

telophase II Of meiosis, final stage when four daughter nuclei form.

temperate pathway A viral infection that enters a latent period; the host is not killed outright.

temporal summation The adding together (summing) of several muscle contractions, resulting in a single, stronger contraction, when stimulatory signals arrive in rapid succession.

tendon A cord or strap of dense, regular connective tissue that attaches a muscle to bone or to another muscle.

test An attempt to produce actual observations that match predicted or expected observations.

testcross Experimental cross to reveal whether an organism is homozygous dominant or heterozygous for a trait. The organism showing dominance is crossed to an individual known to be homozygous recessive for the same trait.

testis, plural testes. Male gonad; primary reproductive organ in which male gametes and sex hormones are produced.

testosterone (tess-TOSS-tuh-rown) In males, a major sex hormone that helps control reproductive functions.

tetany Condition in which a muscle motor unit is maintained in a state of contraction for an extended period.

thalamus Of the forebrain, a coordinating center for sensory input and a relay station for signals to the cerebrum.

theory A testable explanation of a broad range of related phenomena. In modern science, only explanations that have been extensively tested and can be relied on with a very high degree of confidence are accorded the status of theory.

thermal inversion Situation in which a layer of dense, cool air becomes trapped beneath a layer of warm air; can cause air pollutants to accumulate to dangerous levels close to the ground.

thermoreceptor Sensory cell that can detect radiant energy associated with temperature.

thirst center Cluster of nerve cells in the hypothalamus that can inhibit saliva production, resulting in mouth dryness that the brain interprets as thirst and leading a person to seek out drinking fluids.

threshold Of neurons and other excitable cells, a certain minimum amount by which the voltage difference across the plasma membrane must change to produce an action potential.

thymine Nitrogen-containing base in some nucleotides.

thymus A lymphoid organ with endocrine functions; lymphocytes of the immune system multiply, differentiate, and mature in its tissues, and its hormone secretions affect their functions.

thyroid gland Of the endocrine system, a gland that produces hormones that affect overall metabolic rates, growth, and development.

tidal volume Volume of air, about 500 milliliters, that enters or leaves the lungs in a normal breath.

tissue A group of cells and intercellular substances that function together in one or more specialized tasks.

T lymphocyte One of a class of white blood cells that carry out immune responses. The helper T and cytotoxic T cells are examples.

tonicity The relative concentrations of solutes in two fluids, such as inside and outside a cell. When solute concentrations are isotonic (equal in both fluids), water shows no net osmotic movement in either direction. When one fluid is hypotonic (has less solutes than the other), the other is hypertonic (has more solutes) and is the direction in which water tends to move.

touch-killing Mechanism by which cytotoxic T cells directly release perforins and toxins into a target cell and cause its destruction.

toxin A normal metabolic product of one species with chemical effects that can hurt or kill individuals of another species.

trace element Any element that represents less than 0.01 percent of body weight.

tracer A radioisotope used to label a substance so that its pathway or destination in a cell, organism, ecosystem, or some other system can be tracked, as by scintillation counters that detect its emissions.

trachea (TRAY-kee-uh) The windpipe, which carries air between the larynx and bronchi.

transcription [L. *trans*, across, and *scribere*, to write] Of protein synthesis, the assembly of an RNA strand on one of the two strands of a DNA double helix; the base sequence of the resulting transcript is complementary to the DNA region on which it was assembled.

transfer RNA (tRNA) Of protein synthesis, any of the type of RNA molecules that bind and deliver specific amino acids to ribosomes and pair with mRNA code words for those amino acids.

translation In protein synthesis, the conversion of the coded sequence of information in mRNA into a particular sequence of amino acids to form a polypeptide chain; depends on interactions of rRNA, tRNA, and mRNA.

translocation Of cells, a change in a chromosome's structure following the insertion of part of a nonhomologous chromosome into it.

transport protein One of many kind of membrane proteins involves in active or passive transport of water-soluble substances across the lipid bilayer of a plasma membrane. Solutes on one side of the membrane pass through the protein's interior to the other side.

transposable element DNA element that can spontaneously "jump" to new locations in the same DNA molecule or a different one. Such elements often inactivate the genes into which they become inserted and give rise to observable changes in phenotype.

triglyceride (neutral fat) A lipid having three fatty acid tails attached to a glycerol backbone. Triglycerides are the body's most abundant lipids and richest energy source.

trisomy (TRY-so-mee) Of diploid cells, the abnormal presence of three of one type of chromosome.

trophic level (TROE-fik) [Gk. *trophos*, feeder] All the organisms in an ecosystem that are the same number of transfer steps away from the energy input into the system.

trophoblast Surface layer of cells of the blastocyst that secrete enzymes that break down the uterine lining where the forthcoming embryo will implant.

tumor A tissue mass composed of cells that are dividing at an abnormally high rate.

tumor marker A substance that is produced by a specific type of cancer cell or by normal cells in response to cancer.

tumor suppressor gene A gene whose protein product operates to keep cell growth and division within normal bounds, or whose product has a role in keeping cells anchored in place within a tissue.

tympanic membrane The eardrum, which vibrates when struck by sound waves.

ulna One of two long bones of the forearm; the ulna extends along the "little finger" side of the forearm, parallel to the radius on the "thumb" side.

ultrafiltration Bulk flow of a small amount of protein-free plasma from a blood capillary when the outward-directed force of blood pressure is greater than the inward-directed osmotic force of interstitial fluid.

umbilical cord Structure containing blood vessels that connect a fetus to its mother's circulatory system.

uracil (YUR-uh-sill) Nitrogen-containing base found in RNA molecules; can base-pair with adenine.

ureter Tubular channel that carries urine from each kidney to the urinary bladder.

urethra Tube that carries urine from the bladder to the body surface.

urinary bladder Storage organ for urine.

urinary excretion A mechanism by which excess water and solutes are removed by way of the urinary system.

urinary system An organ system that adjusts the volume and composition of blood and so helps maintain extracellular fluid.

urine Fluid formed by filtration, reabsorption, and secretion in kidneys; consists of wastes, excess water, and solutes.

uterus (YOU-tur-us) [L. *uterus*, womb] Chamber in which the developing embryo is contained and nurtured during pregnancy.

vaccine Antigen-containing preparation injected into the body or taken orally; it elicits an immune response leading to the proliferation of memory cells that offer long-lasting protection against a particular pathogen.

vagina Part of a female reproductive system that receives sperm, forms part of the birth canal, and channels menstrual flow to the exterior.

variable Of a scientific experiment, the only factor that is not exactly the same in the experimental group as it is in the control group.

vas deferens Tube leading to the ejaculatory duct; one of several tubes through which sperm move after they leave the testes just prior to ejaculation.

vasoconstriction Decrease in the diameter of an arteriole, so that blood pressure rises; may be triggered by the hormones epinephrine and angiotensin.

vasodilation Enlargement of arteriole diameter, so that blood pressure falls; may be triggered by hormones including epinephrine and angiotensin.

vein Of the circulatory system, any of the large-diameter vessels that lead back to the heart.

ventricle (VEN-tri-kul) Of the heart, one of two chambers from which blood is pumped out. Compare atrium.

venule Small blood vessel that receives blood from tissue capillaries and merges into larger-diameter veins; a limited amount of diffusion occurs across venule walls.

vertebra, plural vertebrae. One of a series of hard bones arranged with intervertebral disks into a backbone.

vertebrate Animal having a backbone of bony segments, the vertebrae.

vesicle (VESS-ih-kul) [L. *vesicula,* little bladder] Within the cytoplasm of cells, one of a variety of small membrane-bound sacs that function in the transport, storage, or digestion of substances or in some other activity.

vestibular apparatus A closed system of fluid-filled canals and sacs in the inner ear that functions in the sense of balance. Compare semicircular canals.

vestigial Applies to a small body part, tissue, or organ abnormally developed or degenerated and unable to function like its normal counterpart (e.g., "tail bones" of humans).

villus (VIL-us), plural villi. Any of several types of absorptive structures projecting from the free surface of an epithelium.

viroid An infectious particle consisting only of very short, tightly folded strands or circles of RNA. Viroids might have evolved from introns, which they resemble.

virus A noncellular infectious agent consisting of DNA or RNA and a protein coat; can replicate only after its genetic material enters a host cell and subverts its metabolic machinery.

vision Precise light focusing onto a layer of photoreceptive cells that is dense enough to sample details concerning a given light stimulus, followed by image formation in the brain.

vital capacity Maximum volume of air that can move out of the lungs after a person inhales as deeply as possible.

vitamin Any of more than a dozen organic substances that the body requires in small amounts for normal cell metabolism but generally cannot synthesize for itself.

vocal cords A pair of elastic ligaments on either side of the larynx wall. Air forced between them causes the cords to vibrate and produce sounds.

water table The upper limit at which the ground in a specified region is fully saturated with water.

watershed Any specified region in which all precipitation drains into a single stream or river.

wax A type of lipid with long-chain fatty acid tails that help form protective, lubricating, or water-repellent coatings.

white blood cell Leukocyte; any of the macrophages, eosinophils, neutrophils, and other cells that, together with their products, make up the immune system.

white matter Of the spinal cord, major nerve tracts so named because of the glistening myelin sheaths of their axons.

X chromosome A sex chromosome with genes that cause an embryo to develop into a female, provided that it inherits a pair of these.

X-linked gene Any gene on an X chromosome.

X-linked recessive inheritance Recessive condition in which the responsible, mutated gene occurs on the X chromosome.

Y chromosome A sex chromosome with genes that cause the embryo that inherited it to develop into a male.

Y-linked gene Any gene on a Y chromosome.

yellow marrow Bone marrow that consists mainly of fat and hence appears yellow. It can convert to red marrow and produce red blood cells if the need arises.

yolk sac Of land vertebrates, one of four extraembryonic membranes. In humans, part becomes a site of blood cell formation and some of its cells give rise to the forerunners of gametes.

zero population growth A population for which the number of births is balanced by the number of deaths over a specified period, assuming immigration and emigration also are balanced.

zygote (ZYE-goat) The first cell of a new individual, formed by the fusion of a sperm nucleus with the nucleus of an egg (fertilization).

CREDITS AND ACKNOWLEDGMENTS

This page constitutes an extension of the copyright page. We have made every effort to trace the ownership of all copyrighted material and to secure permission from copyright holders. In the event of any question arising as to the use of any material, we will be pleased to make the necessary corrections in future printings. Thanks are due to the following authors, publishers, and agents for permission to use the material indicated.

Page 1 Eric Gravé/Science Source/Photo Researchers

CHAPTER 1 **1.1** Frank Kaczmarek / **1.2** Art by Carlyn Iverson / **1.3** Rich Buzzelli/Tom Stack & Associates / **1.5** Daniel McDonald/ The Stock Shop / **Page 5** Art by Precision Graphics / **1.6 Jon** Feingersh/Tom Stack & Associates / **1.10** (a, b, c, d) Photographs Gary Head / **1.11** © Peter Menzel/Stock Boston / **1.12** © 2000 Julian Calder/Stone

CHAPTER 2 **2.1** Photograph Gary Head / **2.2** Photograph Jack Carey / **2.4** Photograph Gary Head / **2.5** (a) Hank Morgan/Rainbow; (b) Art by Raychel Ciemma; (c) Dr. Harry T. Chugani, M.D., UCLA School of Medicine / **2.9** © Y. Arthus-Bertrand / **2.10** Micrograph © Bruce Iverson / **2.11 Vandystadt**/Photo Researchers / **2.12** Art by Palay/Beaubois / **2.13** © Alan Craft/Photofile / **2.14** Photograph Richard Riley/FPG / **2.15** Art by Raychel Ciemma / **2.17** Michael Grecco/ Picture Group / **2.18** (b) Art by Raychel Ciemma / **2.19** (b) © 2000 Liz Von Hoene/ Stone / **2.20** (left) David M. Phillips/Visuals Unlimited; (right) Richard H. Gross / **2.21** Nancy J. Pierce/Photo Researchers; (a, b) Art by Lisa Starr / **2.22** Art by Precision Graphics / **2.23** Micrograph Lewis Lainey; (a, b) Art by Precision Graphics / **2.24** (a) Ron Davis/Shooting Star; (b) Frank Trapper/ Sygma / **2.26** Art by Precision Graphics / **2.29** Art by Palay/Beaubois / **2.30 CNRI**/ SPL/Photo Researchers; (b) Art by Robert Demarest; (c) Art by Lisa Starr / **2.31** Model A. Lesk/SPL/Photo Researchers / **2.32** (left) © Inga Spence/ Tom Stack & Associates; (right) Photograph Gary Head

CHAPTER 3 **3.1** (a) National Library of Medicine; (b) George Musli/Visuals Unlimited / **3.2** (a) Lennart Nilsson from *Behold Man,* © 1974 by Albert Bonniers. Forlag & Little Brown & Company, Boston; (b) David M. Phillips/Visuals Unlimited; (c) CNRI/SPL/Photo Researchers / **3.3** (a) Micrograph G. Cohen-Bazire; (b) Manfred Cage/Bruce Coleman Ltd. / **3.4** (a) Photograph Ed Reschke; (b) Photograph Carolina Biological Supply Company / **3.5** Art by Raychel Ciemma / **3.6** Art by Raychel Ciemma and American Composition and Graphics / **3.7** Micrograph G. L. Decker; Art by Raychel Ciemma and American Composition and Graphics / **3.8** J. Victor Smal & Gottfried Rinnerthaler / **3.9** Art by Lisa Starr / **3.10** Art by Precision Graphics after *Molecular and Cellular Biology,* by

Stephen L. Wolfe, Wadsworth, 1993 / **3.11** Art by Raychel Ciemma / **3.12** Art by Raychel Ciemma and American Composition and Graphics / **3.13** (a, b) Micrographs Don W. Fawcett/Visuals Unlimited; Art by Lisa Starr / **3.14** Micrograph Gary W. Grimes; Art by Robert Demarest after a model by J. Kephart / **3.15** Art by Raychel Ciemma / **3.17** Micrographs M. Sheetz, R. Painter and S. Singer, *Journal of Cell Biology,* 70:193 (1976), by copyright permission of the Rockefeller University Press / **3.18** Art by Lisa Starr / **3.19** Art by Lisa Starr / **3.20** Art by Leonard Morgan / **3.21** Frank L. Lambrecht/Visuals Unlimited / **3.22** D. Fawcett, *The Cell,* Philadelphia: W. B. Saunders Co. / **3.23** Micrograph Keith R. Porter / **3.24** Enzyme model by Thomas A. Steitz; (c) Art by Palay/ Beaubois / **3.27** Micrograph Keith R. Porter; (b) Art by Lisa Starr / **3.28** Photograph Gary Head; Art by Precision Graphics / **3.29** Photograph Gary Head

Pages 66–67 CNRI / SPL / Photo Researchers

CHAPTER 4 **4.1** © Richard B. Levine / **4.2** (a) Focus on Sports; (a-inset) Manfred Cage/ Bruce Coleman Ltd.; (b) Lennart Nilsson from *Behold Man,* © 1974 by Albert Bonniers. Forlag & Little Brown & Company, Boston; (c) Manfred Cage/Bruce Coleman Ltd.; (d) Ed Reschke/Peter Arnold, Inc.; Art by Palay/Beaubois / **4.3** (a–c, e, f-below) Micrograph Ed Reschke; (d) Micrograph Fred Hossler; (f-above) P. Motta/Dept of Anatomy/University "La Sapienza," Rome/ SPL/Photo Researchers / **4.4** (left) Art by Joel Ito; (right) Art by L. Calver / **4.5** Micrograph Ed Reschke / **4.6** (a, b, c) Micrograph Ed Reschke / **4.7** Art by Robert Demarest / **4.8** (a) Lennart Nilsson from *Behold Man,* © 1974 by Albert Bonniers. Forlag & Little Brown & Company, Boston; (b) © Nancy Kedersha/Harvard Medical School / **4.9** © Joseph P. Vacanti, M.D. / **4.10** Photograph NASA/Johnson Space Center **4.11** Art by Raychel Ciemma / **4.12** (a) Fabian/Sygma; Art by Kevin Somerville; (c) Art by Raychel Ciemma; (d) From *Perspectives in Human Biology,* by L. Knapp, Wadsworth, 1998 / **4.13** Art by L. Calver / **4.14** Art by Palay/Beaubois / **4.15** (a) Micrograph Ed Reschke; (b) NRI/SPL/Photo Researchers; Art by Robert Demarest / **Page 86** (above left) Lennart Nilsson from *Behold Man,* © 1974 by Albert Bonniers. Forlag & Little Brown & Company, Boston; (above right, below left and right) Micrographs Ed Reschke; (below center) Lennart Nilsson from *Behold Man,* © 1974 by Albert Bonniers. Forlag & Little Brown & Company, Boston

CHAPTER 5 **5.1** © Kevin Regan/Camera 5 / **5.2** Art by L. Calver / **5.3** Micrograph Ed Reschke; (a-left) Art by Raychel Ciemma; (a-right; b) Art by Joel Ito / **5.4** Art by K. Kasnot / **5.5** National Osteoporosis Foundation / **5.6** Art by Raychel Ciemma / **5.7** Art by John W. Karapelou / **5.8** Art by Ron Ervin / **5.9** Art by John W. Karapelou /

5.10 Art by John W. Karapelou / **5.11** Photograph C. Yokochi and J. Rohen, *Photographic Anatomy of the Human Body, 2/E.* Igaku-Shoin, Ltd.; Art by Precision Graphics /**5.12** Mike Devlin/SPL/ Photo Researchers / **5.13** Art by Raychel Ciemma / **5.14** Art by Robert Demarest / **5.15** Art by Raychel Ciemma / **5.16** (b) Micrograph John D. Cunningham; (c) Micrograph D. W. Fawcett, *The Cell,* Philadelphia: W. B. Saunders Co., 1966; Art by Robert Demarest / **5.17** Art by Nadine Sokol / **5.18** Art by Robert Demarest / **5.19** After *Molecular and Cellular Biology,* by Stephen L. Wolfe, Wadsworth, 1993. / **5.21** Micrograph by Ed Reschke; Art by Kevin Sommerville / **5.23** Tim Davis/Photo Researchers / **5.24** Photograph Michale Neveux

CHAPTER 6 **6.1** Hulton Getty Collection/ Stone / **6.3** Art by Kevin Sommerville / **6.5** (a-left) Art by Kevin Sommerville / **6.5** (a-right; b) Art after *Human Physiology: Mechanisms of Body Function,* Fifth Edition, by A. Vander et al. Copyright © 1990 McGraw-Hill. Used by permission of The McGraw-Hill Companies. / **6.6** (a, b) Art by Carlyn Iverson (c) Art by Precision Graphics / **6.7** Art by Carlyn Iverson / **6.9** (a) Art after *Human Physiology: Mechanisms of Body Function,* Fifth Edition, by A. Vander et al. Copyright © 1990 McGraw-Hill. Used by permission of The McGraw-Hill Companies; (b) Redrawn from *Human Anatomy and Physiology,* Fourth Edition, by A. Spence and E. Mason, p. 763, Brooks/Cole, 1992. / **6.10** (b-above) Microslide courtesy of Mark Nielsen, University of Utah; (b-below) D. W. Fawcett/Photo Researchers; Art by Lisa Starr / **6.11** Art by Raychel Ciemma / **Page 119** (in-text art) After *Human Physiology: Mechanisms of Body Function,* Fifth Edition, by A. Vander et al. Copyright © 1990 McGraw-Hill. Used by permission of The McGraw-Hill Companies / **6.12** Art by Carlyn Iverson / **6.15** Photograph Ralph Pleasant/FPG; Art modified after *Human Physiology: Mechanisms of Body Function,* Fourth Edition, by A. Vander et al., McGraw-Hill, 1985. / **6.16** Elizabeth Hathon/The Stock Market / **6.18** Photograph Gary Head / **6.19** AP/Wide World Photos

CHAPTER 7 **7.1** (a) From A. D. Waller, *Physiology, The Servant of Medicine,* Hitchcock Lectures, University of London Press, 1910; (b) Photograph courtesy of The New York Academy of Medicine Library / **7.2** Art by Palay/Beaubois / **7.3** CNRI/SPL/Photo Researchers; Art by Raychel Ciemma / **7.4** Art by Palay/Beaubois / **7.5** Art by John W. Karapelou / **7.6** (a) After *Modern Genetics,* by F. Ayala and J. Kiger, Benjamin-Cummings, 1980.; (b-both) Lester V. Bergman & Associates, Inc. / **7.7** After *Principles of Anatomy and Physiology,* Sixth Edition, by Gerard J. Tortora and Nicholas P. Anagnos-takos. Copyright © 1990 Biological Sciences Textbooks, Inc., A & P Textbooks, Inc., and Elia-Sparta, Inc. Reprinted by permission of John Wiley & Sons, Inc. / **7.9** Art by Kevin

Sommerville / **7.10** (a) C. Yokochi and J. Rohen, *Photographic Anatomy of the Human Body,* Second edition, Igaku-Shoin Ltd, 1979; (b, c) Art by Raychel Ciemma / **7.11** Art by John W. Karapelou / **7.12** Art by John W. Karapelou **7.14** Art by Kevin Somerville / **7.15** Lewis L. Lainey / **7.16** Micrograph Ed Reschke; Art by Raychel Ciemma / **7.17** Art by Raychel Ciemma / **7.18** Sheila Terry / SPL/Photo Researchers / **7.20** Art by Robert Demarest based on *Basic Human Anatomy,* by A. Spence, Benjamin-Cummings, 1982. / **7.21** Art by John W. Karapelou / **7.22** Art by Kevin Somerville / **7.23** Art by Raychel Ciemma / **7.24** Photograph Lennart Nilsson, © Boehringer Ingelheim International GmbH / **7.25** (a) Ed Reschke; (b) F. Sloop and W. Ober/Visuals Unlimited / **7.27** Art by Raychel Ciemma / **7.28** Art by Raychel Ciemma / **7.29** Lennart Nilsson from *Behold Man,* © 1974 by Albert Bonniers Forlag and Little, Brown and Company, Boston

CHAPTER 8 **8.1** (a) The Granger Collection, New York; (b) Lennart Nilsson © Boehringer Ingelheim International GmbH / **8.2** Robert R. Dourmashkin, courtesy of Clinical Research Centre, Harrow, England / **8.3** Art by Lisa Starr / **8.4** (a) Biology Media/Photo Researchers, Inc.; (b) NSIBC/SPL/Photo Researchers, Inc. / **8.5** Art by Raychel Ciemma / **8.6** Art by Lisa Starr / **8.8** Art by Raychel Ciemma / **8.9** (b) Micrograph Dr. A. Liepins/SPL/Photo Researchers, Inc.; Art by Lisa Starr / **8.10** Computer model antigen courtesy Don C. Wiley, Harvard University; Art by Lisa Starr / **8.12** Art by Palay/Beaubois after *Molecular Biology of the Cell,* by B. Alberts et al., Garland Publishing Company, 1983. / **8.13** Art by Palay/Beaubois after *Molecular Biology of the Cell,* by B. Alberts et al., Garland Publishing Company, 1983. / **8.15** Matt Meadows/Peter Arnold, Inc. / **8.16** K. G. Murti/ Visuals Unlimited / **8.17** David Scharf/Peter Arnold, Inc.

CHAPTER 9 **9.1** Galen Rowell/Peter Arnold, Inc. / **9.3** Art by Kevin Somerville / **9.4** CNRI/SPL/Photo Researchers / **9.5** Modified from *Human Anatomy and Physiology,* Fourth Edition, by A. Spence and E. Mason, Brooks/Cole,1992. / **9.7** (a, b) O. Auerbach/Visuals Unlimited / **9.8** Larry Mulvehill/Photo Researchers / **9.9** (a) © 2000 PhotoDisc, Inc.; (b, c) SIU/Visuals Unlimited; Art by Lisa Starr / **9.10** From *Zoology* by L. G. Mitchell, J. A. Mutchmor, and W. D. Dophin, Fig. 5.15, p. 111. Copyright © 1988 The Benjamin-Cummings Publishing Company, Inc. Reprinted by permission of Addison Wesley Longman Publishers, Inc. / **9.11** (a) © R. Kessel/Visuals Unlimited; Art by Lisa Starr / **9.12** Art by Leonard Morgan / **9.13** (a-below, b) Art by Carlyn Iverson / **9.14** Lennart Nilsson from *Behold Man,* © 1974 by Albert Bonniers Forlag and Little Brown and Company, Boston / **Page 197** Steve Lissau/Rainbow / **9.15** Image Bank

CHAPTER 10 **10.1** (left) National Park Service; (right) © Greg Mancuso/Jeroboam / **Page 200** Photograph Gary Head / **10.3** U.S. Department of Agriculture / **10.4** Art by Robert Demarest / **10.5** Art by Robert Demarest / **10.6** Art by Robert Demarest / **10.7** Art by Precision Graphics / **10.8** Art by Robert Demarest / **10.9** Photograph Gary Head / **10.11** (a) Art by Joel Ito / **10.12** Art by Precision Graphics / **10.13** © 2000 David Joel/Stone; (a) Art from *Perspectives in Human Biology* by L. Knapp, Wadsworth, 1998. / **10.14** Art by (left) Robert Demarest; (right) Kevin Somerville / **10.14** (above) Photograph Colin Monteath, Hedgehog House, New Zealand; Art (left) Robert Demarest, (right) Kevin Somerville; (below) Evan Cerasoli / **10.15** © Corbis-Bettmann

CHAPTER 11 **11.1** (a) © Corbis/Bettmann; (b) Manfred Kage/Peter Arnold, Inc. / **11.2** (a) From *Perspectives in Human Biology,* by L. Knapp, p. 402, Wadsworth, 1998. / **11.3** Art by Raychel Ciemma / **11.4** Art by Raychel Ciemma / **11.5** Art by Raychel Ciemma / **11.7** (b) Micrograph Dr. Constantino Sotelo; (a, c) Art by Lisa Starr / **11.8** Art by D. & V. Hennings / **11.10** Micrograph from *Tissues and Organs: A Text-Atlas of Scanning Electron Microscopy,* by R. G. Kessel and R. H. Kardon. Copyright © 1979 by W. H. Freeman and Company. Reprinted with permission; Art by Robert Demarest / **11.11** (c) © C. Raines/Visuals Unlimited / **11.12** (b) Art by Robert Demarest / **11.13** Art by Robert Demarest / **Page 227** p. 227 © Wiley/Wales/ProFiles West / **11.15** (b) Micrograph Manfred Kage/Peter Arnold, Inc.; Art by Robert Demarest / **11.16** (a) Kenneth Garret/National Geographic Image Collection; (b) Art by Kevin Somerville / **11.17** Art by Kevin Somerville / **11.18** C. Yokochi and J. Rohen, *Photographic Anatomy of the Human Body,* Second edition, Igaku-Shoin Ltd., 1979 / **11.19** (a) From James W. Kalat, *Introduction to Psychology,* Third edition, Brooks/Cole Publishing Company, 1993; (b) Marcus Raichle, Washington University School of Medicine / **11.20** (left) Art by Palay/Beaubois after Wilder Penfield and Theodore Rasmussen, *The Cerebral Cortex of Man,* © 1950 Macmillan Publishing Company; copyright renewed © 1978 by Theodore Rasmussen. Reprinted with permission of Macmillan Publishing.; photograph Colin Chumbley/Science Source/ Photo Researchers **11.21** Art by Lisa Starr / **11.22** Art by Palay/Beaubois / **11.26** (a, b) From Edythe D. London et al., *Archives of General Psychiatry,* 47:567–574 (1990); (b) photograph Ogden Gigli/Photo Researchers

CHAPTER 12 **12.1** (a) F. Wood/Superstock; (b) Comstock, Inc. / **12.2** Art by (above) Kevin Somerville; (below) Lisa Starr / **12.3** From Hensel and Bowman, *Journal of Physiology,* 23: 564–568, 1960. / **12.4** Photograph Colin Chumbley/ Science Source/ Photo Researchers; (left) Art by Palay/Beaubois after Wilder Penfield and Theodore Rasmussen, *The Cerebral Cortex of Man,* © 1950 Macmillan Publishing Company; copyright renewed © 1978 by Theodore Rasmussen. Reprinted with permission of Macmillan Publishing. / **12.5** Art by Raychel Ciemma / **12.6** Art by Gary Head / **12.7** Micrograph Omikron/SPL/Photo Researchers; Art by Robert Demarest / **12.9** Art by Robert Demarest / **12.10** (a) Photograph Edward R. Bower © 1991 TIB/West; (left) Art by Kevin Somerville; (b) Art by Betsy Palay/Artemis / **12.11** (a, b) Robert E. Preston, courtesy Joseph E. Hawkins, Kresge Hearing Research Institute, University of Michigan Medical School / **12.12** Ted Beaudin/FPG; Art by Robert Demarest / **12.13** Art by Kevin Somerville / **12.14** Art by Kevin Somerville / **12.15** Micrograph Lennart Nilsson © Boehringer Ingelheim International GmbH / **12.17** After *From Neuron to Brain,* by S. Kuffler and J. Nicholls, Sinauer, 1977. / **12.18** (a, b) Photographs Gerry Ellis/The Wildlife Collection; Art by Kevin Somerville /

CHAPTER 13 **13.1** Evan Cerasoli / **13.2** Art by Kevin Somerville / **13.3** Art by Lisa Starr / **13.4** Art by Lisa Starr / **13.5, 13.6** Art by Robert Demarest / **13.7** (a) Mitchell Layton; (b) Syndication International (1986) Ltd.; (c) Photographs courtesy of Dr. William H. Daughaday, Washington University School of Medicine, from A. I. Mendelhoff and D. E. Smith, eds., *American Journal of Medicine,* 20:133 (1956) / **13.8** Art by L. Morgan / **13.9** (a, b) Art by Raychel Ciemma; (c) © Corbis-Bettmann / **13.10** Biophoto Associates/SPL/Photo Researchers / **13.11** Art by Leonard Morgan / **Page 275** Thomas Zimmermann/FPG / **13.12** The Mark Shaw Collection/Photo Researchers

Pages 278–279 From Fran Heyl Associates, © Jaques Cohen, Computer enhanced by © Pix Elation

CHAPTER 14 **14.1** (left) Evan Cerasoli; (right) Art by Robert Demarest after Patten, Carlson, and others / **14.2** Art by Raychel Ciemma / **14.3** (b) Micrograph Ed Reschke; Art by Raychel Ciemma / **14.4, 14.5, 14.6, 14.7** Art by Raychel Ciemma / **14.8** Photograph Lennart Nilsson from *A Child Is Born,* © 1966, 1977 Dell Publishing Company. Inc.; Art by Raychel Ciemma / **14.9** Art by Robert Demarest / **14.10** © Lennart Nilsson / **14.12** Adapted from *Human Heredity,* Fifth Edition, by M. Cummings, p. 183, Brooks/Cole, 2000 / **14.13** David Frazier/Photo Researchers / **14.14** Sandy Roessler/FPG

CHAPTER 15 **15.1** (a) Lennart Nilsson from *A Child Is Born,* © 1966, 1977 Dell Publishing Company. Inc.; (b) Carolina Biological Supply; (c) Photograph Gary Head / **15.3** Art by Palay/Beaubois adapted from *Mechanisms of Development,* by R. G. Ham and M. J. Veomett, C. V. Mosby Co., 1980. / **15.4** Art by Raychel Ciemma / **15.5** Photograph Gary Head; Art by Raychel Ciemma / **15.6** Art by Raychel Ciemma / **15.7** Art by Raychel Ciemma / **15.8** Art by Raychel Ciemma / **15.9** (a) Art by Raychel Ciemma after figure from B. Burnside, *Developmental Biology,* 1971, 26:416–441. Copyright © 1971 by Academic Press, reproduced by permission of the publisher; (b) art by Lisa Starr / **15.10**

(a, b, c) Photographs from Lennart Nilsson, *A Child Is Born*, © 1966, 1977 Dell Publishing Company, Inc.; Art by Raychel Ciemma / **15.11** Photographs from Lennart Nilsson, *A Child Is Born*, © 1966, 1977 Dell Publishing Company, Inc.; Art by Raychel Ciemma / **15.12** Art by Carlyn Iverson / **15.13** Art by Robert Demarest / **15.14** Art by Raychel Ciemma / **Page 313** Evan Cerasoli / **15.15** Art by Raychel Ciemma modified after *The Developing Human: Clinically Oriented Embryology,* Fourth Edition, by Keith L. Moore, W. B. Saunders Co., 1988. / **15.16** From Lennart Nilsson, *A Child Is Born*, © 1966, 1977 Dell Publishing Company, Inc.; (b) James W. Hanson, M.D. / **15.17** (inset) From Fran Heyl Associates, © Jaques Cohen, Computer enhanced by © Pix Elation; Art by Raychel Ciemma / **15.18** From Lennart Nilsson, *A Child Is Born*, © 1966, 1977 Dell Publishing Company, Inc.; Art by Lisa Starr / **15.19** Art by Raychel Ciemma adapted from *Developmental Anatomy,* by L. B. Arey, W. B. Saunders Co., 1965. / **15.20** Photograph Gary Head / **15.21** Jose Carillo/Photophile / **15.22** (a) G. Musil/Visuals Unlimited; (b) Cecil Fox/Science Source/Photo Researchers / **15.23** M. Gadomski/Photo Researchers

CHAPTER 16 **16.1** (a) Lennart Nilsson © Boehringer Ingelheim International GmbH; (b) R. Regnery, Centers for Disease Control and Prevention, Atlanta / **16.2** Lee D. Simon/Photo Researchers / **16.3** Art by Palay/Beaubois / **16.4** Art by Raychel Ciemma / **16.5** (a) P. Hawtin, University of Southampton/SPL/Photo Researchers; (b) Stanley Flegler/Visuals Unlimited / **16.6** Art by Raychel Ciemma / **16.7** (above) Centers for Disease Control; (below) photograph Edward S. Ross / **16.8** (a) © Larry Jensen/Visuals Unlimited; (b) Jerome Paulin/Visuals Unlimited; (c) John D. Cunningham/Visuals Unlimited / **16.9** Art by Leonard Morgan / **16.10** Kent Wood/Photo Researchers / **16.11** Oliver Meckes/Ottawa/Photo Researchers / **16.12** CAMR/A.B. Dowsett/Science Photo Library / **16.13** (a) After *Molecular and Cellular Biology,* by Stephen L. Wolfe, Wadsworth, 1993; (b) Z. Salahuddin, National Institutes of Health / **16.15** (a-left) Allen Russell/ProFiles West; (a-right) © M. Richards/PhotoEdit / **16.16** David M. Phillips/Visuals Unlimited / **16.17** (a) CNRI/SPL/Photo Researchers; (b) Biophoto Associates/Photo Researchers; (c) Joel B. Baseman / **16.18** George Musil/Visuals Unlimited / **16.19** E. Grau/SPL/Photo Researchers / **16.20** David M. Phillips/Visuals Unlimited / **16.21** Tony Brain and David Parker/SPL/Photo Researchers

Pages 344–345 David M. Phillips/Science Source/Photo Researchers

CHAPTER 17 **17.1** (a-c),(e) Lennart Nilsson from *A Child Is Born* © 1966, 1967 Dell Publishing Company, Inc.; (d) Lennart Nilsson from *Behold Man*, © 1974 by Albert Bonniers Forlag and Little, Brown and Company, Boston / **Page 348** C. J. Harrison et al., *Cytogenetics and Cell Genetics* 35:21–27, copyright 1983 S. Karger A.G., Basel / **17.2** C. J. Harri-

son et al., *Cytogenetics and Cell Genetics* 35:21–27, copyright 1983 S. Karger A.G., Basel / **17.3** CNRI/Photo Researchers / **17.5** Micrographs Andrew S. Bajer, University of Oregon / **17.6** (a) Courtesy of the family of Henrietta Lacks; (b) Dr. Pascal Madaule, France / **17.7** Micrographs Ed Reschke; Art by Raychel Ciemma / **17.8** Micrographs H. Beams and R. G. Kessel, *American Scientist*, 64:279–290, 1976 / **17.10** (a) C. J. Harrison et al., *Cytogenetics and Cell Genetics* 35:21–27, copyright 1983 S. Karger A.G., Basel; (b) B. Hamkalo; (c) O.L. Miller, Jr., Steve L. McKnight; Art by Lisa Starr / **17.11** Art by Lisa Starr / **17.13, 17.14** Art by Lisa Starr / **17.15** Art by Raychel Ciemma / **17.16** Art by Raychel Ciemma and American Composition and Graphics / **17.17, 17.18** Art by Raychel Ciemma

CHAPTER 18 **18.1** (a) Frank Trapper/Sygma; (b) Focus on Sports; (c) Fabian/Sygma; (d) Moravian Museum, Brno / **18.3** Richard Corman/Outline / **18.9** Art by Raychel Ciemma / **18.12** (a, b) Micrographs Stanley Flegler/Visuals Unlimited / **18.13** Jim Stevenson/SPL/Photo Researchers / **18.14** (top to bottom) Frank Cezus; Frank Cezus; Michael Keller; Ted Beaudin; Stan Sholik/all FPG / **18.15** Dan Fairbanks, Brigham Young University / **Page 379** C. J. Harrison et al., *Cytogenetics and Cell Genetics* 35:21–27, copyright 1983 S. Karger A.G., Basel / **Page 381** Evan Cerasoli

CHAPTER 19 **19.1** (a) © Abraham Menashe, Inc.; (b) From "Multicolor Spectral Karyotyping of Human Chromosomes" by E. Schröck, T. Ried, et. al. *Science*, 26 July 1996, Volume 273, pp. 495. Used by permission of E. Schröck and T. Ried and the American Association for the Advancement of Science / **19.2** (a) Photograph Omikron/Photo Researchers; (f) Art by Palay/Beaubois; photograph Omikron/Photo Researchers / **19.3** (b) Redrawn by Robert Demarest after *Human Heredity: Principles and Issues*, Third Edition, by M. Cummings, p. 126, Brooks/Cole, 1994. / **19.4** (a), (b) Stuart Kenter Associates / **19.7** Photograph Dr. Victor A. McCusick / **Page 389** Fran Heyl Associates, © Jacques Cohen, computer enhanced by © Pix Elation / **19.10** Steve Uzzell / **19.11** Giraudon/Art Resource / **19.13** (a) After *Human Genetics*, Second Edition, by Victor A. McKusick, Copyright © 1969. Reprinted by permission of Prentice-Hall, Inc., Englewood Cliffs, NJ; (b) photograph Corbis-Bettmann / **19.14** W.M. Carpenter / **19.15** Rivera Collection/Superstock, Inc. / **19.17** (a, b) Courtesy of G. H. Valentine / **19.21** Art by Raychel Ciemma / **19.22** (a) Cytogenetics Laboratory, University of California, San Francisco; (b) after Collman and Stoller, *American Journal of Public Health,* 52, 1962.; (c) Courtesy of Peninsula Association for Retarded Children and Adults, San Mateo Special Olympics, Burlingame, CA

CHAPTER 20 **20.1** (a) Photograph A. C. Barrington Brown © 1968 J. D. Watson; model A. Lesk/SPL/Photo Researchers / **20.2** Micrograph Biophoto Associates/SPL/

Photo Researchers /**20.4** Damon Biotech, Inc. / **20.8** From *Molecular and Cellular Biology*, by Stephen L. Wolfe, Wadsworth, 1993. / **20.12** (a) 3-D model of tRNA by David B. Goodin, Ph.D.; Art by Lisa Starr / **20.13** Art by Lisa Starr / **20.14** Art by Lisa Starr / **20.17** Photograph courtesy of the National Neurofibromatosis Foundation / **20.18** Micrograph C. J. Harrison / **20.19** Art by Lisa Starr

CHAPTER 21 **21.1** (a) © Corbis/Bettmann / **21.2** (a) Dr. Huntington Potter and Dr. David Dressler; (b) Stanley N. Cohen/Science Source/Photo Researchers / **21.3, 21.4** Art by Lisa Starr / **21.6** Cellmark Diagnostics, Abingdon, U.K. / **21.7, 21.8** Art by Lisa Starr / **21.9** NCI/Science Source/Photo Researchers / **21.10** (a) Dr. Vincent Chiang, School of Forestry & Wood Products, Michigan Technology University; (b) Courtesy Calgene LLC / **21.11** Courtesy of Exxon Corporation / **21.12** (a) R. Brinster and R. E. Hammer, School of Veterinary Medicine, University of Pennsylvania; (b) PA News Photo Library / **21.13** Courtesy of Victor A. McKusick and Joanna Strayer Amberger, Johns Hopkins University. **21.14** After *Human Heredity*, Fifth Edition by M. Cummings, p. 332, Brooks/Cole, 2000 / **21.15** © Adrian Arbib / Still Pictures

CHAPTER 22 **22.1** Billy E. Barnes/Jeroboam / **22.2** Art by Betsy Palay/Artemis / **22.3** Lennart Nilsson © Boehringer Ingelheim GmbH / **22.4** Art by Betsy Palay/Artemis / **22.6** © Tony Freeman/PhotoEdit / **22.7** Paul Shambroom/Photo Researchers / **22.8** Michael Newman/PhotoEdit / **22.9** From American Cancer Society's Cancer Facts and Figures–2000. Reprinted by permission of the American Cancer Society, Inc. / **22.10** (a) Photograph Mills-Peninsula Hospitals / **22.12** (a) John D. Cunningham/Visuals Unlimited; (b) Alfred Pasieka/Peter Arnold, Inc. / **22.13** (a) Biophoto Associates/Science Source/Photo Researchers; (b) James Stevenson/ SPL/Photo Researchers; (c) Biophoto Associates/Science Source/Photo Researchers; Art by Betsy Palay/Artemis

Pages 448–449 Photograph NASA

CHAPTER 23 **23.1** Douglas Mazonowicz/Gallery of Prehistoric Art / **23.2** (a) Christopher Ralling; (b) Courtesy George P. Darwin, Darwin Museum, Down House; (c-right) Heather Angel / **23.3** (above) Photograph Bruce Coleman Ltd.; (below) Stephen Rapley / **23.4** (a) After *Evolving*, by F. Ayala and J. Valentine, Benjamin-Cummings, 1979.; (b) Art by Jean Paul Tibbles / **23.5** (a) Patricia G. Gensel; (b) Donald Baird, Princeton Museum of Natural History; (c) Art by D. and V. Hennings / **23.6** Art by (a) Leonard Morgan; (b) Lloyd K. Townsend / **23.7** Art by Raychel Ciemma / **23.8** (b) From *General Zoology*, Sixth Edition, by T. Storer et al. Copyright © 1979 McGraw-Hill. Used by permission of the California Academy of Sciences. / **23.9** Art by Victor Royer / **23.10** Art by Raychel Ciemma / **23.12** Art by Raychel Ciemma / **23.14** Art by D. and V. Hennings / **23.17** Kjell B. Sandved/Visuals Unlimited / **23.18** Painting by Chesley Bonestell / **23.19** Stanley W. Awramik / **23.21** (a) Sidney W. Fox; (b) W. Hargreaves and D. Deamer

CHAPTER 24 **24.1** (a, b) Wolfgang Kaehler / **24.2** (a) Harlo H. Hadow; (b) Jack Carey / **24.3** Photographs Roger K. Burnard / **24.4** (a, b) Roger K. Burnard; (c, d) E. R. Degginger / **24.6** Photograph Sharon R. Chester / **24.7** (a) Gene C. Feldman and Compton J. Tucker/ NASA, Goddard Space Flight Center / **24.9** Gerry Ellis/The Wildlife Collection / **24.10** Courtesy United States Department of Agriculture / **24.12** (a) Photograph by Gene E. Likens from G. E. Likens and F. H. Bormann, *Proceedings First International Congress of Ecology*, pp. 330–335, September 1974, Centre Agric. Publ. Doc. Wagenigen, The Hague, The Netherlands; (c) Photograph by Gene E. Likens from G. E. Likens et al., *Ecology Monograph*, 40(1):23–47, 1970; (d) after G. E. Likens and F. H. Bormann, "An Experimental Approach to New England Landscapes" in A. D. Hasler (ed.), *Coupling of Land and Water Systems*, Chapman & Hall, 1975 / **24.13** (b) Photograph John Lawler/ FPG; Art by Lisa Starr and P. Hertz / **24.14, 24.15, 24.16** Art by Lisa Starr / **24.17** (c, d) PhotoDisc; Art by Lisa Starr / **24.18** Photo-Disc; Art by Lisa Starr / **24.19** Heather Angel / **24.20** PhotoDisc; Art by Lisa Starr / **24.21** (above) Peter Scoones/Planet Earth Pictures; (below) Dennis Brokaw

CHAPTER 25 **25.1** (a) Antoinette Jongen/ FPG; (b) Photograph United Nations / **25.3** Data from Population Reference Bureau after *Living in the Environment*, Eighth Edition, by G. T. Miller, Jr., Brooks/Cole, 1993. / **25.6** Photograph NASA / **25.8**: After *Environmental Science*, Sixth Edition, by G. T. Miller, Jr., Brooks/Cole, 1997. / **25.9** Photograph United Nations / **25.10** (a) Photograph National Science Foundation; (b, c, d) Photographs NASA / **25.11** Dr. Charles Henneghiem/Bruce Coleman Ltd. / **25.12** From Water Resources Council / **25.13** Agency for International Development / **25.14** (above) After *Living in the Environment*, Eighth Edition, by G. T. Miller, Jr., Brooks/ Cole, 1993.; Photograph R. Bieregaard/Photo Researchers / **25.15** Photograph NASA / **25.16** William Campbell/*Time* Magazine / **25.17** After *Living in the Environment*, Tenth Edition, by G. T. Miller, Jr., Brooks/Cole, 1997 / **25.18** (a, b) Alex MacLean/Landslides / **25.19** Erwin Christian/ZEFA / **25.20** Richard Parker/Photo Researchers / **25.21** World Bank / **25.22** © 1983 Billy Grimes

Appendix I Evan Cerasoli

Egg. *See also* Ovum; chromosomes in, 284; development of, 300; embryo formation and, 80; fertilization of, 302–303, *302–303;* genetic instructions in, 363; meiosis and, *358*

Eicosanoids, 30

Ejaculation, 291

Ejaculatory ducts: function of, 282–283; male reproductive system and, 283

El Tor, 53

Elastic cartilage, 73

Elastin: aging and, 319; in connective tissues, 72; structure of skin and, 82

Electric gradients, 50

Electrocardiogram (ECG): arrhythmias, 157, *157;* heartbeat and, *136,* 136

Electrolytes: defined, 201; malabsorption disorders and, 125

Electromagnetic energy, 355

Electron microscopy, 40–41

Electron transfers: ATP and, 60, *60;* metabolic pathways and, 56

Electron transport phosphorylation, *63*

Electronegative atoms, 21

Electrons: atomic structure and, *16,* 18; ATP and, 59; carbon's bonding behavior and, 26; chemical bonds and, 20–21, *20–21;* coenzymes and transfer of, 36; defined, 16; distribution of in atoms, *18,* 18–19, *19;* electron microscopes and, 40–41; free radicals and, 22; Krebs cycle and, 62–63

Electroporation, 428

Elements: atomic numbers of, *16,* 16; atoms and, 19; defined, 16; in Earth's crust, *15;* in human body, *15;* major elements in human body and, 14; mass number and, *16,* 16; overview of, 14; trace elements in human body and, 14

Elimination: body water lost through, 200; defined, 115

Elongation stage of protein synthesis, 410

Embolism, 155, 157

Embolus, 155

Embryology, *459*

Embryonic disk, *306;* formation of, 304; function of, 306

Embryo: development and, 322; early stages of, 298–299, *298–299;* emergence of human features in, 308–309; formation of, *306,* 306–307; genetic screening and, 389; genomic imprinting and, 379; hyaline cartilage in, 73; maternal nutrition and, *314;* reproductive organ development in, *280,* 280–281; screening for birth defects, 316; sex determination in, *386,* 386

Emerging diseases, 325

Emigration, 495

Emotions, 112–113

Emphysema: lungs and, *187;* respiratory disorders and, 195; smoking and, 194

Emulsification, *121*

Enamel, *116*

Encapsulated receptors, 244

Encephalitis: bovine spongiform, 77; disorders of the nervous system, 227

Endangered species, 464

Endemic diseases, 332

Endocardium, 146

Endocrine glands: functions of, 71; hormone production and, 269; hormones of, 272

Endocrine system, 260–277. *See also* Hormones; abnormal pituitary output and, 268; amine, peptide, and protein hormones of, 265; blood vessels and, 153; control of, 269–270; function of, 80; growth factors and, 272–273; hormone sources in, 269; hypothalamus and pituitary glands of, 266–267; integration and control, 272–273; other organ systems and, *80;* overview of, 262–263, *263;* signaling mechanisms of, 264–265; steroid hormones and, 264–265

Endocytic vesicles, *48*

Endocytosis, 53, *53,* 154

Endoderm: organs and tissues from, 300; role of, 80–81

Endogenous pyrogen, 167

Endometriosis, 287

Endometrium: cancer of, 443; location and function of, *287;* structure of, 286

Endoplasmic reticulum, *44,* 44, 45. *See* ER

Endorphins, 223

Endothelium: in blood vessels, 152, *152;* in cardiovascular disorders, 157; of the heart, 146, *146*

Endotherms, 212

Energy carriers. *See* ATP (adenosine triphosphate)

Energy, cellular: ATP and, 36, 52, *52,* 55, 56; carbohydrates and, 28–29; carbon atoms and, 26–27; cells' use of, 4, 41; defined, 4; interdependent organisms and, 7, *7;* lipids and, 30–31; metabolism and, 4, 5, 56–57; organization levels in nature and, 7, *7;* sources for muscle contraction, 105, *105;* transfers, 4

Energy, environmental: consumption of, *506,* 506–507; energy flows, 476, *477, 478;* energy inputs, 474; fossil fuels, 506–507; nuclear energy, 506–507; sources of, 507

Energy levels (in atoms), 18

Energy pyramids: energy flow and, *478;* function of, 476

Entamoeba histolytica, 330, *330*

Enveloped virus, *326*

Environment: adaptive radiations and, 6, 461; aging and, 318; air pollution and, 500–501, *500–501;* Alzheimer's disease and, 320; biogeochemical cycles and, 479; cancer risks and, 438, 440; carcinogens and, 436; changes in, 3, 4–5; chemicals and, 14; evolutionary change and, 455; extinction and, 461, 508; fossil record and, 457; genetically engineered bacteria and, 424, 425; infectious protozoa and, 330; noise pollution and, 251; nuclear energy concerns and, 506–507; population control and, 496; sustainable societies and, 509; TB and, 333

Environment, internal: defined, 69; extracellular, *47;* homeostasis and systems control, 84–85; pH scale and, 24–25, *25*

Environmental chemicals. *See* Chemicals, agricultural; Chemicals, industrial

Environmentalists, 445

Enzymes: ATP and, 56, 58–61; ATP synthases and, 60, *60;* body temperature and, 57;

clotting and, 155; coenzymes and, 57; control mechanisms and, 57, 64; defined, 64; digestion and, 119; DNA repair and, 405; DNA replication and, 404; features of, 56–57, *57;* functional groups and, 27; globular proteins and, 33, 35; Golgi bodies and, 48; lysosomes and, 49; metabolic pathways and, 56; mitochondrion and, 55; protein synthesis and, *411;* proteins and, *27,* 32; use in recombinant DNA technology, 418–420; reproduction and, 467

Eosinophils: defined, 139, *139;* role in immunity, 166, 180

Epidemics: disease transmission and, 333; genes of endangered species and, 464

Epidermal growth factor (EGF), 275

Epidermis: structure of skin, *82,* 82–83; UV radiation and, 83, *83*

Epididymis: function of, 282; male reproductive tract and, *284;* reproductive system and, 283

Epiglottis: breathing and, 185; defined, 117; elastic cartilage and, *73;* respiratory system and, *184;* swallowing and, *117*

Epilepsy: disorders of the nervous system, 227; genetic markers and, 427; split brain experiments and, 234

Epinephrine: feedback control and, 271; hormone production and, 269; signaling mechanisms and, 265

Epiphyseal plate, 91

Epiphysis, 91

Epithelial cells, *120, 121*

Epithelial membranes, 79

Epithelial tissue: bones and, 90; cell-to-cell contacts in, 78, *78;* function of, 86; glands derived from, 70–71; lab-grown, 76; loose connective tissues and, 72, *72;* major types of, *70,* 70; membranes and, 79; structure of, *71*

EPSPs (excitatory postsynaptic potentials), 223

Epstein-Barr virus (EBV): lymphatic cancer and, 444; viruses and, 327

Flatulence, 124

Flavin adenine dinucleotide. *See* FAD

Fleming, Alexander, 8

Flexion (of joints), *98*

Flu. *See* Influenza

Fluid pressure, 51

Fluorine, *129*

Fluorouracil, 440

Folic (folate) acid, *128*, 157

Follicles: cyclic changes in ovaries and, *289*; fertilization and, *302*; function of, 288, *288*; menstrual cycle and, *290*

Follicular phase (of menstrual cycle), 286, *287*, *290*

Fontanel, 99

Food: allergies to, 125; cancer risk from chemicals in, 438; chemicals and food production and, 37; energy from, 130–131; food guide pyramid and, 126; genetic engineering and, 430; use of chemicals in production of, 37

Food chain: biological magnification of DDT in, 489; ecosystems and, 474

Food digestion. *See also* Digestion; Nutrition; bacterial fermentation and, 61; body's energy reserves and, *63*; carbohydrates and, 28–29; energy and, 4–5; enzyme activity and, *57*; glucose metabolism and, 273; homeostasis and, 5, *5*; hydrolysis reactions and, *27*; irradiation and, 355; lipoproteins and, 35; organic metabolism pathways and, *125*; pH and, 24

Food supply: contamination of, 45; global, 133; plants and, 7

Food webs: in Antarctica, *475*; ecosystems and, 474; ; effects of DDT on, 489; energy flows through ecosystems and, *476*, *477*; global carbon cycle and, *482–483*, 483; nitrogen cycle and, 486

Foot bones, 97

Foramen ovale: at birth, 312; defined, 311; fetal circulation and, *311*

Foramen magnum, 94, *94*

Forebrain, 230–231

Forests. *See* Deforestation

Formulas, writing, 19

Fossil fuels: alternative energy and, 507; consumption of, 506; fertilizer production and, 487; global carbon cycle and, *482*, 482, *483*, 483; green revolution and, 502; greenhouse effect and global warming and, 485; nitrogen scarcity and, 487; sustainable societies and, 509

Fossil record: evidence for evolution, 455; evolutionary trees and, 460; interpreting, 456–457

Fossilization, 456

Fossils: defined, 456, *456*; origin of life and, *466*

Founder effect, 454

Fovea, 255

Fragile X syndrome, 413, *413*

Frameshift mutation, 412, *412*

Frankenfoods, 430

Fraternal twins, 302

Free nerve endings: defined, 244; pain and, 244

Free radicals: aging and, 318; oxidation and, 22, *22*; photoaging and, 10, *10*

Frontal bones, 94

Frontal sinus, *95*

Frostbite, 212

Fructose, *28*

FSH (follicle-stimulating hormone): aging and, 321; changes in ovaries and, 288, *289*; changes in uterus and, 289; function of, *266*, 266–267, *267*; in males, 285, *285*; menstrual cycle and, 286, *290*

Functional groups, *26*, 26–27, 38

Fungi, *3*, 7

Fungicides: cancer risk from chemicals in, 438; chemical toxins in food and, 37; ozone layer and, 501

Fusion power, 507

GABA (gamma aminobutyric acid), 222

Galilei, Galileo, 40

Gallbladder: function of, *114*, 119, *134*; liver and, 122, *122*

Gallstones, 119, 130

Gamete intrafallopian transfer (GIFT), 294

Gametes: changes in chromosome number and, 396–397; chromosome variations in, *387*; development and, 300, *301*; genetic crosses and, 372–373; independent assortment and, 374–375, 376; male and female, 284; meiosis and, *358*, 358–359; segregation and, 371; spermatogenesis and, *359*

Gametocytes, *330*, 330

Gamma aminobutyric acid (GABA), 222

Gamma globulin, 143

Gamma interferon, 177

Ganglion cells: function of, 254, *254*; neurons and, 226; response to light, 255

Gap junctions, 78, *78*

Gas exchange: alveoli and, 185, 190; carbon dioxide transport in, 191; disrupted controls, 193; extreme environments and, 186; factors influencing, 186; local chemical controls over, 193; matching air flow and blood flow, 192; oxygen transport and, 190–191; pressure gradients and, 186

Gas transport, 190–191; carbon dioxide, 191; oxygen, 190–191; pressure gradients and, *191*

Gastric fluid, 57; cell-to-cell contacts and, 78; function of, 118

Gastric mucosal barrier, 118

Gastrin: defined, 124; function of, 118, 274; production of, 269

Gastrocnemius, *101*

Gastrointestinal tract. *See* GI tract

Gastrulation: in development, 300, *301*; formation of early embryo and, *306*, 306–307

Gel electrophoresis: automated DNA sequencing and, *422*; DNA fingerprints and, 421, *421*; DNA research and, 403, *403*

Gene flow, 454

Gene libraries, 423, *423*

Gene locus, *370*

Gene mutations: cancer and, 433; chromosomes and, 385; diagnosing cancer and, 439; evolutionary change and, 455; types and causes of, 412–413

Gene therapy: defined, 428; early results of, 429; strategies for transferring genes and, *428*, 428–429

Gene transfers: altering gene expression and, 427; animals and, 426–427; mapping and using the human genome and, 426–427

Gene vaccines, 405

Generalist species, 472

Genes. *See also* Chromosome variations; ABO markers, 143; defined, 370; DNA structure and, 174, 402, 403; evaluating evolutionary relationships and, 460; genetically engineered bacteria, 424; independent assortment and, 374–375, *374–375*; isolating specific genes, 423; linked, 387; meiosis I events and, 363; multiple effects of single genes, 377; penetrance and expressivity, 378–379; populations and, 453; programmed behavior and, 379; proteins from, *408*; regulating, 413; weight and, 131

Genetic analysis, 393

Genetic codes, 408, *408*

Genetic counseling, 389

Genetic crosses: rules of probability and, *373*; segregation and, 372–373

Genetic disorders: defined, 389; genetic counseling and, 389; genetic crosses and, *372*; pedigrees and, 388–389

Genetic drift, 454

Genetic engineering, 416–431; bacteria that clean up pollutants, 425; DNA fingerprints and, 421; DNA sequencing and, 422; DNA vaccines and, 405; endangered species and, 464; engineering bacteria and plants, 424–425; gene therapy and, 428–429; gene transfers in animals, 426–427; greenhouse effect and, 485; isolating specific genes for, 423; PCR and, 420; process of, 418–419; recombinant DNA and, 418; societal issues, 430

Genetic markers: DNA fingerprints and, 417; Human Genome Project and, 427

Genetic recombination: defined, 363; locations of genes and, 384

Genetic screening: defined, 389; genetic engineering issues and, 430; Huntington disorder and, 391; promise and

HCV. *See* Hepatitis (types A–E)

HDLs (high-density lipoproteins), 157

Hearing: amplitude and frequency of sound waves, 248, *248*; ear structure and, 248; mechanism of, 248–249, *248–249*; noise pollution and, 251; overview of, 258

Heart: blood vessels, 153; cardiovascular system and, 144; contraction of, 150, *150*; development of, *308*; electrocardiograms and, 136; fetal circulation and, 311; hormones and, 269, 274; muscle tissue of, 74; pacemaker for, 136; photo and location of, *146*; pulmonary and systemic circuits, *148*; valves of, 147, *147*

Heart disease: aging and, 321; atherosclerotic plaques and, *31*, 31; blood circulation and, 156; cholesterol and, *31*, 31, 157; edema and, 154; hydrogenated fats and, 30; obesity and, 112, 130; treatments for, 149

Heartbeat: cardiac muscle and, 146; defining death and, 11; ECG of, *157*; fetal, 308; function of, 147; response to stimuli, 4

Heartburn, 118

Heartstrings (chordae tendineae), 146

Heat capacity of water, 23

Heat exhaustion, 213

Heat stress, 212–213, *213*

Heat stroke, 213

Heimlich maneuver, 116, *116*

HeLa cells, 351, *351*

Helical virus, *326*

Helicobacter pylori: defined, 118; structure of, *328*

Helium, *16*, 18–19

Helper T cells: differentiation in thymus, 170; function of, *172*; HIV attack on, 335; role in immunity, 168–169, 180

Heme groups: hemoglobin and, 34, *34*, 140; red blood cells and, 141

Hemings, Sally, *416*, 416–417

Hemodialysis, 211, *211*

Hemoglobin: altered, in sickle-cell anemia, 377, 412; fetal circulation and, 310–311; function of, 138; gas

exchange and, 186; globular proteins of, *34*, 35; oxygen transport and, 140; structure and function of, *140*, 140; synthetic blood, 77

Hemolytic disease of the newborn, 143

Hemophilia: DNA mutation and, 6; HIV and, 334

Hemophilia A: drugs for, 426; X-linked inheritance and, 392

Hemorrhage, 141

Hemorrhagic fevers, 325

Hemorrhoids, 123

Hepatic artery, 149

Hepatic portal system, *122*

Hepatic portal vein: blood circulation and, *145*; cardiovascular system and, 149; defined, 122; fetal circulation and, 311; in hepatic portal system, *122*

Hepatic vein, *122*

Hepatitis (types A–E): Hepatitis B virus (HBV), 340, 444; Hepatitis C (HCV), 324; symptoms of, *122*

Herbicides: cancer risk from, 438; chemical toxins in food and, 37; in hydrologic cycle, *481*

Herbivores: as consumers, *475*; ecological pyramids and, *476*; energy flows and, *478*; phosphorus cycle and, 488

Herceptin, 443

Heredity: breast cancer and, 442; cancer and, 436; cardiovascular disorders and, 156; ovarian cancer and, 443

Herniated disks, 95

Herpes infections, 427

Herpes simplex (HSV): about, 340; eye disease and, 256; latency of, 327; as a sexually transmitted disease, *340*

Heterotrophs, 474, *474*

Heterozygotes, 377

Heterozygous individuals: allele segregation and, 371; autosomal dominant inheritance in, 390–391; autosomal recessive inheritance in, 390, *390*; defined, 370; independent assortment and, 374–375, *376*; recessive alleles in, 389; testcrosses with, 374

Hexokinase, *57*, 57

Hexosaminidase A, 390

High-density lipoproteins (HDLs), 157

Hindbrain, 230

Hip (coxal) bones, 97

Hip replacement, 99

Histamine: allergies and, *178*; role in immunity, 166–167; somatic sensations and, 245

Histones, 356

Histoplasmosis, 195, 256

HIV (human immunodeficiency virus). *See also* AIDS (acquired immunodeficiency syndrome); blood transfusions and, 77; deficient immune response and, 179; drugs for, 334; infection cycle of, 335, *335*; infectious diseases and, 324; reverse transcriptase and, 327; structure and replication of, *326*, 334; TB and, 333; treating and preventing, 336

HIV protease, 336

HMS *Beagle*, 452, *452*

Hodgkin disease, 444

Homeostasis: acid-base balance and, 210; blood pH and, 191; blood vessels and, *155*; body core temperature, *213*; bone characteristics and, 90, *90*; cardiovascular system and, 144; defined, 5, *5*; disruption by toxins, 53; glucose metabolism and, *273*, 273; glucose sparing and, 270; life and, 12; negative feedback in, *84*, 84–86; neurons and, 75; overview of, 86, 155; positive feedback in, 85; role of liver in, 122; systems control and, 84–85; water-salt balance and, 198

Hominids, 463

Hominoids, 463

Homo (genus), 455

Homo erectus, 461

Homo habilis: adaptive radiations of, 461; speciation and, 455, *455*

Homo sapiens: adaptive radiations and, 461; emergence of, *450*, 450; evolution, 2, 463; speciation and, 455

Homocysteine, 157

Homologous chromosomes: defined, 349; in genetics, *370*; independent assortment

and, 375; locations of genes on, 384; meiosis I events and, 362–363; meiosis II events and, 363, *363*; segregation of, 371

Homologous structures, 458

Homozygotes: autosomal dominant inheritance in, 391; sickle-cell anemia and, 377

Homozygous dominant individuals: allele segregation and, 371, *371*; defined, 370; rules of probability and, *373*; testcrosses and, 374

Homozygous recessive individuals: allele segregation and, 371, *371*; autosomal dominant inheritance in, 390–391; autosomal recessive inheritance in, 390, *390*; defined, 370; effects of single genes, 377; rules of probability and, *373*; testcrosses and, 374

Hooke, Robert, 40

Hookworms, 330

Horizontal canal, *250*

Horizontal cells, 254, *254*

Hormone replacement therapy (HRT): abnormal pituitary output and, 268; Addison's disease and, 271; aging and, 321; use of synthetic testosterone, 108

Hormones, 260–277. *See also* Reproductive system; action of, 264; aging and, 321; amine, peptide and protein, 265; birth and fetal, 312–313; in blood plasma, 138; bone remodeling and, 92; chemical changes and, 272–273; childbirth and, 85; discovery and function of, 262–263; endocrine system and, *263*; enzyme activity and, 57; feedback control of, 269–270; functional groups and, *26*; functions of, 71, 262; gastrointestinal, 124; glucose regulation and, 62; growth factors, 272–273; homeostasis and, 5; integration and control of, 272–273; interactions of, 262; liver and, 122; male hormone secretions, 284–285; menstrual cycle and, 286, 290; negative feedback and, 270–271; obesity and, 131; precursors

Monosaccharides, 28, *121*, 134
Monosomic, 396
Monotremes, 457
Morning-after pill, fertility control and, 293
Morphogenesis: development and, 301, 322; of early embryo, 307, *307*; emergence of human features and, 308
Morphology, comparative: defined, 458–459; evolutionary trees and, 460; morphological convergence, 458; morphological divergence, *458*; morphological traits, 453
Morula: implantation and, *303*; stage of development, 300, *301*
Mosquitoes (*Anopheles*), 330
Motion sickness, 251
Motor areas, cerebral cortex, 232
Motor neurons: action potentials, *218*, *220*; cell bodies of, *75*; defined, 110; function of, 218; information pathways and, *225*; motor units and, *106*; as nerve cells, *42*
Motor proteins, 356
Motor units: defined, 106, *106*; functioning of, 106–107
Mouth: function of, *114*, 116, *134*; taste receptors (of tongue) and, 246
MRI (Magnetic resonance imaging), 355
mRNA. *See* Messenger RNA (mRNA)
MS. *See* Multiple sclerosis (MS)
MTBE (methyl tertiary butyl ether), 201
Mucins, 116
Mucosa: defined, 115; digestion in stomach and, *118*; mucous membranes and, 79, *79*; peristalsis and, *115*; of small intestine, 120, *120*
Mucous membranes: of digestive system, 114; function of, 79, *79*
Mucus, 70
Multicellular organisms, 7
Multiple allele system, 377
Multiple sclerosis (MS): degeneration of myelin sheath in, 224; disorders of the nervous system, 227; treating with interferons, 177
Multiplication cycle of a virus, 326, 327, *327*

Muscle-building substances: anabolic steroids, 107; androstenedione and creatine, 109; uses and abuses of, 108–109
Muscle cells: cardiac, 149; controlling contraction of, *104*, 104–105; glucose metabolism and, *273*; signaling mechanisms and, 265; skeletal-muscular system and, 100; structure of, 102–103, *102–103*
Muscle fatigue, 105, 107
Muscle fibers, 100
Muscle mass, 131
Muscle sense, 244
Muscle spindles, 244
Muscle tension: defined, 106; exercise and, 108
Muscle tissue: function and types of, 74, *74*, 86; neurons and, 75
Muscle tone, 107
Muscle twitch, 107, *107*
Muscles: anaerobic pathways in, 61; antagonistic pairs, 100, *100*; development from mesoderm, 81; ER in, 48; exercise and, 29, *108*, 108–109; glycogen storage and, 62; homeostasis and, 84; lactate fermentation and, 61; motor units of, *106*, 106–107; muscle tissue of, 74; nervous tissue and, 75; structure and function of, 100–101; synergistic pairs, 100
Muscular dystrophy, 392–393
Musculoskeletal system, 88–111; appendicular skeleton, 96–98; axial skeleton, 94–95; characteristics of bone, 90–91; defined, 89; exercise, muscle-building substances and, 108–109; function of, 80; how skeleton grows, 92–93; interaction with skeleton and, 100–101; joints and, 98–99; muscle contraction, 104–105; muscles of, 102–103; as organ systems, 80; overview, *89*; properties of whole muscles, 106–107
Mutagens, 413
Mutations: cancer and, 433; chemicals causing, 438; in chromosomes, 385; defined, 6; DNA repair and, 405;

endangered species and, 464; evolutionary relationships and, 460; gene therapy and, 428; genetic engineering and, 430; genetic triggers for cancer and, 436–437, *437*; microevolution and, 454; types and causes of gene, 412–413; vertebrate development and, 459
Myasthenia gravis, 222
Mycobacterium tuberculosis, 333, 333
Myelin sheath: effect on speed of action potentials, 224; spinal cord and, 229
Myocardium, 146, *146*
Myofibrils: defined, 110; muscle contraction and, 104, *104*; of skeletal muscle cells, 102, *102*
Myoglobin: aerobic exercise and, 108; in skeletal muscle cells, 102; in slow muscle, 107
Myometrium, *287*
Myopia, 256
Myosin: muscle contraction and, *103*; muscles and, 46; skeletal muscle cells and, 102–103, 110

N

Na. *See* Sodium
NaCl (sodium chloride), 23
NAD⁺ (nicotinamide adenine dinucleotide): ATP formation and, 61; as a coenzyme, 36, 57; glycolysis and, 59, 59; Krebs cycle and, 59, 59
NADH: aerobic respiration and, 63; ATP formation and, 61; in glycolysis, 59, 59; role in Krebs cycle, 59, 59, 60
Nails: dermis and, 83; structure of skin and, 82
Nasal cavity: location, *184*; role in breathing, 184
Nasal passages, 70
Natural disasters, ecosystems and, 472
Natural gas reserves, 506
Natural killer cells (NK): DNA vaccines and, 405; response to cancer, 437; role in immunity, 171, 180; use in immunotherapy, 440
Natural opiates, 245
Natural selection: Darwin's concept of, 452; evolutionary

change and, 455; microevolution and, 454
Neandertals, adaptive radiation and, 461
Nearsightedness, 256
Negative feedback, homeostasis and, *84*, 85, 86, 270–271
Neisseria gonorrhoeae, 338, *339*
Neoplasms, 434
Nephritis, 211
Nephrons: ADH effects on, *208*; blood capillaries around, *203*; as components of urinary system, 202–203; reabsorption in, 206
Nerve cells, *42*. *See also* Neurons
Nerve growth factor (NGF), 275
Nerve tissue: cells in, *75*; overview of, 75, 86
Nerves: communication by, 224; functions of, 75; structure of, *224*
Nervous system, 216–239; action potentials and, 220–221; aging of, 320; blood vessels and, 153; brain, 230–231; cardiovascular system and, 150; cerebrum, 232–233; chemical synapses in, 222–223; components of, *226*; controls over muscle contraction, 104–105, *104–105*; development of, 81; disorders of, 227; divisions of, *226*; drug effects on, 236–237; eicosanoids and, 30; environmental assault on, 227; function of, 80; information pathways of, 224–225; memory, 235; neurons and, 218–219; as an organ system, *80*; overview of, 5, 226–227; peripheral nervous system, 228–229; reciprocal innervation, 100; sexual activity and, 85; spinal cord, 229; split brain experiments, 234; states of consciousness, 236
Net charges: defined, 20–21, 38; water and, 23
Net energy, 506
Netherlands, solid waste disposal in, 502
Neural plate, *307*, 307
Neural tube: birth defects and, 307; of early embryo, *306*; function of, 306; prevention of defects, 314
Neuralgia, 227
Neurofibrillary tangles, 320
Neurofibromatosis, 413, *413*

Tarsal bones: location, *93*; pelvic girdle and lower limbs, 97, *97*

Taste: sense of, 246; sensory receptors for, 241

Taste buds, 246, *246*, 247

Taste receptors, 246, *246*

Taxol, 443

Tay-Sachs disease: autosomal recessive inheritance of, 390; genetic screening for, 389, 390

TCRs. *See* T cell receptors (TCRs)

Tectorial membrane, 249, *249*

Teeth: fibrous joints and, 99; primate evolution and, 462; structure and location of, 116, *116*

Telomeres, 318

Telophase: function of, 353; in mitosis and meiosis, *364–365*; stages of mitosis and, *350*, *353*; visual tour of, *360–361*

Temperature, body: age-related changes, 319; feedback control of, 84–85, *85*; maintaining, 212–213; oxygen transport and, 140; protein structural changes and, 35; scrotum and, 282; somatic sensations and, 244; stabilizing, 23; thermoreceptors and, 242, 244

Temperature, global, 484, *484*, 485

Temporal bones, 94, *94*

Temporal summation: defined, *107*, 107; synaptic integration and, 223

Tendon sheath, *100*, 100

Tendons: defined, 92; dense connective tissues of, 72; structure and function of, 100, *100*

Teratogens, 315

Termination stage, 410, *411*

Tertiary structure of proteins, 34

Testcross, 374, *374*

Testes, 284; cancer of, 283, 443; development of, 308; effects of muscle-building substances, 109; as endocrine organs, *263*; hormone production by, 269; location of, 282, 284; self-examination for cancer, *443*; sperm formation, *284*, 284; steroid hormones and, 264–265

Testicular feminization syndrome, 264, 393

Testosterone: chemical structure, *26*; effects of, *264*, 264–265; as a male hormone secretion, 284–285; production of, 269; role in sperm formation, *285*; from sterols, 31; synthetic, 108; X-linked abnormalities and, 393

Tests: experimental designs and, 9; sampling error and, 9, *9*; scientific method and, 8, *8*, 12

Tetanus, 107

Tetany: defined, 107, 110; shifts in body pH and, 25

Tetracycline, 315

TGFs (transforming growth factors), 275

Thalamus, 231

Thalidomide, 315

Theory, scientific, 9

Thermal inversion, 500

Thermoreceptors, 242, 244

Thermus aquaticus, 420

Thirst center, 209

Thoracic cavity: location, *80*, 81; serous membranes of, 79, *79*; volume and pressure changes in, 188–189

Thoracic duct, 158

Thoracic vertebrae, *95*

Thoroughfare channels, 149

Threonine, 127, *127*

Threshold level of stimulation, 220

Throat, 70. *See also* Pharynx

Thrombin, 155, *155*

Thrombocytes, 139

Thrombosis, 155

Thrombus, 155

Thymine, 402, *402*

Thymine dimers, 405, *405*

Thymosins, 269, 274

Thymus gland, *263*; hormone production and, 269; hormones of, 274; lymphatic system and, 158, 159

Thyroid disease, 17

Thyroid gland, 71, *263*; hormones of, 269, 276; iodide deficiency, 133; radiation disease and, 507; skewed feedback from, 271, *271*

Thyroid hormones, 269, 276

Thyrotropin. *See* TSH

Thyrotropin-releasing hormones. *See* TRH

Thyroxine, 269; skewed feedback from thyroid and, 271

Ti plasmid, 425

Tibia: location, *93*; pelvic girdle and lower limbs, 97, *97*

Tidal volume, 189

Tight junctions, 78, *78*

TILs (tumor-infiltrating lymphocytes), 429

Tissue engineering, 76

Tissue plasminogen activator (tPA), 426

Tissues, 68–79; biological magnification of DDT in, 489; categorizing cancers, 441; connective, 72–73, *72–73*; defined, 86; embryonic, 80–81; epithelial, 70–71, *71*; membranes, 79–80; muscle, 74, *74*; nervous, 74, 75; oxygen transport in blood and, *140*, 140; replacement, 76

TNF (tumor necrosis factor), 429

Tobacco. *See also* Smoking; cancer risk and, 440; carcinogens and, 436; cardiovascular disorders and, 156; cause of gene mutations, 413; effect on respiratory health, 194; high blood pressure and, 153; lung cancers and, 444; oral cancers and, 444

Tonicity, 51, *51*

Tonsils: lymphatic system and, 158; role in immunity, 170

Top carnivores, *478*

Touch-killing, 169, *169*, 170, 171, *171*

Toxic waste dumps, 425

Toxins: antibody-mediated responses to, 173; cleaning up pollution with bacteria, 425; DDT and, 489; designer plants and cleaning up, 424; genetically engineered food and, 430; natural plant, 37

tPA (tissue plasminogen activator), 426

Trace elements, 14

Tracers, radioactive, 17

Trachea: as airway, 185; Heimlich maneuver, 117; hyaline cartilage of, 73; pseudostratified epithelium in, 70; swallowing and, *117*

Trachoma, 257

Trans fatty acids, 30

Trans-retinal, 255

Transcription: defined, *406*; RNA and, 406–407, *407*; visual summary, *414*

Transcription factors, 427

Transfection, 428

Transfer RNA (tRNA): role in protein synthesis, 406, 408;

structural features of, *409*; translation and, 410

Transforming growth factors (TGFs), 275

Transgenic organisms, 424

Transitional stage, population growth rates and, 498

Translation: RNA and, 406; stages of, *410*, 410–411; visual summary of, *414*

Translocations, chromosome, *395*, 395

Transmission electron micrographs: components of skeletal muscles, *102*; endoplasmic reticulum, *49*; liver cell components and, *45*; mitochondrion and, *55*

Transmission electron microscopes, 41, *41*

Transmission, infectious disease: disease vectors and, 332–333; of sexually transmitted diseases, 337–341; of TB, 333

Transport proteins: active transport by, *52*, 52; ATP and, 56, 60; passive transport and, *52*, 52; in plasma membrane, 47, *47*; vesicles and, 53

Transposable element, 413

Transverse colon, 123

Trapezius, *101*

Treponema pallidum, 339, *339*

TRH (thyrotropin-releasing hormone): amount in body, 268; function of, 267

Triceps, 100, *100*

Triceps brachii, *101*

Trichomonas vaginalis, 341, *341*

Tricuspid valve, 146

Triglycerides: defined, 30; digestion of, 119, *121*, 121, 134; energy from, 62; heart disease and, 149, 157; structure, *30*

Triiodothyronine: production of, 269; skewed feedback from thyroid and, 271

Triple covalent bond, 20

Trisomic individual, 396

Trisomy 21, 396–397

tRNA. *See* Transfer RNA

Trophic levels, 474

Trophic structure, 476

Trophoblast, 304

Tropomyosin, 104, *105*

Troponin, 104, *105*

True capillaries, 148–149

Trypanosoma brucei, 330, *330*

Trypsin, 119

APPLICATIONS INDEX   *Numbers in italic indicate pages that contain illustrations.*

females, 285; menopausal symptoms and, 287
Eugenic screening, 430
Evolution, 450–469; basic principles, 453; biological, 2; defined, 450; Earth history and, 464–465; endangered species and, 464; evolutionary trees, 460–461; form and development of body parts and, 458–459; fossil record, 456–457; of humans, 462–463
Exercise: aerobic, *108*, 108; aging and, 319, 321; building muscle and, 108; cardiovascular disorders and, 157; dieting and, 112, 131; effect on bone mass, 89; energy sources for, 105; muscle fatigue and, 105; muscle mass and, 107
Exocrine glands, cystic fibrosis and, 382
Experiments, applied scientific method, 10, 12
Exponential population growth, *495*, 495, 496
Expressed sequence tags (EST), 427
Extinction: endangered species and, 464; human role in, 508; process of, 461
Eye: disorders of, 256–257; genetics and, 379; split brain experiments and visual perception, 234

Fertilizers: chemicals and, 14, *14*; green revolution and, 502; greenhouse effect and global warming and, *485*; nitrogen scarcity and, 487
Fetal alcohol syndrome (FAS), 315
Fetal development: malnutrition and undernutrition, 133; maternal nutrition and, *314*; Rh blood typing and, 142–143, *143*
Fetoscopy, 316, *316*
Fingers, joint replacement, 99
Fish, mercury poisoning and, 227
Flu. *See* Influenza
Fluorouracil, 440
Focusing problems (eye), 256
Folic acid (folate), *128*, 157
Food: allergies to, 125; cancer risk from chemicals in, 438; chemicals and food production, 37; energy from, 130–131; food guide pyramid, 126; genetic engineering and, 430; glucose metabolism and, 273; use of chemicals in production, 37
Food chain: biological magnification of DDT in, 489; in ecosystems, 474
Food supply: contamination of, 45; global, 133; plants and, 7
Food webs, use of DDT and, 489
Fossil fuels: alternatives to, 507; consumption of, 506; fertilizer production and, 487; global carbon cycle and, 482, *482*, 483, *483*; green revolution and, 502; greenhouse effect and global warming and, 485; nitrogen scarcity and, 487; sustainable societies and, 509
Fragile X syndrome, 413, *413*
Frankenfoods, 430
Frostbite, 212
Fruits, *126*
Fungicides: cancer risk from, 438; ozone layer and, 501; toxins in food and, 37
Fusion power, 507

Gel electrophoresis: for automated DNA sequencing, *422*; DNA fingerprints and, *421*, 421; in DNA research, *403*, 403
Gene libraries, *423*
Gene mutations: cancer and, 433; in diagnosing cancer, 439; types and causes of, 412–413
Gene therapy: defined, 428; early results of, 429; strategies for transferring genes, *428*, 428–429
Gene transfers: altering gene expression and, 427; in animals, 426–427; mapping and using the human genome, 426–427
Gene vaccines, 405
Genes: and behavior, 379; coding for ABO markers, 143; fat storage, 131; of genetically engineered bacteria, 424; penetrance and expressivity, 378–379; regulating, 413
Genetic analysis, 393
Genetic counseling, 389
Genetic disorders: defined, 389; genetic counseling, 389; genetic crosses, 372; pedigrees and, 388–389
Genetic engineering, 416–431; in animals, 426–427; bacteria that clean up pollutants, 425; DNA fingerprints and, 421; DNA sequencing, 422; DNA vaccines, 405; endangered species and, 464; engineering bacteria and plants, 424–425; in gene therapy, 428–429; isolating specific genes, 423; pedigrees and, *416*, 416–417; process of, 418–419; recombinant DNA, 418; societal issues, 430; using PCR, 420
Genetic markers: DNA fingerprints and, 417; use in Human Genome Project, 427
Genetic screening: defined, 389; genetic engineering issues and, 430; for Huntington disorder, 391; promise and problems of, 389; for Tay-Sachs disease, 390
Genital herpes, *340*; about, 340; causative virus, 327
Genital warts, 340
Genomic imprinting, 378, 379

Genomic libraries, 423
Genotype, 389
German measles, 315
GH (Growth hormone): abnormal effects of, 268, *268*; from genetically engineered bacteria, 424
Giardiasis, 330
Gigantism, 268, *268*
Gingivitis, 116
Glaucoma, 257
Global health: infectious disease, 332; undernutrition, 132
Global population, 295
Global warming: defined, 484; extinction risk and, 508; fossil fuels and, 483
Glomerulonephritis, 211
Glucagon, 272
Gluconeogenesis, 270
Glucose sparing, 270
Glycogen, nutritional requirements for, 126
Glycolysis: exercise and, 105; strength training and, 108
Goiter, 271
Gonorrhea: infection cycle of, 338–339, *339*; pelvic inflammatory disease and, 338; virulence of, 333
Gout: obesity and, 130; uric acid and, 201
Graves' disease, 271
Green Belt movement, *505*, 505
Green revolution, 502
Greenhouse effect. *See also* Global warming; burning of fossil fuels and, 483; defined, 484; global temperature and, *484*
Growth factors, injury repair and, 275
Growth hormone (somatotropin), abnormal effects of, 268, *268*